W9-AYU-760

Symmetry and Spectroscopy

Symmetry and Spectroscopy

AN INTRODUCTION TO VIBRATIONAL AND ELECTRONIC SPECTROSCOPY

BY

DANIEL C. HARRIS

Michelson Laboratory
China Lake, California

AND

MICHAEL D. BERTOLUCCI

DOVER PUBLICATIONS, INC., New York

This Dover edition, first published in 1989, is an unabridged, corrected republication of the work first published by Oxford University Press, New York, 1978. It is reprinted by special arrangement with Oxford University Press, 200 Madison Avenue, New York, N.Y. 10016.

Library of Congress Cataloging-in-Publication Data

Harris, Daniel C., 1948–
Symmetry and spectroscopy : an introduction to vibrational and electronic spectroscopy / by Daniel C. Harris and Michael D. Bertolucci.
p. cm.
Reprint. Originally published: New York : Oxford University Press, 1978.
Includes bibliographies and index.
ISBN-13: 978-0-486-66144-5
ISBN-10: 0-486-66144-X
1. Vibrational spectra. 2. Electron spectroscopy. 3. Molecular orbitals. 4. Symmetry (Physics) I. Bertolucci, Michael D. II. Title.
QD96.V53H37 1989
543′.0858—dc20 89-16810
CIP

Manufactured in the United States by Courier Corporation
66144X11
www.doverpublications.com

Contents

Preface

Late in the afternoon on a hot, smoggy, Los Angeles, September day in 1970, a well-publicized contest pitting the most pollution-free vehicles in the country against each other was coming to an end. It was the second Clean Air Car Race from M.I.T. to Caltech. Among the throngs of dignitaries and spectators at the finish line on Greasy Street that day were two graduate students, George Rossman and myself. We hoped to capture a sample of Clean Air Car exhaust in an evacuated glass cylinder I had wrapped in some rags and carefully protected from the surging crowd. When the third car came in, and the crowd around the second car (an entry from the University of California at Berkeley) began to wane, I pushed my way up to the Berkeley driver's window, still clinging the precious evacuated cylinder. I explained to the driver that I wanted a sample of his exhaust and asked if he would please start the car as I crawled under his tail pipe. Not only did he start the car for me, but when I crawled underneath and positioned the mouth of the cylinder in the exhaust pipe, he drove away! The next time I caught that rascal from Berkeley I did manage to bottle his exhaust. We scurried off to the lab where George produced the gas phase infrared spectra shown below. These spectra pit the Clean Air Car against the Harris Smogmobile and show that the carbon monoxide emission of my car (two humps near 2150 cm^{-1}) is absent in the Clean Air Car exhaust. This little episode marked the start of the section of this book dealing with the carbon monoxide rotation-vibration spectrum. Not every part of this text has such a colorful history, but most have benefited from a similar degree of personal

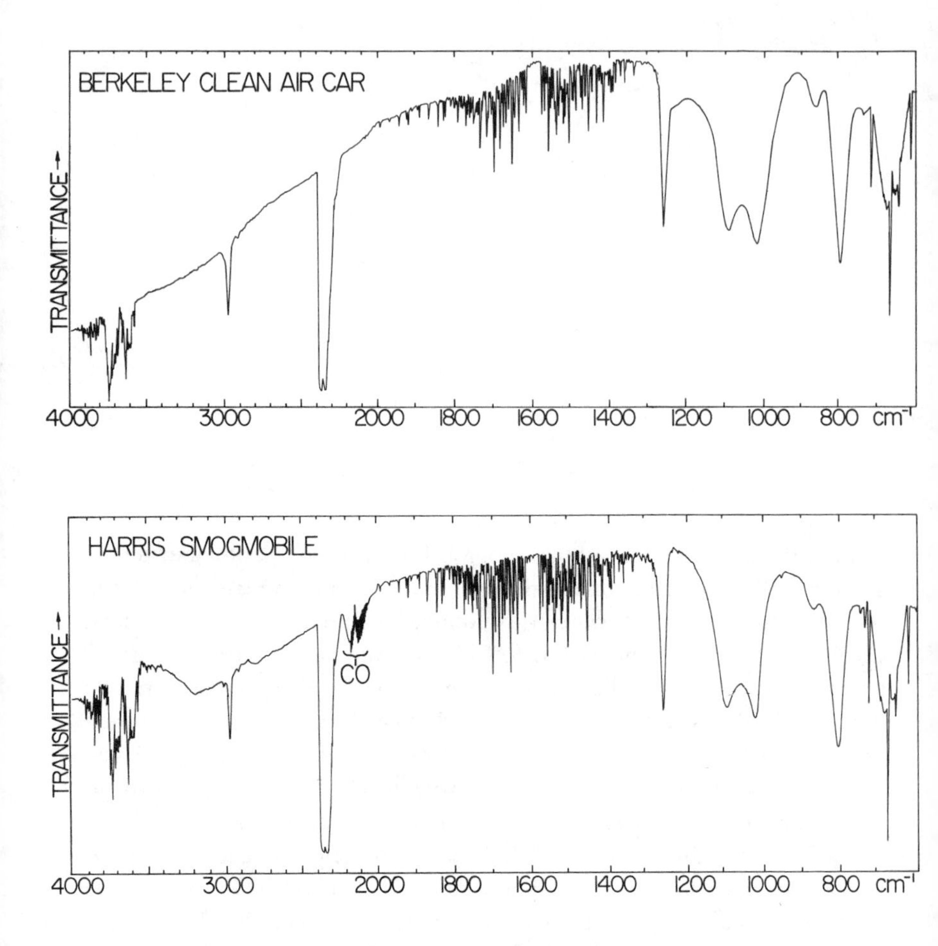

involvement on the part of myself and my students during the three years the manuscript was used for part of an undergraduate spectroscopy course at Caltech.

This book was written with the goal of introducing the student to vibrational and electronic spectroscopy and taking her or him to a rather sophisticated (albeit qualitative) level in some areas. We have tried to write a text most suitable for use on the junior to beginning graduate levels. Taking the approach that group theory is essential to the modern practice of spectroscopy, we devote the first chapter to group theory and then make extensive use of it throughout the text. For this reason we believe that this book may be used as the primary text for a course on the applications of symmetry in chemistry, as well as for a course in spectroscopy. We cover most of the topics in Cotton's fine book, *Chemical Applications of Group Theory,* but do so in the process of teaching vibrational and electronic spectroscopy and molecular orbital theory.

First and foremost this is a textbook. We have taken great pains not to assume very much background knowledge on the part of the reader. To make the exposition clear and meaningful, each new concept is applied or illustrated with experimental results as quickly as possible. The text includes some 200 problems with solutions in Appendix G. We consider these problems to be an integral part of the text and sometimes introduce new material in them. The student is urged to work through as many as time permits.

The present version of this book was written during a two-year period of postdoctoral research in the laboratory of Phil Aisen at the Albert Einstein College of Medicine in New York. The original text was written in collaboration with Mike Bertolucci who taught the course with me for a year at Caltech. Don Titus, Benes Trus, and Harry Gray have made invaluable contributions subsequently. Harry Gray and George Hammond were instrumental in initiating the course and capturing my interest in it (which was similar to capturing the interest of a hungry monkey in a banana). To keep the price of this volume to a level that students can afford, my wife Sally devoted more than half a year of effort to the production of line drawings. Finally, I cannot overestimate the role my students played in the development of this book. Comments on ways to improve the book or on errors are solicited from all readers and will be greatly appreciated.

I dedicate this book to the student who is willing to take it to bed with him at night, along with a pencil and occasionally a calculator, and who falls asleep with a smile on his face.

Dan Harris

NOTE (1989): Daniel C. Harris may be reached at Chemistry Division, Research Department, Michelson Laboratory, China Lake, CA 93555.

Note on units and conventions

Although the Système International d'Unités (SI units) are the primary units in this book,† a number of other standard units are also used. It is suggested that the student become familiar with the different units we employ because all are encountered in practice. Unless otherwise stated, however, all calculations and all equations here employ SI units. When we wish to convert an answer to a unit other than an SI unit, the conversion is the last step of the calculation.

In the SI system, the units of mass, length, time, and charge are the kilogram (kg), meter (m), second (s), and coulomb (C), respectively. Force is expressed in newtons ($1\ \text{N} = 1\ \text{kg m s}^{-2}$) and energy in joules ($1\ \text{J} = 1\ \text{N m} = 1\ \text{kg m}^2\ \text{s}^{-2}$). Coulomb's law is generally written

$$F = k\,\frac{q_1 q_2}{r^2}$$

where F is force, q is charge, r is distance, and k is a constant. In the centimeter-gram-second (cgs) system, the unit of charge, the electrostatic unit (esu), is such that k is dimensionless and has the magnitude unity. In SI units, k is written $1/4\pi\epsilon_0$, where $\epsilon_0 (= 8.85419 \times 10^{-12}\ \text{C}^2\text{N}^{-1}\text{m}^{-2})$ is called the dielectric constant (or permittivity) of free space. The factor $4\pi\epsilon_0$ appears in several equations in this book and should alert you to the fact that coulombs, meters, and kilograms are used in such equations.

† For discussions of SI units, see A.C. Norris, *J. Chem. Ed.*, *48*, 797 (1971); G. Socrates, *Ibid.*, *46*, 711 (1969); T.I. Quickenden and R.C. Marshall, *Ibid.*, *49*, 114 (1972); J.I. Hoppeé, *Ibid.*, *49*, 505 (1972); and G. Pass and H. Sutcliffe, *Ibid.*, *48*, 180 (1971).

We use the largest variety of units in measurements of energy. Chemists familiar with calories will find the use of joules not too difficult because the conversion is simple:

$$1 \text{ calorie} = 4.184 \text{ joules}$$

In the cgs system, 1 erg = 10^{-7} joules. The electron volt (eV) is the kinetic energy of an electron accelerated through one volt: 1 electron volt = 1.602×10^{-19} joules. The corresponding molar energy is 1.602×10^{-19} J $\times$ 6.022×10^{23} mol^{-1} = 96.49 kJ mol^{-1} = 23.06 kcal mol^{-1}.

We very frequently express "energy" as wave numbers. The relation between wave number and energy, in cgs units, is

$$\bar{\nu} = \frac{E}{hc}$$

where $\bar{\nu}$ is wave number (cm^{-1}), E is energy (erg), h is Planck's constant (erg s), and c is the speed of light (cm s^{-1}). The SI units are, respectively, $\bar{\nu}$ (m^{-1}), E (J), h (J s), and c (m s^{-1}). To convert wave number in reciprocal meters to wave number in recriprocal centimeters, we divide by 100

$$\bar{\nu}\ (\text{cm}^{-1}) = \frac{\bar{\nu}(\text{m}^{-1})}{100(\text{cm m}^{-1})} = \frac{E(\text{J})}{100(\text{cm m}^{-1})\, h(\text{J s})\, c(\text{m s}^{-1})}$$

Since "*wave number*" *is universally expressed in reciprocal centimeters,* and since any equation for energy in this book is in joules, unless otherwise specified, the conversion $\bar{\nu} = E/100\,hc$ must be used to too obtain wave numbers in reciprocal centimeters. When we want to emphasize that a quantity is in cm^{-1} units, we will write a bar over the symbol (e.g., $\bar{E}$ and $\bar{\omega}_e$). The use of cm^{-1} units is discussed further in Chapter Three. The 1,000 cm^{-1} unit is a kilokayser (kK).

Units of length often encountered are centimeters and Angstroms (1 Å = 10^{-10} m). Millimicrons (mμ) have been largely replaced by nanometers (1 mμ = 1 nm = 10^{-9} m). Temperature is generally expressed in Kelvins, written K (not °K). Concentrations are always expressed as moles per liter (1 mol l^{-1} $\equiv$ 1 M). The molar extinction coefficient (ϵ) used in spectroscopy is universally expressed in M^{-1}cm^{-1} units, and we dare not tamper with them.

The choice of coordinate systems and symmetry elements can be a major problem in the literature. We recommend that coordinate systems and symmetry elements always be defined at the outset of a research paper, homework problem, blackboard example, or anything else. We adhere to this policy faithfully. Conventions on coordinate systems and symmetry elements have been recommended by the Joint Commission for Spectroscopy of the International Astronomical Union and the International Union of Pure and Applied Physics [*J. Chem. Phys.*, *23*, 1997 (1955)]. We generally adhere to these conventions, with some notable exceptions. We accidentally adopted a coordinate system for ethylene that differs from the system in common use. We apologize for this but chose not to change our coordinate system for fear of introducing errors into the text in the process of making changes. We intentionally disregarded the recommended choice of symmetry elements for XeF_4 (D_{4h} symmetry), but not for benzene (D_{6h}). We recommend that C_2' and σ_v axes always be colinear and C_2'' and σ_d axes always be colinear and that C_2' and σ_v go through as many atoms as possible and that C_2'' and σ_d go through as few atoms as possible. We further recommend the use of our convention for all point groups, instead of adopting different conventions for each point group.

With regard to the naming of symmetry operations (e.g., C_2, C_2', and C_2'') and other conventions in character tables, we adopted the widely used tables of Cotton (F.A. Cotton, *Chemical Applications of Group Theory*, John Wiley & Sons, New York, 1971); we hope that this set of tables will become standard.

Symmetry and Spectroscopy

0 · *Opening remarks*

Spectroscopy is the study of the interaction of electromagnetic radiation (light, radio waves, x-rays, *etc.*) with matter. In this book we will deal with a central portion of the electromagnetic spectrum (Fig. 0-1), spanning the infrared (ir), visible (vis), and ultraviolet (uv) wavelengths.

Molecules, consisting of electrically charged nuclei and electrons, may interact with the oscillating electric and magnetic fields of light and absorb the energy carried by the light. The molecule does not interact with all light that comes its way, but only with light that carries the right amount of energy to promote the molecule from one discrete energy level to another. For example, the diatomic molecule $^{127}I^{79}Br$ in its lowest vibrational state (*ground state*) vibrates with an energy of 2.662×10^{-21} J. The next lowest vibrational energy available to the molecule is 7.961×10^{-21} J. Suppose that far infrared light of energy $(7.961 - 2.662) \times 10^{-21}$ J $= 5.299 \times 10^{-21}$ J ($= 266.8$ cm^{-1}) is shined on a sample of $^{127}I^{79}Br$. The light can be absorbed and a ground state molecule can be promoted to its first excited vibrational state. When this happens we say that the molecule has made a *transition* between the ground state and the first excited state. The two energy levels we have been discussing and the absorption spectrum of $^{127}I^{79}Br$ in the far infrared are shown schematically in Fig. 0-2. Light of energy other than 5.299×10^{-21} J would not be absorbed by the sample because the energy carried by such light does not precisely span two energy levels of the molecule.

Light of infrared frequencies can generally promote molecules from one vibrational energy level to another. Hence, we call infrared spectroscopy

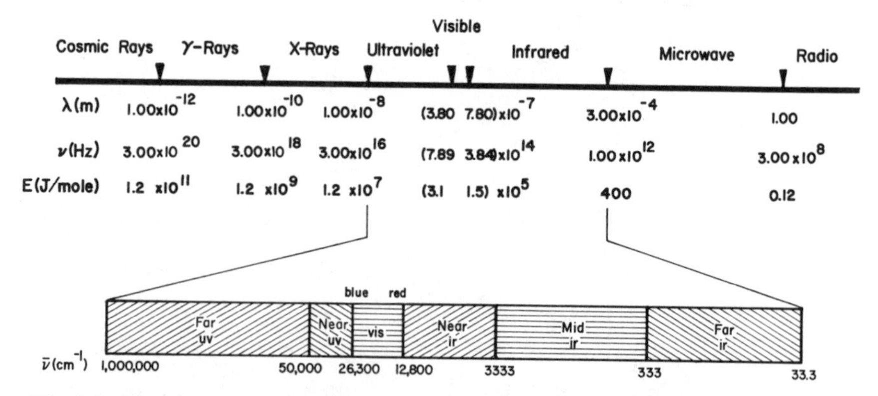

Fig. 0-1. The electromagnetic spectrum. The wavelength, λ, is given in units of meters; the frequency, ν, is given in units of hertz (1 Hz = 1 oscillation per second); and the energy, E, carried by a *mole* of photons is given in joules (4.184 J = 1 calorie). The wave number, $\bar{\nu}$, is expressed in units of cm^{-1} (read "reciprocal centimeters" or "wave numbers"). All of these properties of light are discussed in Section 2-2.

vibrational spectroscopy. Visible and ultraviolet light are much more energetic and can promote the redistribution of electrons in a molecule such that the electronic potential energy of the molecule is changed. Hence, we call visible and ultraviolet spectroscopy *electronic spectroscopy*. In order to treat electronic spectroscopy in a rational way, we will need to study molecular orbital theory and gain some understanding of the distribution of electrons in molecules.

Our study will make extensive use of symmetry. As one simple way to illustrate the relation of symmetry to energy levels, consider the three

Fig. 0-2. (a) A molecule of $^{127}I^{79}Br$ can be promoted from its ground vibrational state to its first excited vibrational state by light of energy 5.299×10^{-21} J per photon ($= 3.191 \times 10^3$ J mol^{-1}). **(b)** The schematic far infrared absorption spectrum of $^{127}I^{79}Br$ shows that only light of this energy is absorbed by the sample. Some reasons why absorption lines are not infinitely sharp are mentioned in Chapter Three.

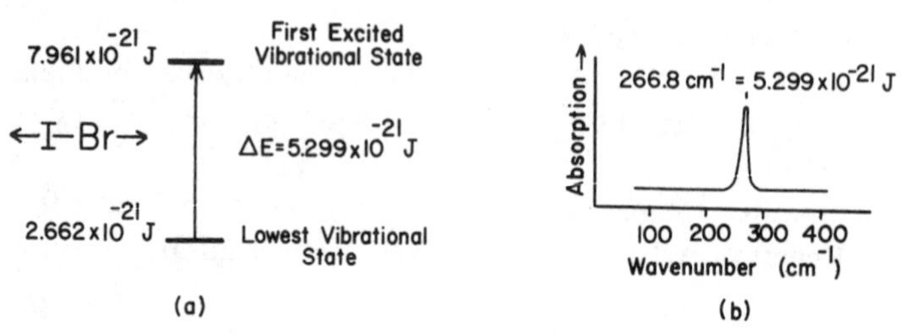

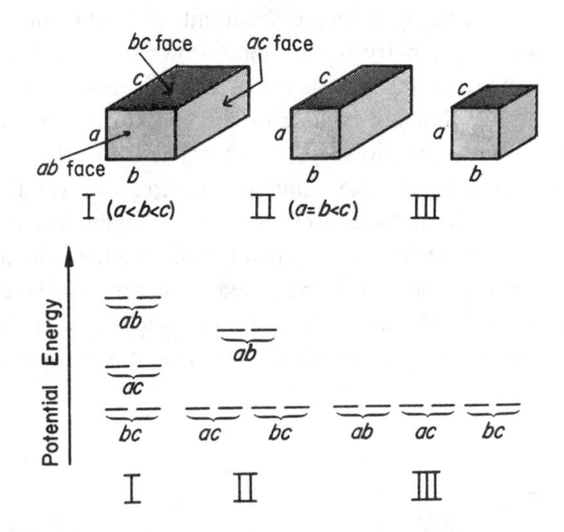

Fig. 0-3. Three parallelepipeds. The length of side a is the same for all three figures. The potential energy of each parallelepiped depends on which face it rests. The higher the center of mass, the greater the energy. The number of resting positions having the same energy for each figure is the degeneracy of that energy level. For example, the degeneracy of the lowest energy level of parallelepiped II is 4.

parallelepipeds in Fig. 0-3. Each of these solid figures has six stable resting positions in the earth's gravitational field, when it is resting on its two *ab*, two *ac*, or two *bc* faces. The potential energy, V, associated with each position can be calculated from the formula $V = mgh$, where m is the mass of the parallelepiped, g is the acceleration of gravity, and h is the height of the center of mass above the surface on which the object rests. The potential energies of the six resting positions of parallelepiped I are divided into three groups, depending on which face it rests. The potential energy is greatest when its center of mass is highest, *i.e.*, when it rests on its *ab* faces (Fig. 0-3). The parallelepiped II has only two energetically distinct resting positions because the lengths of sides a and b are equal. All six faces of the cube, III, are equal and hence only a single potential energy is obtained, regardless of which face it rests on. In this example, as the symmetry of the solids increases in the order I < II < III, the number of different energy levels decreases. The *degeneracy* (number of states which have the same energy) of each level increases with increasing symmetry.

The same trend is generally true for molecules also. The more symmetric the molecule, the fewer different energy levels it has, and the greater the degeneracies of those levels. The study of symmetry, therefore, helps us simplify some problems by reducing the number of different energy levels we must deal with. Symmetry is even more powerful than that, because it helps us decide which transitions between energy levels are possible. That is to say, a molecule may not be able to absorb light even if that light has precisely the correct energy to span two energy levels of the molecule. The symmetries of the states involved must be "compatible" (in a way we will discuss later) in order that the molecule may absorb light. The *selection rules* which tell us which transitions are possible will be one of the most important uses of symmetry and will be explained as we proceed. Since symmetry will be used throughout our studies, we will begin with a discussion of molecular symmetry and the tool which makes use of such symmetry, group theory.

1 · *A chemist's view of group theory*

1-1. Introduction

Our treatment of vibrational spectroscopy, molecular orbital theory, and electronic spectroscopy will make extensive use of molecular symmetry. One can reasonably ask, "Why is symmetry so important when the vast majority of molecules have no symmetry at all?" While it is true that most molecules considered in their entirety don't possess any symmetry, many molecules do have *local symmetry*. Consider the "representative" large molecule in Fig. 1-1. The molecule as a whole is not only devoid of symmetry, it is a mess! But if we focus our attention on the iron surrounded by four nitrogen atoms in an approximately square arrangement, we have found a region possessing useful local symmetry. For many purposes, only the iron atom and its four nearest neighbors need be considered to understand the spectroscopic behavior of this region of the molecule.

Many small molecules do possess real symmetry. We must understand the behavior of these small molecules before we try to understand big ones.

Fig. 1-1. A representative large molecule.

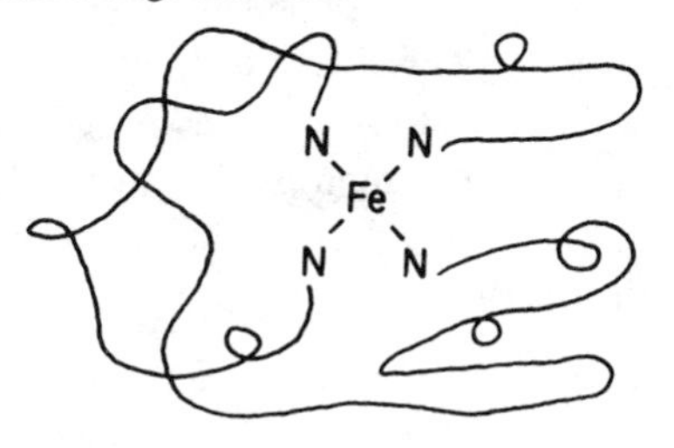

Many of the properties of, say, a benzene ring are nearly the same whether we are looking at free benzene or a phenyl ring bound to a large molecule. The properties of free benzene are easier to understand if we make use of its symmetry. What we learn about benzene can then be transferred to the phenyl substituent.

The mathematical tool which makes use of symmetry is group theory. Since our purpose is to apply group theory to chemistry, our treatment will be aimed toward this goal and will be anything but mathematically rigorous.

1-2. Symmetry Operations and Molecules

An operator is a symbol which tells you to do something to whatever follows it. For example, "d/dx" tells you to take the derivative with respect to x of some mathematical expression. In this book, we will deal with operators which tell us to do such things as "rotate a molecule by 180°."

A *symmetry operation* is an operation which moves a molecule into a new orientation equivalent to its original one. For example, consider the threefold rotation of the planar molecule boron trifluoride in Fig. 1-2. If we could label the flourine atoms we could tell that the molecule had been moved. Since we cannot label the atoms, the second configuration is entirely equivalent to the first one.

A *symmetry element* is a point, line, or plane with respect to which a symmetry operation is performed. In the BF_3 example, the element we used was the axis passing through boron perpendicular to the molecular plane. We performed the threefold rotation about this axis.

There are five kinds of symmetry operations we will use:

1. The simplest operation is the *identity operation*, usually given the symbol "E". This symbol tells you to do nothing to the molecule. We need the identity operation only to satisfy certain mathematical requirements of groups.

2. *Reflection through a plane* is denoted by the Greek letter "σ". For example, the pyramidal molecule F_2SO in Fig. 1-3 has one σ plane, also referred to as a "mirror plane." It passes through the oxygen and sulfur atoms and bisects the FSF angle. Reflection through this plane interchanges the two fluorine atoms. If we perform a second σ operation on the resulting figure, we get back the original figure. This means that two successive reflections have no net effect. We denote the two successive operations as $\sigma \cdot \sigma$ or σ^2 and write the operator equation 1-1:

$$\sigma \cdot \sigma = \sigma^2 = \mathrm{E} \tag{1-1}$$

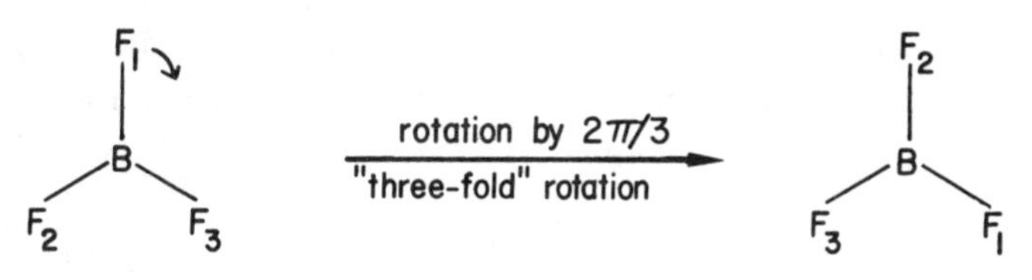

Fig. 1-2. Threefold rotation of BF_3.

It is understood when we write the product of two operations that the operation on the right is performed first. In the case of eq. 1-1 it makes no difference since both operations are the same. But for two different operations, $X \cdot Y$ means that Y operates first. The product $X \cdot Y$ is not necessarily equal to $Y \cdot X$. Two operations whose product does not depend on the order of mulitplication (*i.e.*, $X \cdot Y = Y \cdot X$) are said to *commute*. Two operations whose product is the identity operation are said to be each other's *inverse*. Thus σ is its own inverse.

3. *Rotation about an axis* is denoted C_n. C simply means that a rotation is involved and the subscript n tells us what fraction of a complete rotation through 2π we are to perform. A rotation of 120° ($2\pi/3$) on BF_3 in Fig. 1-2 is called a C_3 rotation. A 90° (or fourfold) rotation of the square planar molecule $PtCl_4^{2-}$ in Fig. 1-4 is written C_4. It makes no difference which

Fig. 1-3. F_2SO has a single mirror plane.

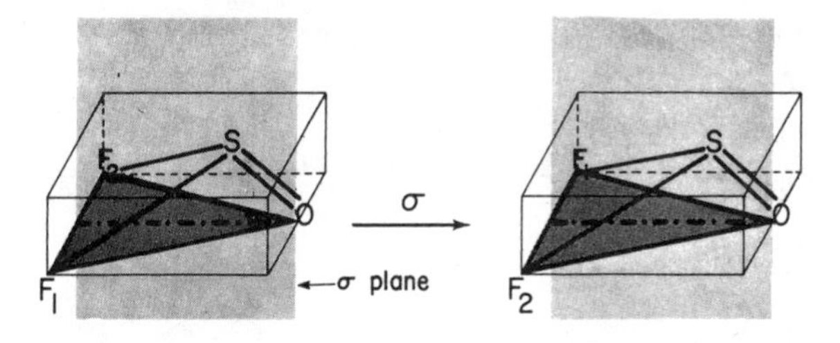

Fig. 1-4. Fourfold rotation of $PtCl_4^{2-}$.

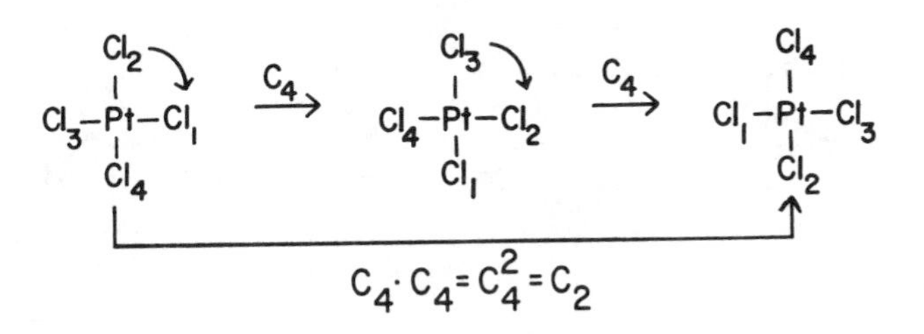

direction you use for the rotation, so long as successive rotations are performed in the same direction. For consistency, we choose the clockwise direction for C_n rotation in this book. Two C_4 operations in a row are the same as a C_2 operation (Fig. 1-4).

$$C_4 \cdot C_4 = C_4{}^2 = C_2 \tag{1-2}$$

Four C_4 operations produce the original configuration.

$$C_4{}^4 = E \tag{1-3}$$

In general, $C_n{}^n = E$. What is the inverse of C_4? It must be $C_4{}^3$ since

$$C_4{}^3 \cdot C_4 = C_4{}^4 = E \tag{1-4}$$

The net effect of $C_4{}^3$ is just rotation in the opposite sense as C_4 (Fig. 1-5). In general, the inverse of $C_n{}^m$ is C_n^{n-m}. For example,

$$\begin{aligned} (C_4)^{-1} &= C_4{}^3 \\ (C_7{}^2)^{-1} &= C_7{}^5 \end{aligned} \tag{1-5}$$

Here we use the superscript -1 to denote the inverse of the operation.

4. An *improper rotation* consists of a rotation followed by a reflection through the plane perpendicular to the axis of rotation. Look at the molecule allene, C_3H_4, in Fig. 1-6. This molecule can be inscribed in a rectangular solid such that each hydrogen atom comes at a vertex and the two ends are rotated 90° from each other. In Fig. 1-7 we rotate the molecule by $2\pi/4$

Fig. 1-5. $C_4{}^3$ is a fourfold rotation in the opposite sense of C_4.

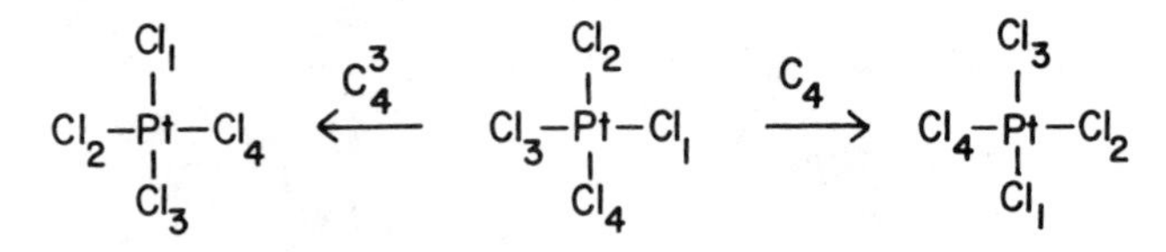

Fig. 1-6. **(a)** A molecule of allene inscribed in a square prism. **(b)** View down the S_4 axis.

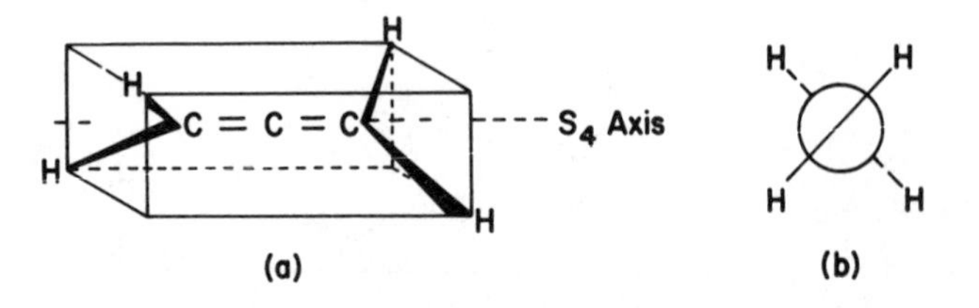

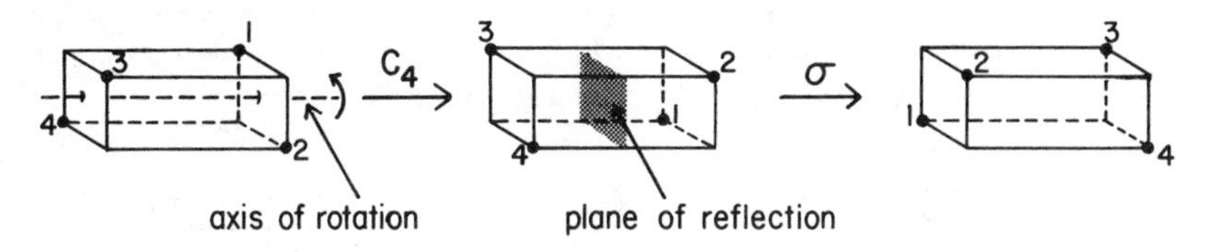

Fig. 1-7. The effect of S_4 on allene. Each solid circle represents a hydrogen atom.

about the axis of the three carbon atoms and then reflect through a plane perpendicular to this containing the central carbon atom. The result, called an S_4 improper rotation, leaves an equivalent configuration of the molecule. The inverse of S_n^m is S_n^{n-m} if n is even and S_n^{2n-m} if n is odd.

5. *Inversion* involves passing each atom through the center of the molecule and placing it on the opposite side of the molecule. Shown in Fig. 1-8 is an example of inversion of the molecule $Mo(CO)_6$. Inversion is equivalent to the operation S_2, but is always given the separate symbol "i".

Fig. 1-8. The effect of inversion on $Mo(CO)_6$.

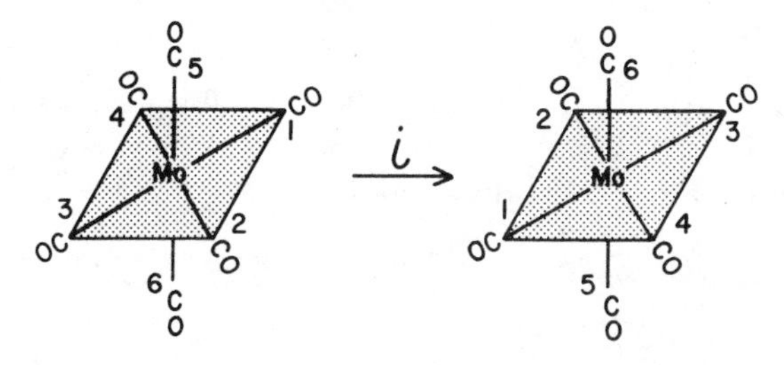

Problems

1-1. Find all of the symmetry elements present in the following species: $PtCl_4^{2-}$ (square planar), ethylene, SF_4 (structure below), cyclopropane, p-difluorobenzene, and tetraphenylcyclobutadiene in the conformation below.

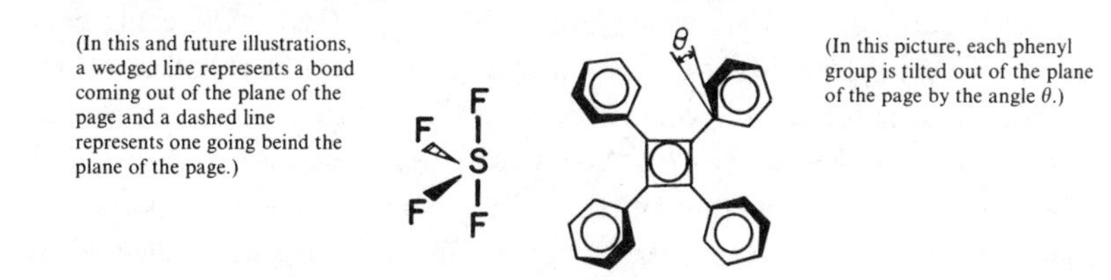

(In this and future illustrations, a wedged line represents a bond coming out of the plane of the page and a dashed line represents one going beind the plane of the page.)

(In this picture, each phenyl group is tilted out of the plane of the page by the angle θ.)

1-2. Molecules with a mirror plane, center of inversion, or improper axis of rotation (*in any accessible conformation*) cannot be optically active. Molecules without such symmetry elements can be active. Using these criteria, which of the molecules below are optically active? For those which are inactive, state which elements are present which tell you that the molecule is inactive.

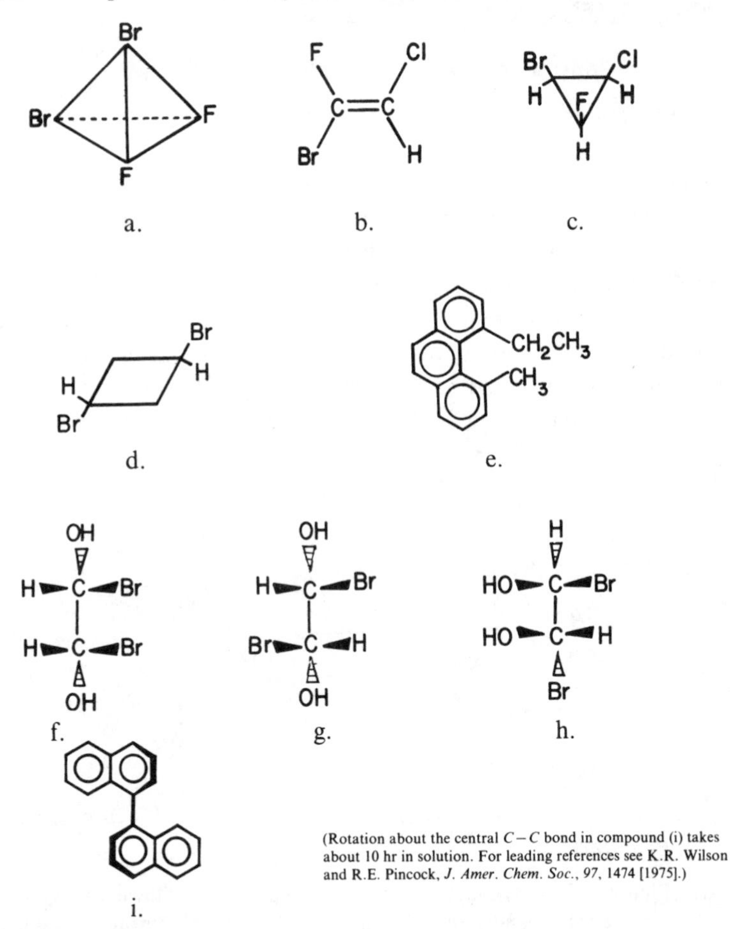

(Rotation about the central $C-C$ bond in compound (i) takes about 10 hr in solution. For leading references see K.R. Wilson and R.E. Pincock, *J. Amer. Chem. Soc.*, *97*, 1474 [1975].)

1-3. Groups

The symmetry operations which apply to any particular molecule collectively possess the properties of a mathematical group. We shall now define a group and some of its properties using the abstract operators Big Bear (B), Cable (C), Temecula (T), Arrowhead (A), Gorgonio (G), and Juliet (J). Whenever they operate on something they obey the properties of a mathematical group. Whenever Gorgonio operates on a molecule and then Temecula operates on it, the result is the same as when Cable operates alone on the same molecule. We express this by the equation 1-6.

$$T \cdot G = C \tag{1-6}$$

If Arrowhead operates on a molecule and then Juliet does, the result is the same as an operation by Temecula alone:

$$J \cdot A = T \tag{1-7}$$

This equation says that Arrowhead does his job *before* Juliet does hers. Table 1-1 governs the operations of this group. We read this table by finding the first operator in the top row and the second in the left column. Thus, the product of an operation first by Juliet and then by Temecula is the same as one by Arrowhead alone. In the table look under J and across from T and find A.

Table 1-1. Group Multiplication Table

	B	C	T	A	G	J
B	B	C	T	A	G	J
C	C	B	J	G	A	T
T	T	G	B	J	C	A
A	A	J	G	B	T	C
G	G	T	A	C	J	B
J	J	A	C	T	B	G

Every group has four properties:

1. There must exist an identity operator (E) which commutes with all other operators and leaves them unchanged. The term "commute" means that the order of multiplication doesn't matter. Big Bear is our identity operator since the product of Big Bear and any other operator is just that other operator. For example, $B \cdot C = C$. Also, $C \cdot B = C$. Although Big Bear doesn't do much, he is a necessary part of the group.

2. The product of any two operators must also be a member of the group. Thus $C \cdot T = J$, and J is indeed a member of the group. Also $J \cdot J = G$, and G is also a member of the group.

3. Multiplication is associative, which means that you may group the operations as you please, so long as you don't reverse the order of operations: That is, $A \cdot (B \cdot C) = (A \cdot B) \cdot C = J$. To work out the first product, for example, we evaluate the product in parentheses first:

$$A \cdot (B \cdot C) = A \cdot (C) = J \tag{1-8}$$

You should use Table 1-1 to satisfy yourself that the second grouping also gives J as the final result. Although multiplication must be associative, it need not be commutative. Thus $A \cdot C \neq C \cdot A$. If we had a group in which multiplication were always commutative, it would be called an *Abelian group*.

4. There must exist in the group an inverse (also called the reciprocal) for each operator. The product of an operator and its inverse is the identity operator. Any operator and its inverse must commute.

$$Z \cdot Z^{-1} = Z^{-1} \cdot Z = E \tag{1-9}$$

In our group, Gorgonio and Juliet are each other's inverse:

$$G \cdot J = J \cdot G = B \equiv E \tag{1-10}$$

Every operator has an inverse and the identity operator is always its own inverse.

A *subgroup* is just a group within a group, such as Big Bear and Cable. Their multiplication table looks like this:

	B	C
B	B	C
C	C	B

Big Bear and Cable possess the four properties which define a group. Big Bear and Temecula also form a subgroup. So do Big Bear and Arrowhead. These are all said to be subgroups of *order* two, since they contain two members. A subgroup of order three is Big Bear, Gorgonio, and Juliet.

	B	G	J
B	B	G	J
G	G	J	B
J	J	B	G

Note that Big Bear, the identity operator, is a member of all subgroups, as property (1) says he must be. Big Bear, alone, constitutes a subgroup of order one. The order of any subgroup must be an integral divisor of the order of the main group. Our group of order six can have subgroups of order $6/2 = 3$, $6/3 = 2$, or $6/6 = 1$. Although these are the only possible orders of subgroups, there need not be such subgroups for all groups of order six.

A *similarity transformation* is defined by the consecutive application of the three operations Z, X, and Z^{-1}, where X and Z are any operations.

$$Z^{-1} \cdot X \cdot Z = Y \tag{1-11}$$

Here X and Y are said to be related by a similarity transformation. They are therefore said to be *conjugate*. Gorgonio and Juliet are conjugate because

$$C^{-1} \cdot G \cdot C = J \tag{1-12}$$

To see this, first note from Table 1-1 that $C^{-1} = C$. Then

$$\begin{aligned} C^{-1} \cdot (G \cdot C) &= C \cdot (G \cdot C) \\ &= C \cdot (T) \\ &= J \end{aligned} \tag{1-13}$$

Whenever $Z^{-1} \cdot X \cdot Z = Y$, we should be able to find another operator, W, such that $W^{-1} \cdot Y \cdot W = X$. To find W, let's try some algebra. First we left-multiply each side of eq. 1-11 by Z:

$$(Z \cdot Z^{-1}) \cdot X \cdot Z = Z \cdot Y \tag{1-14}$$

Then we right-multiply each side of eq. 1-14 by Z^{-1}:

$$(Z \cdot Z^{-1}) \cdot X \cdot (Z \cdot Z^{-1}) = Z \cdot Y \cdot Z^{-1} \tag{1-15}$$

or

$$X = Z \cdot Y \cdot Z^{-1} \tag{1-16}$$

So the operator W is just equal to Z^{-1}. We had to keep track of the order of mutiplication (right- or left-multiplication) because the operators do not necessarily commute.

We are now in a position to define a *class*. A class is a complete set of operators which are conjugate to one another. Gorgonio and Juliet form a class because they are conjugate to each other and to no other operators in the group. Cable, Temecula, and Arrowhead also form a class, as any similarity transformation performed on C, T, or A only generates a member of the class C, T, and A. For example,

$$\begin{aligned} J^{-1} \cdot C \cdot J &= G \cdot (C \cdot J) = G \cdot T = A \\ J^{-1} \cdot T \cdot J &= G \cdot (T \cdot J) = G \cdot A = C \\ J^{-1} \cdot A \cdot J &= G \cdot (A \cdot J) = G \cdot C = T \\ &\vdots \end{aligned} \tag{1-17}$$

The identity operator is always in a class by itself. As with subgroups, the order of a class must be an integral divisor of the order of the group, but not all integral divisors must exist as classes.

Problems

1-3. In the group below, what is the identity operator? Show that the following multiplications are associative: $J \cdot L \cdot M$; $K \cdot H \cdot P$. What is the inverse of O? Of K? What are the possible orders of the subgroups of this group? Find subgroups of order two and four. Find the products $H \cdot K \cdot H^{-1}$, $L \cdot O \cdot L^{-1}$, and $P^{-1} \cdot P \cdot P$. Show that H and P are in the same class.

	G	*K*	*H*	*J*	*L*	*M*	*O*	*P*
G	*G*	*K*	*H*	*J*	*L*	*M*	*O*	*P*
K	*K*	*M*	*J*	*P*	*G*	*L*	*H*	*O*
H	*H*	*O*	*G*	*L*	*J*	*P*	*K*	*M*
J	*J*	*H*	*K*	*G*	*P*	*O*	*M*	*L*
L	*L*	*G*	*O*	*H*	*M*	*K*	*P*	*J*
M	*M*	*L*	*P*	*O*	*K*	*G*	*J*	*H*
O	*O*	*P*	*L*	*M*	*H*	*J*	*G*	*K*
P	*P*	*J*	*M*	*K*	*O*	*H*	*L*	*G*

1-4. Point Groups

It is possible to assemble symmetry operations into groups that obey multiplication tables just as Big Bear and his friends did. These groups satisfy all the properties of mathematical groups and are called *point groups.* It will be possible to assign any molecule to one of these groups depending on what symmetry elements are present. Point groups are so named because at least one point in space is invariant (unchanged) to all operations in the group. The *space groups* used in crystallography contain translation operations which we will not treat in this book.

Point group C_1. This is the trivial group which contains all molecules having no symmetry. Such a molecule is HNClF in Fig. 1-9. There is no center of inversion, no plane of reflection, and no proper or improper axes of rotation. The only operation we can perform on HNClF is E, the identity operation.

Point group C_s. To this point group belong molecules whose only symmetry element is a plane of reflection. The very unstable formyl chloride,

Fig. 1-9. HNClF belongs to the point group C_1, since it has no symmetry.

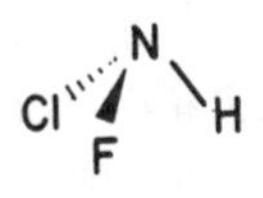

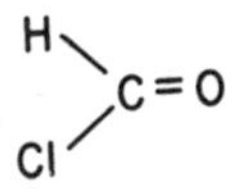

Fig. 1-10. Formyl chloride belongs to the point group C_s.

HCOCl (Fig. 1-10), possesses only its molecular plane as a symmetry element. This σ operation doesn't even interchange any of the atoms. The reflection plane in F_2SO (Fig. 1-3), which also belongs to the point group C_s, would interchange the two fluorine atoms.

Point group C_i. Molecules possessing only a center of inversion belong to the point group C_i. In the particular rotamer (rotational isomer) of 1,2-dichloro-1,2-difluoroethane shown in Fig. 1-11, there is only a center of inversion.

Point group C_n. A molecule in the point group C_n possesses only an n-fold axis of rotation. A molecule of boric acid with the three hydrogen atoms bent up above the BO_3 plane, as shown in Fig. 1-12, has C_3 symmetry.

Before proceeding to the more complicated point groups, we ought to see that the symmetry operations applicable to a molecule such as $B(OH)_3$ do indeed form a mathematical group. To form a group, the operations must satisfy four requirements. First, there must exist the identity operation,

Fig. 1-11. A rotamer of 1,2-dichloro-1,2-difluoroethane having a center of inversion as its only symmetry element.

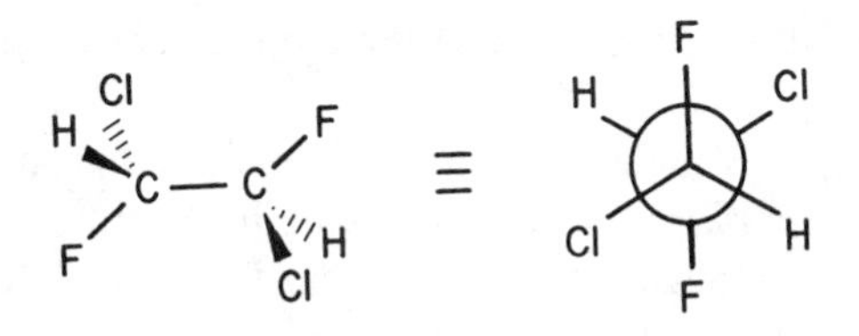

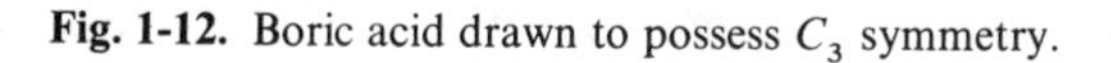
Fig. 1-12. Boric acid drawn to possess C_3 symmetry.

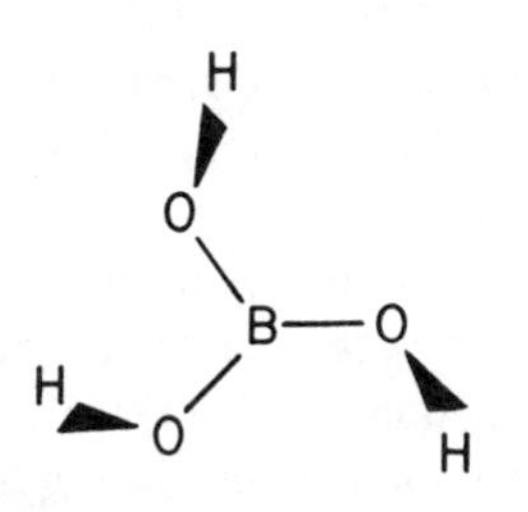

and this operation must commute with all other operations. Second, the product of any two operations in the group must also be a member of the group. What operations are generated by E and C_3?

$$\begin{aligned} E \cdot E &= E \\ E \cdot C_3 &= C_3 \\ C_3 \cdot C_3 &= C_3^2 \end{aligned} \tag{1-18}$$

The product of C_3 times itself gives C_3^2; so this must also be a member of the group. Does C_3^2 generate any more operations?

$$\begin{aligned} E \cdot C_3^2 &= C_3^2 \\ C_3^2 \cdot C_3 &= E \\ C_3^2 \cdot C_3^2 &= C_3^4 = C_3 \end{aligned} \tag{1-19}$$

No it does not. Let's write out a multiplication table for this group:

C_3	E	C_3	C_3^2
E	E	C_3	C_3^2
C_3	C_3	C_3^2	E
C_3^2	C_3^2	E	C_3

The third property is association. Is the product $C_3 \cdot (C_3 \cdot C_3^2)$ the same as $(C_3 \cdot C_3) \cdot C_3^2$? Yes it is. The first product is $C_3 \cdot (C_3 \cdot C_3^2) = C_3 \cdot (C_3^3) = C_3 \cdot E = C_3$. The second one is $(C_3 \cdot C_3) \cdot C_3^2 = (C_3^2) \cdot C_3^2 = C_3$. The fourth property requires an inverse for each operation. The inverse of E is E; of C_3 is C_3^2; of C_3^2 is C_3. These three operations do indeed form a mathematical group.

Point groups C_{nv}. A molecule in one of these point groups possesses a C_n axis and n *vertical* mirror planes which are, by definition, colinear with the C_n axis. H_2O belongs to the point group C_{2v} (Fig. 1-13). It has one twofold axis of rotation and two vertical mirror planes, σ_v and σ_v'. We distinguish these two planes by writing a prime on one of them, and where possible, draw the molecule to be contained in the σ_v plane. If you find a C_n axis and any σ_v plane in a molecule, you are guaranteed that you will find n σ_v planes if you look hard enough. This assertion will be justified shortly.

At this point it will be advantageous to introduce the notion of a stereographic projection, without bothering to go into the derivation of such projections. We start by defining a "working area" with a dashed circle, and marking some "general point" with a little circle, as in Fig. 1-14. Dashed

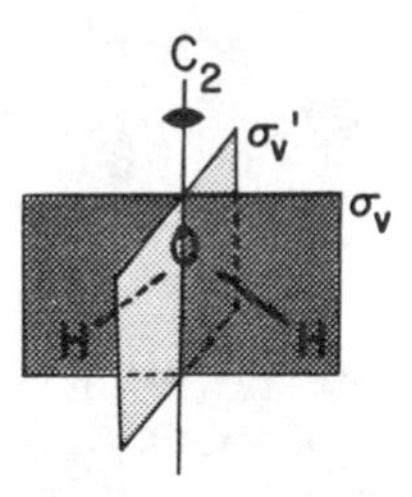

Fig. 1-13. H_2O belongs to the point group C_{2v}.

lines may be used anywhere in the diagram for our own reference, but will not represent symmetry elements. We then use the set of symbols in Table 1-2 to mark symmetry elements in the working area. To generate the point group C_{2v}, we will first apply a C_2 operation to the general point, as shown in Fig. 1-15 (a). The C_2 operation generates the point 2 from the point 1. The vertical dashed line is just used as a reference for our 180° rotation. But now let's introduce a σ_v plane colinear with the vertical dashed line, as shown at the left of Fig. 1-15 (b). Reflection of points 1 and 2 across the mirror plane generates the points 1′ and 2′, respectively. But points 1′ and 2′ could just as well have been generated from points 2 and 1, respectively, by reflection across the plane labelled σ_v' at the right of Fig. 1-15 (b). The

Table 1-2. Symbols for Stereographic Projections

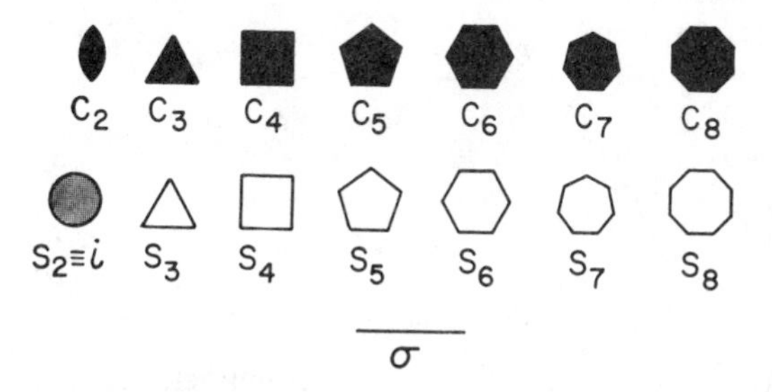

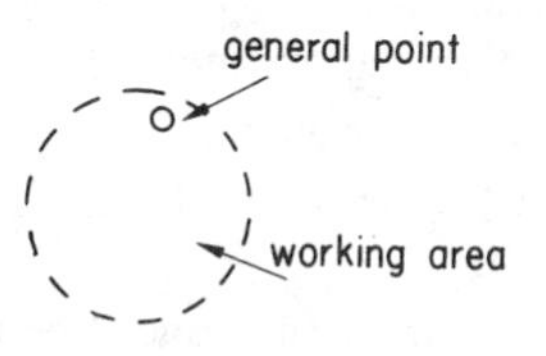

Fig. 1-14. The raw material for a stereographic projection.

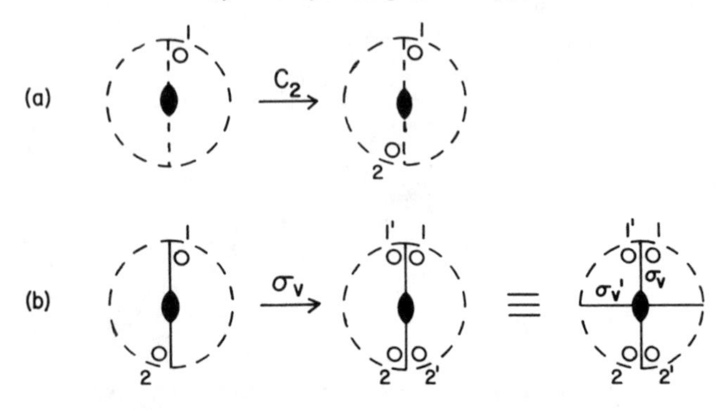

Fig. 1-15. Generation of the point group C_{2v}.

presence of the C_2 axis and one σ_v plane therefore implies that a second σ_v plane must exist perpendicular to the first one. In a similar manner, one can shown that a C_n axis and one σ_v plane guarantee that there must be a total of n σ_v planes present. The C_n axis and one σ_v plane are therefore called "generating elements" for the point groups C_{nv}, since they generate the remaining elements.

Now let's generate a group multiplication table for the point group C_{2v} with the aid of stereographic projections. The four operations of the group are E, C_2, σ_v, and σ_v' and we know that E commutes with all operations and leaves them unchanged. This gives us the first row and column of the multiplication table:

C_{2v}	E	C_2	σ_v	σ_v'
E	E	C_2	σ_v	σ_v'
C_2	C_2			
σ_v	σ_v			
σ_v'	σ_v'			

In this particular group, all of the diagonal elements are E, because $C_2 \cdot C_2 = \sigma_v \cdot \sigma_v = \sigma_v' \cdot \sigma_v' = E$. We demonstrate this for the case $\sigma_v' \cdot \sigma_v'$ in Fig. 1-16 (a). We say that $\sigma_v' \cdot \sigma_v' = E$ because the net result of these two operations is just what would have resulted from the operation E alone, namely nothing. The off diagonal elements of the table can be worked out similarly; two examples are given in Fig. 1-16 (b) and (c). The final result is Table 1-3. You might try to verify that $\sigma_v' \cdot C_2 = \sigma_v$ to be sure that you understand what we just did.

A point of nomenclature can be introduced before we leave the C_{nv} point groups. The group C_{4v}, exemplified by the molecule $BrMn(CO)_5$ (Fig. 1-17)

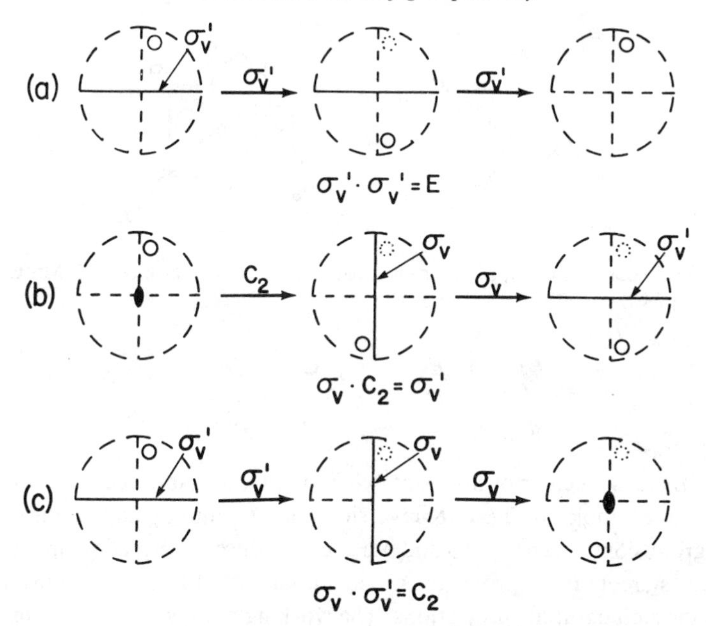

Fig. 1-16. Operator products for the point group C_{2v} worked out with stereographic projections. The small dotted circle is drawn for reference only.

contains two kinds of vertical mirror planes. One kind goes through the central atoms and the carbonyl groups and is still designated σ_v. The other kind passes through the central atoms but bisects the equatorial C-Mn-C angles. It is called a σ_d plane, the "*d*" standing for "dihedral".

Point groups C_{nh}. Such a point group is generated by a C_n axis and a *horizontal* mirror plane, σ_h. By definition, a horizontal mirror plane is one which is perpendicular to the axis of rotation. Butadiene, in its *trans*-planar conformation (Fig. 1-18) belongs to the group C_{2h}. The σ_h plane is the plane of the molecule. The C_2 axis bisects the central C-C bond and is perpendicular to the molecular plane. If these are all the symmetry elements present, their operations ought to form a complete set. By examining the six products in eq. 1-20, however, we discover that we have overlooked the operation i.

Table 1-3. Group Multiplication Table for the Point Group C_{2v}

C_{2v}	E	C_2	σ_v	σ_v'
E	E	C_2	σ_v	σ_v'
C_2	C_2	E	σ_v'	σ_v
σ_v	σ_v	σ_v'	E	C_2
σ_v'	σ_v'	σ_v	C_2	E

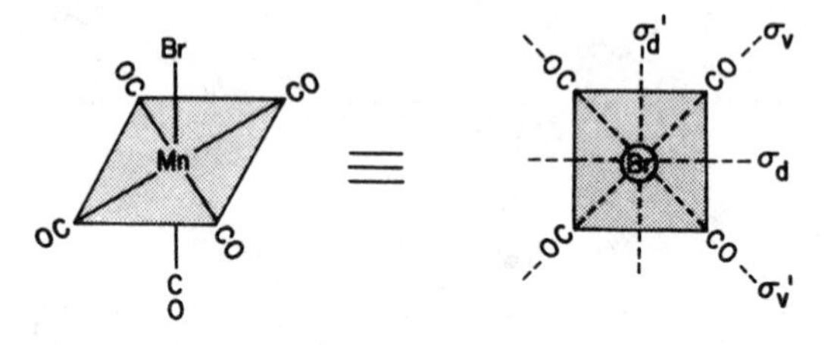

Fig. 1-17. By convention, σ_v planes pass through the carbonyl groups of $BrMn(CO)_5$ and σ_d planes bisect C−Mn−C angles.

$$\begin{aligned} E\cdot E &= E & C_2\cdot C_2 &= E \\ E\cdot C_2 &= C_2 & C_2\cdot\sigma_h &= S_2 = i \\ E\cdot\sigma_h &= \sigma_h & \sigma_h\cdot\sigma_h &= E \end{aligned} \tag{1-20}$$

Indeed there is a center of inversion where the C_2 axis intersects the mirror plane. For example, in Fig. 1-18, inversion takes C_1 into C_4 and H_a into H_f. Is the group complete now? To find out, we introduce the inversion operation into our stereographic projections, and we have to disclose one more fact about the stereographic projections. The working point we have been using, the little circle, represents a point *above* the plane of the page. Inversion will take the point from above the plane of the page, move it through the center of the working area, and push it down below the plane of the page. Such a point below the plane of the page will be designated by the letter "x". So inversion accomplishes the operation in Fig. 1-19. Finally, Fig. 1-20 shows that the products $E\cdot i$, $C_2\cdot i$, $\sigma_h\cdot i$, and $i\cdot i$ are all members of the group; so the group is now complete. The final set of multiplication rules for the point group C_{2h} is given in Table 1-4.

Point groups D_n. A D_n point group is generated by a C_n axis and a C_2 axis perpendicular to the C_n axis. A cation of idealized D_3 symmetry is $Co(H_2NCH_2CH_2NH_2)_3{}^{3+}$. The $-CH_2CH_2-$ groups of the ethylenediamine ligands are relatively free to move and we represent them by curved

Fig. 1-18. *Trans*-planar butadiene belongs to the point group C_{2h}.

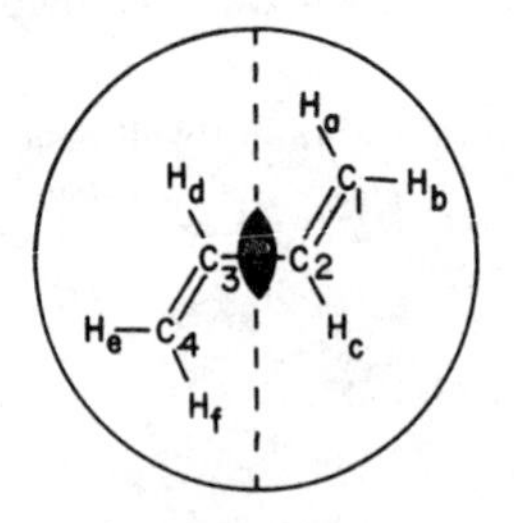

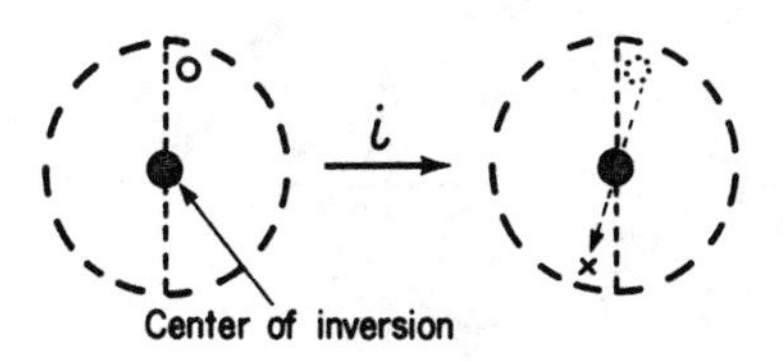

Fig. 1-19. Inversion takes a point above the page (*o*) to one below the page (*x*).

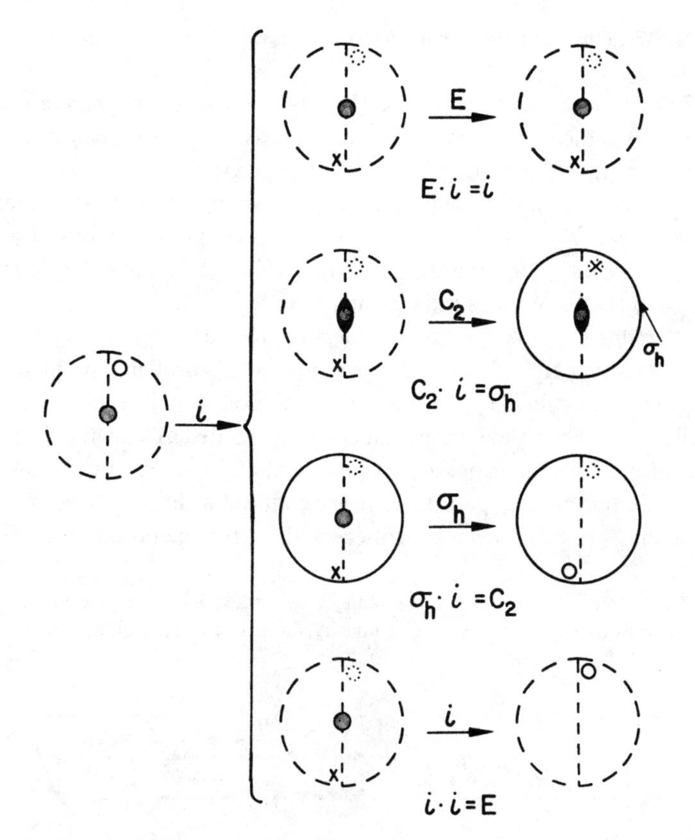

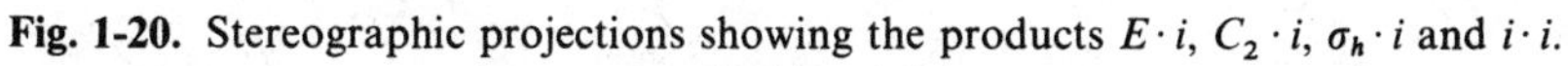

Fig. 1-20. Stereographic projections showing the products $E \cdot i$, $C_2 \cdot i$, $\sigma_h \cdot i$ and $i \cdot i$.

Table 1-4. Group Multiplication Table for the Point Group C_{2h}

C_{2h}	E	C_2	σ_h	i
E	E	C_2	σ_h	i
C_2	C_2	E	i	σ_h
σ_h	σ_h	i	E	C_2
i	i	σ_h	C_2	E

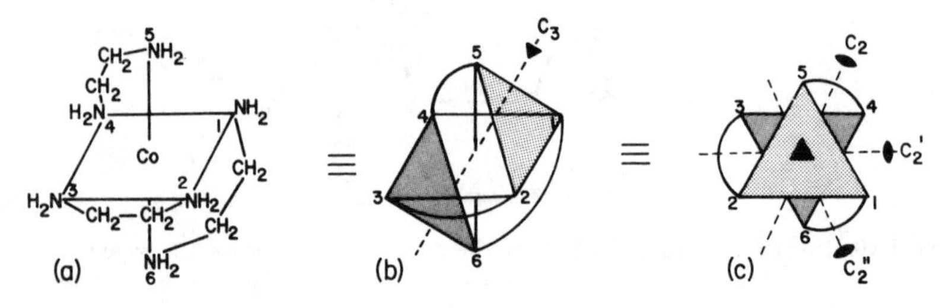

Fig. 1-21. Tris(ethylenediamine)cobalt(III) possesses idealized D_3 symmetry.

lines connecting the nitrogen atoms (Fig. 1-21). Notice that there are *three* C_2 axes perpendicular to the C_3 axis. A C_n axis and one perpendicular C_2 axis will generate the other $(n - 1)C_2$ axes perpendicular to the C_n axis. To be sure that you understand Fig. 1-21, you should see that, for example, C_3 takes N_5 to N_1 and C_2' takes N_5 to N_6. As a point of nomenclature, the axis of highest order in a point group is called the *principal axis*. In the point group D_3, the C_3 axis is the principal axis.

Point groups D_{nd}. As you might guess by now, a D_{nd} group is generated by a C_n axis, a perpendicular C_2 axis, and a dihedral mirror plane, σ_d. The σ_d plane is colinear with the principal axis, but bisects $\perp C_2$ axes. The full D_{4d} stereographic projection can be generated as shown in Fig. 1-22 by the successive operations C_4, C_4, C_4, $\perp C_2$, and σ_d. In the finished diagram, we find that σ_d planes are never colinear with $\perp C_2$ axes. We also find that an S_8 axis has been generated by the other operations.

Fig. 1-22. The full D_{4d} representation can be generated by successive applications of the operations C_4, C_4, C_4, $\perp C_2$, and σ_d. Dihedral mirror planes bisect $\perp C_2$ axes in the D_{4d} point group.

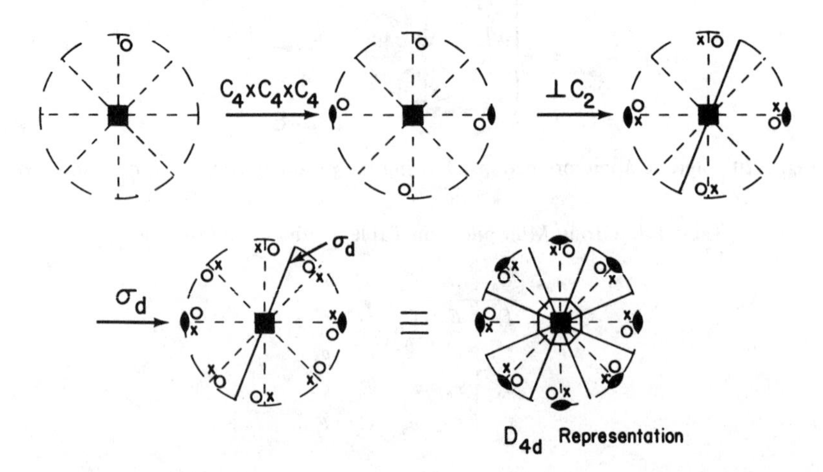

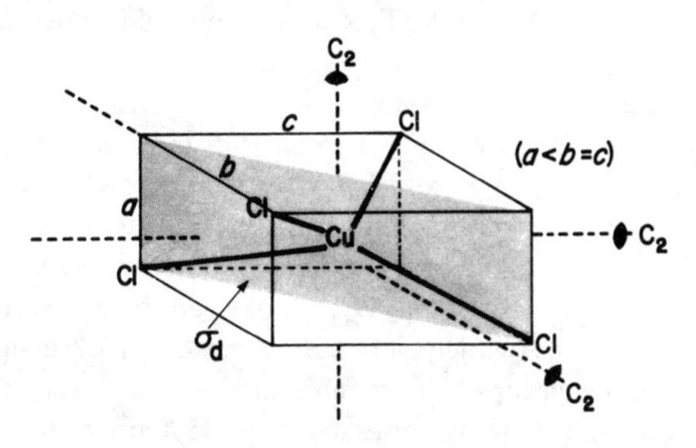

Fig. 1-23. In crystals of Cs_2CuCl_4, the anion has a squashed tetrahedral D_{2d} structure.

In the special case D_{2d}, of which the squashed tetrahedral structure in Fig. 1-23 is an example, there are three C_2 axes. One of the two σ_d planes is shown in this figure. Can you find the second σ_d plane and can you decide which of the C_2 axes is unique?

Point groups D_{nh}. The symbol "D_{nh}" tells us that we need a C_n axis, n $\perp C_2$ axes, and a horizontal mirror plane which, by definition, is *perpendicular to the principal axis.* These symmetry elements will then generate n vertical mirror planes. The planar molecule *trans*-$PtCl_2Br_2{}^{2-}$ in Fig. 1-24 has D_{2h} symmetry. In groups of higher order, the principal axis is always designated as the z axis. In the group D_{2h}, the choice of axes is arbitrary.

Benzene has D_{6h} symmetry. The six $\perp C_2$ axes fall into two classes, designated C_2' and C_2''. The six vertical mirror planes also fall into two classes, designated σ_v and σ_d. By convention, we will always define σ_v and C_2' to be colinear, and σ_d and C_2'' to be colinear. Further, σ_v and C_2' will pass through as many atoms as possible, whereas σ_d and C_2'' will pass

Fig. 1-24. *trans*-$PtCl_2Br_2{}^{2-}$ is planar and possesses D_{2h} symmetry.

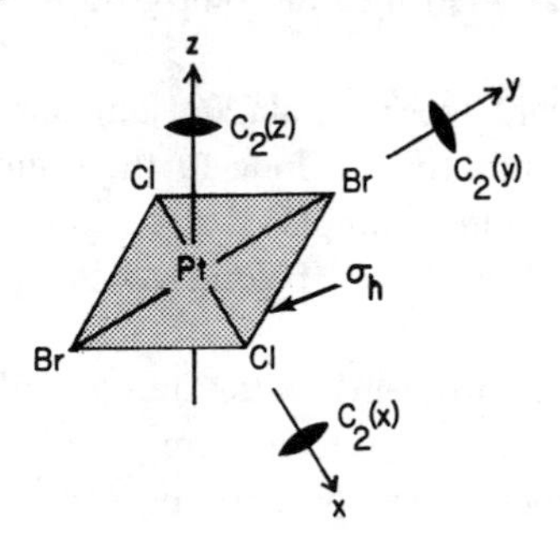

through as few atoms as possible. This convention is illustrated for benzene in Fig. 1-25.

Point groups S_n. A point group of this type is generated by an S_n axis. The substituted spirononane in Fig. 1-26 has *only* the elements generated by S_4, *viz.*, S_4, $S_4^2 = C_2$, S_4^3, and $S_4^4 = E$. The S_2 point group is really just C_i since $S_2 = i$. Therefore, we do not designate anything S_2. When n is *odd*, the S_n point groups are just the same as the C_{nh} point groups and are designated C_{nh}. Thus only S_4, S_6, S_8, ... have a separate existence.

Special point groups. A variety of molecules can be rapidly assigned to one of the special point groups. Linear molecules with a center of inversion, such as homonuclear diatomic molecules (*e.g.*, H_2) are in the point group $D_{\infty h}$. These molecules have a C axis of infinite order and an infinite number of perpendicular C_2 axes, one of which is shown in Fig. 1-27. They also have an infinite number of vertical mirror planes. The center of the molecule is an inversion center, and the C_∞ axis is also an S_∞ axis. Linear molecules lacking a center of inversion, such as heteronuclear diatomic molecules, belong to the point group $C_{\infty v}$. An example would be HCl.

Tetrahedral molecules belong to the point group T_d. There are 24 operations in this group, making use of the symmetry elements E, C_3, C_2, S_4, and σ_d. Some of these symmetry elements are illustrated in Fig. 1-28. Note that there is no center of inversion in this point group.

The anion $Cu(NO_2)_6^{4-}$ in the salt $K_2PbCu(NO_2)_6$ has T_h symmetry (Fig. 1-29). The cube with bars drawn on its face in this figure also has T_h symmetry. Four C_3 axes pass through opposite corners and three C_2 axes bisect opposite faces. Colinear with the C_3 axes are S_6 axes. Also present are three mirror planes and a center of inversion.

Octahedral molecules, such as AlF_6^{3-} (Fig. 1-30) belong to the point group O_h. This group contains C_4, C_3, C_2, S_6, and S_4 axes, as well as σ_h and σ_d mirror planes, and a center of inversion. Subgroups of O_h and T_h possessing only the proper rotations are called O and T, but are not common. All of the octahedral and tetrahedral point groups are also called the *cubic groups*.

The rare molecules with icosahedral (20 triangular faces) or dodecahedral (12 pentagonal faces) structures belong to the point group I_h (Fig. 1-31). The $B_{12}H_{12}^{2-}$ anion is an example of the regular icosahedral structure. Atoms, with *spherical symmetry* belong to the point group K_h whose properties we will explore later.

To help you work problems with stereographic projections, some representations of selected point groups are given in Fig. 1-32. A summary of some of the properties of the point groups is given in Table 1-5.

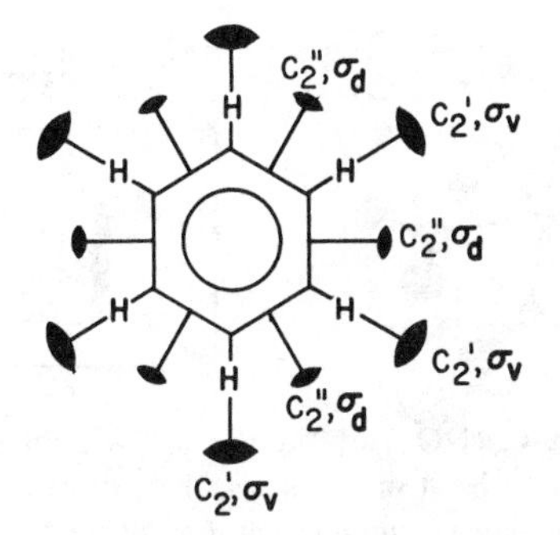

Fig. 1-25. Designation of C_2', C_2'', σ_v and σ_d symmetry elements of benzene.

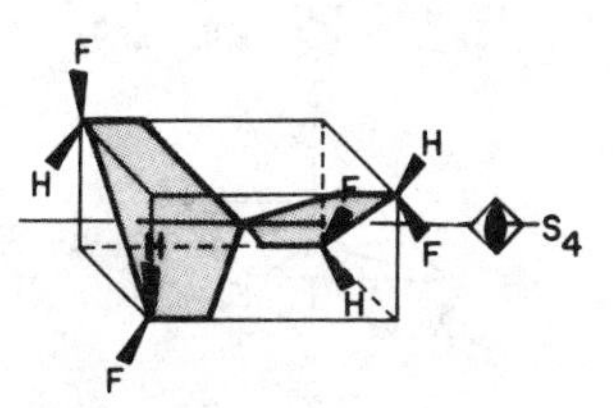

Fig. 1-26. A tetrafluorospirononane having S_4 symmetry.

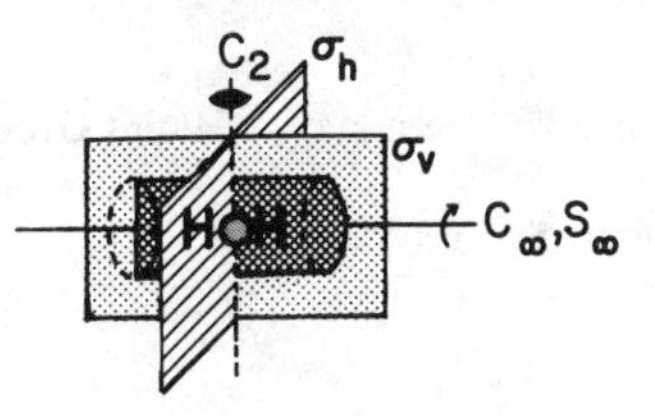

Fig. 1-27. H_2 has $D_{\infty h}$ symmetry. The notation "C_∞" means that a rotation through any arbitrary angle about this axis leaves the molecule unchanged.

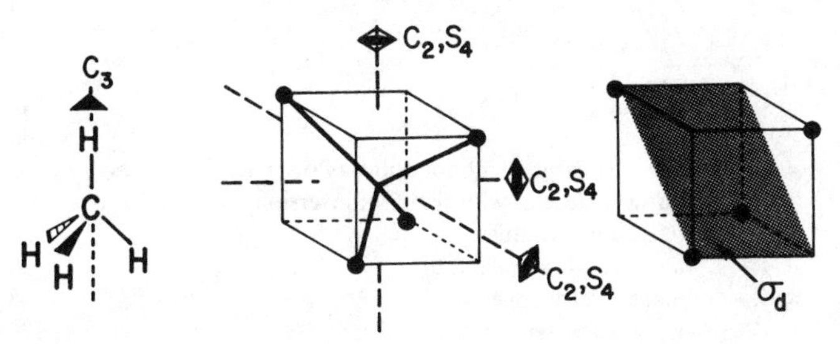

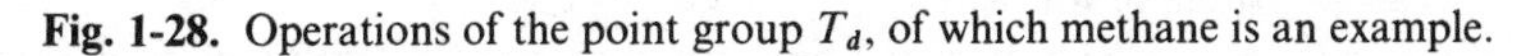
Fig. 1-28. Operations of the point group T_d, of which methane is an example.

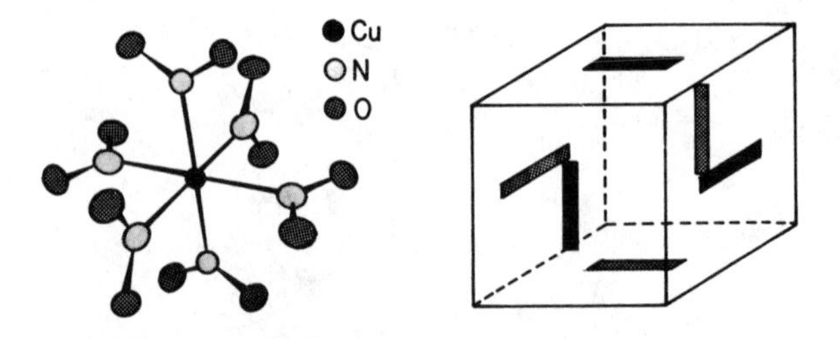

Fig. 1-29. In the anion $Cu(NO_2)_6^{4-}$ in the salt $K_2PbCu(NO_2)_6$, all six Cu-N distances are 2.11 Å. The cube at the right with bars drawn on its faces (solid bars on the front faces) has the same symmetry as $Cu(NO_2)_6^{4-}$, which is T_h. Structure redrawn from D.L. Cullen and E.C. Lingafelter, *Inorg. Chem., 10*, 1264 (1971).

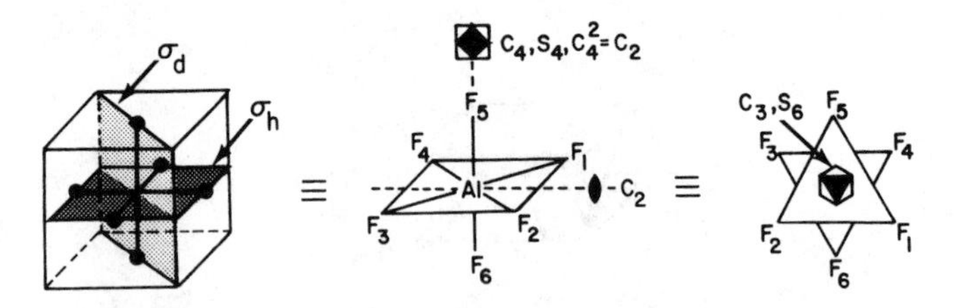

Fig. 1-30. AlF_6^{3-} has O_h (octahedral) symmetry, which includes four C_3 axes, three C_4 axes, three C_4^2 $(=C_2)$ axes, six C_2 axes, four S_6 axes, three S_4 axes, three σ_h planes, six σ_d planes, and a center of inversion.

Table 1-5. Summary of Point Groups

Point Group	Important Symmetry Elements	Order of the Group
C_1	E	1
C_i	i	2
C_s	σ	2
C_n	C_n	n
S_n†	S_n	n
C_{nv}	C_n, σ_v	$2n$
C_{nh}	C_n, σ_h	$2n$
D_n	$C_n, \perp C_2$	$2n$
D_{nd}	$C_n, \perp C_2, \sigma_d$	$4n$
D_{nh}	$C_n, \perp C_2, \sigma_h$	$4n$
$C_{\infty v}$	linear molecules without center of inversion	∞
$D_{\infty h}$	linear molecules with center of inversion	∞
T_d	tetrahedral symmetry	24
T_h	tetrahedral symmetry, σ_h	24
O_h	octahedral symmetry	48
I_h	icosahedral symmetry	120
K_h	spherical symmetry	∞

† n must be even, or else $S_n = C_{nh}$.

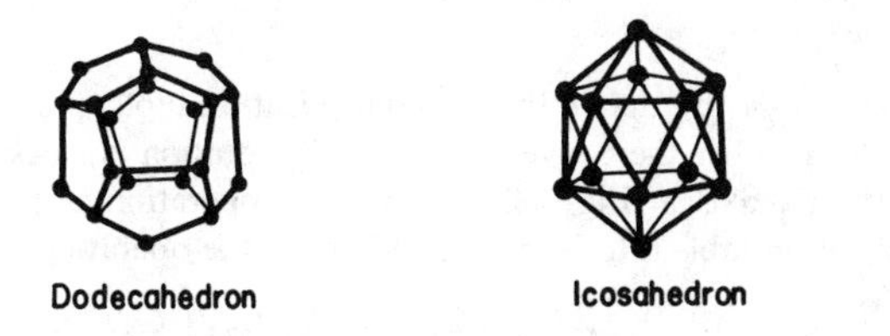

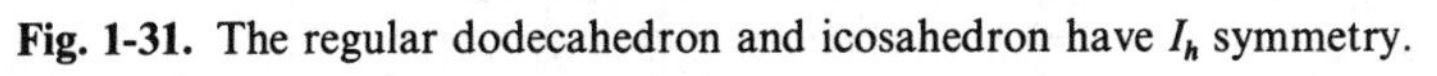

Fig. 1-31. The regular dodecahedron and icosahedron have I_h symmetry.

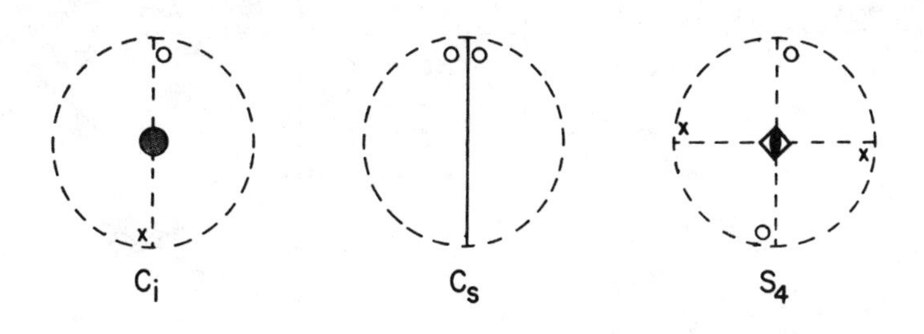

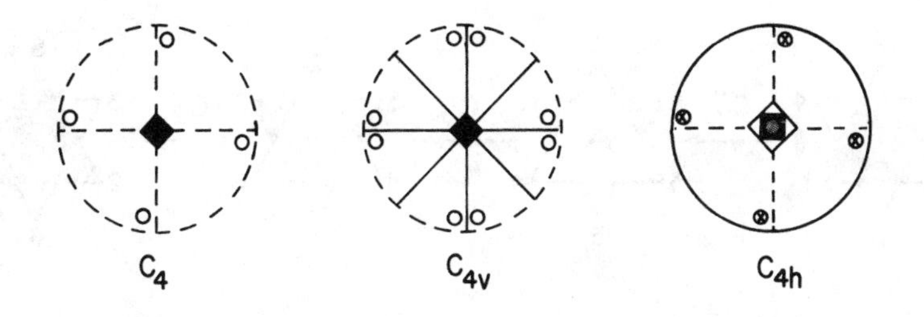

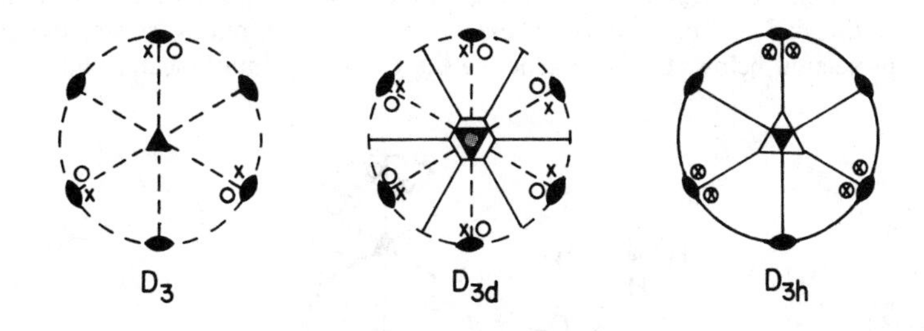

Fig. 1-32. A selection of stereographic projections.

Problems

1-4. Using Fig. 1-21 (c), derive the D_3 multiplication table, part of which is given below. Remember the conventions that C_3 rotation is *clockwise* as you look down the C_3 axis in Fig. 1-21 (c), and the operation in the top row of the multiplication table is to be performed *before* the operation in the left column.

D_3	E	C_3	C_3^2	C_2	C_2'	C_2''
E	E					C_2''
C_3		C_3^2			C_2''	
C_3^2			C_3	C_2''		
C_2			C_2'	E		
C_2'						
C_2''	C_2''					E

You need only manipulate the numbers of the ligands to get the products. For example, we demonstrate below how to determine the product $C_3 \cdot C_2' = C_2''$.

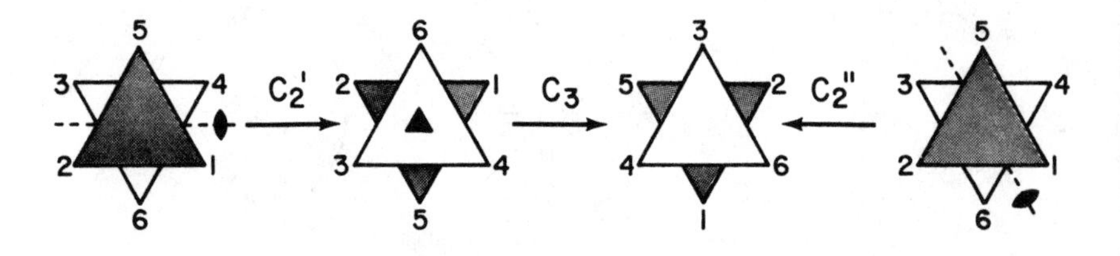

Determine subgroups of order one, two, and three. Divide the operations of D_3 into classes using a few choice similarity transformations.

1-5. Ammonia belongs to the point group C_{3v} which includes the operations E, C_3, C_3^2, σ_v, σ_v', and σ_v''. These operations are set out in the stereographic projection below. Can you generate the C_{3v} group multiplication table?

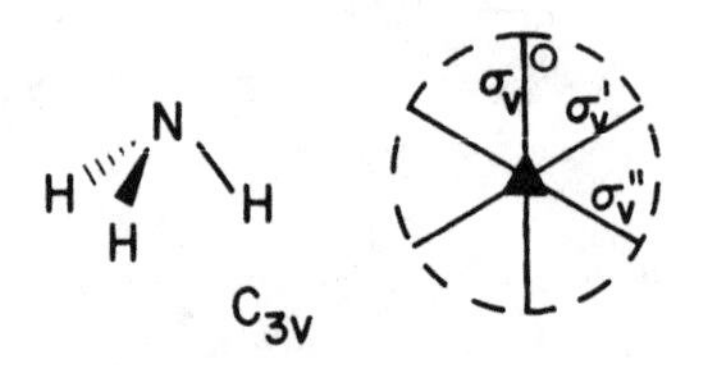

1-6. Starting with the stereographic projection below, show that a C_5 axis and a perpendicular C_2 axis will generate four more C_2 axes perpendicular to the C_5 axis. What point group do these operations define?

1-7. Using stereographic projections, or the drawing of $PtCl_2Br_2^{2-}$ in Fig. 1-24, generate a group multiplication table for the point group D_{2h} which contains the symmetry elements E, $C_2(z)$, $C_2(x)$, $C_2(y)$, i, $\sigma(xy)$, $\sigma(xz)$, and $\sigma(yz)$.

1-8. Determine which point groups are generated by the symmetry elements shown in the partial stereographic projections below by completing each figure.

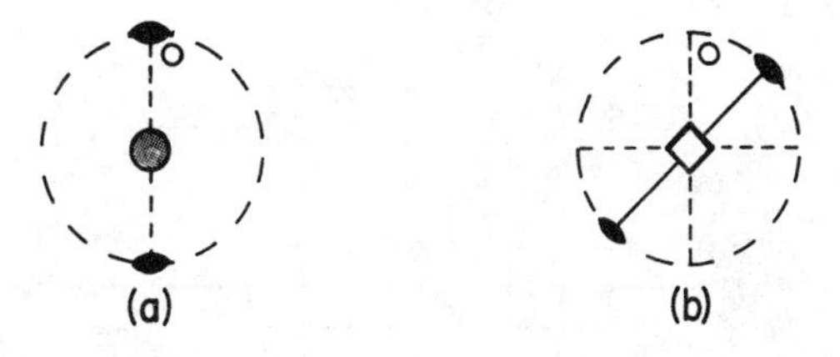

1-9. Using a stereographic projection, verify that the S_5 point group is just one of the point groups which we call by a different name. Which point group is it?

1-5. Classification of Molecules into Point Groups

There exists a systematic procedure for determining the point group of any molecule. The success of this procedure depends on your ability to find symmetry elements present in a molecule, a skill which can only be developed by continued practice. In Fig. 1-33 there is a flow chart which will lead us through the decision making process necessary for the proper assignment of a molecule to a point group. In each box is a question which must be answered YES or NO. The answer to each question leads us to the next question until a point group is finally determined. The first question, "Does the molecule belong to one of the special point groups?" may be challenging for highly symmetric species. The presence of each and every

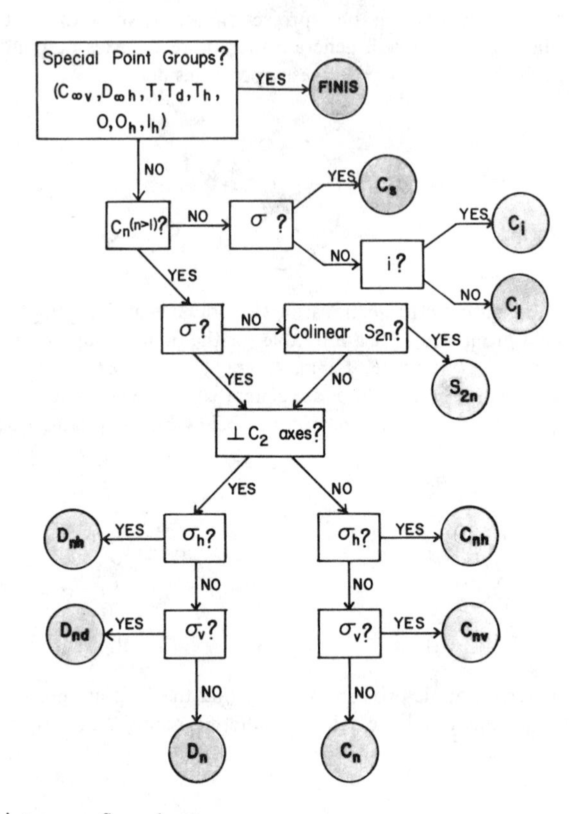

Fig. 1-33. Point group flow chart.

symmetry element should be verified before you can be certain of the point group of such a symmetric species as $Cu(NO_2)_6^{4-}$ with T_h symmetry in Fig. 1-29.

Let's work through some examples:

A. 1. *Trans*-1-bromo-2-chloroethylene (Fig. 1-34) does not belong to the linear, cubic, or icosahedral point groups.
   2. The molecule doesn't have any proper axes of rotation.
   3. The plane of the molecule is a mirror plane, so the point group is C_s.

B. 1. Dichloromethane (Fig. 1-35) does not belong to any of the special point groups.
   2. There is a C_2 axis passing through the carbon atom and bisecting the Cl–C–Cl and H–C–H angles.
   3. There are planes of symmetry present.

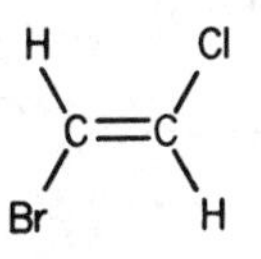

Fig. 1-34. *Trans*-1-bromo-2-chloroethylene.

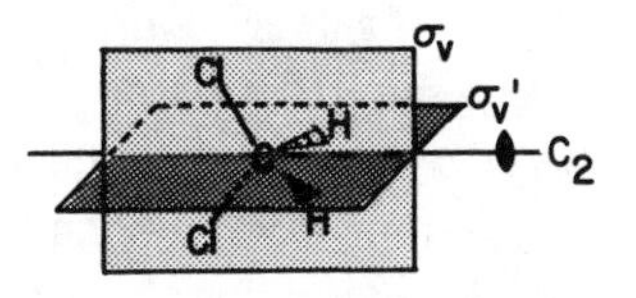

Fig. 1-35. Dichloromethane.

4. There are no C_2 axes perpendicular to the one we already found; so the molecule does not belong to one of the D point groups.
5. There is no horizontal mirror plane.
6. Colinear with the C_2 axis are two mirror planes. One contains the two Cl atoms and the C atom and the other contains the H atoms and the C atom. So the point group is C_{2v}.

C. 1. PF_5 (Fig. 1-36) does not belong to one of the special point groups.
2. There is a C_3 axis passing through the two axial F atoms and a C_2 axis passing through an equatorial F atom and the P atom.
3. There are planes of symmetry.
4. A C_2 axis perpendicular to the C_3 axis guarantees that there are three $\perp C_2$ axes and that the molecule belongs to a D point group.
5. The equatorial plane containing three F atoms and the P atom is a horizontal mirror plane. The point group must be D_{3h}.

Fig. 1-36. Phosphorus pentafluoride.

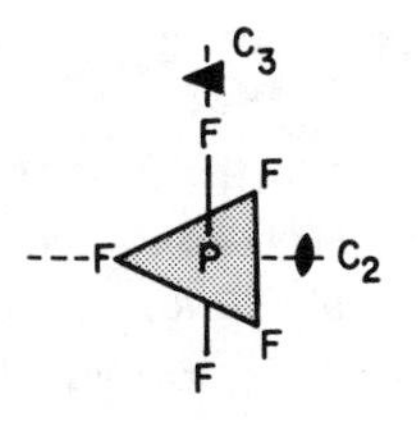

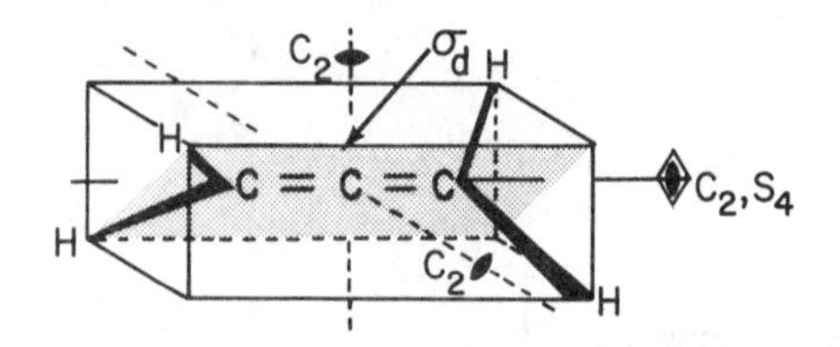

Fig. 1-37. Allene.

D. 1. Allene (Fig. 1-37) does not belong to any of the special point groups.
2. There are three C_2 axes.
3. There are two vertical mirror planes colinear with one of the C_2 axes. One is shown in Fig. 1-37 and the other includes the three C atoms and the two H atoms not in the plane shown in the drawing.
4. Three perpendicular C_2 axes must put us in the D point groups.
5. There is no horizontal mirror plane (which, by definition, would be perpendicular to one of the C_2 axes).
6. The vertical mirror planes already noted must put us in the group D_{2d}.

If you had missed the two perpendicular C_2 axes of allene, you might have concluded that the point group was C_{2v}. The S_4 axis should tell you that something is wrong, though, since the C_{2v} point group doesn't have an S_4 operation. This should lead you to look for more symmetry in the molecule. There is no substitute for experience in finding all of the symmetry elements present in a molecule.

E. 1. *Fac*-trichlorotricyanoferrate(III) (Fig. 1-38) does not belong to one of the special point groups.
2. When viewed properly, a C_3 axis is evident. It passes through the Fe atom and is perpendicular to the plane of the page in Fig. 1-38(b).
3. There are a few mirror planes present.
4. The absence of a $\perp C_2$ axis eliminates the D point groups.
5. No σ_h planes can be found perpendicular to the C_3 (principal) axis.
6. The σ_v planes already noted make the point group C_{3v}.

F. 1. In solution, octacyanomolybdate(IV) appears to have the square antiprismatic structure shown in Fig. 1-39, and does not belong to one of the special groups.
2. A C_4 axis perpendicular to the page passes through the Mo atom.
3. Several planes of symmetry are evident.
4. Four C_2 axes perpendicular to the principal axis are evident.
5. There is no horizontal mirror plane.
6. The presence of vertical mirror planes establishes the point group as D_{4d}.

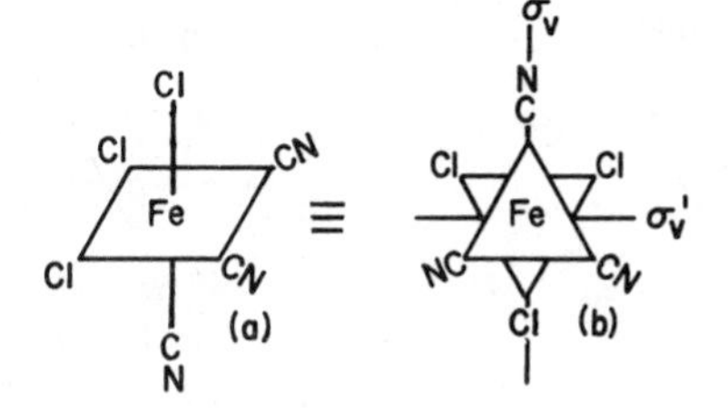

Fig. 1-38. *Mer*-trichlorotricyanoferrate(III).

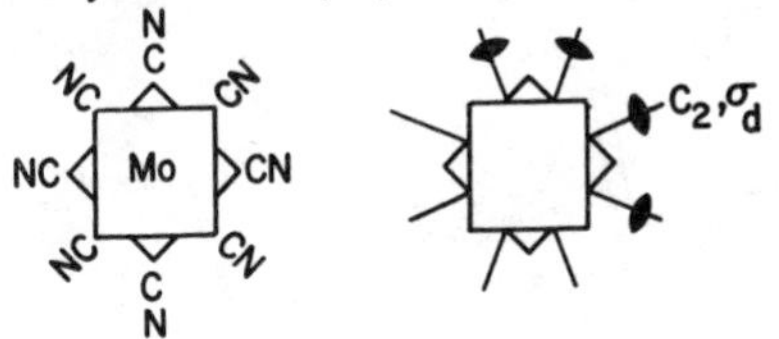

Fig. 1-39. Octacyanomolybdate(IV).

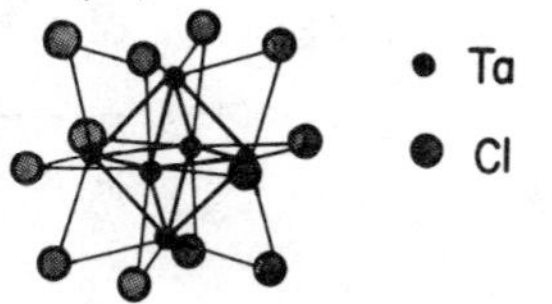

Fig. 1-40. $Ta_6Cl_{12}^{2+}$. (Redrawn from L. Pauling, *The Chemical Bond*, Cornell University Press, Ithaca, New York, 1967.)

G. 1. $Ta_6Cl_{12}^{2+}$ looks like it has an awful lot of symmetry (Fig. 1-40). The Ta atoms are at the vertices of a regular octahedron and the Cl atoms are distributed very symmetrically. One can verify the presence of each of the elements of the point group O_h.

Problems

1-10. Assign each molecule below to the proper point group.

a. O=C=C=C=O (linear) b. HF

c. IF_7 d. XeO_2F_2

e. $TeCl_4$ f. Cl/Sb=O

g. *trans*-dichloroethylene

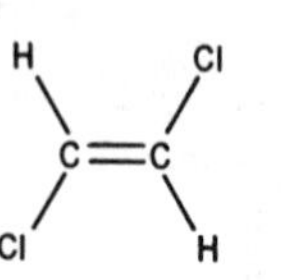

h. cyclopropane

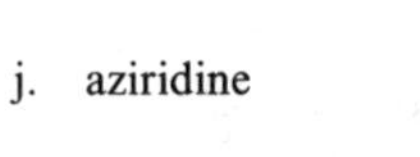

i. cyclopropene

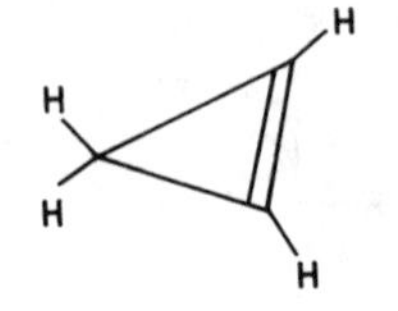

j. aziridine

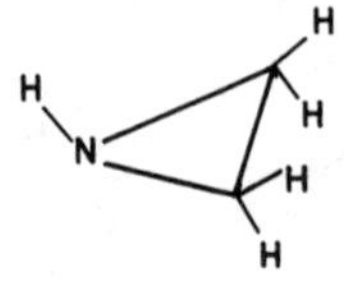

k. $Cr_2(CO)_{10}{}^{2-}$

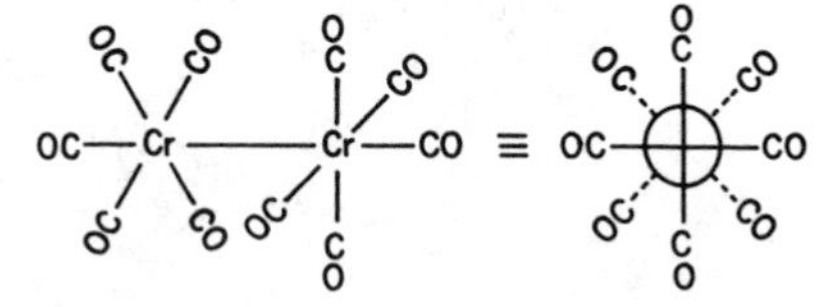

(Staggered Carbonyls)

l. $HCr_2(CO)_{10}{}^{-}$

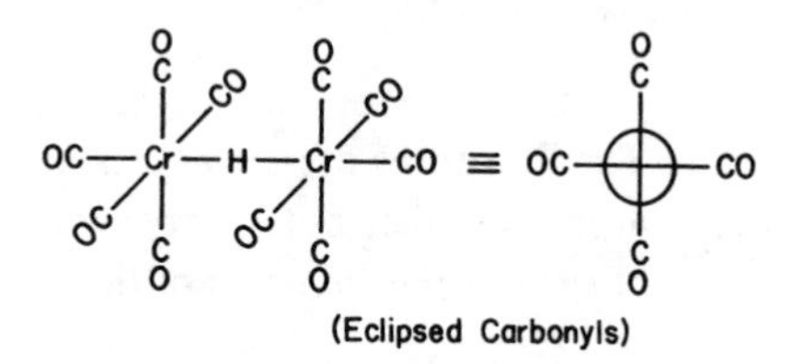

(Eclipsed Carbonyls)

m. $Pt_2Cl_6{}^{2-}$

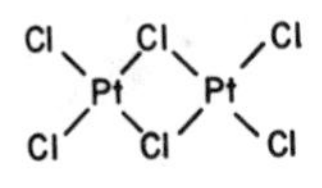

n. white phosphorus, P_4

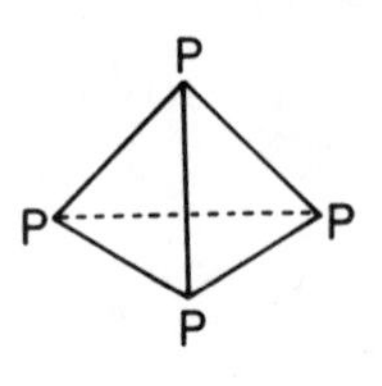

o. cubane, C_8H_8

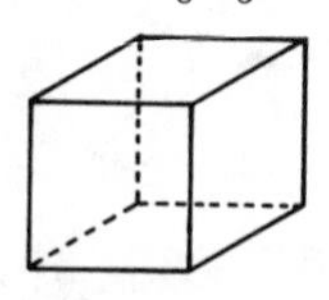

p. tetrafluorocubane

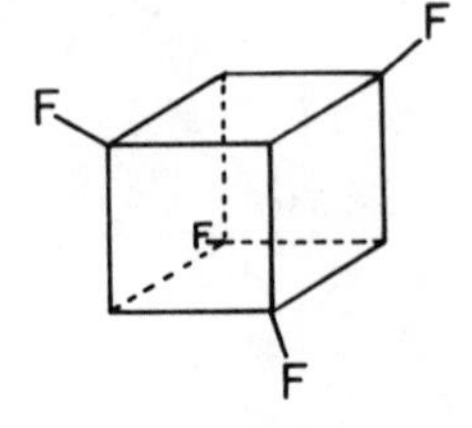

q. $Cu(NO_2)_6^{4-}$ in $Rb_2PbCu(NO_2)_6$
(not the same structure as Fig. 1-29)

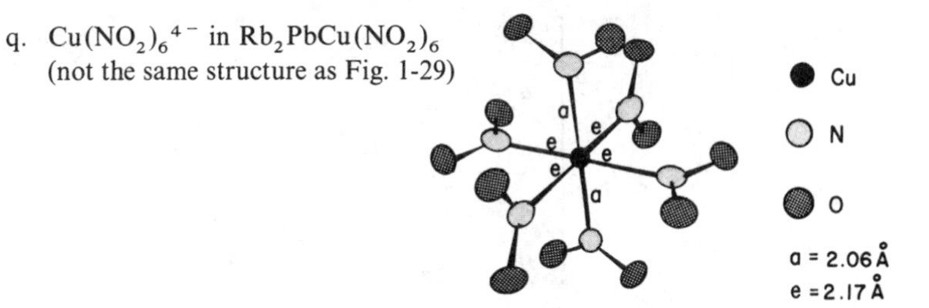

(Structure redrawn from S. Takagi, M.D. Joesten, and P.G. Lenhert, *J. Amer. Chem. Soc.*, *97*, 444 [1975].)

r. $[(CH_3)_6C_6]_3Nb_3Cl_6^+$

C
Cl
Nb

(Structure redrawn from M.R. Churchill and S.W.-Y. Chang, *Chem. Commun.*, *1974*, 248.)

1-6. Matrix Representation of Symmetry Operations

A *vector* is a series of numbers which we will write in a row or a column. We will denote the jth element of the vector, $\vec{x}$, as x_j.

$$(\; 1 \;\; 3 \;\; 12 \;\; e \;\; \cos 2\pi/17 \;) \qquad \begin{bmatrix} 1 \\ 3 \\ 12 \\ e \\ \cos 2\pi/17 \end{bmatrix}$$

a row vector — a column vector

A *matrix* is any rectangular array of numbers set between two brackets. We will only be using *square matrices* with equal numbers of rows and columns. Equation 1-21 gives the matrix A, with the general element a_{ij}. The subscript "i" gives the row number and the subscript "j" the column.

$$A = \begin{bmatrix} a_{11} & a_{12} & a_{13} & \cdots & a_{1n} \\ a_{21} & a_{22} & a_{23} & \cdots & \\ a_{31} & a_{32} & a_{33} & \cdots & \vdots \\ \vdots & & \vdots & & \\ a_{n1} & a_{n2} & \cdots & & a_{nn} \end{bmatrix} \tag{1-21}$$

We will be concerned with two kinds of multiplications, *viz.*, that of a square matrix times a square matrix and that of a square matrix times a vector. The product of a matrix and a vector is easy. If the components of the vector are x_j and the matrix elements are a_{ij}, the product *vector* has components y_m given by eq. 1-22:

$$y_m = \sum_j a_{mj} x_j \tag{1-22}$$

Let's see what this means. Using eq. 1-22, y_1 in eq. 1-23 is just given by the sum of products in eq. 1-24.

$$\begin{bmatrix} a_{11} & a_{12} & \cdots & a_{1n} \\ a_{21} & & & \\ \vdots & & & \vdots \\ a_{n1} & & \cdots & a_{nn} \end{bmatrix} \cdot \begin{bmatrix} x_1 \\ x_2 \\ \vdots \\ x_n \end{bmatrix} = \begin{bmatrix} y_1 \\ y_2 \\ \vdots \\ y_n \end{bmatrix} \tag{1-23}$$

$$A \cdot \vec{x} = \vec{y}$$

$$y_1 = \sum_j a_{1j} x_j = a_{11}x_1 + a_{12}x_2 + a_{13}x_3 + \ldots + a_{1n}x_n \tag{1-24}$$

That is, y_1 is the sum of the products of the first row of the matrix times the components of the column vector. Remember "row times column" and it is simple. Here is an example:

$$\begin{bmatrix} 1 & 2 & 3 \\ 0 & 2 & 1 \\ 1 & 0 & 2 \end{bmatrix} \cdot \begin{bmatrix} 4 \\ 5 \\ 6 \end{bmatrix} = \begin{bmatrix} 1\cdot 4 + 2\cdot 5 + 3\cdot 6 \\ 0\cdot 4 + 2\cdot 5 + 1\cdot 6 \\ 1\cdot 4 + 0\cdot 5 + 2\cdot 6 \end{bmatrix} = \begin{bmatrix} 32 \\ 16 \\ 16 \end{bmatrix} \tag{1-25}$$

Now the product of two matrices is a simple extension of this. Just consider the second matrix to be a series of vectors lined up side-by-side and multiply the first matrix times each vector:

$$\begin{matrix} & \vec{x}_1 \quad \vec{x}_2 \quad\quad \vec{x}_n & & \vec{y}_1 \quad \vec{y}_2 \quad\quad \vec{y}_n \\ \begin{bmatrix} a_{11} & a_{12} & \cdots & a_{1n} \\ a_{21} & & & \\ \vdots & & & \\ a_{n1} & & \cdots & a_{nn} \end{bmatrix} \cdot & \begin{bmatrix} x_{11} & x_{12} & \cdots & x_{1n} \\ x_{21} & & & \\ \vdots & & & \\ x_{n1} & \cdot & \cdots & x_{nn} \end{bmatrix} & = & \begin{bmatrix} y_{11} & y_{12} & \cdots & y_{1n} \\ y_{21} & & & \\ \vdots & & & \\ y_{n1} & \cdot & \cdots & y_{nn} \end{bmatrix} \\ A \quad\quad \cdot & X & = & Y \end{matrix} \tag{1-26}$$

The column vector, $\vec{y}_2$, for example, is obtained by multiplying A times the corresponding column vector, $\vec{x}_2$. Here is an example using numbers. See if you get the same answer.

$$\begin{bmatrix} 0 & 1 & 2 \\ 2 & 1 & 3 \\ 1 & 0 & 2 \end{bmatrix} \cdot \begin{bmatrix} 1 & 2 & 3 \\ 3 & 2 & 1 \\ 0 & 1 & 2 \end{bmatrix} = \begin{bmatrix} 3 & 4 & 5 \\ 5 & 9 & 13 \\ 1 & 4 & 7 \end{bmatrix} \tag{1-27}$$

What does a matrix have to do with a symmetry operation? We can use matrices as *representations* of symmetry operations. Let's consider the point group C_{2h}. We will take a vector, $\vec{v}_1$, which starts at the origin and terminates at the point (x_1, y_1, z_1) and use a 3×3 matrix to represent the effect of the symmetry operations on this vector (Fig. 1-41). Let the C_2 axis be the z axis. The other operations in the group are E, i, and $\sigma_h = \sigma(xz)$. The operation E does nothing to the vector, so we write a representation of E as follows:

$$\begin{matrix} \begin{bmatrix} 1 & 0 & 0 \\ 0 & 1 & 0 \\ 0 & 0 & 1 \end{bmatrix} & \cdot & \begin{bmatrix} x_1 \\ y_1 \\ z_1 \end{bmatrix} & = & \begin{bmatrix} x_1 \\ y_1 \\ z_1 \end{bmatrix} \\ \mathbf{E} & \cdot & \vec{v}_1 & = & \vec{v}_1 \end{matrix} \tag{1-28}$$

Inversion changes each coordinate into minus itself:

$$\begin{matrix} \begin{bmatrix} -1 & 0 & 0 \\ 0 & -1 & 0 \\ 0 & 0 & -1 \end{bmatrix} & \cdot & \begin{bmatrix} x_1 \\ y_1 \\ z_1 \end{bmatrix} & = & \begin{bmatrix} -x_1 \\ -y_1 \\ -z_1 \end{bmatrix} \\ \boldsymbol{i} & \cdot & \vec{v}_1 & = & -\vec{v}_1 \end{matrix} \tag{1-29}$$

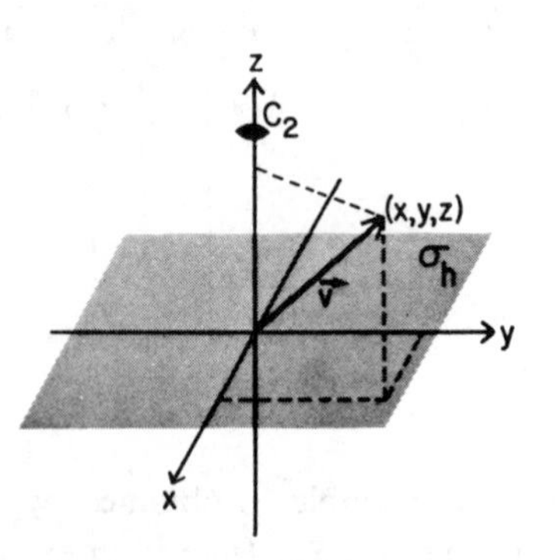

Fig. 1-41. A coordinate system showing the vector $\vec{v}$ and the location of the symmetry operations $C_2(z)$ and $\sigma_h(xy)$.

σ_h leaves the x and y coordinates unchanged but changes z to $-z$.

$$\begin{bmatrix} 1 & 0 & 0 \\ 0 & 1 & 0 \\ 0 & 0 & -1 \end{bmatrix} \cdot \begin{bmatrix} x_1 \\ y_1 \\ z_1 \end{bmatrix} = \begin{bmatrix} x_1 \\ y_1 \\ -z_1 \end{bmatrix} \tag{1-30}$$

$$\sigma_h \cdot \vec{v}_1 = \vec{v}_2$$

Finally, C_2 leaves z alone but changes x and y. We will derive the new x and y coordinates in a general way which will be applicable to rotation through any angle θ. Consider the rotation of the vector $\vec{r}_1$ in Fig. 1-42 through the angle θ to give the vector $\vec{r}_2$. Calling the length of each vector ℓ, the values of x_1 and y_1 are just

$$\begin{aligned} x_1 &= \ell \cos \alpha \\ y_1 &= \ell \sin \alpha \end{aligned} \tag{1-31}$$

The rotated vector has coordinates

$$\begin{aligned} x_2 &= \ell \cos (\theta - \alpha) \\ y_2 &= -\ell \sin (\theta - \alpha) \end{aligned} \tag{1-32}$$

But you may recall that

$$\begin{aligned} \cos (\theta - \alpha) &= \cos \theta \cos \alpha + \sin \theta \sin \alpha \\ \sin (\theta - \alpha) &= \sin \theta \cos \alpha - \cos \theta \sin \alpha \end{aligned} \tag{1-33}$$

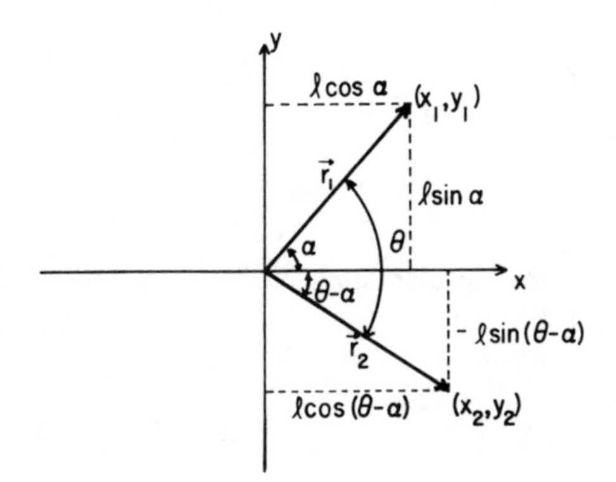

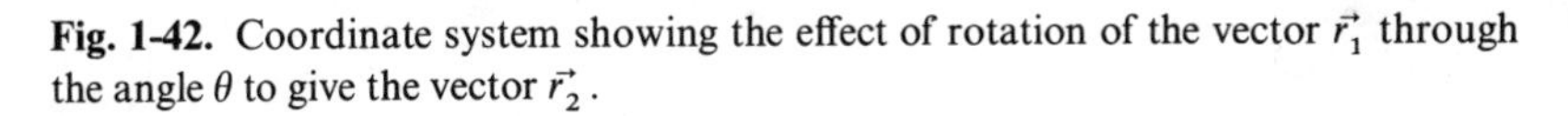

Fig. 1-42. Coordinate system showing the effect of rotation of the vector $\vec{r}_1$ through the angle θ to give the vector $\vec{r}_2$.

Therefore,

$$\begin{aligned} x_2 &= \underbrace{\ell \cos\theta \cos\alpha}_{x_1 \cos\theta} + \underbrace{\ell \sin\theta \sin\alpha}_{y_1 \sin\theta} \\ y_2 &= \underbrace{-\ell \sin\theta \cos\alpha}_{-x_1 \sin\theta} + \underbrace{\ell \cos\theta \sin\alpha}_{y_1 \cos\theta} \end{aligned} \tag{1-34}$$

Or,

$$\begin{aligned} x_2 &= x_1 \cos\theta + y_1 \sin\theta \\ y_2 &= -x_1 \sin\theta + y_1 \cos\theta \end{aligned} \tag{1-35}$$

In matrix-vector language, we say†

$$\boxed{\begin{array}{ccccc} \begin{bmatrix} \cos\theta & \sin\theta \\ -\sin\theta & \cos\theta \end{bmatrix} & \begin{bmatrix} x_1 \\ y_1 \end{bmatrix} & = & \begin{bmatrix} x_2 \\ y_2 \end{bmatrix} \\ \boldsymbol{R}_\theta & \vec{r}_1 & = & \vec{r}_2 \end{array}} \tag{1-36}$$

The matrix $\boldsymbol{R}_\theta$ is a *representation* of a rotation through the angle θ. A C_2 rotation is a rotation through $\theta = 2\pi/2 = \pi$; so the $\boldsymbol{C}_2$ matrix is

$$\boldsymbol{C}_2 = R_\pi = \begin{bmatrix} \cos\pi & \sin\pi \\ -\sin\pi & \cos\pi \end{bmatrix} = \begin{bmatrix} -1 & 0 \\ 0 & -1 \end{bmatrix} \tag{1-37}$$

†A box around an equation should alert you to its importance.

Our final matrix equation for the C_{2h} point group is then

$$\begin{bmatrix} -1 & 0 & 0 \\ 0 & -1 & 0 \\ 0 & 0 & +1 \end{bmatrix} \cdot \begin{bmatrix} x_1 \\ y_1 \\ z_1 \end{bmatrix} = \begin{bmatrix} -x_1 \\ -y_1 \\ z_1 \end{bmatrix} \tag{1-38}$$

$$\boldsymbol{C}_2 \cdot \vec{v}_1 = \vec{v}_3$$

To recapitulate, one representation of the operations of the point group C_{2h} is

$$\boldsymbol{E} = \begin{bmatrix} 1 & 0 & 0 \\ 0 & 1 & 0 \\ 0 & 0 & 1 \end{bmatrix} \qquad \sigma_h = \begin{bmatrix} 1 & 0 & 0 \\ 0 & 1 & 0 \\ 0 & 0 & -1 \end{bmatrix}$$

$$\boldsymbol{C}_2 = \begin{bmatrix} -1 & 0 & 0 \\ 0 & -1 & 0 \\ 0 & 0 & 1 \end{bmatrix} \qquad \boldsymbol{i} = \begin{bmatrix} -1 & 0 & 0 \\ 0 & -1 & 0 \\ 0 & 0 & -1 \end{bmatrix} \tag{1-39}$$

Believe it or not, these four matrices form a mathematical group which obeys the same multiplication table as the operations. The group multiplication table has already been set out in Table 1-4, and one example is now presented showing that the matrices obey the same table:

$$\begin{bmatrix} -1 & 0 & 0 \\ 0 & -1 & 0 \\ 0 & 0 & 1 \end{bmatrix} \cdot \begin{bmatrix} 1 & 0 & 0 \\ 0 & 1 & 0 \\ 0 & 0 & -1 \end{bmatrix} = \begin{bmatrix} -1 & 0 & 0 \\ 0 & -1 & 0 \\ 0 & 0 & -1 \end{bmatrix} \tag{1-40}$$

$$\boldsymbol{C}_2 \cdot \sigma_h = \boldsymbol{i}$$

Now let's consider an interesting property of *block diagonal matrices.* A block diagonal matrix is one which has blocks of numbers along the top-to-bottom left-to-right diagonal, and zeros elsewhere:

$$\left[\begin{array}{ccc:cc:c} 1 & 2 & 3 & 0 & 0 & 0 \\ 3 & 2 & 1 & 0 & 0 & 0 \\ 0 & 1 & 2 & 0 & 0 & 0 \\ \hdashline 0 & 0 & 0 & 2 & 3 & 0 \\ 0 & 0 & 0 & 1 & 2 & 0 \\ \hdashline 0 & 0 & 0 & 0 & 0 & 1 \end{array}\right]$$

a block diagonal matrix.

The product of two block diagonal matrices with similar arrays of blocks is another matrix obtained by multiplying the individual blocks:

$$\left[\begin{array}{ccc:cc:c} 1 & 2 & 3 & 0 & 0 & 0 \\ 3 & 2 & 1 & 0 & 0 & 0 \\ 0 & 1 & 2 & 0 & 0 & 0 \\ \hdashline 0 & 0 & 0 & 2 & 3 & 0 \\ 0 & 0 & 0 & 1 & 2 & 0 \\ \hdashline 0 & 0 & 0 & 0 & 0 & 1 \end{array}\right] \cdot \left[\begin{array}{ccc:cc:c} 2 & 0 & 0 & 0 & 0 & 0 \\ 1 & 0 & 2 & 0 & 0 & 0 \\ 0 & 1 & 1 & 0 & 0 & 0 \\ \hdashline 0 & 0 & 0 & 0 & 2 & 0 \\ 0 & 0 & 0 & 1 & 2 & 0 \\ \hdashline 0 & 0 & 0 & 0 & 0 & 3 \end{array}\right] = \left[\begin{array}{ccc:cc:c} 4 & 3 & 7 & 0 & 0 & 0 \\ 8 & 1 & 5 & 0 & 0 & 0 \\ 1 & 2 & 4 & 0 & 0 & 0 \\ \hdashline 0 & 0 & 0 & 3 & 10 & 0 \\ 0 & 0 & 0 & 2 & 6 & 0 \\ \hdashline 0 & 0 & 0 & 0 & 0 & 3 \end{array}\right] \tag{1-41}$$

That is

$$\begin{gathered} \begin{bmatrix} 1 & 2 & 3 \\ 3 & 2 & 1 \\ 0 & 1 & 2 \end{bmatrix} \cdot \begin{bmatrix} 2 & 0 & 0 \\ 1 & 0 & 2 \\ 0 & 1 & 1 \end{bmatrix} = \begin{bmatrix} 4 & 3 & 7 \\ 8 & 1 & 5 \\ 1 & 2 & 4 \end{bmatrix} \\ \begin{bmatrix} 2 & 3 \\ 1 & 2 \end{bmatrix} \cdot \begin{bmatrix} 0 & 2 \\ 1 & 2 \end{bmatrix} = \begin{bmatrix} 3 & 10 \\ 2 & 6 \end{bmatrix} \\ [1] \cdot [3] = [3] \end{gathered} \tag{1-42}$$

Each matrix has an inverse matrix just as each operation of a group has an inverse operation.† Just as an operation times its inverse gives the identity operation, E, a matrix times its inverse gives the identity matrix, $\boldsymbol{I}$, which has ones on the diagonal and zeros elsewhere. Furthermore, a matrix and its inverse must commute:

$$\boldsymbol{Q} \cdot \boldsymbol{Q}^{-1} = \boldsymbol{Q}^{-1} \cdot \boldsymbol{Q} = \boldsymbol{I} \tag{1-43}$$

Using a matrix and its inverse, we can perform similarity transformations with matrices:

$$\boldsymbol{B} = \boldsymbol{Q}^{-1} \cdot \boldsymbol{A} \cdot \boldsymbol{Q} \tag{1-44}$$

Here $\boldsymbol{A}$ and $\boldsymbol{B}$ are said to be conjugate just as symmetry operations related by similarity transformations are said to be conjugate.

Now suppose we had an expensive machine that takes a matrix, $\boldsymbol{X}$, and acts on it with a similarity transformation and then spits out a conjugate matrix, $\boldsymbol{Y}$, *in block diagonal form.* Behold! We have a bunch of matrices

†A matrix has no inverse only if the determinant of that matrix is zero.

representing the various operations of a point group. Take these matrices, $\boldsymbol{A}, \boldsymbol{B}, \boldsymbol{C}, \ldots$ and feed them into the expensive machine. Out comes a new set of matrices, $\boldsymbol{A}', \boldsymbol{B}', \boldsymbol{C}', \ldots$ in block diagonal form:

$$\begin{aligned} \boldsymbol{A}' &= \boldsymbol{Q}^{-1} \cdot \boldsymbol{A} \cdot \boldsymbol{Q} \\ \boldsymbol{B}' &= \boldsymbol{Q}^{-1} \cdot \boldsymbol{B} \cdot \boldsymbol{Q} \\ \boldsymbol{C}' &= \boldsymbol{Q}^{-1} \cdot \boldsymbol{C} \cdot \boldsymbol{Q} \\ &\vdots \end{aligned} \tag{1-45}$$

These new matrices are still representations of the same operations, because they still obey the same multiplication table as $\boldsymbol{A}, \boldsymbol{B}, \boldsymbol{C}, \ldots$. For example, if $\boldsymbol{A} \cdot \boldsymbol{B} = \boldsymbol{C}$, we can show that $\boldsymbol{A}' \cdot \boldsymbol{B}' = \boldsymbol{C}'$:

$$\begin{aligned} \boldsymbol{A}' \cdot \boldsymbol{B}' &= (\boldsymbol{Q}^{-1} \cdot \boldsymbol{A} \cdot \boldsymbol{Q})(\boldsymbol{Q}^{-1} \cdot \boldsymbol{B} \cdot \boldsymbol{Q}) \\ &= \boldsymbol{Q}^{-1} \cdot \boldsymbol{A} \cdot (\boldsymbol{Q} \cdot \boldsymbol{Q}^{-1}) \cdot \boldsymbol{B} \cdot \boldsymbol{Q} \\ &= \boldsymbol{Q}^{-1} \cdot \boldsymbol{A} \cdot \boldsymbol{I} \cdot \boldsymbol{B} \cdot \boldsymbol{Q} \\ &= \boldsymbol{Q}^{-1} \cdot \boldsymbol{A} \cdot \boldsymbol{B} \cdot \boldsymbol{Q} \\ &= \boldsymbol{Q}^{-1} \cdot \boldsymbol{C} \cdot \boldsymbol{Q} \\ \boldsymbol{A}' \cdot \boldsymbol{B}' &= \boldsymbol{C}' \end{aligned} \tag{1-46}$$

Here we have used the fact that the identity matrix, $\boldsymbol{I}$, has no effect on any matrix it multiplies ($\boldsymbol{I} \cdot \boldsymbol{X} = \boldsymbol{X} \cdot \boldsymbol{I} = \boldsymbol{X}$).

But look at the product $\boldsymbol{A}' \cdot \boldsymbol{B}' = \boldsymbol{C}'$.

$$\begin{bmatrix} A1 & & \\ & A2 & \\ & & A3 \end{bmatrix} \cdot \begin{bmatrix} B1 & & \\ & B2 & \\ & & B3 \end{bmatrix} = \begin{bmatrix} C1 & & \\ & C2 & \\ & & C3 \end{bmatrix} \tag{1-47}$$

The small matrix $\boldsymbol{C1}$ must be the product $\boldsymbol{A1} \cdot \boldsymbol{B1}$. Similarly, $\boldsymbol{C2} = \boldsymbol{A2} \cdot \boldsymbol{B2}$ and $\boldsymbol{C3} = \boldsymbol{A3} \cdot \boldsymbol{B3}$. What this means is that *each little block of each big matrix, $\boldsymbol{A}', \boldsymbol{B}', \boldsymbol{C}', \ldots$ is a new representation of the operation* since the little blocks obey the same multiplication table that the big matrices obey.

The point is that if you have such a matrix-reducing machine, you can feed in a big *reducible matrix representation* and reduce it to block diagonal form. Then take each little block and feed it back into the machine again and again until you have the simplest representation possible. When you have either a set of 1×1 matrices or a set of matrices which cannot be further reduced by similarity transformations, each member of the set is called an *IRREDUCIBLE REPRESENTATION*. Although the number of reducible representations of any operation is infinite, all of these representations reduce to just a small, finite number of irreducible representations for most point groups. These are the simplest representations of the operations of the point group and we will use irreducible representations throughout the remainder of this book.

An Example.

"Stop! Wait!" you are probably screaming by now. If you feel totally confused by irreducible representations at this point, you are probably in good company. To help clarify what we just did, let's look at a specific example outlined below:

Objective 1. Generate reducible matrix representations of the operations C_5, C_5^2, and C_5^3.
Objective 2. Verify by matrix multiplication that $C_5 \cdot C_5^2 = C_5^3$.
Objective 3. Block diagonalize each of the reducible representations using a similarity transformation.
Objective 4. Show that the individual blocks of the diagonalized matrices still obey the same multiplication rules as the reducible representations.

To generate a reducible representation of the operation C_5, consider the pentagon in Fig. 1-43. The C_5 rotation interchanges the vertices as follows:

$$
\begin{aligned}
1 &\rightarrow 2 \\
2 &\rightarrow 3 \\
3 &\rightarrow 4 \\
4 &\rightarrow 5 \\
5 &\rightarrow 1
\end{aligned}
$$

Fig. 1-43. A scheme for generating reducible representations of the operations C_5, C_5^2, and C_5^3.

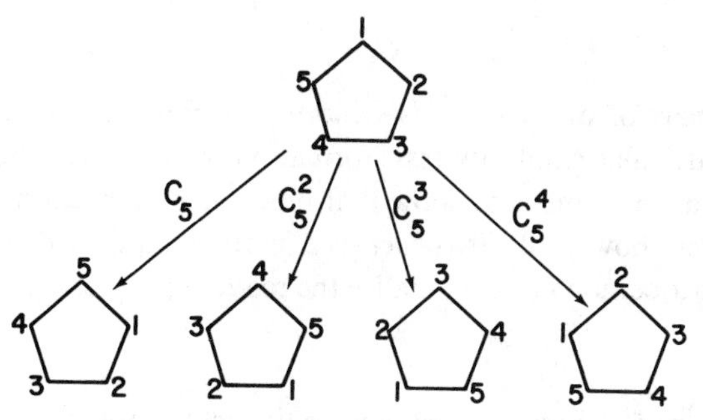

This transformation can be represented by the matrix equation 1-48.

$$C_5 \cdot \begin{bmatrix} 1 \\ 2 \\ 3 \\ 4 \\ 5 \end{bmatrix} = \underset{C_5}{\begin{bmatrix} 0 & 1 & 0 & 0 & 0 \\ 0 & 0 & 1 & 0 & 0 \\ 0 & 0 & 0 & 1 & 0 \\ 0 & 0 & 0 & 0 & 1 \\ 1 & 0 & 0 & 0 & 0 \end{bmatrix}} \cdot \begin{bmatrix} 1 \\ 2 \\ 3 \\ 4 \\ 5 \end{bmatrix} = \begin{bmatrix} 2 \\ 3 \\ 4 \\ 5 \\ 1 \end{bmatrix} \tag{1-48}$$

In a completely analagous manner, we can generate representations of the operations $C_5{}^2$ and $C_5{}^3$ using Fig. 1-42. These come out as follows:

$$C_5{}^2 = \begin{bmatrix} 0 & 0 & 1 & 0 & 0 \\ 0 & 0 & 0 & 1 & 0 \\ 0 & 0 & 0 & 0 & 1 \\ 1 & 0 & 0 & 0 & 0 \\ 0 & 1 & 0 & 0 & 0 \end{bmatrix} \qquad C_5{}^3 = \begin{bmatrix} 0 & 0 & 0 & 1 & 0 \\ 0 & 0 & 0 & 0 & 1 \\ 1 & 0 & 0 & 0 & 0 \\ 0 & 1 & 0 & 0 & 0 \\ 0 & 0 & 1 & 0 & 0 \end{bmatrix} \tag{1-49}$$

Objective 1 of our outline above is attained. You should be able to satisfy yourself that the product in Objective 2 is indeed valid, *viz*.,

$$\underset{C_5}{\begin{bmatrix} 0 & 1 & 0 & 0 & 0 \\ 0 & 0 & 1 & 0 & 0 \\ 0 & 0 & 0 & 1 & 0 \\ 0 & 0 & 0 & 0 & 1 \\ 1 & 0 & 0 & 0 & 0 \end{bmatrix}} \cdot \underset{C_5^2}{\begin{bmatrix} 0 & 0 & 1 & 0 & 0 \\ 0 & 0 & 0 & 1 & 0 \\ 0 & 0 & 0 & 0 & 1 \\ 1 & 0 & 0 & 0 & 0 \\ 0 & 1 & 0 & 0 & 0 \end{bmatrix}} = \underset{C_5^3}{\begin{bmatrix} 0 & 0 & 0 & 1 & 0 \\ 0 & 0 & 0 & 0 & 1 \\ 1 & 0 & 0 & 0 & 0 \\ 0 & 1 & 0 & 0 & 0 \\ 0 & 0 & 1 & 0 & 0 \end{bmatrix}} \tag{1-50}$$

The tricky part of our scheme is Objective 3. For this we are going to introduce a particular similarity transformation with no justification given. Our objective at this time is to show that block diagonalization is possible, and not to show how to do it (which is a more complicated matter). We will call the matrices used to diagonalize the reducible representations $\boldsymbol{Q}$ and $\boldsymbol{Q}^{-1}$†:

† The matrices $\boldsymbol{Q}$ and $\boldsymbol{Q}^{-1}$ have many special properties, not the least of which is that $\boldsymbol{Q}^{-1}$ is the *transpose* of $\boldsymbol{Q}$. The transpose of a matrix is obtained by interchanging the rows and columns. That is, the first row of $\boldsymbol{Q}$ is the first column of $\boldsymbol{Q}^{-1}$.

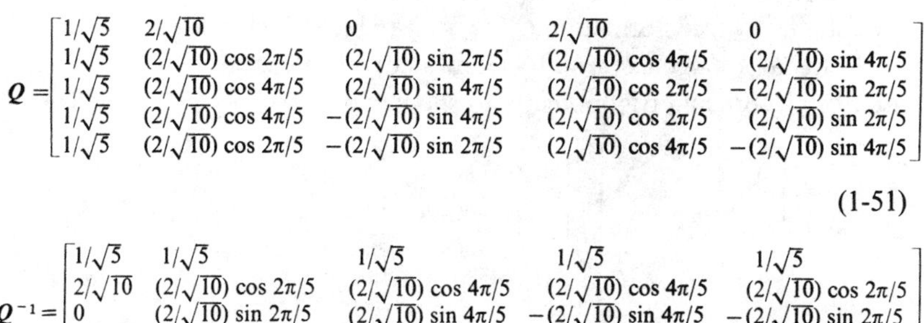

$$Q = \begin{bmatrix} 1/\sqrt{5} & 2/\sqrt{10} & 0 & 2/\sqrt{10} & 0 \\ 1/\sqrt{5} & (2/\sqrt{10})\cos 2\pi/5 & (2/\sqrt{10})\sin 2\pi/5 & (2/\sqrt{10})\cos 4\pi/5 & (2/\sqrt{10})\sin 4\pi/5 \\ 1/\sqrt{5} & (2/\sqrt{10})\cos 4\pi/5 & (2/\sqrt{10})\sin 4\pi/5 & (2/\sqrt{10})\cos 2\pi/5 & -(2/\sqrt{10})\sin 2\pi/5 \\ 1/\sqrt{5} & (2/\sqrt{10})\cos 4\pi/5 & -(2/\sqrt{10})\sin 4\pi/5 & (2/\sqrt{10})\cos 2\pi/5 & (2/\sqrt{10})\sin 2\pi/5 \\ 1/\sqrt{5} & (2/\sqrt{10})\cos 2\pi/5 & -(2/\sqrt{10})\sin 2\pi/5 & (2/\sqrt{10})\cos 4\pi/5 & -(2/\sqrt{10})\sin 4\pi/5 \end{bmatrix}$$

(1-51)

$$Q^{-1} = \begin{bmatrix} 1/\sqrt{5} & 1/\sqrt{5} & 1/\sqrt{5} & 1/\sqrt{5} & 1/\sqrt{5} \\ 2/\sqrt{10} & (2/\sqrt{10})\cos 2\pi/5 & (2/\sqrt{10})\cos 4\pi/5 & (2/\sqrt{10})\cos 4\pi/5 & (2/\sqrt{10})\cos 2\pi/5 \\ 0 & (2/\sqrt{10})\sin 2\pi/5 & (2/\sqrt{10})\sin 4\pi/5 & -(2/\sqrt{10})\sin 4\pi/5 & -(2/\sqrt{10})\sin 2\pi/5 \\ 2/\sqrt{10} & (2/\sqrt{10})\cos 4\pi/5 & (2/\sqrt{10})\cos 2\pi/5 & (2/\sqrt{10})\cos 2\pi/5 & (2/\sqrt{10})\cos 4\pi/5 \\ 0 & (2/\sqrt{10})\sin 4\pi/5 & -(2/\sqrt{10})\sin 2\pi/5 & (2/\sqrt{10})\sin 2\pi/5 & -(2/\sqrt{10})\sin 4\pi/5 \end{bmatrix}$$

Three similarity transformations then produce the *block diagonalized matrices*, C_5', $C_5^{2\prime}$, and $C_5^{3\prime}$:

$$C_5' = Q^{-1} \cdot C_5 \cdot Q = \begin{bmatrix} 1 & 0 & 0 & 0 & 0 \\ 0 & \cos 2\pi/5 & \sin 2\pi/5 & 0 & 0 \\ 0 & -\sin 2\pi/5 & \cos 2\pi/5 & 0 & 0 \\ 0 & 0 & 0 & \cos 4\pi/5 & \sin 4\pi/5 \\ 0 & 0 & 0 & -\sin 4\pi/5 & \cos 4\pi/5 \end{bmatrix}$$

(1-52)

$$C_5^{2\prime} = Q^{-1} \cdot C_5^2 \cdot Q = \begin{bmatrix} 1 & 0 & 0 & 0 & 0 \\ 0 & \cos 4\pi/5 & \sin 4\pi/5 & 0 & 0 \\ 0 & -\sin 4\pi/5 & \cos 4\pi/5 & 0 & 0 \\ 0 & 0 & 0 & \cos 2\pi/5 & -\sin 2\pi/5 \\ 0 & 0 & 0 & \sin 2\pi/5 & \cos 2\pi/5 \end{bmatrix}$$

(1-53)

$$C_5^{3\prime} = Q^{-1} \cdot C_5^3 \cdot Q = \begin{bmatrix} 1 & 0 & 0 & 0 & 0 \\ 0 & \cos 4\pi/5 & -\sin 4\pi/5 & 0 & 0 \\ 0 & \sin 4\pi/5 & \cos 4\pi/5 & 0 & 0 \\ 0 & 0 & 0 & \cos 2\pi/5 & \sin 2\pi/5 \\ 0 & 0 & 0 & -\sin 2\pi/5 & \cos 2\pi/5 \end{bmatrix}$$

(1-54)

The similarity transformation $Q^{-1} \cdot X \cdot Q$ reduced each 5×5 representation, X, to *block diagonal form.* It is now a complicated, but not impossible, matter of algebra and trigonometry to show that Objective 4 is indeed met:

$$\begin{bmatrix} A1 & & \\ & B1 & \\ & & C1 \end{bmatrix} \cdot \begin{bmatrix} A2 & & \\ & B2 & \\ & & C2 \end{bmatrix} = \begin{bmatrix} A3 & & \\ & B3 & \\ & & C3 \end{bmatrix} \qquad (1\text{-}55)$$

$$C_5{}' \qquad \cdot \qquad C_5{}^{2\prime} \qquad = \qquad C_5{}^{3\prime}$$

The remarkable thing about eq. 1-55 is that each small block of each block diagonalized representation still obeys the product rule $C_5 \cdot C_5^2 = C_5^3$. For example, matrix multiplication would show that the lower right block in each matrix is such that $C1 \cdot C2 = C3$. The three blocks, $C1$, $C2$, and $C3$, are *irreducible representations* as there is no similarity transformation which can be performed on them to further diagonalize them.

$$\begin{bmatrix} \cos 4\pi/5 & \sin 4\pi/5 \\ -\sin 4\pi/5 & \cos 4\pi/5 \end{bmatrix} \cdot \begin{bmatrix} \cos 2\pi/5 & -\sin 2\pi/5 \\ \sin 2\pi/5 & \cos 2\pi/5 \end{bmatrix} = \begin{bmatrix} \cos 2\pi/5 & \sin 2\pi/5 \\ -\sin 2\pi/5 & \cos 2\pi/5 \end{bmatrix} \qquad (1\text{-}56)$$

$C1$	$\cdot$	$C2$	$=$	$C3$
an irreducible representation of C_5		an irreducible representation of C_5^2		an irreducible representation of C_5^3

This rather complicated example was chosen to show that it is possible to take a large (in this case 5×5) reducible representation and find smaller irreducible representations by performing similarity transformations. The smaller irreducible representations still obey the same multiplication rules as do the large reducible representations.

Problems

1-11. Determine the products:

a. $\begin{bmatrix} 4 & 1 & 4 \\ -1 & -2 & -1 \\ 2 & -1 & 1 \end{bmatrix} \cdot \begin{bmatrix} 3 \\ -1 \\ 0 \end{bmatrix}$ b. $\begin{bmatrix} 4 & 1 & 4 \\ -1 & -2 & -1 \\ 2 & -1 & 1 \end{bmatrix} \cdot \begin{bmatrix} 3 & 0 & 1 \\ -1 & 2 & 0 \\ 0 & 0 & 0 \end{bmatrix}$

1-12. Write a three dimensional (*i.e.*, 3×3) matrix representation of each of the operations of the point group D_{2h} which are: $C_2(z)$, $C_2(x)$, $C_2(y)$, E, i, $\sigma(xz)$, $\sigma(yz)$, and $\sigma(xy)$.

1-13. Using the coordinate system below, write a 3×3 representation of each of the operations of the point group C_{3v}: E, C_3, $C_3{}^2$, σ_v, $\sigma_v{}'$, and $\sigma_v{}''$. The representations of $\sigma_v{}'$ and $\sigma_v{}''$ are not trivially set up. Show by matrix multiplication that $C_3{}^2$ is the inverse of C_3.

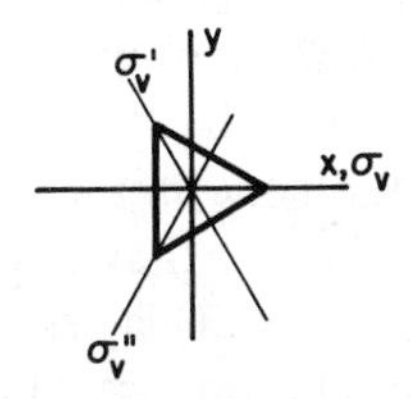

1-14. An easier and completely different way to generate a set of 3×3 matrix representations of the operations of C_{3v} is to use the figure below. To generate a representation of C_3, for example, we note that C_3 takes 1 to 2, 2 to 3, and 3 to 1. We write this in a matrix equation as follows:

$$C_3 \cdot \begin{bmatrix} 1 \\ 2 \\ 3 \end{bmatrix} = \begin{bmatrix} 0 & 1 & 0 \\ 0 & 0 & 1 \\ 1 & 0 & 0 \end{bmatrix} \cdot \begin{bmatrix} 1 \\ 2 \\ 3 \end{bmatrix} = \begin{bmatrix} 2 \\ 3 \\ 1 \end{bmatrix}$$

Write matrix representations of the other operations of C_{3v}. Show that the similarity transformation below will block diagonalize C_3.

$$\underset{Q^{-1}}{\begin{bmatrix} 1/\sqrt{3} & 1/\sqrt{3} & 1/\sqrt{3} \\ 2/\sqrt{6} & -1/\sqrt{6} & -1/\sqrt{6} \\ 0 & 1/\sqrt{2} & -1/\sqrt{2} \end{bmatrix}} \cdot \underset{C_3}{\begin{bmatrix} 0 & 1 & 0 \\ 0 & 0 & 1 \\ 1 & 0 & 0 \end{bmatrix}} \cdot \underset{Q}{\begin{bmatrix} 1/\sqrt{3} & 2/\sqrt{6} & 0 \\ 1/\sqrt{3} & -1/\sqrt{6} & 1/\sqrt{2} \\ 1/\sqrt{3} & -1/\sqrt{6} & -1/\sqrt{2} \end{bmatrix}}$$

1-7. Characters and Character Tables

Instead of working with irreducible representations, we are going to make things even simpler and work with *characters* of irreducible representations. The character of a matrix is the sum of its diagonal elements. The character is also called the *trace* of the matrix. The character of the matrix below is 21.

$$\begin{bmatrix} 1 & 2 & 3 & 4 & 5 \\ 6 & 7 & 8 & 9 & 10 \\ 9 & 8 & 7 & 6 & 5 \\ 4 & 3 & 2 & 1 & 0 \\ 1 & 2 & 3 & 4 & 5 \end{bmatrix}$$

Character = trace =
$1 + 7 + 7 + 1 + 5 = 21$

There are theorems which tell how to find the characters for any irreducible representation of any point group, but we shall not be concerned with deriving these characters. All of the *character tables* needed for common chemical problems are given in Appendix A.†

Let's look at Table 1-6, a rudimentary character table for the point group C_{3v}. Across the top row is the complete set of operations of the group. Down the left column are the names of the irreducible representations, which we are calling Γ_1, Γ_2, and Γ_3 for now. The number of irreducible representations is equal to the number of classes of operations of the point group. By constructing a C_{3v} multiplication table (Table 1-7), noting the reciprocal of each operation,

$$\begin{aligned} E^{-1} &= E & \sigma_v^{-1} &= \sigma_v \\ C_3^{-1} &= C_3^2 & \sigma_v'^{-1} &= \sigma_v' \\ (C_3^2)^{-1} &= C_3 & \sigma_v''^{-1} &= \sigma_v'' \end{aligned} \tag{1-57}$$

and trying some similarity transformations,

$$\begin{aligned} \sigma_v \cdot C_3 \cdot \sigma_v &= C_3^2 & C_3^2 \cdot \sigma_v \cdot C_3 &= \sigma_v'' & C_3^2 \cdot E \cdot C_3 &= E \\ \sigma_v'' \cdot C_3 \cdot \sigma_v'' &= C_3^2 & \sigma_v \cdot \sigma_v \cdot \sigma_v &= \sigma_v & \sigma_v \cdot E \cdot \sigma_v &= E \\ C_3^2 \cdot C_3^2 \cdot C_3 &= C_3^2 & C_3 \cdot \sigma_v \cdot C_3^2 &= \sigma_v' & \sigma_v'' \cdot E \cdot \sigma_v'' &= E \\ \sigma_v' \cdot C_3^2 \cdot \sigma_v' &= C_3 & \sigma_v'' \cdot \sigma_v' \cdot \sigma_v'' &= \sigma_v' & & \\ & & \sigma_v' \cdot \sigma_v'' \cdot \sigma_v' &= \sigma_v & & \end{aligned} \tag{1-58}$$

we find that C_3 and C_3^2 form one class. σ_v, σ_v', and σ_v'' form another class and E, as always, is in a class by itself. The number of classes of C_{3v} is three so the number of irreducible representations is also three.

In most, but not all, point groups, operations which produce the same effect are in the same class. C_4 and C_4^3, for example, both rotate the molecule by 90° and are in the same class in most point groups. Various σ_v operations

Table 1-6. A Rudimentary C_{3v} Character Table

C_{3v}	E	C_3	C_3^2	σ_v	σ_v'	σ_v''
Γ_1	1	1	1	1	1	1
Γ_2	1	1	1	−1	−1	−1
Γ_3	2	−1	−1	0	0	0

†The knowledgeable reader will note that we include character tables up through tenfold axes for each kind of point group in Appendix A. We hope that this will stimulate synthetic chemists to prepare new compounds with such high symmetry!

Table 1-7. C_{3v} Group Multiplication Table

C_{3v}	E	C_3	C_3^2	σ_v	σ_v'	σ_v''
E	E	C_3	C_3^2	σ_v	σ_v'	σ_v''
C_3	C_3	C_3^2	E	σ_v''	σ_v	σ_v'
C_3^2	C_3^2	E	C_3	σ_v'	σ_v''	σ_v
σ_v	σ_v	σ_v'	σ_v''	E	C_3	C_3^2
σ_v'	σ_v'	σ_v''	σ_v	C_3^2	E	C_3
σ_v''	σ_v''	σ_v	σ_v'	C_3	C_3^2	E

are generally in the same class and σ_d operations form a different class. Inversion is always in a class by itself because its function is unique. The same is true for σ_h. In general, you will not have to determine which operations form each class because the character tables in Appendix A contain this information.

Getting back to our C_{3v} character table, the characters for any given irreducible representation must be the same for operations in the same class. That is, the characters under C_3 and C_3^2 must be the same in any row of the table. Hence the character tables are not written as in Table 1-6, but rather, as in Table 1-8. The symbols "$2C_3$" and "$3\sigma_v$" mean that there are two operations in the class containing C_3 and three operations in the class containing σ_v. That the unwritten operations are C_3^2 in the first instance and σ_v' and σ_v'' in the second is understood.

The character tables of Appendix A are more complicated than what we have just seen. The full C_{4v} table (Table 1-9), for example, is divided into several areas. The main part contains the characters. On the left are the *names* of the irreducible representations, known as Mulliken symbols. Conventionally, we use the letters A, B, E, and T (or F in some tables). A and B are one-dimensional. E is two-dimensional and T is three-dimensional. The dimension of an irreducible representation is the dimension of any of its matrices (the number of rows or columns in the matrix). Since the representation of the operation E is always the identity matrix, the

Table 1-8. Condensed C_{3v} Character Table

C_{3v}	E	$2C_3$	$3\sigma_v$
Γ_1	1	1	1
Γ_2	1	1	−1
Γ_3	2	−1	0

Table 1-9. Complete C_{4v} Character Table

C_{4v}	E	$2C_4$	C_2	$2\sigma_v$	$2\sigma_d$			
A_1	1	1	1	1	1	z	$x^2 + y^2, z^2$	z^3
A_2	1	1	1	-1	-1	R_z		
B_1	1	-1	1	1	-1		$x^2 - y^2$	$z(x^2 - y^2)$
B_2	1	-1	1	-1	1		xy	xyz
E	2	0	-2	0	0	$(x, y), (R_x, R_y)$	(xz, yz)	$(xz^2, yz^2), [x(x^2 - 3y^2), y(3x^2 - y^2)]$
$\Gamma_{x,y,z}$	3	1	-1	1	1			

character of E is always the dimension of the irreducible representation. For example, for a three-dimensional irreducible representation, whose symbol would be T, the matrix for E (the identity operation, not to be confused with E, the Mulliken symbol) is equation 1-59 and the trace of this matrix is 3. The difference between A and B is that the character under the principal rotational operation, C_n, is always $+1$ for A and -1 for B representations. The subscripts 1, 2, 3, *etc.*, which may appear with A, B, E, or T can be considered arbitrary labels. The subscripts g and u and the superscripts $'$ and $''$ are worth knowing about. A g representation is symmetric with respect to inversion and a u representation is antisymmetric to inversion. The meaning of symmetry with respect to inversion can be illustrated with the atomic orbitals.† Any p or f orbital is transformed into minus itself upon inversion, and is therefore a u function (Fig. 1-44). A d orbital is transformed into itself upon inversion and is a g function. The symbols g and u stand for the German words *gerade* and *ungerade*, meaning even and odd. In a similar way, the superscripts $'$ and $''$ denote irreducible representations which are, respectively, symmetric and antisymmetric with respect to reflection through a horizontal mirror plane.

$$\boldsymbol{E} = \begin{bmatrix} 1 & 0 & 0 \\ 0 & 1 & 0 \\ 0 & 0 & 1 \end{bmatrix} \tag{1-59}$$

The three columns on the right side of the table contain basis functions for the irreducible representations. These basis functions have the same symmetry properties as the atomic orbitals which bear the same names. To understand what a basis function is, let's write matrix representations for the operations of C_{4v}. The C_4 operation about the z axis leaves the z co-

†Pictures showing the shapes of the s, p, d, and f orbitals appear in Appendix D.

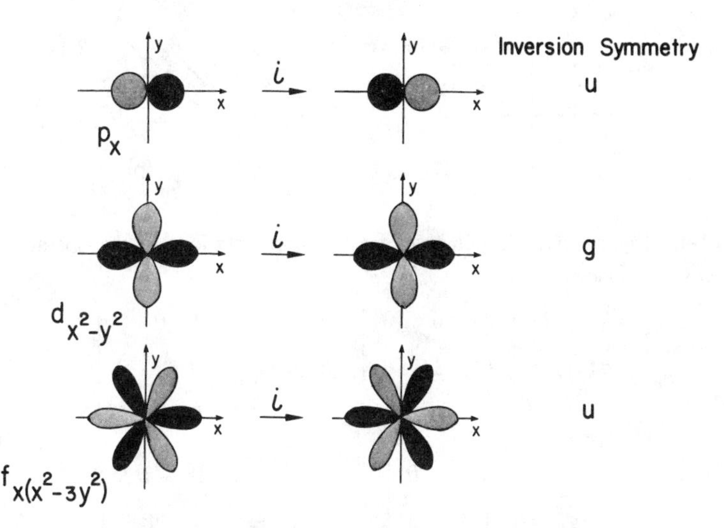

Fig. 1-44. The p and f atomic orbitals go into minus themselves upon inversion, and so have *u* symmetry. The d orbitals go into plus themselves and have *g* symmetry. Dark and light shaded lobes are of opposite sign.

ordinate of any point unchanged, but changes the x and y coordinates according to the $\boldsymbol{R}_\theta$ matrix in eq. 1-36, in which we use the value $2\pi/4$ for θ. Therefore the matrix representation of C_4 would be as follows:

$$\boldsymbol{C}_4 = \begin{bmatrix} \cos 2\pi/4 & \sin 2\pi/4 & 0 \\ -\sin 2\pi/4 & \cos 2\pi/4 & 0 \\ 0 & 0 & 1 \end{bmatrix} = \begin{bmatrix} 0 & 1 & 0 \\ -1 & 0 & 0 \\ 0 & 0 & 1 \end{bmatrix} \quad (1\text{-}60)$$

Another way to derive the same matrix is to note that $C_4(z)$ takes x to y and y to $-x$. Using the same reasoning, the C_2 representation is then

$$\boldsymbol{C}_2 = \begin{bmatrix} \cos\pi & \sin\pi & 0 \\ -\sin\pi & \cos\pi & 0 \\ 0 & 0 & 1 \end{bmatrix} = \begin{bmatrix} -1 & 0 & 0 \\ 0 & -1 & 0 \\ 0 & 0 & 1 \end{bmatrix} \quad (1\text{-}61)$$

There are two σ_v operations, one utilizing the xz and the other the yz plane. $\sigma(xz)$ leaves x and z unchanged but changes y to $-y$ (Fig. 1-45[a]). $\sigma(yz)$ leaves y and z unchanged but takes x to $-x$. Similarly, Fig. 1-45 (b) shows that σ_d' takes the point (x, y, z) to $(-y, -x, z)$. The four σ matrices come out as shown in eq. 1-62. You should be able to see from what we have done

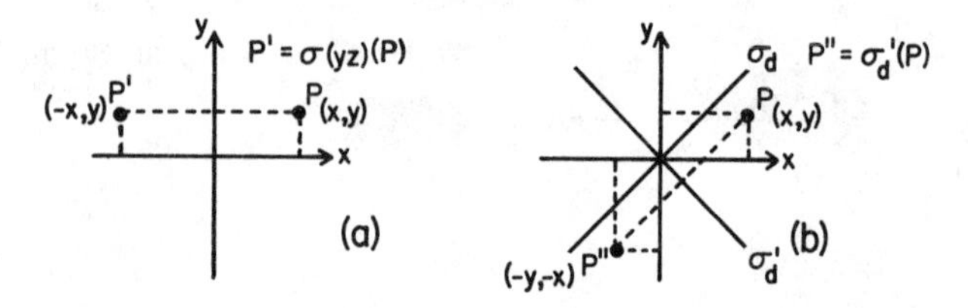

Fig. 1-45. Figures used to construct matrix representations of the operations (a) $\sigma(yz)$ and (b) σ_d'.

$$\sigma(xz) = \begin{bmatrix} 1 & 0 & 0 \\ 0 & -1 & 0 \\ 0 & 0 & 1 \end{bmatrix} \quad \sigma(yz) = \begin{bmatrix} -1 & 0 & 0 \\ 0 & 1 & 0 \\ 0 & 0 & 1 \end{bmatrix}$$

$$\sigma_d = \begin{bmatrix} 0 & 1 & 0 \\ 1 & 0 & 0 \\ 0 & 0 & 1 \end{bmatrix} \quad \sigma_d' = \begin{bmatrix} 0 & -1 & 0 \\ -1 & 0 & 0 \\ 0 & 0 & 1 \end{bmatrix} \tag{1-62}$$

that z is never mixed with x or y by these operations. But x and y are mixed with each other by the C_4 rotation and two of the σ operations. The matrices are all blocked out with x and y in one block and z in another:

$$\begin{array}{ccccc} & E & C_4 & C_4{}^3 & C_2 \\ \Gamma_{x,y} & \begin{bmatrix} 1 & 0 \\ 0 & 1 \end{bmatrix} & \begin{bmatrix} 0 & 1 \\ -1 & 0 \end{bmatrix} & \begin{bmatrix} 0 & -1 \\ 1 & 0 \end{bmatrix} & \begin{bmatrix} -1 & 0 \\ 0 & -1 \end{bmatrix} \\ \Gamma_z & [1] & [1] & [1] & [1] \end{array}$$

$$\begin{array}{ccccc} & \sigma(xz) & \sigma(yz) & \sigma_d & \sigma_d' \\ \Gamma_{x,y} & \begin{bmatrix} 1 & 0 \\ 0 & -1 \end{bmatrix} & \begin{bmatrix} -1 & 0 \\ 0 & 1 \end{bmatrix} & \begin{bmatrix} 0 & 1 \\ 1 & 0 \end{bmatrix} & \begin{bmatrix} 0 & -1 \\ -1 & 0 \end{bmatrix} \\ \Gamma_z & [1] & [1] & [1] & [1] \end{array}$$

The *characters* of z are all ones while the characters of x and y, *together*, are:

	E	$2C_4$	C_2	$2\sigma_v$	$2\sigma_d$
$\Gamma_{x,y}$	2	0	-2	0	0

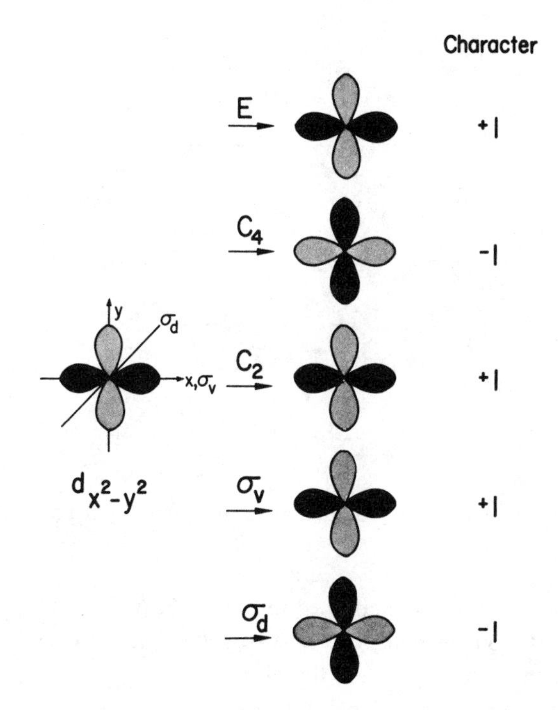

Fig. 1-46. Transformation of the $d_{x^2-y^2}$ orbital by the operations of the point group C_{4v}.

We say that z *forms a basis* for the irreducible representation A_1 in the group C_{4v} because the characters of both Γ_z and A_1 are all ones. We say that x and y *together* form a basis for the two-dimensional irreducible representation, E, whose characters are 2 0 −2 0 0.

The function $x^2 - y^2$ has the symmetry of a $d_{x^2-y^2}$ orbital. Figure 1-46 shows how this orbital transforms under the symmetry operations of C_{4v}. The characters of the representation whose basis function is $x^2 - y^2$ is therefore:

	E	$2C_4$	C_2	$2\sigma_v$	$2\sigma_d$
$\Gamma_{x^2-y^2}$	1	−1	1	1	−1

But these are just the characters of the irreducible representation B_1. We therefore say that the function $x^2 - y^2$ *forms a basis* for the representation B_1, or *transforms* as B_1.

Other symbols in the character table are the rotations about the x, y, and z axes (called R_x, R_y, and R_z) and the cubic (or f orbital) basis functions.

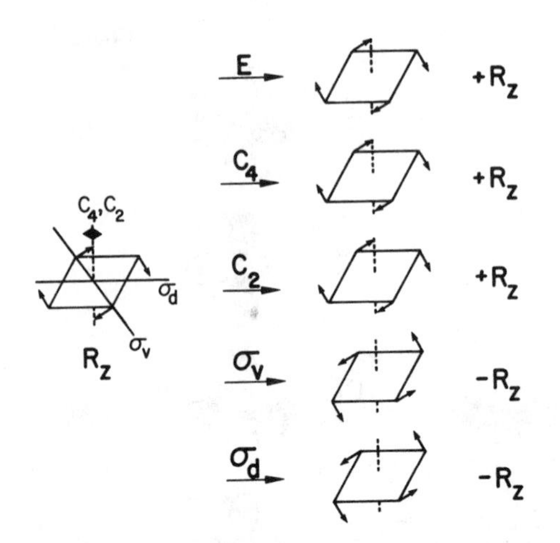

Fig. 1-47. Transformation of R_z (rotation about the z axis) by the operations of the point group C_{4v}.

Rotation about the z axis is examined in Fig. 1-47. E, C_4, and C_2 leave the direction of rotation unchanged. The two reflections reverse the sense of rotation. The characters of R_z are therefore

	E	$2C_4$	C_2	$2\sigma_v$	$2\sigma_d$
Γ_{R_z}.	1	1	1	-1	-1

R_z, therefore, forms a basis for, or transforms as, the A_2 irreducible representation. R_x and R_y, together, form a basis for the E representation in C_{4v}. The importance of the representation $\Gamma_{x, y, z}$ at the bottom of the character table will be explained in Chapter Three.

Many point groups have imaginary characters given either by $\pm i(\pm\sqrt{-1})$ or the symbols "ε" and "ε*". For a group with a principal axis of rotation, C_n, $\varepsilon = e^{2\pi i/n}$. Noting the identity $e^{i\theta} = \cos\theta + i\sin\theta$, we find that

$$\varepsilon = e^{2\pi i/n} = \cos 2\pi/n + i \sin 2\pi/n \tag{1-63}$$

The imaginary characters always appear in pairs of complex conjugates. Their application to physical problems requiring real characters will be described as needed.

The tables for the *infinite* point groups $C_{\infty v}$ and $D_{\infty h}$ (each of which has an infinite number of operations) have different names for the irreducible representations. Instead of A, B, E, and T, these tables use Greek letters. Σ representations are one-dimensional and all others are two-dimensional.

Problems

1-15. A C_{2v} character table without the characters is given below. Using the basis functions at the right side of the table, fill in the characters.

C_{2v}	E	C_2	$\sigma_v(xz)$	$\sigma_v'(yz)$			
A_1					z	x^2, y^2, z^2	$z^3, z(x^2-y^2)$
A_2					R_z	xy	xyz
B_1					x, R_y	xz	$xz^2, x(x^2-3y^2)$
B_2					y, R_x	yz	$yz^2, y(3x^2-y^2)$

1-16. A partial D_3 character table is given below. Using the basis functions at the right side of the table, determine the missing characters.

D_3	E	$2C_3$	$3C_2$			
A_1	1				x^2+y^2, z^2	$x(x^2-3y^2)$
A_2	1			z, R_z		$z^3, y(3x^2-y^2)$
E	2			$(x, y), (R_x, R_y)$	$(x^2-y^2, xy), (xz, yz)$	$(xz^2, yz^2), [xyz, z(x^2-y^2)]$

When two functions are grouped inside of parentheses or brackets (*e.g.*, "(x, y)"), they must be considered *together* as bases of a two dimensional representation.

1-17. Write as many of the basis functions [$x, y, z, R_x, R_y, R_z, x^2+y^2, z^2, xy, xz, yz, x^2-y^2, z^3, xz^2, yz^2, xyz, z(x^2-y^2), x(x^2-3y^2)$, and $y(3x^2-y^2)$] as you can along side the proper irreducible representations of the D_4 character table below. By convention the C_2' axes are colinear with the x and y axes and the C_2'' axes bisect the x and y axes.

D_4	E	$2C_4$	$C_2(=C_4^2)$	$2C_2'$	$2C_2''$			
A_1	1	1	1	1	1			
A_2	1	1	1	−1	−1			
B_1	1	−1	1	1	−1			
B_2	1	−1	1	−1	1			
E	2	0	−2	0	0			

1-8. Decomposition of Reducible Representations and the Direct Product

When we begin to apply group theory to chemical problems, we will have to deal with characters of *reducible* representations. The task will be to find the irreducible representations whose sum is our reducible representation. Table 1-10 presents an example. In the point group C_{2v}, we come across a

Table 1-10. An Example of a Reducible Representation in the Point Group C_{2v}

C_{2v}	E	C_2	$\sigma_v(xz)$	$\sigma_v(yz)$
A_1	1	1	1	1
A_2	1	1	−1	−1
B_1	1	−1	1	−1
B_2	1	−1	−1	1
Γ_{red}	3	1	3	1

reducible representation, Γ_{red}, with characters 3 1 3 1. It can be found that Γ_{red} is just the sum $2A_1 + B_1$:

$$\begin{array}{rrrrr} 2A_1 & = 2 & 2 & 2 & 2 \\ B_1 & = 1 & -1 & 1 & -1 \\ \hline 2A_1 + B_1 & = 3 & 1 & 3 & 1 \end{array}$$

How can you determine that $\Gamma_{\text{red}} = 2A_1 + B_1$? There are two ways. The decomposition of any reducible representation is *unique*; so if you find any combination of irreducible representations whose sum is Γ_{red}, you have the answer. Sometimes this process of inspection is the fastest way to decompose a reducible representation.

The second, longer and infallible† way of decomposing a reducible representation is to use eq. 1-64:

$$a_i = (1/h) \sum_R (\chi^R \cdot \chi_i^R \cdot C^R) \tag{1-64}$$

where a_i = the number of times the irreducible representation, Γ_i, appears in the reducible representation, Γ_{red}

h = order of the point group

R = an operation of the group

χ^R = character of the operation R in the reducible representation, Γ_{red}

χ_i^R = character of the operation R in the irreducible representation, Γ_i

C^R = the number of members in the class to which R belongs

†Infallible if the order of the group is not infinite. Equation 1-64 will not work for the point groups $C_{\infty v}$ or $D_{\infty h}$, for which cases the process of inspection is described in detail in Chapter Three. Although we find the method of inspection to be the easiest, alternate methods have been described by L. Schäfer and S.J. Cyvin (*J. Chem. Ed.*, *48*, 295 [1971]) and by D.P. Strommen and E.R. Lippincott (*J. Chem. Ed.*, *49*, 341 [1972]).

This formula may appear more formidable than it actually is. To see how it works, we go back to the example in Table 1-10. The order of the group is 4. The number of times the irreducible representation A_1 appears in the reducible representation, Γ_{red}, is given by:

$$\begin{aligned} a_{A_1} &= (1/4) \sum_R (\chi^R \cdot \chi_{A_1}^R \cdot C^R) \\ &= (1/4)[3 \times 1 \times 1 + 1 \times 1 \times 1 + 3 \times 1 \times 1 + 1 \times 1 \times 1] \\ &= 2 \end{aligned} \tag{1-65}$$

Similarly,

$$\begin{aligned} a_{A_2} &= (1/4)[3 \times 1 \times 1 + 1 \times 1 \times 1 + 3 \times (-1) \times 1 + 1 \times (-1) \times 1] = 0 \\ a_{B_1} &= (1/4)[3 \times 1 \times 1 + 1 \times (-1) \times 1 + 3 \times 1 \times 1 + 1 \times (-1) \times 1] = 1 \\ a_{B_2} &= (1/4)[3 \times 1 \times 1 + 1 \times (-1) \times 1 + 3 \times (-1) \times 1 + 1 \times 1 \times 1] = 0 \end{aligned} \tag{1-66}$$

Table 1-11 gives an example emphasizing that you need to remember to include the number of members of each class (C^R) in these computations.

$$\begin{aligned} a_{A_1} &= (1/6)[12 \times 1 \times 1 + 0 \times 1 \times 2 + 2 \times 1 \times 3] = 3 \\ a_{A_2} &= (1/6)[12 \times 1 \times 1 + 0 \times 1 \times 2 + 2 \times (-1) \times 3] = 1 \\ a_E &= (1/6)[12 \times 2 \times 1 + 0 \times (-1) \times 2 + 2 \times 0 \times 3] = 4 \\ \Gamma_{\text{red}} &= 3A_1 + A_2 + 4E \end{aligned} \tag{1-67}$$

Our final topic concerns the *direct product* of the representations Γ_i and Γ_j. It is obtained by multiplying the characters of the two representations under each operation. Table 1-12 gives some examples of direct products in the point group D_3. The characters of the direct product $A_2 \times E$, for example, were obtained by multiplying the characters of A_2 and E under each operation. The coefficients C^R do not come into the direct product. Notice that the direct product of two irreducible representations is a new representation which is either irreducible itself, or which can be reduced.

Table 1-11. A Reducible Representation in the Point Group C_{3v}

C_{3v}	E	$2C_3$	$3\sigma_v$
A_1	1	1	1
A_2	1	1	−1
E	2	−1	0
Γ_{red}	12	0	2

Table 1-12. Examples of Direct Products in the Point Group D_3

D_3	E	$2C_3$	$3C_2$	
A_1	1	1	1	
A_2	1	1	-1	
E	2	-1	0	
$A_1 \times E$	2	-1	0	$= E$
$A_2 \times E$	2	-1	0	$= E$
$E \times E$	4	1	0	$= A_1 + A_2 + E$
$A_2 \times A_2$	1	1	1	$= A_1$

Problems

1-18. Decompose the reducible representation 6 0 0 in the point group C_{3v} into a sum of irreducible representations. Write the product $E \times E$ in this group as a sum of irreducible representations.

1-19. Decompose the following reducible representations:

D_4	E	$2C_4$	C_2	$2C_2'$	$2C_2''$
Γ_1	3	-1	-1	1	-1
Γ_2	2	2	2	0	0
Γ_3	8	0	0	0	0
Γ_4	4	-2	0	-2	2

1-20. In the point group D_{3h}, find the following direct products in terms of irreducible representations: $A_1' \times A_2'$, $A_2' \times E'$, $E' \times E''$, $E'' \times E''$.

1-21. In the point group $C_{\infty v}$, decompose Γ_{red} by inspection:

$C_{\infty v}$	E	$2C_\infty^\phi$	$\cdots$	$\infty\sigma_v$
Γ_{red}	6	$2+4\cos\phi$		2

Additional Problems

1-22. Assign each molecule below to the proper point group.

a. o-phenanthroline

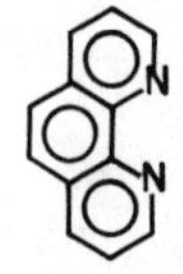

b. stannous chloride

Sn
Cl Cl

c. cyclopentadienide anion

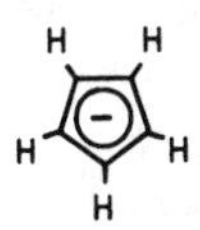

d. phosphorus oxychloride

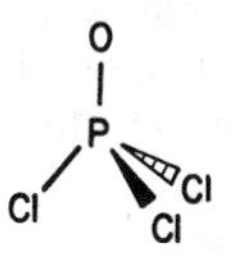

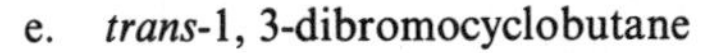
e. *trans*-1, 3-dibromocyclobutane

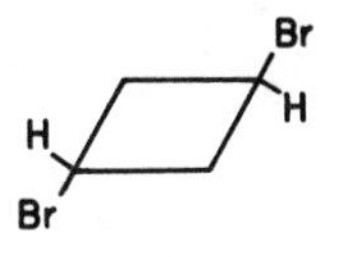

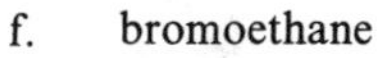
f. bromoethane

H
Br H
H H
H

g. NbF_7^{2-} (monocapped trigonal prism)

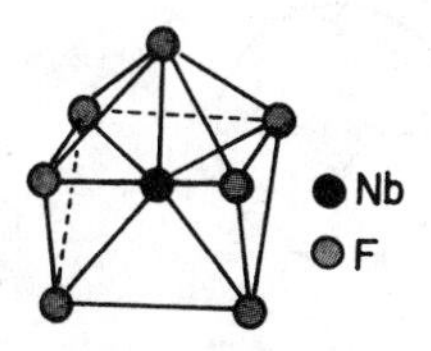

Redrawn from J.L. Hoard, *J. Amer. Chem. Soc., 61*, 1252 (1939).

h. ReH_9^{2-} (tricapped trigonal prism)

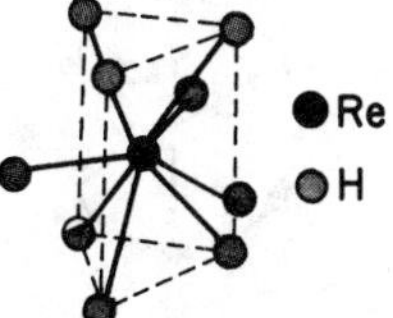

Redrawn from C.S. Abrahams, *et al., Inorg. Chem., 3*, 558 (1964).

i. I—I—I
3.10Å 2.82Å

(found in $NH_4^+ I_3^-$)

j. I—I—I
2.90Å

(found in $\phi_4As^+ I_3^-$)

k. staggered ethane

l. eclipsed ethane

m. cyclobutane

n. naphthalene

1-23. In the drawing of bis[heptanetrionato(2−)]bispyridinedicopper(II) below, the six oxygen atoms define a plane. Each Cu atom is displaced 0.24 Å from this plane toward the pyridine ligand to which it is attached. What is the point group of the ten atoms of the unit $Cu_2O_6N_2$?

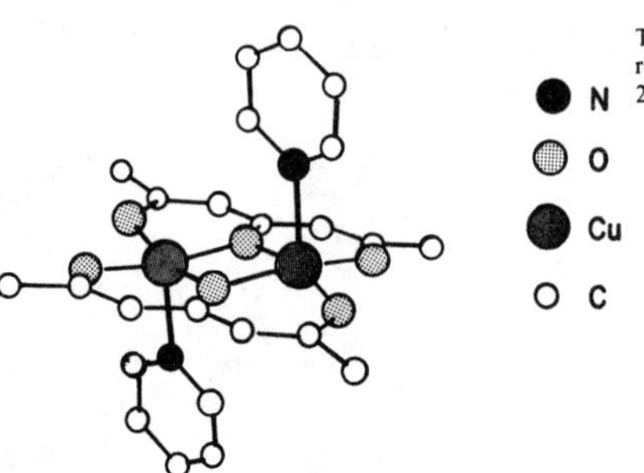

The structure of bis[heptanetrionato(2-)]bispyridinedicopper(II) redrawn from A.B. Blake and L.R. Fraser, *J. Chem. Soc. Dalton*, 2554 (1974).

1-24. Complete the following stereographic projections and label each with the proper point group.

1-25. Show that the matrix U and its transpose (obtained by interchanging rows and columns of U) are each other's inverse. (A matrix whose transpose is its inverse is said to be orthogonal.) Show that the similarity transformation $U \cdot B \cdot U^{-1}$ will block diagonalize B.

$$U = \begin{bmatrix} 1/\sqrt{2} & 1/\sqrt{2} & 0 \\ 0 & 0 & 1 \\ 1/\sqrt{2} & -1/\sqrt{2} & 0 \end{bmatrix} \qquad B = \begin{bmatrix} 1 & 2 & 3 \\ 2 & 1 & 3 \\ 3 & 3 & 9 \end{bmatrix}$$

1-26. In the point group D_2, whose characters are given below, to which irreducible representations do the functions x, y, z, R_x, R_y, R_z, $x^2 - y^2$, $x^2 + y^2$, z^2, xz, xy, yz, z^3, xz^2, yz^2, xyz, $z(x^2 - y^2)$, $x(x^2 - 3y^2)$, and $y(3x^2 - y^2)$ belong?

D_2	E	$C_2(z)$	$C_2(y)$	$C_2(x)$			
A_1	1	1	1	1			
B_1	1	1	−1	−1			
B_2	1	−1	1	−1			
B_3	1	−1	−1	1			

1-27. Using the basis functions below, determine the characters of the irreducible representations of the point group C_{5v}.

C_{5v}	E	$2C_5$	$2C_5{}^2$	$5\sigma_v$	
A_1					z
A_2					R_z
E_1					(x, y)
E_2					$(x^2 - y^2, xy)$

To determine the effect of a rotation, R_θ, on a quadratic function such as xy, we let R_θ operate on each component (x and y) separately, and then take the product. For example, using eq. 1-35,

$$R_\theta \cdot x = x \cos\theta + y \sin\theta$$
$$R_\theta \cdot y = -x \sin\theta + y \cos\theta$$

Therefore,

$$\begin{aligned} R_\theta \cdot xy &= (x\cos\theta + y\sin\theta)(-x\sin\theta + y\cos\theta) \\ &= xy(\cos^2\theta - \sin^2\theta) - (x^2 - y^2)(\cos\theta\sin\theta) \\ &= xy\cos 2\theta - (1/2)(x^2 - y^2)\sin 2\theta \end{aligned}$$

1-28. Decompose the following direct products in the indicated point groups. Do not use Appendix B.

Point Group	*Direct Products*
C_{4v}	$A_1 \times A_1, A_2 \times B_1, A_2 \times B_2, B_2 \times E, E \times E$
C_{6v}	$A_1 \times A_1, A_2 \times B_1, A_2 \times B_2, B_2 \times E_1, B_2 \times E_2,$ $E_1 \times E_1, E_1 \times E_2, E_2 \times E_2$
D_2	$B_1 \times B_2, B_2 \times B_3, B_1 \times B_3, B_3 \times A$
O_h	$E_g \times T_{2g}$
C_4	$B \times E, E \times E$ (see text of Appendix B for handling imaginary characters)

Related Reading

D.M. Bishop, *Group Theory and Chemistry,* Oxford University Press, London, 1973.
F.A. Cotton, *Chemical Applications of Group Theory,* John Wiley & Sons, N.Y., 1971.
J.P. Fackler, Jr., *Symmetry in Coordination Chemistry,* Academic Press, N.Y., 1971.
L.H. Hall, *Group Theory and Symmetry in Chemistry,* McGraw-Hill, N.Y., 1969.
R.M. Hochstrasser, *Molecular Aspects of Symmetry,* Benjamin, N.Y., 1966.
H.H. Jaffé and M. Orchin, *Symmetry in Chemistry*, John Wiley & Sons, N.Y., 1965.
A. Nussbaun, *Applied Group Theory for Chemists, Physicists and Engineers,* Prentice-Hall, Englewood Cliffs, New Jersey, 1971.
M. Tinkham, *Group Theory and Quantum Mechanics,* McGraw-Hill, N.Y., 1964.

2 · *A skirmish with quantum mechanics*

2-1. Introduction

Since spectroscopy deals with transitions of molecules from one *state* (or energy level) to another, we need some machinery to describe these states. Quantum mechanics provides that machinery. This chapter is intended to either introduce some of the ideas and terminology of quantum mechanics to a reader who has not seen them in an undergraduate physical chemistry course, or to reinforce the experience of one who has. Readers familiar with these topics might choose to skim this chapter and proceed to Chapter Three.

2-2. Light

A. *Some Properties of Light.* It is convenient to think of light as having the properties of both particles and waves. Light waves consist of perpendicular oscillating electric (ε) and magnetic (B) fields. To simplify our discussion, consider *plane polarized light* in which each field oscillates in only one plane, as in Fig. 2-1. It is the oscillating field that interacts with the charged nuclei and electrons of a molecule and results in energy transfer to the molecule. The *wavelength*, λ, is the crest-to-crest distance shown in Fig. 2-1. The *frequency*, ν, is the number of times per second that a crest passes a given point on the x axis. The relation between frequency and wavelength is

$$\boxed{\lambda\nu = c} \qquad (2\text{-}1)$$

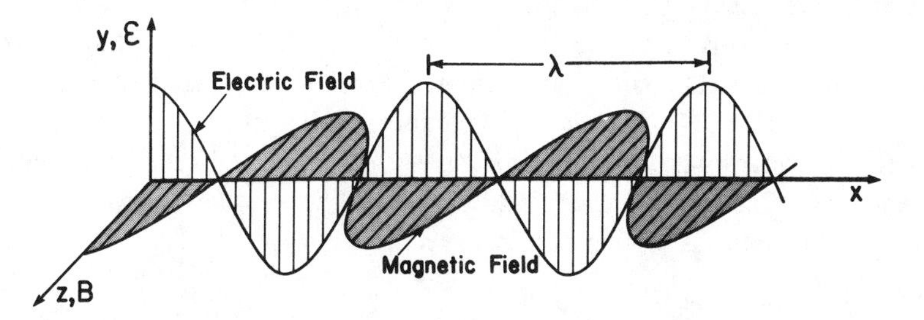

Fig. 2-1. Plane polarized electromagnetic radiation of wavelength λ propagating along the x axis. The electric field is confined to the xy plane and the magnetic field is confined to the xz plane. The maximum value of ε is ε_0.

where c is the speed of light. The speed of light in a vacuum is 2.9979×10^8 m s^{-1}, but in any other medium the speed of light is c/n, where n is the *refractive index* of that medium. When light passes from one medium to another of different refractive index, the value of ν is unchanged, but the value of λ changes to accommodate the different value of c.

If light is viewed as discrete particles called *photons*, each photon carries the energy E, given by†

$$\boxed{E = h\nu} \tag{2-2}$$

where h is Planck's constant. The *wave number*, $\bar{\nu}$, is defined as

$$\boxed{\bar{\nu} \equiv 1/\lambda} \tag{2-3}$$

Wave number is almost always used with the unit cm^{-1}, read "reciprocal centimeters" or "wave numbers." Since energy is directly proportional to wave number,

$$E = h\nu = hc/\lambda = hc\bar{\nu} \tag{2-4}$$

† We will gloss over an awful lot of profound statements, such as $E = h\nu$, in this chapter as if they are trivial or obvious facts of nature. Among all the spectacular achievements of Albert Einstein, it was the application of this statement to light for which he was awarded the Nobel Prize in 1921. Planck first proposed this relationship in connection with black body emission, but Einstein was the first to apply it to photons.

we will often speak of "energy" in wave number units. An "energy" of 12,000 cm^{-1} really means an energy of $hc\bar{\nu}$ = $(6.6262 \times 10^{-34}$ J s) (2.9979 $\times 10^8$ m s^{-1}) (12,000 cm^{-1}) (100 cm m^{-1}) = 2.3838×10^{-19} J or 143.56 kJ mol^{-1} or 34.311 kcal mol^{-1}. The energy of a mole of photons of a particular wavelength is called an *einstein* of energy. For example, one einstein is equal to 176 kJ for 680 nm red light and 249 kJ for 480 nm blue light. Light has momentum, p, as well as energy†:

$$\boxed{p = E/c} \tag{2-5}$$

It will be of interest later to know that each photon has one unit (1 unit $\equiv h/2\pi \equiv \hbar$) of angular momentum, as well as whatever linear momentum it carries.

Viewing light as a wave again, the *energy density* (energy per unit volume), ρ, stored in an electromagnetic field is

$$\rho = \epsilon_0 \langle \varepsilon^2 \rangle \tag{2-6}$$

where ϵ_0 is the dielectric constant of free space (see note on units at the beginning of the book) and the symbol $\langle\ \rangle$ denotes the average of a quantity. For the plane wave in Fig. 2-1, $\langle \varepsilon^2 \rangle = (1/2)\varepsilon_0{}^2$; so the energy density is $(1/2)\epsilon_0\varepsilon_0{}^2$. The *intensity, I,* of an electromagnetic wave is the energy per unit time crossing a unit square surface (J s^{-1} m^{-2}), which would be the energy density times the speed of the wave:

$$I = c\epsilon_0 \langle \varepsilon^2 \rangle = (1/2)c\epsilon_0\varepsilon_0{}^2 \tag{2-7}$$

We will make use of these equations later in the discussion of the absorption of light by molecules.

B. *Equations of waves.* The oscillating electric field in Fig. 2-1 is described by the equation††

$$\varepsilon = \varepsilon_0 \cos\left(\frac{2\pi x}{\lambda} - 2\pi\nu t\right) \tag{2-8}$$

†The relativistic expression for the energy of a particle of rest mass m_0 and momentum p is: $E = (m_0^2 c^4 + p^2 c^2)^{1/2}$. This reduces to the famous equation $E = m_0 c^2$ when $p = 0$. For a photon, $m_0 = 0$ and this reduces to $E = pc$.

††Several potentially confusing uses of the Greek letter epsilon appear in this chapter:

ε – electric field
ε_0 – maximum value of electric field
ϵ – extinction coefficient (Section 2-2-C)
ϵ_0 – dielectric constant of free space

This equation defines a wave of maximum amplitude ε_0, wavelength λ, and frequency ν moving in the $+x$ direction. The factor $2\pi x/\lambda$ assures that the wave completes one full oscillation in the interval $\Delta x = \lambda$, and the factor $2\pi\nu t$ fixes the speed of the wave at $c = \lambda\nu$. For example, at the time $t = 1/4\nu$,

$$\varepsilon = \varepsilon_0 \cos\left(\frac{2\pi x}{\lambda} - \frac{2\pi\nu}{4\nu}\right) = \varepsilon_0 \cos\left(\frac{2\pi x}{\lambda} - \frac{\pi}{2}\right) = \varepsilon_0 \sin\frac{2\pi x}{\lambda} \quad (2\text{-}9)$$

This says that at time $t = 1/4\nu$, the wave has moved a distance $\lambda/4$ to the right (Fig. 2-2). To see that this is consistent with the speed being $\lambda\nu$, we compute the ratio $\Delta x/\Delta t$:

$$\text{speed} = \frac{\Delta x}{\Delta t} = \frac{\lambda/4}{1/4\nu} = \lambda\nu \quad (2\text{-}10)$$

It is often remarkably convenient to use a *complex* wave equation containing the imaginary number i. Making use of Euler's identity,

$$\boxed{e^{i\theta} = \cos\theta + i\sin\theta} \quad (2\text{-}11)$$

we can write a complex wave equation of the form

$$\psi(x,t) = \varepsilon_0 e^{2\pi i(x/\lambda - \nu t)} = \underbrace{\varepsilon_0 \cos\left[2\pi\left(\frac{x}{\lambda} - \nu t\right)\right]}_{\varepsilon} + i\varepsilon_0 \sin\left[2\pi\left(\frac{x}{\lambda} - \nu t\right)\right] \quad (2\text{-}12)$$

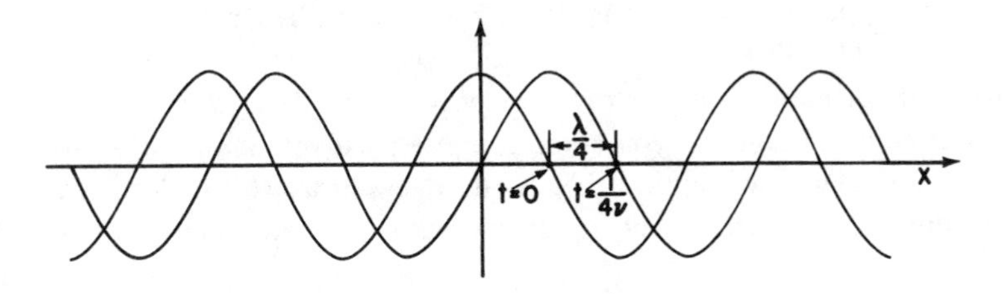

Fig. 2-2. The wave moves a distance $\lambda/4$ in the time $1/4\nu$.

The real wave form, eq. 2-8, is just the real part of the complex equation 2-12.

C. *Absorption of Light.* In the most common forms of spectroscopy, a sample is illuminated with light and we measure how much light is absorbed by the sample as a function of the energy of the light. Two ways are commonly used to express how much light is absorbed by a sample. If the intensity of light striking a sample is I_0, and if the intensity of light which passes out of the other side of the sample is I, the *transmittance*, T, is the ratio

$$\boxed{T = I/I_0} \qquad (2\text{-}13)$$

Percent transmittance (= 100 T) is also often encountered. More commonly, we are interested in the *absorbance, A,* defined as

$$\boxed{A = \log_{10} \frac{I_0}{I} = \log_{10} (1/T)} \qquad (2\text{-}14)$$

The absorbance is often called the *optical density*, or just OD. When $I = I_0$, no light is absorbed and $A = 0$. When $I = 0.01\ I_0$, the fraction of light passing through the sample is $1/10^2$ and the absorbance is 2.

The reason A is the most useful measure of light absorption is that it is directly proportional to the concentration of the sample, C, as well as the length of the light path through the sample cell, ℓ. This is expressed by *Beer's Law*,

$$\boxed{A = \epsilon C \ell} \qquad (2\text{-}15)$$

in which the proportionality constant, ϵ, is called the *extinction coefficient.* If C is in moles per liter, and ℓ is in cm, ϵ has the units liter mole^{-1} cm^{-1} (M^{-1} cm^{-1}) and is often referred to as the *molar absorptivity.* When ϵ is reported without units, as it often is, *the units just mentioned are assumed.* The extinction coefficient tells you how strongly a particular sample absorbs light at a particular wavelength. It is almost always measured in electronic spectroscopy, but, unfortunately, rarely measured in vibrational spectroscopy.

A summary of the important equations of Section 2-1 appears in Table 2-1.

Table 2-1. Important Properties of Light

Relationship	Symbols		Frequently Used Units
$\lambda\nu = c$	λ	= wavelength	nm (nanometers, 10^{-9} m) Å (Ångstrom units, 10^{-10} m) μ (microns, 10^{-6} m) mμ (millimicrons, 10^{-9} m)
	ν	= frequency	Hz (hertz, 1 Hz = 1 oscillation per second) s^{-1}
	c	= speed of light	2.9979×10^8 m s^{-1} in vacuum
$\bar{\nu} = 1/\lambda$	$\bar{\nu}$	= wave number	cm^{-1} kK (kilokayser, 1 kK = 1000 cm^{-1})
$E = h\nu$ $= hc/\lambda$ $= hc\bar{\nu}$	E	= energy	J (1 joule = 1 kg m^2 s^{-2}) cal (calorie, 1 cal = 4.184 J) erg (1 erg = 10^{-7} J) eV (1 electron volt = 1.6022×10^{-19} J)
	h	= Planck's constant	6.6262×10^{-34} J s
	$\hbar$	= $h/2\pi$	1.0546×10^{-34} J s
$p = E/c$	p	= momentum	kg m s^{-1}
$A = \epsilon C\ell$	A	= absorbance = optical density (OD)	dimensionless
	ϵ	= extinction coefficient	liter mol^{-1} cm^{-1} (M^{-1} cm^{-1})
	C	= concentration	moles per liter (M)
	ℓ	= light path length	cm
$A = \log_{10}(I_0/I)$	I_0	= incident light intensity	einstein m^{-2} s^{-1} J m^{-2} s^{-1}
	I	= emergent light intensity	
$T = I/I_0$	T	= transmittance	
$\%T = 100\ T$	$\%T$	= percent transmittance	

Problems

2-1. Fill in the following table and label the direction of increasing energy.

	x-ray	vacuum uv	uv	visible	near ir	mid ir	far ir	microwave
λ (nm)	10	200	380	780				
$\bar{\nu}$ (cm^{-1})					3333	333	33.3	
ν (s^{-1})								

uv = ultraviolet ir = infrared

The term "vacuum uv" refers to a region where air (N_2, O_2, H_2O, *etc.*) absorbs uv light. In order to study a sample at these wavelengths, one evacuates one's spectrometer.

2-2. The absorbance of a 2.31×10^{-5} M solution of some compound is 0.822 at 266 nm. What is the value of ϵ at this wavelength if $\ell = 1$ cm?

2-3. What value of absorbance corresponds to 40% transmittance? If a 0.0100 M solution shows 40% transmittance at some wavelength, what will be the percent transmittance at this same wavelength for a 0.0200 M solution? If $\ell = 1$ cm, what is the optical density of each solution?

2-4. If all the energy is dissipated as light, how many photons per second are emitted by a 366 nm uv lamp rated at a power of 150 watts (1 watt = 1 J s^{-1})? Would a 150 watt tungsten lamp, which emits mainly visible light, produce more or less photons per second than the 366 nm lamp?

2-5. Write an equation analagous to eq. 2-8 to describe a wave of wavelength λ and frequency ν moving to the *left* along the x axis. A *standing wave*, one which oscillates in time but not in space, can be written as the sum of a wave moving to the left plus one moving to the right. Write an equation for a standing wave as the sum of eq. 2-8 plus the answer to the first part of this problem. Using the formulas $\cos(A + B) = \cos A \cos B - \sin A \sin B$ and $\cos(A - B) = \cos A \cos B + \sin A \sin B$, rewrite the equation of the standing wave as a product of two cosine terms, one of which depends only on x and one only on t. Show that the positions of the maxima and minima of this standing wave remain fixed in space, even though the amplitude oscillates in time.

2-3. The Postulates of Quantum Mechanics[†]

A. *Wave-like Properties of Matter*. Just as $E = h\nu$ established the particle-like nature of light, eq. 2-16 establishes the wave-like properties of matter:

$$\boxed{\lambda = h/p} \tag{2-16}$$

This relationship, postulated by de Broglie and named after him,[††] relates the *wavelength of a particle* to its momentum, p. The implications of eq. 2-16 are profound. For example, an electron accelerated through a potential of 100 V will have a kinetic energy of

$$\frac{mv^2}{2} = e\,V = (1.602 \times 10^{-19} \text{ coulomb})\,(100 \text{ V}) = 1.602 \times 10^{-17} \text{ J} \tag{2-17}$$

[†] We don't have the space in this volume to examine the fascinating history of the development of quantum mechanics in the first third of this century. But a clear, readable, and historically oriented discussion of the foundations of quantum mechanics can be found in Vol. 4 of the Berkeley Physics Series: E. H. Wichmann, *Quantum Physics*, McGraw-Hill, New York, 1971.

[††] This equation is another of the Gargantuan relationships on which quantum mechanics was built. De Broglie received the Nobel Prize in 1929. Equation 2-16 can be motivated in a very simple manner from the relation which applies to photons, $p = E/c$, by substituting $h\nu$ for E: $p = h\nu/c = h/\lambda$.

The momentum is obtained by rearranging eq. 2-17:

$$m^2 v^2 = 2meV$$
$$p = mv = \sqrt{2meV} = 5.403 \times 10^{-24} \text{ kg m s}^{-1} \qquad (2\text{-}18)$$

The *wavelength* of a 100 V electron will therefore be

$$\lambda = h/p = h/\sqrt{2meV} = 1.226 \times 10^{-10}\text{m} = 1.226 \text{ Å} \qquad (2\text{-}19)$$

That is to say, an electron accelerated through a potential of 100 V will behave like a *wave* with $\lambda = 1.226$ Å. Electromagnetic radiation of this wavelength is in the x-ray region of the spectrum. Experimental confirmation of de Broglie's hypothesis came almost immediately, when it was shown that electrons are diffracted by crystals just as x-rays are. It is therefore clear that not only does light have particle-like properties, but "particles" must have wave-like properties.

Problems

2-6. What is the de Broglie wavelength of an electron with a kinetic energy of 10 eV? What would your answer be if you used the correct relativistic expression $p^2c^2 = E^2 - m_0{}^2c^4$, where m_0 is the rest mass and E is the total energy (= kinetic energy + m_0c^2)? Calculate the de Broglie wavelength of a 1 MeV electron using the Newtonian and relativistic expressions.

2-7. What is the energy and momentum (in units of J mol^{-1} and kg m s^{-1} mol^{-1}) of 250 nm uv light? Of 500 nm visible light? Of 2000 cm^{-1} ir light? What is the ratio of the momenta of an electron and a 500 nm photon, if the electron has the same kinetic energy as the photon?

2-8. In the Davisson-Germer experiment, which confirmed de Broglie's hypothesis, electrons accelerated through a potential of 54 V were scattered off a crystal of Ni. Essentially only electrons scattered from the first layer of atoms on the crystal were detected. Under these conditions, the electron "waves" will

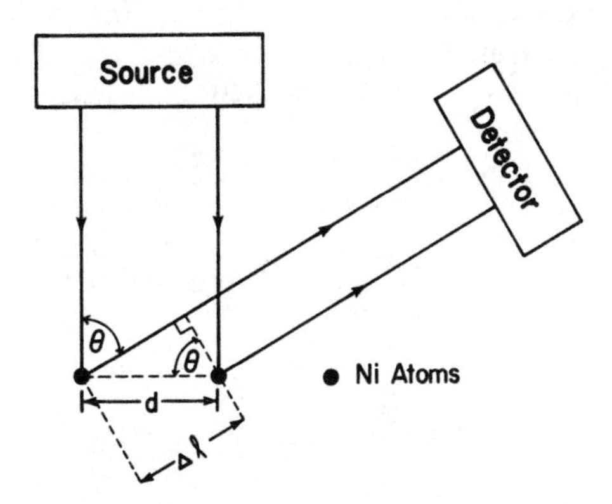

reinforce each other to produce maxima in the diffraction pattern at angles, θ, such that the distance $\Delta\ell\,(= d\sin\theta$, where d is the spacing between Ni atoms) is an integral multiple of λ: $n\lambda = d\sin\theta$. At what angle should the first diffraction maximum ($n = 1$) occur if $d = 2.15$ Å (from x-ray diffraction)? The observed angle was 50°.

2-9. The work function, ϕ, of a substance is defined as the minimum energy needed to free an electron from the surface of that substance. The values of ϕ for elemental Ni and Na are, respectively, 5.0 and 2.3 eV. What is the maximum wavelength of light sufficient to knock an electron off of each of these metals? Such light loosens an electron which then has zero kinetic energy. If light of energy $E > \phi$ is used, the kinetic energy of the free electron will be $K = E - \phi$. What is the kinetic energy of an electron loosened from each metal by 230 nm uv light?

B. *The Schrödinger Equation.* For most applications in this volume, the Schrödinger equation is the fundamental postulate of quantum mechanics.† The Schrödinger equation is an assumption, with no more derivation than $E = h\nu$. Each of these equations has withstood the test of time, so far, so we use them with confidence. But no derivation exists from more fundamental principles. What follows is a motivation, not a derivation, of the Schrödinger equation.

We start with the desire of writing a wave equation for a particle such as an electron; since we know that an electron does possess wave-like properties. We call the *wave function* Ψ, and choose to write it in a complex form analagous to eq. 2-12:

$$\Psi(x,t) = a_0 e^{2\pi i(x/\lambda - \nu t)} \tag{2-20}$$

where a_0 is a constant and λ and ν are the wavelength and frequency of the wave which describes the behavior of the particle. The wave function eq. 2-20 depends on one space coordinate and one time coordinate. Later we will extend the equation to include three space coordinates. To develop the Schrödinger equation, which is just a differential equation describing a function such as eq. 2-20, we start with some partial derivatives of $\Psi(x,t)$:

$$\frac{\partial^2\Psi(x,t)}{\partial x^2} = -\left(\frac{2\pi}{\lambda}\right)^2 \Psi(x,t) \tag{2-21}$$

$$\frac{\partial\Psi(x,t)}{\partial t} = -2\pi i\nu\Psi(x,t) \tag{2-22}$$

† Erwin Schrödinger published his equation in 1926 and received the Nobel Prize in 1933.

But we can rewrite eq. 2-21 using the successive substitutions $\lambda = h/p$ and $p^2 = 2mK$ (where K is kinetic energy):

$$\frac{\partial^2 \Psi}{\partial x^2} = -\left(\frac{2\pi}{\lambda}\right)^2 \Psi = -\left(\frac{p^2}{\hbar^2}\right)\Psi = -\left(\frac{2mK}{\hbar^2}\right)\Psi \tag{2-23}$$

where $\hbar \equiv h/2\pi$. But the total energy, E, is the sum of K plus $V(x,t)$, the potential energy. (For the moment we consider V as a function of both x and t. If V is independent of t, the system is said to be *conservative*, which just means that no energy is being added or taken out.) Writing $K = E - V(x,t)$, we find that eq. 2-23 becomes

$$\frac{\partial^2 \Psi(x,t)}{\partial x^2} = -\left(\frac{2mE}{\hbar^2}\right)\Psi(x,t) + \left(\frac{2mV(x,t)}{\hbar^2}\right)\Psi(x,t) \tag{2-24}$$

Returning to eq. 2-22, we make the substitution $\nu = E/2\pi\hbar$:

$$\frac{\partial \Psi(x,t)}{\partial t} = -\left(\frac{iE}{\hbar}\right)\Psi(x,t) \tag{2-25}$$

The Schrödinger equation (2-27) results from solving eqs. 2-24 and 2-25 for $E\Psi(x,t)$ and equating the results:

$$\underbrace{-\frac{\hbar^2}{2m}\frac{\partial^2 \Psi(x,t)}{\partial x^2} + V(x,t)\,\Psi(x,t)} = E\Psi(x,t)$$
$$= -\frac{\hbar}{i}\frac{\partial \Psi(x,t)}{\partial t} = \underbrace{i\hbar\frac{\partial \Psi(x,t)}{\partial t}} \tag{2-26}$$

or

$$-\frac{\hbar^2}{2m}\frac{\partial^2 \Psi(x,t)}{\partial x^2} + V(x,t)\,\Psi(x,t) = i\hbar\frac{\partial \Psi(x,t)}{\partial t} \tag{2-27}$$

Equation 2-27 can be extended to three dimensions in a straightforward manner, the result being

$$-\frac{\hbar^2}{2m}\left[\frac{\partial^2}{\partial x^2} + \frac{\partial^2}{\partial y^2} + \frac{\partial^2}{\partial z^2}\right]\Psi(x,y,z,t) + V(x,y,z,t)\Psi(x,y,z,t)$$
$$= i\hbar\frac{\partial \Psi(x,y,z,t)}{\partial t} \tag{2-28}$$

The three derivatives in brackets are usually combined in the symbol "∇^2," read "del squared," to give

$$\boxed{\left[-\frac{\hbar^2}{2m}\nabla^2 + V(x,y,z,t)\right]\Psi(x,y,z,t) = i\hbar\frac{\partial \Psi(x,y,z,t)}{\partial t}} \tag{2-29}$$

The term in brackets on the left side of eq. 2-29 is called the *hamiltonian operator*, $\mathscr{H}$.

$$\mathscr{H}(x,y,z,t) = -\frac{\hbar^2}{2m}\nabla^2 + V(x,y,z,t) \; . \tag{2-30}$$

The *operator* on the right side of eq. 2-29 is called the *energy operator*, $\hat{E}$.

$$\hat{E} = i\hbar\frac{\partial}{\partial t} \tag{2-31}$$

(Do not confuse the energy operator, $\hat{E}$, with the scalar quantity energy, E.) The *time-dependent Schrödinger equation* can therefore be written in the famous compact form

$$\boxed{\mathscr{H}\Psi = \hat{E}\Psi} \tag{2-32}$$

What have we done so far? We started by *assuming* that a particle could be described by the wave function eq. 2-20; and we derived the differential equation 2-32 relating Ψ to the particle's kinetic, potential, and total energy. Most of our applications of the Schrödinger equation will be to problems in which only the potential energy term varies.

For simplicity, we will confine our discussion to the one-dimensional equation 2-27. Suppose that the system (a molecule, for example) is conservative: energy is neither being added to the molecule nor emitted by the molecule. If V does not vary with time, the molecule has no way of ever changing its *state* which is characterized by some distribution of atomic positions, momenta, *etc.* It is said to be in a *stationary state*. It is precisely the stationary states of molecules in which we are generally interested. In the lowest energy stationary state of a hydrogen atom, for example, the electron is in the 1s orbital and will remain there forever unless we change the potential. Spectroscopy deals with transitions *between* different stationary states. These transitions can occur only when V varies with time, as when an oscillating electromagnetic field interacts with the molecule or when the molecule spontaneously spits out a photon and goes to a lower energy state. Until we get to the interaction of molecules with light in Chapter Three, we will be concerned only with conservative systems, whose wave functions are fully characterized by their spatial dependence only. It is our immediate task, then, to get rid of time from the Schrödinger equation.

The form of the wave function we have assumed, eq. 2-20, can be factored into separate functions of position and time:

$$\Psi(x,t) = a_0 e^{2\pi i(x/\lambda - \nu t)} = \underbrace{a_0 e^{2\pi i x/\lambda}}\ \underbrace{e^{-iEt/\hbar}}$$
$$\equiv \psi(x) \cdot \phi(t) \qquad (2\text{-}33)$$

To characterize the stationary states of a system, we consider what happens to the Schrödinger equation if V is a function of x but not t. Equation 2-26 allows us to solve for $\psi(x)$ by writing $\Psi(x,t) = \psi(x)\phi(t)$:

$$-\frac{\hbar^2}{2m}\frac{\partial^2[\psi(x)\phi(t)]}{\partial x^2} + V(x)\psi(x)\phi(t) = E\psi(x)\phi(t) \qquad (2\text{-}34)$$

Since $\partial^2/\partial x^2$ does nothing to $\phi(t)$, we can divide both sides of eq. 2-34 by $\phi(t)$ to get the desired result:

$$\boxed{-\frac{\hbar^2}{2m}\frac{\partial^2\psi(x)}{\partial x^2} + V(x)\psi(x) = E\psi(x)} \qquad (2\text{-}35)$$

Equation 2-35 is the *time-independent* form of the Schrödinger equation. By plugging in a potential function, $V(x)$, which pertains to a particular physical system, we can, in principle, solve for the wave functions, $\psi(x)$, which characterize the stationary states of that system. Equation 2-35 can be written in the more compact form

$$\boxed{\mathscr{H}\psi = E\psi} \qquad (2\text{-}36)$$

which differs from eq. 2-32 in that ψ is just a function of position while Ψ is a function of position and time. Furthermore, the quantity E in eq. 2-36 is the *scalar quantity,* energy, while the symbol $\hat{E}$, in eq. 2-32, is an *operator*. The hamiltonian has the same form in both the time-dependent and time-independent equations.

The solutions of the differential equation 2-35 (or any other form of the Schrödinger equation) have several noteworthy properties:

(1) In general, there will be an infinite number of wave functions, ψ_i, each with a corresponding energy, E_i, which satisfy eq. 2-35. The functions, ψ, are called *eigenfunctions* and the energies, E, are called *eigenvalues* of the hamiltonian operator. In general, any function which satisfies an

equation of the form 2-36 (in which $\mathscr{H}$ may be replaced by any operator) is called an eigenfunction and the corresponding scalar is called an eigenvalue.

(2) Sometimes two or more eigenfunctions (ψ_i, ψ_j, ψ_k, for example) have the *same* eigenvalues ($E_i = E_j = E_k$). In such a case, the functions (ψ_i, ψ_j, ψ_k) are said to be *degenerate.*

(3) If ψ is a solution of eq. 2-35, any constant multiple of ψ, $c\psi$ is also a solution.

(4) If ψ_i and ψ_j are degenerate solutions of eq. 2-35, having the same eigenvalue, any linear combination, $a\psi_i + b\psi_j$, is also a solution.

(5) Any two functions ψ_i and ψ_j which are not degenerate are *orthogonal,* which means that

$$\boxed{\int_{-\infty}^{\infty} \psi_i^* \psi_j dx = \int_{-\infty}^{\infty} \psi_j^* \psi_i dx = 0} \qquad (2\text{-}37)$$

where ψ^* is the complex conjugate of ψ, in case ψ is imaginary. If ψ_i and ψ_j are degenerate, eq. 2-37 does not necessarily hold, but it will always be possible to find a linear combination, $\psi_i + c\psi_j$, which is orthogonal to ψ_i.

Problems

2-10. If $V(x)$ is a constant, show that the functions $\psi = e^{i\alpha x}$, $\cos \beta x$ and $\sin \beta x$ all satisfy eq. 2-35. What are the values of α and β?

2-11. Verify that if ψ is a solution of eq. 2-35 ($V(x) \neq$ constant), any constant multiple of ψ is also a solution. Show that if ψ_1 and ψ_2 are degenerate, $\psi_1 + c\psi_2$ is also a solution of eq. 2-35, where c is any constant.

2-12. Show that $\cos 4x$ is an eigenfunction of the operator d^2/dx^2 and find the corresponding eigenvalue. Show that $xe^{-x^2/2}$ is an eigenfunction of the operator $(-d^2/dx^2 + x^2)$ and find the corresponding eigenvalue.

2-13. Using the right hand side of eq. 2-26 and making the substitution $\Psi(x,t) = \psi(x)\phi(t)$, derive a differential equation analagous to eq. 2-35, but which is satisfied by $\phi(t)$. Show that the form of $\phi(t)$ in eq. 2-33 satisfies your differential equation.

C. *The Meaning of* Ψ. We have so far *assumed* that the behavior of a particle can be described by a wave function which satisfies the Schrödinger equation. But what is the meaning of this wave function? We would like to postulate that the wave function Ψ (or ψ if we are dealing with stationary states, as we will in the remainder of this section) contains all of the information about a particle that one can know—its position, momentum, energy, *etc.* We will now examine the machinery used to extract this information from ψ.

First, it is postulated that the product $\psi^*(x)\psi(x)$ gives the *probability* that the particle can be found in the infinitesimal length interval, dx, at the point x. If we are dealing in three dimensions, the product $\psi^*(x,y,z)\psi(x,y,z)$ gives the probability that the particle can be found in the infinitesimal volume element $d\tau$ ($\equiv dxdydz$) at the point (x,y,z). For example, if the product $\psi^*(x)\psi(x)$ has the gaussian shape in Fig. 2-3, given by the equation

$$\psi^*\psi = \frac{1}{\sqrt{2\pi}}e^{-x^2/2} \tag{2-38}$$

the particle has a maximum probability of being found at the point $x = 0$. The probability that the particle can be found at $x = 1$ is just 0.607 times the probability at $x = 0$. There is less liklihood that the particle will be at $x = 1$ than at $x = 0$. The gaussian function in Fig. 2-3 never goes to zero for finite values of x, though it gets exponentially smaller as $|x|$ increases. For example, at $x = 10$, $\psi^*\psi = 7.69 \times 10^{-23}$, compared to a value of 0.399 at $x = 0$.

Since the sum of the probabilities of finding the particle at each point in space must be unity, the sum (or integral) of $\psi^*\psi$ over all space must be unity:

$$\boxed{\int_{-\infty}^{\infty} \psi^*\psi dx = 1} \tag{2-39}$$

Equation 2-39 is the *normalization condition* which we will demand of all wave functions. It was noted in the last section that if ψ is a solution of the Schrödinger equation, any multiple if ψ, $c\psi$, is also a solution. We will therefore choose c (which we assume to be real) such that the function $c\psi$ is normalized:

$$1 = \int_{-\infty}^{\infty} c^*\psi^* c\psi dx = c^2\int_{-\infty}^{\infty} \psi^*\psi dx$$

$$c = \left(\int_{-\infty}^{\infty} \psi^*\psi dx\right)^{-1/2} \tag{2-40}$$

As a further interpretation of ψ, the integral $\int_a^b \psi^*\psi dx$ gives the probability that the particle can be found somewhere in the interval between $x = a$ and $x = b$. For example, in Fig. 2-3, the probability of finding the particle between $x = -2$ and $x = -1$ is

$$\int_{-2}^{-1} \psi^*\psi dx = \int_{-2}^{-1} (1/\sqrt{2\pi})e^{-x^2/2}\, dx = 0.136 \tag{2-41}$$

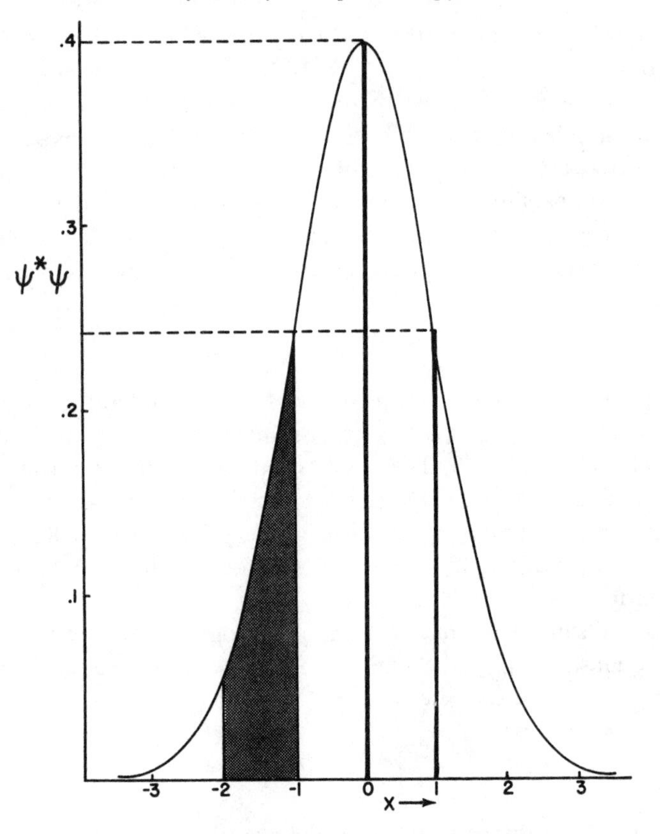

Fig. 2-3. A gaussian probability distribution, $\psi^*\psi = (1/\sqrt{2\pi})e^{-x^2/2}$. The maximum probability of finding the particle occurs at $x = 0$. The probability of finding the particle at $x = 1$ is $0.242/0.399 = 0.607$ times the probability of finding the particle at $x = 0$. The probability of finding the particle in the interval from $x = -2$ to $x = -1$ is equal to the hatched area, 0.136. The total area under the curve from $x = -\infty$ to $x = +\infty$ is unity.

(The integral was evaluated numerically.) That is, there is a 13.6% chance that if you look for the particle between $x = -2$ and $x = -1$, you will find it.

If a particle may be found in any of, say, three states, its wave function will be of the form

$$\psi = a_1\psi_1 + a_2\psi_2 + a_3\psi_3 \tag{2-42}$$

The probability that the particle is in state number 2, for example, will be given by the magnitude of $a_2{}^2$. This interpretation requires that $a_1{}^2 + a_2{}^2 + a_3{}^2 = 1$, since the sum of the three probabilities must be unity.

To extract other physical quantities, such as momentum, from ψ, it is postulated that there are quantum mechanical operators, $\hat{g}$, that correspond to each physical quantity such that if a large number of measurements of the property G are made, the average value found, $\langle G \rangle$, will be given by

$$\langle G \rangle = \frac{\int_{-\infty}^{\infty} \psi^* \hat{g} \psi \, dx}{\int_{-\infty}^{\infty} \psi^* \psi \, dx} \tag{2-43}$$

If ψ is normalized, the denominator will be unity. $\langle G \rangle$ is also called the *expectation value* of G. Some of the most common operators are listed in Table 2-2. The standard deviation, σ, of some measurable quantity is given by

$$\sigma = \sqrt{\langle G^2 \rangle - \langle G \rangle^2} \tag{2-44}$$

where $\langle G^2 \rangle$ is the average value of G^2. If $\sigma = 0$, G is precisely determined. If $\sigma \neq 0$, there is some distribution of measured values of G, a distribution which might have the shape of the curve in Fig. 2-3. Let's try to understand eqs. 2-43 and 2-44 with an example.

Suppose we have a particle whose wave function is

$$\psi = [(1/\sqrt{2\pi})e^{-x^2/2}]^{1/2} \tag{2-45}$$

We have already plotted the probability distribution, $\psi^*\psi$ in Fig. 2-3. What is the *average value* of the position of the particle if we measure x a large number of times? Equation 2-43 says that we get

$$\langle x \rangle = \frac{\int_{-\infty}^{\infty} \psi^* \hat{x} \psi \, dx}{\int_{-\infty}^{\infty} \psi^* \psi \, dx} = \frac{(1/\sqrt{2\pi}) \int_{-\infty}^{\infty} x e^{-x^2/2} \, dx}{(1/\sqrt{2\pi}) \int_{-\infty}^{\infty} e^{-x^2/2} \, dx} \tag{2-46}$$

To evaluate these integrals, you need to be aware of *even* and *odd* functions. If you plot $y = f(x)$ in the xy plane, an even function is one for which the yz plane is a mirror plane. An odd function is one for which the origin is a center of inversion. For example, in Fig. 2-4 the function $y = (1/\sqrt{2\pi})e^{-x^2/2}$ is an even function. It is symmetric to reflection across the y axis. The integral of this function over all space (*i.e.*, from $x = -\infty$ to $x = +\infty$) has some finite value equal to the area under the curve. The function $y = xe^{-x^2}$, on the other hand, is an odd function. The integral of this odd function over all space is exactly zero, because the positive

Table 2-2. Some One-Dimensional Quantum Mechanical Operators

Variable	Name	Symbol	Operation
x	position	$\hat{x}$	multiplication by x
x^2	—	$\hat{x}^2$	multiplication by x^2
p	momentum	$\hat{p}$	$-i\hbar(\partial/\partial x)$
p^2	—	$\hat{p}^2$	$[-i\hbar(\partial/\partial x)][-i\hbar(\partial/\partial x)] = -\hbar^2(\partial^2/\partial x^2)$
t	time	$\hat{t}$	multiplication by t
E†	energy	$\hat{E}$	$i\hbar(\partial/\partial t)$
E†	hamiltonian	$\mathscr{H}$	$-(\hbar^2/2m)(\partial^2/\partial x^2) + V(x)$

† Both the energy and hamiltonian operators give the energy as their eigenvalue. $\hat{E}$ operates on $\phi(t)$ while $\mathscr{H}$ operates on $\psi(x)$.

area is cancelled by the negative area (Fig. 2-4). In general, any function involving only even powers of x is even, and any function involving only odd powers of x is odd.

Observing that the integrand in the denominator of eq. 2-46 is an even function, we can break up the integral from $-\infty$ to $+\infty$ into a sum of two integrals from 0 to $+\infty$:

$$(1/\sqrt{2\pi})\int_{-\infty}^{\infty} e^{-x^2/2}\,dx = 2(1/\sqrt{2\pi})\int_{0}^{\infty} e^{-x^2/2}\,dx = 1 \tag{2-47}$$

The integral in eq. 2-47 was evaluated with the help of the definite integral $\int_0^\infty e^{-ax^2}dx = (1/2)\sqrt{\pi/a}$ which can be found in the *Handbook of Chemistry and Physics*, or elsewhere. The numerator of eq. 2-46 is the integral of an odd function, so it must be zero. Therefore $\langle x\rangle = 0/1 = 0$, which is certainly not too surprising, since the probability distribution is symmetric about $x = 0$. All we have said so far is that if you measure the position many times of a particle whose wave function is eq. 2-45, the average position is zero.

But what if you measure x^2 instead of x? x^2 is always positive, so the average value, $\langle x^2\rangle$, will be a positive number:

$$\langle x^2\rangle = \frac{(1/\sqrt{2\pi})\int_{-\infty}^{\infty} x^2 e^{-x^2/2}\,dx}{(1/\sqrt{2\pi})\int_{-\infty}^{\infty} e^{-x^2/2}\,dx} = \frac{1}{1} = 1 \tag{2-48}$$

To evaluate the numerator of eq. 2-48, we made use of the fact that the integrand is even, and used the definite integral $\int_0^\infty x^2 e^{-ax^2}dx = (1/4)\sqrt{\pi/a^3}$. The *standard deviation* of a large number of measurements of x would therefore be:

$$\sigma_x = \sqrt{\langle x^2\rangle - \langle x\rangle^2} = \sqrt{1-0} = 1 \tag{2-49}$$

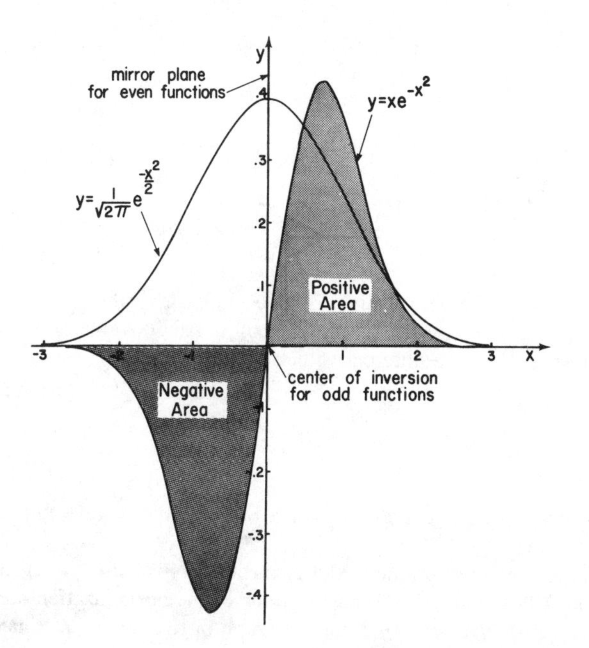

Fig. 2-4. Graphs of $y = (1/\sqrt{2\pi})e^{-x^2/2}$, an even function, and $y = xe^{-x^2}$, an odd function. The area under an odd function from $-\infty$ to $+\infty$ must be zero.

This is not overly surprising, either, since the probability distribution eq. 2-38 is the gaussian error function, in which x is the quantity known as the standard deviation.

Before beginning to apply the Schrödinger equation to some simple systems, we should make a few more points about wave functions:

(1) ψ must be finite, or the probability of finding the particle at a point where ψ is infinite would be infinite. (An exception to this would be the "wave function" for a particle at rest.)

(2) Since the probability of finding a particle in the interval dx at the point x is $\psi^*(x)\psi(x)$, we will require all wave functions to be single-valued (Fig. 2-5). A many-valued wave function would be unacceptable; for which value of the function would you choose to calculate the probability?

(3) ψ must be continuous since a real, physical quantity such as probability cannot be discontinuous.

(4) We will also require the first derivative, $d\psi/dx$, to be continuous unless $V(x)$ becomes infinite.

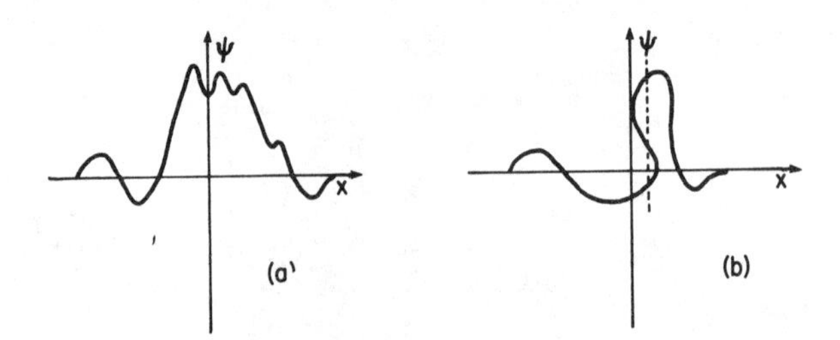

Fig. 2-5. (a) A single-valued function of x. For each value of x, there is only one value of ψ. (b) A many-valued function of x. At the value of x indicated by the dashed line, ψ has three different values.

Problems

2-14. Calculate $\langle p \rangle$, $\langle p^2 \rangle$ and σ_p for a particle whose wave function is given by eq. 2-45.

2-15. Suppose the wave function of a particle is $\psi = A/(1 + x)$ for $0 \leqslant x \leqslant 10$ and $\psi = 0$ elsewhere. Find the value of the normalization constant, A, such that $\int_0^{10} \psi^* \psi dx = 1$ and make a graph of this function. What are the values of $\langle x \rangle$, $\langle x^2 \rangle$, $\langle x \rangle^2$ and σ_x?

2-16. A linear operator, $\hat{g}$, has the property that $\hat{g}[r(x) + s(x)] = \hat{g}r(x) + \hat{g}s(x)$, where r and s are functions of x. Which of the following operators are linear: x, d/dx, d^2/dx^2, $\sqrt{}$, $(\ )^2$? Are the operators in Table 2-2 linear?

2-4. Some Simple Illustrations from Quantum Mechanics

A. *The Particle in a One-Dimensional Box.* Imagine a particle constrained to move along the x axis in the interval $0 \leqslant x \leqslant a$. In the language of potential energy diagrams, this corresponds to the infinitely deep, infinitely steep potential well shown in Fig. 2-6. Between the points $x = 0$ and $x = a$, the potential is zero and there are no constraints on the particle's motion. At the points $x = 0$ and $x = a$ the potential suddenly goes to infinity and remains at infinity outside of this interval. By definition, an infinite potential means that the particle cannot exist in the region outside $0 \leqslant x \leqslant a$. If a particle of mass m lived in such a well, what would its life be like?

Well, first of all, its wave function must be zero for $x < 0$ and $x > a$. Inside of this interval, the particle should be governed by the Schrödinger equation with $V = 0$:

$$-\frac{\hbar^2}{2m}\frac{\partial^2 \psi}{\partial x^2} = E\psi \tag{2-50}$$

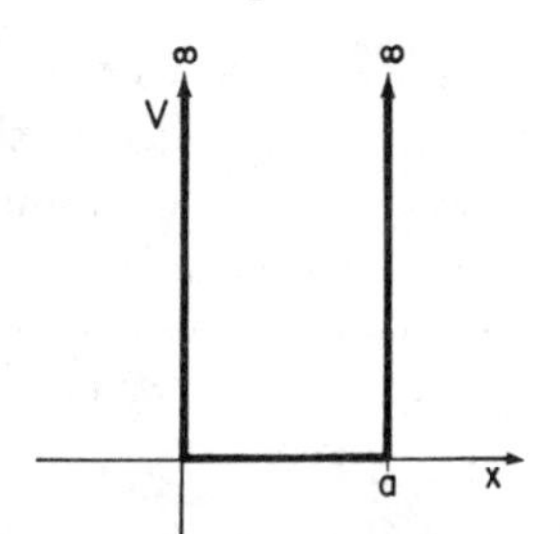

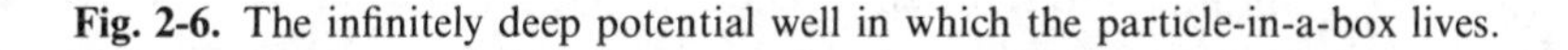

Fig. 2-6. The infinitely deep potential well in which the particle-in-a-box lives.

Equation 2-50 is satisfied, in general, by a solution of the form

$$\psi = A\cos\alpha x + B\sin\alpha x \tag{2-51}$$

where A and B are constants. The condition that $\psi = 0$ at $x = 0$ requires that the cosine term vanish; A must therefore be zero. The condition that $\psi = 0$ at $x = a$ demands that $\alpha = n\pi/a$, where n is an *integer* called a *quantum number*. Finally, the requirement that ψ be normalized gives us the value of B:

$$\int_0^a [B\sin(n\pi x/a)][B\sin(n\pi x/a)]dx = B^2 a/2 \equiv 1$$

$$B = \sqrt{2/a} \tag{2-52}$$

The various conditions we just used to find A, B and, α are called *boundary conditions* and are common in many quantum mechanical problems. Our final solution is therefore

$$\psi_n = (\sqrt{2/a})\sin(n\pi x/a) \tag{2-53}$$

We put a subscript "n" on ψ to emphasize that the exact form of the wave function depends on the integer quantum number, n. What would the particle's energy be? We can find it in two ways. One way is to use equation 2-43 with $\hat{g} = \mathscr{H}$. The other, which we use here, is to plug eq. 2-53 back into eq. 2-50 and solve for the eigenvalue, E:

$$-\frac{\hbar^2}{2m}\frac{d^2[(\sqrt{2/a})\sin(n\pi x/a)]}{dx^2} = E[(\sqrt{2/a})\sin(n\pi x/a)]$$

$$+\frac{\hbar^2}{2m}\frac{n^2\pi^2}{a^2}\psi_n = E\psi_n$$

$$E_n = \frac{n^2\pi^2\hbar^2}{2ma^2} \tag{2-54}$$

In a very natural way we have discovered that the particle's behavior is *quantized*. Depending on the value of the integer n, the particle can have only certain discrete wave functions and energies. The four lowest energy wave functions and probability distributions are shown in Fig. 2-7. As the probability distributions show, the particle does not have a uniform probability of being found at any value of x in the lower energy states. For $n = 1$, the distribution is symmetrical about $x = a/2$. For $n = 2$, there is a *node* (a point at which $\psi = 0$) at $x = a/2$. In this state, the particle lives on either side of $a/2$, but not at $a/2$. As $n \rightarrow \infty$, the wave function becomes a smoother and smoother oscillating curve until the probability distribution is essentially uniform across the entire interval. This would be the classical behavior of a particle in such a potential field; *viz.*, it could be found anywhere in the interval $0 \leqslant x \leqslant a$ with equal probability. It is a general feature of quantum mechanical problems that as energies, masses, and lengths approach macroscopic values, the wave functions approach the classical limit. They had better, because very few macroscopic phenomena exhibit quantization. This trend toward the classical limit is called the Bohr correspondence principle.

A qualitative result which will be of importance to molecular spectroscopy is that, in general, a particle confined by a potential well of any form has quantized energies (Fig. 2-8). A particle with energy greater than the height of the well is *not* limited to discrete energies, and may possess any energy greater than the height of the well.

Problems

2-17. Use eq. 2-43 with $\hat{g} = \mathscr{H}$ to find the energy of the particle in a box.

2-18. Show that ψ_1 and ψ_2 of the particle in a box are orthogonal, *i.e.*, $\int_0^a \psi_1{}^*\psi_2 dx = 0$. Do the same for ψ_1 and ψ_3.

2-19. What is the probability that the particle in a box will be found in the region $0.9a \leqslant x \leqslant a$ in the states ψ_1, ψ_8, and ψ_{100}? What is the classical value expected?

2-20. What are the values of $\langle p \rangle$ and $\langle p^2 \rangle$ for ψ_1? For ψ_{100}?

B. *The Particle in a Three-Dimensional Box: Degeneracy of States.* Imagine now a three-dimensional parallelepiped with sides of length $a \neq b \neq c$, as in Fig. 2-9. If the potential is zero everywhere inside this box and infinity everywhere outside of it, we can write a three-dimensional Schrödinger equation which must be satisfied by the particle's wave functions as follows:

$$-\frac{\hbar^2}{2m}\left[\frac{\partial^2}{\partial x^2} + \frac{\partial^2}{\partial y^2} + \frac{\partial^2}{\partial z^2}\right]\psi(x,y,z) = E\psi(x,y,z) \qquad (2\text{-}55)$$

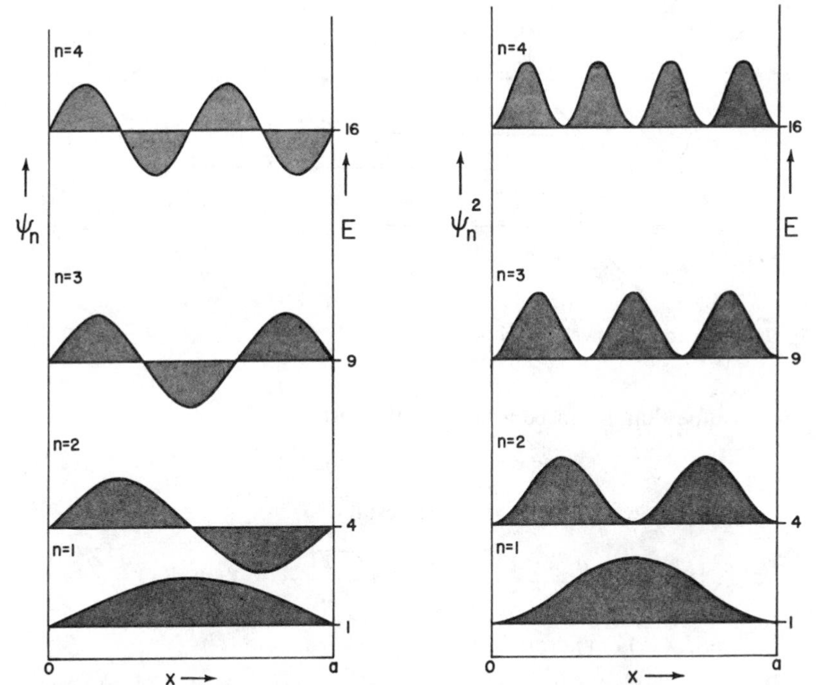

Fig. 2-7. Wave functions (left) and probability distributions (right) for the four lowest energy levels of the particle in a one-dimensional box. Energies are expressed in units of $\pi^2\hbar^2/2ma^2$.

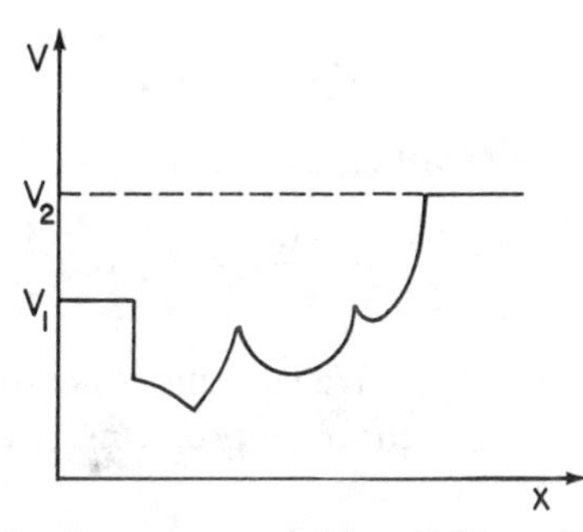

Fig. 2-8. A particle confined to any potential well ($E < V_1$) will have quantized energy levels. When $E > V_1$, any value of the energy is allowed.

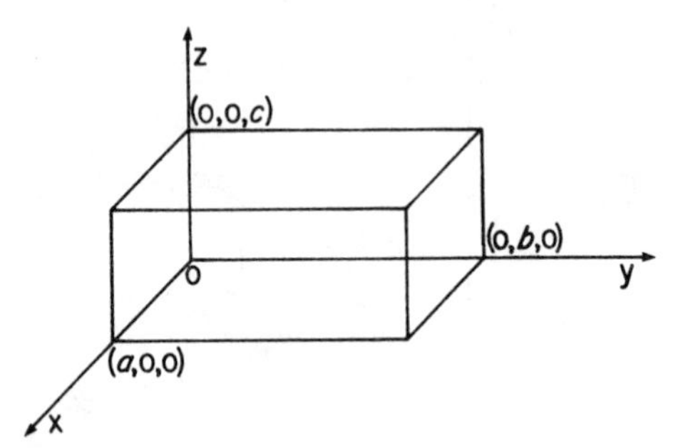

Fig. 2-9. The three-dimensional box.

If we assume that ψ can be written in the form

$$\psi(x,y,z) = X(x)Y(y)Z(z) \tag{2-56}$$

and plug eq. 2-56 into eq. 2-55, there results

$$-\frac{\hbar^2}{2m}\left[Y(y)Z(z)\frac{d^2X(x)}{dx^2} + X(x)Z(z)\frac{d^2Y(y)}{dy^2} + X(x)Y(y)\frac{d^2Z(z)}{dz^2}\right]$$
$$= EX(x)Y(y)Z(z) \tag{2-57}$$

After dividing both sides by $X(x)\,Y(y)\,Z(z)$ and rearranging, this becomes

$$\frac{1}{X(x)}\frac{d^2X(x)}{dx^2} + \frac{1}{Y(y)}\frac{d^2Y(y)}{dy^2} + \frac{1}{Z(z)}\frac{d^2Z(z)}{dz^2} = -\frac{2mE}{\hbar^2} \tag{2-58}$$

But in eq. 2-58 each term on the left is independent of the others, and the term on the right is a constant. The only way such an equation can hold is if each term on the left is separately equal to a constant. This implies that the energy can be expressed as a sum of energies due to motion along each coordinate,

$$E = E_x + E_y + E_z \tag{2-59}$$

and that eq. 2-58 can be rewritten as a sum of three equations of the form

$$-\frac{\hbar^2}{2m}\frac{d^2X(x)}{dx^2} = E_x X(x) \tag{2-60}$$

and similarly for Y(y) and Z(z). We have just *separated* a three-dimensional problem into three one-dimensional problems, which we have already solved in Section 2-4-A. So

$$X(x) = (\sqrt{2/a})\sin(n_x\pi x/a) \tag{2-61}$$

and
$$E_x = \frac{n_x^2 \pi^2 \hbar^2}{2ma^2} \tag{2-62}$$

and the three-dimensional wave function $\psi = X(x)Y(y)Z(z)$ must be

$$\psi(x,y,z) = \sqrt{\frac{2}{a}}\sqrt{\frac{2}{b}}\sqrt{\frac{2}{c}} \sin(n_x \pi x/a) \sin(n_y \pi y/b) \sin(n_z \pi z/c) \tag{2-63}$$

The allowed energy levels are

$$E_{n_x, n_y, n_z} = \frac{\pi^2 \hbar^2}{2m}\left[\frac{n_x^2}{a^2} + \frac{n_y^2}{b^2} + \frac{n_z^2}{c^2}\right] \tag{2-64}$$

where each quantum number, n_x, n_y, and n_z, can have the values 1, 2, 3, ...

We are about to discover for the second time (*cf.* Chapter Zero, Fig. 0-3) that the greater the symmetry of a problem, the more *degeneracy* is associated with the solutions. Suppose that $a \neq b \neq c$ and neither a nor b nor c is an integer multiple of any other side of the box. The symmetry of such a box (Fig. 2-9) is D_{2h}. For a specific example, let $a = 1.0$, $b = 1.1$, and $c = 1.8$ arbitrary units. The energies of the lowest few stationary states of a particle inside this box are shown at the left of Fig. 2-10. None of the states are degenerate (*i.e.*, none have the same energy), though some, such as (121) and (113) are very close in energy. The notation (121) means that $n_x = 1$, $n_y = 2$, and $n_z = 1$. When symmetry does not require any degeneracy, as in the present case, but two states have the same energy, they are said to be *accidentally degenerate*. Sometimes degeneracy is a relative matter. The separation of the states (121) and (113) is 0.010 energy units in our example. If we performed an experiment with a resolution of only 0.1 energy units, these two states would appear to be degenerate. If our resolution were 0.001 units, they would not.

At the center of Fig. 2-10 are shown the energy levels for the case $a = b = 1.0$ and $c = 1.8$ units (symmetry D_{4h}). Now the two states (121) and (211) are rigorously degenerate (symmetry required) and no experiment, no matter how high the resolution, would ever tell us otherwise. At the right of Fig. 2-10 are the energy levels of a particle in a cube ($a = b = c = 1.3$ units), the symmetry of which is O_h. As you can see, a great deal of symmetry required degeneracy is found. It will be found again and again that the greater the symmetry of a molecule, the more degeneracy will be associated with its rotational, vibrational, and electronic states.

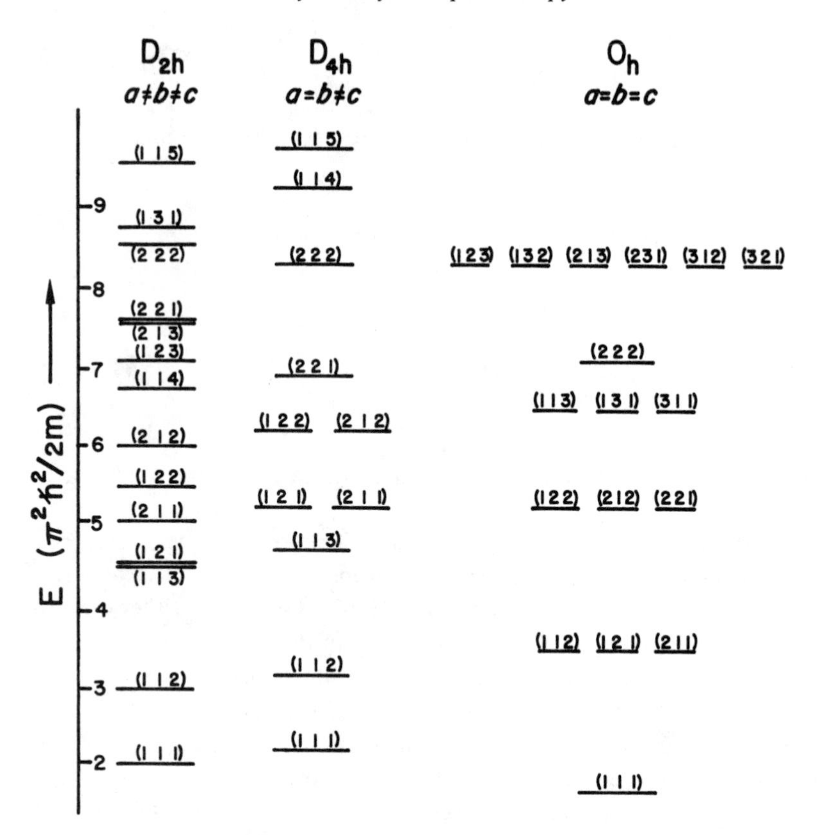

Fig. 2-10. Low lying energy levels of a particle in a three-dimensional box. Values of a, b, and c for each case are given in the text.

Problems

2-21. Show that the particle in a three-dimensional box has degenerate states if b equals an integer multiple of a, and c can have any value.

2-22. Write the Schrödinger equation and the solutions for a particle in a two-dimensional box with sides of lengths a and b. Is there any degeneracy if b is not an integer multiple of a? What if $b = a$?

C. *The Uncertainty Principle.* It should be clear by now that a main feature of quantum mechanics is that the properties of a particle are generally not precisely determined, but are described by probability distributions. Heisenberg made an observation which puts conditions on certain pairs of probability distributions. According to the *uncertainty principle*, the results of any experiment designed to simultaneously measure the position and momentum of a particle will be such that the product of the uncertainties in each measurement can never be smaller than the order of magnitude of $\hbar$:

$$\boxed{\Delta x \cdot \Delta p \gtrsim \hbar} \tag{2-65}$$

This statement is "fuzzy" in that we have no precise definition of the uncertainties, Δx and Δp.

For the sake of an illustration, let's take Δx and Δp as the standard deviations of each quantity, given by eq. 2-44. What would be the value of $\sigma_x \sigma_p$ for the particle in a one-dimensional box? Using eq. 2-43 and the operators in Table 2-2, we calculate the following:

$$\langle x \rangle = (2/a)\int_0^a x \sin^2(n\pi x/a)dx = a/2 \tag{2-66}$$

$$\langle x^2 \rangle = (2/a)\int_0^a x^2 \sin^2(n\pi x/a)dx = \frac{a^2}{3} - \frac{a^2}{2n^2\pi^2} \tag{2-67}$$

$$\sigma_x = \frac{a\sqrt{n^2\pi^2 - 6}}{2n\pi\sqrt{3}} \tag{2-68}$$

$$\langle p \rangle = (2/a)\int_0^a \sin(n\pi x/a)\left(-i\hbar\frac{d}{dx}\right)\sin(n\pi x/a)dx = 0 \tag{2-69}$$

$$\langle p^2 \rangle = (2/a)\int_0^a \sin(n\pi x/a)\left(-\hbar^2\frac{d^2}{dx^2}\right)\sin(n\pi x/a)dx = \frac{n^2\pi^2\hbar^2}{a^2} \tag{2-70}$$

$$\sigma_p = \frac{n\pi\hbar}{a} \tag{2-71}$$

$$\sigma_x\sigma_p = \frac{\hbar\sqrt{n^2\pi^2 - 6}}{2\sqrt{3}} \tag{2-72}$$

For the lowest energy state ($n = 1$), $\sigma_x\sigma_p = 0.57\hbar$. For $n = 2$, $\sigma_x\sigma_p = 1.67\hbar$, and the product increases as n increases. One can do the same kind of calculation for other quantum mechanical systems to see that the product $\Delta x\ \Delta p$ never gets much smaller than $\hbar$.

A corollary of the Heisenberg uncertainty principle is that if we could measure either Δx or Δp with very great precision, the other measurement would necessarily have very low precision. If, for example, we measured the position of an electron very accurately by bouncing a short wavelength γ-ray off of it, the photon would transfer considerable momentum to the electron, leading to a large uncertainty in the momentum of the electron.

The uncertainty principle also applies to the product $\Delta E \Delta t$, which has a direct implication in spectroscopy. Many molecules *fluoresce* after absorbing a photon and being promoted to an excited state. In fluorescence, the molecule emits a new photon, typically about 10^{-8} s after the absorption, and returns to the ground state. This means that the *lifetime*, Δt, of the excited state is of the order of 10^{-8} s. What therefore is the uncertainty in the energy of the excited state?

$$\Delta E \gtrsim \hbar/\Delta t = 1.0 \times 10^{-26}\text{J} = 0.0005 \text{ cm}^{-1} \tag{2-73}$$

That is, the width of (or uncertainty in) the absorption or emission bands cannot be less than about 0.0005 cm^{-1} if the lifetime of the excited state is 10^{-8} s. Usually this is not an important limitation in vibrational or electronic spectroscopy, where the energies of the transitions are measured in hundreds and thousands of cm^{-1} and where line broadening due to other causes is much greater than this *uncertainty broadening*. However, in magnetic resonance spectroscopy (which uses radio or microwave energies), uncertainty broadening is often the main determinant of linewidth when the lifetime of the excited state is short.

Problems

2-23. The first two wave functions of a harmonic oscillator (a particle governed by the potential $V = \frac{1}{2}kx^2$, where k is a constant) are $\psi_o = (\alpha/\pi)^{1/4} e^{-\alpha x^2/2}$ and $\psi_1 = (4\alpha^3/\pi)^{1/4} x e^{-\alpha x^2/2}$, where $\alpha = \sqrt{mk}/\hbar$. What is the value of $\Delta x \Delta p$ for a particle in each state?

D. *Potential Barriers*. What happens when a particle of mass m and energy E, moving from left to right in the x direction, suddenly encounters an increased potential, V_o, as shown in Fig. 2-11? The applicable wave equation is

$$-\frac{\hbar^2}{2m}\frac{d^2\psi}{dx^2} + [V(x) - E]\psi = 0 \tag{2-74}$$

where $V(x) = V_0$ for $x \geqslant 0$ and $V(x) = 0$ for $x < 0$. Classically, the behavior you would predict is very simple. If $E > V_0$, the particle continues on toward the right, but with decreased kinetic energy equal to $E - V_0$. If $E < V_0$, the particle is reflected at the barrier, and returns to the left. But the solution of eq. 2-74 is not as simple as this, and is much more interesting.

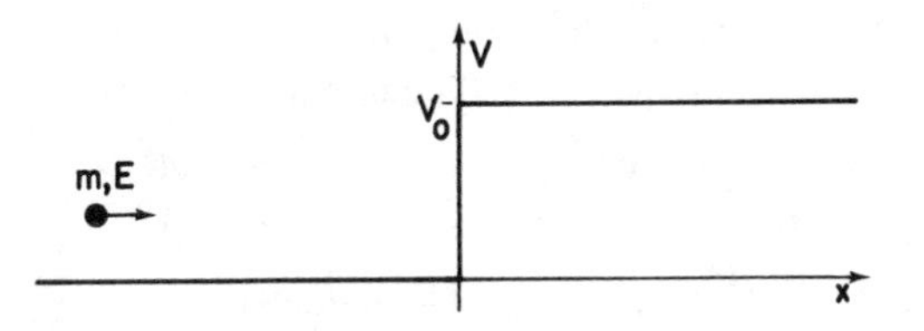

Fig. 2-11. The step-potential barrier. A particle of mass m and energy E is initially coming from the left toward the barrier.

Since V is a constant, the solutions of the differential equation are generally of the form $\psi = e^{\pm\sqrt{2m(V-E)}x/\hbar}$. If $V < E$, the quantity $\sqrt{2m(V-E)}$ is imaginary and the solution is a complex wave of the form $e^{\pm ix\sqrt{2m(E-V)}/\hbar}$. If $E < V$, the quantity $\sqrt{2m(V-E)}$ is real and the wave function is an exponential. For the particle in Fig. 2-11 the wave function must be

$$\psi_1 = Ae^{i\alpha x} + Be^{-i\alpha x} \qquad (x < 0) \tag{2-75}$$

$$\psi_2 = Ce^{\beta x} + De^{-\beta x} \qquad (x \geqslant 0) \tag{2-76}$$

where $\alpha = \sqrt{2mE}/\hbar$ and $\beta = \sqrt{2m(V_0 - E)}/\hbar$ and A, B, C, and D are constants.

We first consider the case where $E < V_0$; so β is real. This means that we must set $C = 0$ or else ψ_2 goes to infinity for large values of x. At the barrier ($x = 0$), continuity requires that $\psi_1 = \psi_2$:

$$\begin{aligned} Ae^{i\alpha 0} + Be^{-i\alpha 0} &= De^{-\beta 0} \\ A + B &= D \end{aligned} \tag{2-77}$$

Continuity of the first derivative at $x = 0$ gives

$$\frac{d\psi_1}{dx} = i\alpha A - i\alpha B = -\beta D = \frac{d\psi_2}{dx} \tag{2-78}$$

Combining eq. 2-77 with eq. 2-78 yields the ratios B/A and D/A:

$$\frac{B}{A} = \frac{i\alpha + \beta}{i\alpha - \beta} \qquad \frac{D}{A} = \frac{2i\alpha}{i\alpha - \beta} \tag{2-79}$$

But we have said in Section 2.3-C that if a wave function is written in the form $A\psi_A + B\psi_B + D\psi_D$, the probability that the particle is found in each state ψ_A, ψ_B, or ψ_D is proportional to A^2, B^2, and D^2, respectively. This has a simple interpretation in this case. ψ_A $(= e^{i\alpha x})$ corresponds to a particle moving to the right in the region $x < 0$. ψ_B $(= e^{-i\alpha x})$ corresponds to a particle moving to the left in this region. ψ_D $(= e^{-\beta x})$ corresponds to a particle inside the potential barrier ($x \geqslant 0$). Since the coefficients A, B, and D are imaginary, the probabilities are proportional to the squares of the *magnitudes* $|A|^2$, $|B|^2$, and $|D|^2$, given, for example by $|A|^2 \equiv |A^*A|$.

$|B|^2/|A|^2$ gives the ratio of the probabilities of finding a particle moving to the left or right in the region $x < 0$, and is just

$$\frac{|B^* B|}{|A^* A|} = \frac{(-i\alpha + \beta)(i\alpha + \beta)}{(-i\alpha - \beta)(i\alpha - \beta)} = \frac{\alpha^2 + \beta^2}{\alpha^2 + \beta^2} = 1 \tag{2-80}$$

All particles striking the barrier are (eventually) reflected. But, there is also a finite, if very small, probability that the particle can be found inside the barrier, since $D \neq 0$. This penetration of the barrier is a strictly non-classical form of behavior, referred to as *tunneling*. In general, in quantum mechanics, unless the potential barrier is infinite, there is a finite probability that it will be penetrated.

Let's go back and consider the case where $E > V_0$. This means that β is imaginary and we find oscillatory behavior of the wave function in both regions ($x < 0$ and $x \geqslant 0$). If we consider the case where the particle comes initially from the left, any particle passing the point $x = 0$ will continue indefinitely to the right. This means that the coefficient D in eq. 2-76 must be zero, since $De^{-\beta x}$ corresponds to a particle moving to the left when $x \geqslant 0$. Solving for A, B, and C, as above, we find

$$\frac{B}{A} = \frac{i\alpha - \beta}{i\alpha + \beta} = \frac{\sqrt{E} - \sqrt{E - V_0}}{\sqrt{E} + \sqrt{E - V_0}} \tag{2-81}$$

$$R \equiv \frac{|B|^2}{|A|^2} = \frac{2E - V_0 - 2\sqrt{E(E - V_0)}}{2E - V_0 + 2\sqrt{E(E - V_0)}} \tag{2-82}$$

$$\frac{C}{A} = \frac{2i\alpha}{i\alpha + \beta} = \frac{2\sqrt{E}}{\sqrt{E} + \sqrt{E - V_0}} \tag{2-83}$$

$$\frac{|C|^2}{|A|^2} = \frac{4E}{2E - V_0 + 2\sqrt{E(E - V_0)}} \tag{2-84}$$

Classically, a particle with $E > V_0$ is not reflected by the barrier. It just continues on to the right with reduced kinetic energy. Our quantum mechanical solution, however, gives a finite value for the *reflection coefficient*, R ($\equiv |B|^2/|A|^2$). This means that there is a finite probability that the particle will be reflected even though it has a greater energy than the height of the potential barrier. A real example of this effect is that of light striking a window. Even though a window is "transparent" and you can see through it, some light is also reflected and you can see your reflection in a window as well. Since the probability that a particle will be reflected (R) plus the probability that it will be transmitted (T) must be unity, the *transmission coefficient* will be given by

$$T = 1 - R = 1 - \frac{|B|^2}{|A|^2} = \frac{\beta |C|^2}{\alpha |A|^2} \tag{2-85}$$

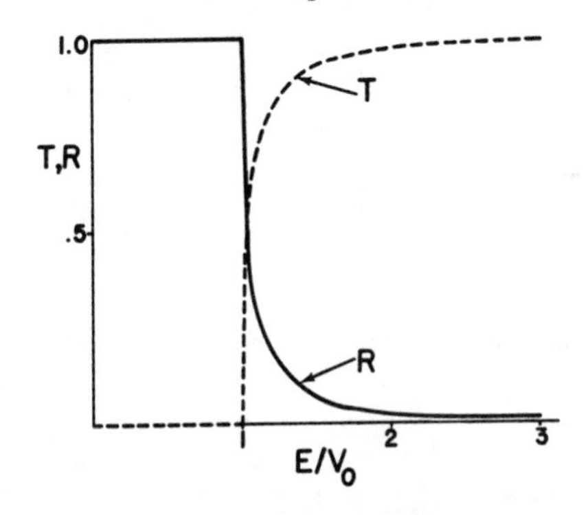

Fig. 2-12. Plot of transmission and reflection coefficients *vs.* E/V_o.

Graphs of T and R vs. the ratio E/V_0 are shown in Fig. 2-12.

Finally we outline, briefly, the case of a particle coming from the left and striking the *limited barrier* shown in Fig. 2-13. The allowed solutions must be

$$\psi_1 = Ae^{i\alpha x} + Be^{-i\alpha x} \qquad (x < 0) \tag{2-86}$$

$$\psi_2 = Ce^{\beta x} + De^{-\beta x} \qquad (0 \leqslant x \leqslant a) \tag{2-87}$$

$$\psi_3 = Fe^{-i\alpha x} + Ge^{-i\alpha x} \qquad (x > a) \tag{2-88}$$

We will throw out the term $Ge^{-i\alpha x}$ since no particle can travel to the left in the region $x > a$ because the particle originally came from the left and cannot be reflected after passing $x = a$. We allow for reflection at the points $x = 0$ and $x = a$ with the terms $Be^{-i\alpha x}$ and $De^{-\beta x}$. Now we will further restrict our attention only to the case $E < V_0$; so β is real.

The continuity conditions are, at $x = 0$:

$$(\psi_1 = \psi_2) : A + B = C + D \tag{2-89}$$

$$\left(\frac{d\psi_1}{dx} = \frac{d\psi_2}{dx}\right) : i\alpha A - i\alpha B = \beta C - \beta D \tag{2-90}$$

and at $x = a$:

$$(\psi_2 = \psi_3) : Ce^{\beta a} + De^{-\beta a} = Fe^{i\alpha a} \tag{2-91}$$

$$\left(\frac{d\psi_2}{dx} = \frac{d\psi_3}{dx}\right) : \beta Ce^{\beta a} - \beta De^{-\beta a} = i\alpha Fe^{i\alpha a} \tag{2-92}$$

which, after a ton of algebra, yield a transmission coefficient of the form

$$T = \frac{|F|^2}{|A|^2} = [\cosh^2 \beta a + \frac{(\alpha^2 - \beta^2)^2}{4\alpha^2\beta^2} \sinh^2 \beta a]^{-1} \tag{2-93}$$

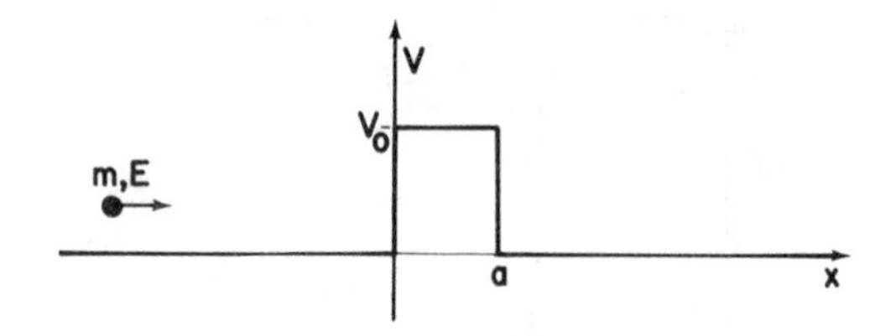

Fig. 2-13. The limited barrier of height V_o which extends over the region $0 \leqslant x \leqslant a$.

in which $\cosh \beta a = (1/2)(e^{\beta a} + e^{-\beta a})$ and $\sinh \beta a = (1/2)(e^{\beta a} - e^{-\beta a})$. We find, therefore, that a particle with $E < V_0$ striking the limited barrier from the left has a finite chance of tunneling through the barrier and continuing on its merry way to infinity.

Problems

2-24. Verify that $e^{\pm\sqrt{2m(V-E)}x/\hbar}$ is a solution of eq. 2-74.

2-25. Find the reflection coefficient for a particle of energy $E > 0$ impinging on the energy *well* (a negative barrier) below.

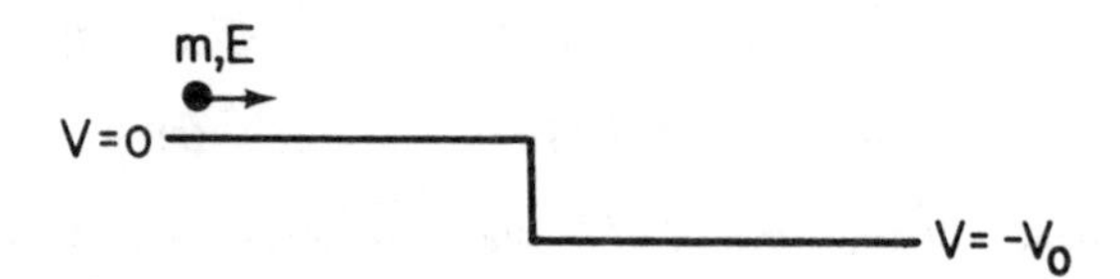

2-26. Calculate the transmission coefficient in eq. 2-93 for an electron passing through a barrier of dimensions $V_o = 0.01$ eV and $a = 100$ Å if $E/V_o = 0.9$. Do the same for $V_o = 0.02$ eV, $a = 50$ Å, and $E/V_o = 0.9$. Show that for a given value of E/V_o, the transmission coefficient is constant if $V_o a^2$ is constant.

Related Reading

J.C. Davis, *Advanced Physical Chemistry*, Ronald Press Co., New York, 1965.
H. Eyring, J. Walter, and G.E. Kimball, *Quantum Chemistry*, John Wiley & Sons, New York, 1944.
W.M. Hanna, *Quantum Mechanics In Chemistry*, W.A. Benjamin, New York, 1965.
M. Karplus and R.N. Porter, *Atoms and Molecules*, W.A. Benjamin, Menlo Park, California, 1970.
W. Kauzmann, *Quantum Chemistry*, Academic Press, New York, 1957.
E. Merzbacher, *Quantum Mechanics*, John Wiley & Sons, New York, 1961.
A. Messiah, *Quantum Mechanics*, North-Holland Publishing Co., Amsterdam, 1961.
L. Pauling and E.B. Wilson, Jr., *Introduction to Quantum Mechanics*, McGraw-Hill, New York, 1935.
E.H. Wichmann, *Quantum Physics*, McGraw-Hill, New York, 1971.

3 · *Vibrational spectroscopy*

3-1. Introduction

The job of the spectroscopist is the interpretation—and sometimes even the prediction—of the interaction of light with matter. In this chapter we will be chiefly concerned with the infrared region of the spectrum, typically confining ourselves to the region 100–5000 cm^{-1}. The energy of such light, 1.2–60 kJ mol^{-1} (0.3–15 kcal mol^{-1}), is sufficient to excite vibrations of the molecules which absorb it. Rotational energies of molecules are even smaller than vibrational energies; so light energetic enough to excite vibrations usually simultaneously excites rotations. By the "excitation" of rotation or vibration, we mean that the molecule is promoted to a state of higher energy in which its rotational frequency or vibrational amplitude is increased.

3-2. Infrared and Raman Spectra

A. *The Phenomena.* Vibrational spectra are ordinarily measured by two very different techniques. In *infrared* (ir) *spectroscopy* light of all different frequencies is passed through a sample and the intensity of the transmitted light is measured at each frequency. Typically the sample is a solution in a cell transparent to the ir radiation. NaCl and KBr windows are most frequently used for this purpose. In the far ir region (Fig. 0-1) these salts are opaque but polyethylene becomes transparent. Alternatively, a solid sample is mixed with KBr or some other pelleting agent, ground to a fine powder, and fused into a transparent disc under high pressure. At frequencies corresponding to vibrational energies of the sample, some light is absorbed and less light is transmitted than at frequencies which do not correspond to vibrational energies of the molecule. In order to compensate for absorption and scattering of the light by the solvent and sample cell, the incident light is split into two beams, one of which goes through the sample, and the other is passed through a reference usually consisting of pure solvent in a cell identical to the sample cell. Transmittance is then defined as I_s/I_r where I_s is the intensity of light passing through (that is, emerging from) the sample cell, and I_r is the intensity of light passing through the reference cell (Fig. 3-1).

In *Raman spectroscopy* we do not observe transmitted light but light scattered by the sample. Typically the sample is a solution in a glass capillary tube and the incident light passes through the length of the capillary. Sometimes a solid sample is simply pressed into a depression in a metal plate and is irradiated directly. The scattered light may be observed from any convenient direction with respect to the incident light. Light of a single frequency, monochromatic light, must be used for a Raman experiment. Ordinarily if you shine light of frequency ν_0 through, say, a homogeneous solution, most of the light will merely pass directly through the sample. Some of the light (about 1/1000 of the incident intensity) is scattered in all directions and can be seen from the side of the sample (Fig. 3-1). This phenomenon, in which light of frequency ν_0 is scattered in all directions, is called *Rayleigh scattering*. A very small fraction of the scattered light (about 1/1000 of it) does not have a frequency ν_0. Instead, this light has frequencies ν_i such that $\Delta E = h|\nu_0 - \nu_i|$ corresponds to energies that are absorbed by the sample. The frequency ν_0 may be in any region of the spectrum (with visible light the most typical), but the difference, $|\nu_0 - \nu_i|$, is an infrared (vibrational) frequency. The process which produces light of frequency other than ν_0 is called *Raman scattering*. ν_i may be greater or less than ν_0, but the amount of light with frequency $\nu_i < \nu_0$ is much greater than that with frequency $\nu_i > \nu_0$. The former, Raman scattered radiation, is called *Stokes* radiation, and the latter is called *anti-Stokes* radiation.

The Stokes and anti-Stokes portions of the Raman spectrum of CCl_4 are shown in Fig. 3-2. The Rayleigh scattering is centered at $\nu_0 = 22{,}938\ \text{cm}^{-1}$, the position of a blue emission line of a mercury arc lamp used as the light source in this experiment. The Stokes lines, to the low energy side, are more intense than the anti-Stokes lines, whose intensity decreases approximately exponentially with increasing energy from the exciting line (see Section 3-3-E for an explanation of this).

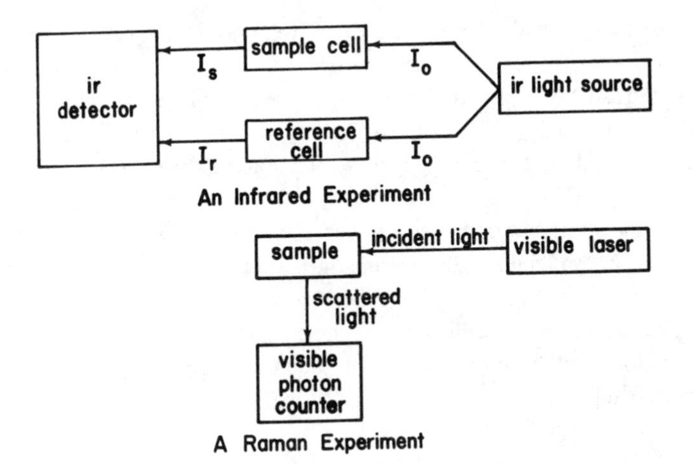

Fig. 3-1. Schematic representation of infrared and Raman experiments. In infrared spectroscopy the incident light contains all different infrared wavelengths. In Raman spectroscopy the incident light is monochromatic and of higher energy than infrared light. Visible laser light is typically used.

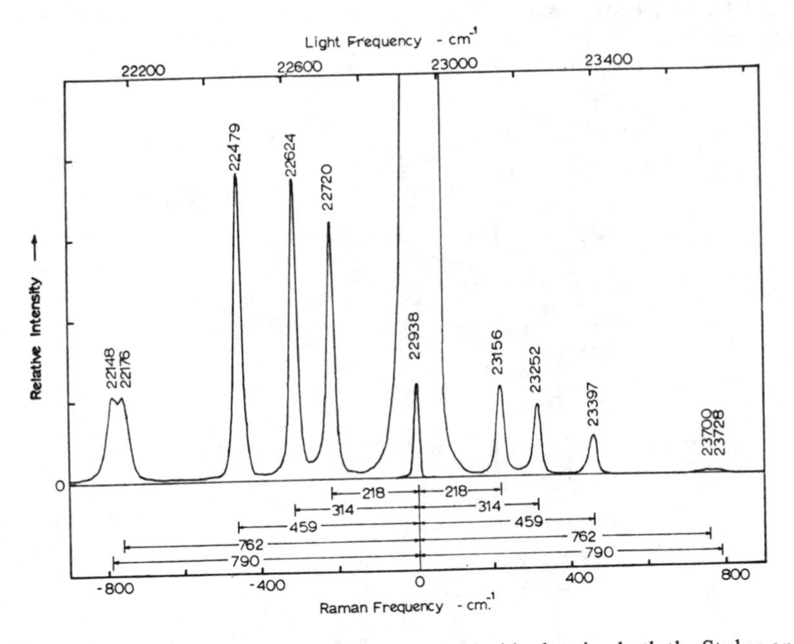

Fig. 3-2. The Raman spectrum of carbon tetrachloride showing both the Stokes and anti-Stokes portions of the spectrum. Reproduced from R.S. Tobias, *J. Chem. Ed.*, *44*, 2 (1967).

Comparing the two techniques, then, we find in ir spectroscopy that light of infrared frequencies must be used. We simply measure the wavelengths of absorbed light. In Raman spectroscopy, light of greater than ir frequencies is used and we measure the differences, $|\nu_0 - \nu_i|$, of the Raman scattered light. Figure 3-3 shows the ir and Raman spectra of the salt $ClO_2F_2{}^+BF_4{}^-$. You should notice that some bands are common to both spectra while some bands which appear in one spectrum do not appear in the other. The molecular vibrations stimulated in the Raman process are not necessarily the same as those excited by the absorption of infrared light. Therefore, the ir and Raman spectra of a particular sample will generally look different and will complement each other. The *selection rules* which tell us which absorptions to expect in each spectrum will be

Fig. 3-3. Vibrational spectra of $ClO_2F_2{}^+BF_4{}^-$. Notice that in an infrared spectrum absorption corresponds to a *decrease* in transmittance (Section 2-2-C). Hence in this figure ir absorption goes *down.* Raman bands are displayed as *upward* peaks in a plot of scattered light intensity *vs.* wave number from the exciting line. The markings P and DP on the lowest spectrum stand for polarized and depolarized and will be discussed in Section 3-6-D. Reproduced from K.O. Christe, R.D. Wilson, and E.C. Curtis, *Inorg. Chem.*, *12*, 1358 (1973).

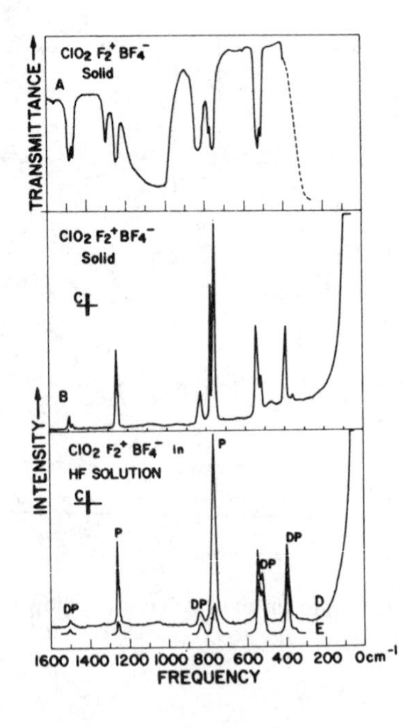

discussed in Section 3-6. At present, instrumentation of ir spectroscopy is generally more sensitive than that of Raman spectroscopy in terms of the amount of signal one can get from a given amount of sample. This was not the case before about 1950 when Raman spectroscopy was the better developed of the two techniques, and it may not be the case too long in the future. The resonance Raman effect, to be discussed in Section 3-10, sometimes enhances the sensitivity of Raman spectroscopy by orders of magnitude and may result in greater-than-infrared sensitivity. It should be emphasized that the two techniques are complementary and both are needed for a complete picture of the vibrational states of a compound.

Problems

3-1. For an incident frequency, ν_0, a Stokes line occurs when the molecule absorbs light of frequency ν, and light of frequency $\nu_i = \nu_0 - \nu$ is emitted. In the process of absorbing light of frequency ν, the molecule is excited to a higher energy vibrational state. What process occurs to account for the anti-Stokes line of frequency $\nu_0 + \nu$?

B. *A Molecular Explanation.* The preceding description of ir and Raman spectroscopy was strictly phenomenological. We have merely noted that matter is observed to absorb certain frequencies of infrared light and to produce Stokes and anti-Stokes scattering. Now we shall try to understand on a molecular level why these phenomena occur.

First we need to describe the permanent dipole moment of a molecule. If two particles of charges $+e$ and $-e$ are separated by a distance r, the permanent electric dipole moment, μ, is given by

$$\boxed{\mu = er} \tag{3-1}$$

To see what we might expect for the magnitude of molecular dipoles, consider a proton and electron separated by one Ångstrom, a typical molecular dimension. The dipole moment would be $\mu = (1.602 \times 10^{-19}$ coulomb) $\times$ (10^{-10}m) $= 1.602 \times 10^{-29}$ coulomb-m. In cgs units, on which most chemical literature and terminology is based, this corresponds to 4.803×10^{-18} esu-cm. The unit 10^{-18} esu-cm is called one *Debye* (1 D) and molecular dipoles are of the order of one Debye. Heteronuclear diatomic molecules must have a permanent dipole moment since one atom will be more electronegative than the other and will have a net negative charge. Homonuclear diatomic molecules cannot have permanent dipole moments since both nuclei attract the electrons equally. Polyatomic molecules with a center of inversion will not have a dipole moment whereas noncentrosymmetric molecules (those without a center of inversion) will usually have $\mu \neq 0$ (Fig. 3-4). An exception would be a highly symmetric molecule such as CCl_4 which has no center of inversion and also no permanent dipole. The direction of the dipole is indicated by an arrow with a "+" at the positive end of the dipole. In molecules like ethylene and hexafluorobenzene, each C–H and each C–F bond has a dipole moment, even though the molecule as a whole has no net moment because the vector sum of the individual bond dipoles is zero.

Now consider a heteronuclear diatomic molecule vibrating at a particular frequency. The molecular dipole moment also oscillates about its equilibrium value as the two atoms (with their net negative and positive charges) move back and forth. This oscillating dipole can absorb energy from an oscillating electric field only if the field also oscillates at the same frequency. If you have ever pushed someone on a swing you know how easy it is to push the swing at its natural frequency and how difficult it is to try to push the swing either out of phase with its natural motion or at a different frequency. In the case of an oscillating permanent dipole, the "push" is achieved by, say, the repulsion between the net negative charge on one end of the molecule and the "negative sign" of the electric field of the light wave. If the frequencies of the light and the vibrations are not the same, this *resonance* energy transfer does not occur. The absorption of energy from the light wave by the oscillating permanent dipole is a molecular explanation of ir spectroscopy.

If a molecule is placed in an electric field, ε, a dipole moment, μ_{ind}, is *induced* in the molecule because the nuclei are attracted toward the negative pole of the field, and the electrons are attracted the opposite way. The induced dipole is proportional to the field strength, the proportionality constant, α, being called the *polarizability* of the molecule.

$$\boxed{\mu_{\text{ind}} = \alpha\varepsilon} \tag{3-2}$$

The dimensions of α are μ/ε = charge · length/(charge/length^2) = length^3 = volume. Typical α values are in units of cubic Ångstroms. All atoms and molecules will have non-zero polarizability even if they have no permanent dipole moment (Fig. 3-5).

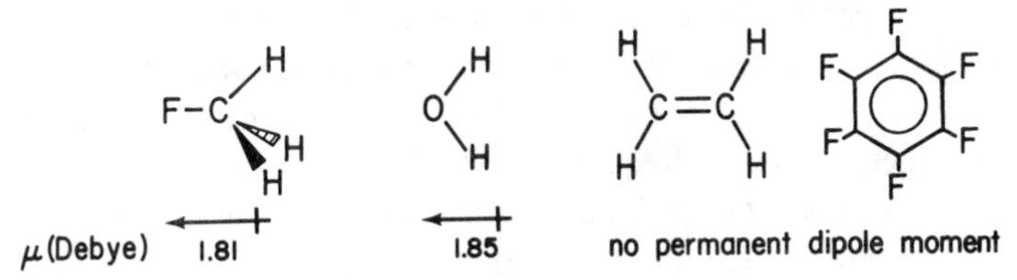

Fig. 3-4. Dipole moments of some small molecules.

Now we come to a classical explanation of the Raman process. Consider a light wave whose electric field oscillates at a certain point in space according to the equation

$$\varepsilon = \varepsilon_0 \cos 2\pi \nu t \tag{3-3}$$

where ε_0 is the maximum value of the field, ν is the frequency, and t is time. The induced dipole moment of a molecule in this oscillating field is obtained from eqs. 3-2 and 3-3:

$$\mu_{\text{ind}} = \alpha\varepsilon_0 \cos 2\pi\nu t \tag{3-4}$$

But α is a molecular property whose magnitude should vary as the molecule oscillates. The polarizability of a bond stretched 0.1 Å from its equilibrium position ought to be different from the polarizability of that same bond compressed 0.1 Å from the equilibrium position. Hence α will vary at the natural vibrational frequency of the bond:

$$\alpha = \alpha_0 + (\Delta\alpha)\cos 2\pi\nu_0 t \tag{3-5}$$

where α_0 is the equilibrium polarizability, $\Delta\alpha$ is its maximum variation, and ν_0 is the natural vibrational frequency. Putting eq. 3-5 into eq. 3-4,

$$\mu_{\text{ind}} = [\alpha_0 + (\Delta\alpha)\cos 2\pi\nu_0 t][\varepsilon_0 \cos 2\pi\nu t] \tag{3-6}$$

which can be rearranged to

$$\mu_{\text{ind}} = \alpha_0\varepsilon_0 \cos 2\pi\underline{\underline{\nu}}t + (1/2)\Delta\alpha\varepsilon_0\,[\cos 2\pi(\underline{\underline{\nu + \nu_0}})t + \cos 2\pi(\underline{\underline{\nu - \nu_0}})t] \tag{3-7}$$

Fig. 3-5. Polarizabilities of some representative species.

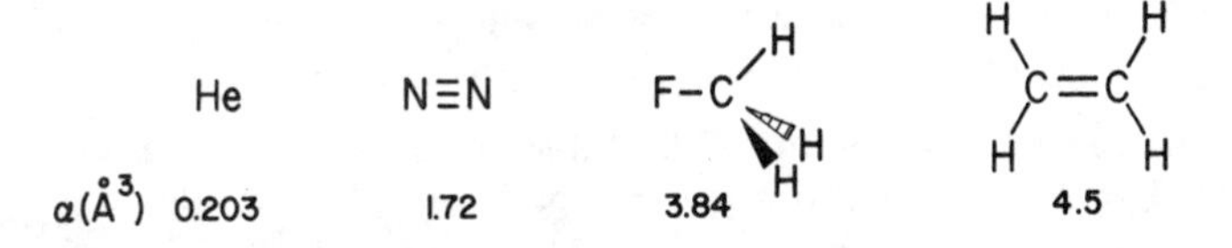

because $\cos\theta \cos\phi = (1/2) [\cos(\theta + \phi) + \cos(\theta - \phi)]$. Equation 3-7 predicts that the induced dipole moment will oscillate with components of frequency ν, $\nu + \nu_0$, and $\nu - \nu_0$. The oscillating electric dipole radiates electromagnetic waves of frequency ν (Rayleigh scattering), $\nu - \nu_0$ (Stokes radiation) and $\nu + \nu_0$ (anti-Stokes radiation). This simple model serves to rationalize Raman scattering on a molecular level.†

Problems

3-2. In order that a molecular vibration may give rise to ir absorption, the dipole moment of the molecule must change during the vibration. Which of the following molecular vibrations can result in ir absorption?

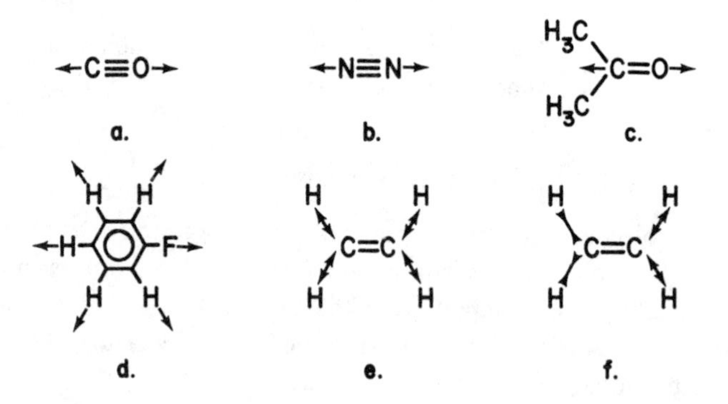

(In parts e and f the symbol ↔ denotes bond stretching and ⤚ denotes bond compression.)

3-3. Diatomic Molecules

The vibrational spectra of diatomic molecules illustrate most of the fundamental principles which apply to complicated polyatomic molecules. Since ir radiation will generally excite not only molecular vibrations but also rotations, we will need models for both rotation and vibration of diatomic molecules in order to analyze their spectra.

A. *Rotation.* For the simplest model of a rotating diatomic molecule, suppose the two nuclei are simply fixed at their equilibrium separation, r_e (Fig. 3-6). If the nuclei have masses m_1 and m_2, the molecule will rotate

†This phenomenon was predicted by A. Smekal five years before it was observed by C.V. Raman in 1928. For his discovery, Raman received the Nobel Prize in 1930.

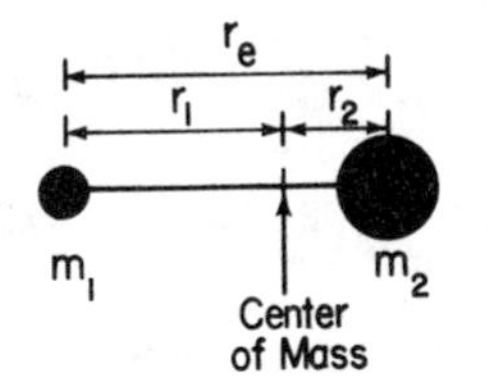

Fig. 3-6. The rigid rotor.

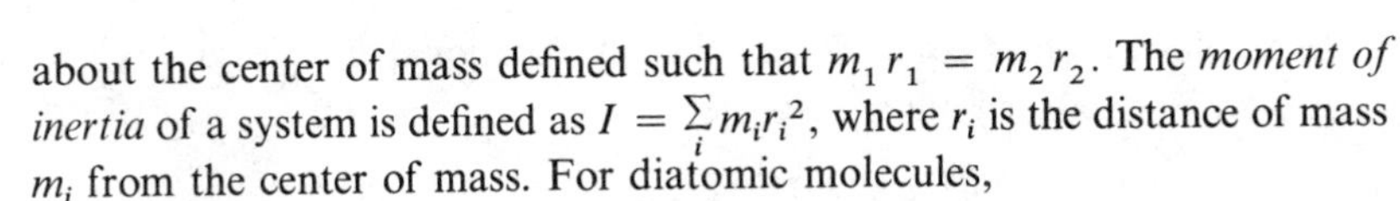

about the center of mass defined such that $m_1 r_1 = m_2 r_2$. The *moment of inertia* of a system is defined as $I = \sum_i m_i r_i^2$, where r_i is the distance of mass m_i from the center of mass. For diatomic molecules,

$$I = \frac{m_1 m_2}{m_1 + m_2} r_e^2 \equiv \mu r_e^2 \tag{3-8}$$

where

$$\mu \equiv \frac{m_1 m_2}{m_1 + m_2} \tag{3-9}$$

The quantity μ is called the *reduced mass* and should not be confused with the dipole moment which has the same symbol.

The quantum mechanical treatment of the rigid rotor (Section 3-3-D) leads to the important result that the molecule can only rotate with discrete energies:

$$E_J = \frac{\hbar^2}{2I} J(J + 1)$$

$$\boxed{E_J \equiv BJ(J + 1)} \tag{3-10}$$

Here E_J is the energy of the Jth rotational energy level, and J is the *rotational* quantum number with integer values 0, 1, 2, 3, ... In spectroscopy we often refer to "energy" and "frequency" in wave number (cm^{-1}) units. To convert a true energy in joules to wave numbers in cm^{-1}, we divide by $100hc$:†

$$\frac{E\,(\text{joules})}{h\,(\text{joule-s}) \cdot c\,(\text{m/s}) \cdot 100\,(\text{cm/m})} = \bar{E}\,(\text{cm}^{-1}) \tag{3-11}$$

We will often write a bar over a symbol in this chapter to emphasize that its units are wave numbers. Thus the number B in eq. 3-10 has units of joules but the number $\bar{B}$ would have units of cm^{-1}.

† Old folks, used to cgs units, will wince at the factor of 100 in eq. 3-11. We apologize for this, but it is an unavoidable consequence of using SI units for calculations which must finally be converted to units of cm^{-1}.

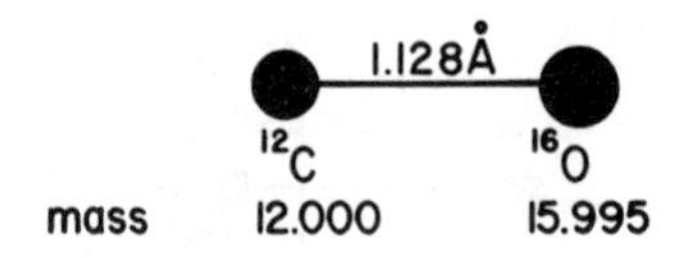

Fig. 3-7. The dimensions of $^{12}C^{16}O$.

As an illustration of eq. 3-10, let's find the rotational energies of $^{12}C^{16}O$ (Fig. 3-7). The reduced mass is

$$\mu = \frac{(12.000)(15.995)}{12.000 + 15.995} = 6.856 \text{ a.u.} = \frac{6.856}{6.022 \times 10^{26}}$$

$$= 1.138 \times 10^{-26} \text{ kg} \tag{3-12}$$

The number 6.022×10^{26} is Avogadro's number in SI units. The moment of inertia is

$$\begin{aligned} I = \mu r_e^2 &= (1.138 \times 10^{-26})(1.128 \times 10^{-10})^2 \\ &= 1.449 \times 10^{-46} \text{ kg m}^2 \end{aligned} \tag{3-13}$$

and the possible rotational energies will be

$$\begin{aligned} E_J &= \frac{(1.055 \times 10^{-34} \text{ kg m}^2 \text{ s}^{-1})^2}{2(1.449 \times 10^{-46} \text{ kg m}^2)} J(J+1) \\ &= 3.84 \times 10^{-23} J(J+1) \text{ joule} \\ &= 1.93\, J(J+1) \text{ cm}^{-1} \end{aligned} \tag{3-14}$$

The calculated values of the energy levels of carbon monoxide are shown in Table 3-1. This energy must be kinetic as we have not introduced any potential energy in our rigid rotor model. For a transition from the $J = 1$ to the $J = 2$ levels, CO must absorb a photon of wave number $11.6 - 3.9 = 7.7 \text{ cm}^{-1}$. This radiation is in the microwave portion of the spectrum (Fig. 0-1). Pure rotational spectroscopy is generally microwave or far ir spectroscopy.

Table 3-1. Rotational Energies of $^{12}C^{16}O$ Calculated with Eq. 3-14

J	$J(J+1)$	$\bar{E}_J$ (cm^{-1})	$\Delta\bar{E}(J \rightarrow J+1)$
0	0	0	$2\bar{B}$
1	2	3.9	$4\bar{B}$
2	6	11.6	$6\bar{B}$
3	12	23.2	$8\bar{B}$
4	20	38.6	$10\bar{B}$

The selection rules for the rigid rotor tell us that the transitions we may expect to see in the rotational absorption spectrum correspond to $J \rightarrow J + 1$. Other transitions, such as $J \rightarrow J + 2$ are "forbidden" and will be very weak. The energy difference between levels J and $J + 1$ is

$$\Delta\overline{E} = \overline{B}(J + 1)(J + 2) - \overline{B}(J)(J + 1) \\ = 2\overline{B}(J + 1) \tag{3-15}$$

That is, each absorption in the spectrum is separated from its neighbors by the energy $2\overline{B}$ for a diatomic molecule.

Real molecules do not behave as rigid rotors, though the approximation is good. A model for a nonrigid rotor assumes that the bond length increases as the rate of rotation of the molecule increases, *i.e.*, as J increases. Such a model leads to an expression like eq. 3-16 for the energy levels.

$$E_J = BJ(J + 1) - DJ^2(J + 1)^2 \tag{3-16}$$

D is called the centrifugal distortion constant because it is a measure of the increased bond length resulting from the rotation of the molecule. The application of equations 3-10 and 3-16 to HCl is compared in Table 3-2.

Table 3-2. Comparison of Rigid Rotor and Nonrigid Rotor Models for the Rotational Energies of HCl[a]

Transition $J \rightarrow J + 1$	$\bar{\nu}_{obs}$ (cm^{-1})	$\bar{\nu}_{calc} = 2\overline{B}(J + 1)$ (with $\overline{B} = 10.34\ cm^{-1}$)	$\bar{\nu}_{calc} = 2\overline{B}(J + 1) - 4\overline{D}(J + 1)^3$ ($\overline{B} = 10.395$, $\overline{D} = 0.0004\ cm^{-1}$)
3 → 4	83.03	82.72	83.06
4 → 5	104.1	103.40	103.75
5 → 6	124.30	124.08	124.39
6 → 7	145.03	144.76	144.98
7 → 8	165.51	165.44	165.50
8 → 9	185.86	186.12	185.94
9 → 10	206.38	206.80	206.30
10 → 11	226.50	227.48	226.55

[a]From G. Herzberg, *Spectra of Diatomic Molecules*, Van Nostrand Reinhold, N.Y., 1950.

Problems

3-3. Predict the positions in cm^{-1} of the first three absorption bands in the rotational spectrum of $^{14}N_2$ ($r_e = 1.094$ Å, mass = 14.00 a.u.) and $^1H_2$ ($r_e = 0.7417$ Å, mass = 1.008 a.u.) using the rigid rotor model.

3-4. The microwave spectrum of $^{79}Br^{19}F$ exhibits a series of absorptions 0.714 cm^{-1} apart. Using the masses ^{79}Br = 78.92 a.u. and ^{19}F = 19.00 a.u., calculate r_e.

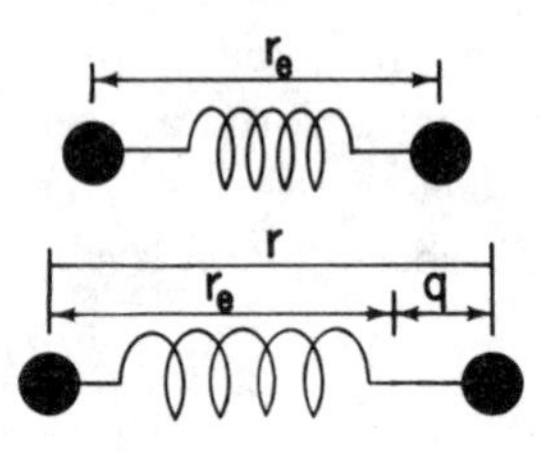

Fig. 3-8. A model for the vibration of a diatomic molecule. An oscillator which obeys Hooke's Law is said to be a harmonic oscillator.

B. *The Harmonic Oscillator.* As a simple model for the vibration of a diatomic molecule (Fig. 3-8), suppose the bond between the two nuclei behaves as a spring which obeys Hooke's Law:

$$\boxed{\text{restoring force} = f = -kq} \tag{3-17}$$

The constant, k, is called the *force constant*. The SI unit of k is newtons per meter, which is 100 times smaller than the most commonly encountered unit, millidyne per Ångstrom. The vibrational coordinate, $r - r_e$ will be called q. The potential energy is then

$$dV = -f\,dq = +k\,q\,dq$$

$$\boxed{V = (1/2)kq^2} \tag{3-18}$$

The quantum mechanical treatment of the harmonic oscillator (Section 3-3-D) tells us that the molecule can have only discrete vibrational energy levels characterized by the quantum number v:

$$\boxed{E_v = (v + 1/2)\frac{h}{2\pi}\sqrt{k/\mu} \equiv (v + 1/2)h\omega} \tag{3-19}$$

where
$$\omega = \frac{1}{2\pi}\sqrt{k/\mu} \tag{3-20}$$

v can have the values 0, 1, 2, 3.... In the lowest vibrational state ($v = 0$), the molecule still has the *zero point energy*, $E_0 = (1/2)h\omega$. This is different from the rotational energy levels, whose zero point energy is zero. According to the harmonic oscillator model, a diatomic molecule has equally spaced vibrational energy levels starting $(1/2)h\omega$ from the bottom of the potential well with spacings between the levels equal to $h\omega$ (Fig. 3-9). The vibrational spectra of diatomic molecules usually result from excitation from the $v = 0$ to the $v = 1$ energy levels. This gives us the value $\Delta E = h\omega$ from which

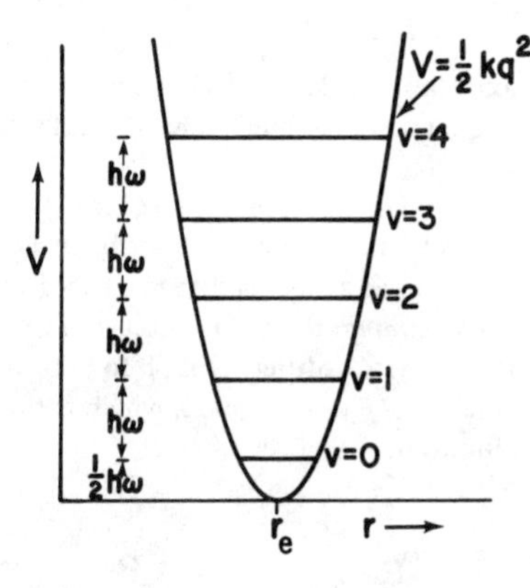

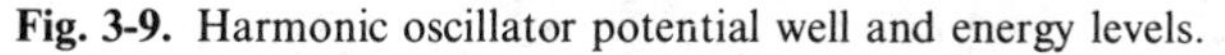

Fig. 3-9. Harmonic oscillator potential well and energy levels.

we can calculate the force constant of the chemical bond. A list of such data is presented in Table 3-3. The precise meaning of the constants in this table will be discussed in the next section. But you should notice that the bonds we consider to be the strongest (such as the triple bond of N_2) have the largest force constants.

Table 3-3. Vibrational Constants of Diatomic Molecules

Molecule[a]	Fundamental Frequency (cm^{-1})	$\bar{\omega}_e(cm^{-1})$[b]	$\bar{\omega}_e x_e(cm^{-1})$[b]	$k_e(N\ m^{-1})$[c]	Bond Enthalpy[d] kJ/mol	Bond Enthalpy[d] kcal/mol	Free Energy[d] kJ/mol	Free Energy[d] kcal/mol
H_2	4159.5	4395.3	117.90	573.4	436	104.2	406	97.2
D_2	2990.3	3118.5	64.10	576.9	—	—	—	—
HF	3958.4	4138.52	90.069	965.5	563	134.6	533	127.5
HCl	2885.7	2988.90	51.60	515.74	432	103.2	404	96.5
HBr	2559.2	2649.67	45.21	411.6	366	87.5	339	81.0
HI	2230.0	2309.5	39.73	314.1	299	71.4	272	65.0
CO	2143.3	2170.21	13.461	1902	1076	257.3	1040	248.6
NO	1876.1	1904.03	13.97	1594	630	150.5	599	143.1
F_2	892	—	—	440	153	36.6	119	28.4
Cl_2	556.9	564.9	4.0	328.6	243	58.0	211	50.4
$^{79}Br^{81}Br$	321	323.2	1.07	245.8	193	46.1	162	38.6
I_2	213.4	214.57	0.6127	172.1	151	36.1	121	28.9
O_2	1556.2	1580.361	12.0730	1177	495	118.4	460	110.0
N_2	2330.7	2359.61	14.456	2297	945	225.9	911	217.7
Li_2	346.3	351.44	2.592	25.5	111	26.5	86.9	20.8
Na_2	157.8	159.23	0.726	17.2	75.3	18.0	52.3	12.5
KCl	278	280	0.9	85.9	427	102.1	402	96.0

[a] Data refer to the most abundant isotope of each element.
[b] Values of $\bar{\omega}_e$ and $\bar{\omega}_e x_e$, defined in Section 3-3-C, were taken from G. Herzberg, *Spectra of Diatomic Molecules*, Van Nostrand Reinhold, N.Y., 1950.
[c] k_e is calculated from $\bar{\omega}_e$ using the relation $k_e = (200\pi c\bar{\omega}_e)^2\mu$. For F_2, $\bar{\omega}$ was used instead of $\bar{\omega}_e$. The almost universal unit of force constant is mdyne/Å. The SI unit, Newton/m, is equal to 0.01 mdyne/Å. The force constant for H_2, for example, is 573.4 N/m = 5.734 mdyne/Å.
[d] Bond enthalpies and free energies were taken from H.A. Bent, *The Second Law*, Oxford University Press, N.Y., 1965.

Problems

3-5. Calculate the force constants of H_2 and D_2 using the $v = 0 \rightarrow v = 1$ fundamental absorption frequencies listed in Table 3-3 while assuming the harmonic oscillator model.

3-6. The Pt complex, a, shows an infrared band at 822 cm^{-1} that has been attributed to an oxygen-oxygen stretching mode. This seems to be a mode in which the two oxygen atoms just move away from each other and Pt-O stretching is insignificant. If you *approximate* the O-O unit as a harmonic oscillator, where would you expect the ^{18}O-substituted compound, b, to absorb? (The band in b is observed within 2 cm^{-1} of this value, which is rather fortuitous in view of our severe approximation.)

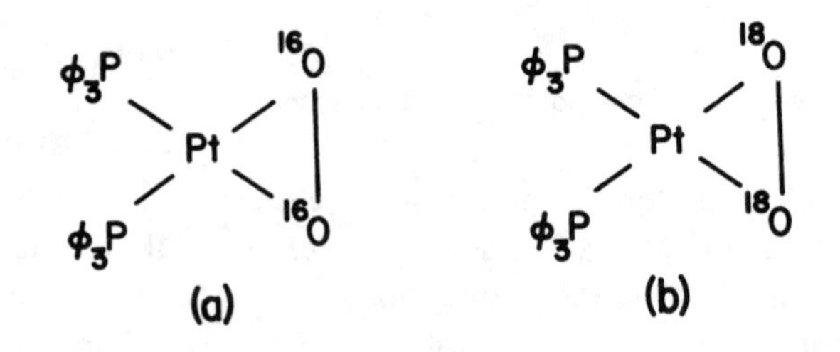

C. *The Anharmonic Oscillator.* The parabolic potential well in Fig. 3-9 is actually a pretty poor representation of the force felt by diatomic molecules. A real molecule ought to have an asymmetric potential well. As q decreases the nuclei come together and repel each other. This repulsion grows very strong at small values of q. On the other hand, as q gets bigger the restoring force eventually levels off and the molecule dissociates. Such a potential well is shown in Fig. 3-10.

An analytical function which has this shape in the region of interest for real molecules is the Morse potential. This purely empirical function has the form

$$\boxed{V = D_e(1 - e^{-\beta q})^2} \tag{3-21}$$

D_e is the depth of the potential well and β is a measure of the curvature at the bottom of the well. When the Morse potential is used to calculate the allowed energy levels of the system (Section 3-3-D), we find

$$\boxed{\bar{E}_v = \bar{\omega}_e(v + 1/2) - \bar{\omega}_e x_e(v + 1/2)^2} \tag{3-22}$$

The quantity $\bar{\omega}_e$ is nearly the same as $\bar{\omega}$ of the harmonic oscillator and the second term, containing the quantity $\bar{\omega}_e x_e$, is a small anharmonic correction.

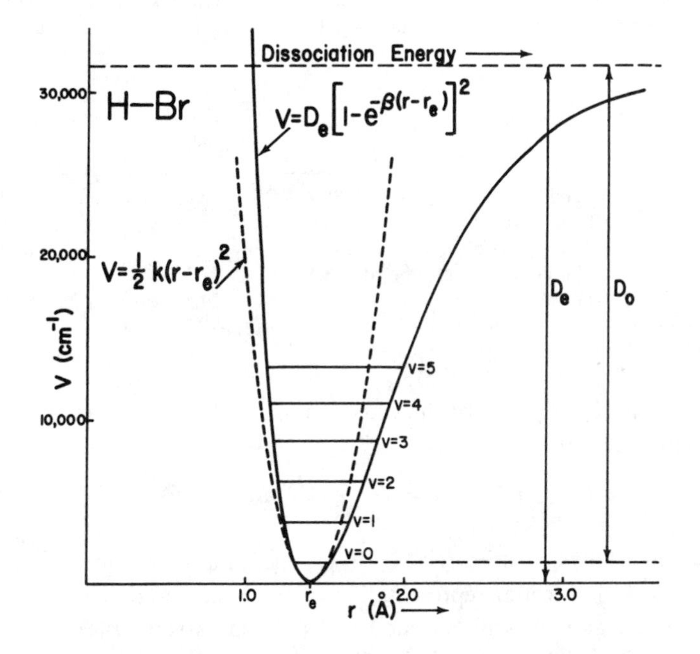

Fig. 3-10. The Morse potential (solid line) and harmonic oscillator potential (dashed line) which best describe the vibrational states of HBr. The parameters used are: $D_e = 31{,}590 \text{ cm}^{-1}$, $\beta = 1.811 \text{ Å}^{-1}$, $k = 103{,}600 \text{ cm}^{-1} \text{ Å}^{-2}$, $r_e = 1.414 \text{ Å}$, $\overline{\omega}_e = 2649.67 \text{ cm}^{-1}$, and $\overline{\omega}_e x_e = 45.21 \text{ cm}^{-1}$.

(Remember that a bar over a symbol implies units of cm^{-1} and an expression for the energy in joules would be analogous to eq. 3-19. That is, $\overline{\omega} = \omega/100c$.) $\overline{\omega}_e$ and x_e may be considered to be experimentally determined constants from which β and D_e can be derived, as shown in Section 3-3-D.

Transitions from the ground state ($v = 0$) to higher energy levels will involve the energies

$$\Delta\overline{E} = \overline{E}_v - \overline{E}_0 = \overline{\omega}_e v - \overline{\omega}_e x_e v(v + 1) \qquad \text{(3-23)}$$

The transition from the $v = 0$ to the $v = 1$ state is called the *fundamental transition*. The transition $v = 0 \rightarrow v = 2$ is the *first overtone*. The transition $v = 0 \rightarrow v = 3$ is the *second overtone, etc.* Table 3-4 illustrates the advantage of the anharmonic oscillator model over the harmonic oscillator in fitting the energies of the vibrational levels of HCl. The harmonic oscillator model predicts equally spaced overtones when, in fact, the vibrational spacing decreases with increasing values of v. As shown in Fig. 3-10, the true dis-

Table 3-4. Observed Vibrational Energies of HCl and Those Calculated Using the Harmonic and Anharmonic Oscillator Models[a]

Transition	Description	$\bar{\nu}_{obs}(cm^{-1})$	$\bar{\nu}_{calc}(cm^{-1})$ Harmonic oscillator	$\bar{\nu}_{calc}(cm^{-1})$ Anharmonic oscillator
$0 \rightarrow 1$	Fundamental	2,885.9	(2,885.9)	2,885.70
$0 \rightarrow 2$	First overtone	5,668.0	5,771.8	5,668.20
$0 \rightarrow 3$	Second overtone	8,347.0	8,657.7	8,347.50
$0 \rightarrow 4$	Third overtone	10,923.1	11,543.6	10,923.6
$0 \rightarrow 5$	Fourth overtone	13,396.5	14,429.5	13,396.5

[a]From G.M. Barrow, *Introduction to Molecular Spectroscopy*, McGraw-Hill, N.Y., 1962.

sociation energy, $\overline{D}_0$, of the molecule is not $\overline{D}_e$, but rather, $\overline{D}_e$ minus the zero point energy:

$$\overline{D}_0 = \overline{D}_e - \overline{E}_0 = \overline{D}_e - (1/2)\overline{\omega}_e + (1/4)\overline{\omega}_e x_e \qquad (3\text{-}24)$$

The value of $\overline{\omega}_e$ can be related to k_e, the anharmonic force constant. In theory, k_e is a better representation of the bond force than k obtained from the harmonic oscillator model. As a comparison, k(HCl) derived from eq. 3-20 is 480 N/m while k_e(HCl) calculated from $\overline{\omega}_e$ is 520 N/m. To get values of $\overline{\omega}_e$ for polyatomic molecules requires vibrational frequencies of isotopically substituted molecules or the observation of overtones and is

Fig. 3-11. Plot of bond enthalpy *vs.* force constant for some diatomic molecules.

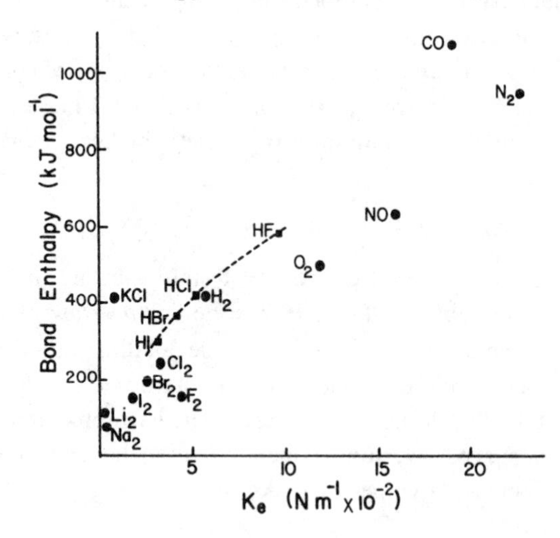

not generally practical. So we ordinarily content ourselves with simple k values.

It is tempting to say that a large value of the force constant indicates a large value of the dissociation energy, D_0. Figure 3-11 shows a plot of the bond enthalpies of the molecules in Table 3-3 *vs.* their force constants. You can see that there is a gross correlation between the force constant and bond energy, but it is no better than just a gross correlation. For a series of closely related compounds, such as HI, HBr, HCl and HF (shown by squares), there is a fairly smooth correlation.

Problems

3-7. Calculate the energies (in cm^{-1}) of the fundamental and first three overtones of N_2 using the anharmonic constants in Table 3-3. Compare these values to those predicted by the harmonic oscillator model using $\overline{\omega} = 2330.7\ \mathrm{cm}^{-1}$.

3-8. The radius of curvature, ρ, of a function $y = f(x)$ is given by

$$\rho = [1 + (dy/dx)^2]^{3/2}/(d^2y/dx^2)$$

Evaluate the radius of curvature of the harmonic oscillator and Morse potentials at their minima. If the best values of k, D_e, and β are such that each curve has the same curvature at the minimum, what is the relation of k to D_e and β? (This is, in fact, the relation obtained from the solutions of the Schrödinger equations for the harmonic and anharmonic oscillators. See Section 3-3-D.)

D. *The Quantum Mechanics of Vibration and Rotation.* As a model for the quantum mechanical treatment of the motions of a diatomic molecule, we consider the molecule to consist of two atoms of masses m_1 and m_2 separated by the (variable) distance r, as in Fig. 3-12. The potential energy does not depend on where the molecule is in space, but only on the internuclear separation. The Schrödinger equation for this two-particle problem will have a kinetic energy term $(-\hbar^2/2m_i)\,\nabla_i^2$ for each atom and a single potential energy term:

$$\left[-\frac{\hbar^2}{2m_1}\nabla_1^2 - \frac{\hbar^2}{2m_2}\nabla_2^2 + V(r)\right]\psi_T = E_T\psi_T \qquad (3\text{-}25)$$

where $\nabla_i^2 = \dfrac{\partial^2}{\partial x_i^2} + \dfrac{\partial^2}{\partial y_i^2} + \dfrac{\partial^2}{\partial z_i^2}$, ψ_T is the total wave function for motion of the two particles in space, and E_T is the total energy of translation, vibration, and rotation.

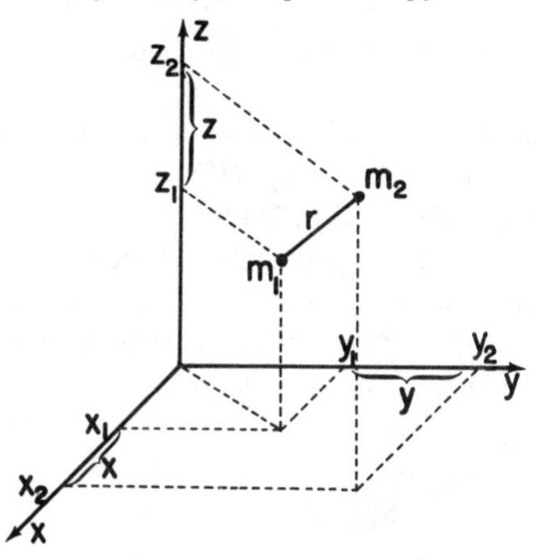

Fig. 3-12. Cartesian coordinates for treatment of a diatomic molecule.

It is desirable to eliminate translation of the entire molecule through space by transforming to center-of-mass coordinates by introducing the substitutions

$$X = \frac{m_1 x_1 + m_2 x_2}{m_1 + m_2} \qquad x = x_2 - x_1$$

$$Y = \frac{m_1 y_1 + m_2 y_2}{m_1 + m_2} \qquad y = y_2 - y_1 \tag{3-26}$$

$$Z = \frac{m_1 z_1 + m_2 z_2}{m_1 + m_2} \qquad z = z_2 - z_1$$

into eq. 3-25. Assuming $\psi_T = \psi_t \psi_{rv}$, where ψ_t is the translational wave function and ψ_{rv} is the wave function for rotation and vibration, it is possible to rearrange eq. 3-25 to the form

$$-\frac{\hbar^2}{2M}\frac{1}{\psi_t}\left[\frac{\partial^2 \psi_t}{\partial X^2} + \frac{\partial^2 \psi_t}{\partial Y^2} + \frac{\partial^2 \psi_t}{\partial Z^2}\right] - \frac{\hbar^2}{2\mu}\frac{1}{\psi_{rv}}\left[\frac{\partial^2 \psi_{rv}}{\partial x^2} + \frac{\partial^2 \psi_{rv}}{\partial y^2} + \frac{\partial^2 \psi_{rv}}{\partial z^2}\right]$$
$$+ V(x,y,z) = E_T \tag{3-27}$$

where $M = m_1 + m_2$ and μ is the reduced mass. Since the first term depends only on the coordinates X, Y, and Z and the next two terms depend only on x, y, and z, the first term must equal a constant and the second two terms together must equal a constant. Further, the sum of these two constants must be E_T. We can therefore rewrite eq. 3-27 as two independent equations, the first of which describes translation and the second of which describes rotation and vibration:

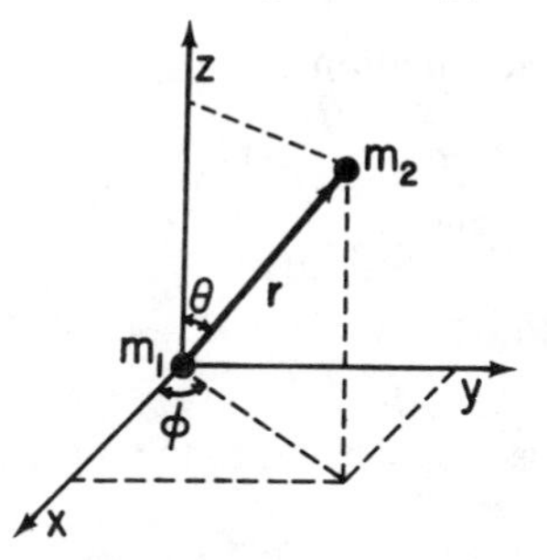

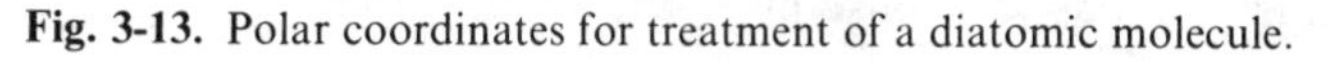

Fig. 3-13. Polar coordinates for treatment of a diatomic molecule.

$$-\frac{\hbar^2}{2M}\left[\frac{\partial^2\psi_t}{\partial X^2}+\frac{\partial^2\psi_t}{\partial Y^2}+\frac{\partial^2\psi_t}{\partial Z^2}\right] = E_t\psi_t \tag{3-28}$$

$$-\frac{\hbar^2}{2\mu}\left[\frac{\partial^2\psi_{rv}}{\partial x^2}+\frac{\partial^2\psi_{rv}}{\partial y^2}+\frac{\partial^2\psi_{rv}}{\partial z^2}\right] + V(x,y,z)\psi_{rv} = E_{rv}\psi_{rv} \tag{3-29}$$

Equation 3-28 is the equation of motion of the molecule in free space and we will not consider it further except to say that the translational energies are not quantized. Equation 3-29 describes the motion of the molecule in a coordinate system fixed at atom number 1. This equation is most conveniently solved by transforming it into spherical polar coordinates, defined in Fig. 3-13, in which

$$x = r\sin\theta\cos\phi$$

$$y = r\sin\theta\sin\phi$$

$$z = r\cos\theta \tag{3-30}$$

$$d\tau = dx\,dy\,dz = r^2\sin\theta\,dr\,d\theta\,d\phi$$

$$\nabla^2 = \frac{1}{r^2}\frac{\partial}{\partial r}\left(r^2\frac{\partial}{\partial r}\right)+\frac{1}{r^2\sin\theta}\frac{\partial}{\partial\theta}\left(\sin\theta\frac{\partial}{\partial\theta}\right)+\frac{1}{r^2\sin^2\theta}\frac{\partial^2}{\partial\phi^2}$$

Writing ψ_{rv} in the form

$$\psi_{rv} = \underbrace{R(r)}_{\psi_{\text{vib}}}\underbrace{\Theta(\theta)\Phi(\phi)}_{\psi_{\text{rot}}} \tag{3-31}$$

eq. 3-29 can be rewritten as

$$-\frac{\hbar^2}{2\mu r^2}\left[\Theta\Phi\frac{d}{dr}\left(r^2\frac{dR}{dr}\right)+\frac{R\Phi}{\sin\theta}\frac{d}{d\theta}\left(\sin\theta\frac{d\Theta}{d\theta}\right)+\frac{R\Theta}{\sin^2\theta}\frac{d^2\Phi}{d\phi^2}\right]$$

$$+\,V(r)R\,\Theta\,\Phi = E_{rv}\,R\,\Theta\,\Phi \tag{3-32}$$

Rearrangement yields the equation

$$\frac{\sin^2\theta}{R}\frac{d}{dr}\left(r^2\frac{dR}{dr}\right)+\frac{\sin\theta}{\Theta}\frac{d}{d\theta}\left(\sin\theta\frac{d\Theta}{d\theta}\right)+\frac{1}{\Phi}\frac{d^2\Phi}{d\phi^2}$$

$$+\frac{2\mu r^2\sin^2\theta}{\hbar^2}[E_{rv}-V(r)]=0 \tag{3-33}$$

in which the term in Φ is independent of the other terms and must therefore be equal to a constant.

Not so arbitrarily calling this constant $-m^2$, we can write

$$\frac{1}{\Phi}\frac{d^2\Phi}{d\phi^2}=-m^2 \tag{3-34}$$

Equation 3-34 has solutions of the form $\Phi_m = Ae^{im\phi} + Be^{-im\phi}$, where A and B are constants. The condition that the wave function be single valued requires that $\Phi_m(\phi) = \Phi_m(\phi + 2\pi)$:

$$\begin{aligned} Ae^{im\phi}+Be^{-im\phi} &= Ae^{im(\phi+2\pi)}+Be^{-im(\phi+2\pi)} \\ &= Ae^{im\phi}e^{im2\pi}+Be^{-im\phi}e^{-im2\pi} \end{aligned} \tag{3-35}$$

This is possible only if $e^{\pm im2\pi} = 1$, which says that m must be an integer. We can rewrite the wave function in the form

$$\Phi_m=\frac{1}{\sqrt{2\pi}}e^{im\phi} \tag{3-36}$$

if we allow m the values $0, \pm 1, \pm 2, \pm 3, \ldots$ The value of A $(= 1/\sqrt{2\pi})$ was obtained by normalizing the function.

We now return to eq. 3-33 in which we substitute $-m^2$ for the third term and divide each term by $\sin^2\theta$:

$$\frac{1}{R}\frac{d}{dr}\left(r^2\frac{dR}{dr}\right)+\frac{2\mu r^2}{\hbar^2}[E_{rv}-V(r)]+\frac{1}{\Theta\sin\theta}\frac{d}{d\theta}\left(\sin\theta\frac{d\Theta}{d\theta}\right)-\frac{m^2}{\sin^2\theta}$$
$$=0 \tag{3-37}$$

In this equation, the first two terms and the last two must separately equal constants. With some foresight we will call the first constant $J(J + 1)$ and we can write

$$\frac{1}{\sin\theta}\frac{d}{d\theta}\left(\sin\theta\frac{d\Theta}{d\theta}\right)-\frac{m^2}{\sin^2\theta}\Theta+J(J+1)\Theta=0 \tag{3-38}$$

$$\frac{d}{dr}\left(r^2\frac{dR}{dr}\right)-J(J+1)R+\frac{2\mu r^2}{\hbar^2}[E_{rv}-V(r)]R=0 \tag{3-39}$$

It is beyond the scope of this text to detail the method of solution of these

Table 3-5. The Spherical Harmonics

J	m	Imaginary Wave Function	Alternative Real Wave Functions
0	0	$(1/4\pi)^{1/2}$	—
1	0	$(3/4\pi)^{1/2}\cos\theta$	—
1	± 1	$(3/8\pi)^{1/2}\sin\theta\, e^{\pm i\phi}$	$(3/4\pi)^{1/2}\sin\theta\cos\phi$ $(3/4\pi)^{1/2}\sin\theta\sin\phi$
2	0	$(5/16\pi)^{1/2}(3\cos^2\theta - 1)$	—
2	± 1	$(15/8\pi)^{1/2}\sin\theta\cos\theta\, e^{\pm i\phi}$	$(15/4\pi)^{1/2}\sin\theta\cos\theta\cos\phi$ $(15/4\pi)^{1/2}\sin\theta\cos\theta\sin\phi$
2	± 2	$(15/32\pi)^{1/2}\sin^2\theta\, e^{\pm 2i\phi}$	$(15/16\pi)^{1/2}\sin^2\theta\cos 2\phi$ $(15/16\pi)^{1/2}\sin^2\theta\sin 2\phi$
3	0	$(63/16\pi)^{1/2}(\frac{5}{3}\cos^3\theta - \cos\theta)$	—
3	± 1	$(21/64\pi)^{1/2}\sin\theta(5\cos^2\theta - 1)\, e^{\pm i\phi}$	$(21/32\pi)^{1/2}\sin\theta(5\cos^2\theta - 1)\cos\phi$ $(21/32\pi)^{1/2}\sin\theta(5\cos^2\theta - 1)\sin\phi$
3	± 2	$(105/32\pi)^{1/2}\sin^2\theta\cos\theta\, e^{\pm 2i\phi}$	$(105/16\pi)^{1/2}\sin^2\theta\cos\theta\cos 2\phi$ $(105/16\pi)^{1/2}\sin^2\theta\cos\theta\sin 2\phi$
3	± 3	$(35/64\pi)^{1/2}\sin^3\theta\, e^{\pm 3i\phi}$	$(35/32\pi)^{1/2}\sin^3\theta\cos 3\phi$ $(35/32\pi)^{1/2}\sin^3\theta\sin 3\phi$

two differential equations.† Rather, we will just state the results. The Θ equation, 3-38, is satisfied by functions of $\cos\theta$ called *Legendre polynomials*. It is found that J must have an integer value (0, 1, 2, 3, . . .) and that for a given value of J, m can only assume the values 0, ± 1, $\pm 2, \ldots \pm J$. The products $\Theta(\theta)\,\Phi(\phi)$, which are the *rotational wave functions* for the diatomic molecule, are known as *spherical harmonics* and some are listed in Table 3-5. If you are familiar with the quantum mechanical treatment of the hydrogen atom, you will recognize the spherical harmonics as the angular parts of the hydrogen atomic orbitals.

The rotational wave functions, $\psi_{\text{rot}} = \Theta\,\Phi$, arose in a natural way by separating the original Schrödinger equation, 3-25, into translational, rotational, and vibrational parts. Since neither Θ nor Φ depends on the form of $V(r)$, the rotational wave functions will be the same regardless of the model we choose for vibration of the molecule. We first sketch the treatment of the harmonic oscillator and then of the anharmonic oscillator.

In the harmonic oscillator model, $V(r) = (1/2)k(r - r_e)^2$ (Fig. 3-8) and eq. 3-39 becomes

$$\frac{d}{dr}\left(r^2\frac{dR}{dr}\right) - J(J+1)R + \frac{2\mu r^2}{\hbar^2}[E_{rv} - (1/2)k(r - r_e)^2]R = 0 \qquad (3\text{-}40)$$

†The general method of solution of eqs. 3-38 and 3-39 can be found in F.B. Hildebrand, *Advanced Calculus for Applications*, Prentice-Hall, Englewood Cliffs, N.J., 1962, Chapter 4. Specific details of these solutions appear in any of the advanced quantum mechanics texts listed at the end of Chapter Two.

This equation is satisfied to a very good approximation by functions of q $(= r - r_e)$ which are products of an exponential term and a polynomial known as a *Hermite polynomial.* These functions have the general form

$$\psi_v = N_v H_v e^{-\alpha q^2/2} \tag{3-41}$$

where N_v is a normalization constant, H_v is the vth Hermite polynomial, $\alpha = \sqrt{\mu k}/\hbar$, and v is the vibrational quantum number. Specifically, the first four vibrational wave functions are

$$\begin{aligned}
\psi_0 &= (\alpha/\pi)^{1/4} e^{-\alpha q^2/2} \\
\psi_1 &= (4\alpha^3/\pi)^{1/4} q e^{-\alpha q^2/2} \\
\psi_2 &= (\alpha/4\pi)^{1/4}(2\alpha q^2 - 1) e^{-\alpha q^2/2} \\
\psi_3 &= (\alpha^3/9\pi)^{1/4}(2\alpha q^3 - 3q) e^{-\alpha q^2/2}
\end{aligned} \tag{3-42}$$

Some of the vibrational wave functions of the harmonic oscillator are shown in Fig. 3-14. These functions have the same qualitative behavior as the solutions of the particle in a box (Section 2-4-A). In particular, for even values of v, ψ_v is an even function of q and for odd values of v, ψ_v is an odd function of q. These properties will be important in the discussion of the selection rules in Section 3-6.

The energy levels of the harmonic oscillator are found to be

$$E_{v,J} = (v + 1/2)h\omega + BJ(J + 1) - DJ^2(J + 1)^2 \tag{3-43}$$

where

$$\begin{aligned}
\omega &= (1/2\pi)\sqrt{k/\mu} \\
B &= \hbar^2/2I = \hbar^2/2\mu r_e^2 \\
D &= \hbar^4/8\pi^2\omega^2 I^3 = B^2/2\pi^2\omega^2 I
\end{aligned} \tag{3-44}$$

This is just the sum of the energies of rotation and vibration given previously in eqs. 3-16 and 3-19. But now we can relate B and D to molecular properties.

A more important problem is to consider the case where $V(r)$ in eq. 3-39 is the Morse potential, eq. 3-21. We will not detail the wave functions in this rather complicated problem, but will state that to a good approximation the energy levels take the form†

$$\begin{aligned}
E_{v,J} = {} & (v + 1/2)h\omega_e - (v + 1/2)^2 h\omega_e x_e + B_e J(J + 1) - DJ^2(J + 1)^2 \\
& - \alpha_e(v + 1/2)J(J + 1)
\end{aligned} \tag{3-45}$$

† In eq. 3-45 we use the subscript e on B_e and α_e to emphasize that these constants correspond to $r = r_e$, the value of r at the minimum of the potential well. In many texts, the constant D also is written with a subscript e but we avoid this since the symbol D_e has already been used for the depth of the Morse potential well.

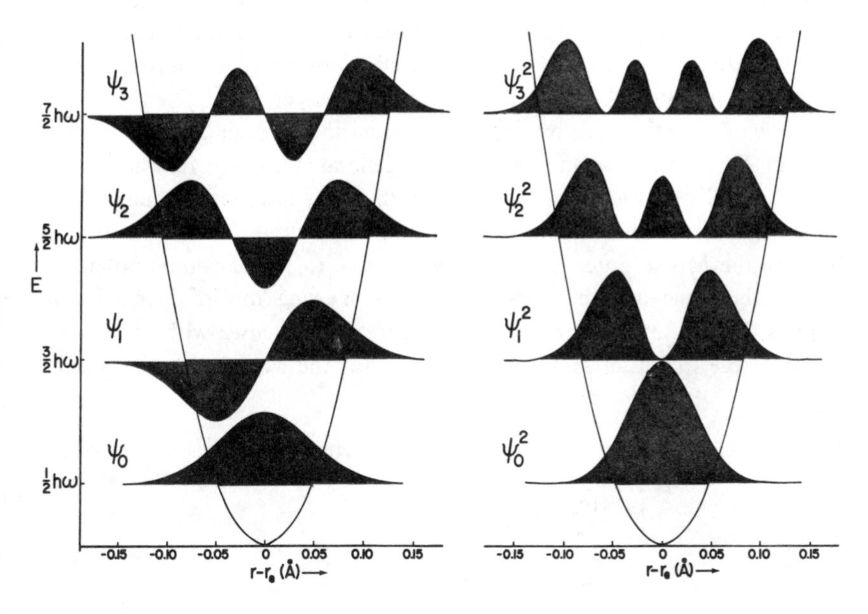

Fig. 3-14. Wave functions (left) and probability distributions (right) for the harmonic oscillator.

where

$$\omega_e = \frac{\beta}{\pi}\sqrt{\frac{D_e}{2\mu}} \qquad x_e = \frac{h\omega_e}{4D_e}$$

$$B_e = \frac{\hbar^2}{2I} = \frac{\hbar^2}{2\mu r_e^2} \qquad D = \frac{\hbar^4}{8\pi^2\omega_e^2\mu^3 r_e^6} \tag{3-46}$$

$$\alpha_e = \frac{3\pi\hbar^3\omega_e}{2\mu r_e^2 D_e}\left(\frac{1}{\beta r_e} - \frac{1}{\beta^2 r_e^2}\right)$$

Equation 3-45 contains something new and useful. The first two terms we have already stated give the vibrational energies of the anharmonic oscillator. The second two terms we have already seen in connection with rotational energies. But the last term is new and represents an interaction between vibration and rotation. We will make use of this term in our treatment of the infrared spectrum of CO in Section 3-3-F. Using the relations 3-46, we can, in principle, use the experimental values of ω_e and x_e to determine the molecular dissociation energy. In practice, dissociation energies obtained in this manner are not in very close agreement with dissociation energies obtained by more direct means. This should alert you to the fact that although the Morse potential is a pretty good approximation of the intramolecular force, it is by no means perfect.

The wave functions for one particular anharmonic oscillator, an excited state of H_2, are shown in Fig. 3-15. In general, the wave functions and probability distributions are skewed such that they have greater magnitude on the side of the well corresponding to bond stretching. This is reasonable because the Morse potential is skewed relative to the harmonic potential in just the same manner (Fig. 3-10). This skewing toward increasing q implies that the average internuclear separation *increases* with increasing value of the quantum number v. Such is not the case for the harmonic

Fig. 3-15. Anharmonic wave functions for some vibrational states of an excited electronic state of H_2 calculated by A.S. Coolidge, H.M. James, and R.D. Present (*J. Chem. Phys.*, *4*, 193 [1936]).

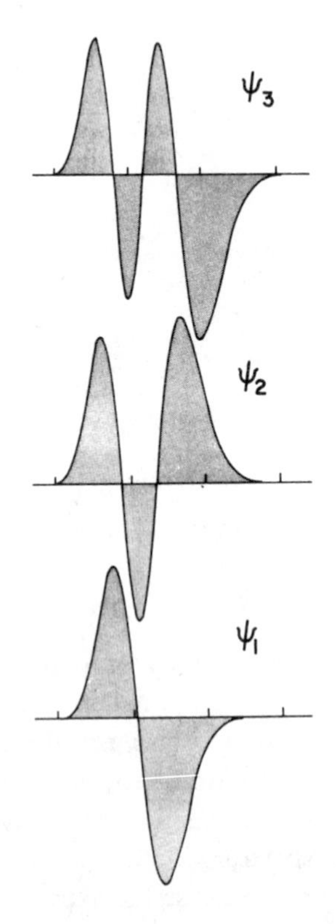

oscillator in which the average value of r is always r_e. This skewing of the wave functions and probability distributions in the anharmonic oscillator is the source of the last term in the energy expression, eq. 3-45, and has significant physical consequences which will be discussed in Section 3-3-F.

Problems

3-9. Verify that ψ_0 of the harmonic oscillator, eq 3-42, is normalized by showing that $\int_{-\infty}^{\infty}\psi_0{}^*\psi_0 dq = 1$. Note that $\int_{-\infty}^{\infty} e^{-ax^2}\,dx = (1/2)\sqrt{\pi/a}$.

3-10. Verify that ψ_0 and ψ_1 of the harmonic oscillator are orthogonal. That is, show that $\int_{-\infty}^{\infty}\psi_0{}^*\psi_1\,dq = 0$.

3-11. Calculate the values of $\langle q\rangle$, $\langle q^2\rangle$, $\langle p\rangle$, $\langle p^2\rangle$ and the product $\sigma_q\sigma_p$ (eq. 2-44) for the harmonic oscillator in the $v = 2$ state (*cf.* Problem 2-23).

3-12. Calculate the value of the constant D (eq. 3-46) for $H^{35}Cl$ ($r_e = 1.2746$ Å) and compare your answer to the experimental value 0.0004 cm^{-1}.

E. *The Population of Energy Levels.* The Maxwell-Boltzmann distribution law states that if molecules in thermal equilibrium occupy two states of energy ϵ_i and ϵ_j (Fig. 3-16[a]), the relative populations of molecules occupying these states will be

$$\frac{n_j}{n_i} = \frac{e^{-\epsilon_j/kT}}{e^{-\epsilon_i/kT}} = e^{-\Delta\epsilon/kT} \tag{3-47}$$

Here n_i and n_j are the number of molecules in each state, k is Boltzmann's constant, and T is the temperature (Kelvin). This applies to the case where there is just one state of energy ϵ_i and one of energy ϵ_j. Suppose instead that there are two states of energy ϵ_i (Fig. 3-16[b]). The distribution law will have an additional factor of 2 in the denominator:

$$\frac{n_j}{n_i} = \frac{e^{-\epsilon_j/kT}}{2e^{-\epsilon_i/kT}} \tag{3-48}$$

In such a case the lower state is said to be doubly degenerate. In general, for two states of degeneracies g_i and g_j,

$$\boxed{\frac{n_j}{n_i} = \frac{g_j e^{-\epsilon_j/kT}}{g_i e^{-\epsilon_i/kT}} = \frac{g_j}{g_i}e^{-\Delta\epsilon/kT}} \tag{3-49}$$

Let's look at the relative populations of the vibrational energy levels of diatomic molecules. We are interested in the relative populations of the

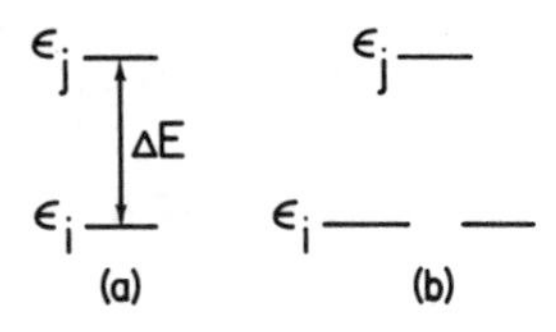

Fig. 3-16. (a) Two nondegenerate energy levels separated by the energy ΔE. (b) Same as (a) except the lower state is doubly degenerate.

vth state and the ground state. The relevant distribution, assuming that the molecule behaves as a harmonic oscillator, is

$$\frac{n_v}{n_0} = \frac{e^{-(v + \frac{1}{2})h\omega/kT}}{e^{-(\frac{1}{2})h\omega/kT}} = e^{-vh\omega/kT} \tag{3-50}$$

The vibrational states are not degenerate so $g_v = g_0 = 1$. Putting actual values of ω into eq. 3-50 shows that for all but the molecules of smallest vibrational frequency the population of any level above the ground state is nil at 300 K (Table 3-6 and Fig. 3-17). Further, the populations of successively higher levels decrease exponentially.

In Section 3-2-A it was stated that the Stokes lines of the Raman spectrum are much more intense than the anti-Stokes lines. The reason for this is simply that Stokes lines originate from the $v = 0$ level while anti-Stokes lines originate from the $v = 1$ level. Therefore only for the lowest vibrational energies will there be any significant population of the $v = 1$ level to give rise to anti-Stokes lines.

The population of rotational levels is another matter altogether. Two factors cause the first few rotational levels of most molecules to be well populated. One is that the values of $\Delta\epsilon$ are much smaller than those for vibrational energy levels. The other is that the rotational levels are degenerate. For each value of J, the rotational quantum number, there are $2J + 1$ wave functions since $m = 0, \pm 1, \pm 2, \ldots \pm J$ (Section 3-3-D). The Jth rotational level is therefore $(2J + 1)$-fold degenerate. This means, for example, that a molecule in the fourth rotational energy level can be in any of $2(4) + 1 = 9$ rotational states, each of the same energy. Hence the populations of the rotational energy levels will be given by

$$\frac{n_J}{n_0} = \frac{(2J + 1)e^{-BJ(J + 1)/kT}}{[2(0) + 1]e^{-B0(0 + 1)/kT}} = (2J + 1)e^{-BJ(J + 1)/kT} \tag{3-51}$$

This function does not decrease monotonically, but rather, rises to a maximum. This is because the exponential part decreases, but the degen-

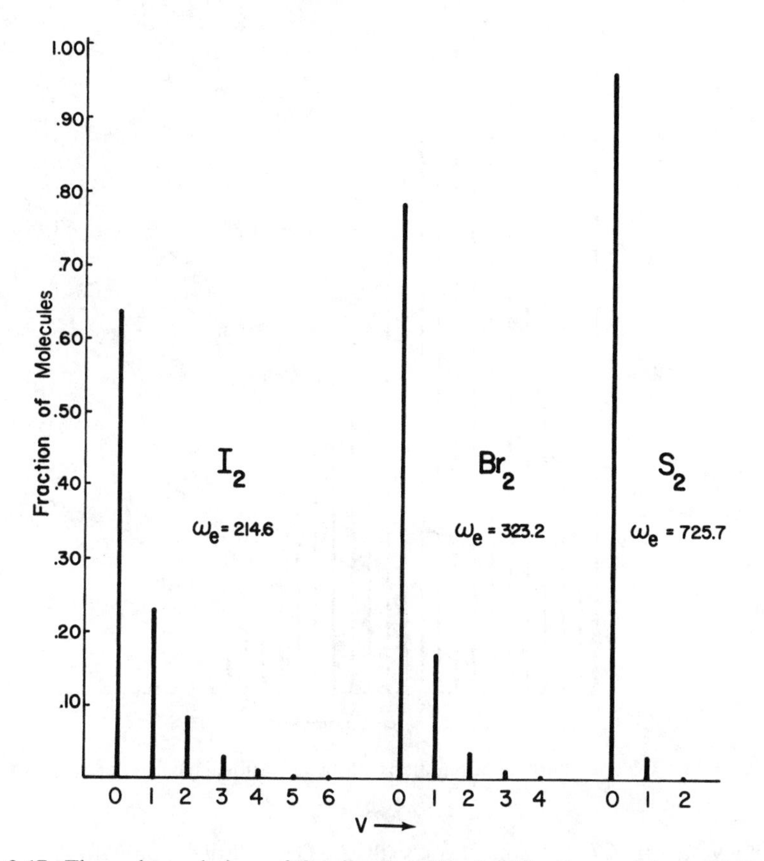

Fig. 3-17. Thermal populations of the vibrational levels of I_2, Br_2, and S_2 at 300 K.

Table 3-6. Populations of the Lowest Vibrationally Excited State for Some Diatomic Molecules

Molecule	$\bar{\omega}$ (cm^{-1})	n_1/n_0 300 K	n_1/n_0 1000 K
H_2	4159.5	2.17×10^{-9}	2.52×10^{-3}
D_2	2990.3	5.91×10^{-7}	1.35×10^{-2}
CO	2143.3	3.43×10^{-5}	4.58×10^{-2}
O_2	1556.2	5.73×10^{-4}	1.07×10^{-1}
S_2	721.6	3.14×10^{-2}	3.54×10^{-1}
Cl_2	556.9	6.92×10^{-2}	4.49×10^{-1}
I_2	213.4	3.59×10^{-1}	7.36×10^{-1}

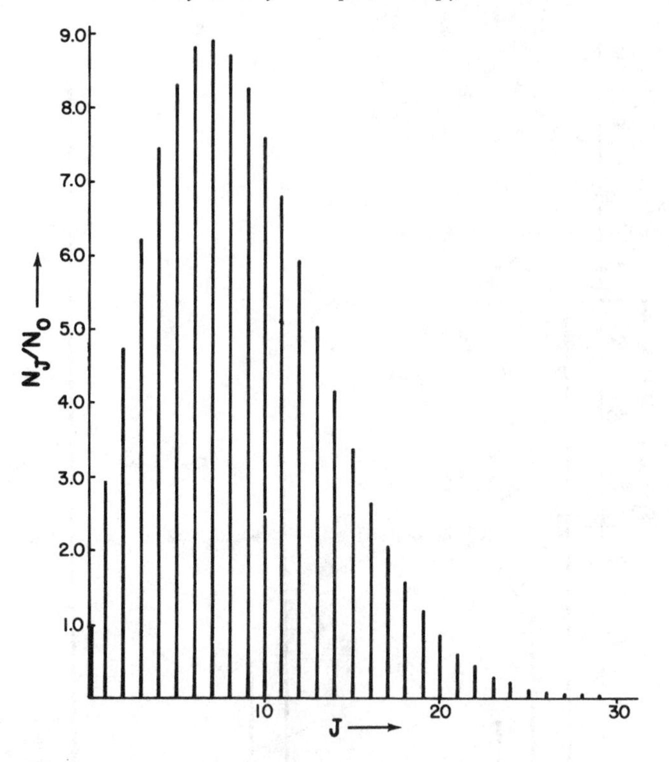

Fig. 3-18. Relative thermal populations of the rotational energy levels of CO at 300 K.

eracy factor, $2J + 1$, increases with J. The relative populations of the rotational levels of CO at 300 K are shown in Fig. 3-18. For this molecule, the $J = 7$ level has the greatest population.

Problems

3-13. If you wish to study the anti-Stokes portion of the Raman spectrum of a sample, should you heat or cool the sample to increase the intensity of the spectrum?

3-14. For which species will the $v = 1$ state have a greater population at thermal equilibrium at 168 K: (a) $^{35}Cl_2$, (b) $^{35}Cl^{37}Cl$; (c) $^{37}Cl_2$?

3-15. What is the value of kT in cm^{-1} at 300 K? If state j is kT cm^{-1} above state i, what is the value of n_j/n_i, if $g_j = g_i = 1$?

3-16. Which rotational state ($J = ?$) of N_2 ($\bar{B} = 2.01$ cm^{-1}) has the greatest population at 500 K? At 100 K? This problem is best approached by finding an expression for dn_J/dJ, setting it equal to zero to find the maximum, and rounding off to the nearest integer value of J at each temperature.

F. *Analysis of the Carbon Monoxide Spectrum.* Before taking the great plunge into the world of polyatomic molecules, we will try to analyze the rotational-vibrational spectrum of CO. The spectrum of CO in solution is not very interesting since it just shows a broad band for the $v = 0 \rightarrow v = 1$ transition at about 2150 cm^{-1}. The effect of the frequent molecular collisions in solution, as well as a variety of intermolecular interactions, is to create a near continuum of very slightly different molecular species (differing, say, in geometry or the arrangement of nearby solvent molecules). The result is a broadening of the absorption bands such that rotational fine structure is generally not seen for liquid or solid samples. However, Fig. 3-19 shows the beautiful structure of the 2150 cm^{-1} band of CO in the gas phase. What we see are two sets of absorptions called the P and R branches. The center of the spectrum (which is called the Q branch when present) is blank. This appearance is determined by the selection rules which govern the possible transitions. The small peaks which overlap the P branch in Fig. 3-19 are due to the 1% natural abundance of $^{13}C^{16}O$. We will examine the CO spectrum in detail and extract information about the bond distance in the ground and first excited states, as well as at the minimum of the Morse potential well.

Let us assume that carbon monoxide has some asymmetric potential well as shown in Fig. 3-20. Because of the asymmetry of the well, the equilibrium internuclear separation in each vibrational state is different, increasing slightly with v (Section 3-3-D, Fig. 3-15). For each value of v, the molecule may be in many rotational levels. The rigid rotor model led us to eq. 3-52 for the rotational energy levels.

$$\bar{E}_J = \bar{B}J(J + 1) \tag{3-52}$$

where

$$\bar{B} = h/800\pi^2 Ic = h/800\pi^2 \mu r^2 c \tag{3-53}$$

Let's account for the anharmonicity by allowing the value of $\bar{B}$ to be different for each vibrational level, since the value of r is changed. Hence, we put subscripts on $\bar{B}$ and r:

$$\bar{E}_J = \bar{B}_v J(J + 1) \tag{3-54}$$

$$\bar{B}_v = h/800\pi^2 \mu r_v{}^2 c \tag{3-55}$$

Each molecule will be characterized by its v and J quantum numbers and its total energy will be the sum of its vibrational and rotational energies:

$$\bar{E}_{v,J} = \bar{\omega}_e(v + 1/2) - \bar{\omega}_e x_e(v + 1/2)^2 + \bar{B}_v J(J + 1) \tag{3-56}$$

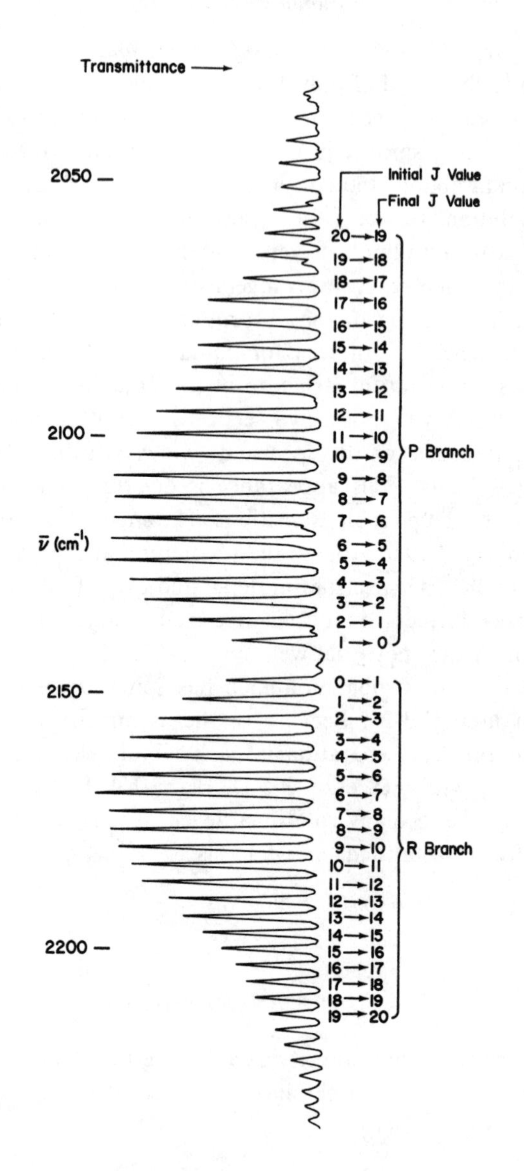

Fig. 3-19. High resolution infrared spectrum of gaseous carbon monoxide.

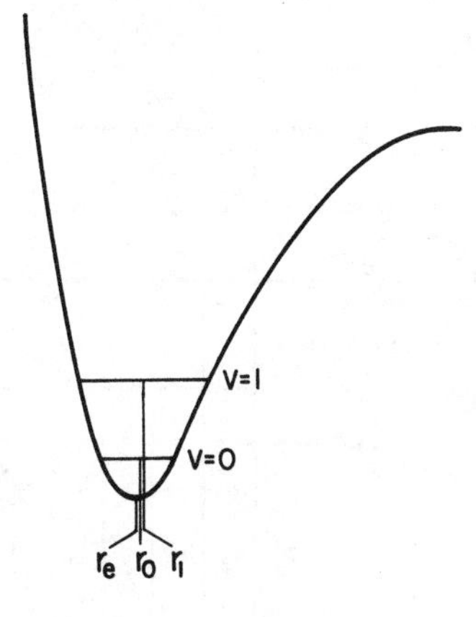

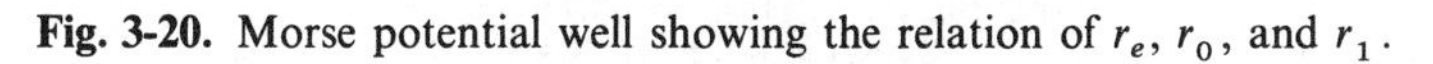

Fig. 3-20. Morse potential well showing the relation of r_e, r_0, and r_1.

For example, a molecule with $v = 1$ and $J = 2$ will have the energy $(3/2)\overline{\omega}_e - (9/4)\overline{\omega}_e x_e + 6\overline{B}_1$ in units of cm^{-1}.

The ir selection rules for diatomic molecules, which we shall not derive yet, state that in a transition from the state (v, J) to the state (v', J') the only allowable changes in the quantum numbers are[†]

$$\boxed{\begin{aligned} \Delta v &= \pm 1 \\ \Delta J &= \pm 1 \end{aligned}} \tag{3-57}$$

Since the infrared spectrum in Fig. 3-19 is that of the $v = 0 \rightarrow v = 1$ transition, all transitions are of the form $(0, J) \rightarrow (1, J \pm 1)$. This is illustrated in Fig. 3-21. The center transition of the spectrum, which is not observed for CO, corresponds to the $(0,0) \rightarrow (1,0)$ transition which violates the ΔJ selection rule. The low energy transitions (the P branch) represent the $\Delta J = -1$ transitions, while the R branch corresponds to the $\Delta J = +1$ transitions.

By simply subtracting the lower energy from the higher energy for each vertical arrow in Fig. 3-21, it is possible to derive the following expressions. For P and R transitions that *start from the same J value* in $v = 0$,

$$\bar{\nu}_{\text{R}}(J) - \bar{\nu}_{\text{P}}(J) = 2\overline{B}_1(2J + 1) \tag{3-58}$$

[†]The Raman selection rules for diatomic molecules are $\Delta V = \pm 1$ and $\Delta J = 0$ or ± 2. Diatomic molecules therefore exhibit rotational Raman fine structure with $\text{O}(\Delta J = -2)$, $\text{Q}(\Delta J = 0)$, and $\text{S}(\Delta J = +2)$ branches.

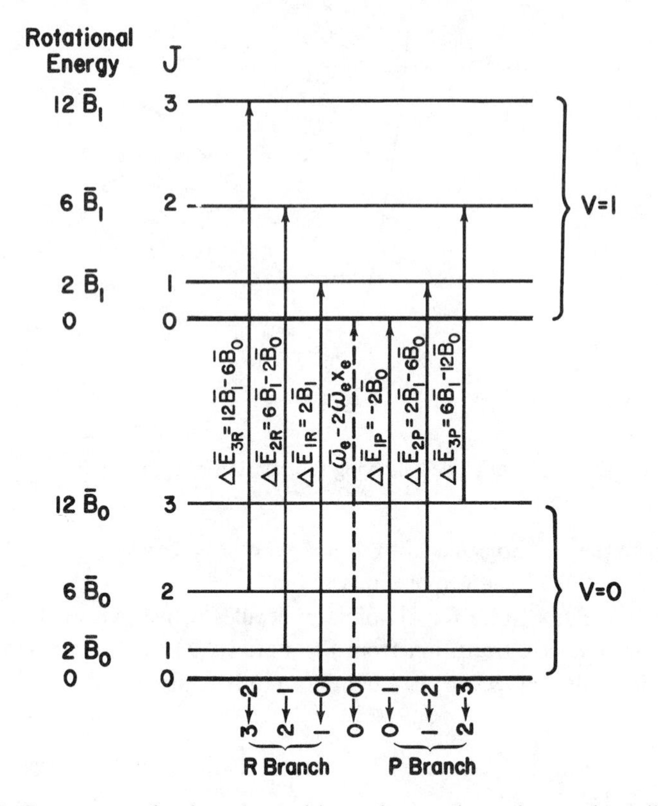

Fig. 3-21. Some energy levels and transitions of a rotating anharmonic oscillator.

An R transition starting at $(0, J)$ goes to $(1, J + 1)$. A P transition starting at $(0, J + 2)$ goes to the same upper level, $(1, J + 1)$. For these transitions,

$$\bar{\nu}_R(J) - \bar{\nu}_P(J + 2) = 2\bar{B}_0(2J + 3) \tag{3-59}$$

By plotting the left hand sides of eqs. 3-58 and 3-59 *vs.* $2J + 1$ and $2J + 3$, respectively, we obtain linear plots whose slopes are $2\bar{B}_1$ and $2\bar{B}_0$. The observed frequencies of the P and R branches of CO are given in Table 3-7. Using these data, one can obtain the values of $\bar{B}_1$ and $\bar{B}_0$ as illustrated in Fig. 3-22. The excellent linearity of this plot attests to the validity of the simple rigid rotor model for rotation. We have made no use of the centrifugal distortion term, $DJ^2(J + 1)^2$, in eq. 3-16.

Once we have $\bar{B}_1$ and $\bar{B}_0$, we can calculate values of r_1 and r_0 with eq. 3-55. In Section 3-3-D we found that the Morse potential led to the prediction that there will be an energy term that represents an interaction of

Table 3-7. Vacuum Wave Numbers for the Rotational Structure of the $v = 0 \rightarrow v = 1$ and $v = 0 \rightarrow v = 2$ Bands of $^{12}C^{16}O$[a]

	$v = 0 \rightarrow v = 1$		$v = 0 \rightarrow v = 2$	
J	$R(J)$	$P(J)$	$R(J)$	$P(J)$
0	2147.0831	—	4263.8396	—
1	2150.8579	2139.4281	4267.5445	4256.2196
2	2154.5975	2135.5482	4271.1790	4252.3047
3	2158.3016	2131.6336	4274.7430	4248.3201
4	2161.9700	2127.6844	4278.2365	4244.2659
5	2165.6028	2123.7008	4281.6592	4240.1423
6	2169.1996	2119.6829	4285.0111	4235.9494
7	2172.7604	2115.6309	4288.2918	4231.6874
8	2176.2850	2111.5449	4291.5014	4227.3564
9	2179.7733	2107.4251	4294.6397	4222.9565
10	2183.2251	2103.2715	4297.7065	4218.4880
11	2186.6403	2099.0845	4300.7018	4213.9509
12	2190.0187	2094.8640	4303.6250	4209.3454
13	2193.3601	2090.6103	4306.4764	4204.6716
14	2196.6645	2086.3234	4309.2558	4199.9298
15	2199.9317	2082.0037	4311.9630	4195.1200
16	2203.1616	2077.6511	4314.5978	4190.2424
17	2206.3540	2073.2658	4317.1601	4185.2971
18	2209.5087	2068.8480	4319.6498	4180.2843
19	2212.6256	2064.3979	4322.0666	4175.2041
20	2215.7045	2059.9155	4324.4106	4170.0568
21	2218.7454	2055.4011	4326.6814	4164.8423
22	2221.7481	2050.8547	4328.8790	4159.5609
23	2224.7124	2046.2766	4331.0033	4154.2128
24	2227.6381	2041.6668	4333.0540	4148.7980
25	2230.5252	2037.0255	4335.0310	4143.3167
26	2233.3735	2032.3529	4336.9342	4137.7691
27	2236.1829	2027.6491	4338.7635	4132.1553
28	2238.9531	2022.9142	4340.5187	4126.4755
29	2241.6841	2018.1484	4342.1996	4120.7297
30	2244.3757	2013.3519	4343.8061	4114.9183

[a] Wave numbers listed in this table have been corrected for the index of refraction of the air and correspond to values which would be observed *in vacuo*. Such correction is not necessary unless energies are measured to an accuracy of tenths of cm^{-1} or better. Data from K.N. Rao and C.W. Mathews, eds., *Molecular Spectroscopy: Modern Research*, Academic Press, N.Y., 1972.

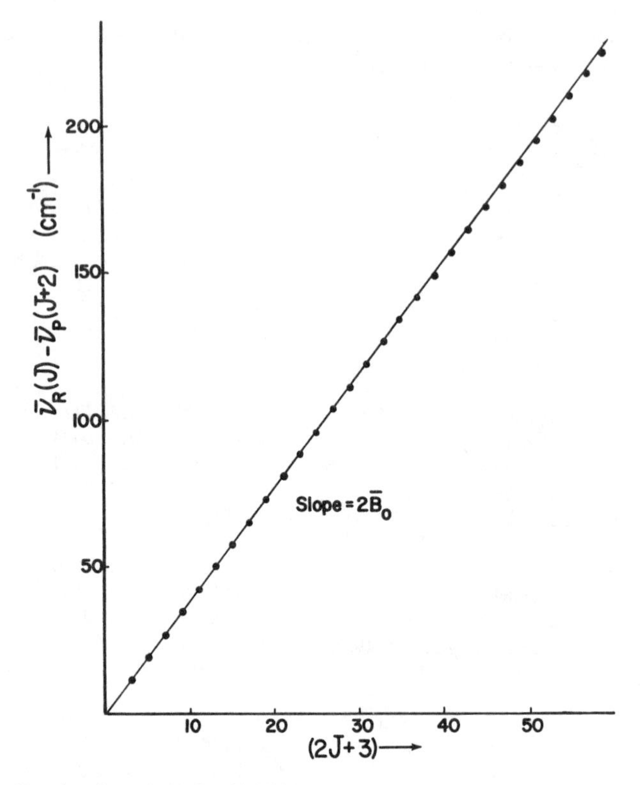

Fig. 3-22. Graph of eq. 3-59 for $^{12}C^{16}O$.

vibration and rotation. Specifically, if we focus on the terms which go as $J(J + 1)$ in eq. 3-45, we see that B_v must have the following form:

$$\begin{aligned}\bar{E}_{\text{rot}} &= \bar{B}_e J(J+1) - \bar{\alpha}_e(v + 1/2)J(J+1) \\ &= \underbrace{[\bar{B}_e - \bar{\alpha}_e(v + 1/2)]}_{\bar{B}_v} J(J+1)\end{aligned}$$

$$\bar{B}_v = \bar{B}_e - \bar{\alpha}_e(v + 1/2) \tag{3-60}$$

$\bar{\alpha}_e$ is quite small compared to $\bar{B}_e$. Having values for $\bar{B}_0$ and $\bar{B}_1$, it is possible to obtain values for $\bar{\alpha}_e$ and $\bar{B}_e$ using eq. 3-60. $\bar{B}_e$ then leads to a value of r_e. Our analysis of the rotational fine structure of the ir spectrum of CO has therefore given us values of r_e, r_0 and r_1. Values of these constants for several diatomic molecules are listed in Table 3-8.

Table 3-8. Values of $\overline{B}_e$, $\overline{\alpha}_e$, r_e, r_0, and r_1 for Selected Diatomic Molecules[a]

Molecule	$\overline{B}_e$ (cm^{-1})	α_e (cm^{-1})	r_e (A)	r_0 (A)	r_1 (A)
H_2	60.809	2.993	0.7417	0.7505	0.7702
HD	45.655	1.993	0.7414	0.7495	0.7668
D_2	30.429	1.049	0.7416	0.7481	0.7616
HCl	10.5909	0.3019	1.27460	1.2838	1.3028
DCl	5.445	0.1118	1.275	1.282	1.295
CO	1.9314	0.01748	1.1282	1.1307	1.1359
N_2	2.010	0.0187	1.094	1.097	1.102

[a] From G.M. Barrow, *Introduction to Molecular Spectroscopy*, McGraw-Hill, N.Y., 1962.

Problems

3-17. Using the data in Table 3-7, prepare graphs of the first overtone data analogous to that in Fig. 3-22. That is, plot the equations

$$\overline{\nu}_R(J) - \overline{\nu}_P(J) = 2\overline{B}_2(2J + 1) \quad (3\text{-}58')$$

$$\overline{\nu}_R(J) - \overline{\nu}_P(J + 2) = 2\overline{B}_0(2J + 3) \quad (3\text{-}59')$$

using the first ten points for each plot. A least squares analysis of these data gives the following slopes and intercepts:

Transition	Equation	Slope (cm^{-1})	Intercept (cm^{-1})
$0 \rightarrow 1$	3-58	3.8070	0.0180
$0 \rightarrow 1$	3-59	3.8420	0.0184
$0 \rightarrow 2$	3-58′	3.7720	0.0180
$0 \rightarrow 2$	3-59′	3.8420	0.0181

From these data, determine $\overline{B}_2$ and $\overline{B}_0$. If you have access to a calculator, work all answers to four significant figures. From $\overline{B}_2$ and $\overline{B}_0$, determine r_2 and r_0. Find $\overline{B}_e$, $\overline{\alpha}_e$, and r_e. Which two slopes in the table above are expected to be the same?

3-18. Using eq. 3-46, calculate the value of $\overline{\alpha}_e$ for $^{12}C^{16}O$ and compare this to the observed value given in Table 3-8.

3-19. The small peaks in Fig. 3-19 which overlap the P branch are due to the 1% natural abundance of $^{13}C^{16}O$. Using the harmonic oscillator approximation and $\overline{\omega}(^{12}C^{16}O) = 2143$ cm^{-1}, at what wave number should the spectrum of $^{13}C^{16}O$ be centered? Which branch of its spectrum appears in Fig. 3-19? The well resolved small peaks near 2060 cm^{-1} have nearly 10% of the height of the corresponding $^{12}C^{16}O$ peaks near 2120 cm^{-1}, instead of 1%. Why? (Hint: What is plotted in Fig. 3-19?)

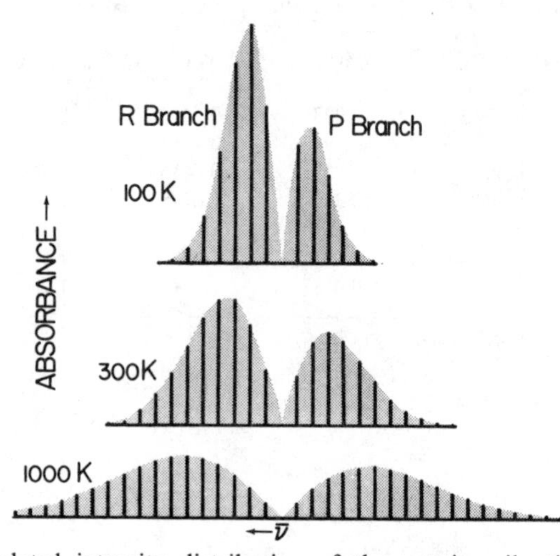

Fig. 3-23. Calculated intensity distribution of the rotation-vibration absorption lines at 100, 300, and 1000 K for $\bar{B} = 10\ \text{cm}^{-1}$. The absorption intensity is not proportional to the population of the ground state, n_J, given by eq. 3-51, but to the closely related expression $\nu(J + J' + 1)e^{-BJ(J+1)/kT}$, where J applies to the ground state and J' to the excited state. For this figure the dependence on ν was neglected, as was the variation in $\bar{B}$ between the upper and lower vibrational states.

G. *The Effect of Temperature.* In Fig. 3-18 it was seen that at 300 K the rotational state of CO with the greatest population of molecules is $J = 7$. We would expect that the absorption intensity of each rotational band in the ir spectrum should closely reflect the fraction of molecules in that rotational level in the ground vibrational state. This is confirmed by the spectrum of CO in Fig. 3-19 in which transitions originating from $J \approx 7$ have the greatest intensity. Since the J value of the state with maximum population decreases with decreasing temperature (at absolute zero all molecules have $J = 0$), we expect the "rotational envelope" of the absorption band to sharpen at lower temperatures and broaden at higher temperatures, as shown in Fig. 3-23. Does this suggest a way to measure temperature using infrared spectroscopy?

The spectra of samples in condensed phases also sharpen at low temperatures, but for reasons less well understood. Broadening can be attributed in a general way to molecular motion and collisions. As molecular motion decreases at low temperatures, spectra can sharpen markedly. Vibrational spectroscopy at liquid nitrogen and lower temperatures is often

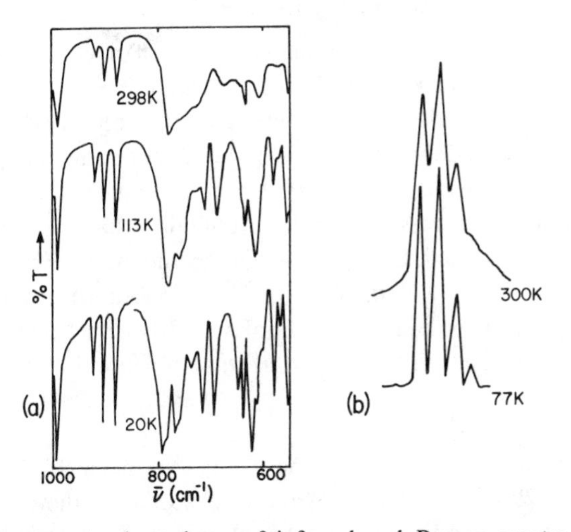

Fig. 3-24. Temperature dependence of infrared and Raman spectra. (a) Infrared spectrum of the sugar lactose reproduced from J.E. Katon, J.T. Miller, Jr., and F.F. Bentley, *Carbohyd. Res.*, *10*, 505 (1969). (b) Raman spectrum showing the structure of the band near 460 cm^{-1} for CCl_4. The four peaks are due to the presence of $C^{35}Cl_4$, $C^{35}Cl_3{}^{37}Cl$, $C^{35}Cl_2{}^{37}Cl_2$, and $C^{35}Cl^{37}Cl_3$. From H.A. Szymanski, ed., *Raman Spectroscopy*, Plenum Press, N.Y., 1970, Vol. 2.

very revealing, as overlapping bands at room temperature are often resolved into distinct bands at low temperatures. Some examples appear in Fig. 3-24.

Problems

3-20. One mechanism which broadens the individual lines of the absorption spectrum of a gas is the *Doppler effect*. Molecules moving away from the light source "see" a lower frequency light beam than is emitted by the source in the laboratory frame of reference. Molecules moving toward the beam "see" higher frequency (energy) light. The average speed of a molecule is given by $v = (8kT/\pi m)^{1/2}$, where k is Boltzmann's constant and m is the mass of the molecule. Estimate the line width of an absorption line at 2000 cm^{-1} as the difference in wave number (in the laboratory frame of reference) absorbed by molecules of CO at 300 K moving toward the source at speed v and molecules moving away from the source at speed v. The difference in wavelength, $\Delta\lambda$, seen by a molecule moving at speed v compared to the wavelength, λ, seen by a molecule at rest in the lab frame is given by $\Delta\lambda/\lambda = v/c$, where c is the speed of light.

3-4. Transitions between Stationary States†

Suppose a one-dimensional harmonic oscillator of mass m and charge e (*i.e.*, a charged particle which feels the potential $V = (1/2)kx^2$) is in its lowest vibrational state described by the wave function Ψ_0. As in Chapter Two, a capital psi, Ψ, denotes the total wave function, including time, and a small psi, ψ, indicates only the spatial part of the wave function. In this section, we seek an expression for the probability that the transition $\Psi_0 \rightarrow \Psi_1$ can be stimulated by electromagnetic radiation. From this one-dimensional model we will then generalize to vibrational transitions of real molecules which contain electric dipoles oscillating in three dimensions.

We adopt the formalism that the total wave function is

$$\Psi(x,t) = c_0 \Psi_0(x,t) + c_1 \Psi_1(x,t) \tag{3-61}$$

where the coefficient c_0 is unity at time $t = 0$, and the coefficient c_1 is zero at time $t = 0$. This just means that the oscillator is in the $v = 0$ state at time $t = 0$. If the Schrödinger equation contains no time dependent potential, c_0 will always be unity and c_1 will always be zero; the oscillator will always be in the ground state. The reason is that we have not provided any mechanism for the molecule to change its state. So we turn on a one-dimensional oscillating electric field, ε, which oscillates along the x axis. The force exerted on a particle of charge e, will be charge times field,

$$F = e\varepsilon \tag{3-62}$$

and the potential energy is

$$V = e\varepsilon x \tag{3-63}$$

So our complete time-dependent Schrödinger equation for the one-dimensional harmonic oscillator of charge e is

$$\left(-\frac{\hbar^2}{2m}\frac{\partial^2}{\partial x^2} + \frac{1}{2}kx^2 + e\varepsilon x\right)\Psi(x,t) = i\hbar\frac{\partial \Psi(x,t)}{\partial t} \tag{3-64}$$

Let us assume a solution of the form

$$\Psi = c_0\Psi_0 + c_1\Psi_1 = c_0\psi_0 e^{-iE_0 t/\hbar} + c_1\psi_1 e^{-iE_1 t/\hbar} \tag{3-65}$$

where E_0 and E_1 are the energies of the two states and the form of the time-dependence comes from eq. 2-33. The coefficients c_0 and c_1 will be functions of time if the oscillator is to change its vibrational state.

† As you thrash your way through the algebra in this section, you should keep in mind that the most important thing to understand when you are finished is the significance of the transition moment integral.

Putting eq. 3-65 into the right side of eq. 3-64 gives

$$i\hbar\frac{\partial}{\partial t}(c_0\psi_0 e^{-iE_0 t/\hbar} + c_1\psi_1 e^{-iE_1 t/\hbar}) = i\hbar\left(\Psi_0\frac{dc_0}{dt} + \Psi_1\frac{dc_1}{dt}\right) + c_0\Psi_0 E_0 + c_1\Psi_1 E_1 \quad (3\text{-}66)$$

Putting eq. 3-65 into the left side of eq. 3-64 gives

$$\left(-\frac{\hbar^2}{2m}\frac{\partial^2}{\partial x^2} + \frac{1}{2}kx^2 + e\varepsilon x\right)(c_0\Psi_0 + c_1\Psi_1)$$

$$= \left(-\frac{\hbar^2}{2m}\frac{\partial^2}{\partial x^2} + \frac{1}{2}kx^2\right)(c_0\Psi_0 + c_1\Psi_1) + e\varepsilon x(c_0\Psi_0 + c_1\Psi_1)$$

$$= c_0\Psi_0 E_0 + c_1\Psi_1 E_1 + e\varepsilon x(c_0\Psi_0 + c_1\Psi_1) \quad (3\text{-}67)$$

because each wave function is an eigenfunction of the time-independent hamiltonian:

$$\left(-\frac{\hbar^2}{2m}\frac{\partial^2}{\partial x^2} + \frac{1}{2}kx^2\right)\Psi_i = E_i\Psi_i \quad (3\text{-}68)$$

Equating eqs. 3-66 and 3-67, and dropping the terms which are the same on both sides, we get

$$e\varepsilon x(c_0\Psi_0 + c_1\Psi_1) = ih\left(\Psi_0\frac{dc_0}{dt} + \Psi_1\frac{dc_1}{dt}\right) \quad (3\text{-}69)$$

Multiplying both sides of eq. 3-69 by $\Psi_1{}^*$, and integrating over all space, we get

$$\varepsilon\int_{-\infty}^{\infty} ex\,(c_0\,\Psi_1{}^*\Psi_0 + c_1\,\Psi_1{}^*\Psi_1)\,dx = i\hbar\left(\frac{dc_0}{dt}\int_{-\infty}^{\infty}\Psi_1{}^*\Psi_0\,dx + \frac{dc_1}{dt}\int_{-\infty}^{\infty}\Psi_1{}^*\Psi_1\,dx\right) \quad (3\text{-}70)$$

But the integral $\int_{-\infty}^{\infty}\Psi_1{}^*\Psi_0\,dx$ vanishes because the wave functions are orthogonal. Further, the integral $\int_{-\infty}^{\infty}\Psi_1{}^*\Psi_1\,dx$ is unity because the wave functions are normalized. So eq. 3-70 reduces to

$$\varepsilon\int_{-\infty}^{\infty} ex(c_0\Psi_1{}^*\Psi_0 + c_1\Psi_1{}^*\Psi_1)dx = i\hbar\frac{dc_1}{dt} \quad (3\text{-}71)$$

Now we are getting to the heart of the matter. Equation 3-71 gives the time dependence of c_1, the coefficient of the excited state. How does c_1 change with time at $t = 0$ when we turn on the electric field? First note that at

$t = 0$, $c_1 = 0$, and $c_0 = 1$; so we throw out the term containing c_1 on the left side of eq. 3-71. For ε we will use the exponential form in eq. 3-72:

$$\varepsilon = \varepsilon_0 \cos 2\pi\nu t = (1/2)\varepsilon_0 \left(e^{2\pi i\nu t} + e^{-2\pi i\nu t}\right) \qquad \text{(3-72)}$$

For Ψ_0 and Ψ_1 we will use the forms in eq. 3-65. Making these substitutions in eq. 3-71 yields

$$\frac{1}{2}\varepsilon_0 \left(e^{2\pi i\nu t} + e^{-2\pi i\nu t}\right) \int_{-\infty}^{\infty} ex\psi_1{}^* e^{iE_1 t/\hbar}\psi_0 e^{-iE_0 t/\hbar}\,dx = i\hbar\frac{dc_1}{dt}$$

or

$$\frac{dc_1}{dt} = \frac{\varepsilon_0}{2i\hbar}\left(e^{2\pi i\nu t} + e^{-2\pi i\nu t}\right)\left(e^{i(E_1 - E_0)t/\hbar}\right) \int_{-\infty}^{\infty} ex\psi_1{}^*\psi_0\,dx$$

or

$$\frac{dc_1}{dt} = \frac{\varepsilon_0}{2i\hbar}\left(e^{i(E_1 - E_0 + h\nu)t/\hbar} + e^{i(E_1 - E_0 - h\nu)t/\hbar}\right) \int_{-\infty}^{\infty} ex\psi_1{}^*\psi_0\,dx \qquad \text{(3-73)}$$

by the judicious substitution $2\pi = h/\hbar$. We are almost there.

We will call the integral $\int_{-\infty}^{\infty} ex\psi_1{}^*\psi_0 dx$ the *transition moment integral* for the transition from the state $v = 0$ to the state $v = 1$ and denote it as M_{01}.† This integral is given the symbol μ with appropriate subscripts in many texts. We can integrate eq. 3-73 with respect to time over the *short* interval 0 to t and find

$$c_1 = \frac{1}{2}\varepsilon_0 M_{01}\left[\frac{1 - e^{i(E_1 - E_0 + h\nu)t/\hbar}}{E_1 - E_0 + h\nu} + \frac{1 - e^{i(E_1 - E_0 - h\nu)t/\hbar}}{E_1 - E_0 - h\nu}\right] \qquad \text{(3-74)}$$

Now we are interested in the case where $h\nu$ approaches $E_1 - E_0$ since we know that the condition for absorption of the light is $\Delta E = h\nu$. In such a case, the denominator of the second term in brackets in eq. 3-74 approaches zero and the second term overwhelms the first term in magnitude. For this reason, we are going to drop the first term of eq. 3-74.

What we really need is a value of $c_1{}^*c_1$ since the probability of being in the excited state goes as $c_1{}^*c_1$. Discarding the first term of eq. 3-74 and multiplying by $c_1{}^*$, we get

$$c_1{}^*c_1 = \frac{\varepsilon_0{}^2}{4} M_{01}{}^2 \left[\frac{2 - e^{i(E_1 - E_0 - h\nu)t/\hbar} - e^{-i(E_1 - E_0 - h\nu)t/\hbar}}{(E_1 - E_0 - h\nu)^2}\right] \qquad \text{(3-75)}$$

†More precisely, M_{01} is the transition moment integral for an electric dipole transition, which is the only kind we have been considering. In the absence of an electric dipole mechanism, integrals involving magnetic dipole or electric quadrupole mechanisms would have to be considered.

Now we make the tricky substitution $\sin^2\theta = \frac{1}{4}(2 - e^{2i\theta} - e^{-2i\theta})$ which yields

$$c_1{}^*c_1 = \frac{\varepsilon_0{}^2 M_{01}{}^2 \sin^2[(E_1 - E_0 - h\nu)t/2\hbar]}{(E_1 - E_0 - h\nu)^2} \tag{3-76}$$

What we want is the factor $c_1{}^*c_1$ for all values of incident radiation frequency since in the ordinary absorption experiment we shine *white light* (light of all different frequencies) on the sample. We recognize that only values of ν near to the resonance frequency will be effective in giving magnitude to $c_1{}^*c_1$. To take into account all frequency values, we integrate eq. 3-76 from $\nu = 0$ to $\nu = \infty$ which gives†

$$c_1{}^*c_1 = \frac{\varepsilon_0{}^2 M_{01}{}^2 t}{4\hbar^2} \tag{3-77}$$

or

$$\boxed{\frac{d(c_1{}^*c_1)}{dt} \propto \varepsilon_0{}^2 M_{01}{}^2} \tag{3-78}$$

Equation 3-78 is the final result we have been seeking. The initial rate at which the excited state is occupied when the light is turned on is proportional to the square of the transition moment integral and the square of the amplitude of the electromagnetic radiation. Because the intensity of light is proportional to $\varepsilon_0{}^2$ (eq. 2-7), the rate of population of the excited state is proportional to the intensity of the incident radiation. The fact that the transition probability, $d(c_1{}^*c_1)/dt$, is zero if the transition moment integral, M_{01}, is zero will be of fundamental significance to the remainder of this book. We will return to the properties of the transition moment integral when we discuss selection rules in Section 3-6.

You may wonder what happens when all the molecules in a given sample have absorbed light and are in the excited state? The qualitative answer to this question is that molecules in the excited state can lose energy by

† To perform the integration, we used the definite integral $\int_{-\infty}^{\infty}(1/x^2)\sin^2 x\, dx = \pi$ and the substitution $x = (h\nu - E_1 + E_0)t/2\hbar$ which converts eq. 3-76 to

$$c_1{}^*c_1 = \frac{\varepsilon_0{}^2 M_{01}{}^2 t}{4\pi\hbar^2}\int_{-\infty}^{\infty}\frac{\sin^2 x}{x^2}\,dx$$

The lower limit of the integration, $-\infty$, is not rigorously correct, but is a very good approximation. The area under the integrand from the true lower limit, $x = (-E_1 + E_0)t/2\hbar$ to $x = -\infty$ is a very small fraction of the total area under the curve from $x = -\infty$ to $x = +\infty$.

The very alert reader may realize that eq. 3-77 does not have the correct units. $c_1{}^*c_1$ should be dimensionless but the right hand side of eq. 3-77 has the units s^{-1}. The way out of this dilemma is to include a spectral "shape factor" as described by A. Abragam and B. Bleaney, *Electron Paramagnetic Resonance of Transition Ions*, Clarendon Press, Oxford, 1970, p 100.

spontaneous emission of light, by *collisional energy transfer*, or by *stimulated emission* caused by the incident radiation. (Light of energy ΔE, where ΔE is the difference in energy between two states, can not only be absorbed but has an equal probability of causing a molecule in the excited state to emit a photon of energy ΔE and return to the ground state. This is called stimulated emission.) These mechanisms, which we combine under the heading *relaxation processes*, allow excited molecules to return to the ground state and be available to absorb more incident radiation. Provided that the incident light intensity is not too great, relaxation can keep up with absorption and most samples do not run out of their supply of ground state molecules. At high incident intensities, relaxation cannot compete with absorption, the ground state population becomes depleted, and absorption of light ceases. This process is called *saturation* and is usually not important in vibrational and electronic spectroscopy unless extremely intense sources of light, such as lasers, are used (Fig. 3-25). Saturation is very often achieved in magnetic resonance spectroscopy where relaxation is not efficient.

Problems

3-21. Evaluate the transition moment integral $e\int_{-\infty}^{\infty}\psi_1{}^*x\psi_0 dx$ for the one-dimensional harmonic oscillator and show that it has a nonzero value. Show that the transition moment integrals M_{02} and M_{03} are identically equal to zero and that the $v = 0 \rightarrow v = 2$ and $v = 0 \rightarrow v = 3$ transitions are therefore "forbidden". Some useful definite integrals are

$$\int_0^{\infty} x^{2n}e^{-ax^2}dx = \frac{1\cdot 3\cdot 5\cdots(2n-1)}{2^{n+1}}\sqrt{\frac{\pi}{a^{2n+1}}} \quad \text{and} \quad \int_0^{\infty} x^{2n+1}e^{-ax^2}dx = \frac{n!}{2a^{n+1}}$$

3-22. Eq. 3-76 tells us that the absorption of radiation for an electric dipole transition has a dependence on ν of the form $f(\nu) = \dfrac{\sin^2[(E_1 - E_0 - h\nu)t/2\hbar]}{(E_1 - E_0 - h\nu)^2}$.

(a) Using L'Hôpital's rule, show that the value of $f(\nu)$ in the limit $h\nu \rightarrow E_1 - E_0$ is $f(\nu) = \pi^2 t^2/h^2$. To refresh your memory, L'Hopital's rule states that

$$\lim_{\nu \rightarrow (E_1 - E_0)/h} \frac{h(\nu)}{g(\nu)} = \frac{h'(\nu)}{g'(\nu)}$$

where h and g are functions of ν, and h' and g' are the derivatives with respect to ν. If the first derivatives of the numerator and denominator still give an indeterminate form, the rule may be applied again.

(b) Show that the absorption falls to half of its maximum value when the condition $\sin[(E_1 - E_0 - h\nu)t\pi/h] = \pi t(E_1 - E_0 - h\nu)/\sqrt{2}h$ is met.

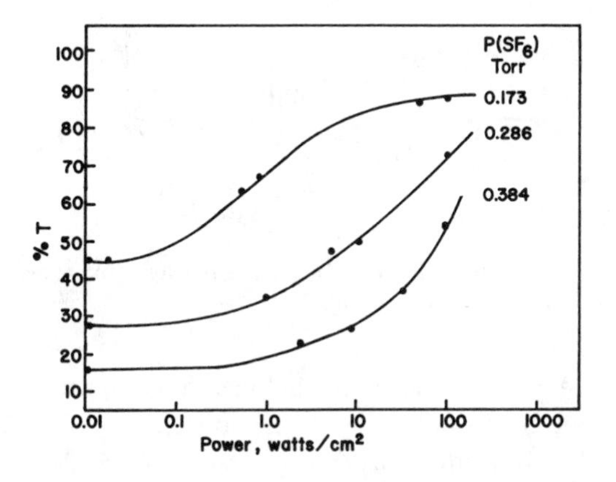

Fig. 3-25. Saturation curves for SF_6 vapor. The higher the concentration, the more efficient is relaxation and the greater is the power needed to induce saturation. Reproduced from I. Burak, J.I. Steinfeld, and D.G. Sutton, *J. Quant. Spec. and Radiative Transfer*, *9*, 959 (1966).

(c) The expression in part (b) is satisfied when $|E_1 - E_0 - h\nu|\pi t/h = 1.39$. This illustrates the Heisenberg uncertainty principle which says that $\Delta E \Delta t \gtrsim \hbar$, where ΔE is the uncertainty of the energy of the state and Δt is the uncertainty of the time for the transition to occur, which is t in the above expression. What value of t gives a line width $(= 2|E_1 - E_0 - h\nu|)$ of 0.01 cm^{-1}?

3-5. The Normal Modes of Vibration of Polyatomic Molecules

A diatomic molecule is simple since it has just a single vibration. Even at absolute zero this vibration occurs because the molecule cannot have less than the zero point energy. Polyatomic molecules undergo much more complex vibrations. However, all of this motion may be resolved into a superposition of a limited number of fundamental motions called *normal modes of vibration*.† We shall be interested in the number, types, and symmetries of these modes.

†This statement is rigorously true only if the vibrations are harmonic in nature, *i.e.*, if the restoring force is proportional to the displacement. Such an approximation is not terrible, however, and without it the study of molecular vibrations would be hopelessly complicated. The statements we will make about the symmetries of vibrational wave functions are rigorously true regardless of anharmonicity.

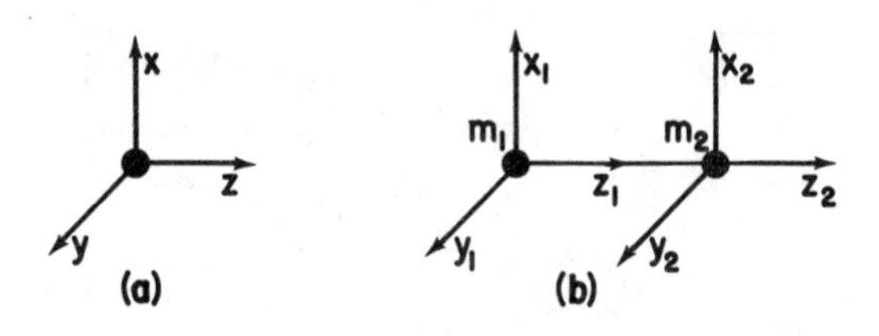

Fig. 3-26. (a) Three degrees of freedom of a single particle. (b) Six degrees of freedom of the two atoms of a diatomic molecule.

Consider a single particle in three-dimensional space. The motion of the particle can be described by three coordinates (Fig. 3-26[a]). The particle is said to have three *degrees of freedom*. Each degree of freedom, in this case, represents a translation of the particle through space. Now consider a set of two particles such as the two nuclei of a diatomic molecule (Fig. 3-26[b]). Since each particle has three degrees of freedom, the system as a whole has six degrees of freedom. Three of these are just translations of the entire unit in the *x*, *y*, or *z* directions (Fig. 3-27). Two degrees of freedom are accounted for by rotation about the center of mass. You might expect a third rotational degree of freedom, but rotation about the molecular axis of a linear molecule is undefined because it does not represent any change of the nuclear coordinates. For nonlinear molecules, there will be a third axis of rotation. Finally, one vibrational degree of freedom is left (Fig. 3-27). Any motion of this diatomic molecule can be obtained by superimposing these six kinds of motions.

Fig. 3-27. The six degrees of freedom of a diatomic molecule. Rotation about the *z* axis does not represent any change of the nuclear coordinates.

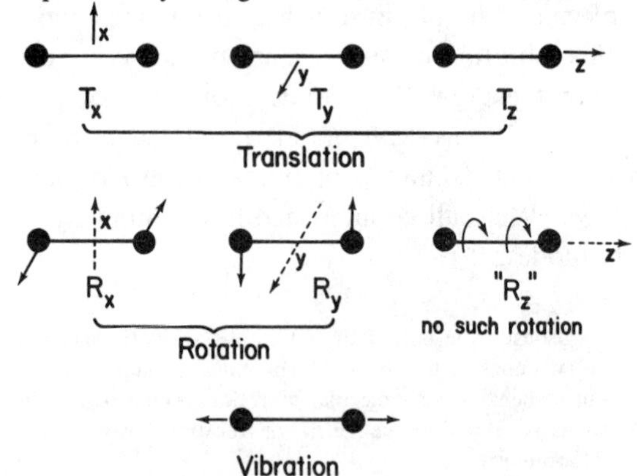

A collection of three particles, such as a molecule of water with three atoms, will have $3 \cdot 3 = 9$ degrees of freedom. Three are again translations of the whole molecule. This time there are three possible rotational degrees of freedom about three mutually perpendicular axes passing through the center of mass of the molecule. This leaves three degrees of freedom for vibrations. These three kinds of vibration are the three normal modes of vibration of the molecule. If you were to sit on one of the atoms and translate and rotate with the molecule, every possible motion you would see could be decomposed into a sum of the three simple normal modes of vibration. In general, a molecule with n atoms will have $3n - 6$ modes of vibration. A linear molecule will have $3n - 5$ modes of vibration because there is no rotation about the molecular axis.

What is a *normal coordinate*, q? A normal coordinate is a single coordinate along which the progress of a single normal mode of vibration can be followed. In $^{12}C^{16}O$, for example, a vibration which does not affect the position of the center of mass, and hence has no component of translation, will occur if the carbon atom moves 15.995/12.000 as far as the oxygen atom (Fig. 3-28). By motion along the normal coordinate we mean that the carbon atom moves a distance $\Delta r(\mathrm{C})$ and the oxygen atom moves a distance $\Delta r(\mathrm{O})$,

$$q = r - r_0 = \Delta r(\mathrm{C}) + \Delta r(\mathrm{O}) \tag{3-79}$$

and

$$m_{\mathrm{C}} \Delta r(\mathrm{C}) = m_{\mathrm{O}} \Delta r(\mathrm{O})$$

or

$$\frac{\Delta r(\mathrm{C})}{\Delta r(\mathrm{O})} = \frac{m_{\mathrm{O}}}{m_{\mathrm{C}}} = \frac{15.995}{12.000} \tag{3-80}$$

Both atomic displacements occur at the same frequency and in phase. Displacement would be measured from the "equilibrium" atomic separation, r_0 in the ground state of CO.

Mathematically, the normal coordinates are defined such that the potential energy of the molecule can be expressed by eq. 3-81 and the kinetic energy by eq. 3-82:

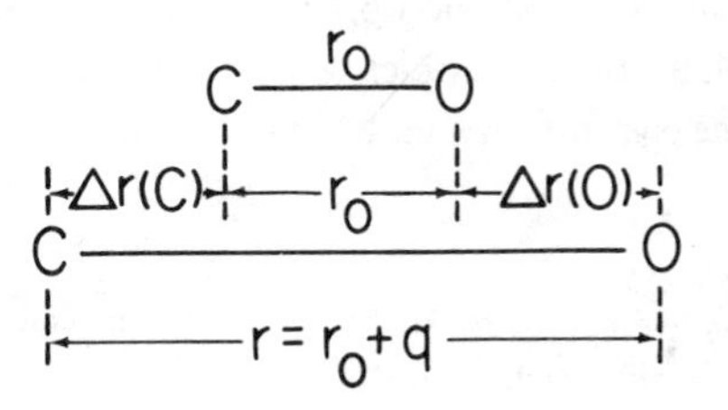

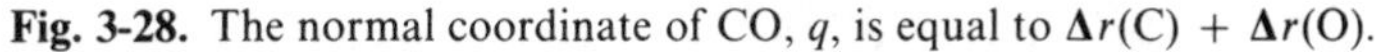

Fig. 3-28. The normal coordinate of CO, q, is equal to $\Delta r(\mathrm{C}) + \Delta r(\mathrm{O})$.

$$V = (1/2)\sum_i \lambda_i q_i^2 \tag{3-81}$$

$$K = (1/2)\sum_i (dq_i/dt)^2 \tag{3-82}$$

where λ_i is a constant. We will concentrate on the qualitative aspects of normal coordinate analysis only.

What we are very much concerned with is the *symmetry* of each normal mode of vibration. *Each normal mode of vibration will form a basis for an irreducible representation of the point group of the molecule.*† Let me say that again: *Each normal mode of vibration will form a basis for an irreducible representation of the point group of the molecule.* We have just made the critical connection between Chapters One, Two, and Three.

Let's illustrate the meaning of this statement with the three normal vibrations of water shown in Fig. 3-29 and customarily labelled ν_1, ν_2, and ν_3. In this figure we have shown large displacements of hydrogen and small displacements of oxygen, because oxygen is 16 times as massive as hydrogen. One often sees normal vibrations drawn without any displacement of the oxygen atom, but this is not strictly correct since such motion would displace the center of mass of the entire molecule and would therefore contain a component of translation as well as vibration. (For simplicity, however, we will occasionally be guilty of this omission.) Calling the C_2 axis of H_2O the z axis, and calling the molecular plane the yz plane, Table 3-9 shows how each normal mode transforms under the symmetry operations of the point group C_{2v}. Let's look at ν_3 in detail. The operation E leaves the vibration unchanged so it has a character of $+1$. The operation C_2 changes the sense of vibration as shown in Fig. 3-30. Each atom moves in the opposite direction after performing the C_2 operation. The vibration has been transformed into *minus* itself, so the character is -1. Similarly, $\sigma(xz)$ changes the vibration into minus itself and $\sigma(yz)$ leaves the vibration unchanged. ν_3 therefore transforms as the irreducible representation of C_{2v} whose characters are $1\ -1\ -1\ 1$, which is b_2. Both ν_1 and ν_2 are unchanged by any of the operations and transform as a_1. The significance of the symmetry of normal vibrations will become apparent when we get to selection rules.

We will now look at some examples which illustrate a systematic procedure for determining the symmetries of the normal modes of vibration of any molecule:

†Proof of this statement is offered by H. Eyring, J. Walter, and G.E. Kimbal, *Quantum Chemistry*, John Wiley & Sons, New York, 1944, Section 10c.

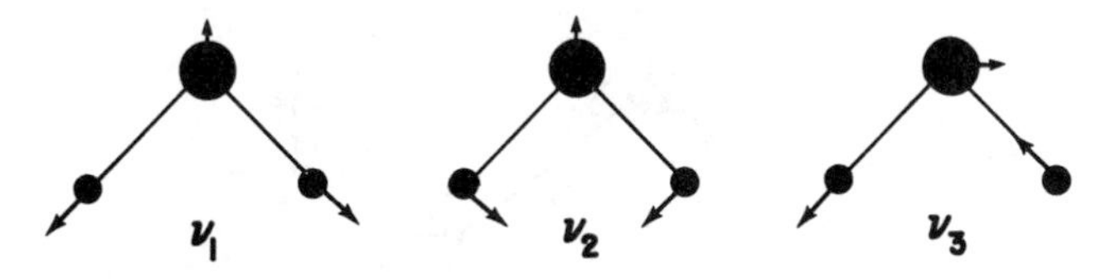

Fig. 3-29. The three normal modes of vibration of H_2O.

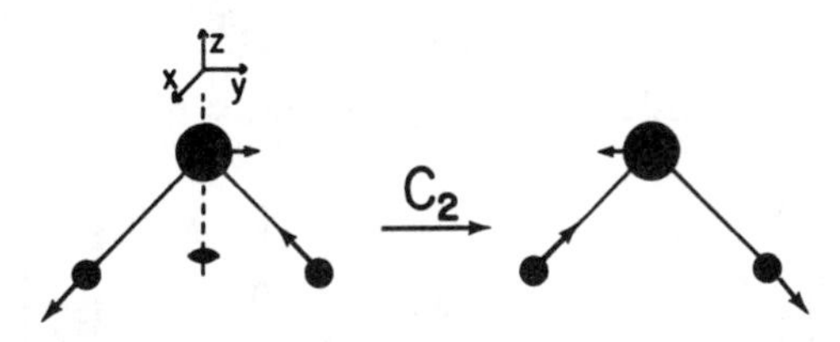

Fig. 3-30. The C_2 operation changes the direction of motion of each atom when the molecule is vibrating in the normal mode ν_3.

Table 3-9. Determining the Symmetries of the Three Modes of Vibration of H_2O

C_{2v}	E	C_2	$\sigma(xz)$	$\sigma(yz)$		
ν_1	1	1	1	1	=	a_1
ν_2	1	1	1	1	=	a_1
ν_3	1	−1	−1	1	=	b_2

H_2O. Water belongs to the point group C_{2v}. In order to use our character table we must assign a coordinate system consistent with the table. Generally, the only restriction is that the principal axis of rotation be the z axis. We can orient x and y as we please. Calling the molecular plane yz, our object is to see how three displacement vectors on each atom, representing the three degrees of freedom of that atom, transform under the operations of the point group (Fig. 3-31). The operation E is trivial. It accomplishes the following transformations:

$$\begin{array}{lll} x_1 \rightarrow x_1 & x_2 \rightarrow x_2 & x_3 \rightarrow x_3 \\ y_1 \rightarrow y_1 & y_2 \rightarrow y_2 & y_3 \rightarrow y_3 \\ z_1 \rightarrow z_1 & z_2 \rightarrow z_2 & z_3 \rightarrow z_3 \end{array} \tag{3-83}$$

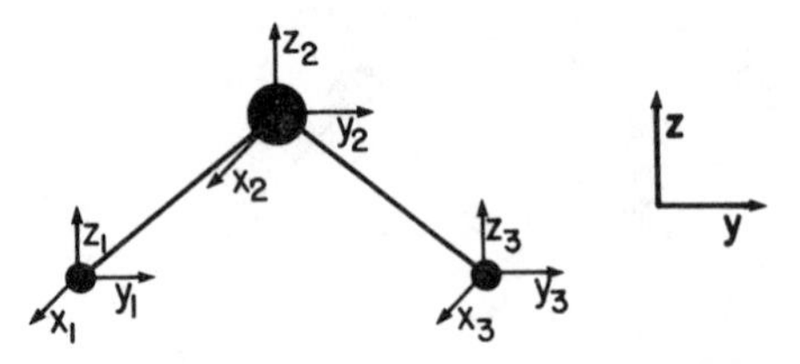

Fig. 3-31. Nine cartesian displacement vectors for H_2O. The x vectors come out of the plane of the page.

We represent these transformations by the matrix equation 3-84:

$$E \cdot \begin{bmatrix} x_1 \\ y_1 \\ z_1 \\ x_2 \\ y_2 \\ z_2 \\ x_3 \\ y_3 \\ z_3 \end{bmatrix} = \begin{bmatrix} 1 & 0 & 0 & 0 & 0 & 0 & 0 & 0 & 0 \\ 0 & 1 & 0 & 0 & 0 & 0 & 0 & 0 & 0 \\ 0 & 0 & 1 & 0 & 0 & 0 & 0 & 0 & 0 \\ 0 & 0 & 0 & 1 & 0 & 0 & 0 & 0 & 0 \\ 0 & 0 & 0 & 0 & 1 & 0 & 0 & 0 & 0 \\ 0 & 0 & 0 & 0 & 0 & 1 & 0 & 0 & 0 \\ 0 & 0 & 0 & 0 & 0 & 0 & 1 & 0 & 0 \\ 0 & 0 & 0 & 0 & 0 & 0 & 0 & 1 & 0 \\ 0 & 0 & 0 & 0 & 0 & 0 & 0 & 0 & 1 \end{bmatrix} \cdot \begin{bmatrix} x_1 \\ y_1 \\ z_1 \\ x_2 \\ y_2 \\ z_2 \\ x_3 \\ y_3 \\ z_3 \end{bmatrix} \tag{3-84}$$

The trace or character of this matrix, the sum of the diagonal elements, is 9.

The operation C_2 is more interesting:

$$\begin{array}{lll} x_1 \to -x_3 & x_2 \to -x_2 & x_3 \to -x_1 \\ y_1 \to -y_3 & y_2 \to -y_2 & y_3 \to -y_1 \\ z_1 \to +z_3 & z_2 \to +z_2 & z_3 \to +z_1 \end{array} \tag{3-85}$$

Table 3-10. Determination of the Reducible Representation Spanned by the $3n = 9$ Degrees of Freedom of H_2O

C_{2v}	E	C_2	$\sigma(xz)$	$\sigma(yz)$	
$\Gamma_{x,y,z}$	3	-1	1	1	
number of unmoved atoms	3	1	1	3	
Γ_{tot}	9	-1	1	3	$= 3a_1 + a_2 + 2b_1 + 3b_2$

The matrix statement of eq. 3-85 is eq. 3-86:

$$C_2 \cdot \begin{bmatrix} x_1 \\ y_1 \\ z_1 \\ x_2 \\ y_2 \\ z_2 \\ x_3 \\ y_3 \\ z_3 \end{bmatrix} = \begin{bmatrix} 0 & 0 & 0 & 0 & 0 & 0 & -1 & 0 & 0 \\ 0 & 0 & 0 & 0 & 0 & 0 & 0 & -1 & 0 \\ 0 & 0 & 0 & 0 & 0 & 0 & 0 & 0 & +1 \\ 0 & 0 & 0 & -1 & 0 & 0 & 0 & 0 & 0 \\ 0 & 0 & 0 & 0 & -1 & 0 & 0 & 0 & 0 \\ 0 & 0 & 0 & 0 & 0 & +1 & 0 & 0 & 0 \\ -1 & 0 & 0 & 0 & 0 & 0 & 0 & 0 & 0 \\ 0 & -1 & 0 & 0 & 0 & 0 & 0 & 0 & 0 \\ 0 & 0 & +1 & 0 & 0 & 0 & 0 & 0 & 0 \end{bmatrix} \cdot \begin{bmatrix} x_1 \\ y_1 \\ z_1 \\ x_2 \\ y_2 \\ z_2 \\ x_3 \\ y_3 \\ z_3 \end{bmatrix} \qquad (3\text{-}86)$$

The character of the C_2 matrix is -1.

By constructing similar matrices for the operations $\sigma(xz)$ and $\sigma(yz)$ we will obtain the remaining characters of the *reducible representation* corresponding to the nine degrees of freedom of water. This procedure is very cumbersome, however, and we can streamline it tremendously. First note that the only way a vector can contribute to the trace of the matrix is if the atom on which it is fixed does not change its location during the symmetry operation. For example, exchange of the two H atoms by the C_2 operation moved all of the vectors fixed on these two atoms off the diagonal of the matrix. So we will confine our attention only to atoms which do not translate during an operation. Here is the streamilined procedure:

(1) Determine the number of atoms which do not change location during each symmetry operation.

(2) For each operation, multiply the number of unmoved atoms by the character of $\Gamma_{x,y,z}$ at the bottom of the character table. This gives the characters of Γ_{tot}, the total representation of all degrees of freedom of the molecule, including translation, rotation, and vibration.

Table 3-10 shows how this procedure works in the case of H_2O. Using the formula 1-64, we can decompose Γ_{tot} to $3a_1 + a_2 + 2b_1 + 3b_2$. Notice that there are *nine* irreducible representations corresponding to the nine degrees of freedom of the three atoms. If you don't come up with $3n$ irreducible representations at this point, you have made a mistake. We now need to subtract translation (Γ_{trans}) and rotation (Γ_{rot}) from Γ_{tot} to obtain the symmetries of the vibrational degrees of freedom (Γ_{vib}). Γ_{trans} is just the set of irreducible representations for which the functions x, y, and z form bases. These are the representations of translation in each cartesian direction. The irreducible representations corresponding to rotation are labelled R_x, R_y, and R_z at the right side of the character table. For H_2O,

$$\begin{aligned} \Gamma_{\text{vib}} &= \Gamma_{\text{tot}} - \Gamma_{\text{trans}} - \Gamma_{\text{rot}} \\ &= (3a_1 + a_2 + 2b_1 + 3b_2) - (a_1 + b_1 + b_2) - (a_2 + b_1 + b_2) \\ \Gamma_{\text{vib}} &= 2a_1 + b_2 \end{aligned} \tag{3-87}$$

So the three normal modes of vibration of water have the symmetries a_1, a_1, and b_2.

By way of explanation, the characters of $\Gamma_{x,y,z}$ at the bottom of each character table are just the sums of the characters of the representations to which x, y, and z belong. For example, in C_{2v}, $\Gamma_{x,y,z} = \Gamma_x + \Gamma_y + \Gamma_z = b_1 + b_2 + a_1$. These characters tell us how each vector is transformed by each operation. For example, under the operation C_2, $x \to -x$, $y \to -y$, and $z \to +z$. So the character of Γ_x (which is b_1) under the operation C_2 is -1; the character of Γ_y (which is b_2) under C_2 is -1; and the character of Γ_z (which is a_1) under C_2 is $+1$. The character of $\Gamma_{x,y,z}$ for a given operation will be the same regardless of the point group. That is, the character of $\Gamma_{x,y,z}$ under C_2 is always -1 no matter which character table you look at.

CO_2. This molecule (Fig. 3-32) introduces the curiosities of the infinite point group, $D_{\infty h}$. Γ_{tot} is derived in Table 3-11 in which the symbols c and s are used for cosine and sine when they appear in characters. To understand this table, let's see what the character $-1 + 2\text{c}\phi$ under S_∞^ϕ means. The operation S_∞^ϕ signifies a rotation through the arbitrary angle ϕ about the molecular axis followed by reflection through the horizontal mirror plane. The two oxygen atoms are exchanged by this process so only the set of vectors on carbon need to be considered. S_∞^ϕ changes z to $-z$ and transforms x and y according to the rotation matrix, eq. 1-36. The total effect is

$$S_\infty^\phi \cdot \begin{bmatrix} x \\ y \\ z \end{bmatrix} = \begin{bmatrix} \text{c}\phi & \text{s}\phi & 0 \\ -\text{s}\phi & \text{c}\phi & 0 \\ 0 & 0 & -1 \end{bmatrix} \cdot \begin{bmatrix} x \\ y \\ z \end{bmatrix} \tag{3-88}$$

The character of the S_∞^ϕ matrix is $-1 + 2\text{c}\phi$.

We cannot use the formula 1-64 to decompose Γ_{tot} because the order of the group is infinite and the term $1/h$ blows up. Instead we resort to the Eyeball Technique.† You can simplify Γ_{tot} by first subtracting Γ_{trans} and

† See the footnote on page 56 for alternatives to the eyeball technique.

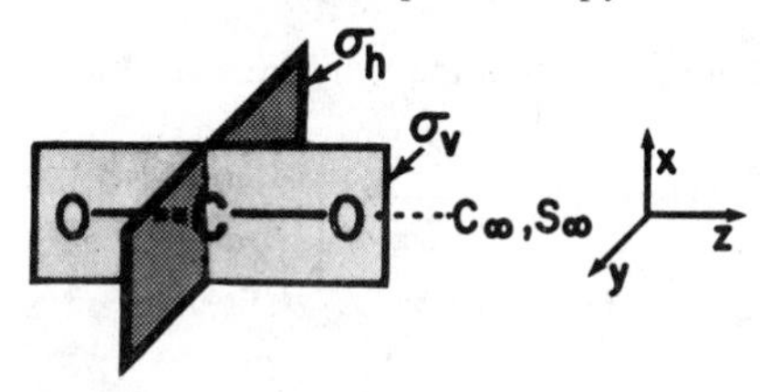

Fig. 3-32. Coordinate system and symmetry elements of CO_2.

Γ_{rot} to obtain Γ_{vib} (Table 3-12). For Γ_{rot} we use only (R_x, R_y) because there is no rotation about the z axis. Now Γ_{vib} in Table 3-12 contains factors of $2c\phi$ under C_∞^ϕ and S_∞^ϕ. Looking at the $D_{\infty h}$ character table, 3-11, you should become suspicious that π_u is involved because it has characters of $2c\phi$ under C_∞^ϕ and S_∞^ϕ. So we just try subtracting π_u from Γ_{vib} as shown at the bottom of Table 3-12. After scratching your head awhile, you recognize the difference at the bottom of the table as the sum $\sigma_g^+ + \sigma_u^+$. The normal modes of vibration of CO_2 therefore transform as $\Gamma_{vib} = \sigma_g^+ + \sigma_u^+ + \pi_u$. Counting π_u as two representations, we find four representations for the four degrees of freedom. When two modes transform as a single degenerate irreducible representation (π or e) they must have the same energy and are degenerate.

Table 3-11. Determination of the Representations Spanned by the Nine Degrees of Freedom of CO_2

$D_{\infty h}$	E	$2C_\infty^\phi$	$\infty\sigma_v$	i	$2S_\infty^\phi$	∞C_2	
Σ_g^+	1	1	1	1	1	1	
Σ_g^-	1	1	-1	1	1	-1	R_z
Π_g	2	$2c\phi$	0	2	$-2c\phi$	0	(R_x, R_y)
Δ_g	2	$2c2\phi$	0	2	$2c2\phi$	0	
Σ_u^+	1	1	1	-1	-1	-1	z
Σ_u^-	1	1	-1	-1	-1	1	
Π_u	2	$2c\phi$	0	-2	$2c\phi$	0	(x, y)
Δ_u	2	$2c2\phi$	0	-2	$-2c2\phi$	0	
$\Gamma_{x,y,z}$	3	$1+2c\phi$	1	-3	$-1+2c\phi$	-1	
unmoved atoms	3	3	3	1	1	1	
Γ_{tot}	9	$3+6c\phi$	3	-3	$-1+2c\phi$	-1	

Table 3-12. Determination of the Symmetries of Vibration of CO_2

$D_{\infty h}$	E	$2C_\infty^\phi$	$\infty\sigma_v$	i	$2S_\infty^\phi$	∞C_2
Γ_{tot}	9	$3+6c\phi$	3	-3	$-1+2c\phi$	-1
Γ_{trans}	3	$1+2c\phi$	1	-3	$-1+2c\phi$	-1
Γ_{rot}	2	$2c\phi$	0	2	$-2c\phi$	0
$\Gamma_{vib} = \Gamma_{tot} - \Gamma_{trans} - \Gamma_{rot}$	4	$2+2c\phi$	2	-2	$2c\phi$	0
π_u	2	$2c\phi$	0	-2	$2c\phi$	0
$\Gamma_{vib} - \pi_u$	2	2	2	0	0	0

Since CO_2 has two chemical bonds, we expect two stretching vibrations. Since there are four normal modes of vibration, two must be stretching modes and two must be bending modes (Fig. 3-33). The symbols used to label these vibrations are set out in Table 3-13. Using this table, we decipher the names of the CO_2 vibrations in Fig. 3-33 as: ν_s—symmetric stretching; ν_{as}—asymmetric stretching; and δ_d—degenerate bending. We will explain in Section 3.7 how to draw these modes, but for now we are more concerned with just understanding the symmetries of the modes. To find the symmetry of a vibration, we apply the operations of the point group to its picture and write down the characters, as shown in Table 3-14. To get these characters you need to consider each whole vibration as a *single* picture. For example, ν_{as} goes to minus itself upon inversion, as shown in Fig. 3-34. The bending modes transform *together* as the degenerate representation π_u. These two vibrations must be considered together because the two bending vibrations (in the xz and yz planes) are mixed by the C_∞^ϕ and S_∞^ϕ operations. You can see that both bending modes do exactly the same thing, physically, and therefore have the same energy. On the other hand, the two stretching modes do not do the same thing and will have different energies. The asymmetric stretch often (but not always) has higher energy than the symmetric stretch. The energies of bending modes generally fall below the energies of stretching modes.

Fig. 3-33. Normal modes of vibration of CO_2. The symbols are defined in Table 3-13. The symmetry of each vibration is given in parentheses.

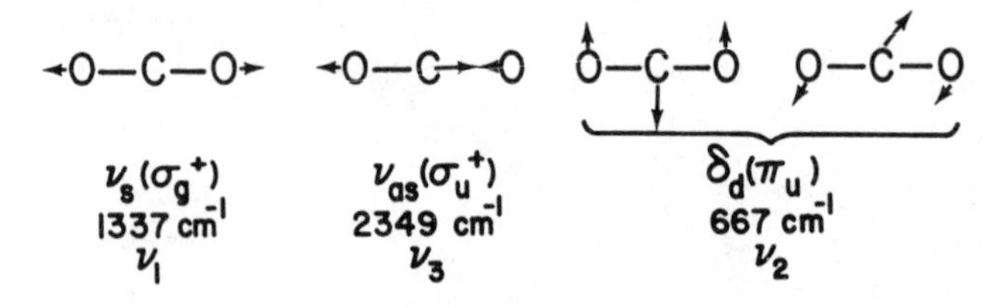

Table 3-13. Abbreviations Used for the Descriptions of Vibrations

ν	— stretching	π	— out of plane
δ	— deformation (bending)	*as*	— asymmetric
ρ_w	— wagging	*s*	— symmetric
ρ_r	— rocking	*d*	— degenerate
ρ_t	— twisting		

Table 3-14. Determining the Symmetries of the Normal Modes in Fig. 3-33

$D_{\infty h}$	E	$2C_\infty^\phi$	$\infty\sigma_v$	i	$2S_\infty^\phi$	∞C_2	
ν_s	1	1	1	1	1	1	$= \sigma_g^+$
ν_{as}	1	1	1	-1	-1	-1	$= \sigma_u^+$
δ_d	2	$2c\phi$	0	-2	$2c\phi$	0	$= \pi_u$

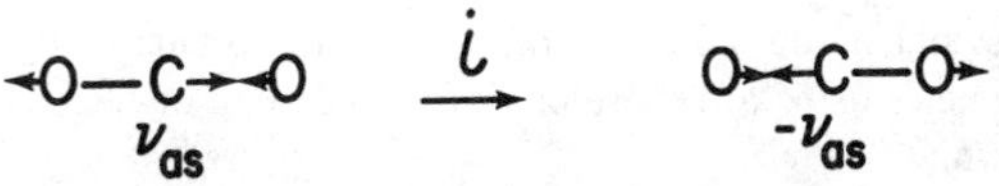

Fig. 3-34. The vibration ν_{as} goes to minus itself upon inversion. That is, the direction of motion of each atom changes sign.

XeF_4. We shall call the C_4 axis of this square planar molecule the z axis (Fig. 3-35). By convention, the σ_v planes include the Xe-F bonds and the σ_d planes bisect the bonds. Similarly, C_2' axes include the bonds and C_2'' axes bisect the bonds. In Table 3-15 Γ_{vib} is found to include nine modes when each e_u mode is counted twice. The entry $2e_u$ implies that there will be two sets of degenerate vibrations, each *set* having its own energy. There will therefore be seven different vibrational energies for the nine normal modes of XeF_4. An analysis which we reserve for Section 3-7 allows us to draw these vibrations as shown in Fig. 3-36. Since there are four chemical bonds, there are four stretching vibrations and $9 - 4 = 5$ bending vibrations.

Fig. 3-35. Coordinate system and symmetry elements of XeF_4.

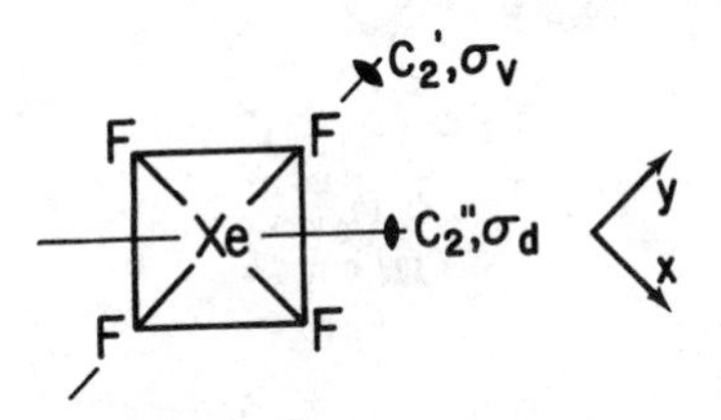

Table 3-15. Determination of Γ_{vib} for XeF_4

D_{4h}	E	$2C_4$	C_2	$2C_2'$	$2C_2''$	i	$2S_4$	σ_h	$2\sigma_v$	$2\sigma_d$
$\Gamma_{x,y,z}$	3	1	−1	−1	−1	−3	−1	1	1	1
unmoved atoms	5	1	1	3	1	1	1	5	3	1
Γ_{tot}	15	1	−1	−3	−1	−3	−1	5	3	1

$$\Gamma_{\text{tot}} = a_{1g} + a_{2g} + b_{1g} + b_{2g} + e_g + 2a_{2u} + b_{2u} + 3e_u$$
$$\Gamma_{\text{trans}} = a_{2u} + e_u$$
$$\Gamma_{\text{rot}} = a_{2g} + e_g$$
$$\Gamma_{\text{vib}} = a_{1g} + b_{1g} + b_{2g} + a_{2u} + b_{2u} + 2e_u$$

Fig. 3-36. The normal modes of vibration of a square planar ML_4 molecule. The vibrational frequencies listed are those of XeF_4. Only one of each set of degenerate vibrations is drawn.

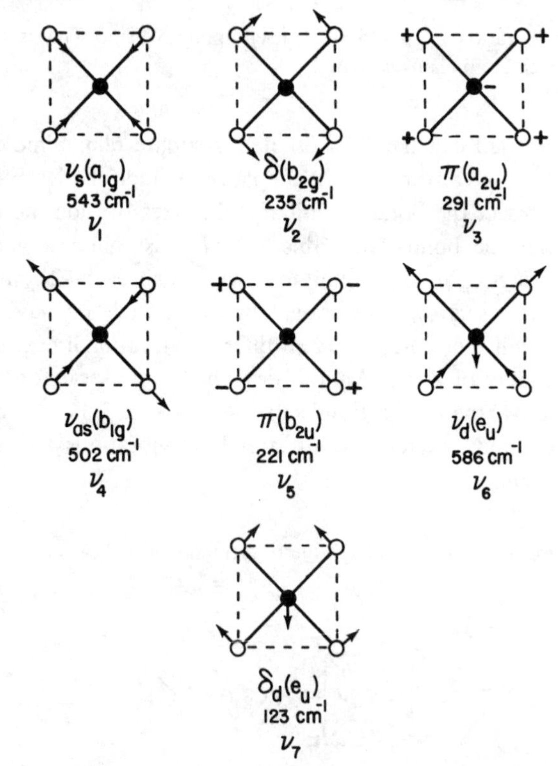

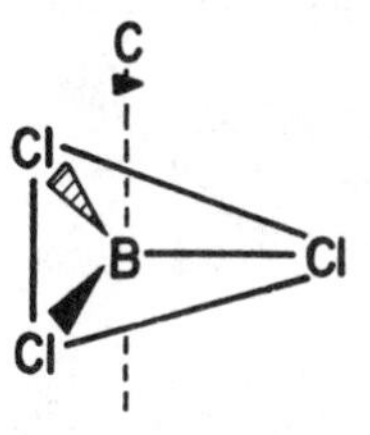

Fig. 3-37. The structure of BCl_3.

BCl_3. This planar molecule has D_{3h} symmetry (Fig. 3-37). As shown in Table 3-16, we obtain six vibrational degrees of freedom spanning four discrete energies. The four kinds of vibrations are shown in Fig. 3-38. Notice that the frequencies assigned to the a_1' modes of both the ^{10}B and ^{11}B species are identical because no B atom motion is involved in the vibration. In all cases with B atom motion involved, the lower frequency is assigned to the heavier species. $^{35}Cl-^{37}Cl$ isotopic splittings are small and not resolved in this work, though such splittings of the bands are observed in some other molecules containing Cl.

Fig. 3-38. Normal modes of vibration of a trigonal planar ML_3 molecule.

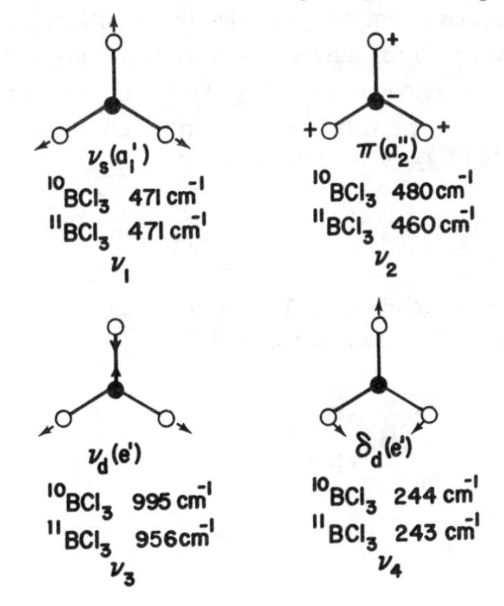

Table 3-16. Determination of Γ_{vib} for BCl_3

D_{3h}	E	$2C_3$	$3C_2$	σ_h	$2S_3$	$3\sigma_v$
$\Gamma_{x,y,z}$	3	0	−1	1	−2	1
unmoved atoms	4	1	2	4	1	2
Γ_{tot}	12	0	−2	4	−2	2

$$\Gamma_{\text{tot}} = a_1' + a_2' + 3e' + 2a_2'' + e''$$
$$\Gamma_{\text{trans}} = e' + a_2''$$
$$\Gamma_{\text{rot}} = a_2' + e''$$
$$\Gamma_{\text{vib}} = a_1' + 2e' + a_2''$$

B_2H_6. As a final example, we will look at diborane (Fig. 3-39, Table 3-17). This molecule is used primarily to illustrate the numbering and descriptive naming of the vibrations (rocking, wagging, *etc.*). The eighteen normal modes are shown in Fig. 3-40. The usual convention for numbering normal modes is to order the modes by descending symmetry with totally symmetric modes at the top of the list. Within a set of modes of the same symmetry, we order the modes by decreasing energy. For example, ν_8, ν_9, and ν_{10} all transform as b_{1u} and the energies decrease in the order $\nu_8 > \nu_9 > \nu_{10}$. It is not obvious, from first principles, which picture describes the vibration of greatest energy. In fact, from one molecule to another of the same type of structure, the energies of two vibrations may be inverted. For many simple molecules the numbering of the vibrations is based solely on pictures agreed upon by convention. Most books and journal articles you encounter dealing with molecules of the symmetries described in this section will use the same numbering systems we have used in Figs. 3-29, 3-33, 3-36, 3-38, and 3-40.

Fig. 3-39. The structure of diborane. The four terminal H atoms lie in the xz plane and the two bridging atoms lie in the yz plane.

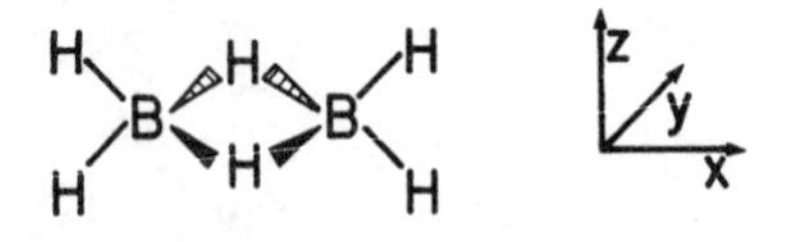

Fig. 3-40. Normal modes of vibration of B_2H_6. Frequencies in parentheses are calculated from a fit of the observed spectrum. As an exercise you might look at the vibrations ν_5, ν_7, and ν_9 and see why they are called twisting (ρ_t), wagging (ρ_w), and rocking (ρ_r), respectively.

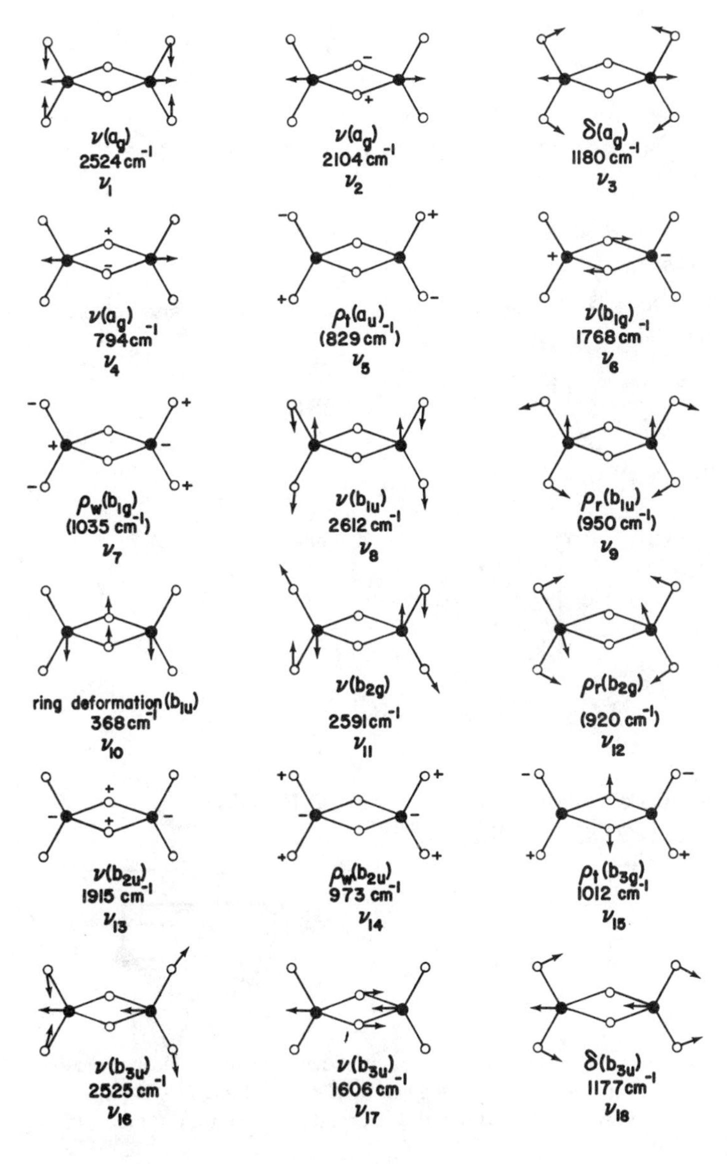

Table 3-17. Determination of Γ_{vib} for B_2H_6

D_{2h}	E	$C_2(z)$	$C_2(y)$	$C_2(x)$	i	$\sigma(xy)$	$\sigma(xz)$	$\sigma(yz)$
$\Gamma_{x,y,z}$	3	-1	-1	-1	-3	1	1	1
unmoved atoms	8	0	2	2	0	4	6	2
Γ_{tot}	24	0	-2	-2	0	4	6	2

$$\Gamma_{\text{tot}} = 4a_g + 3b_{1g} + 3b_{2g} + 2b_{3g} + a_u + 4b_{1u} + 3b_{2u} + 4b_{3u}$$
$$\Gamma_{\text{trans}} = b_{1u} + b_{2u} + b_{3u}$$
$$\Gamma_{\text{rot}} = b_{1g} + b_{2g} + b_{3g}$$
$$\Gamma_{\text{vib}} = 4a_g + 2b_{1g} + 2b_{2g} + b_{3g} + a_u + 3b_{1u} + 2b_{2u} + 3b_{3u}$$

Problems

3-23. Determine the symmetries of the normal modes of vibration of the following molecules:

a. 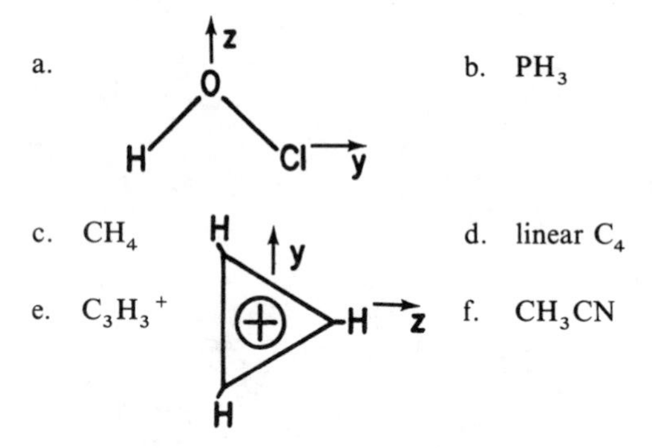

b. PH_3

c. CH_4

d. linear C_4

e. $C_3H_3{}^+$

f. CH_3CN

g. $FeF_6{}^{3-}$ (ligands on cartesian axes)

h. C_3H_4 (allene)

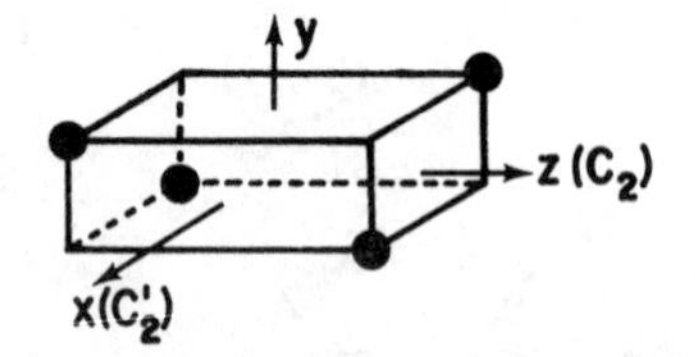

3-24. The quantum mechanical treatment of rotation of a rigid body showed that the allowed energy levels are given by $E = (\hbar^2/2I)J(J + 1)$ (eq. 3-10), where I is the moment of inertia and J is an integer quantum number. In Fig. 3-27

it is stated that rotation about the internuclear axis is undefined because such rotation does not represent any change of the nuclear coordinates. Calculate the energy of rotation of a ^{12}C *nucleus* about its own axis if $J = 1$. For this purpose, consider the nucleus to be a sphere of mass m and uniform density whose moment of inertia is given by $I = (2/5)mr^2$, where the nuclear radius, r, is $2.7 \times 10^{-15}m$ (= 2.7 fermi). What is the frequency of radiation with this energy? In what region of the spectrum does this occur? Will nuclear rotational transitions (which should have energies comparable to the energy of rotation of the nucleus) give rise to ir or microwave absorption?

3-6. Selection Rules and Polarization

A. *What Are Selection Rules*? At room temperature most molecular vibrations will be occurring with their zero point energy. Suppose each mode of vibration of a molecule is described by the wave function $\psi_i(v_i)$.† The subscript i means that we are dealing with the ith normal mode and v_i is the vibrational quantum number of that mode. If we have a molecule with three modes of vibration, its total vibrational wave function would be well approximated by the product of the wave functions describing each normal mode:

$$\psi_{\text{vib}} = \psi_1(v_1)\psi_2(v_2)\psi_3(v_3) \tag{3-89}$$

We could represent an excitation of ν_1 from the ground state to the first excited state as follows:

$$\psi_1(0)\psi_2(0)\psi_3(0) \rightarrow \psi_1(1)\psi_2(0)\psi_3(0) \tag{3-90}$$

If ν_1 and ν_2 were excited simultaneously, we could write

$$\psi_1(0)\psi_2(0)\psi_3(0) \rightarrow \psi_1(1)\psi_2(1)\psi_3(0) \tag{3-91}$$

(In general the probability of two simultaneous transitions is much less than the probability of their happening one at a time.)

The *selection rules* tell us which transitions we might hope to see in the spectrum. In a real spectrum we find band intensities varying from very weak to very strong. The selection rules do not distinguish shades of intensity. They merely tell us which transitions are expected to have *zero* intensity (based on the harmonic oscillator approximation) and which

† Previously we used the symbol ψ_v for the vibrational wave function of a diatomic molecule. Since polyatomic molecules have one wave function for each vibrational degree of freedom, we now switch to the notation $\psi_i(v_i)$. When the value of i, the number of the normal mode, is not important to our discussion, the symbol will simply be $\psi(v)$.

have nonzero intensity. Transitions predicted to have zero intensity are said to be *forbidden* and those predicted to have nonzero intensity are said to be *allowed*. We generally, but by no means always, observe that transitions forbidden by the selection rules are weak or absent and transitions allowed by our simple model are moderate or strong. Transitions predicted to be observed in the ir spectrum are called *infrared active* while those expected in the Raman spectrum are *Raman active*. Depending on the symmetry of the molecule, some transitions may appear in one or the other spectrum, or *both*, or *neither*.

B. *The Physical Basis of Selection Rules for Fundamental Transitions. A vibrational transition is ir active if the dipole moment of the molecule changes during the vibration.*† *A transition is Raman active if the polarizability of the molecule changes during the vibration.* The infrared selection rule follows from the transition moment integral for the one-dimensional harmonic oscillator derived in Section 3.4. Using the definition of the dipole moment, $\mu = ex$, the transition moment integral for the transition $v \rightarrow v'$ is written

$$M_{vv'} = \int_{-\infty}^{\infty} \psi^*(v')\mu\psi(v)dx \tag{3-92}$$

The probability of the transition occurring is proportional to the square of $M_{vv'}$ (eq. 3-78). If μ were a constant, independent of the vibration, it could be factored out of the integral leaving

$$M_{vv'} = \mu \int_{-\infty}^{\infty} \psi^*(v')\psi(v)dx \tag{3-93}$$

which is zero because $\psi(v')$ and $\psi(v)$ are orthogonal. Hence μ must be a function of x and must change during the vibration if the transition is to be allowed. The Raman selection rule follows from eq. 3-7 in which $\Delta\alpha$ must be nonzero in order to produce Raman scattering. As a point of information, the magnitude of $d\mu/dx$ for the stretching of a C–H bond is of the order of 1 Debye/Å and the magnitude of $d\alpha/dx$ is of the order of 1 Å^3/Å.

The physical statement of the infrared selection rule directly implies that the fundamental transitions of all homonuclear diatomic molecules cannot be ir active because the homonuclear diatomic molecule has no

† In this book we only consider electric dipole transitions, as in Section 3-4. In the absence of allowed electric dipole transitions, magnetic dipole or electric quadrupole transitions may become important and transitions forbidden by our simple model may become allowed when considered in greater detail. However, transitions forbidden by the electric dipole selection rules will generally be very weak, even if allowed by other mechanisms.

dipole moment no matter what stage of its vibration you look at. When we derive the formal selection rules, they had better agree with this simple statement. Of the two vibrations of ethylene in Fig. 3-41, the totally symmetric mode does not change the dipole moment and will not be ir active. The asymmetric stretch does change the dipole moment and will be ir active.

In the next few paragraphs we will describe some properties of dipole moments, polarizability, definite integrals, and wave functions needed for a discussion of the selection rules. For a polyatomic molecule, we associate an electric dipole component with each cartesian coordinate. For example, the x component of the moment would be

$$\mu_x = \sum_i e_i x_i \tag{3-94}$$

where e_i is the charge on the ith atom and x_i is its x coordinate. There will be three cartesian components of the dipole moment for any molecule. If any one of these components changes during the vibration, the fundamental transition will be ir active.

Polarizability is complicated by the directional nature of chemical bonds. In general, an electric field in the x direction not only induces a dipole in the x direction, but also in the y and z directions. We represent this by making α a tensor (a square matrix):

$$\begin{aligned} \mu_{\text{ind}}(x) &= \alpha_{xx}E_x + \alpha_{xy}E_y + \alpha_{xz}E_z \\ \mu_{\text{ind}}(y) &= \alpha_{yx}E_x + \alpha_{yy}E_y + \alpha_{yz}E_z \\ \mu_{\text{ind}}(z) &= \alpha_{zx}E_x + \alpha_{zy}E_y + \alpha_{zz}E_z \end{aligned} \tag{3-95}$$

$$\begin{bmatrix} \mu_{\text{ind}}(x) \\ \mu_{\text{ind}}(y) \\ \mu_{\text{ind}}(z) \end{bmatrix} = \underbrace{\begin{bmatrix} \alpha_{xx} & \alpha_{xy} & \alpha_{xz} \\ \alpha_{yx} & \alpha_{yy} & \alpha_{yz} \\ \alpha_{zx} & \alpha_{zy} & \alpha_{zz} \end{bmatrix}}_{\text{polarizability tensor}} \begin{bmatrix} E_x \\ E_y \\ E_z \end{bmatrix} \tag{3-96}$$

The tensor is symmetric, with $\alpha_{xy} = \alpha_{yx}$, $\alpha_{xz} = \alpha_{zx}$, and $\alpha_{yz} = \alpha_{zy}$. A transition will be Raman active if one of the six different components of the polarizability tensor changes during the vibration.

We now proceed to a very useful feature of definite integrals. We shall encounter integrals of the type $\int_{-\infty}^{\infty} \phi_a \phi_b d\tau$ and $\int_{-\infty}^{\infty} \phi_a \phi_b \phi_c d\tau$. Here ϕ_a, ϕ_b, and ϕ_c are just some functions and $d\tau$ is a generalized differential,

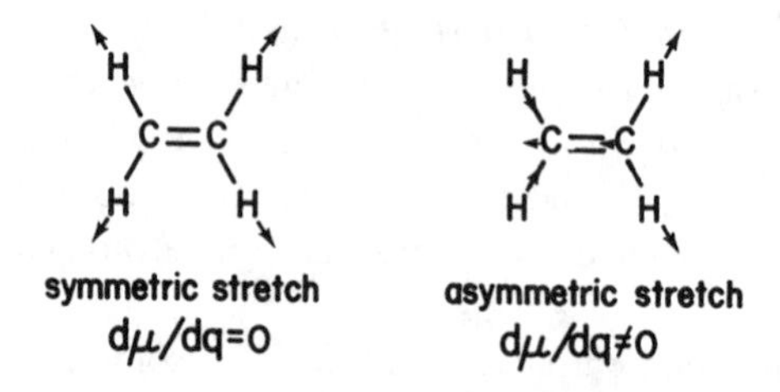

Fig. 3-41. In the symmetric stretching vibration of ethylene any changes in the individual bond dipoles cancel each other. In the asymmetric stretch, changes on the right side of the molecule do not cancel changes on the left side of the molecule, and there is a net change in the molecular dipole.

such as *dxdydz* in cartesian coordinates. For a molecule belonging to a particular point group, the functions ϕ_a, ϕ_b, and ϕ_c will each form a basis for some irreducible representation of the group. Suppose ϕ_a forms a basis for the irreducible representation Γ_a and ϕ_b forms a basis for Γ_b. The direct product $\Gamma_a \times \Gamma_b$ will be, in general, some reducible representation. *The integral* $\int_{-\infty}^{\infty} \phi_a \phi_b d\tau$ *will have a nonzero value only if the direct product* $\Gamma_a \times \Gamma_b$ *contains the totally symmetric irreducible representation of the point group.* For example, in the point group D_{3h} the integral of the product of two functions of symmetry e' and a_2'' will be zero because $e' \times a_2'' = e'' \neq a_1'$. Two functions of symmetry e' will yield a nonzero integral because $e' \times e' = \underline{\underline{a_1}}' + a_2' + e'$. For an integral of three functions we examine the direct product of all three irreducible representations to find out if the integral is nonzero.

The reason for this rule is that the integrand can, in general, be expressed as a sum of even and odd functions (Fig. 2-4) in three-dimensional space. The integrand has an even component only if the direct product $\Gamma_a \times \Gamma_b \times \Gamma_c \ldots$ contains the totally symmetric irreducible representation. If the direct product does not contain the totally symmetric irreducible representation, all components of the integrand are odd and the integral over all space vanishes.

The vibrational wave functions of the harmonic oscillator were given in eq. 3-42. We know that each normal coordinate, q, of a polyatomic molecule forms a basis for an irreducible representation of the point group of the molecule. What is the symmetry of the *wave function* $\psi(0)$? If q forms a basis for a nondegenerate irreducible representation, then all symmetry operations of the point group take q into plus or minus itself. Hence q^2 will always go into plus itself. The function $\psi(0)$ will always remain unchanged under any symmetry operation because it depends on q^2. Hence $\psi(0)$ *transforms as the totally symmetric irreducible representa-*

tion. If q transforms as a degenerate irreducible representation, $\psi(0)$ is still totally symmetric but the analysis is a little more complicated. If q_a and q_b belong to a doubly degenerate irreducible representation, any symmetry operation must take q_a into $\pm q_a$ or some linear combination of q_a and q_b. Suppose the operation R takes q_a into the linear combination in eq. 3-97:

$$Rq_a = q_a' = aq_a + bq_b \tag{3-97}$$

in which a and b are constants. If q_a and q_b were normalized vibrational functions ($q_a^2 = q_b^2 = 1$), the coefficients a and b must be such that q_a' is also a normalized function ($q_a'^2 = 1$). Since $\psi(0)$ depends on the square of the vibrational coordinate, the operation R still has no net effect on $\psi(0)$ even if q_a is degenerate.

To find the symmetry of $\psi(1)$ we need the symmetry of the product $qe^{-(1/2)\alpha q^2}$. The exponential part we just found to be totally symmetric, regardless of the symmetry of q. But the function q has the symmetry of the vibration. So the product $(q)\,(e^{-(1/2)\alpha q^2})$ transforms as q itself:

$$\Gamma[\psi(1)] = \Gamma(q) \tag{3-98}$$

$\psi(2)$ has two components, one of which transforms as q^2 and one of which transforms as unity. Both are totally symmetric. $\psi(3)$ will transform as q^3 and q, both of which have the symmetry of q. We find that even wave functions are totally symmetric and odd wave functions have the symmetry of the vibration. The rules for obtaining the symmetries of higher wave functions if q is degenerate are more complicated and are given in Appendix C.

With the information at hand, we can finally get to the selection rules. We know that the transition probability is proportional to the square of the transition moment integral:

$$\text{transition probability} \propto M_{vv'}^2 = \left[\int_{-\infty}^{\infty} \psi^*(v')\hat{O}\psi(v)d\tau\right]^2 \tag{3-99}$$

A transition is allowed if $M_{vv'} \neq 0$. In eq. 3-99 we have replaced the dipole moment operator, ex, by the more general operator $\hat{O}$. For ir absorption, $\hat{O}$ will be the dipole moment operator in three dimensions given by

$$\hat{O} = \hat{\mu} = \sum_i (e_i x_i + e_i y_i + e_i z_i) \tag{3-100}$$

$\hat{\mu}$ therefore has three components which transform as x, y and z. For Raman transitions, $\hat{O}$ will be the polarizability operator, $\hat{\alpha}$, which has six different components ($\alpha_{xx}, \alpha_{yy}, \alpha_{zz}, \alpha_{xy}, \alpha_{xz}, \alpha_{yz}$). We state without derivation that *the six components of α transform as the binary products x^2, y^2, z^2, xy, xz*

and yz.† These functions (or equally suitable linear combinations such as $x^2 - y^2$) all appear at the right side of each character table.

The selection rules telling which transitions are allowed in the ir or Raman spectra are readily derived when we know the symmetries of $\psi(v')$††, $\psi(v)$, and the operators $\hat{\mu}$ and $\hat{\alpha}$. *If the direct product* $\Gamma[\psi(v')] \times \Gamma(\hat{\mu}) \times \Gamma[\psi(v)]$ *contains the totally symmetric irreducible representation of the point group, the transition* $v \rightarrow v'$ *is ir active. If the direct product* $\Gamma[\psi(v')] \times \Gamma(\hat{\alpha}) \times \Gamma[\psi(v)]$ *contains the totally symmetric irreducible representation of the point group, the transition* $v \rightarrow v'$ *is Raman active.*

C. *The Selection Rules In Action.* Let's try a vibrational analysis for a homonuclear diatomic molecule such as nitrogen (Table 3-18). You should have your character tables handy for this section. The lone stretching mode of N_2 has symmetry σ_g^+. The fundamental transition, $v = 0 \rightarrow v = 1$, is usually an order of magnitude more intense than other kinds of transitions, when it is allowed. To see if the fundamental transition of N_2 is allowed in the ir spectrum, we set up the appropriate transition moment integral:

$$M_{01} = \int_{-\infty}^{\infty} \psi(1)\hat{\mu}\psi(0)d\tau \tag{3-101}$$

The symmetry of the integrand is

$$\psi(1)\hat{\mu}\psi(0) \sim \sigma_g^+ \begin{pmatrix} \sigma_u^+ \\ \pi_u \end{pmatrix} \sigma_g^+ = \begin{pmatrix} \sigma_u^+ \\ \pi_u \end{pmatrix} \tag{3-102}$$

The symbol $\sim$ should be read "transforms as". Let's examine eq. 3-102 carefully. $\psi(0)$ is *always* totally symmetric. $\psi(1)$ has the symmetry of the normal vibration, σ_g^+ in this case. The dipole moment operator has the symmetry of the three functions, x, y, and z. Your $D_{\infty h}$ character table tells you that x and y transform as π_u and z transforms as σ_u^+. The triple direct products in eq. 3-102 are $\sigma_g^+ \times \sigma_u^+ \times \sigma_g^+ = \sigma_u^+$ and $\sigma_g^+ \times \pi_u \times \sigma_g^+ = \pi_u$. Since neither triple product contains the totally symmetric representation, σ_g^+, the integral is necessarily equal to zero. The transition is forbidden and will not be seen in the infrared spectrum. We already knew this had to be so because vibration of N_2 does not change its dipole moment.

What about the Raman spectrum? The polarizability components transform as

$$\begin{aligned} x^2 + y^2 &\sim \sigma_g^+ & xz, yz &\sim \pi_g \\ z^2 &\sim \sigma_g^+ & x^2 - y^2, xy &\sim \delta_g \end{aligned} \tag{3-103}$$

†This is shown in Appendix XV of E.B. Wilson, J.C. Decius, and P.C. Cross, *Molecular Vibrations*, McGraw-Hill, N.Y., 1955.

††Since the spatial part of a vibrational wave function is real, the symmetry of $\psi^*(v')$ is the same as the symmetry of $\psi(v')$.

Table 3-18. Vibrational Analysis of $N \equiv N$

$D_{\infty h}$	E	$2C_{\infty}^{\phi}$	$\infty\sigma_v$	i	$2S_{\infty}^{\phi}$	∞C_2	
$\Gamma_{x,y,z}$	3	$1+2c\phi$	1	-3	$-1+2c\phi$	-1	
unmoved atoms	2	2	2	0	0	0	
Γ_{tot}	6	$2+4c\phi$	2	0	0	0	
$\Gamma_{\text{trans}} = \Sigma_u^+ + \Pi_u$	3	$1+2c\phi$	1	-3	$-1+2c\phi$	-1	
$\Gamma_{\text{rot}} = \Pi_g$	2	$2c\phi$	0	2	$-2c\phi$	0	
$\Gamma_{\text{vib}} = \Gamma_{\text{tot}} - \Gamma_{\text{trans}} - \Gamma_{\text{rot}}$	1	1	1	1	1	1	$= \sigma_g^+$

Two different components transform as σ_g^+ but we need only carry one of them through the calculation. Putting these components into eq. 3-99, we get

$$M_{01} = \int_{-\infty}^{\infty} \psi(1)\hat{\alpha}\psi(0)d\tau \tag{3-104}$$

$$\psi(1)\hat{\alpha}\psi(0) \sim \sigma_g^+ \begin{pmatrix} \sigma_g^+ \\ \pi_g \\ \delta_g \end{pmatrix} \sigma_g^+ = \begin{pmatrix} \sigma_g^+ \\ \pi_g \\ \delta_g \end{pmatrix} \tag{3-105}$$

The direct products in eq. 3-105 *do* contain σ_g^+ so the transition is Raman allowed and we expect to see it in the Raman spectrum. Indeed, the Raman spectrum shows a band centered at 2331 cm^{-1} and the ir spectrum shows nothing.

Symmetry analysis of transition moment integrals tells us only if the integral is necessarily equal to zero. If the integral is nonzero, symmetry considerations alone do not tell us what the value is. The strength of the absorption band cannot be calculated without actually performing the integration.

Let's try the same vibrational analysis for carbon monoxide (Table 3-19). The two integrals are:

$$\text{infrared:} \quad \psi(1)\hat{\mu}\psi(0) \sim \sigma^+ \begin{pmatrix} \sigma^+ \\ \pi \end{pmatrix} \sigma^+ = \begin{pmatrix} \sigma^+ \\ \pi \end{pmatrix} \tag{3-106}$$

$$\text{Raman:} \quad \psi(1)\hat{\alpha}\psi(0) \sim \sigma^+ \begin{pmatrix} \sigma^+ \\ \pi \\ \delta \end{pmatrix} \sigma^+ = \begin{pmatrix} \sigma^+ \\ \pi \\ \delta \end{pmatrix} \tag{3-107}$$

Both direct products contain σ^+ so the transition is both ir and Raman allowed. A band is seen in both spectra centered at 2143 cm^{-1}. Any heteronuclear diatomic molecule will exhibit both ir and Raman spectra.

Table 3-19. Vibrational Analysis of CO

$C_{\infty v}$	E	$2C_{\infty}^{\phi}$	$\infty\sigma_v$	
$\Gamma_{x,y,z}$	3	$1+2c\phi$	1	
unmoved atoms	2	2	2	
Γ_{tot}	6	$2+4c\phi$	2	
Γ_{trans}	3	$1+2c\phi$	1	
Γ_{rot}	2	$2c\phi$	0	
Γ_{vib}	1	1	1	$=\sigma^+$

We can now look at the subject of polarization using CO as an example. The direct product $\psi(1)\hat{\mu}_z\psi(0)$ has the symmetry σ^+. The product $\psi(1)\begin{pmatrix}\mu_x\\ \mu_y\end{pmatrix}\psi(0)$ is π. Only the z component (parallel to the C—O bond axis) is allowed. This means that a molecule of carbon monoxide can absorb z-polarized light but not x- or y-polarized light. By z-polarized light, we mean light whose electric vector oscillates in the z direction.

XeF_4 is an instructive example. We found in Section 3-5 that $\Gamma_{\text{vib}} = a_{1g} + b_{1g} + b_{2g} + a_{2u} + b_{2u} + 2e_u$. There are nine modes of vibration, two pairs of which are degenerate. Hence, there will be a total of seven possible fundamental transitions. Now you could examine the six different kinds of integrals if you wish, but that's too much work. Look at b_{1g}, for example:

$$\text{infrared:}\quad M_{01} \sim b_{1g}\begin{pmatrix}a_{2u}\\ e_u\end{pmatrix}a_{1g} = \begin{pmatrix}b_{2u}\\ e_u\end{pmatrix}\quad \text{forbidden} \tag{3-108}$$

$$\text{Raman:}\quad M_{01} \sim b_{1g}\begin{pmatrix}a_{1g}\\ b_{1g}\\ b_{2g}\\ e_g\end{pmatrix}a_{1g} = \begin{pmatrix}b_{1g}\\ a_{1g}\\ a_{2g}\\ e_g\end{pmatrix}\quad \text{allowed} \tag{3-109}$$

The symmetry of $\psi(0)$ is always a_{1g} for any vibration of XeF_4. Since a_{1g} times any irreducible representation is just equal to that irreducible representation, the only part of each integrand we need to evaluate is the product $\psi(1)\hat{O}$. *The only way to get a_{1g} as a component of the product $\psi(1)\hat{O}$ is if one of the components of O has the same symmetry as* $\psi(1)$. Therefore, *the only possible ir active fundamentals will have the symmetries of the three components of $\hat{\mu}$*. The ir active modes of XeF_4 can have symmetry a_{2u} or e_u only. Inspection of Γ_{vib} immediately tells us that there will be only three fundamental transitions in the ir spectrum of XeF_4. *The only possible Raman active fundamentals will have the symmetries of the six components of* α. The Raman active bands of XeF_4 will have symmetries a_{1g}, b_{1g}, and b_{2g}. There will be only three allowed fundamental transitions in the Raman spectrum of this compound. One of the normal modes of XeF_4, the b_{2u} mode, is both Raman and ir inactive. It should not give rise to absorption in either spectrum.

Let's pause for a moment to recapitulate because we just passed a milestone in this chapter. In evaluating the selection rules for the fundamental transitions of any molecule, regardless of its symmetry, *the only ir active modes will have the symmetries of the three components of* $\hat{\mu}$. *The only Raman active modes will have the symmetries of the six components of* $\hat{\alpha}$. Our procedure for evaluating selection rules henceforth will be one of *inspection.* We merely search the list of vibrational symmetries and pick out those which transform as components of $\hat{\mu}$ as the ir active modes and those which transform as components of $\hat{\alpha}$ as the Raman active modes.

The analysis of XeF_4 illustrates the *mutual exclusion rule*. In a molecule with a center of symmetry, $\hat{\mu}$ will always have u symmetry and $\hat{\alpha}$ will always have g symmetry. Therefore, only u modes can be ir active and only g modes can be Raman active. (But not all modes need be active.) No mode can be both ir and Raman active. If you find strong bands at the same frequency in the ir and Raman spectra of a compound, the chances are that the compound does not have a center of symmetry.

As a final example, we will analyze BCl_3, for which $\Gamma_{\text{vib}} = a_1' + 2e' + a_2''$ and $\hat{\mu} \sim a_2'' + e'$ and $\hat{\alpha} \sim a_1' + e' + e''$. By inspection, the a_2'' and two e' modes will be ir active, and the a_1' and two e' modes will be Raman active. Therefore, two bands will be common to both spectra and each spectrum will have one band not found in the other. The observed spectra of $^{10}B^{35}Cl_3$ show bands at 995, 480, and 244 cm^{-1} in the ir and 995, 471, and 244 cm^{-1} in the Raman. From this we immediately assign the a_1' mode at 471 cm^{-1}, the a_2'' mode at 480 cm^{-1}, and the two e' modes at 995 and 244 cm^{-1}.

Problems

3-25. For each molecule in Problem 3-23, determine whether the possible fundamental transitions are infrared or Raman active or inactive.

3-26. The spectra of HF_2^- and DF_2^-, isolated as alkali metal salts, are listed below. From these data, determine the geometry of these anions, which may be bent or linear. Draw a picture of each normal mode and make an assignment of the frequencies of these vibrations.

HF_2^-	DF_2^-	Activity
1550	1140	ir
1200	860	ir
675	675	Raman

3-27. The ir and Raman spectra of benzene are shown below. If you knew nothing of this compound's structure, what symmetry element would you suspect from these spectra?

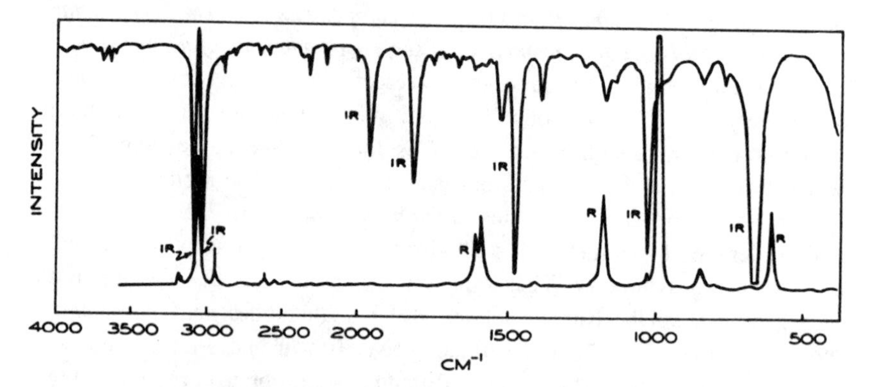

Vibrational spectra of benzene reproduced from S.K. Freeman, *Applications of Laser Raman Spectroscopy*, John Wiley & Sons, N.Y. 1974.

D. *Polarization.* We can illustrate the meaning of polarization quite nicely with H_2O whose three vibrations transform as $\nu_1 \sim a_1$, $\nu_2 \sim a_1$, and $\nu_3 \sim b_2$ (Section 3.4). Since the components of $\hat{\mu}$ are $z \sim a_1$, $x \sim b_1$, and $y \sim b_2$, the three ir transition moment integrals will be

$$\nu_1 \text{ or } \nu_2: \quad \psi(1)\hat{\mu}\psi(0) \sim a_1 \begin{pmatrix} a_1 \\ b_1 \\ b_2 \end{pmatrix} a_1 = \begin{pmatrix} a_1 \\ b_1 \\ b_2 \end{pmatrix} \tag{3-110}$$

$$\nu_3: \quad \psi(1)\hat{\mu}\psi(0) \sim b_2 \begin{pmatrix} a_1 \\ b_1 \\ b_2 \end{pmatrix} a_1 = \begin{pmatrix} b_2 \\ a_2 \\ a_1 \end{pmatrix} \tag{3-111}$$

All three transitions are ir allowed and are observed at 3657, 1595, and 3756 cm^{-1}, respectively, in gaseous water. The effect of using polarized light for the infrared spectrum is shown schematically in Fig. 3-42. ν_1

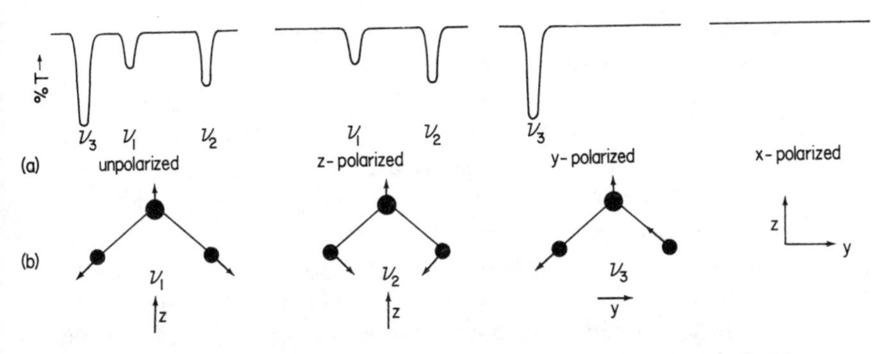

Fig. 3-42. (a) Hypothetical polarized infrared spectra of H_2O. (b) The vibrations ν_1 and ν_2 change only the *z* component of the dipole moment of H_2O while ν_3 changes only the *y* component.

and ν_2 are *z*-polarized while ν_3 is *y*-polarized. Fig. 3-42 is purely fictitious as there is no way to hold the water molecule oriented to obtain the spectrum. It is instructive as a hypothetical case because the physical basis of the polarization rules is easy to see. ν_1 and ν_2 only change the *z* component of the dipole moment of the molecule (Fig. 3-42). ν_3 only changes the *y* component of the dipole.

We now discuss the most common use of polarization in Raman spectroscopy. If the light shining on a sample in solution is plane polarized, it is possible to examine the Raman scattered light through a second polarizer that is either parallel or perpendicular to the plane of polarization of the incident light (Fig. 3-43). The intensity of scattered light polarized in the same plane as the incident light is $I_{\parallel}$, while that of light polarized

Fig. 3-43. To measure the polarization of Raman scattered radiation, a solution is placed at the origin of the coordinate system and *xz* plane polarized light is shined on it. $I_{\parallel}$ is the intensity of the *yz*-polarized scattered light. $I_{\perp}$ is the intensity of the *xy*-polarized scattered light. The polarizer is essentially transparent to light in one polarization and opaque to light in any other polarization.

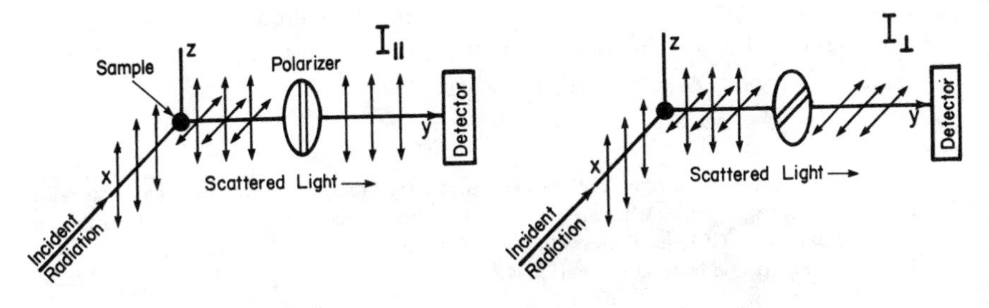

perpendicular to the incident beam is $I_\perp$. The *depolarization ratio*, ρ, is defined as $I_\perp / I_\parallel$. (To measure the intensity of a band, one integrates the area under the band.) For a sample in solution, theory predicts that a *depolarized* (unpolarized) band will exhibit $\rho = 3/4$. Polarized bands exhibit $0 \leqslant \rho \leqslant 3/4$. Polarized bands can only arise from vibrations which are totally symmetric. Therefore the bands which are polarized can be immediately assigned to the totally symmetric representation and the depolarized bands can be assigned to modes of any other symmetry. The more highly symmetric the molecule, the closer to 0 will be ρ for a totally symmetric vibration. It should also be pointed out that the totally symmetric vibrations usually give rise to the strongest Raman bands. In the spectrum of CCl_4 (Fig. 3-44), the band at 459 cm^{-1} almost disappears when the polarizers are perpendicular to each other. The bands at 314 and 218 cm^{-1} exhibit depolarization ratios of 0.75 ± 0.02.† The band at 459 cm^{-1} must have a_1 symmetry and the ones at 314 and 218 cannot have a_1 symmetry. Another example of Raman polarization was seen in Fig. 3-3.

Polarization is nice to deal with in theory but a very difficult tool to use in practice. Generally molecules can only be held fixed in the solid state. If you wish to study a molecule in a crystal you must do an x-ray crystal structure to know how the molecules are oriented in the crystal. Even then, crystals are often opaque to ir radiation and difficult reflectance techniques must be used. The polarized ir spectra in Fig. 3-45 were taken using two different orientations of crystals of dimethyl sulfone, $(CH_3)_2SO_2$, which has local C_{2v} symmetry about the sulfur atom. Analysis of a spectrum of a crystalline sample is complicated by the fact that the *site* that the molecule occupies also has symmetry and can affect the spectrum. The point of this illustration is the drastic change in the spectrum when the sample orientation is changed.

Four polarized Raman spectra of solid pentaerythritol, $C(CH_2OH)_4$, are shown in Fig. 3-46. This molecule has S_4 symmetry, as does its site in the crystal lattice. In this particular case, various combinations of polarization of incident and scattered radiation allow modes of only selected symmetry to be seen in each spectrum. For example, in the bottom spectrum only modes of symmetry *e* are seen.

† Theory and practice are often not well aware of each other. Since the depolarization ratio may not always agree with the theoretical value for a given spectrometer, one usually calibrates one's machine with CCl_4 using the values $\rho = 0.75$ for the 218 cm^{-1} band of CCl_4 and $\rho = 0.005 \pm 0.002$ for the band at 459 cm^{-1}.

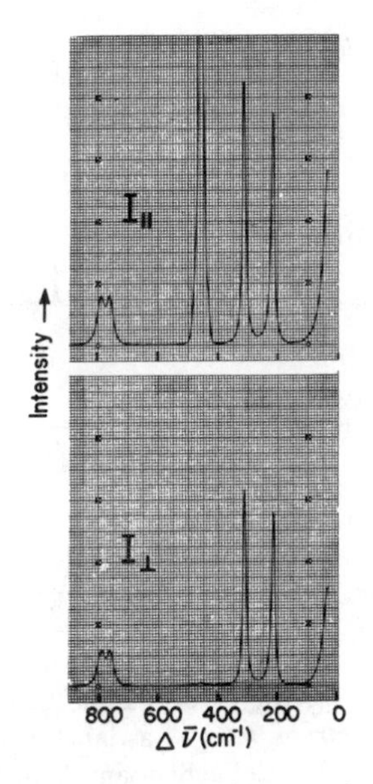

Fig. 3-44. Polarized Raman spectra of CCl_4 reproduced from *Sadtler Standard Raman Spectra*, © Sadtler Research Laboratories, Philadelphia, 1973.

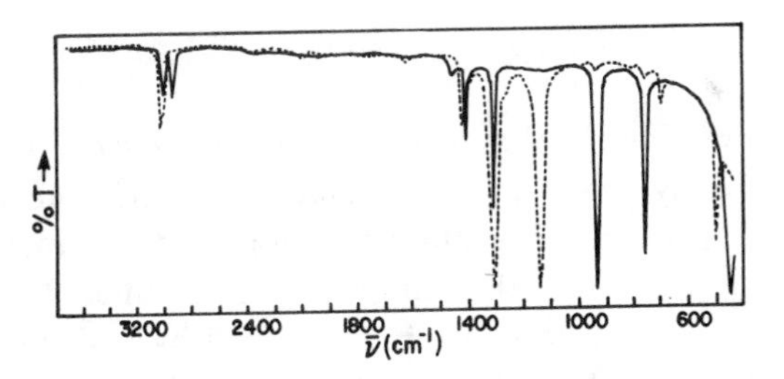

Fig. 3-45. Polarized infrared spectra of dimethyl sulphone. Solid line: Electric vector parallel to the direction of crystal growth. Dashed line: Electric vector perpendicular to the direction of crystal growth. Reproduced from T. Uno, K. Machida, and K. Hanai, *Spectrochim. Acta*, *27A*, 107 (1971).

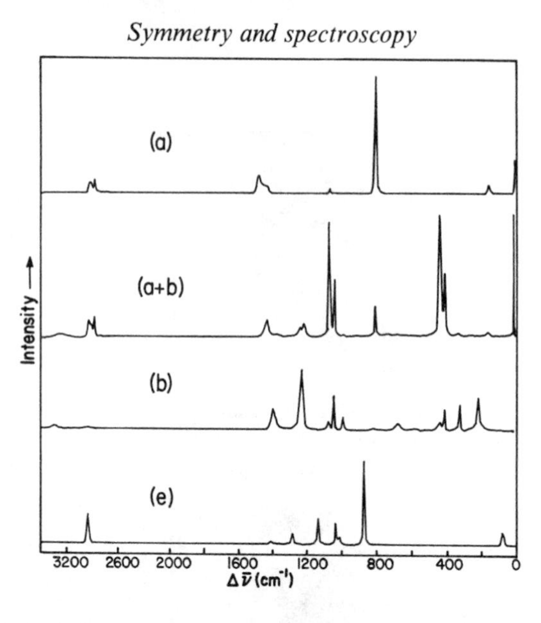

Fig. 3-46. Polarized Raman spectra of crystalline pentaerythritol, $C(CH_2OH)_4$. Only the vibrations of symmetry *a*, *b*, or *e*, as labelled in each trace, are active in a given polarization. From. R.D. McLachlan and V.B. Carter, *Spectrochim. Acta*, *27A*, 853 (1971).

Problems

3-28. For each ir allowed fundamental transition of each molecule in Problem 3-23, determine the direction of polarization of the absorption. Determine which Raman active fundamentals of these compounds are polarized or depolarized.

E. *Overtones, Combination Bands, and Hot Bands.* Most infrared spectra show vastly more than the number of bands predicted for the fundamentals. There are some other common kinds of transitions whose intensities are generally less than those of fundamentals. *Overtones* occur when a mode is excited beyond the $v = 1$ level by a single photon. For example, in the transition $\psi_1(0)\psi_2(0)\psi_3(0) \rightarrow \psi_1(0)\psi_2(3)\psi_3(0)$, ν_2 is excited to its $v = 3$ level. The absorption band would be called the second overtone of ν_2. The harmonic oscillator approximation allows us to estimate that the energy of this transition will be about three times that of the fundamental. A more realistic guess is that the energy will be somewhat less than three times the fundamental energy. A *combination band* is observed when more than one vibration is excited by one photon. Two examples are

$$\psi_1(0)\psi_2(0)\psi_3(0) \rightarrow \psi_1(1)\psi_2(1)\psi_3(0) \tag{3-112}$$

and

$$\psi_1(0)\psi_2(0)\psi_3(0) \rightarrow \psi_1(2)\psi_2(0)\psi_3(1)$$

In the first case the energy of the combination band should be the sum of the two fundamentals, $\nu_1 + \nu_2$. In the second case the energy will be the sum $2\nu_1 + \nu_3$. A *hot band* is observed when an already excited vibration is further excited. For example, the transition $\psi_1(0)\psi_2(1)\psi_3(0) \rightarrow \psi_1(0)\psi_2(2)\psi_3(0)$ gives rise to a hot band, the energy of which is slightly less than ν_2. Since the thermal population of the initial state is probably low, the intensity of the hot band will usually be weak. However, since the population of the initial state increases with increasing temperature, the intensity of the hot band will increase with temperature. Hence the name "hot band."

To decide whether a transition is allowed or not, we evaluate the transition moment integral eq. 3-113,

$$M = \int \psi_{es}\hat{O}\psi_{gs}d\tau \tag{3-113}$$

where ψ_{es} is the excited state wave function and ψ_{gs} is the ground state wave function. Consider an overtone first. Suppose a vibration, ν_i, is excited to its vth level. ψ_{gs} is totally symmetric and only ψ_{es} needs thought. If ν_i is not degenerate, the rule we discovered in Section 3-6-B was that if v is even, $\psi(v)$ is totally symmetric. If v is odd, $\psi(v)$ has the symmetry of ν_i. Now suppose ν_a and ν_b are degenerate. In the fundamental transition, just one mode is excited:

$$\psi_a(0)\psi_b(0) \rightarrow \psi_a(1)\psi_b(0)$$

or

$$\psi_a(0)\psi_b(0) \rightarrow \psi_a(0)\psi_b(1) \tag{3-114}$$

For the first overtone there are three possibilities:

$$\psi_a(0)\psi_b(0) \rightarrow \psi_a(2)\psi_b(0)$$

or

$$\psi_a(0)\psi_b(0) \rightarrow \psi_a(1)\psi_b(1) \tag{3-115}$$

or

$$\psi_a(0)\psi_b(0) \rightarrow \psi_a(0)\psi_b(2)$$

The third overtone would represent five possibilities:

$$
\begin{array}{ll}
 & \psi_a(0)\psi_b(0) \rightarrow \psi_a(4)\psi_b(0) \\
\text{or} & \psi_a(0)\psi_b(0) \rightarrow \psi_a(3)\psi_b(1) \\
\text{or} & \psi_a(0)\psi_b(0) \rightarrow \psi_a(2)\psi_b(2) \\
\text{or} & \psi_a(0)\psi_b(0) \rightarrow \psi_a(1)\psi_b(3) \\
\text{or} & \psi_a(0)\psi_b(0) \rightarrow \psi_a(0)\psi_b(4)
\end{array}
\tag{3-116}
$$

Suppose v_a and v_b transform as e in the point group C_{4v}. The symmetry of $\psi_a(0)\psi_b(0)$ is a_1. The symmetry of $[\psi_a(1)\psi_b(0), \psi_a(0)\psi_b(1)]$ is e. The symmetry of $[\psi_a(2)\psi_b(0), \psi_a(1)\psi_b(1), \psi_a(0)\psi_b(2)]$ is *not* $e \times e = a_1 + a_2 + b_1 + b_2$. It cannot be $e \times e$ since there are three wave functions and $e \times e$ has a dimension of four. A formula which allows us to determine the proper symmetries is given in Appendix C. The excited state vibrational wave functions of b_1 and e vibrations of a C_{4v} molecule are found to transform as follows:

b_1 vibration		e vibration	
.	.	.	.
.	.	.	.
.	.	.	.
$v = 4$	a_1	$v = 4$	$2a_1 + a_2 + b_1 + b_2$
$v = 3$	b_1	$v = 3$	$2e$
$v = 2$	a_1	$v = 2$	$a_1 + b_1 + b_2$
$v = 1$	b_1	$v = 1$	e
$v = 0$	a_1	$v = 0$	a_1

Note that even numbered levels always contain at least one totally symmetric component. Suppose we wish to know if the third overtone of the e vibration is allowed. The transition moment integrals are

$$
\text{ir}: \int \psi_{es}\hat{\mu}\psi_{gs}d\tau \sim \begin{pmatrix} a_1 \\ a_2 \\ b_1 \\ b_2 \end{pmatrix} \begin{pmatrix} a_1 \\ e \end{pmatrix} (a_1) \sim \underline{\underline{a_1}} + a_2 + b_1 + b_2 + e
$$

$$
\text{Raman}: \int \psi_{es}\hat{\alpha}\psi_{gs}d\tau \sim \begin{pmatrix} a_1 \\ a_2 \\ b_1 \\ b_2 \end{pmatrix} \begin{pmatrix} a_1 \\ b_1 \\ b_2 \\ e \end{pmatrix} (a_1) \sim \underline{\underline{a_1}} + a_2 + b_1 + b_2 + e
\tag{3-117}
$$

which show that the third overtone is allowed in both spectra. This does not mean that the third overtone will be strong in both spectra. It only means that it may have some nonzero value.

Combination bands are not much more complicated. Again we seek the symmetry of ψ_{es} to plug into the appropriate transition moment integrals. Suppose we want to know if the transition $\psi_1(0)\psi_2(0)\psi_3(0) \rightarrow \psi_1(2)\psi_2(0)\psi_3(1)$ is allowed. The symmetry of the excited state will be the symmetry of the product $\Gamma[\psi_1(2)] \times \Gamma[\psi_3(1)]$ regardless of whether either or both modes are degenerate. If ν_1 is degenerate in this case, though, you will have to use Appendix C to evaluate $\Gamma[\psi_1(2)]$. If ν_1 has symmetry e and ν_3 has symmetry a_2 in the point group C_{4v}, we get:

$$\Gamma[\psi_{es}] = \Gamma[\psi_1(2)] \times \Gamma[\psi_3(1)] = (a_1 + b_1 + b_2) \times a_2 = a_2 + b_2 + b_1 \tag{3-118}$$

The transition moment integrals for hot bands are slightly different from the examples so far, in that ψ_{gs} will not generally be totally symmetric since one of the vibrations is not in the $v = 0$ state. If ν_i has symmetry b_2, the transition $\psi_i(1) \rightarrow \psi_i(2)$ will be characterized by the symmetries $\psi_{gs} \sim b_2$ and $\psi_{es} \sim a_1$.

Although overtones and combination bands are generally much less intense than fundamental transitions, an "intensity borrowing" mechanism is possible which can add intensity to otherwise weak absorptions. In CO_2, for example (Fig 3-33), ν_1 (σ_g^+) should be found at 1337 cm^{-1} and ν_2 (π_u) occurs at 667 cm^{-1}. The first overtone of ν_2, with symmetry species $\sigma_g^+ + \delta_g$ (by use of Appendix C) is expected near $2 \times 667 = 1334$ cm^{-1}. However, quantum mechanics allows a "mixing" of the two σ_g^+ states expected at 1337 (ν_1) and 1334 ($2\nu_2$) cm^{-1}. This mixing, called *Fermi resonance*, has two effects: (1) The overtone can gain intensity from the nearby fundamental of the same symmetry; and (2) both energy levels are shifted away from each other. Instead of observing a strong band at 1337 cm^{-1} and a weak band at 1334 cm^{-1}, two strong bands are observed in the Raman spectrum at 1388 and 1286 cm^{-1}. Of course the spectrum does not run up to you and say, "Look at this marvelous Fermi resonance!" The interpretation of the two strong bands is just that—interpretation. Such pitfalls in spectroscopy are commonplace.

Problems

3-29. What are the symmetries of the $v = 0, 1, 2,$ and 3 levels of an e' vibration in the point group D_{3h}? Is the first overtone of an e' vibration allowed in the ir spectrum? In the Raman spectrum?

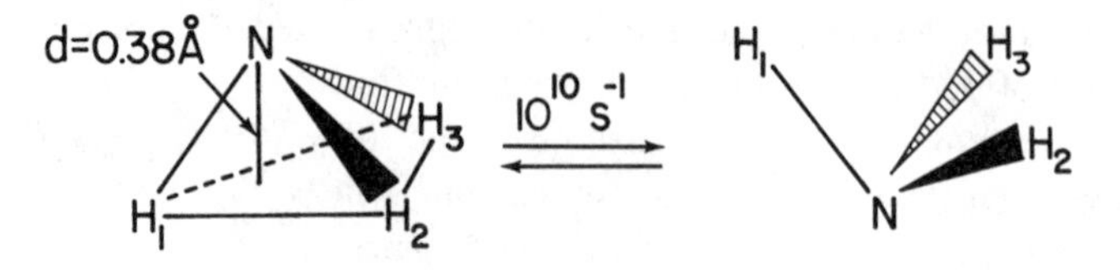

Fig. 3-47. Inversion of gaseous ammonia occurs more than 10^{10} times per second.

F. *Inversion Doubling.*† The ammonia molecule is pyramidal with the nitrogen atom lying 0.38 Å above the plane of the hydrogen atoms. It is known that the barrier to passage of the nitrogen atom through the H_3 plane cannot be too great since such *pyramidal inversion* occurs more than 10^{10} times per second at 300 K (Fig. 3-47). If one considers the potential energy diagram for N-H umbrella-like bending, it ought to be as shown in Fig 3-48.

The quantum mechanical consequence of such a multiple-minimum potential well is that there are two states for every one state of a single-minimum well. This is illustrated for NH_3 in Fig. 3-48. This figure is constructed by superimposing two single-minimum potential wells. The splitting between the states increases with increasing v until each state is split by half the energy separation of successive states of the single-minimum wells. Each splitting of vibrational levels is a result of symmetric (+) and antisymmetric (−) combinations of the wave functions of the single-minimum potential wells (Fig. 3-49). In each case, the symmetric combination is of lower energy than the antisymmetric combination.

For NH_3 the height of the potential barrier is 2076 cm^{-1} and the N-H bending frequency (ν_2) is 950 cm^{-1}. Therefore the $v = 2$ level of ν_2 is just below the potential maximum, and the $v = 3$ level is above the maximum (Fig. 3-48). NH_3 molecules undergoing 10^{10} inversions per second at 300 K are largely in the $v = 0$ state, so they must be *tunneling through the potential barrier* (Section 2-4-D).

The selection rule for infrared transitions in such a case is $+ \leftrightarrow -$. For Raman transitions, the rules are $+ \leftrightarrow +$ and $- \leftrightarrow -$. As a result, the fundamental transition will be split into two components as shown in Fig. 3-50. For NH_3 the largest splitting (37 cm^{-1}) is observed for $\nu_2(\delta_s)$, which is an umbrella-like bending motion. The other vibrations of ammonia do

†Good discussions of this subject appear in G. Herzberg, *Infrared and Raman Spectra of Polyatomic Molecules*, Van Nostrand Reinhold, N.Y., 1945; and in Chapter 12 of C.H. Townes and A.L. Schwalow, *Microwave Spectroscopy*, McGraw-Hill, N.Y., 1955.

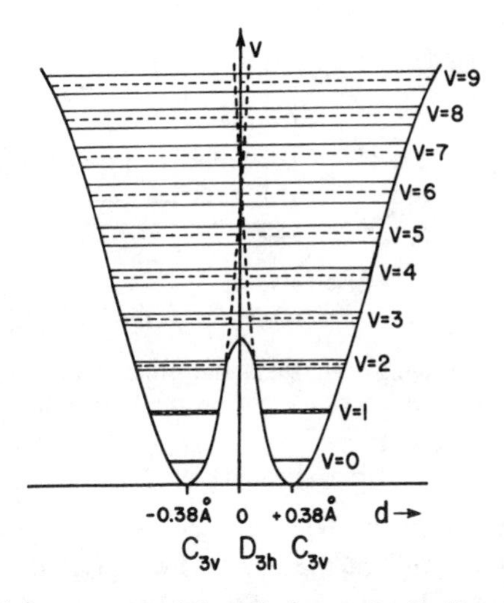

Fig. 3-48. Double-minimum potential well for umbrella-like bending of NH_3 showing the vibrational energy levels. The abscissa is the distance of N from the H_3 plane.

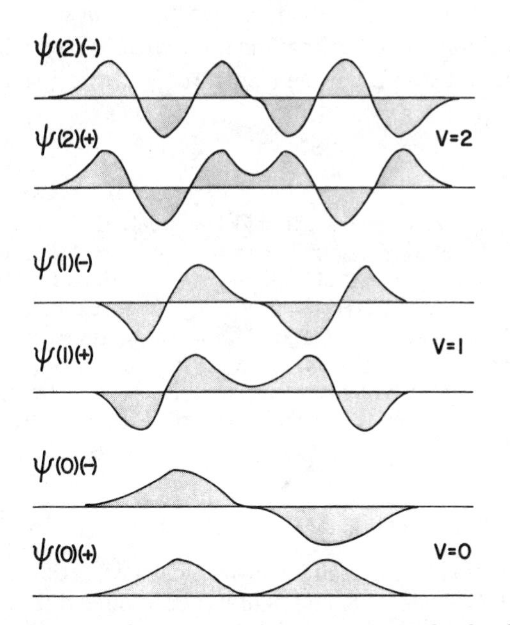

Fig. 3-49. Wave functions for some of the lowest energy levels of a particle in a double-minimum potential well.

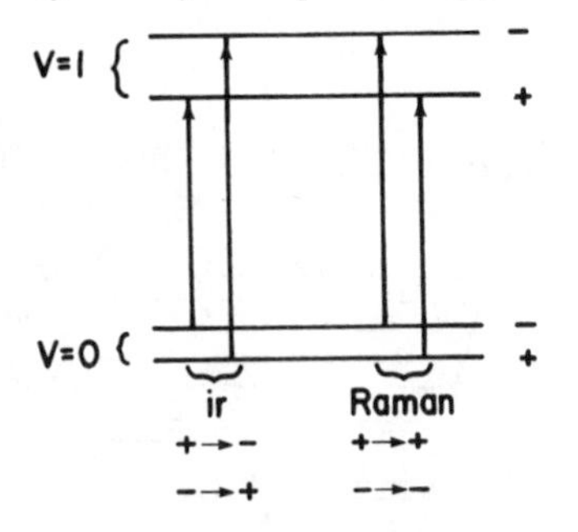

Fig. 3-50. Inversion doubling of the ν_2 fundamental of ammonia. The ir frequencies are observed at 931.58 and 968.08 cm^{-1}, and the Raman transitions occur at 934.0 and 964.3 cm^{-1}.

involve some change in the NH_3 distance and so, in principle, will also exhibit *inversion doubling* of the bands. For $\nu_1(\nu_s)$ the ir splitting is only 2 cm^{-1}, and for ν_3 and ν_4 the splittings are too small to observe. The transition $v = 0(+) \rightarrow v = (0)(-)$, which does not involve a quantum of vibrational energy, is observed at 0.8 cm^{-1}, in the microwave spectrum. By $v = 0$, of course, we really mean $v_1 = v_2 = v_3 = v_4 = 0$.

In general, any nonplanar molecule may exhibit multiple minima in its potential well and inversion doubling in its vibrational spectra. In practice, few molecules exhibit such doubling, and those which do not exhibit doubling can just as well be treated as if their potential wells were single-minimum.

Problems

3-30. The time for inversion of a pyramidal molecule is given by $t = 1/(2\Delta_0 c)$, where Δ_0 is the splitting of $\psi(0)(+)$ and $\psi(0)(-)$ in cm^{-1} (Fig. 3-50) and c is the speed of light in cm s^{-1}. Calculate the inversion times for NH_3, PH_3, and AsH_3, given that $\Delta_0(NH_3) = 0.79$ cm^{-1} (observed); $\Delta_0(PH_3) = 4.8 \times 10^{-4} cm^{-1}$ (calculated); and $\Delta_0(AsH_3) = 3.7 \times 10^{-18}$ cm^{-1} (calculated). Which kinds of compounds, $NR_1R_2R_3$, $PR_1R_2R_3$, or $AsR_1R_2R_3$ (where R_i is an organic substituent) might you expect to exist in optically active forms? Ref.: C.C. Costain and G.B.B.M. Sutherland, *J. Phys. Chem.*, *56*, 321 (1952).

3-7. Symmetry Coordinates and Normal Modes

So far no derivation of the pictures of molecular vibrations, such as those in Figs. 3-29, 3-33, 3-36, 3-38, and 3-40 has been offered. In this section we will consider the problem of drawing pictures which have the required symmetries of the normal modes. Such pictures represent *symmetry coordinates*. In favorable circumstances, the symmetry coordinates are very nearly the normal vibrations. Only a numerical normal coordinate analysis, which is beyond the scope of this text,† making use of force constants derived from vibrational spectra can establish the relation between the symmetry coordinates and the normal modes. We will discuss the factors which determine how closely the symmetry coordinates may resemble the normal vibrations.

To illustrate a case in which the symmetry coordinates are not good approximations of the normal modes, let's consider the totally symmetric (a_{1g}) modes of Au_2Cl_6 which has the planar structure shown in Fig. 3-51. Four independent totally symmetric vibrations can be drawn. By independent, we mean that none are linear combinations of the others. We call these four symmetry coordinates S_1, S_2, S_3, and S_4:

$$\begin{aligned} S_1 &= \Delta t_1 + \Delta t_2 + \Delta t_3 + \Delta t_4 \\ S_2 &= \Delta b_1 + \Delta b_2 + \Delta b_3 + \Delta b_4 \\ S_3 &= \Delta\alpha_1 + \Delta\alpha_2 + \Delta\alpha_3 + \Delta\alpha_4 \\ S_4 &= \Delta\beta_1 + \Delta\beta_2 \end{aligned} \tag{3-119}$$

The symbols in eq. 3-119 are defined in Fig. 3-51 in which Δt_i represents stretching of terminal Au-Cl bonds, Δb_i represents stretching of bridging Au-Cl bonds, and $\Delta\alpha_i$ and $\Delta\beta_i$ represent bending of the indicated bonds.

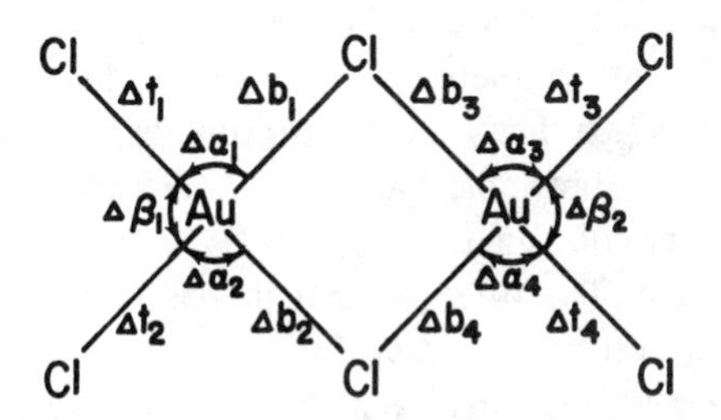

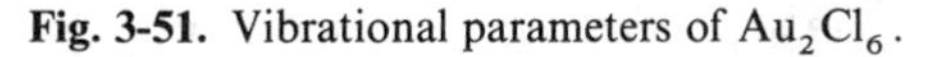
Fig. 3-51. Vibrational parameters of Au_2Cl_6.

†A readable, but elementary, discussion of normal coordinate analysis can be found in K. Nakamoto, *Infrared Spectra of Inorganic and Coordination Compounds*, John Wiley & Sons, N.Y., 1970. A readable, and very complete, discussion of this subject appears in L.A. Woodward, *Introduction to the Theory of Molecular Vibrations and Vibrational Spectroscopy*, Oxford University Press, Oxford, 1972. The most authoritative, but least readable, treatment is found in E.B. Wilson, Jr., J.C. Decius, and P. C. Cross, *Molecular Vibrations*, McGraw-Hill, N.Y., 1955.

The pictures corresponding to S_1-S_4 are shown in Fig. 3-52. These pictures were constructed by drawing arrows at each Cl and Au atom to indicate the proper stretching or bending motions and then adding all arrows on each atom vectorially. It should be clear to you that each vibration is totally symmetric, which is to say that application of each symmetry operation of the group D_{2h} leaves the picture unchanged.

In the remaining examples in this section, the symmetry coordinates will generally be good representations of the normal modes of vibration. In this case, however, they are not. Four totally symmetric (polarized) Raman bands are observed at 380, 331, 168, and 99 cm^{-1}. These are labelled ν_1, ν_2, ν_3, and ν_4, respectively. Normal coordinate analysis indicates the fractional composition of each vibration shown in Table 3-20. An example of what this means is that ν_1 consists of 71% S_1, 28% S_2, and very little of anything else. This mode, consisting of predominantly terminal bond stretching, might be called $\nu(\text{Au-Cl}_t)$, but a more accurate description would be $\nu(\text{Au-Cl}_t) + \nu(\text{Au-Cl}_b)$. Similarly, ν_2 is $\nu(\text{Au-Cl}_b) + \nu(\text{Au-Cl}_t)$. ν_3 and ν_4 are thorough mixtures of bending modes.

The reason the vibrations of this molecule do not correspond more simply to the symmetry coordinates is that the energies of all the "pure symmetry coordinate" vibrations are very close together. Had the modes been separated by greater energies, such as C–H stretching (3000 cm^{-1}) and C=O stretching (1700 cm^{-1}), there would be much less mixing of the symmetry coordinates. Another point is that only symmetry coordinates of the same symmetry can be mixed. Therefore, we did not consider other modes of Au_2Cl_2 which have nearly the same energies as ν_1, ν_2, ν_3, or ν_4, but which have different symmetries. In the remainder of this section, we will consider cases in which the symmetry coordinates more closely resemble the normal modes of vibrations, and we will discuss a method of constructing the symmetry coordinates.

Let's try H_2O again. The three normal modes transform as $2a_1 + b_2$ (Section 3-5). It is easiest to construct the symmetry coordinates if we can separate the stretching vibrations from the bending vibrations. To isolate the stretching modes, draw a two-headed arrow between each set of atoms

Fig. 3-52. Four independent, totally symmetric symmetry coordinates of Au_2Cl_6.

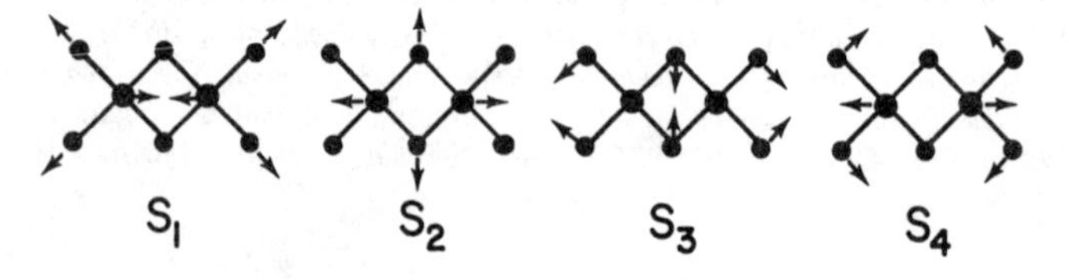

Table 3-20. Fractional Composition of the Normal Modes of Vibration of Au_2Cl_6[a]

	S_1	S_2	S_3	S_4
ν_1	0.71	0.28	0.01	—
ν_2	0.35	0.62	0.03	—
ν_3	—	0.09	0.56	0.35
ν_4	—	0.02	0.55	0.43

[a]From D.M. Adams and R.G. Churchill, *J. Chem. Soc.* A, *1968*, 2141.

in the molecule (Fig. 3-53 [a]), and determine how these arrows transform under each operation of the point group (Table 3-21). The O–H stretching modes of water are classified as symmetric (a_1) and antisymmetric (b_2) because one is symmetric to the C_2 operation and the other is not. A symmetric stretch is easy to construct and is shown in Fig. 3-53 (b). The antisymmetric stretch involves stretching of one O–H bond and compression of the other (Fig. 3-53 [c]). The bending mode of H_2O must have a_1 symmetry since the three vibrations have symmetry $2a_1 + b_2$ and stretching used up $a_1 + b_2$. In Fig. 3-53 (d) the only reasonable way to illustrate bending of the H–O–H bond is shown.

We have just constructed stretching and bending symmetry coordinates for any symmetrical, bent XY_2 molecule. How close are the symmetry coordinates to the actual vibrations? As shown in Fig. 3-54, in the cases of H_2O, NO_2, and F_2O, the symmetry coordinates are good approximations of the normal modes. For Cl_2O, however (which is not shown), the two symmetry coordinates of a_1 symmetry are extensively mixed and ν_1 and ν_2 are difficult to classify as either pure stretching or pure bending modes.

Linear CO_2 has four normal modes of symmetry $\sigma_g^+ + \sigma_u^+ + \pi_u$. The stretching modes transform as $\sigma_g^+ + \sigma_u^+$ (Fig. 3-55 [a] and Table 3-22).

Fig. 3-53. (a) Bond stretching vectors for H_2O. (b–d) Normal modes of vibration of H_2O.

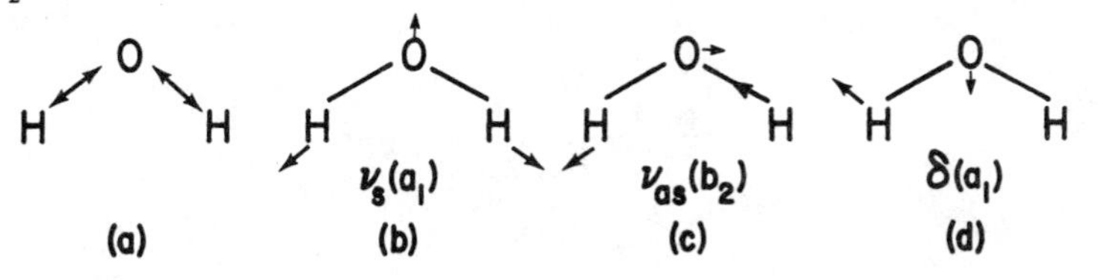

Table 3-21. Stretching Analysis of H_2O

C_{2v}	E	C_2	$\sigma(xz)$	$\sigma(yz)$	
Γ_{stretch}	2	0	0	2	$= a_1 + b_2$

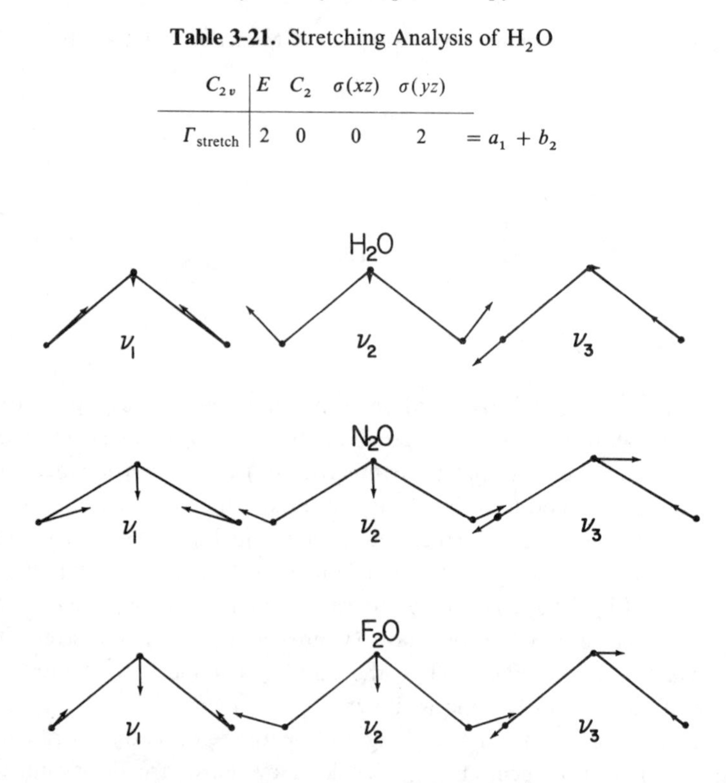

Fig. 3-54. Actual forms of the normal vibrations of H_2O, NO_2, and F_2O. The scale of vibrational amplitudes is much larger than the scale of internuclear distance if the $v = 1$ states are considered. Vibrations have not been normalized, so different pictures are not directly comparable. Calculated from data in G. Herzberg, *Infrared and Raman Spectra of Polyatomic Molecules*, Van Nostrand Reinhold, N.Y., 1945, Section II, 4.

Fig. 3-55. (a) Stretching vectors for CO_2. (b–d) Normal modes of vibration of CO_2.

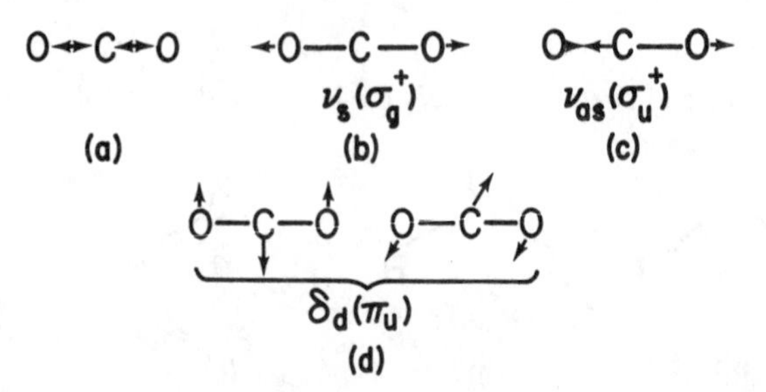

Table 3-22. Stretching Analysis for CO_2

$D_{\infty h}$	E	$2C_\infty^\phi$	$\infty\sigma_v$	i	$2S_\infty^\phi$	∞C_2	
Γ_{stretch}	2	2	2	0	0	0	$= \sigma_g^+ + \sigma_u^+$

Hence the two bending modes must transform as π_u and must be degenerate. Once again we can construct symmetric and antisymmetric stretching vibrations, as shown in Fig. 3-55(b) and (c). You can draw bending in the plane of the page and perpendicular to the page to account for the degenerate π_u modes (Fig. 3-55[d]).

Let's make a generalization. In any molecule with a C_2 axis (and no higher order axis), you can construct symmetric and antisymmetric stretching modes. For example, in dichloromethane (Fig. 3-56[a]), the two C–Cl stretching modes transform as $a_1 + b_1$ and the two C–H stretching modes transform as $a_1 + b_2$ (Table 3-23). Both sets of vibrations just amount to symmetric and antisymmetric combinations (Fig. 3-56[b–e]). While the symmetry coordinates in Fig. 3-56 might be good representations of the normal modes of vibration of CH_2Cl_2, they might not be such good

Fig. 3-56. (a) Coordinate system for CH_2Cl_2 stretching analysis. The Cl atoms are in the *xz* plane and the H atoms are in the *yz* plane. (b) Symmetric C–Cl stretching. (c) Antisymmetric C–Cl stretching. (d) Symmetric C–H stretching. (e) Antisymmetric C–H stretching.

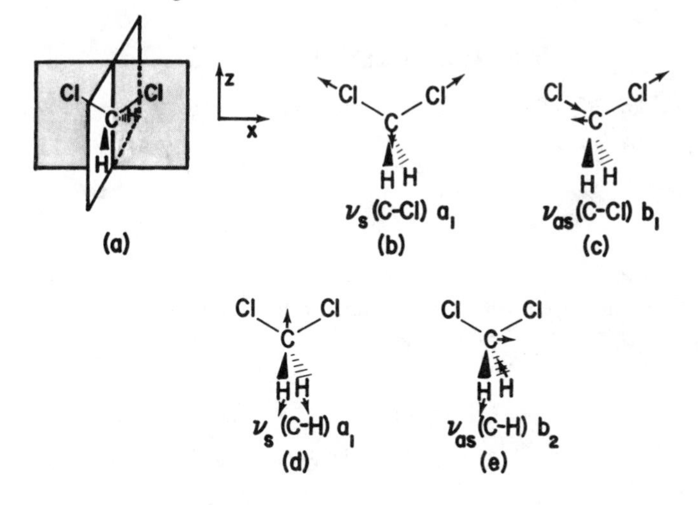

Table 3-23. Stretching Analysis of CH_2Cl_2

C_{2v}	E	C_2	$\sigma(xz)$	$\sigma(yz)$	
Γ(C–Cl)	2	0	2	0	$= a_1 + b_1$
Γ(C–H)	2	0	0	2	$= a_1 + b_2$

representations of the normal modes of CCl_2Br_2. There will still be two a_1 stretching modes in CCl_2Br_2, but neither will be pure C–Cl or C–Br stretching.

To handle molecules with axes of symmetry higher than C_2, such as XeF_4, we will introduce a more systematic procedure for constructing symmetry coordinates. Using the symmetry elements defined in Fig. 3-57, we find that the four Xe–F stretching modes transform as $a_{1g} + b_{1g} + e_u$ (Table 3-24). Since all of the vibrations together transform as $a_{1g} + b_{1g} + b_{2g} + a_{2u} + b_{2u} + 2e_u$, we deduce that the bending vibrations transform as $\Gamma_{\text{bend}} = \Gamma_{\text{vib}} - \Gamma_{\text{stretch}} = b_{2g} + a_{2u} + b_{2u} + e_u$.

For a molecule with a principal axis C_n, the character table for the pure rotational point group C_n usually contains all of the information we need to construct symmetry coordinates. (Complications may arise in the cubic or icosahedral point groups where there are more than one axes of symmetry > C_2.) Since the C_n character tables contain imaginary characters for the degenerate representations, we will transform the pairs of imaginary

Table 3-24. Stretching Analysis of XeF_4

D_{4h}	E	$2C_4$	C_2	$2C_2'$	$2C_2''$	i	$2S_4$	σ_h	$2\sigma_v$	$2\sigma_d$	
Γ_{stretch}	4	0	0	2	0	0	0	4	2	0	$= a_{1g} + b_{1g} + e_u$

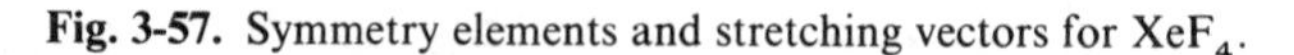

Fig. 3-57. Symmetry elements and stretching vectors for XeF_4.

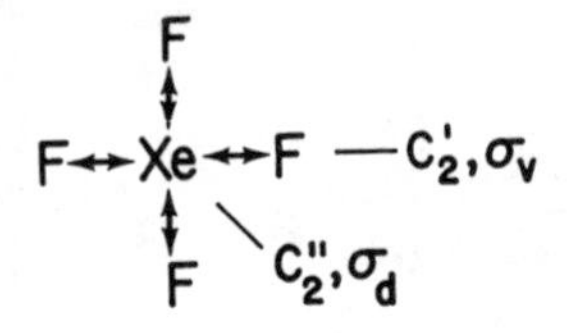

characters into pairs of real characters. This is done by adding and subtracting the two lines for each degenerate representation and dividing the resulting characters by the greatest common denominator. For the E representation of C_4, this recipe translates as follows:

C_4:	E	1	i	-1	$-i$	line a
		1	$-i$	-1	i	line b
add $(a + b)$:		2	0	-2	0	line c
subtract $(a - b)$:		0	$2i$	0	$-2i$	line d
divide c by 2:		1	0	-1	0	real characters for the E representation
divide d by $2i$:		0	1	0	-1	

We therefore convert the true C_4 character table into a more useful one of the following form:

C_4	E	C_4	C_2	C_4^3
A	1	1	1	1
B	1	-1	1	-1
E	1	i	-1	$-$i
	1	$-$i	-1	i

$\Rightarrow$

C_4	E	C_4	C_2	C_4^3
A	1	1	1	1
B	1	-1	1	-1
E	1	0	-1	0
	0	1	0	-1

Now consider the four stretching modes of XeF_4 which transform as $a_{1g} + b_{1g} + e_u$. To generate a symmetry coordinate, S_i, from a single Xe$-$F stretching vector, $\boldsymbol{\Delta}s$, we use the formula 3-120:

$$S_i \propto \sum_R R \cdot \chi_i{}^R \cdot \boldsymbol{\Delta}s \quad (3\text{-}120)$$

in which R is an operation of the point group C_4, $\chi_i{}^R$ is the character under that operation in the ith irreducible representation, and the sum is extended over all operations of the group. To construct the a_{1g} stretching symmetry coordinate, we apply eq. 3-120 to $\boldsymbol{\Delta}s$ using the characters of the A irreducible representation of the C_4 character table, as shown in Fig. 3-58. When we add the four components in Fig. 3-58, we obtain the totally symmetric stretching coordinate $S_{a_{1g}}$:

$$S_{a_{1g}} \propto \boldsymbol{\Delta}s_1 + \boldsymbol{\Delta}s_2 + \boldsymbol{\Delta}s_3 + \boldsymbol{\Delta}s_4 \quad (3\text{-}121)$$

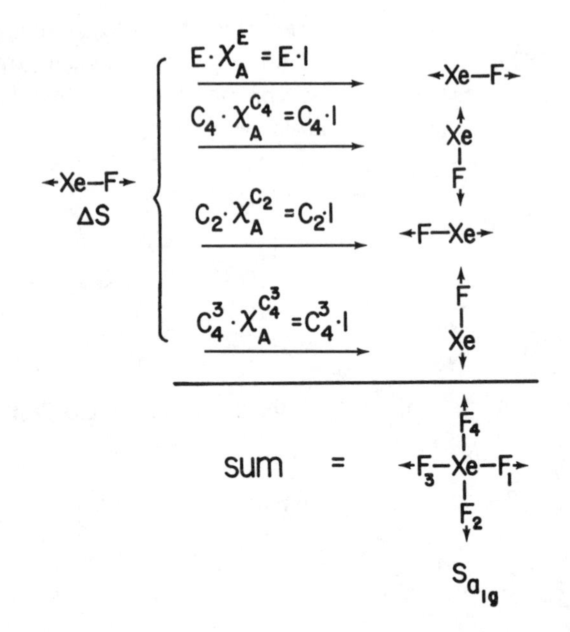

Fig. 3-58. Application of eq. 3-120 to a single Xe−F stretching vector to produce the totally symmetric stretching symmetry coordinate. The four vectors on Xe add up to zero.

We write a proportional symbol ($\propto$) instead of an equality because the symmetry coordinates we will derive are not normalized. Since normalization serves no purpose in this chapter, we will not introduce it. In deriving $S_{a_{1g}}$, we used the A irreducible representation of the C_4 character table. Since this really only guarantees us that we have constructed a stretching coordinate of symmetry a, you should check that the function in eq. 3-121 does indeed possess the full symmetry a_{1g} in the point group D_{4h}.

In Fig. 3-59 we use the same procedure with the characters of the B representation of C_4 to generate the symmetry coordinate $S_{b_{1g}}$:

$$S_{b_{1g}} \propto \Delta s_1 - \Delta s_2 + \Delta s_3 - \Delta s_4 \tag{3-122}$$

Finally, we apply both sets of real characters of the irreducible representation E to produce the two vibrations which together transform as e_u and are shown in Fig. 3-60:

$$\nu_d(e_u)\begin{cases} S_{e_u}(1) \propto \Delta s_1 - \Delta s_3 & (3\text{-}123) \\ S_{e_u}(2) \propto \Delta s_2 - \Delta s_4 & (3\text{-}124) \end{cases}$$

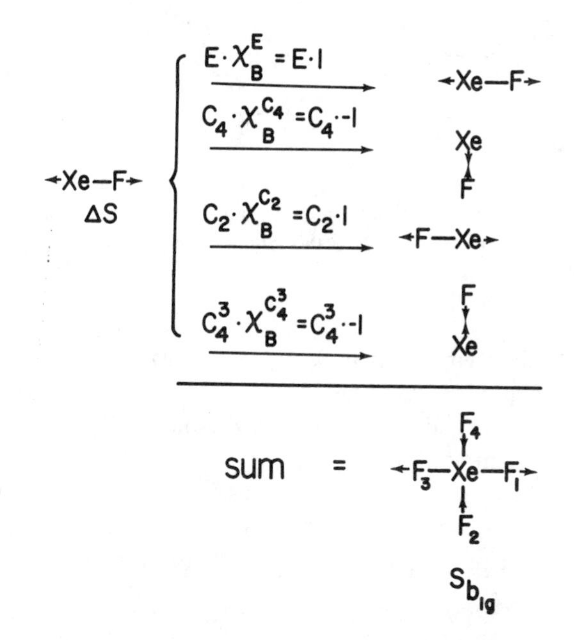

Fig. 3-59. Application of eq. 3-120 to a single Xe–F stretching vector to produce the symmetry coordinate $S_{b_{1g}}$.

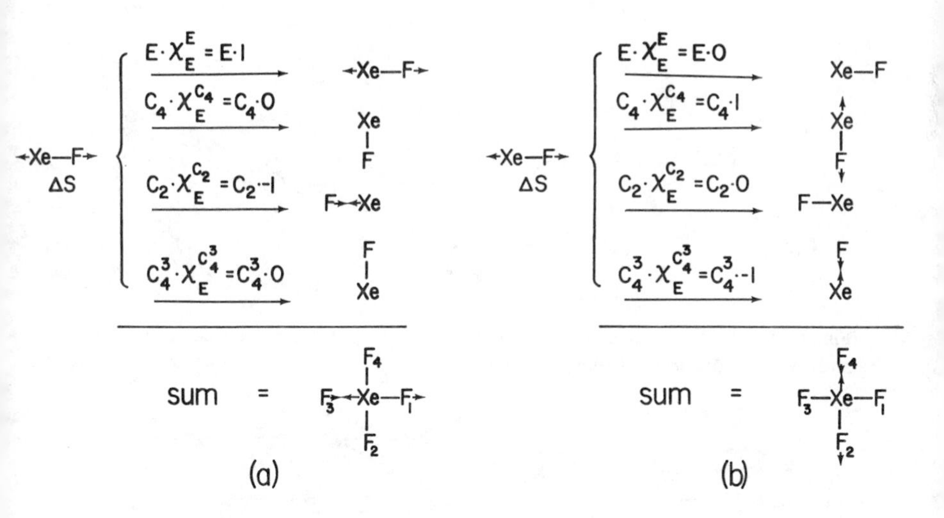

Fig. 3-60. Application of eq. 3-120 using the two different sets of real characters for the E irreducible representation to generate the two degenerate stretching modes of XeF_4.

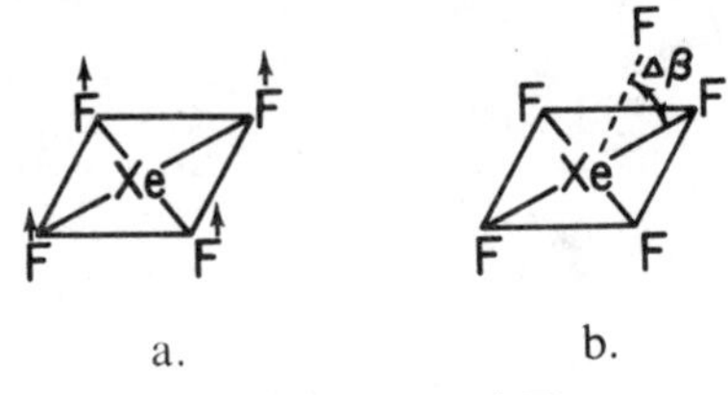

Fig. 3-61. (a) Vectors in the z direction on each F atom are used as bases for out-of-plane bending. (b) Out-of-plane bending of a single Xe–F bond is represented by changes in the angle β.

To handle bending vibrations of XeF_4, we consider first out-of-plane bending and then in-plane bending. As a basis for out-of-plane motion, consider a vector in the z direction on each F atom in Fig. 3-61(a). These four vectors transform as shown in the top line of Table 3-25. Using the bending motion $\Delta\beta$ defined in Fig. 3-61(b) as a basis function, and applying an equation analagous to eq. 3-120, in which we substitute $\Delta\beta$ for Δs, we generate the four out-of-plane motions shown in Fig. 3-62:

$$\pi(a_{2u}) \propto \Delta\beta_1 + \Delta\beta_2 + \Delta\beta_3 + \Delta\beta_4 \tag{3-125}$$

$$\pi(b_{2u}) \propto \Delta\beta_1 - \Delta\beta_2 + \Delta\beta_3 - \Delta\beta_4 \tag{3-126}$$

$$e_g \begin{cases} R_x \propto \Delta\beta_1 - \Delta\beta_3 & (3\text{-}127) \\ R_y \propto \Delta\beta_2 - \Delta\beta_4 & (3\text{-}128) \end{cases}$$

Table 3-25. Bending Analysis of XeF_4

D_{4h}	E	$2C_4$	C_2	$2C_2'$	$2C_2''$	i	$2S_4$	σ_h	$2\sigma_v$	$2\sigma_d$	
$\Gamma_{\text{out-of-plane}}$	4	0	0	−2	0	0	0	−4	2	0	$= a_{2u} + b_{2u} + e_g$
$\Gamma_{\text{in-plane}}$	4	0	0	−2	0	0	0	4	−2	0	$= a_{2g} + b_{2g} + e_u$

Fig. 3-62. The four out-of-plane motions of XeF_4. Two are true bending motions and two are rotations of the whole molecule.

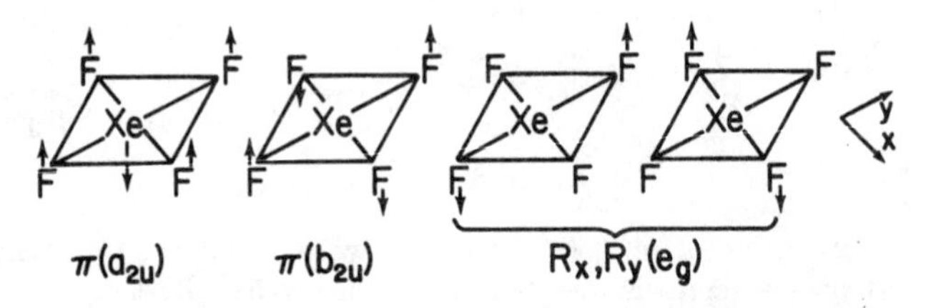

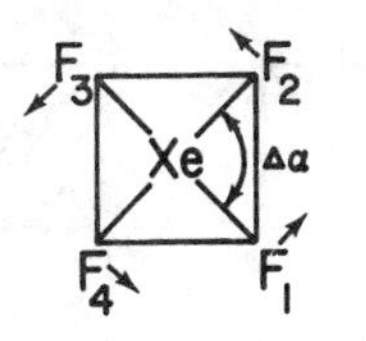

Fig. 3-63. In-plane Xe–F bending motion can be represented by four arrows in the plane of the molecule. This picture is not meant to indicate rotation of the whole molecule about the C_4 axis, which would be drawn the same way. Rather, it is supposed to indicate bending of each Xe–F bond while the other three remain fixed. In-plane bending of a single Xe–F bond results in a change in the angle α.

Two are true out-of-plane bending motions, designated by the symbol π (Table 3-13), and two are degenerate rotational motions (R_x, R_y). Finally we come to in-plane bending which can be represented by the four arrows in Fig. 3-63. As shown in Table 3-25, in-plane bending transforms as $a_{2g} + b_{2g} + e_u$. Applying an equation analogous to eq. 3-120 to in-plane bending, $\Delta\alpha$, generates the three true bending modes and one rotational mode shown in Fig. 3-64:

$$R_z(a_{2g}) \propto \Delta\alpha_1 + \Delta\alpha_2 + \Delta\alpha_3 + \Delta\alpha_4 \qquad (3\text{-}129)$$

$$\delta(b_{2g}) \propto \Delta\alpha_1 - \Delta\alpha_2 + \Delta\alpha_3 - \Delta\alpha_4 \qquad (3\text{-}130)$$

$$\delta_d(e_u) \propto \begin{cases} \Delta\alpha_1 - \Delta\alpha_3 & (3\text{-}131) \\ \Delta\alpha_2 - \Delta\alpha_4 & (3\text{-}132) \end{cases}$$

The four Xe–F stretching motions, two out-of-plane bending motions, and three in-plane bending motions account for all nine vibrational degrees of freedom. The pictures of these modes in Figs. 3-58, 3-59, 3-60, 3-62, and 3-64 agree with those in Fig. 3-36, except that the degenerate vibrations in Fig. 3-36 are drawn as linear combinations of the degenerate vibrations in Figs. 3-60 and 3-64. For example, $\nu_6(e_u)$ in Fig. 3-36 is the sum (eq. 3-123) + (eq. 3-124). The other linear combination, (eq. 3-123) − (eq. 3-124) is not shown.

We have now invested some effort in using eq. 3-120 to generate symmetry coordinates. This is justified because there are other places in this book, notably in the study of molecular orbital theory, where procedures directly analogous to the use of eq. 3-120 will be of great importance. The operator in eq. 3-120, $\sum_R R \cdot \chi_i{}^R$, is a rudimentary form of a *projection*

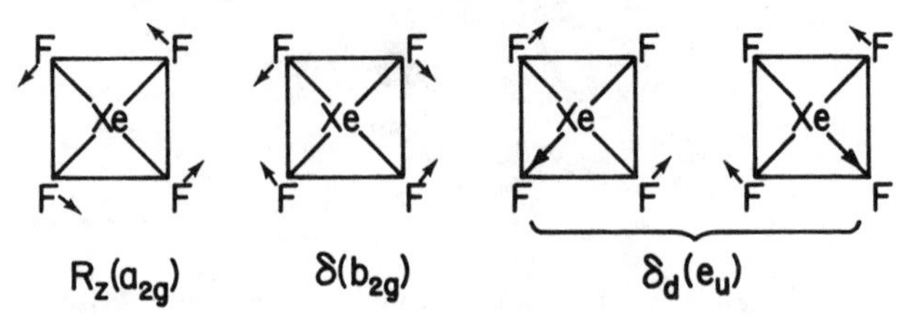

Fig. 3-64. The four in-plane bending motions of XeF_4. Three are true bending modes and one is rotation about the C_4 axis.

operator.† This operator has also been called a "function generating machine" since it generates functions of the desired symmetry starting with a single basis function, such as a stretching or bending motion.

As a further illustration of the use of eq. 3-120, as well as an important example of handling imaginary characters, we will construct the B–Cl stretching coordinates of BCl_3. This molecule belongs to the point group D_{3h}, but we will be most successful in treating it with the character table for the point group C_3. The imaginary characters of C_3 are transformed into real characters as follows, where $\varepsilon = e^{2\pi i/3}$:

$$C_3:\ E\left\{\begin{matrix} 1 & \varepsilon & \varepsilon^* \\ 1 & \varepsilon^* & \varepsilon \end{matrix}\right\} = \left\{\begin{matrix} 1 & -\frac{1}{2}+i\frac{\sqrt{3}}{2} & -\frac{1}{2}-i\frac{\sqrt{3}}{2} \\ 1 & -\frac{1}{2}-i\frac{\sqrt{3}}{2} & -\frac{1}{2}+i\frac{\sqrt{3}}{2} \end{matrix}\right\} \quad \begin{matrix}\text{line a}\\ \text{line b}\end{matrix}$$

$$\begin{array}{lccccl} \text{add (a + b):} & 2 & -1 & -1 & \text{line c} \\ \text{subtract (a − b):} & 0 & i\sqrt{3} & -i\sqrt{3} & \text{line d} \\ \text{divide d by } i\sqrt{3}\text{:} & 0 & 1 & -1 & \end{array}$$

C_3	E	C_3	C_3^2
A	1	1	1
E	$\left\{\begin{matrix}1\\1\end{matrix}\right.$	$\begin{matrix}\varepsilon\\ \varepsilon^*\end{matrix}$	$\left.\begin{matrix}\varepsilon^*\\ \varepsilon\end{matrix}\right\}$

$\Rightarrow$

C_3	E	C_3	C_3^2
A	1	1	1
E	$\left\{\begin{matrix}2\\0\end{matrix}\right.$	$\begin{matrix}-1\\1\end{matrix}$	$\left.\begin{matrix}-1\\-1\end{matrix}\right\}$

The three B–Cl stretching vectors (Fig. 3-65) transform as $a_1' + e'$ (Table 3-26). To construct the totally symmetric mode, $v_s(a_1')$, we use the

† For a derivation and other applications of the projection operator, see A.B. Sannigrahi, *J. Chem. Ed.*, *52*, 307 (1975).

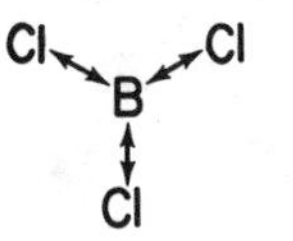

Fig. 3-65. Three B – Cl stretching vectors.

Table 3-26. Stretching Analysis of BCl_3

D_{3h}	E	$2C_3$	$3C_2$	σ_h	$2S_3$	$3\sigma_v$	
Γ_{stretch}	3	0	1	3	0	1	$= a_1' + e'$

characters of the irreducible representation A of the point group C_3. These are all $+1$ so the totally symmetric mode is simply simultaneous stretching of all three B – Cl bonds.

$$\nu_s(a_1') \propto \Delta s_1 + \Delta s_2 + \Delta s_3 \tag{3-133}$$

The construction of the degenerate stretching modes, $\nu_d(e')$, using the real characters of the modified C_3 character table, is illustrated in Fig. 3-66.

$$\nu_d(e') \propto \begin{cases} 2\Delta s_1 - \Delta s_2 - \Delta s_3 \\ \qquad\quad \Delta s_2 - \Delta s_3 \end{cases} \tag{3-134}$$

You should realize that the coefficients in eq. 3-134 are just the real forms of the characters of the E irreducible representation in the C_3 character table.

So far we have been simplifying our work by using the pure rotational groups as guides for constructing symmetry coordinates. In Fig. 3-67 we consider an example in which we use the full D_{2h} character table to construct the b_{3u} C – H vibrational symmetry coordinate of ethylene. You can see that the first four operations, E, $C_2(z)$, $C_2(y)$, and $C_2(x)$, are sufficient to generate the symmetry coordinate and that the next four operations just duplicate the effects of the first four. This means that we could just as well have used the D_2 point group and could have cut our work in half. The symmetry coordinate constructed in Fig. 3-67 would be written

$$\nu(b_{3u}) \propto \Delta s_1 + \Delta s_2 - \Delta s_3 - \Delta s_4 \tag{3-135}$$

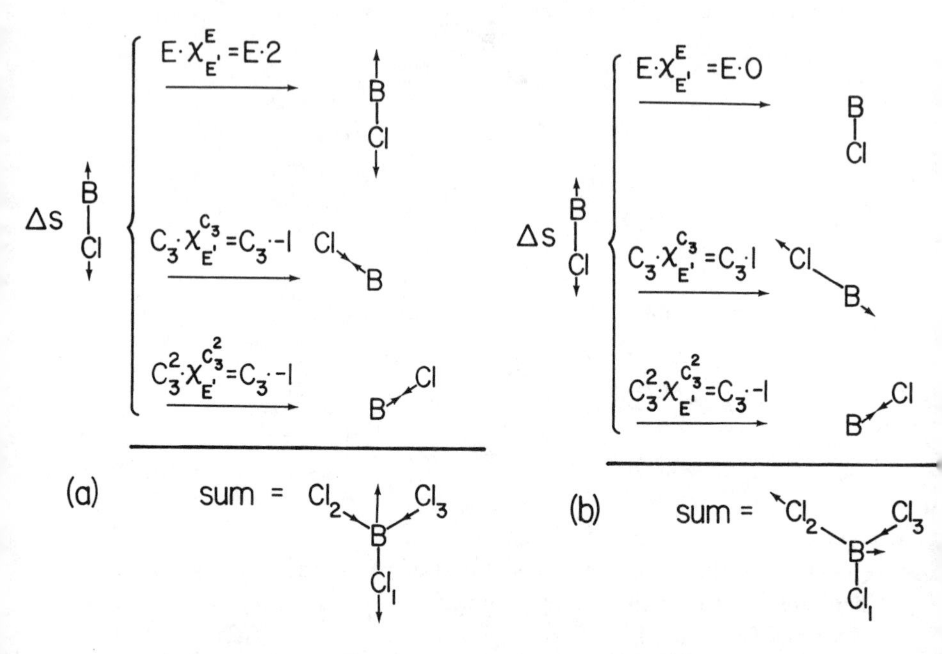

Fig. 3-66. Generation of the two degenerate B—Cl stretching modes of BCl_3 using the two different sets of real characters for the E representation.

The fact that the operations performed in Fig. 3-67 add up to $2\Delta s_1 + 2\Delta s_2 - 2\Delta s_3 - 2\Delta s_4$ is of no consequence because we are free to divide through by 2 and reduce the symmetry coordinate to its simplest form. For quantitative work, the simplest form of the symmetry coordinate would then be multiplied by a normalization coefficient.

Problems

3-31. PH_3 (pyramidal, C_{3v} symmetry) has ir and Raman active vibrations at 2421, 2327, 1121, and 991 cm^{-1}. PD_3 exhibits its fundamentals at 1698, 1694 806, and 730 cm^{-1}. Suggest an assignment of these frequencies and draw a picture of each normal mode.

3-32. Construct symmetry coordinates for the bending modes of BCl_3.

3-33. Construct symmetry coordinates for all of the stretching modes of a hypothetical planar ML_5 molecule of symmetry D_{5h}.

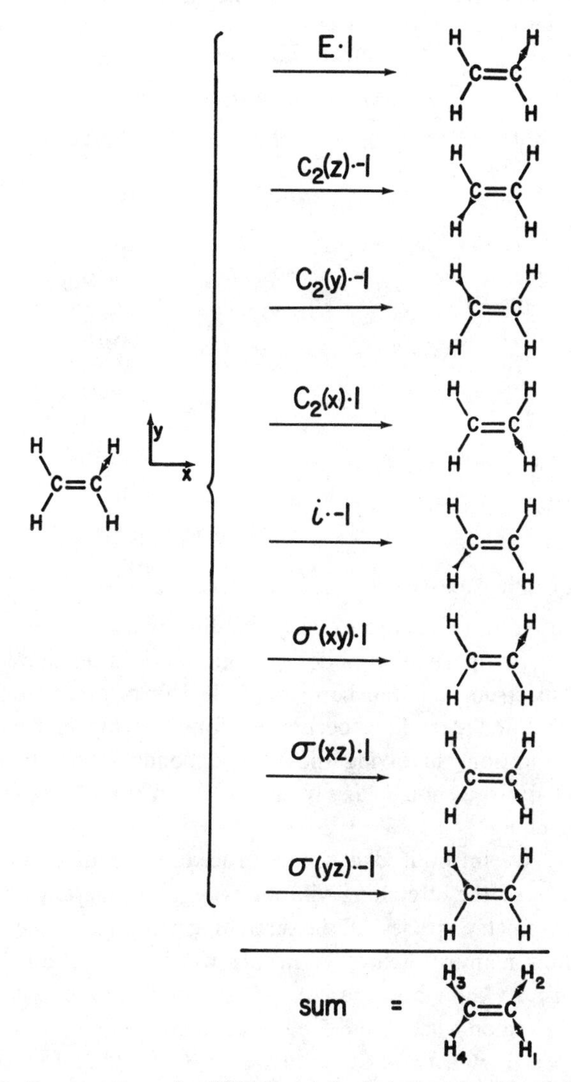

Fig. 3-67. Construction of $\nu(\mathrm{CH})(b_{3u})$ of ethylene using the full D_{2h} character table. In this figure ↔ denotes bond stretching and ↣ denotes bond compression.

3-34. The fundamental frequencies of ethylene and ethylene-d_4 are listed below. Draw a picture of each normal mode using the diborane vibrations (Fig. 3-40) and the D_{2h} character table as guides. Use the coordinate system of Fig. 3-67.

C_2H_4	C_2D_4	number	symmetry	description	activity
3108	2304	ν_5	b_{1g}	ν(CH)	R
3106	2345	ν_9	b_{2u}	ν(CH)	ir
3019	2251	ν_1	a_g	ν(CH)	R
2990	2200	ν_{11}	b_{3u}	ν(CH)	ir
1623	1515	ν_2	a_g	ν(CC)	R
1444	1078	ν_{12}	b_{3u}	$\delta_{as}(CH_2)$	ir
1342	981	ν_3	a_g	$\delta_s(CH_2)$	R
1236	1009	ν_6	b_{1g}	$\rho_r(CH_2)$	R
1007	726	ν_4	a_u	$\rho_t(CH_2)$	inactive
949	721	ν_8	b_{2g}	$\rho_w(CH_2)$	R
943	780	ν_7	b_{1u}	$\rho_w(CH_2)$	ir
810	586	ν_{10}	b_{2u}	$\rho_r(CH_2)$	ir

3-8. Stretching Mode Analysis

A frequent use of vibrational spectroscopy is to distinguish possible isomers of a compound. Often particular modes of vibration occur in characteristic regions of the spectrum and can be distinguished from other transitions. Modes in which bond stretching occurs are found at higher frequencies than bending vibrations involving those same bonds. The difference in frequency is often large enough to say clearly which transitions represent nearly pure stretching.

In Section 3-7 we introduced a simple procedure for determining the symmetry species of the stretching vibrations (Γ_{stretch}) for any molecule. Knowing the symmetry species of the stretching vibrations, one can tell by inspection how many stretching vibrations will be ir or Raman active. Suppose you wish to know how many C–H stretching modes will be seen for methane. The region where these occur is 2800–3100 cm^{-1}. To represent the stretching, we draw a two-headed arrow in the middle of each C–H bond (Fig. 3-68). Table 3-27 shows how these four arrows transform under the operations of T_d. We find only two distinct vibrational energy levels, one of which is threefold degenerate. The T_d character table tells us that both possible transitions are Raman active, but only the t_2 transition is ir active. Therefore, we expect only one band in the ir spectrum in the C–H stretching region and two bands in the Raman spectrum.

Carbonyl stretching in transition metal complexes typically occurs in the region 1700–2100 cm^{-1}. How many bands would one expect for the

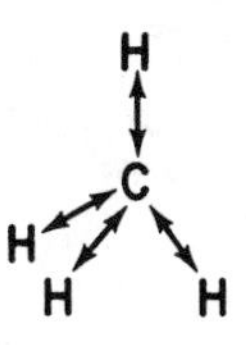

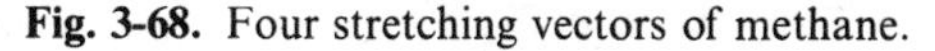

Fig. 3-68. Four stretching vectors of methane.

Table 3-27. Stretching Analysis of CH_4

T_d	E	$8C_3$	$3C_2$	$6S_4$	$6\sigma_d$	
$\Gamma_{stretch}$	4	1	0	0	2	$= a_1 + t_2$

complex $Mn(CO)_5I$ (Fig. 3-69[a]) in this region of the spectrum? The analysis in Table 3-28 indicates that there should be three ir bands (Fig. 3-69[b]) and four Raman bands.

What if you replace the iodine atom by a more complex ligand such as a phosphine? The true symmetry of the molecule may be C_1, or C_s at best, but the carbonyl region of the spectrum does not suddenly exhibit five bands. The effective symmetry of the complex is still close to C_{4v} and most likely the pattern of the bands will not be very different from the true C_{4v} case.

Fig. 3-69. (a) Five C–O stretching vectors of $Mn(CO)_5I$. (b) Carbonyl stretching region infrared spectrum of $Mn(CO)_5I$ in CCl_4. From M.A. El-Sayed and H.D. Kaesz, *J. Mol. Spec.*, *9*, 310 (1962).

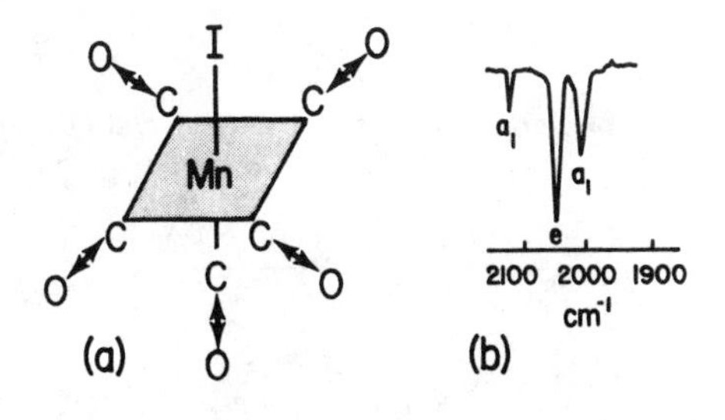

Table 3-28. Stretching Analysis of $Mn(CO)_5I$

C_{4v}	E	$2C_4$	C_2	$2\sigma_v$	$2\sigma_d$	
$\Gamma_{stretch}$	5	1	1	3	1	$= 2a_1 + b_1 + e$

ir: $2a_1 + e$

Raman: $2a_1 + b_1 + e$

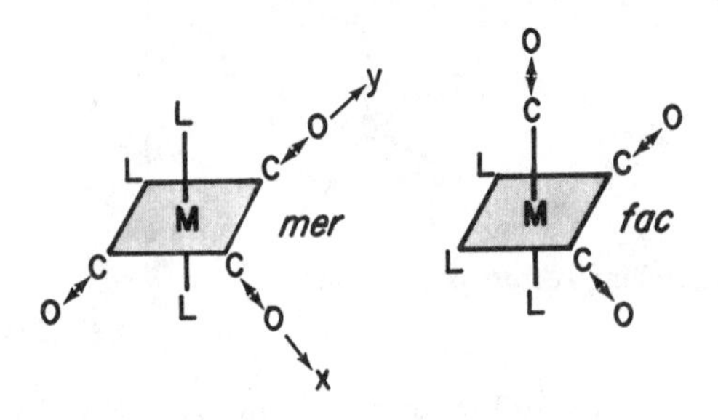

Fig. 3-70. *Mer* and *fac* isomers of an octahedral $ML_3(CO)_3$ complex showing carbonyl stretching vectors. The prefix *mer* stands for meridional since similar ligands are arranged on a meridian of the octahedron. The prefix *fac* stands for facial since similar ligands are situated on the same face of the octahedron.

What then, would you expect for the *mer* and *fac* isomers of an $ML_3(CO)_3$ complex (Fig. 3-70)? Table 3-29 indicates that we can expect two bands for the *fac* isomer and three bands for the *mer* isomer in both the ir and Raman spectra. Thus, the compound $Mo(CO)_3[P(OCH_3)_3]_3$, whose ir spectrum exhibits bands at 1993, 1919, and 1890 cm^{-1}, is assigned the *mer* structure; and the complex $Cr(CO)_3(CNCH_3)_3$, with bands at 1942 and 1860 cm^{-1}, is assigned the *fac* structure. You might also wish to use the metal-carbon stretching vibrations to tell the same thing since the $M-C$ stretching vectors have the same symmetries as the $C-O$ stretching vectors. Unfortunately, the energies of the $M-C$ stretching vibrations are low enough so that these modes are extensively mixed with the $M-C-O$ bending modes.

Table 3-29. Carbonyl Stretching Analysis of the *mer* and *fac* Isomers of $ML_3(CO)_3$ Complexes

	C_{2v}	E	C_2	$\sigma(xz)$	$\sigma(yz)$	
mer:	$\Gamma_{stretch}$	3	1	1	3	$= 2a_1 + b_2$

ir: $2a_1 + b_2$

Raman: $2a_1 + b_2$

	C_{3v}	E	$2C_3$	$3\sigma_v$	
fac:	$\Gamma_{stretch}$	3	0	1	$= a_1 + e$

ir: $a_1 + e$

Raman: $a_1 + e$

Problems

3-35. Two important idealized geometries for seven coordination are the mono-capped trigonal prism (a) and the pentagonal bipyramid (b). The seven-coordinate complex $Mo(CN)_7{}^{4-}$ has been examined both as solid $K_4Mo(CN)_7 \cdot 2H_2O$ and in aqueous solution. In the C–N stretching region the ir spectrum has bands at 2119, 2115, 2090, 2080, 2074, and 2059 cm^{-1} for the solid and at 2080 and 2040 cm^{-1} for solutions. How many Raman and infrared bands would you expect for each geometry above and how many coincidences (bands present in both the ir and Raman spectra) should there be for each geometry? How do you interpret these data?

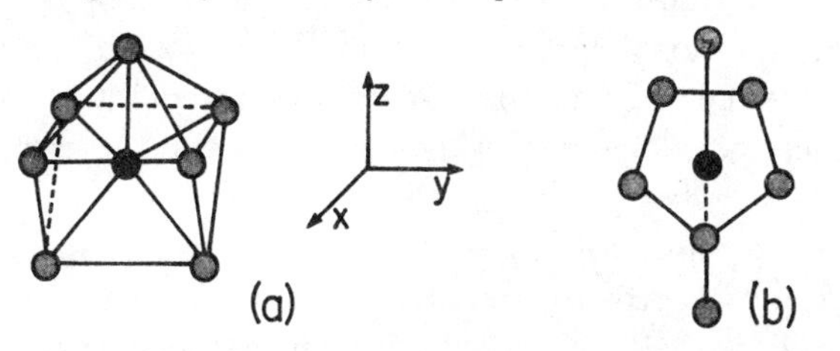

3-36. (a) Determine the number and activities of the carbonyl stretching modes of both the *cis* and *trans* isomers of $L_2M(CO)_4$. (b) $Fe(CO)_4Cl_2$ has ir bands at 2167, 2126, and 2082 cm^{-1} in $CHCl_3$ solution. How would you interpret this spectrum? (c) Draw a picture of each carbonyl stretching symmetry coordinate of each isomer of $L_2M(CO)_4$. As an example of what we are looking for, the modes of *mer*-$L_3M(CO)_3$ are drawn below. The symbol ↔ stands for bond stretching and the symbol ↣ denotes bond compression. Note that the stretching of a unique chemical bond, as in the center diagram below, is always a legitimate basis for a symmetry coordinate.

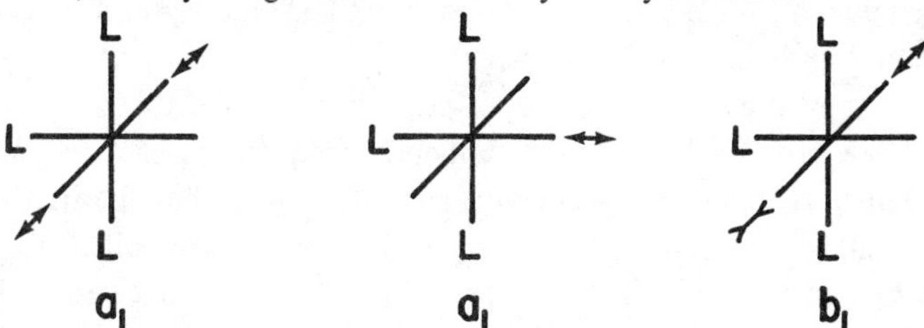

3-37. Show that all of the C_nH_n hydrocarbons below have just a single ir active C–H stretching mode.

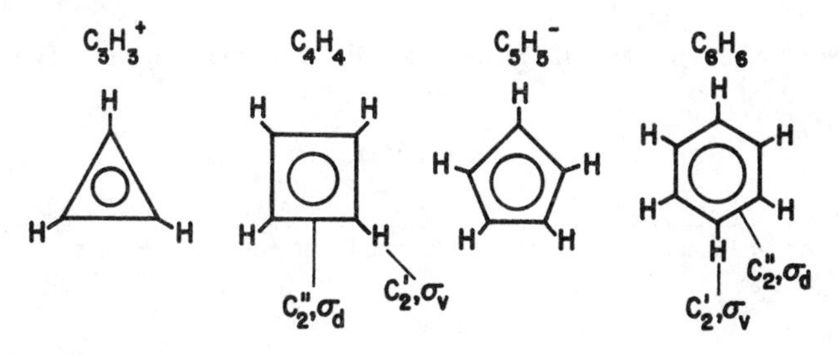

3-9. Assignment of Real Spectra

Lest the reader be led to believe that spectra are very much more straightforward than they actually are, we feel obliged to present some typical vibrational analyses. Most spectra do not exhibit as many strong bands as one would predict based on the number of allowed fundamentals. "Fine," you respond, "let's count the weaker bands." The trouble now is that if you count the weaker bands you have many more bands than you expect for fundamentals. The reason is that the weaker bands include combination bands and overtones as well. One cannot generalize on the method of assignment of vibrational spectra. Each molecule is a unique case which requires resourcefulness on the part of the spectroscopist. In this section, we will examine the assignment of the spectra of two rather simple compounds.

XeF_4. Two possible geometries would be square planar (D_{4h}) and tetrahedral (T_d). The predicted activities of the fundamentals in each case are shown in Table 3-30. It should certainly be possible to distinguish these two geometries spectroscopically since the predictions are different. For example, in the Xe−F stretching region (500–700 cm^{-1}) we expect one allowed ir band and two allowed Raman bands for both D_{4h} and T_d geometry. However, if the molecule is tetrahedral the ir absorption will coincide with one of the Raman bands. In planar geometry this is forbidden by the mutual exclusion rule (Section 3-6-C). That is, there will be no ir-Raman coincidences. In the Xe−F bending region (< 500 cm^{-1}) we expect two ir and one Raman band in D_{4h} symmetry and one ir and two Raman bands in T_d symmetry.

We now reproduce an entire paper from the Journal of the American Chemical Society in which the fundamental frequencies of XeF_4 are assigned. Before reading this paper you should be told that the irreducible representations which we call b_{1g}, b_{2g}, b_{1u}, and b_{2u} are called b_{2g}, b_{1g}, b_{2u}, and b_{1u}, respectively, in this paper. That is, the 1 and 2 subscripts on the b representations have been reversed. (This happens if you reverse the conventions we have adopted on the definitions of C_2', C_2'', σ_v and σ_d.)

Table 3-30. Predictions About the Spectra of XeF_4 with Square Planar or Tetrahedral Geometry

	Number of Active Fundamentals								
	All Vibrations			Xe–F Stretching			Xe–F Bending		
	ir	Raman	Coincidences	ir	Raman	Coincidences	ir	Raman	Coincidences
D_{4h} Structure	3	3	0	1	2	0	2	1	0
T_d Structure	2	4	2	1	2	1	1	2	1

(Reproduced from *J. Am. Chem. Soc.*, *85*, 1927 [1963].)

Vibrational Spectra and Structure of Xenon Tetrafluoride[1]

By Howard H. Claassen,[2] Cedric L. Chernick, and John G. Malm

(Received March 18, 1963)

The infrared spectrum of XeF_4 vapor has strong bands at 123, 291, and 586 cm^{-1}. The Raman spectrum of the solid has very intense peaks at 502 and 543 cm^{-1} and weaker ones at 235 and 442 cm^{-1}. These data show that the molecule is planar and of symmetry D_{4h}. The seven fundamental frequencies have been assigned as 543 (a_{1g}), 291 (a_{2u}), 235(b_{1g}), 221 (b_{1u}), 502(b_{2g}), 586(e_u), and 123 (e_u). The (b_{1u}) frequency value is quite uncertain.

Introduction

The preparation of XeF_4 has been described previously[3] and the results of a preliminary study of its vibrational spectra reported briefly.[4] We report here the results of a more complete study of the Raman spectrum of the solid phase and the infrared spectrum of the vapor.

Experimental Procedures

Preparation The purity of the XeF_4, prepared as described elsewhere,[4] was checked by infrared analysis. The probable impurities are XeF_6 and XeF_2 which have absorption peaks at 612 and 566 cm^{-1}, respectively. The sample was found to contain small amounts of the more volatile XeF_6 but this was easily removed since its vapor pressure is higher by a factor of 10. Pumping the equilibrium vapor rapidly out of the storage can several times removed the XeF_6 so that none of the 612 cm^{-1} absorption could be detected in the bulk of sample remaining.

Infrared Spectra The vapor pressure of XeF_4 (approximately 2 mm at 20°) was sufficient to allow the observation of the fundamentals at or slightly above room temperature in a 10-cm cell. For weaker bands a 60-cm absorbing path was obtained by use of the mirror cell designed at this Laboratory and previously described.[5] The cells were made of nickel and were used with either AgCl or polyethylene windows. The spectra were obtained with a Beckman IR-7 with CsI prism and Perkin-Elmer 421 and 301 spectrophotometers. We are indebted to the Perkin-Elmer Corporation for the opportunity to use the 301 instrument at Norwalk, Conn., and to Charles Helms and Robert Anacreon for their help with the operation of that spectrophotometer.

The reproducibility of the spectrum and uniform composition of the sample were established by scanning several samples. A nickel can containing about 1 g

[1] Based on work performed under the auspices of the U. S. Atomic Energy Commission.

[2] Permanent address: Wheaton College, Wheaton, Ill.

[3] H.H. Claassen, H. Selig and J.G. Malm, *J. Am. Chem. Soc.*, **84,** 3593 (1962).

[4] C.L. Chernick, *el al.*, *Science*, **138,** 136 (1962).

[5] B. Weinstock, H.H. Claassen and C.L. Chernick, *J, Chem. Phys.*, **38,** 1470 (1963).

of XeF_4 was connected to the cell and to a similar can. The whole sample was transferred batchwise to the second can and vapor samples of each batch were taken into the cell. In most cases just the two most intense bands were examined to look for possible changes which would be indicative of impurities, but several complete spectra were also observed.

Raman Spectra The sample used was approximately 1 g of XeF_4 that had grown to a single crystal in a sealed quartz tube. The spectrum was obtained using a Cary 81 photoelectric instrument with the lens system designed for solids.

Results and Interpretation

Figure 1 shows tracings of the regions of the infrared spectrum where bands were observed, and Fig. 2 is a tracing of the Raman spectrum. Judging from their positions and intensities the three infrared bands at 123, 291, and 586 cm^{-1} are probably fundamentals. Of the four bands observed in the Raman spectrum the one at 442 cm^{-1} is the least intense and may not represent a fundamental. In fact, the reality of the 442 cm^{-1} frequency is doubtful since the 543 cm^{-1} vibration excited by 4339 Å and the 502 cm^{-1} one excited by 4337 Å would occur at apparent shifts of 442 and 445 cm^{-1}, respectively, from 4358 Å.

In considering the information the spectral data furnish on the molecular symmetry, it must be noted that the Raman measurements are for the solid compound and the infrared ones are for the vapor. Some solid-vapor shifts in frequencies are to be expected.

From the infrared spectrum alone one can conclude that there is high symmetry in the XeF_4 molecule. Only one band is observed in the region where bond stretching motions occur (500–700 cm^{-1}). Of all the symmetries possible for a YZ_4-molecule, only for T_d(tetrahedral) and D_{4h} (square-planar) would there be just one infrared-active bond stretching fundamental. The infrared spectrum also allows the distinction to be made between these two symmetries since a T_d molecule would have one bending mode that would be infrared active while a D_{4h} molecule would have two. As two are observed for XeF_4 the D_{4h} model is the preferred one. Strong support for this is provided in the Raman spectrum, also.

The fundamental vibrations of a D_{4h}, YZ_4, molecule are described in Fig. 3 as to their symmetries, numbering, spectral activity and modes of atomic motions. The assignment of ν_2 is definite from the band contours expected according to Gerhard and Dennison.[6] Only for the out-of-plane motion, ν_2, should there be a very intense Q-branch and this is observed at 291 cm^{-1}. The other two infrared fundamentals are then assigned without ambiguity.

The Raman spectrum of the solid fits very well and lends strong support for the planar model. The two very intense bands at 543 and 502 cm^{-1} must be due to the two stretching vibrations. Although polarization measurements could not be made it is quite certain that the symmetric vibration is the higher one because any significant repulsion between fluorines would almost require this. Further support for this interpretation of the Raman spectrum of the solid has recently been obtained

[6] S.L. Gerhard and D.M. Dennison, *Phys. Rev.*, **43**, 197 (1933).

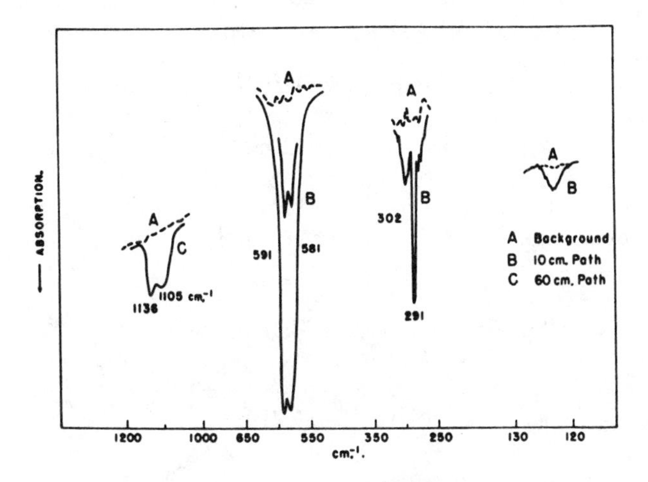

Fig. 1. Infrared spectrum of XeF_4 vapor.

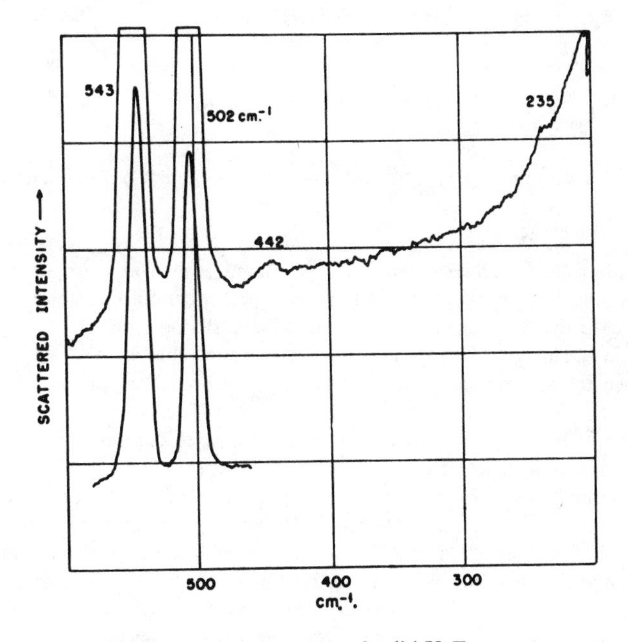

Fig. 2. Raman spectrum of solid XeF_4.

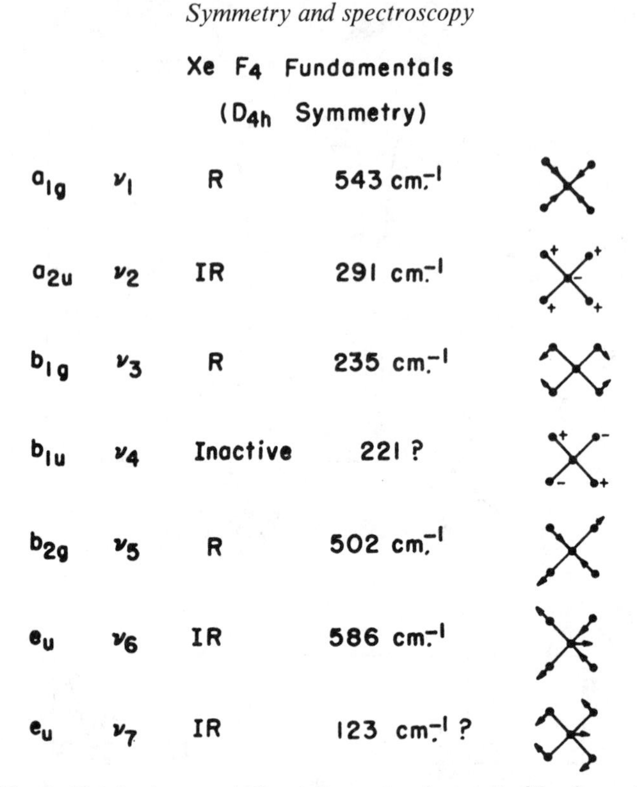

Fig. 3. Numbering, spectral activity, and assignment of fundamentals.

in this Laboratory.[7] The 235 cm^{-1} Raman band is then assigned to ν_3 and this leaves 442 cm^{-1} to be assigned. If the band is real it cannot be a fundamental and it must be an overtone or combination band and the only plausible assignment is that it is $2\nu_4$. This gives the value of 221 cm^{-1} for ν_4 that is listed with a question mark since the assignment is not certain. The infrared absorption peaks at 1105 and 1136 cm^{-1} may be assigned as $\nu_5 + \nu_6 = 1088$ cm^{-1} and $\nu_1 + \nu_6 = 1129$ cm^{-1}. The fit is satisfactory when account is taken of corrections needed due to vapor to solid shift of frequencies.

There is one feature of the infrared spectrum that we do not understand and that is the doublet appearance of ν_6. This has been traced many times and the peaks reproducibly found at 581 and 591 cm^{-1}. Expected is a triplet band with all three peaks of about equal intensity and with a P-R separation of approximately 14

[7] H.H. Hyman and L.A. Quarterman observed a Raman band at 553 cm^{-1} of XeF_4 in HF solution. This must correspond to the 543 cm^{-1} band for the solid. The ν_5 band was so broadened, however, that it was not definitely observed in the very dilute solution. That the higher frequency band remained sharp is good indication that it represents the totally symmetric vibration.

cm^{-1}.[6] The observed splitting is much too large to ascribe to isotopes of xenon, and may be due to a Coriolis coupling between the doubly degenerate vibration and rotation.

The structure of XeF_4 has been obtained by X-ray diffraction by Ibers and Hamilton.[8] They find that for the solid, also, the molecule is square planar within experimental error.

Several theoretical discussions[9–11] have stated that the square planar model best fits the theory and one of them[9] suggests that the molecule could possibly be distorted by coulomb repulsion. Therefore, it seems interesting to question whether the vibrational data require an exactly planar molecule or whether the "ring" of fluorines might be slightly puckered. If the latter were true the Raman-active ν_5 would be infrared active, but a slight distortion would, of course, result in a very weak infrared band. One can set a rough upper limit to the amount of possible puckering if one looks at the infrared spectrum in the region of 502 cm^{-1} and makes the plausible assumption that the rate of change of bond moment with stretching is approximately the same for ν_5 and ν_6. The result is that an upper limit can be set for deviation of the Xe–F bond from the plane of about 0.5 degree, or fluorine distances of 0.02 Å from the plane.

The Q–R separation of 11 ± 1 cm^{-1} in the 291 cm^{-1} band can be used to calculate a bond length. This gives 1.85 ± 0.2 Å for the Xe–F bond, in good agreement with the value of 1.92 Å for the solid obtained from X-ray diffraction.[8] Since a more precise value of the bond length for the vapor molecule will probably be available soon from electron diffraction studies, we have not calculated thermodynamic functions.

Preliminary force constant calculations using a valence plus interaction terms type of potential function similar to that used by Claassen for hexafluorides[12] gave a value of 3.00 mdynes/Å for the bond stretching constant and 0.12 for the interaction constant between bonds at right angles. The interaction constant for opposite bonds cannot be determined accurately, but is approximately 0.06 mdyne/Å. These may be compared with values given by Smith[13] for XeF_2 and with those for PuF_6,[12] a molecule that also has fluorine bonds at right angles and a comparable bond length.

Molecule	XeF_4	XeF_2	PuF_6
Bond length, Å	1.92[8]	2.00[14]	1.972
Stretching force constant, mdynes/Å	3.00	2.85	3.59
Interaction constant for perpendicular bonds	0.12	—	0.22
Interaction constant for opposite bonds	~0.06	0.11	−0.08

[8] J.A. Ibers and W.C. Hamilton, *Science*, **139,** 106 (1963).
[9] R.E. Rundle, *J. Am. Chem. Soc.*, **85,** 112 (1963).
[10] L.L. Lohr and W.N. Lipscomb, *ibid.*, **85,** 240 (1963).
[11] L.C. Allen, *Science*, **138,** 892 (1962).
[12] H.H. Claassen, *J, Chem. Phys.*, **30,** 968 (1959).
[13] D.F. Smith, *ibid.*, **38,** 276 (1963).
[14] P.A. Agron, G.M. Begun, H.A. Levy, A.A. Mason, C.F. Jones and D.F. Smith, *Science*, **139,** 842 (1963).

Since the publication of the paper which was just reproduced, the same research group has succeeded in recording the laser Raman spectrum of gaseous XeF_4 (as opposed to the solid state spectrum in the 1963 paper). The following passage, which serves to update the XeF_4 results, is reproduced from P. Tsao, C.C. Cobb, and H.A. Claassen, *J. Chem. Phys.*, *54*, 5247 (1971).

"Figure 2 is the Raman spectrum of XeF_4 in the gaseous state. The three strongest bands at 554.3, 524, and 218 cm^{-1} are shifted by 11, 22, and -17 cm^{-1}, respectively, from corresponding bands for solid XeF_4. In addition to these fundamental frequencies, two very weak bands are observed at 322 and 433 cm^{-1}. The polarized band at 554.3 cm^{-1} is overlapped by the depolarized band centered at 524 cm^{-1}, hence its depolarization factor cannot be measured accurately. It is very probably below 0.05.

"Although the gas phase frequencies are appreciably shifted from those observed earlier for condensed phases, as noted above, the present results suggest no changes in assignments from the earlier work. The fundamental frequencies for both XeF_4 and $XeOF_4$, from present and earlier work, are listed in Table 1, together with the symmetry designations and spectral activity. For XeF_4 two of the seven fundamentals are not observed directly, but they can be estimated rather closely. The out-of-plane puckering motion, ν_4, is inactive. The force constant for this motion must be very close to that for the other out-of-plane fundamental, ν_2, observed at 291 cm^{-1}. If one assumes the same force constant, the calculated value for ν_4 is 232 cm^{-1}. In the octahedral hexafluorides, ν_6 is an exactly identical motion, and its overtone usually shows up as a surprisingly intense band. Thus, it seems very probable that our observed Raman band at 433 is the overtone of the inactive ν_4, and therefore we list 216 cm^{-1} in parentheses for ν_4. The other fundamental that has not been observed is the infrared-active ν_7. The value of 123 cm^{-1} listed for it (in the previous work) with a question mark is not correct, but was due to some HF impurity. The band must be very weak, for we have not been able to observe it in several attempts. Its approximate value can be obtained from the essentially identical motion in $XeOF_4$, there labelled ν_9, and observed at 161 cm^{-1} in the Raman effect. This means that the Raman band of XeF_4 at 322 cm^{-1} must be the overtone of ν_7, and we list 161 cm^{-1} as the fundamental frequency. It is rather remarkable that the only two overtones observed in the Raman effect happen to be those for the only two fundamentals not directly observed."

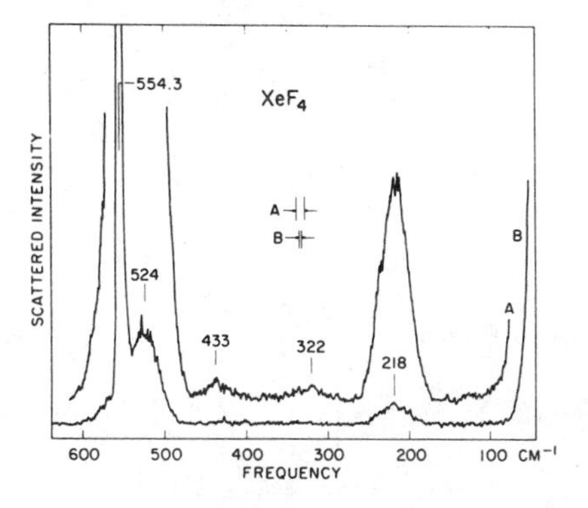

Fig. 2 From P. Tsao, C.C. Cobb, and H.A. Claassen, *J. Chem. Phys.*, *54*, 5247 (1971). Raman spectrum of gaseous XeF_4 at 0.9 atm and 105°C. Excitation line at 6471 Å. Spectral slitwidths: A — 10 cm^{-1} and B — 3 cm^{-1}.

ClF_3. The ir and Raman spectra of this compound are shown in Fig. 3-71. Since ClF_3 is known to have C_{2v} symmetry (roughly a T-shaped structure with a fluorine at the end of each line of the T and chlorine at the vertex), we predict that all six fundamentals will be both ir and Raman active. The Raman spectrum shows only two strong bands and the ir spectrum exhibits a multitude of bands. It should be pointed out that the Raman spectrum in Fig. 3-71 predates laser Raman spectroscopy and a modern instrument would probably reveal more bands. But the ir spectrum recorded today would probably be quite similar to that shown in Fig.

Table I. Fundamental Frequencies for XeF_4 and $XeOF_4$[a]

XeF_4			$XeOF_4$			
Frequency (cm^{-1})	Designation	Spectrum observed	Frequency (cm^{-1})	Designation	Spectrum observed	Description
			926.3	$\nu_1(a_1)$	R, ir	Xe–O stretch
554.3	$\nu_1(a_{1g})$	R	576.9	$\nu_2(a_1)$	R, ir	XeF_4 in-phase stretch
291	$\nu_2(a_{2u})$	ir	285.9	$\nu_3(a_1)$	R, ir	XeF_4 out-of-plane bend
218	$\nu_3(b_{1g})$	R	225	$\nu_4(b_1)$	R	XeF_4 in-plane scissor
524	$\nu_5(b_{2g})$	R	543	$\nu_5(b_2)$	R	XeF_4 out-of-phase stretch
(216)	$\nu_4(b_{1u})$	inactive	(219)	$\nu_6(b_2)$	none observed	F_4 out-of-plane pucker
586	$\nu_6(e_u)$	ir	609	$\nu_7(e)$	R, ir	F_4 degenerate stretch
			362	$\nu_8(e)$	R, ir	Xe–O bend
(161)	$\nu_7(e_u)$	none observed	161	$\nu_9(e)$	R, ir	F_4 degenerate bend

[a] From P. Tsao, C.C. Cobb, and H.A. Claassen, *J. Chem. Phys.* *54*, 5247 (1971).

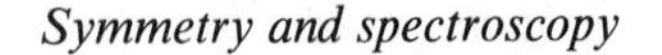

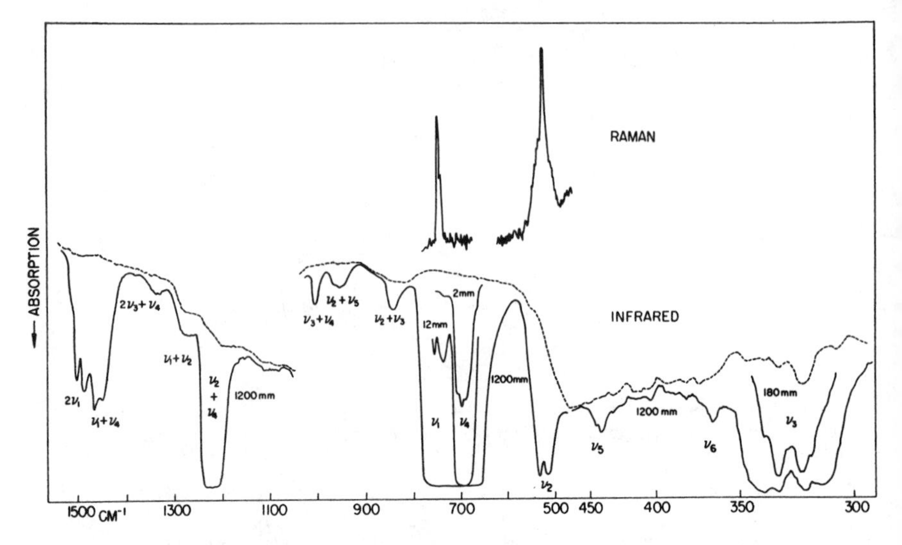

Fig. 3-71. Infrared and Raman spectra of gaseous ClF_3. Reproduced from H.H. Claassen, B. Weinstock, and J.G. Malm, *J. Chem. Phys.*, *28*, 285 (1958).

3-71. Notice that some of the bands assigned as overtones and combination bands are more intense than some of the bands assigned as fundamentals. The assignment of vibrational spectra is generally done to give the greatest overall consistency of the observed frequencies. Figure 3-71 should dispel any illusions you may have had about the simplicity of vibrational spectroscopy.

Problems

3-38. Derive the results in Table 3-30.

3-10. The Resonance Raman Effect

If a compound is irradiated at a frequency within an electronic absorption band, the Raman emission may be greatly enhanced. This effect, known as the *resonance Raman effect* is illustrated in Fig. 3-72 in which we see the spectrum of $Cs^+Cl_2^-$ produced by codeposition of Cl_2 mixed with Ar (Ar/Cl = 100) and Cs vapor on a surface cooled to 15 K. The fundamental appears at 259 cm^{-1}, and the first seven overtones are clearly seen. Each

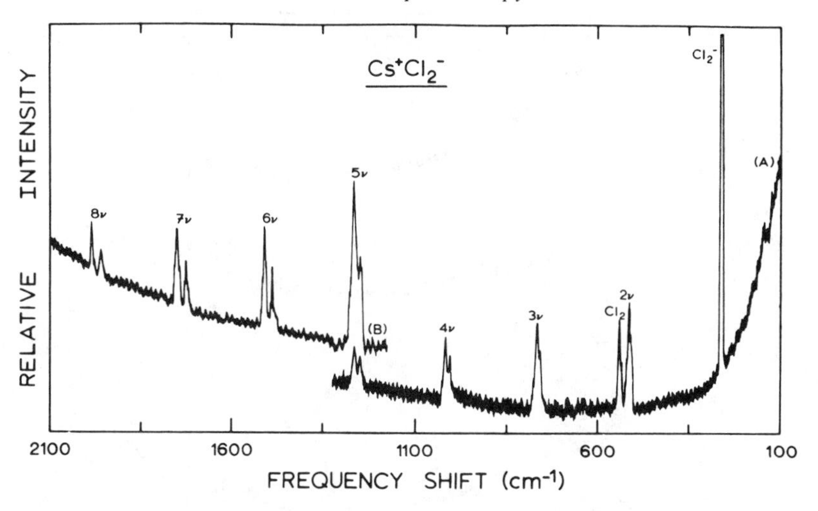

Fig. 3-72. Resonance Raman spectrum of the product of chlorine molecule-cesium atom matrix reaction. Laser exciting line is at 457.9 nm. Receiver gain is increased in trace B. From W.F. Howard, Jr. and L. Andrews, *Inorg. Chem.*, *14*, 767 (1975).

band consists of a doublet due to $^{35}Cl-^{35}Cl$ and $^{35}Cl-^{37}Cl$. Also seen in this spectrum is some unreacted Cl_2 at 530 cm^{-1}. As the laser wavelength is increased (Fig. 3-73), fewer and fewer overtones are seen and the intensity of the fundamental decreases as the laser frequency becomes more and more removed from the Cl_2^- visible absorption band.

The resonance Raman effect is generally observed when the frequency of the incident light is such that the molar extinction coefficient of the sample is $\gtrsim 10^3$. A lesser effect ("pre-resonance Raman") is observed when $\epsilon \approx 10^0 - 10^2$, and ordinary Raman spectra are observed if $\epsilon \lesssim 10^0$. Since a sample in a resonance Raman experiment is actually absorbing the intense laser light, it can be quickly burned to a crisp if no precautions are taken. To counter this problem, the sample is spun at about 3000 revolutions per minute and the laser is aimed at some point other than the axis of rotation. This way no part of the sample is seriously overheated. For the ordinary Raman spectrum it was stated that $\rho = I_\perp/I_\parallel \leqslant 3/4$ for solution spectra (Section 3-6-D). In resonance Raman spectra this is no longer always so and *anomalous polarization* ($\rho > 3/4$) may be observed. In fact, complete *inverse polarization* ($\rho \approx \infty$) is possible for some nontotally symmetric vibrations.

The resonance Raman effect is of potential value in biology for the study of colored metalloproteins and other colored macromolecules. Since the

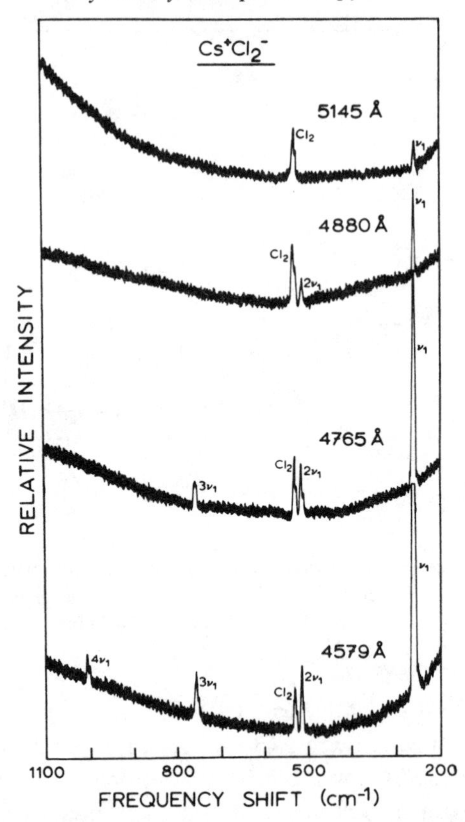

Fig. 3-73. Wavelength dependence of laser excitation on the resonance Raman spectrum of $Cs^+Cl_2^-$. All spectra were recorded at the same receiver gain but the laser power was 150 mW at 514.5 and 488.0 nm, 100 mW at 476.5 nm, and 75 mW at 457.9 nm. Reproduced from W.F. Howard, Jr. and L. Andrews, *Inorg. Chem.*, *14*, 767 (1975).

emission intensity is greatly enhanced, the available low concentrations of macromolecules do not prohibit observing a spectrum. Further, the resonance Raman spectrum is generally just associated with the region of the molecule responsible for the electronic absorption band. For this reason the immediate environment of colored transition metal ions may be especially amenable to study by resonance Raman spectroscopy.† When uv lasers become available the potential of resonance Raman spectroscopy as a probe of colorless compounds, as well, will be enormous.

† For a review of this subject, see T.G. Spiro, *Acc. Chem. Res.*, 7, 339 (1974).

Problems

3-39. The observed frequencies of the first eight vibrational transitions of $Cs^+Cl_2^-$ in Fig. 3-72 are listed below. Show that the anharmonic oscillator model leads to an expression of the form $\bar{\nu}_v/v = (\bar{\omega}_e - \bar{\omega}_e x_e) - \bar{\omega}_e x_e v$, where $\bar{\nu}_v$ is the wave number of the vth transition. From a plot of $\bar{\nu}_v/v$ *vs.* v, find the values of $\bar{\omega}_e$ and $\bar{\omega}_e x_e$ for $^{35}Cl_2^-$ and $^{35}Cl^{37}Cl^-$. Why don't we observe the spectrum of $^{37}Cl_2^-$ in Fig. 3-72?

$v = 0 \rightarrow v =$	$Cs^{+\,35}Cl_2^-$	$Cs^{+\,35}Cl^{37}Cl^-$
1	259.0	——
2	515.0	508.4
3	767.8	758.4
4	1017.8	1005.7
5	1264.2	1249.3
6	1508.1	1489.1
7	1749.3	1726.2
8	1984.5	1960.0

3-11. Functional Group Analysis

Probably the most routine use of vibrational spectroscopy is to ascertain which functional groups are present in a compound. If you had no idea of the composition of the compound, this would probably be impossible because more than one functional group absorbs in most regions of the spectrum. But with a knowledge of the compound's origin, its elemental composition, its mass spectrum, and its other spectral properties (nuclear magnetic resonance, uv, *etc.*), vibrational spectra can be very useful. We have already alluded to some inorganic functional groups. For example, C–O stretching of terminal metal carbonyl groups comes in the region 1800–2100 cm^{-1}, and Xe–F stretching frequencies come in the range 500–700 cm^{-1}. In this section we will examine the infrared spectra of some small organic compounds—and a few more exotic samples—with special reference to characteristic frequencies of the functional groups present.

When scanning a spectrum for the purpose of identifying functional groups present, it is most useful to begin at the high frequency side of the spectrum ($\gtrsim$ 1500 cm^{-1}) and work down to lower frequencies. This is because the stretching frequencies are usually much more useful than bending frequencies for preliminary identification of functional groups. In Table 3-31 are listed some of the more useful stretching frequencies.

Table 3-31. Characteristic Stretching Frequencies

Bond	Type of Compound	Frequency (cm^{-1})	Intensity
$C-H$	alkane	2800–3000	strong
$=C-H$	alkene or arene	3000–3100	medium
$\equiv C-H$	alkyne	3300	strong
$C=C$	alkene	1620–1680	variable
$C\equiv C$	alkyne	2100–2260	variable
$C\equiv N$	nitrile	2200–2300	variable
$C=O$	ketones, aldehydes acids, esters	1700–1750	strong
$O-H$	alcohols	3590–3650	variable, sharp
	$H-$bonded alcohols	3200–3400	strong, broad
	$H-$bonded acids	2500–3000	variable, broad
$N-H$	amines	3300–3500	medium

Once you suspect the presence of a functional group, use Table 3-32 to locate its other characteristic absorption frequencies. These tables are by no means infallible, but they provide a good start in identifying a functional group.

The spectrum of hexane, Fig. 3-74, exhibits an intense set of absorptions just below 3000 cm^{-1} characteristic of $C-H$ stretching of saturated hydrocarbons. According to Table 3-32, there should be bending frequencies of methyl groups in the region 1350–1400 and 1430–1480 cm^{-1}. Overlapping the 1430–1480 absorption should be the methylene (CH_2) bending mode. In the spectrum we observe strong bands near 1450 and 1375 cm^{-1}. Table 3-32 also leads us to expect a medium intensity band near 720 cm^{-1} due to CH_2 rocking when several adjacent methylene groups occur in the same compound. Indeed, this is the next most intense band in the spectrum. The vibrations of a methylene group are shown in Fig. 3-75. Figure

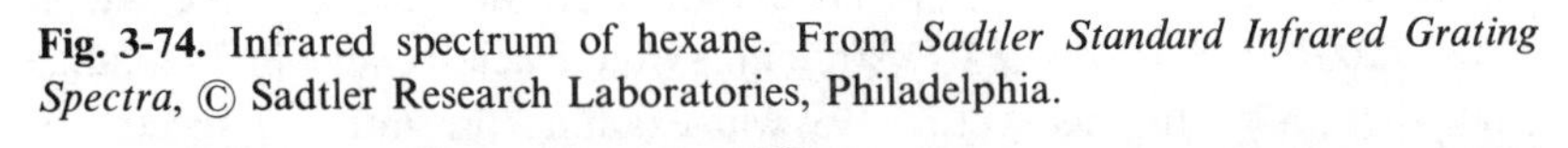

Fig. 3-74. Infrared spectrum of hexane. From *Sadtler Standard Infrared Grating Spectra*, © Sadtler Research Laboratories, Philadelphia.

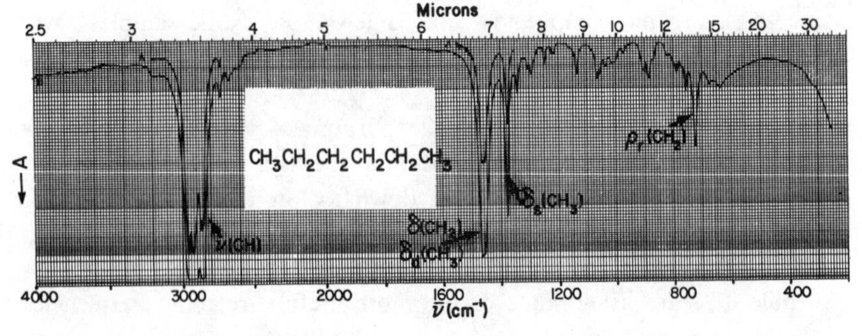

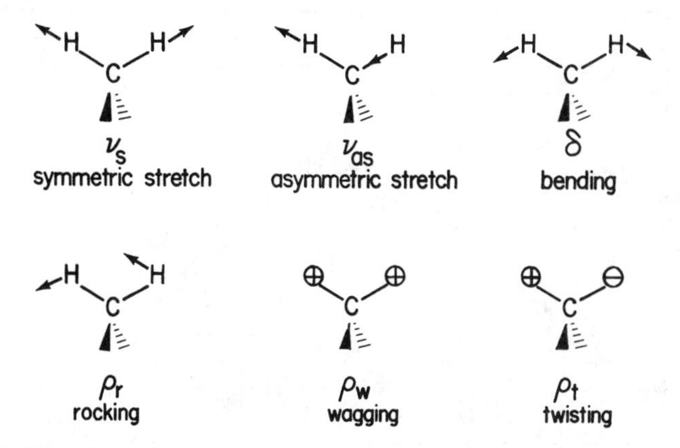

Fig. 3-75. Vibrations of a methylene group.

3-76 illustrates how the absorption bands due to CH_3 groups can be distinguished empirically from those of CH_2 groups. In the series of compounds whose spectra are shown, the bands whose intensity grows with an increasing number of CH_2 groups are logically assigned to CH_2 absorptions.

The spectrum of 1-butene in Fig. 3-77 illustrates the bands characteristic of a vinyl ($-CH{=}CH_2$) group. We observe $=C-H$ stretching near 3100 cm^{-1}, $C{=}C$ stretching near 1650 cm^{-1}, and a combination band that is often present in compounds with $C{=}C$ double bonds near 1900 cm^{-1}. Table 3-32 indicates that strong bands, due to bending vibrations of the vinyl group, should be found near 1425, 975, and 925 cm^{-1}, and these bands are all seen.

Toluene (Fig. 3-78) exhibits aromatic $C-H$ stretching near 3030 cm^{-1} as well as the asymmetric and symmetric methyl stretching below 3000 cm^{-1}. Variable intensity $C{=}C$ stretching modes (in this case medium to strong) of a benzene ring are seen at 1600, 1500, and 1450 cm^{-1}. Benzene rings also have a relatively rich spectrum of overtones and combination bands in the region 1650–2000 cm^{-1} whose pattern is characteristic of the geometry of substituents on the benzene ring. Strong bands near 700 cm^{-1} are due to $C-H$ out-of-plane bending and are also characteristic of the pattern of ring substitution.

Noteworthy features of the spectrum of phenylacetylene (Fig. 3-79) are alkyne $C-H$ stretching near 3300 cm^{-1} and sharp, weak $C{\equiv}C$ stretching near 2100 cm^{-1}. Although the ir absorption of $C{\equiv}C$ triple bonds is usually weak, the Raman spectrum is usually quite strong. (The same is true of symmetrically substituted alkenes.) In Fig. 3-80 we see the $C{\equiv}C$ stretching

Table 3-32. Characteristic Infrared Frequencies of Common Functional Groups†

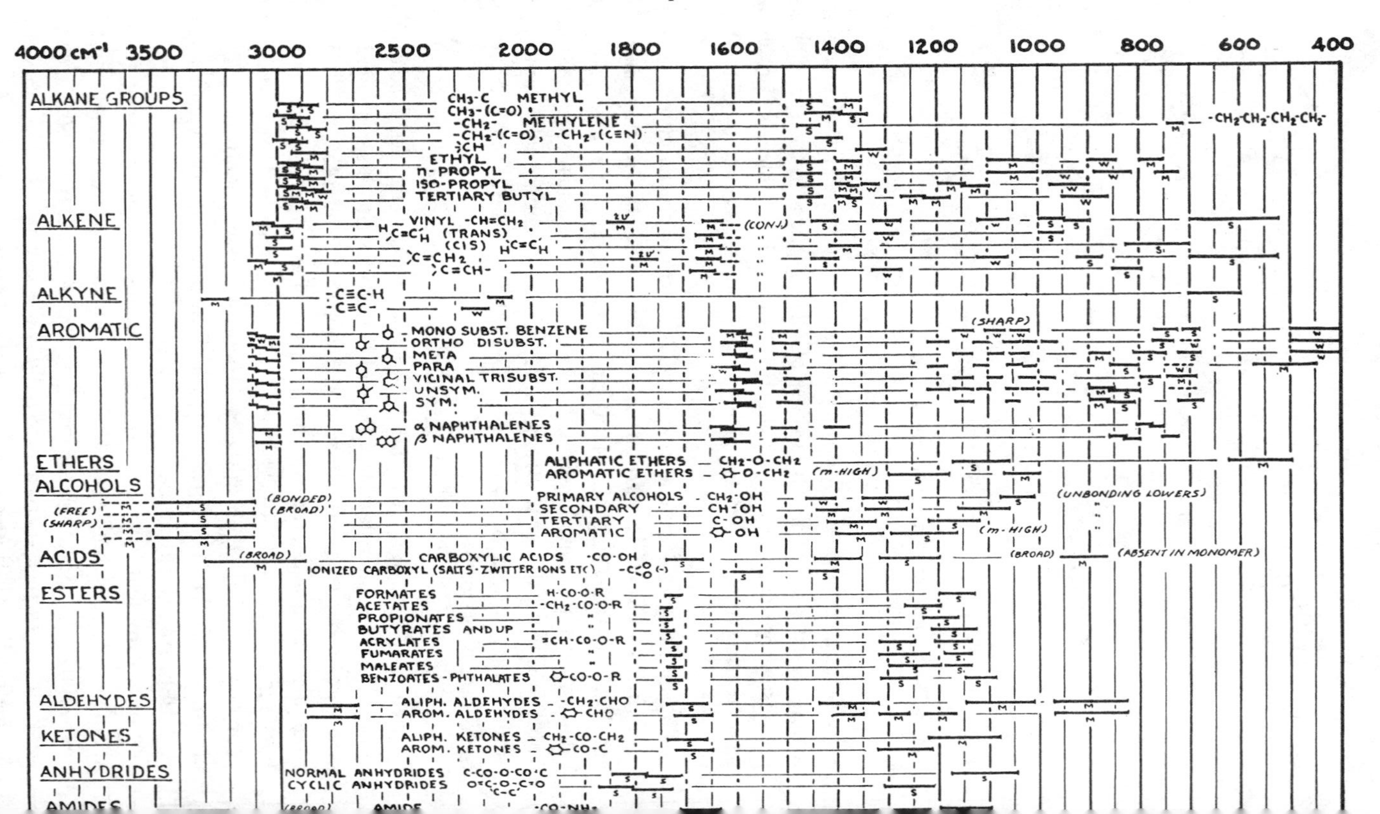

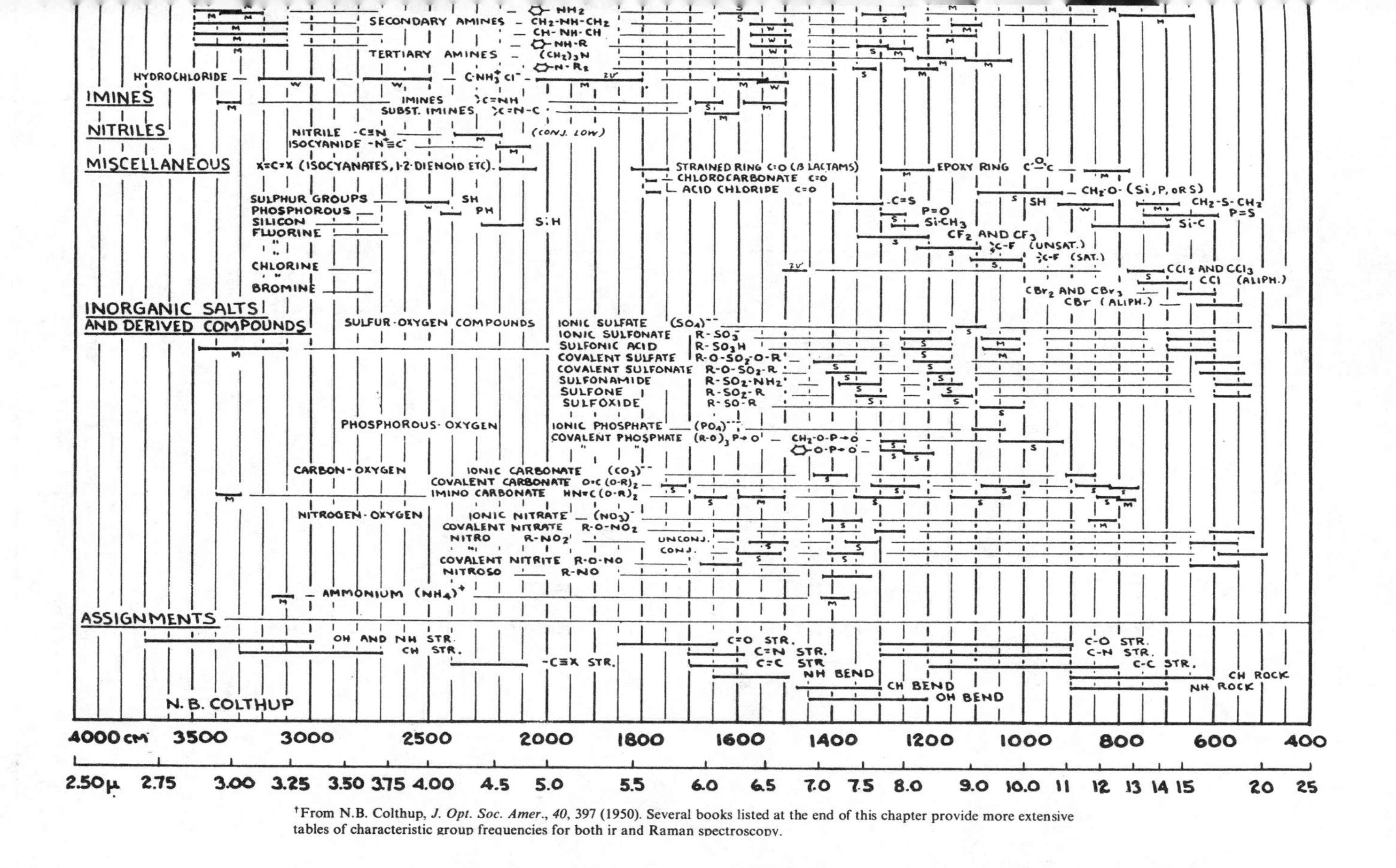

†From N.B. Colthup, *J. Opt. Soc. Amer.*, *40*, 397 (1950). Several books listed at the end of this chapter provide more extensive tables of characteristic group frequencies for both ir and Raman spectroscopy.

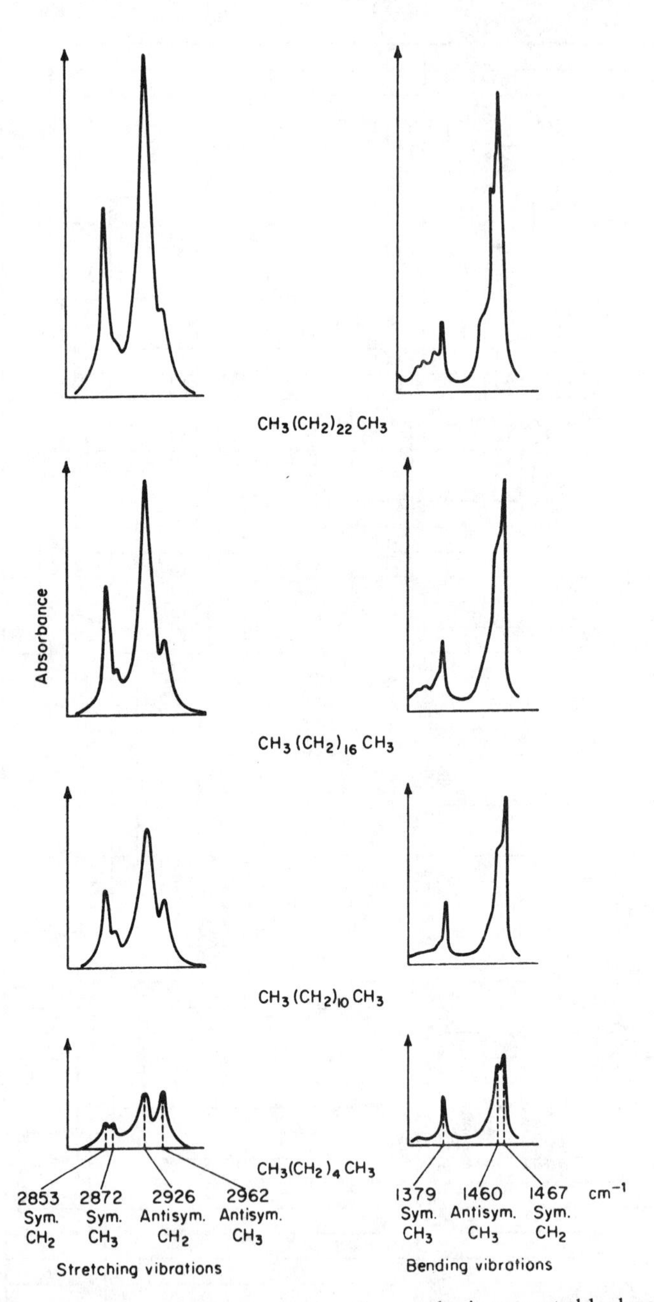

Fig. 3-76. CH_2 and CH_3 stretching and bending modes in saturated hydrocarbons. In this figure absorption goes *up*. From E.F.H. Brittain, W.O. George, and C.H.J. Well, *Introduction to Molecular Spectroscopy*, Academic Press, London, 1970.

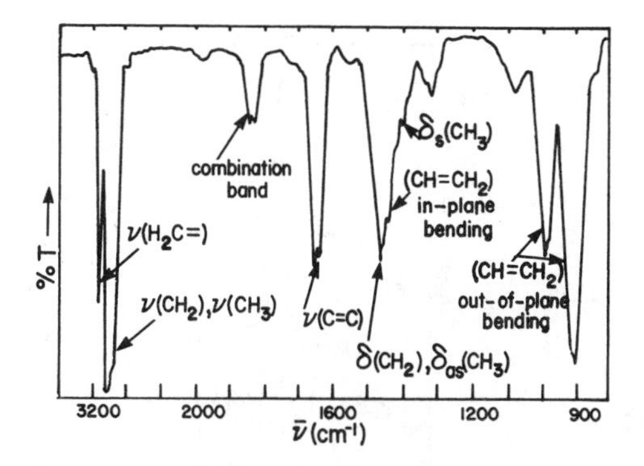

Fig. 3-77. Infrared spectrum of 1-butene. From J.D. Roberts and M.C. Caserio, *Basic Principles of Organic Chemistry*, W.A. Benjamin, N.Y., 1964.

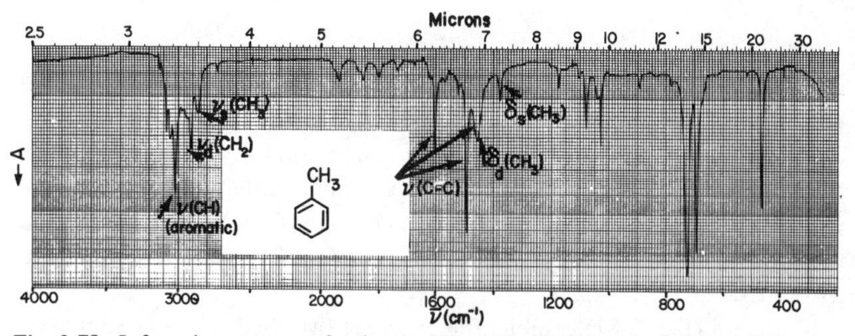

Fig. 3-78. Infrared spectrum of toluene. From *Sadtler Standard Infrared Grating Spectra*, © Sadtler Research Laboratories, Philadelphia.

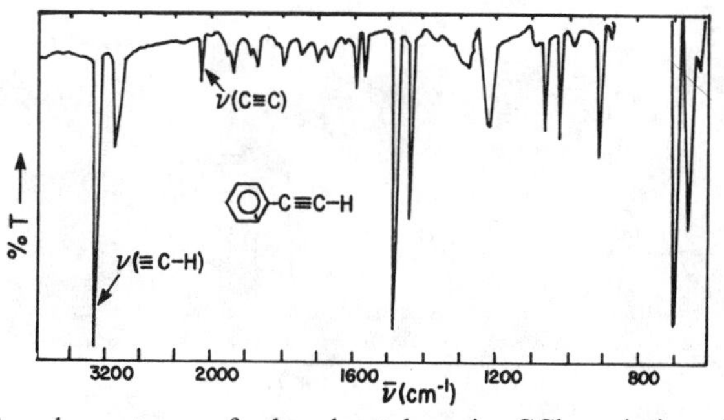

Fig. 3-79. Infrared spectrum of phenylacetylene in CCl_4 solution. From J.D. Roberts and M.C. Caserio, *Basic Principles of Organic Chemistry*, W.A. Benjamin, N.Y., 1964.

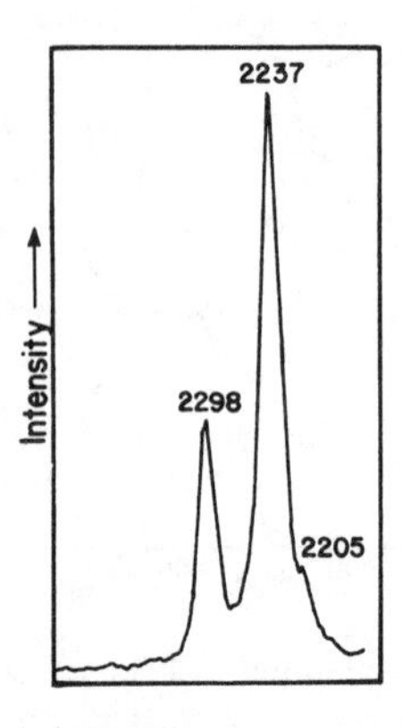

Fig. 3-80. Raman spectrum of the C≡C stretching region of 3-hexyn-1-ol. From S.K. Freeman, *Applications of Laser Raman Spectroscopy*, John Wiley & Sons, N.Y., 1974.

region of the Raman spectrum of 3-hexyn-1-ol. The strong band at 2237 is $\nu(C\equiv C)$ and the moderate band at 2298 cm^{-1} is attributed to an overtone or combination band which has "borrowed intensity" (Section 3-6-E) from the fundamental at 2237 cm^{-1}. A similar pattern of bands in the C≡N stretching region of nitriles is commonly found and explained in the same manner. The weak band in Fig. 3-80 at 2205 cm^{-1} is attributed to $\nu({}^{13}C\equiv{}^{12}C)$.

Methanol (Fig. 3-81) exhibits the expected two methyl C–H stretching absorptions near 2900 cm^{-1} and bending absorptions near 1400 cm^{-1}. In addition, we find strong O–H stretching near 3300 cm^{-1}, C–O stretch-

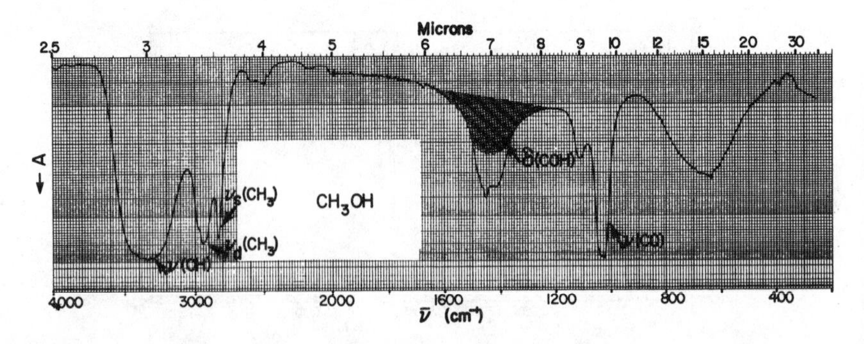

Fig. 3-81. Infrared spectrum of methanol. From *Sadtler Standard Infrared Grating Spectra*, © Sadtler Research Laboratories, Philadelphia.

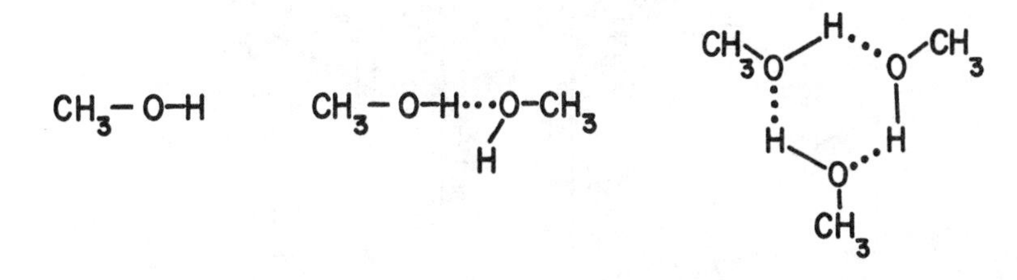

Fig. 3-82. Methanol and possible structures of hydrogen-bonded aggregates.

ing near 1050 cm^{-1}, and a broad C–O–H bending absorption underlying the sharper C–H bands in the region 1400–1500 cm^{-1}. In dilute solutions alcohols exhibit a relatively sharp O–H stretch near 3600 cm^{-1} due to monomeric species (Figs. 3-82 and 3-83). In more concentrated solutions, broad O–H bands at frequencies down to 3200 cm^{-1} due to dimers and polymers appear. In carboxylic acids (Fig. 3-84) extensive association in solution can lower the very broad O–H absorption to 2600 cm^{-1}.

A primary amine (RNH_2, Fig. 3-85[a]) exhibits two sharp N–H stretching absorptions (ν_{as} and ν_s) in the region 3500–3300 cm^{-1}. A secondary amine (R_2NH, Fig. 3-85[b]) has only one such absorption.

The spectrum of heroin hydrochloride (Fig. 3-86) (yes, I said heroin) exhibits two C=O absorptions near 1750 cm^{-1} due to the two different acetate groups present. The strong, broad band at 3400 cm^{-1} is due to water present in the sample. (Most spectra obtained by grinding a compound in KBr and pressing the finely ground mixture into a translucent pellet exhibit strong H_2O absorptions due to water adhering to the KBr.) The ammonium N–H stretch gives rise to the broad band at 2600 cm^{-1}.

The drug LSD (Fig. 3-87) exhibits the amide carbonyl absorption just above 1600 cm^{-1}. C=C stretching bands are weaker than carbonyl absorptions and are probably buried under the carbonyl peak. Primary ($CONH_2$) and secondary (CONH–) amides exhibit several characteristic bands in the region 1250–1650 shown in detail for N-methylacetamide in Fig. 3-88.

The drug STP gives a very intense, broad band at 3000 cm^{-1} due to the $-NH_3^+$ stretching vibrations (Fig. 3-89). Sharp bands near 1600 and 1510 cm^{-1} are due to δ_d and δ_s of the ammonium group. $R_2NH_2^+$ salts show corresponding stretching in the region 2700–2250 (broad) and bending occurs near 1600 cm^{-1}. The R_3NH^+ group, already seen in heroin hydrochloride (Fig. 3-86) exhibits N–H stretching also in the region 2700–2250 cm^{-1}.

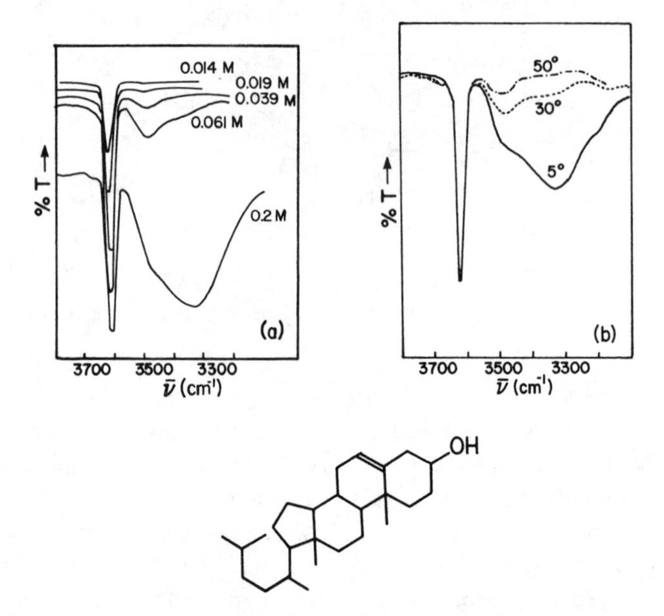

Fig. 3-83. Infrared spectrum of cholesterol as a function of (a) concentration and (b) temperature in CCl_4 solution. The sharp band near 3620 cm^{-1} is assigned to monomeric species while broad bands near 3500 and 3350 cm^{-1} are assigned to H-bonded dimers and trimers, respectively. Reproduced from F.S. Parker and K.R. Bhaskar, *Biochem.*, 7, 1286 (1968).

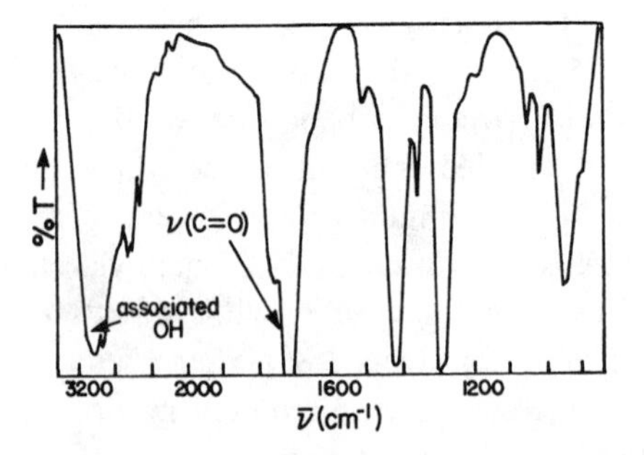

Fig. 3-84. Infrared spectrum of acetic acid. From J.D. Roberts and M.C. Caserio, *Basic Principles of Organic Chemistry*, W.A. Benjamin, N.Y., 1964.

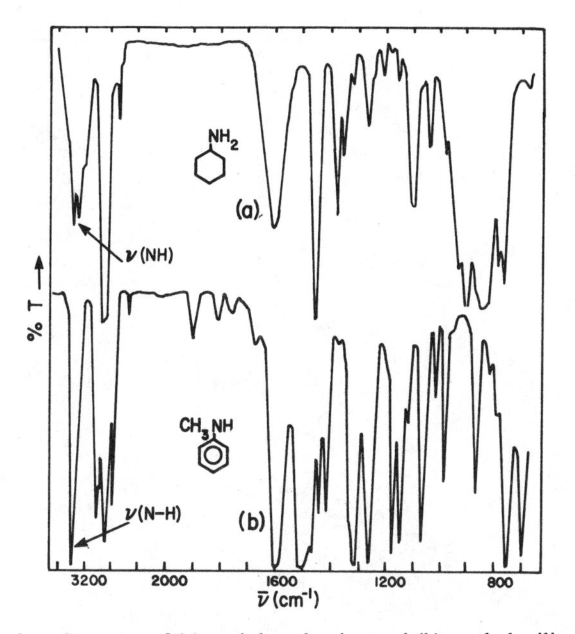

Fig. 3-85. Infrared spectra of (a) cyclohexylamine and (b) methylaniline. From J.D. Roberts and M.C. Caserio *Basic Principles of Organic Chemistry*, W.A. Benjamin, N.Y., 1964.

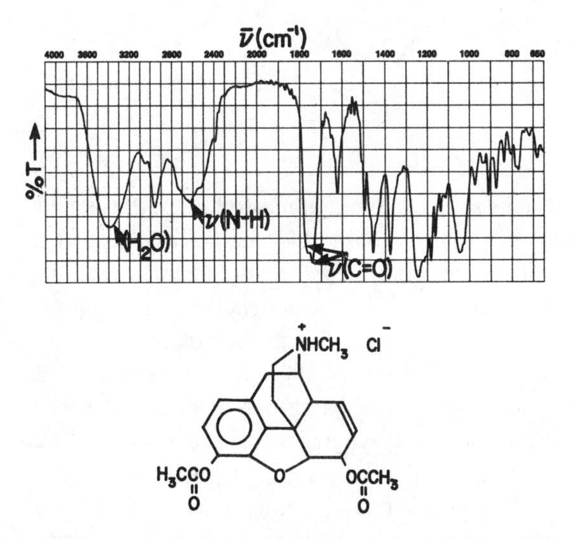

Fig. 3-86. Infrared spectrum of heroin hydrochloride in a KBr pellet. From J.S. Swinehart and R.C. Gore, *Perkin-Elmer Infrared Applications Study No. 6*, Perkin-Elmer Corp., Norwalk, Connecticut, 1968.

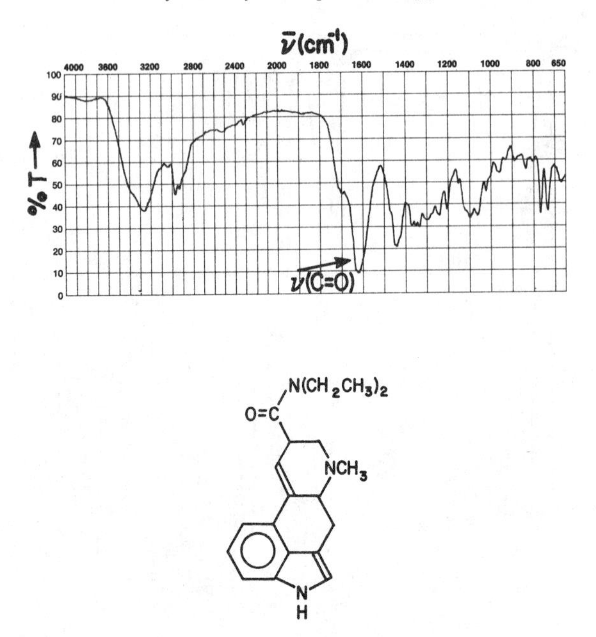

Fig. 3-87. Infrared spectrum of LSD in a KBr pellet. From J.S. Swinehart and R.C. Gore, *Perkin-Elmer Infrared Applications Study No. 6*, Perkin-Elmer Corp., Norwalk, Connecticut, 1968.

Figure 3-90 shows the spectrum of mastodon tooth enamel. The spectrum and elemental analysis establish this enamel as the mineral Dahllite, a calcium phosphate carbonate. Besides the characteristic H_2O bands, we see a few relatively broad bands at lower energy than we have been observing so far for stretching vibrations. The phosphate anions give rise to ν_d near 1050 cm^{-1}, ν_s as a doublet near 600 cm^{-1}, and δ_d near 300 cm^{-1}. The asymmetry or splitting of these bands can be due to several effects, such as lower than T_d crystal site symmetry or chemically different phosphate groups present in the crystal lattice. $\nu_d(CO_3)$ is similarly split into a doublet near 1500 cm^{-1}. The point of this illustration is that vibrational spectra are of importance to inorganic functional group analysis as well as organic functional group analysis.

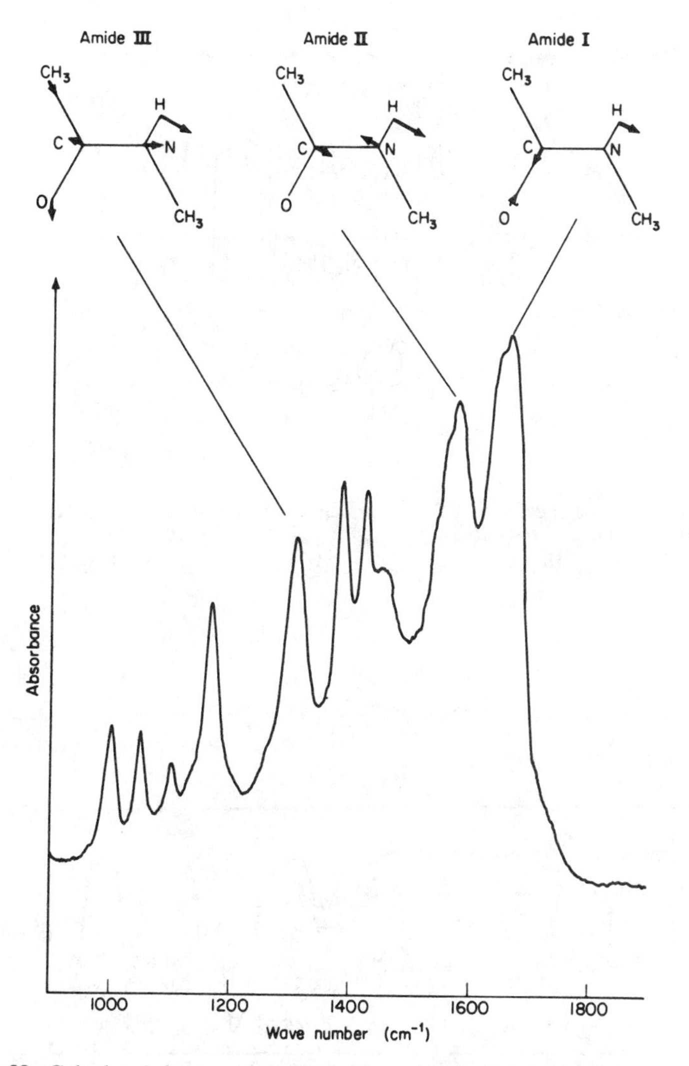

Fig. 3-88. Calculated forms of amide I, II, and III vibrations and spectrum of N-methylacetamide. Note reversed absorbance scale. From E.F.H. Brittain, W.O. George, and C.H.J. Wells, *Introduction to Molecular Spectroscopy*, Academic Press, London, 1970.

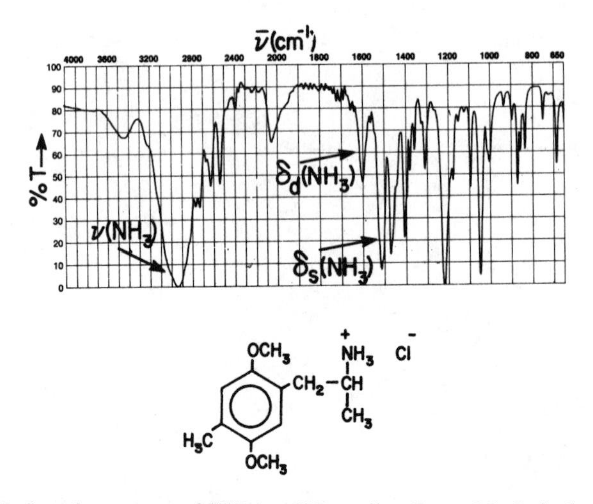

Fig. 3-89. Infrared spectrum of STP in a KBr pellet. From J.S. Swinehart and R.C. Gore, *Pèrkin-Elmer Infrared Applications Study No. 6*, Perkin-Elmer Corp., Norwalk, Connecticut, 1968.

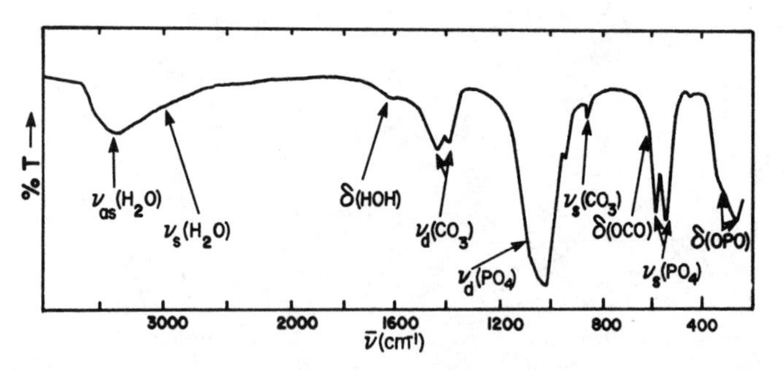

Fig. 3-90. Infrared spectrum of mastodon tooth enamel ground in a KBr pellet. Courtesy George Rossman, California Institute of Technology.

Problems

3-40. Suggest a structure consistent with the molecular formula and the ir spectrum for each compound below. A compound with the composition $C_xH_yN_zO_n$ will have $(x + 1 - \frac{1}{2}y + \frac{1}{2}z)$ (rings plus double bonds). For example, ethylene, C_2H_4, has $(2 + 1 - 2 + 0) = 1$ (rings plus double bonds). Ethylene has one double bond. A triple bond counts as two double bonds, and multiple bonds may include atoms other than carbon (*e.g.*, C≡N and C=O). Toluene ($C_7H_8 \Rightarrow 7 + 1 - 4 + 0 = 4$) has one ring and three double bonds. For compounds with other elements present, monovalent elements count in y and trivalent elements count in z. Divalent elements do not contribute to the total count of rings and double bonds. Spectra a–c below are from *Sadtler Standard Infrared Grating Spectra*. © Sadtler Research Laboratories, Philadelphia; and spectra d–f are from F. Scheinmann, ed., *An Introduction to Spectroscopic Methods for the Indentification of Organic Compounds*, Vol. I, Pergamon Press, Oxford, 1970.

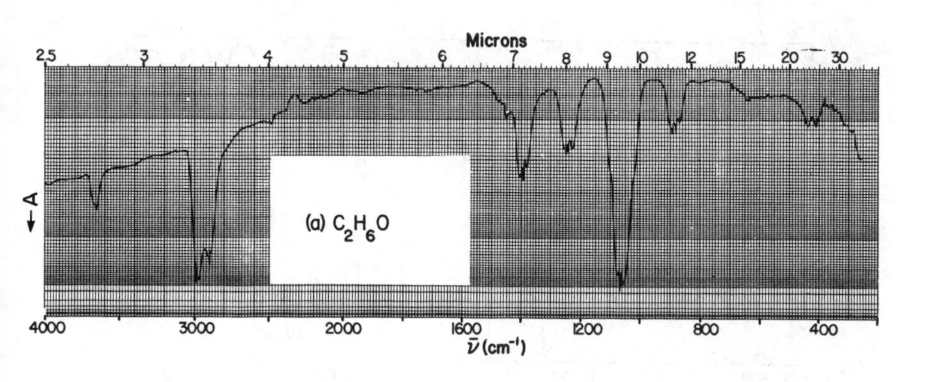

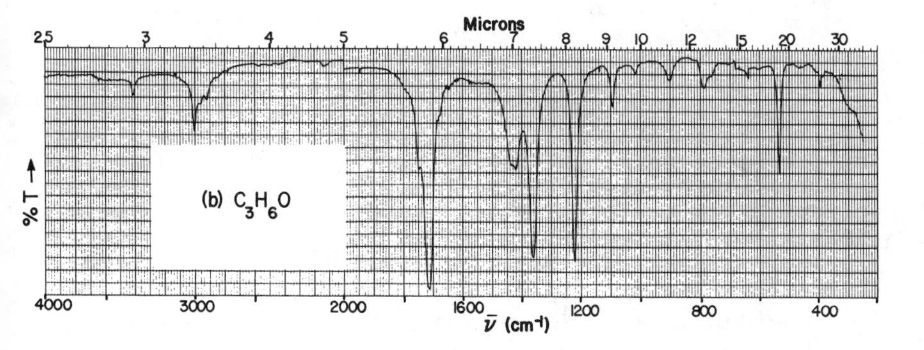

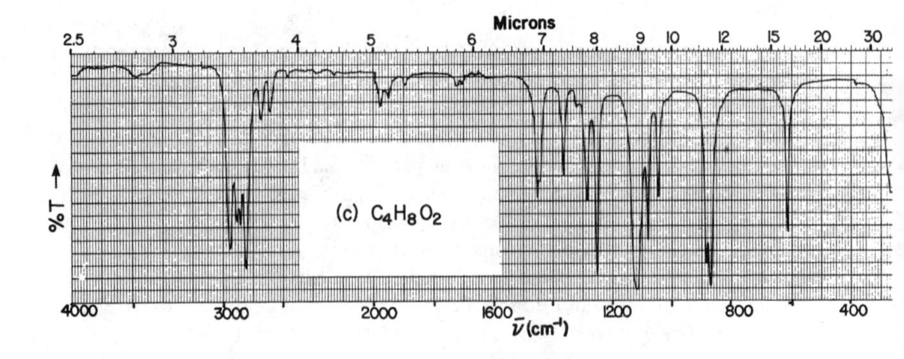
Microns
2.5
3
4
5
6
7
8
9
10
12
15
20
30
%T →
(c) C4H8O2
4000
3000
2000
1600
1200
800
400
ν̄ (cm⁻¹)

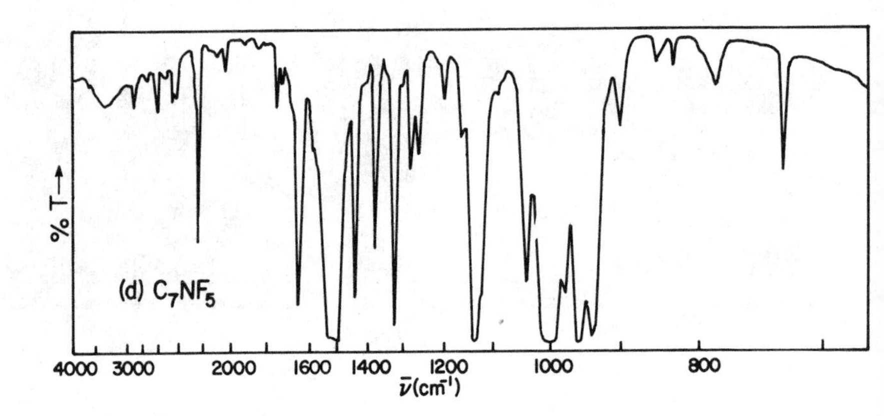
% T →
(d) C7NF5
4000
3000
2000
1600
1400
1200
1000
800
ν̄ (cm⁻¹)

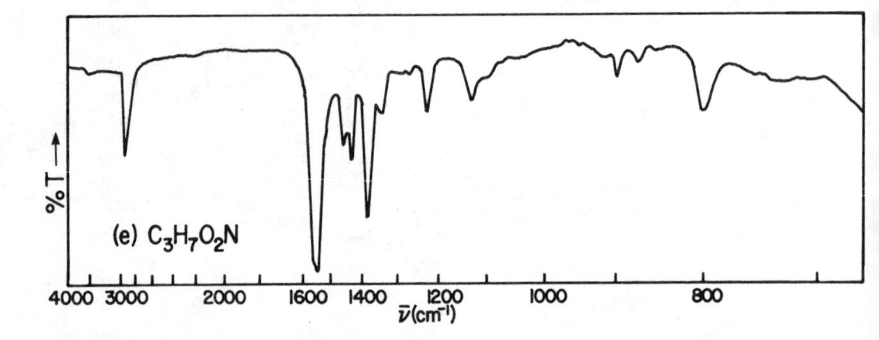
% T →
(e) C3H7O2N
4000
3000
2000
1600
1400
1200
1000
800
ν̄ (cm⁻¹)

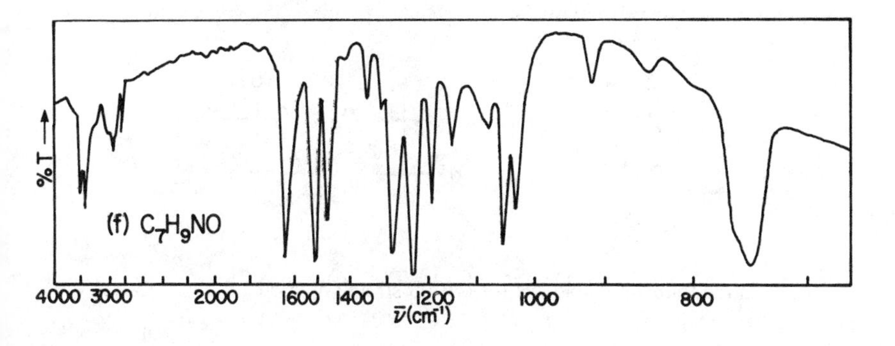

Additional Problems

3-41. The Raman spectrum of the pyramidal (C_{3v}) anion, SO_3^{2-}, in aqueous solution, exhibits four bands: 966 (strong, p), 933 (shoulder, dp), 620 (weak, p), and 473 cm^{-1} (medium, dp) (p = polarized, dp = depolarized). Make an assignment of these bands, labelling each with the proper symmetry and descriptive symbol from Table 3-13.

3-42. The linear $X-C-N$ molecules below exhibit the indicated ir bands. (a) Determine the number, symmetries, and activities of the fundamentals for the $X-C-N$ structure. (b) Draw a picture of each ir active mode and label it with its symmetry and proper symbol from Table 3-13. (c) Fill in the table below and reorganize the frequencies such that each column of the table has the same kind of vibration for all six molecules. Pay attention to the effect of increasing mass when you do this.

HCN	3311	2097	712
DCN	2630	1925	569
FCN	2290	1077	449
ClCN	2219	714	380
BrCN	2200	574	342
ICN	2158	470	321

	ν	ν	ν
description			
symmetry			
HCN			
DCN			
FCN			
ClCN			
BrCN			
ICN			

3-43. Four reasonable structures for OsO_4N (N=pyridine) are given below:

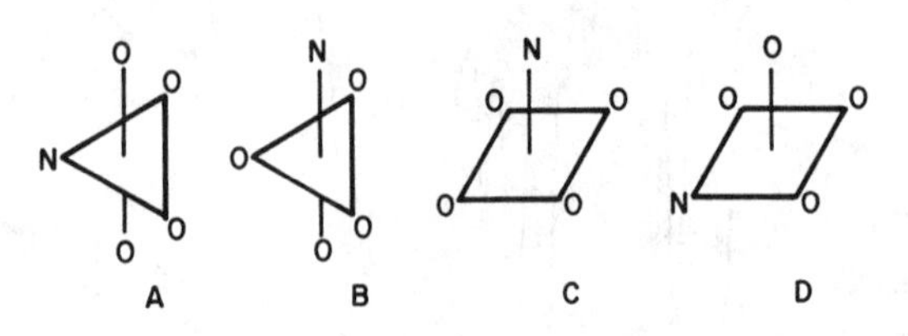

Fill in the following table just for Os–O stretching modes:

structure	point group	number of ir bands	number of Raman bands	number of coincidences	number of polarized Raman bands
A					
B					
C					
D					

The observed spectra in the Os–O stretching region are as follows:

ir: 926, 915, 908, 885 cm^{-1}.

Raman: 928 (p), 916 (p), 907 (p), 886 (dp) cm^{-1}.

With which structure are these data consistent?

3-44. Determine the number, activities, and polarizations of all of the normal modes of vibration of $Pd_2Cl_6^{2-}$ which has the planar structure below. Determine the number and symmetries of the stretches of *terminal* Pd–Cl bonds. Try the same for *bridge* Pd–Cl bonds. Draw a picture of each terminal and bridge stretching mode.

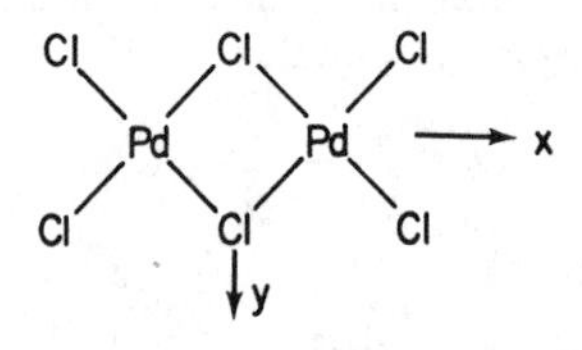

3-45. The ir spectrum of $Os(NH_3)_4(N_2)_2^{2+}$ is shown in (a) below. Based on the appearance of the N ≡ N stretching region (~2000 cm^{-1}), is this a *cis* or *trans* isomer? The ir spectra of I and II below are shown in (b). What is the formal oxidation state of Os in I and II? Can you interpret the spectrum of the N ≡ N stretching region in terms of *local* symmetry about the N_2 groups? Local symmetry means just considering nearest neighbors, *i.e.*, $X-N\equiv N-Y$.

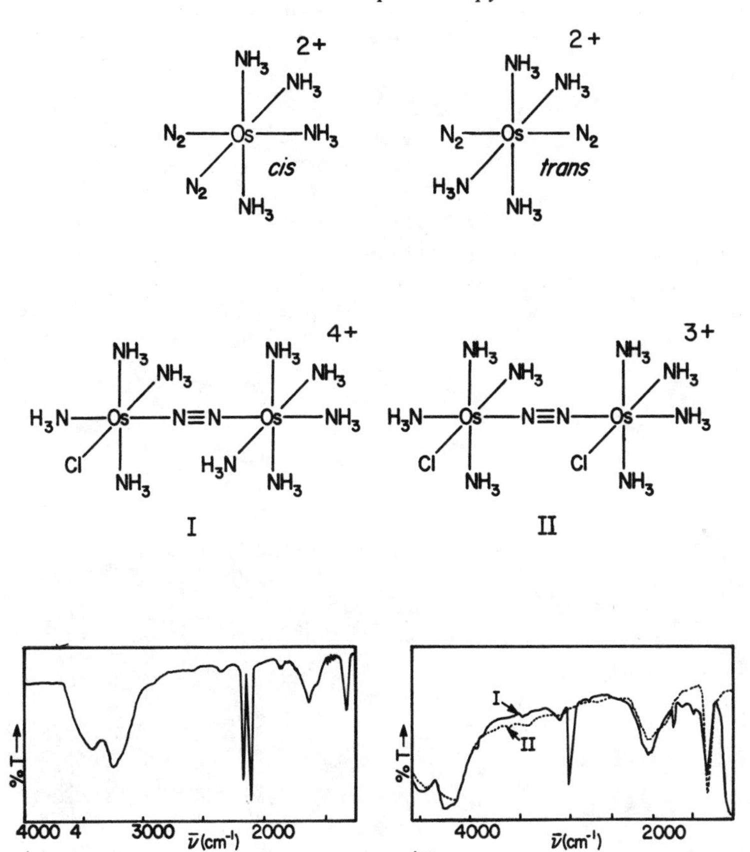

Reproduced from (a) H.A. Scheidegger, J.N. Armor, and H. Taube, *J. Amer. Chem. Soc.*, *90*, 3263 (1968) and (b) R.H. Magnuson and H. Taube, *J. Amer. Chem. Soc.*, *94*, 7213 (1972).

3-46. Construct a picture of each of the symmetry coordinates of a linear molecule of C_4 which exists to a small extent in the vapors of molten carbon.

3-47. Suggest a structure consistent with the molecular formula and the ir spectrum of each compound below. Spectra *a* and *b* are from *Frequently used Spectra for the Infrared Spectroscopist*, © Sadtler Research Laboratories, Philadelphia. Spectra c–f are reproduced from *Sadtler Standard Infrared Grating Spectra*, © Sadtler Research Laboratories. Spectrum g is from F. Scheinmann, ed., *An Introduction to Spectroscopic Methods for the Identification of Organic Compounds*, Vol. I, Pergamon Press, Oxford, 1970.

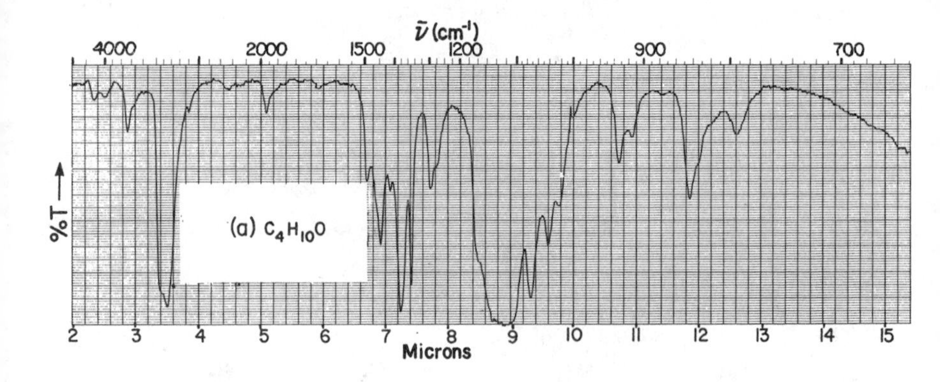
ν̃ (cm⁻¹)
4000
2000
1500
1200
900
700
%T
(a) $C_4H_{10}O$
2
3
4
5
6
7
8
9
10
11
12
13
14
15
Microns

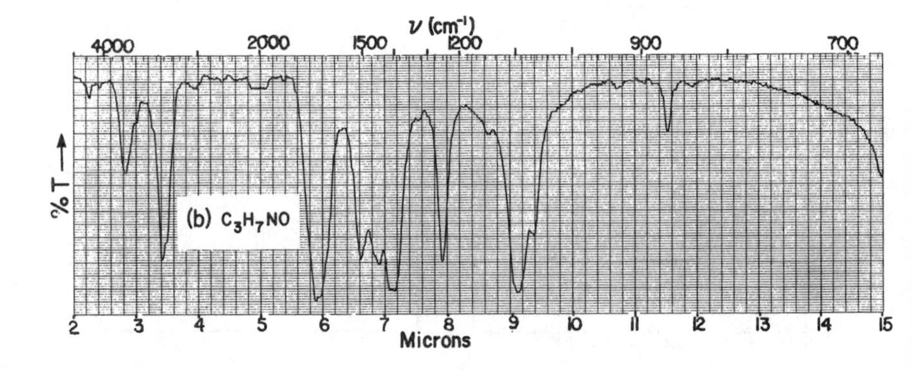
ν (cm⁻¹)
4000
2000
1500
1200
900
700
%T
(b) C_3H_7NO
2
3
4
5
6
7
8
9
10
11
12
13
14
15
Microns

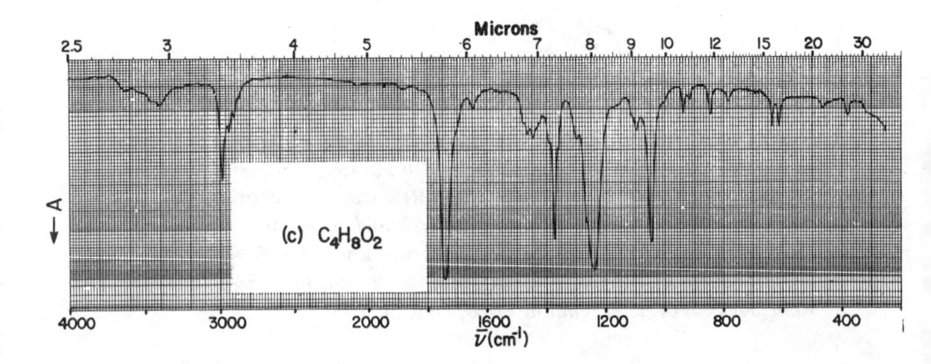
Microns
2.5
3
4
5
6
7
8
9
10
12
15
20
30
A
(c) $C_4H_8O_2$
4000
3000
2000
1600
1200
800
400
ν̃ (cm⁻¹)

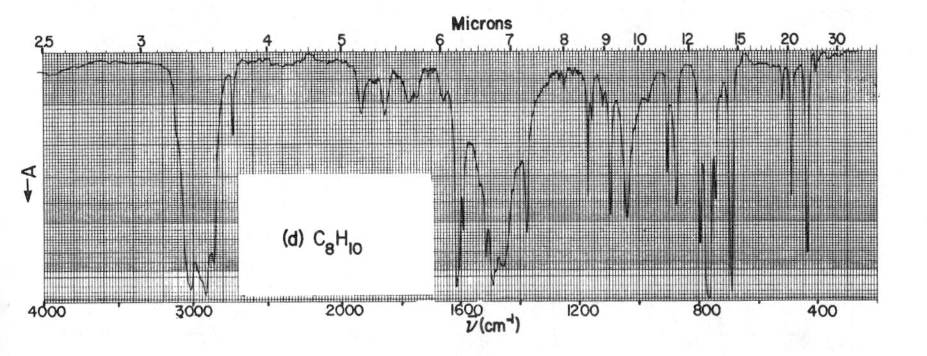
Microns
2.5
3
4
5
6
7
8
9
10
12
15
20
30
← A
(d) C_8H_{10}
4000
3000
2000
1600
1200
800
400
$\bar{\nu}$ (cm⁻¹)

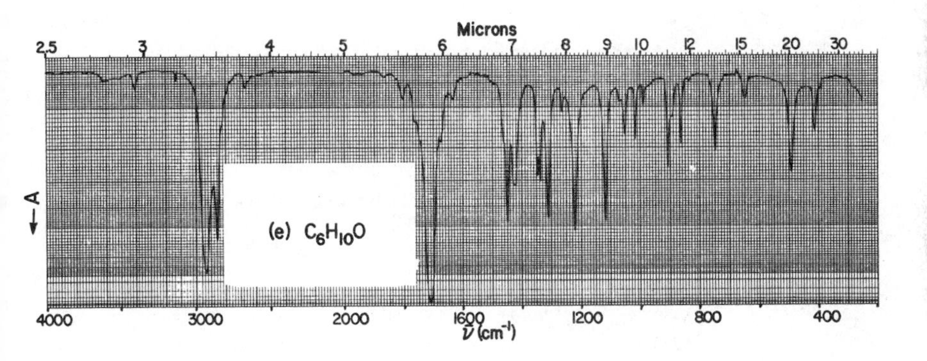
Microns
2.5
3
4
5
6
7
8
9
10
12
15
20
30
← A
(e) $C_6H_{10}O$
4000
3000
2000
1600
1200
800
400
$\bar{\nu}$ (cm⁻¹)

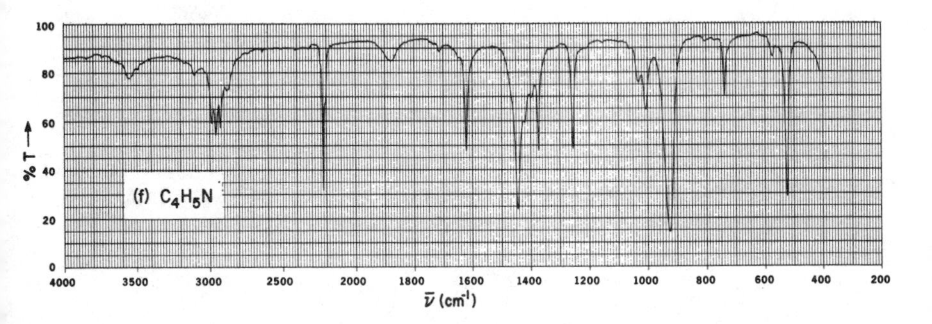
100
80
60
40
20
0
% T →
(f) C_4H_5N
4000
3500
3000
2500
2000
1800
1600
1400
1200
1000
800
600
400
200
$\bar{\nu}$ (cm⁻¹)

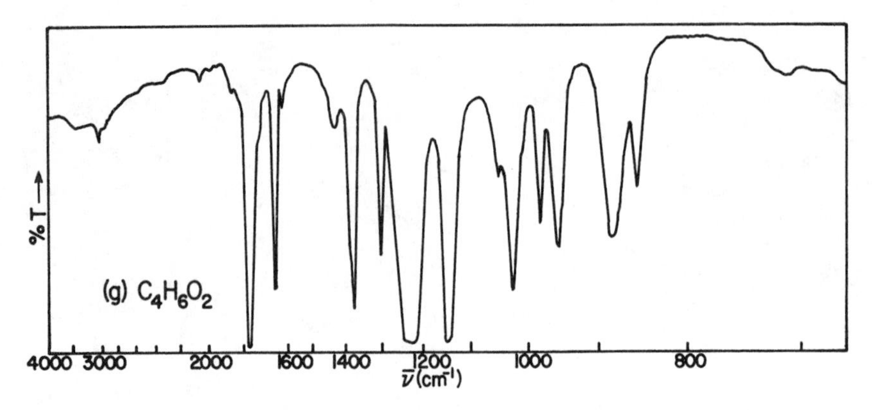

3-48. $H^{35}Cl$ exhibits the vibrational spectrum tabulated below. From this spectrum, calculate the values of $\bar{B}_e$, $\bar{\alpha}_e$, r_e, r_0, r_1, r_2, and r_3.

Absorption Lines of $H^{35}Cl$[a]

	Fundamental		
3059.32	2981.00	2863.02	2750.13
3045.06	2963.29	2841.58	2725.92
3030.09	2944.90	2819.56	2701.18
3014.41	2925.90	2796.97	2675.94
2998.04	2906.24	2773.82	2650.22
	First Overtone		
5739.29	5706.21	5647.03	5602.05
5723.29	5687.81	5624.81	5577.25
	Second Overtone		
8412.25	8383.29	8326.10	8278.99
8398.70	8366.02	8303.39	

[a] Report of the Commission on Molecular Structure and Spectroscopy, IUPAC, *Pure and Applied Chemistry*, 1, 573 (1961).

3-49. Assign symmetry labels and descriptive symbols (Table 3-13) to the fundamental frequencies of CH_4 which appear at 3019 (ir, R), 2917 (R), 1534 (R), and 1306 (ir, R) cm^{-1}.

3-50. Suggest an assignment with symmetry labels and descriptive symbols for the fundamentals of CH_3CN which appear at 2999, 2942, 2249, 1440, 1376, 1124, 918, and 380 cm^{-1}.

3-51. The compounds $Ru(CO)_3(P\phi_3)_2$, $Fe(CO)_3(P\phi_3)_2$, and $Ru(CO)_3(As\phi_3)_2$ (ϕ = phenyl) each exhibit just a single ir active CO stretching mode. Six plausible structures for these compounds are drawn below, in which CO ligands are indicated by dark circles. How many ir and Raman bands are expected in the carbonyl stretching region for each structure? With which structure is the spectrum consistent? Draw a picture of this ir active CO stretching mode.

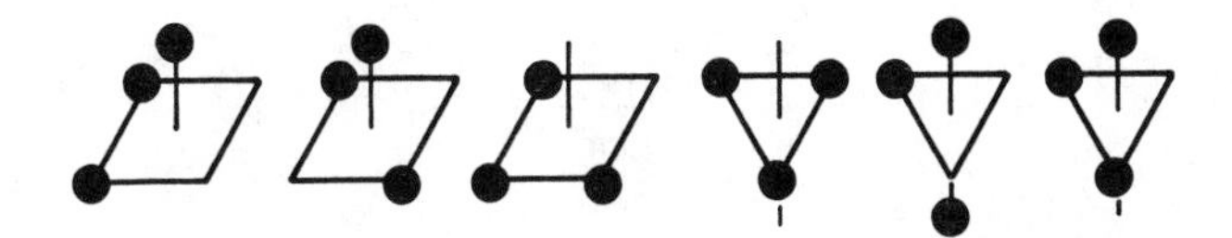

3-52. The compound $Mn_2(CO)_{10}$ has the same staggered D_{4d} structure shown for $Cr_2(CO)_{10}^{2-}$ in Problem 1-10(k). An interesting kind of polarization study has been done (R.A. Levenson, H.B. Gray, and G.P. Ceasar, *J. Amer. Chem. Soc.*, *92*, 3653 [1970]) in which $Mn_2(CO)_{10}$ was dissolved in an oriented nematic liquid crystal. Such a solution consists of long, thin, solvent molecules aligned so that their long axes generally point in the same direction. Since $Mn_2(CO)_{10}$ also has one unique long axis, it is presumed that the solute molecules align themselves parallel to the solvent molecules. The ir spectrum of the carbonyl stretching region with the electric vector parallel and perpendicular to the long axis of the solvent molecules is listed below. (a) What are the symmetries and activities of the ν(CO) vibrations of $Mn_2(CO)_{10}$? (b) Which absorptions will be polarized parallel and perpendicular to the long axis of the molecule? (c) Assign the proper symmetries to the three observed absorptions.

	2045	2009	1980 cm^{-1}
$A_{\parallel}$*	0.62	0.81	0.21
$A_{\perp}$**	0.54	0.92	0.19

*$A_{\parallel}$ is arbitrary relative absorbance of light polarized parallel to the long solvent axis.
**$A_{\perp}$ is relative absorbance of light polarized perpendicular to the long solvent axis. Relative absorbance is to the same scale as the $A_{\parallel}$ data.

3-53. The fundamentals, $\nu_1 - \nu_6$, of CH_3Cl are observed at 2966, 1355, 732, 3042, 1455, and 1015 cm^{-1}, respectively, and are labelled on the spectrum below. Assign a symmetry label and descriptive symbol (Table 3-13) to each frequency. Notice that the bands fall into two distinct classes based on their appearance. On this basis, what is the symmetry of the absorption centered at 2879 cm^{-1}, which is not a fundamental? Suggest an assignment of this band consistent with its bandshape and position. Spectrum reproduced from *Sadtler Reference Spectra Collections*, © Sadtler Research Laboratories, Philadelphia.

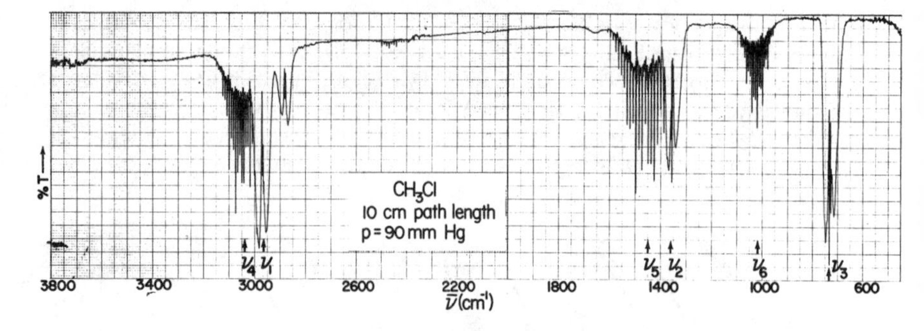

Related Reading

D.M. Adams, *Metal-Ligand and Related Vibrations*, St. Martin's Press, N.Y., 1968.

N.L. Alpert, W.E. Keiser, and H.A. Szymanski, *IR-Theory and Practice of Infrared Spectroscopy*, Plenum Press, N.Y., 1970.

G.M. Barrow, *Introduction to Molecular Spectroscopy*, McGraw-Hill, N.Y., 1962.

L.J. Bellamy, *The Infrared Spectra of Complex Molecules*, Methuen, London, 1958.

L.J. Bellamy, *Advances in Infrared Group Frequencies*, Methuen, London, 1968.

R.T. Conley, *Infrared Spectroscopy*, Allyn and Bacon, Boston, 1972.

F.R. Dollish, W.G. Fately, and F.F. Bentley, *Characteristic Raman Frequencies of Organic Compounds*, John Wiley & Sons, N.Y., 1974.

J.R. Ferraro, *Low Frequency Vibrations of Inorganic and Coordination Compounds*, Plenum Press, N.Y., 1971.

S.K. Freeman, *Applications of Laser Raman Spectroscopy*, John Wiley & Sons, N.Y., 1974.

G. Herzberg, *Infrared and Raman Spectra of Polyatomic Molecules*, Van Nostrand Reinhold, N.Y., 1945.

G. Herzberg, *Spectra of Diatomic Molecules*, Van Nostrand Reinhold, N.Y., 1950.

K. Nakamoto, *Infrared Spectra of Inorganic and Coordination Compounds*, John Wiley & Sons, N.Y., 1970.

F. Scheinmann, ed., *An Introduction to Spectroscopic Methods for the Identification of Organic Compounds*, Vol. I., Pergamon Press, Oxford, 1970.

R.M. Silverstein, G.C. Bassler, and T.C. Morrill, *Spectrometric Identification of Organic Compounds*, John Wiley & Sons, N.Y., 1974.

D. Steele, *Theory of Vibrational Spectroscopy*, W.B. Saunders, Philadelphia, 1971.

H.A., Szymanski, *Interpreted Infrared Spectra*, Vols. I-III, Plenum Press, N.Y., 1964, 1966, 1967.

C.H. Townes and A. L. Schawlow, *Microwave Spectroscopy*, McGraw-Hill, N.Y., 1955.

S. Walker and H. Straw, *Spectroscopy*, Vol. II, Chapman and Hall, Great Britain, 1967.

E.B. Wilson, Jr., J.C. Decius, and P.C. Cross, *Molecular Vibrations*, McGraw-Hill N.Y., 1955.

L.A. Woodward, *Introduction to the Theory of Molecular Vibrations and Vibrational Spectroscopy*, Oxford University Press, London, 1972.

4 · *Molecular orbital theory*

4-1. Introduction

The qualitative molecular orbital (mo) theory developed in this chapter will form a foundation for our study of electronic spectroscopy. It should be stressed that there are a variety of ways to treat the electronic structure of molecules and we are taking a rather naive qualitative approach chosen to emphasize the symmetry of molecular orbitals. What you should get out of this chapter is an appreciation of the distribution of electrons in some selected small molecules, and some idea of the relative energies of the molecular orbitals of those molecules. You should also have some appreciation of how the electron distribution changes upon going to some low-lying excited electronic states. Electronic spectroscopy deals with changes in the distribution of electrons within a molecule. The symmetries of molecular orbitals are important to electronic spectroscopy in the same way that the symmetries of vibrational wave functions were important to vibrational spectroscopy. The experimental technique most closely related to molecular orbital theory, photoelectron spectroscopy, will also be introduced in this chapter.

4-2. Atoms

A. *Atomic Orbitals.* Let's see how the time-independent form of the Schrödinger equation (2-36) applies to an atom with a nucleus of charge $+Ze$ and mass M and a single electron of charge $-e$ and mass m. We can make best use of the inherent spherical symmetry of this problem by using the spherical polar coordinate system defined in Fig. 4-1. The relations between

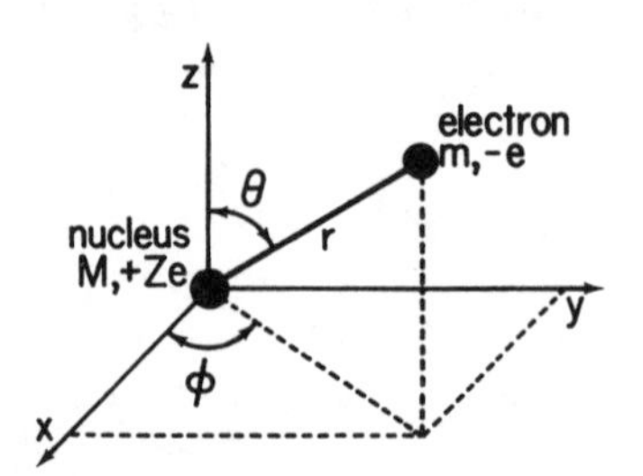

Fig. 4-1. A spherical polar coordinate system fixed at the nucleus of a one-electron atom.

spherical polar and cartesian coordinates were given in eq. 3-30. The hamiltonian will have kinetic energy terms for both the nucleus and the electron, as well as a potential energy term describing the electrostatic interaction between them:

$$\mathscr{H} = \underbrace{-\frac{\hbar^2}{2M}\nabla_n^2 - \frac{\hbar^2}{2m}\nabla_e^2}_{\text{kinetic energy terms}} \underbrace{- \frac{Ze^2}{4\pi\epsilon_0 r}}_{\text{potential energy term}} \tag{4-1}$$

where ∇_n^2 applies to the nuclear coordinates, ∇_e^2 applies to electron coordinates, and $-Ze^2/4\pi\epsilon_0 r$ is the Coulombic energy of attraction of the two particles. In this expression the charge on the electron is given in SI units. (See note on units at the beginning of the book.) In a manner analagous to the treatment of diatomic molecules in Section 3-3-D, it can be shown that the solution of the Schrödinger equation for the one-electron atom can be factored into a product of two functions. One gives the electronic energy of the atom and the other gives the translational kinetic energy. We are not at all concerned with translation of the entire atom through space and will disregard the translational equation. The "electronic" Schrödinger equation which is left has the form

$$\left(-\frac{\hbar^2}{2\mu}\nabla^2 - \frac{Ze^2}{4\pi\epsilon_0 r}\right)\psi_e = E\psi_e \tag{4-2}$$

where ∇^2 has the polar form in eq. 3-30, ψ_e is the electronic wave function, and μ is the reduced mass of the atom $(= Mm/(M + m))$. The reduced mass is very nearly equal to the mass of the electron because M is thousands of times larger than m. The term $-\hbar^2\nabla^2/2\mu$ gives the kinetic energy of a particle of mass μ moving about the origin of the coordinate system. Since μ is so nearly equal to m, we are effectively describing an electron moving about a nucleus fixed at the center of the coordinate system. Since we will only deal with electronic wave functions throughout the remainder of this chapter, we will drop the subscript on ψ_e and it will be understood that any wave function is an electronic wave function.

The process of solving eq. 4-2 is complicated and can be found in the quantum mechanics texts mentioned at the end of Chapter Two. We will only look at the final result. The electronic wave function, ψ, can be factored into a product of functions of the individual coordinates, r, θ, and ϕ:

$$\psi(r,\theta,\phi) = \underbrace{R_{n\ell}(r)}\,\underbrace{\Theta_{\ell m_\ell}(\theta)}\,\underbrace{\Phi_{m_\ell}(\phi)} \tag{4-3}$$

The total wave function, $\psi(r, \theta, \phi)$, is called an *atomic orbital* and describes the behavior of the electron with respect to the nucleus. The function $R_{n\ell}(r)$ depends on two integer quantum numbers: n, the *principal quantum number*, and ℓ, the *azimuthal quantum number*. n can have the values 1, 2, 3, 4 . . . and ℓ ranges from 0 to $n - 1$. For example, for $n = 4$, ℓ can be 0, 1, 2, or 3. $\Theta_{\ell m_\ell}(\theta)$ depends on ℓ and m_ℓ, the *magnetic quantum number*, which can have the values $-\ell$, $-\ell + 1$, .. 0, 1, .. $+\ell$. For $\ell = 3$, for example, m_ℓ can be -3, -2, -1, 0, 1, 2, or 3. $\Phi_{m_\ell}(\phi)$ depends only on the magnetic quantum number. A list of some one-electron wave functions is given in Table 4-1. In this table the constant a_0 is called the *Bohr radius*. It is the most probable distance of the electron from the nucleus in the lowest energy (1s) orbital and is equal to the electron-nuclear separation predicted by the "old quantum theory" of Bohr. You may recognize the angular factors in Table 4-1 as the same spherical harmonics of Table 3-5, which are the rotational wave functions of diatomic molecules. In Table 4-1 we show only the real wave functions of Table 3-5. Direct solution of the Φ part of the Schrödinger equation gives imaginary wave functions (eq. 3-36). The real functions are linear combinations of the imaginary functions and, therefore, also satisfy the Schrödinger equation.

The value of n determines the *shell*. Values of $n = 1, 2, 3, 4 \ldots$ are called K, L, M, N . . . shells, respectively. The value of ℓ determines the name of the orbital. $\ell = 0, 1, 2, 3 \ldots$ corresponds to s, p, d, f . . . orbitals. A complete set of each of these types of orbitals (such as p_x, p_y, and p_z) is called a *subshell*. For the one-electron atom, the energy of the orbital is completely determined by n:

$$E_n = -\frac{Z^2 e^4 \mu}{8\epsilon_0{}^2 h^2 n^2} \tag{4-4}$$

Table 4-1. Wave Functions for the One-Electron Atom[a]

n	ℓ	m_ℓ	Name	Wave Function
				K Shell
1	0	0	$\psi(1s)$	$\frac{1}{\sqrt{\pi}}\left(\frac{Z}{a_0}\right)^{3/2} e^{-\sigma}$
				L Shell
2	0	0	$\psi(2s)$	$\frac{1}{4\sqrt{2\pi}}\left(\frac{Z}{a_0}\right)^{3/2} (2-\sigma)e^{-\sigma/2}$
2	1	0	$\psi(2p_z)$	$\frac{1}{4\sqrt{2\pi}}\left(\frac{Z}{a_0}\right)^{3/2} \sigma e^{-\sigma/2}\cos\theta$
2	1	± 1[b]	$\psi(2p_x)$	$\frac{1}{4\sqrt{2\pi}}\left(\frac{Z}{a_0}\right)^{3/2} \sigma e^{-\sigma/2}\sin\theta\cos\phi$
			$\psi(2p_y)$	$\frac{1}{4\sqrt{2\pi}}\left(\frac{Z}{a_0}\right)^{3/2} \sigma e^{-\sigma/2}\sin\theta\sin\phi$
				M Shell
3	0	0	$\psi(3s)$	$\frac{1}{81\sqrt{3\pi}}\left(\frac{Z}{a_0}\right)^{3/2} (27-18\sigma+2\sigma^2)e^{-\sigma/3}$
3	1	0	$\psi(3p_z)$	$\frac{\sqrt{2}}{81\sqrt{\pi}}\left(\frac{Z}{a_0}\right)^{3/2} (6-\sigma)e^{-\sigma/3}\cos\theta$
3	1	± 1[b]	$\psi(3p_x)$	$\frac{\sqrt{2}}{81\sqrt{\pi}}\left(\frac{Z}{a_0}\right)^{3/2} (6-\sigma)\sigma e^{-\sigma/3}\sin\theta\cos\phi$
			$\psi(3p_y)$	$\frac{\sqrt{2}}{81\sqrt{\pi}}\left(\frac{Z}{a_0}\right)^{3/2} (6-\sigma)\sigma e^{-\sigma/3}\sin\theta\sin\phi$
3	2	0	$\psi(3d_{z^2})$	$\frac{1}{81\sqrt{6\pi}}\left(\frac{Z}{a_0}\right)^{3/2} \sigma^2 e^{-\sigma/3}(3\cos^2\theta-1)$
3	2	± 1[b]	$\psi(3d_{xz})$	$\frac{\sqrt{2}}{81\sqrt{\pi}}\left(\frac{Z}{a_0}\right)^{3/2} \sigma^2 e^{-\sigma/3}\sin\theta\cos\theta\cos\phi$
			$\psi(3d_{yz})$	$\frac{\sqrt{2}}{81\sqrt{\pi}}\left(\frac{Z}{a_0}\right)^{3/2} \sigma^2 e^{-\sigma/3}\sin\theta\cos\theta\sin\phi$
3	2	± 2[b]	$\psi(3d_{x^2-y^2})$	$\frac{1}{81\sqrt{2\pi}}\left(\frac{Z}{a_0}\right)^{3/2} \sigma^2 e^{-\sigma/3}\sin^2\theta\cos 2\phi$
			$\psi(3d_{xy})$	$\frac{1}{81\sqrt{2\pi}}\left(\frac{Z}{a_0}\right)^{3/2} \sigma^2 e^{-\sigma/3}\sin^2\theta\sin 2\phi$

[a] $\sigma = Zr/a_0$; $a_0 = 4\pi\epsilon_0\hbar^2/\mu e^2 = 0.5292$ Å

[b] The atomic orbitals having m_ℓ equal to $+1$ or -1 (or $+2$ or -2) are imaginary. The real functions given in this table are linear combinations of those imaginary functions. For example, $2p_x \propto \phi_+ + \phi_-$, where ϕ_+ is the imaginary function having $m_\ell = +1$, and ϕ_- is the imaginary function having $m_\ell = -1$.

The values of ℓ and m_ℓ determine the orbital angular momentum of the electron.†

If the atomic orbitals are factored into radial (R) and angular ($\Theta\Phi$) parts, the radial functions are as shown (by dashed lines) in Fig. 4-2. The 1s radial wave function goes to a maximum at $r = 0$. But the maximum probability of finding the electron is not at $r = 0$, but at $r = a_0$ if $Z = 1$. The reason is that although the value of $R^2_{n\ell}(r)$ is proportional to the probability of finding the electron in a given infinitesimal volume element at a distance r from the nucleus, there are more such volume elements for each value of r as r increases. In fact, the number of volume elements increases as r^2 since the area of a sphere is $4\pi r^2$ and the number of volume elements is proportional to the surface area of the sphere. Hence the *radial distribution function* of interest is proportional to $r^2 R^2_{n\ell}(r)$. These functions are given as solid lines in Fig. 4-2. We see in this figure that there are some points where the wave functions change sign and the probabilities go to zero. Such points are called *nodes*.

The angular parts of some of the wave functions are shown in Appendix D. The sign of each lobe represents the sign of the wave function in that region of space. The probability of finding the electron in a region of space does not depend on the sign of the wave function, but on its square. However, the bonding interactions between different orbitals will depend very much on the signs of the individual wave functions. This will become clearer later in this chapter.

Problems

4-1. Fill in the table below.

orbital	quantum numbers			nodes in radial function
	n	ℓ	m_ℓ	
2s				
3s	3	0	0	2
2p			0, ± 1	
3d				
4f				0
	4			1

†The magnitude of the orbital angular momentum of the electron of a hydrogen atom is given by $\sqrt{\ell(\ell + 1)}\hbar$. The only measurable component of angular momentum along a given direction is $m_\ell \hbar$.

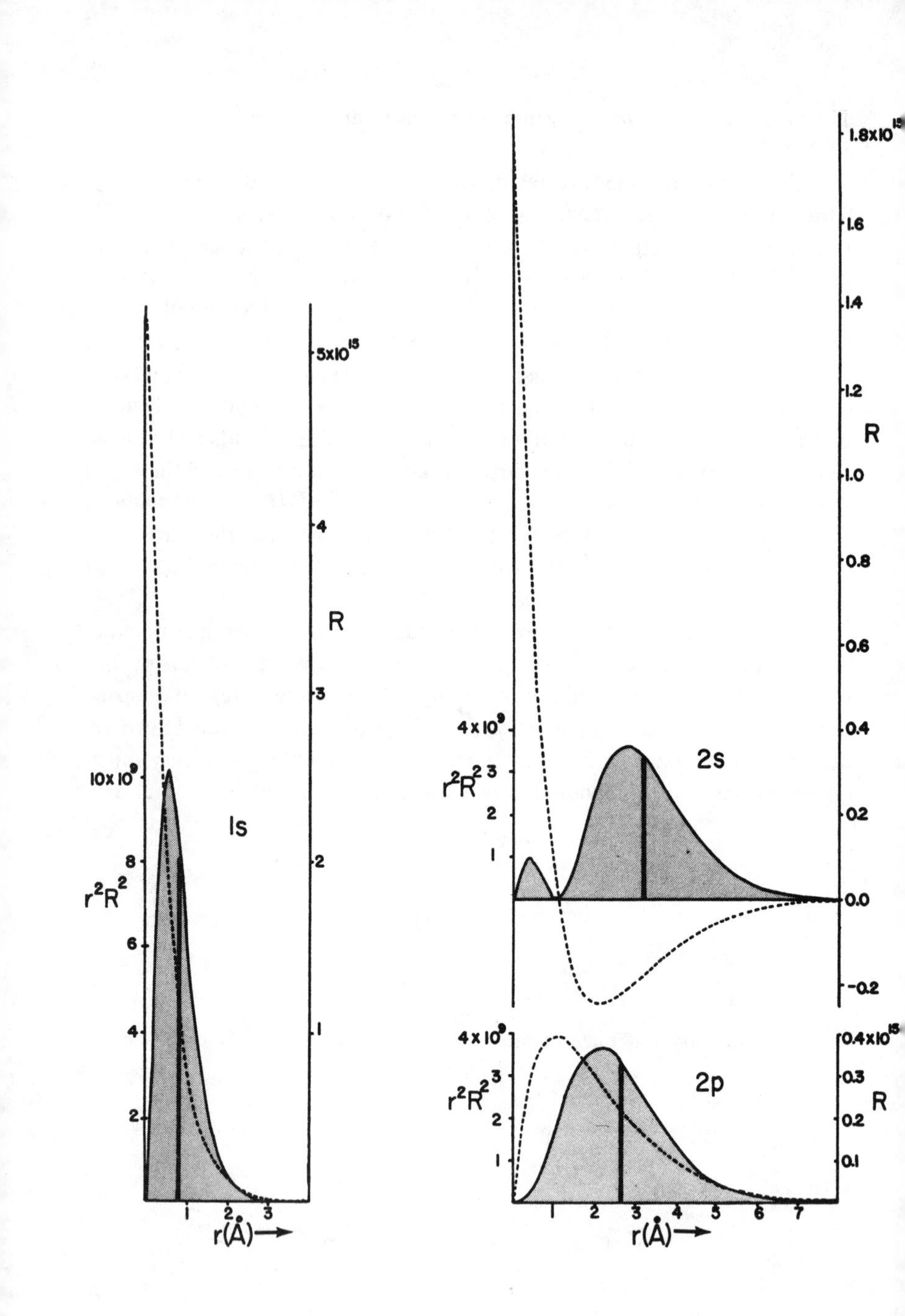

5x10^15
4
R
3
10x10^9
8
r^2R^2
6
4
2
2
1
ls
1
2
3
r(Å)
1.8x10^15
1.6
1.4
1.2
R
1.0
0.8
0.6
4x10^9
0.4
r^2R^2
3
2
1
2s
0.2
0.0
-0.2
4x10^9
0.4x10^15
r^2R^2
3
2
1
2p
0.3
R
0.2
0.1
1
2
3
4
5
6
7
r(Å)

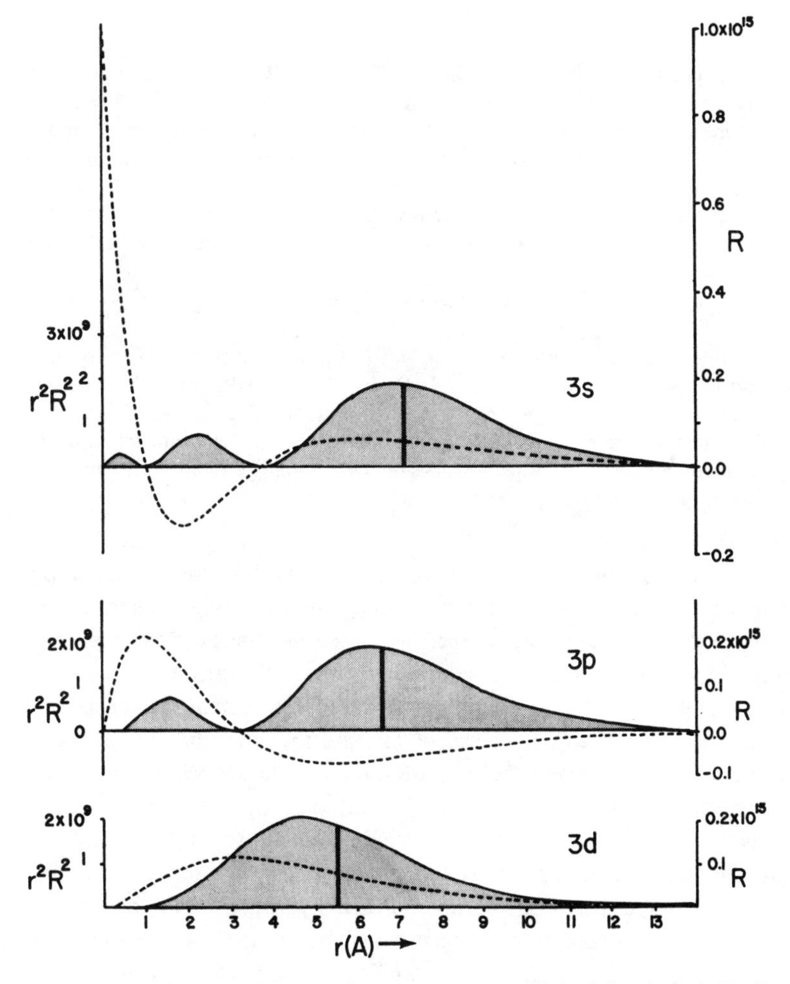

Fig. 4-2. Graphs of radial wave functions, $R_{n\ell}(r)$ (dashed lines), and distribution functions, $r^2R^2_{n\ell}(r)$ (solid lines), for the hydrogen atom. Units of R are $m^{-3/2}$ and units of r^2R^2 are m^{-1}. Vertical lines mark the average value of r for an electron in each orbital.

4-2. Atomic orbitals are often named after cartesian functions. The cartesian description of, say, a $3d_{xy}$ orbital comes from transforming the spherical polar form of the wave function in Table 4-1:

$$\begin{aligned}\psi(3d_{xy}) &= (81\sqrt{2\pi})^{-1}\,(Z/a_0)^{3/2}\sigma^2 e^{-\sigma/3}\sin^2\theta\sin 2\phi\\ &\propto r^2\sin^2\theta\sin 2\phi\\ &\propto r^2\sin^2\theta\sin\phi\cos\phi \quad (\text{since } \sin 2\phi = 2\sin\phi\cos\phi)\\ &= \underbrace{r\sin\theta\cos\phi}_{x}\cdot\underbrace{r\sin\theta\sin\phi}_{y}\end{aligned}$$

$$\psi(3d_{xy}) \propto xy$$

Using polar graph paper, plot the angular part of the $3d_{xy}$ function in the xy plane (*i.e.*, $\theta = 90°$). What you are really plotting is the function $\psi = \sin^2\theta \times \sin 2\phi$, where ψ will be the distance from the origin of the graph and ϕ determines which radial you are on. Do the same for the $3d_{z^2}$ function in the xz plane. Write the angular part (just the angular part) of the $4f_{y(z^2-x^2)}$ orbital.

4-3. Long before the advent of quantum mechanics, it was known that very hot atoms of hydrogen in the sun emit light of discrete energies given by the formula

$$E = 100hc\bar{v} = 100hc\bar{R}_H\left(\frac{1}{n_1^{\,2}} - \frac{1}{n_2^{\,2}}\right)$$

where n_1 and n_2 are integers and $n_2 > n_1$. $\bar{R}_H$ is called the Rydberg constant and is equal to 1.097×10^5 cm^{-1}. Derive this equation using the quantum mechanical result, eq. 4-4, and express $\bar{R}_H$ in terms of fundamental constants. Calculate the wave number of the highest energy hydrogen atom emission. To what atomic electronic transition does this correspond? Calculate the ionization energy of a hydrogen atom in its ground state.

B. *Atomic Configurations.* The wave functions discussed so far apply only to atoms with just one electron. To calculate wave functions for atoms with more than one electron is exceedingly complex because the repulsion between electrons must be considered as well as the attraction between electrons and the nucleus. The problem of many-electron atoms is well beyond the scope of this text, except to state that the qualitative shapes of the many-electron wave functions are similar to the shapes of the orbitals we have already encountered. We will call these many-electron atomic orbitals "hydrogen-like" and will use such orbitals through the remainder of this chapter. The energy of a one-electron atomic orbital depends only on n, the principle quantum number. For a hydrogen atom the orbitals 3s, 3p, and 3d, for example, are degenerate. The energy of a many-electron atomic orbital depends on both n and ℓ. For many-electron atoms, the order of orbital energies is generally 1s < 2s < 2p < 3s < 3p < 4s < 3d < 4p < 5s < 4d < 5p < 6s < 4f < 5d < 6p < 7s < 5f < 6d, but there are atoms and ions for which some levels are reversed.

It was known before the Schrödinger equation was postulated that electrons behave as if they have an intrinsic angular momentum, called *spin*. Spin is quantized and an electron may have either $+1$ or -1 unit of angular momentum. We speak of the *spin quantum number*, m_s, as being $+1/2$ or $-1/2$ and represent these two numbers by the symbols ↑ and ↓, respectively.[†] In Chapter Five the two spin wave functions corresponding to $m_s = +1/2$ and $m_s = -1/2$ will be called α and β. For the present purposes, the arrows will serve well. Electron spin did not come out of the quantum mechanical treatment of the one-electron atom outlined above. It also does not come out of similar treatment of the many-electron atom. Spin can be derived if relativistic effects are taken into account, or it can just be tacked on to the results we already have in a completely *ad hoc* manner. We will take the latter approach.

Putting all of this information together, then, what is the *electronic configuration* of a given atom? That is, what are the values of n, ℓ, m_ℓ, and m_s for each of its electrons? We need two more guiding principles, one of which is the *Pauli exclusion rule* (or *Pauli principle*), which states that no two electrons in an atom can have the same set of quantum numbers. Consider the He atom with two electrons, e_1 and e_2. If e_1 has the quantum numbers $n = 1$, $\ell = 0$, $m_\ell = 0$, and $m_s = +1/2$, e_2 can have the quantum numbers $n = 1$, $\ell = 0$, $m_\ell = 0$, and $m_s = -1/2$. e_2 could not have had $m_s = +1/2$ since we have already assigned this spin quantum number to e_1; and n, ℓ, and m_ℓ are all the same for both electrons. These values of n, ℓ, and m_ℓ describe a 1s orbital, and we represent the electronic configuration of He as follows:

He ↓↑ 1s

For Li we would have

Li ↓↑ ↑ or ↓↑ ↓
1s 2s 1s 2s

These configurations are degenerate in the absence of an external magnetic field so we will only write one of them from now on.[††] Continuing, we draw Be and B as follows:

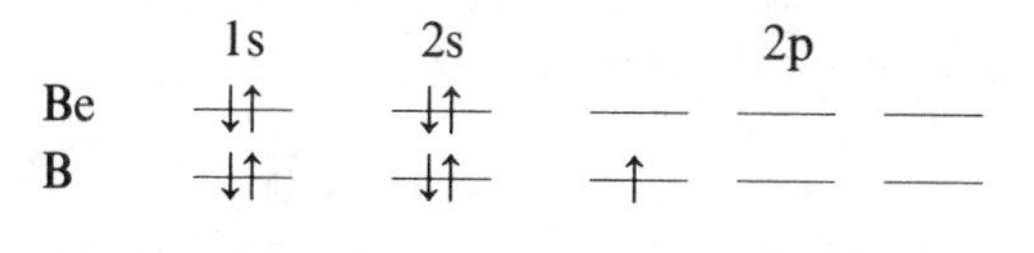

[†]The magnitude of the angular momentum of an electron is given by $\sqrt{m_s(m_s + 1)}\hbar$. The measurable component of angular momentum along a given direction, such as that of an external magnetic field, is $m_s\hbar = \pm(1/2)\hbar$.

[††]In the presence of an external magnetic field, the energy is lower if the electron angular momentum is parallel to the field than if it is antiparallel.

For carbon we could write

C ⇅ ⇅ ↑ ↑ —
1s 2s 2p
two unpaired spins

or ⇅ ⇅ ⇅ — —
1s 2s 2p
no unpaired spins

or ⇅ ⇅ ↑ ↓ —
1s 2s 2p

no unpaired spins

all of which do not violate the Pauli principle. We now introduce *Hund's first rule* which states that, in a case such as this, the configuration with the greatest number of unpaired spins has the lowest energy. So the first configuration above is correct for the ground state of carbon, and we can fill out the rest of the first row of the periodic table:

	1s	2s		2p		
N	⇅	⇅	↑	↑	↑	= [He]$2s^2 2p^3$
O	⇅	⇅	⇅	↑	↑	= [He]$2s^2 2p^4$
F	⇅	⇅	⇅	⇅	↑	= [He]$2s^2 2p^5$
Ne	⇅	⇅	⇅	⇅	⇅	= [He]$2s^2 2p^6$ ≡ [Ne]

The electronic configurations of the elements are listed in Table 4-2.

Problems

4-4. What is the first atom of Table 4-2 which violates the simple rules for electronic configurations we have just stated? What would its configuration be if it did not violate our rules?

C. *Atomic States.* In Section 4-2-B we found that we could draw three different arrangements of the electrons of carbon:

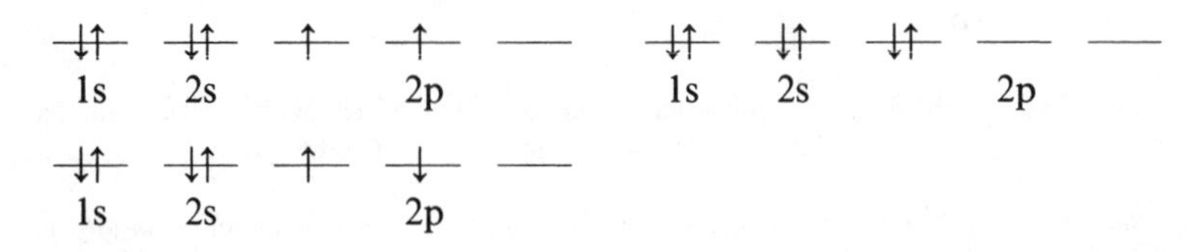

By Hund's first rule, we concluded that the first arrangement represents the ground state since it has the greatest number of unpaired electrons. But all three of these diagrams represent energetically distinct *states* of the carbon atom, each of which comes from the same configuration, $1s^2 2s^2 2p^2$. *Electronic transitions of atoms or molecules represent changes in the state of the atom or molecule, and not necessarily the configuration.* For this reason, we will introduce a systematic means of determining the states which arise from a given atomic configuration. In Chapter Five we will make use of both atomic and molecular states.

For the lighter atoms of the periodic table it is most useful to classify the state of an atom according to its total orbital angular momentum, L, and total spin angular momentum, S. We find that we can ignore all filled subshells (*e.g.*, $2s^2$ of carbon) because both the total spin and total orbital angular momentum of the electrons of a filled subshell are zero. For carbon we therefore consider only the $2p^2$ partially filled subshell. When considering several electrons, we define

$$M_L = \sum_{i=1}^{n} (m_\ell)_i \tag{4-5}$$

$$M_S = \sum_{i=1}^{n} (m_s)_i \tag{4-6}$$

where n is the total number of electrons. These equations say that M_L, for an atom, is the sum of the values of m_ℓ of all of its electrons, and similarly for M_S. It is also true that

$$M_L = L, L-1, L-2, \ldots, -L \tag{4-7}$$

$$M_S = S, S-1, S-2, \ldots -S \tag{4-8}$$

Equations 4-7 and 4-8 are analagous to the relation given in Section 4-2-A between m_ℓ and ℓ for an individual electron.

To classify the states of carbon, we set up Table 4-3 in which each set of parentheses contains a *microstate* of the p^2 configuration. The symbols 1, 0, and -1 give the value of m_ℓ for each electron. Since each electron is in a p orbital, $\ell = 1$ and m_ℓ may be only 1, 0, or -1. The symbols + and $-$ give the value of m_s ($= +1/2$ or $-1/2$) for each electron. Therefore the symbol $(\overset{+}{1},\overset{-}{1})$ at the top of the table means that electron 1 has the quantum numbers ($m_\ell = 1$, $m_s = +1/2$) and electron 2 has quantum numbers ($m_\ell = 1$, $m_s = -1/2$). Table 4-3 contains all possible microstates that can be written for the configuration p^2 which do not violate the Pauli principle. For example, in the box ($M_L = 0$, $M_S = 1$) we might have written

Table 4-2. Atomic Configurations and States

Z	Element	Electron Configuration	Ground State
1	H	$1s$	$^2S_{1/2}$
2	He	$1s^2$	$^1S_0$
3	Li	$[He]2s$	$^2S_{1/2}$
4	Be	$[He]2s^2$	$^1S_0$
5	B	$[He]2s^2 2p$	$^2P_{1/2}$
6	C	$[He]2s^2 2p^2$	$^3P_0$
7	N	$[He]2s^2 2p^3$	$^4S_{3/2}$
8	O	$[He]2s^2 2p^4$	$^3P_2$
9	F	$[He]2s^2 2p^5$	$^2P_{3/2}$
10	Ne	$[He]2s^2 2p^6$	$^1S_0$
11	Na	$[Ne]3s$	$^2S_{1/2}$
12	Mg	$[Ne]3s^2$	$^1S_0$
13	Al	$[Ne]3s^2 3p$	$^2P_{1/2}$
14	Si	$[Ne]3s^2 2p^2$	$^3P_0$
15	P	$[Ne]3s^2 3p^3$	$^4S_{3/2}$
16	S	$[Ne]3s^2 3p^4$	$^3P_2$
17	Cl	$[Ne]3s^2 3p^5$	$^2P_{3/2}$
18	Ar	$[Ne]3s^2 3p^6$	$^1S_0$
19	K	$[Ar]4s$	$^2S_{1/2}$
20	Ca	$[Ar]4s^2$	$^1S_0$
21	Sc	$[Ar]3d4s^2$	$^2D_{3/2}$
22	Ti	$[Ar]3d^2 4s^2$	$^3F_2$
23	V	$[Ar]3d^3 4s^2$	$^4F_{3/2}$
24	Cr	$[Ar]3d^5 4s$	$^7S_3$
25	Mn	$[Ar]3d^5 4s^2$	$^6S_{5/2}$
26	Fe	$[Ar]3d^6 4s^2$	$^5D_4$
27	Co	$[Ar]3d^7 4s^2$	$^4F_{9/2}$
28	Ni	$[Ar]3d^8 4s^2$	$^3F_4$
29	Cu	$[Ar]3d^{10} 4s$	$^2S_{1/2}$
30	Zn	$[Ar]3d^{10} 4s^2$	$^1S_0$
31	Ga	$[Ar]3d^{10} 4s^2 4p$	$^2P_{1/2}$
32	Ge	$[Ar]3d^{10} 4s^2 4p^2$	$^3P_0$
33	As	$[Ar]3d^{10} 4s^2 4p^3$	$^4S_{3/2}$
34	Se	$[Ar]3d^{10} 4s^2 4p^4$	$^3P_2$
35	Br	$[Ar]3d^{10} 4s^2 4p^5$	$^2P_{3/2}$
36	Kr	$[Ar]3d^{10} 4s^2 4p^6$	$^1S_0$
37	Rb	$[Kr]5s$	$^2S_{1/2}$
38	Sr	$[Kr]5s^2$	$^1S_0$
39	Y	$[Kr]4d5s^2$	$^2D_{3/2}$
40	Zr	$[Kr]4d^2 5s^2$	$^3F_2$
41	Nb	$[Kr]4d^4 5s$	$^6D_{1/2}$
42	Mo	$[Kr]4d^5 5s$	$^7S_3$
43	Tc	$[Kr]4d^5 5s^2$	$^6S_{5/2}$
44	Ru	$[Kr]4d^7 5s$	$^5F_5$
45	Rh	$[Kr]4d^8 5s$	$^4F_{9/2}$
46	Pd	$[Kr]4d^{10}$	$^1S_0$
47	Ag	$[Kr]4d^{10} 5s$	$^2S_{1/2}$
48	Cd	$[Kr]4d^{10} 5s^2$	$^1S_0$
49	In	$[Kr]4d^{10} 5s^2 5p$	$^2P_{1/2}$
50	Sn	$[Kr]4d^{10} 5s^2 5p^2$	$^3P_0$
51	Sb	$[Kr]4d^{10} 5s^2 5p^3$	$^4S_{3/2}$
52	Te	$[Kr]4d^{10} 5s^2 5p^4$	$^3P_2$
53	I	$[Kr]4d^{10} 5s^2 5p^5$	$^2P_{3/2}$

Z	Element	Electron Configuration	Ground State
54	Xe	$[Kr]4d^{10}5s^2 5p^6$	$^1S_0$
55	Cs	$[Xe]6s$	$^2S_{1/2}$
56	Ba	$[Xe]6s^2$	$^1S_0$
57	La	$[Xe]5d6s^2$	$^2D_{3/2}$
58	Ce	$[Xe]4f5d6s^2$	$(^3H_5)$
59	Pr	$[Xe]4f^3 6s^2$	$(^4I_{9/2})$
60	Nd	$[Xe]4f^4 6s^2$	$^5I_4$
61	Pm	$[Xe]4f^5 6s^2$	$(^6H_{5/2})$
62	Sm	$[Xe]4f^6 6s^2$	$^7F_0$
63	Eu	$[Xe]4f^7 6s^2$	$^8S_{7/2}$
64	Gd	$[Xe]4f^7 5d6s^2$	$^9D_2$
65	Tb	$[Xe]4f^9 6s^2$	$(^6H_{15/2})$
66	Dy	$[Xe]4f^{10}6s^2$	$(^5I_8)$
67	Ho	$[Xe]4f^{11}6s^2$	$(^4I_{15/2})$
68	Er	$[Xe]4f^{12}6s^2$	$(^3H_6)$
69	Tm	$[Xe]4f^{13}6s^2$	$^2F_{7/2}$
70	Yb	$[Xe]4f^{14}6s^2$	$^1S_0$
71	Lu	$[Xe]4f^{14}5d6s^2$	$^2D_{3/2}$
72	Hf	$[Xe]4f^{14}5d^2 6s^2$	$^3F_2$
73	Ta	$[Xe]4f^{14}5d^3 6s^2$	$^4F_{3/2}$
74	W	$[Xe]4f^{14}5d^4 6s^2$	$^5D_0$
75	Re	$[Xe]4f^{14}5d^5 6s^2$	$^6S_{5/2}$
76	Os	$[Xe]4f^{14}5d^6 6s^2$	$^5D_4$
77	Ir	$[Xe]4f^{14}5d^7 6s^2$	$^4F_{9/2}$
78	Pt	$[Xe]4f^{14}5d^9 6s$	$^3D_3$
79	Au	$[Xe]4f^{14}5d^{10}6s$	$^2S_{1/2}$
80	Hg	$[Xe]4f^{14}5d^{10}6s^2$	$^1S_0$
81	Tl	$[Xe]4f^{14}5d^{10}6s^2 6p$	$^2P_{1/2}$
82	Pb	$[Xe]4f^{14}5d^{10}6s^2 6p^2$	$^3P_0$
83	Bi	$[Xe]4f^{14}5d^{10}6s^2 6p^3$	$^4S_{3/2}$
84	Po	$[Xe]4f^{14}5d^{10}6s^2 6p^4$	$^3P_2$
85	At	$[Xe]4f^{14}5d^{10}6s^2 6p^5$	$(^2P_{3/2})$
86	Rn	$[Xe]4f^{14}5d^{10}6s^2 6p^6$	$^1S_0$
87	Fr	$[Rn]7s$	$(^2S_{1/2})$
88	Ra	$[Rn]7s^2$	$^1S_0$
89	Ac	$[Rn]6d7s^2$	$^2D_{3/2}$
90	Th	$[Rn]6d^2 7s^2$	$^3F_2$
91	Pa	$[Rn]5f^2 6d7s^2$	$(^4K_{11/2})$
92	U	$[Rn]5f^3 6d7s^2$	$^5L_6$
93	Np	$[Rn]5f^4 6d7s^2$	$(^6L_{11/2})$
94	Pu	$[Rn]5f^6 7s^2$	$(^7F_0)$
95	Am	$[Rn]5f^7 7s^2$	$(^8S_{7/2})$
96	Cm	$[Rn]5f^7 6d7s^2$	$(^9D_2)$
97	Bk	$[Rn]5f^9 7s^2$	$(^6H_{15/2})$
98	Cf	$[Rn]5f^{10}7s^2$	$(^5I_8)$
99	Es	$[Rn]5f^{11}7s^2$	$(^4I_{15/2})$
100	Fm	$[Rn]5f^{12}7s^2$	$(^3H_6)$
101	Md	$[Rn]5f^{13}7s^2$	$(^2F_{7/2})$
102	No	$[Rn]5f^{14}7s^2$	$(^1S_0)$
103	Lr	$[Rn]5f^{14}6d7s^2$	$(^2D_{3/2})$

Data from C.E. Moore, *Atomic Energy Levels*, National Bureau of Standards Circular No. 467, Washington, D.C., 1949, 1952, and 1958. Data for actinides are from G.T. Seaborg, *Ann. Rev. Nucl. Sci.*, *18*, 53 (1968) and references therein. Terms in parentheses are not established experimentally.

Table 4-3. Microstates which Arise from the Configuration p²

$M_L \backslash M_S$	1	0	−1
2		$(\overset{+}{1},\overset{-}{1})$	
1	$(\overset{+}{1},\overset{+}{0})$	$(\overset{+}{1},\overset{-}{0})\ (\overset{+}{0},\overset{-}{1})$	$(\overset{-}{1},\overset{-}{0})$
0	$(\overset{+}{1},-\overset{+}{1})$	$(\overset{+}{1},-\overset{-}{1})\ (-\overset{+}{1},\overset{-}{1})$ $(\overset{+}{0},\overset{-}{0})$	$(\overset{-}{1},-\overset{-}{1})$
−1	$(-\overset{+}{1},\overset{+}{0})$	$(-\overset{+}{1},\overset{-}{0})\ (\overset{+}{0},-\overset{-}{1})$	$(-\overset{-}{1},\overset{-}{0})$
−2		$(-\overset{+}{1},-\overset{-}{1})$	

≡

$M_L \backslash M_S$	1	0	−1
2		1	
1	1	2	1
0	1	3	1
−1	1	2	1
−2		1	

$(\overset{+}{0},\overset{+}{0})$ since $M_L = 0 + 0 = 0$ and $M_S = +1/2 + 1/2 = 1$. But the two electrons in the microstate $(\overset{+}{0},\overset{+}{0})$ have the same set of quantum numbers, which violates the Pauli principle. The simplified form of the microstate table, also shown under the heading of Table 4-3, simply lists the number of microstates for each value of M_L and M_S.

In the *Russell-Saunders* (or *LS*) *coupling scheme*, each atomic state is given the symbol ^{2S+1}L. Starting at the top of the simplified Table 4-3, we find a microstate for which ($M_L = 2$, $M_S = 0$). This must be part of an atomic state for which $L = 2$ and $S = 0$, which is called a ^1D state (read "singlet d"). The superscript, called the *spin multiplicity*, is $2S + 1 = 2(0) + 1 = 1$. The value $L = 2$ is written as a capital D according to the convention: $S = 0$, $P = 1$, $D = 2$, $F = 3$, $G = 4$, $H = 5$, $I = 6$, $K = 7$, $L = 8$, *etc.* (Note that J is omitted.) Equation 4-7 tells us that for a state with $L = 2$, M_L can be 2, 1, 0, −1, or −2. Equation 4-8 tells us that for $S = 0$, the only possible value of M_S is 0. The ^1D state, therefore, accounts for one microstate from each box for which $M_L = 2, 1, 0, -1$, and -2, and $M_S = 0$. We therefore subtract one (arbitrary) microstate from each of these boxes of the table. This transforms the simplified table as follows:

$M_L \backslash M_S$	1	0	−1
2		1	
1	1	2	1
0	1	3	1
−1	1	2	1
−2		1	

remove ^1D microstates ⟶

$M_L \backslash M_S$	1	0	−1
2			
1	1	1	1
0	1	2	1
−1	1	1	1
−2			

The top left microstate, ($M_L = 1$, $M_S = 1$), in the resulting table must be part of a ^3P (read "triplet p") atomic state, characterized by $L = 1$ and $S = 1$. This state must have the components $M_L = 1$, 0, and -1, and $M_S = 1$, 0, and -1. Subtracting these *nine* microstates, we are left with a single microstate:

$M_L \backslash M_S$	1	0	−1
2			
1	1	1	1
0	1	2	1
−1	1	1	1
−2			

remove ^3P microstates $\longrightarrow$

$M_L \backslash M_S$	1	0	−1
2			
1			
0		1	
−1			
−2			

The remaining microstate defines a ^1S ($L = 0$, $S = 0$) atomic state.†

The fifteen microstates of carbon in Table 4-3 therefore define the three energetically distinct atomic states ^1D, ^3P, and ^1S. Hund's first rule says that the state with the maximum spin multiplicity, which is ^3P in this case, will be lowest in energy. *Hund's second rule* says that for states of the same spin multiplicity (^1D, ^1S) the state with greater orbital angular momentum will usually be lower in energy. Since for a D state $L = 2$ and for an S state $L = 0$, we expect ^1D to be of lower energy than ^1S. We therefore arrange the three states of carbon which arise from the ground state configuration, $1s^2 2s^2 2p^2$, by order of increasing energy as follows: $^3P < {}^1D < {}^1S$. The splitting between these states is considerable. ^1D lies some 10,200 cm^{-1} (122 kJ/mol, 29.2 kcal/mol) above ^3P, and ^1S is 21,600 cm^{-1} above ^3P.

Table 4-4 summarizes the atomic states which arise from a variety of configurations. The *LS* coupling scheme works well for light atoms but breaks down for heavier atoms. In a heavy atom the total of spin plus orbital angular momentum ($\equiv J = L + S, L + S - 1, L + S - 2, ..0, ..L - S$) becomes quite significant in determining the energy of the state.†† The ^3P state of carbon has three *levels*, designated $^3P_0$, $^3P_1$, and $^3P_2$, in which the subscript

†The method of determining Russell-Saunders terms is based mostly on H.B. Gray, *Electrons and Chemical Bonding*, W.A. Benjamin, N.Y., 1965; and partly on K.E. Hyde, *J. Chem. Ed.*, *52*, 87 (1975). The reader is referred to either of these sources for additional examples. A completely different method of deriving Russell-Saunders terms is given in Problem 5-54, which should not be attempted until you have read Section 5-3-D.

††Do not confuse the total angular momentum, J, with the rotational quantum number for which we use the same symbol. The meaning of J will usually be clear in context. Also do not confuse S with the multiplicity, which is equal to $2S + 1$.

Table 4-4. States which Arise from Common Configurations

Equivalent Electrons

p^1, p^5	^2P
p^2, p^4	^3P, ^1D, ^1S
p^3	^4S, ^2D, ^2P
d^1, d^9	^2D
d^2, d^8	^3F, ^3P, ^1G, ^1D, ^1S
d^3, d^7	^4F, ^4P, ^2H, ^2G, ^2F, two ^2D's, ^2P
d^4, d^6	^5D, ^3H, ^3G, two ^3F's, ^3D, two ^3P's, ^1I, two ^1G's, ^1F, two ^1D's, two ^1S's
d^5	^6S, ^4G, ^4F, ^4D, ^4P, ^2I, ^2H, two ^2G's, two ^2F's, three ^2D's, ^2P, ^2S
f^1, f^{13}	^2F
f^2, f^{12}	^3H, ^3F, ^3P, ^1I, ^1G, ^1D, ^1S
f^3, f^{11}	^4I, ^4G, ^4F, ^4D, ^4S, ^2L, ^2K, ^2I, two ^2H's, two ^2G's, two ^2F's, two ^2D's, ^2P
f^4, f^{10}	^5I, ^5G, ^5F, ^5D, ^5S, ^3M, ^3L, two ^3K's, two ^3Is, four ^3H's, three ^3G's, four ^3F's, two ^3D's, three ^3P's, ^1N, two ^1L's, ^1K, three ^1I's, two ^1H's, four ^1G's, ^1F, four ^1D's, two ^1S's
f^5, f^9	^6H, ^6F, ^6P, ^4M, ^4L, two ^4K's, three ^4I's, three ^4H's, four ^4G's, four ^4F's, three ^4D's, two ^4P's, ^4S, ^2O, ^2N, two ^2M's, three ^2L's, five ^2K's, five ^2I's, seven ^2H's, six ^2G's, seven ^2F's, five ^2D's, four ^2P's,
f^6, f^8	^7F, ^5L, ^5K, two ^5I's, two ^5H's, three ^5G's, two ^5F's, three ^5D's, ^5P, ^5S, ^3O, ^3N, three ^3M's, three ^3L's, six ^3K's, six ^3I's, nine ^3H's, seven ^3G's, nine ^3F's, five ^3D's, six ^3P's, ^1Q, two ^1N's, two ^1M's, four ^1L's, three ^1K's, seven ^1I's, four ^1H's, eight ^1G's, four ^1F's, six ^1D's, ^1P, four ^1S's
f^7	^8S, ^6I, ^6H, ^6G, ^6F, ^6D, ^6P, ^4N, ^4M, three ^4L's, three ^4K's, five ^4I's, five ^4H's, seven ^4G's, five ^4F's, six ^4D's, two ^4P's, two ^4S's, ^2Q, ^2O, two ^2N's, four ^2M's, five ^2L's, seven ^2K's, nine ^2I's, nine ^2H's, ten ^2G's, ten ^2F's, seven ^2D's, five ^2P's, two ^2S's

Nonequivalent Electrons[a]

ss	^1S, ^3S
sp	^1P, ^3P
sd	^1D, ^3D
pp	^3D, ^1D, ^3P, ^1P, ^3S, ^1S
pd	^3F, ^1F, ^3D, ^1D, ^3P, ^1P
dd	^3G, ^1G, ^3F, ^1F, ^3D, ^1D, ^3P, ^1P, ^3S, ^1S
sss	^4S, two ^2S's
ssp	^4P, two ^2P's
spp	^4D, two ^2D's, ^4P, two ^2P's, ^4S, two ^2S's
spd	^4F, two ^2F's, ^4D, two ^2D's, ^4P, two ^2P's

[a]The notation ss, for example, implies that the two s electrons are in different shells.

gives the value of J ($= 1 + 1 = 2$; $= 1 + 1 - 1 = 1$; $= 1 - 1 = 0$). For carbon these three levels are only split by a total of 43.5 cm^{-1}, which is insignificant compared to the 10,200 cm^{-1} splitting between Russell-Saunders terms. But in Pb, which also has the p^2 configuration, the splitting between these three levels is a total of 10,651 cm^{-1}. *Hund's third rule* states that in atoms with a less-than-half-filled subshell (such as p^2), the level of lowest J will generally be lowest in energy ($^3P_0 < {}^3P_1 < {}^3P_2$). In atoms with a greater-than-half-filled subshell (such as p^4) the level of highest J will be lowest in energy. The trend of the splitting of the levels of the Group IV atoms is illustrated in Table 4-5. In the limit in which the value of J is the most important determinant of the energy of a level, the energy order would be $^3P_0 < {}^1S_0 < {}^3P_1 < {}^3P_2 < {}^1D_2$ (Fig 4-3). Clearly this is not achieved, even for Pb. This extreme limit in which the value of J is most critical is called the *jj* (or *spin-orbit*) coupling scheme. In practice, most atoms are closer to the *LS* coupling limit than the *jj* coupling limit. The ground state levels for the elements were given in Table 4-2.

Table 4-5. Energies (cm^{-1}) of Group IV Atomic Levels

Atom	$^3P_0$	$^3P_1$	$^3P_2$	$^1D_2$	$^1S_0$
C	0	16.4	43.5	10,193.7	21,648.4
Si	0	77.15	223.31	6,298.81	15,394.24
Ge	0	557.10	1,409.90	7,125.26	16,367.14
Sn	0	1,691.8	3,427.7	8,613.0	17,162.6
Pb	0	7,819.35	10,650.47	21,457.90	29,466.81

From C.E. Moore, *Atomic Energy Levels*, National Bureau of Standards Circular No. 467, Washington, D.C., 1949, 1952, and 1958.

Problems

4-5. Explain why the Russell-Saunders term for He, with the configuration $1s^2$ is ^1S. Explain why the configuration d^1 gives the state ^2D. What levels are contained in the ^2D term?

4-6. Using a table of microstates, derive the Russell-Saunders terms for the configurations (a) sp, (b) d^2, and (c) d^3 (a long problem).

4-7. What is the ground state atomic *configuration* of Na? What is the first excited *configuration*? To what atomic *levels* do these configurations give rise? The yellow Na "*D*" line represents emission from the first excited configuration to the ground configuration. The *D* line is actually a doublet, with components at 589.15788 and 589.75537 nm. Label each transition with the appropriate symbol indicating initial and final atomic levels. What is the energy separation in cm^{-1} between the two levels of the excited configuration?

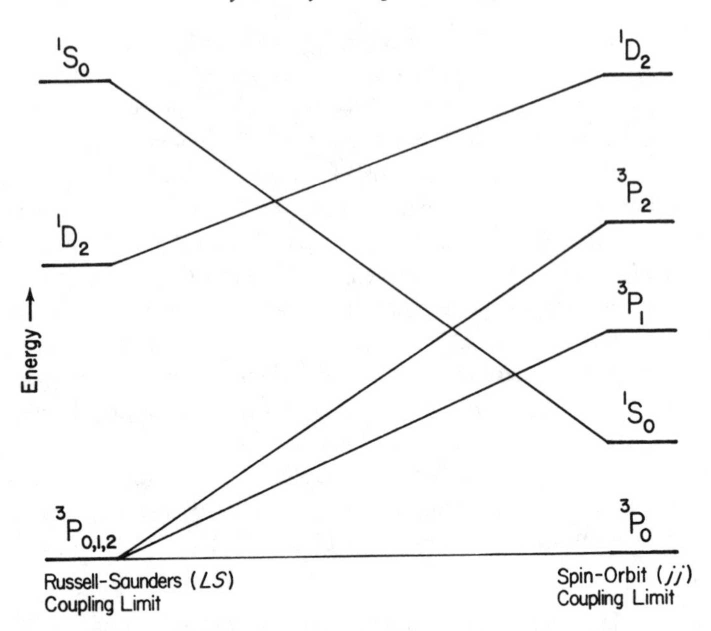

Fig. 4-3. Schematic illustration of the variation in energy of each atomic state of the p^2 configuration upon going from the Russell-Saunders (LS) coupling limit to the spin-orbit (jj) coupling limit. In the former case, spin and orbital angular momentum separately determine the energy of the state. In the latter case, the total angular momentum (spin + orbital) is most important.

4-3. Photoelectron Spectroscopy

When a high energy photon strikes an electron in an atom or molecule, ionization may result. The kinetic energy of the ejected electron is equal to the photon energy minus the ionization energy (IE):

$$K = h\nu - IE \tag{4-9}$$

In *photoelectron spectroscopy* (PES) high energy monochromatic radiation, typically the 58.4 nm (He I) emission of a He discharge lamp, is shined on a gaseous sample to be studied. Electrons ejected from the sample are counted as a function of their kinetic energy. That is, the photoelectron spectrometer produces a graph of electrons ejected *vs.* electron kinetic energy. Knowing the values of $h\nu$ and K in eq. 4-9, one can plot the data as electrons ejected *vs.* ionization energy. Such a plot is called a photoelectron spectrum. The 58.4 nm photon carries an energy of 21.2 eV (1 eV = 96.49 kJ/mol =

23.06 kcal/mol) and so can ionize electrons with an ionization energy less than 21.2 eV. Typically, these include just the more loosely bound valence electrons. By using a higher energy source, more tightly held electrons can be ionized. For example, the Cr K_α x-ray emission line (5414.7 eV) is commonly employed in a method known as ESCA (Electron Spectroscopy for Chemical Analysis) which is analogous to PES.

As an instructive example, consider the complete photoelectron spectrum of Ar shown in Fig. 4-4(a), a composite of PES and ESCA results. The electronic configuration of Ar is $1s^2 2s^2 2p^6 3s^2 3p^6$, which defines a $^1S_0$ ground state. That is, there is zero orbital and zero spin angular momentum. When an electron is ejected from the outermost orbital (3p), the atom is left with the configuration $1s^2 2s^2 2p^6 3s^2 3p^5$, which defines $^2P_{3/2}$ and $^2P_{1/2}$ levels,† of which the $^2P_{3/2}$ level is lower in energy according to Hund's third rule (Section 4-2-C). In the photoelectron spectrum of Ar, we assign the lowest *IE* doublet near 16 eV to the transitions $^1S_{1/2} \rightarrow {}^2P_{3/2}$ and $^1S_{1/2} \rightarrow {}^2P_{1/2}$ (transitions a and b, Fig. 4-4[b]). The next lowest energy ionization occurs if a 3s electron is removed from the neutral atom, yielding a $^2S_{1/2}$ level (transition c, Fig. 4-4[b]). This gives rise to the peak near 29 eV in Fig. 4-4(a). In a similar manner, the complete spectrum of Ar can be readily and satisfyingly explained.

Koopmans' theorem, which states that the *IE* is equal in magnitude to the orbital energy (with electron-electron repulsions included), is implicit in the preceding discussion. Koopmans' "theorem" is really an approximation. For example, although there is only one "3p orbital energy" for Ar, there are two levels resulting from the two physically distinguishable ways that the orbital and spin angular momenta may be arranged in the configuration $3p^5$. The orbital and spin angular momenta may be either parallel or antiparallel, resulting in a total momentum, J, of $L + S$ or $L - S$ in this case:

↑	+	↑	⇒	↑		↑	+	↓	⇒	↑
L		S		J		L		S		J
1		1/2		3/2		1		−1/2		1/2

To extract the energy of an electron in the 3p orbital from the two peaks near 16 eV would require some additional theory. (It is also an approximation, albeit a good one, to associate a particular ^2P state with the 3p orbital.

† You might try to convince yourself of this by making a table of microstates. It is very useful to know that a more-than-half-filled shell with n vacancies gives rise to the same terms as a less-than-half-filled shell with n electrons. That is, the configurations p^5 and p^1 give rise to the same terms. So, for example, do the pairs of configurations p^2 and p^4, and d^3 and d^7.

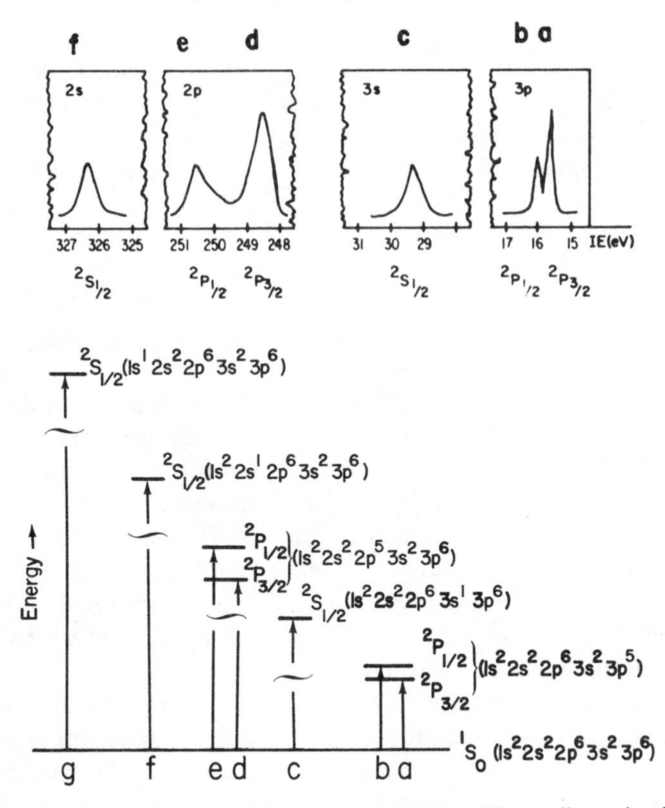

Fig. 4-4. (a) Complete photoelectron spectrum of Ar. The ordinate is electron counts per second. Reproduced from H. Bock and P.D. Mollère, *J. Chem. Ed.*, *51*, 506 (1974). (b) Ground state and singly ionized states of the Ar atom showing transitions observed in the photoelectron spectrum.

Any other configuration which gives rise to a ^2P state will contribute to some extent to the ^2P state in question.)

One other feature of photoelectron spectra worth noting is that under ideal circumstances† the area under a peak is proportional to the degeneracy of the orbital from which ionization occurs. For example, in comparing the areas under the 2s and 2p peaks of Ar at 326 and ~249 eV, the two peaks resulting from 2p ionization together have about three times the area of the 2s peak. (The 2p orbital is threefold degenerate and the 2s orbital is non-

†For a discussion of the limitations of using peak areas in PES, as well as other limitations of PES in general, see R.L. DeKock and D.R. Lloyd, *Adv. Inorg. Chem. Radiochem.*, *16*, 65 (1974).

degenerate.) Further, within the 2p region, the $^2P_{3/2}$ peak has twice the area of the $^2P_{1/2}$ peak because the degeneracy of a state characterized by a certain value of J is $2J + 1$. (For $^2P_{3/2}$, $J = 3/2$, and $2J + 1 = 4$. For $^2P_{1/2}$, $J = 1/2$, and $2J + 1 = 2$.) This is the most direct way to assign the order of the two levels $^2P_{3/2}$ and $^2P_{1/2}$ and is experimental confirmation of Hund's third rule.†

Problems

4-8. Make a list showing the order of increasing energy of all of the transitions expected in the photoelectron spectrum of nitrogen (ground state $^4S_{3/2}$) in which N^+ is the product. (Note that for $L = 0$ [*i.e.*, S terms], the total angular momentum is not, $L + S, L + S - 1 \ldots L - S$, but just S. For a ^4S term, therefore, the only state is $^4S_{3/2}$).

4-4. The LCAO Molecular Orbital Method

In this section we are going to outline one of the simplest molecular orbital theories in which molecular wave functions are written as linear combinations of atomic wave functions. *It must be emphasized that molecular electronic wave functions are not really linear combinations of atomic electronic wave functions* (*orbitals*). This is just an approximation commonly chosen for its simplicity and is adequate for our purposes. Before beginning, we need to introduce the *variation theorem.* From eq. 2-43 and Table 2-2 we know that the energy of a particle whose wave function is ψ will be given by eq. 4-10:

$$E = \frac{\int_{-\infty}^{\infty} \psi^* \mathscr{H} \psi \, d\tau}{\int_{-\infty}^{\infty} \psi^* \psi \, d\tau} \tag{4-10}$$

If ψ is normalized, the denominator is unity. Now suppose we have a hamiltonian but the Schrödinger equation cannot be solved. What happens if we *guess* a function, ψ_g, and put it into eq. 4-10? This will give us an energy,

†In Fig. 4-4 the area under the pair of 3p peaks cannot be compared to the rest of the spectrum because the 3p region was recorded on a different spectrometer than was used for the higher energy peaks.

E_g, based on a fictional wave function plus the actual hamiltonian describing the system of interest:

$$E_g = \frac{\int_{-\infty}^{\infty} \psi_g^* \mathscr{H} \psi_g d\tau}{\int_{-\infty}^{\infty} \psi_g^* \psi_g d\tau} \tag{4-11}$$

The variation theorem tells us that the energy, E_g, that we get by guessing *any* function, ψ_g, will be *greater* than the true energy, E, if we could actually solve the differential equation. This theorem is the basis of an important approximation method called the variation method. We are going to guess a wave function and give it some adjustable parameters. We will then minimize the energy calculated with eq. 4-11 by varying the parameters. If the resulting energy is close to the real energy of the molecule (determined by some kind of an experiment) we can hope that the guessed wave function is a good approximation to the actual wave function. The variation theorem allows us to be certain that the guessed wave function can never be "too good" in that it can never predict a more stable molecule than actually exists.

With this meager background, we are ready to look at some molecular problems. The simplest molecule is the H_2^+ cation, containing two nuclei and one electron. With the aid of Fig. 4-5 one can construct a suitable hamiltonian for the slightly more general case of one electron and two arbitrary nuclei of charges $+Z_A e$ and $+Z_B e$. The hamiltonian must contain a kinetic energy term for each of the three particles as well as a potential energy term including the three different electrostatic interactions. Thus, we get

$$\mathscr{H} = \underbrace{-\frac{\hbar^2}{2m_e}\nabla_e^2 - \frac{\hbar^2}{2m_A}\nabla_A^2 - \frac{\hbar^2}{2m_B}\nabla_B^2}_{\text{kinetic energy terms}} \underbrace{- \frac{Z_A e^2}{4\pi\epsilon_0 R_A} - \frac{Z_B e^2}{4\pi\epsilon_0 R_B} + \frac{Z_A Z_B e^2}{4\pi\epsilon_0 R_{AB}}}_{\text{potential energy terms}} \tag{4-12}$$

As in the treatment of atoms (Section 4-2-A) and diatomic molecules (Section 3-3-D), the Schrödinger equation can be divided into one equation describing translation of the entire unit through space and another equation describing the internal motions of the system. We will neglect the translation equation. A common and very useful approximation at this point is the *Born-Oppenheimer approximation.* Since the nuclei are several thousand times more massive than the electrons and therefore will move much more slowly than the electrons, we will consider the problem in which the nuclei are fixed at

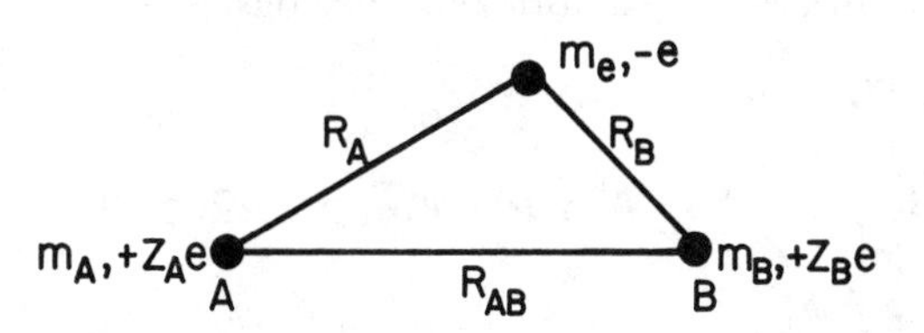

Fig. 4-5. A system of two nuclei of masses m_A and m_B, and charges $+Z_A$ and $+Z_B$, and an electron of mass m_e and charge $-e$. The distances between particles are R_{AB}, R_A, and R_B.

the separation R_{AB} and only the electron will be allowed to move. This eliminates the second and third terms of the Hamiltonian since the ∇^2 terms are zero for stationary particles. We will solve the problem for several values of R_{AB} and the equilibrium internuclear distance will be that corresponding to the lowest energy of the system.

Although it is easy to write the hamiltonian eq. 4-12, it is another matter to solve the corresponding Schrödinger equation. In fact, this equation can be solved exactly, if you are clever enough. But as soon as a second electron is added the problem becomes hopelessly complicated. A more generally useful procedure is to guess a reasonable form of the solution and to minimize the energy of the guessed wave function. The variation theorem assures us that our guess cannot be "too good." Let us guess a wave function that is some linear combination of the 1s atomic orbitals of the two isolated atoms whose nuclei are A and B.

$$\psi = c_1\phi_{1s_A} + c_2\phi_{1s_B} \tag{4-13}$$

Our wave function, ψ, is said to be a molecular orbital. The game we are playing is called the LCAO-MO (Linear Combination of Atomic Orbitals-Molecular Orbital) method.

The functions ϕ_{1s_A} and ϕ_{1s_B} are already normalized and each is a solution of the Schrödinger equation for atoms A or B. For atom A we would have

$$\mathscr{H}_A = -\frac{\hbar^2}{2m_e}\nabla_e^2 - \frac{Z_A e^2}{4\pi\epsilon_0 R_A} \tag{4-14}$$

$$H_A\phi_{1s_A} = E_A\phi_{1s_A} \tag{4-15}$$

$$\int_{-\infty}^{\infty}\phi_{1s_A}^*\,\phi_{1s_A}\,d\tau = 1 \tag{4-16}$$

Now let us normalize ψ using a normalization constant N:

$$1 = \int_{-\infty}^{\infty} (N\psi)^*(N\psi)d\tau \tag{4-17}$$

$$1 = N^2 \int_{-\infty}^{\infty} (c_1 \phi^*_{1s_A} + c_2 \phi^*_{1s_B})(c_1 \phi_{1s_A} + c_2 \phi_{1s_B})d\tau$$

$$1 = N^2 \int_{-\infty}^{\infty} (c_1^2 \phi^2_{1s_A} \, d\tau + 2c_1 c_2 \phi_{1s_A} \phi_{1s_B} \, d\tau + c_2^2 \phi^2_{1s_B} \, d\tau) \tag{4-18}$$

$$1 = N^2 (c_1^2 \cdot 1 + 2c_1 c_2 \int_{-\infty}^{\infty} \phi_{1s_A} \phi_{1s_B} \, d\tau + c_2^2 \cdot 1)$$

$$1 = N^2 (c_1^2 + c_2^2 + 2c_1 c_2 S_{1s_A \, 1s_B})$$

$$N = (c_1^2 + c_2^2 + 2c_1 c_2 S_{1s_A \, 1s_B})^{-\frac{1}{2}} \tag{4-19}$$

where
$$S_{1s_A \, 1s_B} \equiv \int_{-\infty}^{\infty} \phi_{1s_A} \phi_{1s_B} \, d\tau \tag{4-20}$$

and we assume that the wave functions are real. The first and third terms of eq. 4-18 were readily evaluated because each atomic wave function is already normalized (eq. 4-16). The middle term of eq. 4-18 involves the *overlap integral* $S_{1s_A \, 1s_B}$. In general, the overlap integral of any two orbitals ϕ_i and ϕ_j is denoted S_{ij}. Physically, the overlap integral is a measure of the volume in which there is electron density from both atoms (Fig. 4-6). The maximum value of any overlap integral is unity, when the two orbitals completely coincide.

We now have a normalized LCAO wave function. The variation method can next be used to determine the values of c_1 and c_2 that give the lowest energy. Using eq. 4-11 we find that the energy of the molecular orbital is

$$E = \int_{-\infty}^{\infty} (N\psi)^* \mathscr{H} (N\psi)d\tau \tag{4-21}$$

$$= N^2 \int (c_1 \phi_{1s_A} + c_2 \phi_{1s_B}) \mathscr{H} (c_1 \phi_{1s_A} + c_2 \phi_{1s_B})d\tau$$

$$= N^2 [c_1^2 \underbrace{\int \phi_{1s_A} \mathscr{H} \phi_{1s_A} \, d\tau}_{H_{AA}} + c_1 c_2 \underbrace{\int \phi_{1s_A} \mathscr{H} \phi_{1s_B} \, d\tau}_{H_{AB}}$$

$$+ c_2 c_1 \underbrace{\int \phi_{1s_B} \mathscr{H} \phi_{1s_A} \, d\tau}_{H_{BA}} + c_2^2 \underbrace{\int \phi_{1s_B} \mathscr{H} \phi_{1s_B} \, d\tau}_{H_{BB}}]$$

$$E = N^2 [c_1^2 H_{AA} + c_1 c_2 H_{AB} + c_2 c_1 H_{BA} + c_2^2 H_{BB}] \tag{4-22}$$

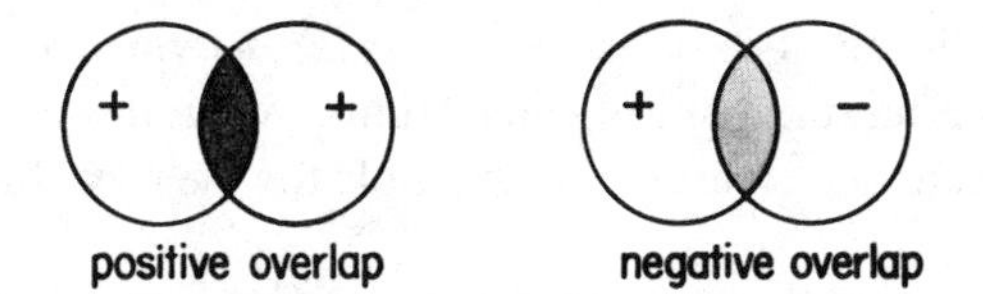

Fig. 4-6. The overlap integral is a measure of the volume in which orbitals overlap.

Here we have assumed that the coefficients are real and we have just given names to the various integrals. H_{AA} and H_{BB} are called *Coulomb integrals* and H_{AB} and H_{BA} are called *exchange integrals.* They are definite integrals and hence merely numbers. One more simplification of the problem is to note that all hamiltonians are Hermitian. This is a big word meaning that $H_{ij} = H_{ji}$. Making this simplification, and combining eq. 4-22 with eq. 4-19, we find

$$E = \frac{c_1^2 H_{AA} + c_2^2 H_{BB} + 2c_1 c_2 H_{AB}}{c_1^2 + c_2^2 + 2c_1 c_2 S} \tag{4-23}$$

Here we have abbreviated $S_{1s_A\, 1s_B}$ to just plain S. What we want to do is to differentiate E with respect to c_1 and c_2 and find the coefficients resulting in the most stable molecular orbital.

$$E(c_1^2 + c_2^2 + 2c_1c_2S) = c_1^2 H_{AA} + c_2^2 H_{BB} + 2c_1c_2H_{AB} \tag{4-24}$$

$$\frac{\partial E}{\partial c_1}(c_1^2 + c_2^2 + 2c_1c_2S) + E(2c_1 + 2c_2S) = 2c_1H_{AA} + 2c_2H_{AB} \tag{4-25}$$

$$\frac{\partial E}{\partial c_2}(c_1^2 + c_2^2 + 2c_1c_2S) + E(2c_2 + 2c_1S) = 2c_2H_{BB} + 2c_1H_{AB} \tag{4-26}$$

Setting each derivative equal to zero, there results

$$c_1(H_{AA} - E) + c_2(H_{AB} - ES) = 0 \tag{4-27}$$

$$c_1(H_{AB} - ES) + c_2(H_{BB} - E) = 0 \tag{4-28}$$

Now eqs. 4-27 and 4-28 can be solved only if[†]

$$\begin{vmatrix} H_{AA} - E & H_{AB} - ES \\ H_{AB} - ES & H_{BB} - E \end{vmatrix} = 0 \tag{4-29}$$

[†]To see why eq. 4-29 must hold, let us solve for c_1 in eqs. 4-27 and 4-28 using determinants:

$$c_1 = \begin{vmatrix} 0 & H_{AB} - ES \\ 0 & H_{BB} - E \end{vmatrix} \div \begin{vmatrix} H_{AA} - E & H_{AB} - ES \\ H_{AB} - ES & H_{BB} - E \end{vmatrix}$$

Since the determinant in the numerator is zero, c_1 can have a nonzero value only if the denominator is zero as well, making the whole expression indeterminate.

The determinant in eq. 4-29 is called a *secular determinant* and appears frequently in quantum mechanical calculations. When the condition in eq. 4-29 is met, we have the lowest energy possible for the wave function in eq. 4-13.

Let's look at the case of a homonuclear diatomic molecule. That means that A and B are identical and $H_{AA} = H_{BB}$. It does not imply that $H_{AB} = H_{AA}$ because the exchange integral, H_{AB}, involves wave functions at two centers (A and B) and the Coulomb integral, H_{AA}, involves functions at only one center. The Hermitian property of the hamiltonian implies that $H_{BA} = H_{AB}$ regardless of whether A and B are the same. Equation 4-29 then becomes

$$(H_{AA} - E)^2 = (H_{AB} - ES)^2 = 0 \tag{4-30}$$

$$H_{AA} - E = \pm(H_{AB} - ES)$$

$$E = \frac{H_{AA} \pm H_{AB}}{1 \pm S} \tag{4-31}$$

When we evaluate the Coulomb and exchange integrals we find bonding (ψ_b) and antibonding (ψ^*)[†] energy levels (Fig. 4-7). The antibonding level is destabilized somewhat more than the bonding level is stabilized, relative to the energy of the 1s orbital. A plot of energy *vs.* internuclear distance is given in Fig. 4-8. What is actually plotted here is the energy of the electron in each molecular orbital minus the energy of the electron in a hydrogen atom. We find that the lowest energy configuration is 1.76 eV (170 kJ/mol) more stable than a single hydrogen atom, and that the internuclear distance, R_{AB}, corresponding to this minimum energy is 1.32 Å. These values compare to experimental values of 2.79 eV and 1.06 Å. You can see that the wave function eq. 4-13 does not get us very close to the true results, but it takes us in the right direction, giving correct qualitative results. The mo theory we use in this book will never be more sophisticated than this. What is important is that the symmetries of the wave functions we derive by our method and the symmetries of the more sophisticated wave functions must be the same.

We need, finally, to plug the values of E for the bonding and antibonding orbitals into eqs. 4-27 and 4-28 to find c_1 and c_2. When this is done, we find $c_2 = c_1$ for ψ_b and $c_2 = -c_1$ for ψ^*. Thus we get

[†]The symbol ψ^* will generally be used from now on in this chapter to denote an antibonding orbital, and not the complex conjugate of ψ. In Chapter Five the meaning of the asterisk should be clear in context.

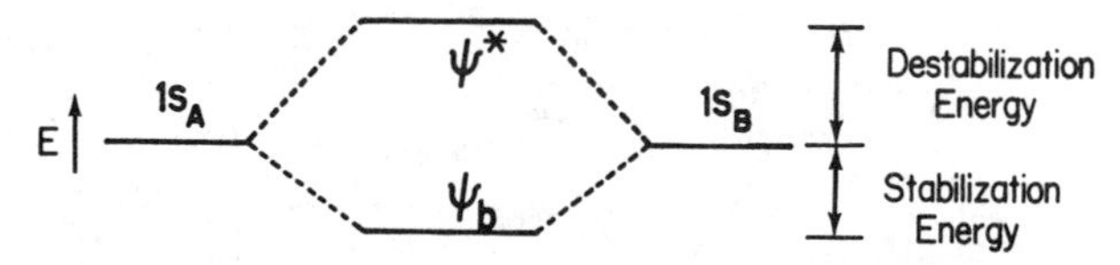

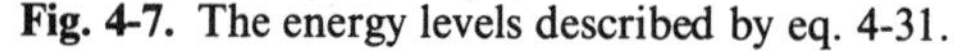

Fig. 4-7. The energy levels described by eq. 4-31.

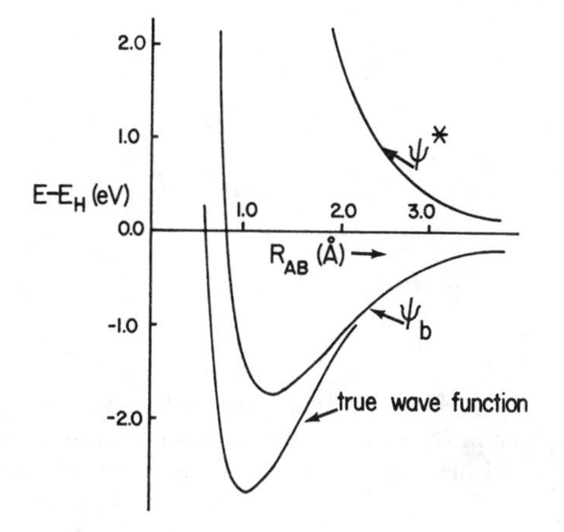

Fig. 4-8. Comparison of LCAO and true orbital energies for the $H_2{}^+$ molecule as a function of R_{AB}. Adapted from H. Eyring, J. Walter, and G.E. Kimball, *Quantum Chemistry*, John Wiley & Sons, N.Y., 1944.

$$\psi_b = \frac{1}{\sqrt{2+2S}}(\phi_{1s_A} + \phi_{1s_B}) \tag{4-32}$$

$$\psi^* = \frac{1}{\sqrt{2-2S}}(\phi_{1s_A} - \phi_{1s_B}) \tag{4-33}$$

In the case where atoms A and B are not the same, it is found that the plus and minus combinations of the orbitals will produce the energy levels shown in Fig. 4-9. Again bonding and antibonding combinations are produced, but the stabilization of ψ_b relative to the energy of the $1s_B$ orbital is not as great as in the homonuclear case. The coefficients c_1 and c_2 are not the same either. In the bonding molecular orbital, the coefficient will be larger for the more electronegative atom. In the antibonding orbital, this will be reversed (Fig. 4-10). These qualitative features of bonding and antibonding molecular orbitals are true of more complex polyatomic molecules as well.

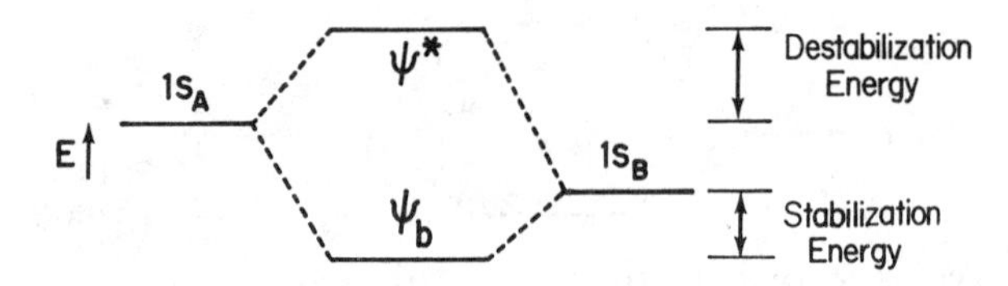

Fig. 4-9. Energy levels obtained when $A \neq B$.

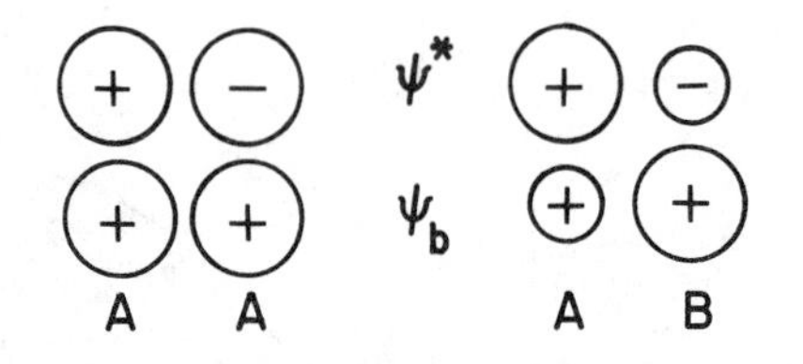

Fig. 4-10. Schematic LCAO wave functions for homonuclear (left) and heteronuclear (right) diatomic molecules, assuming atom B is more electronegative than A. We will generally call pictures such as these molecular orbitals, even though they just show in a schematic manner the proper linear combinations of atomic orbitals. Better representations of molecular orbitals are those such as in Fig. 4-27.

Problems

4-9. Using eq. 4-10, find the energy of ψ_0 of the harmonic oscillator (Section 3-3-D). You will need the integral $\int_0^\infty x^2 e^{-ax^2}\,dx = (1/4)\sqrt{\pi/a^3}$ and your answer should come out to $(1/2)\hbar\sqrt{k/m}$.

4-10. Construct a hamiltonian for a system of two nuclei and two electrons. Remember to include the electrostatic interaction of each electron with the other. Which terms are eliminated by the Born-Oppenheimer approximation?

4-11. Suggest a reason why both ψ_b and ψ^*, in Fig. 4-8, approach the exact solutions of the Schrödinger equation as R_{AB} increases.

4-12. The two sp hybrid orbitals, often used by organic chemists, are made by adding and subtracting s and p atomic orbitals:

$$\psi_a = A(\phi_s + \phi_p)$$
$$\psi_b = B(\phi_s - \phi_p)$$

Determine the values of the normalization constants, A and B. Using eq. 4-10, determine the energies of the two sp hybrids in terms of the energies of the component atomic orbitals.

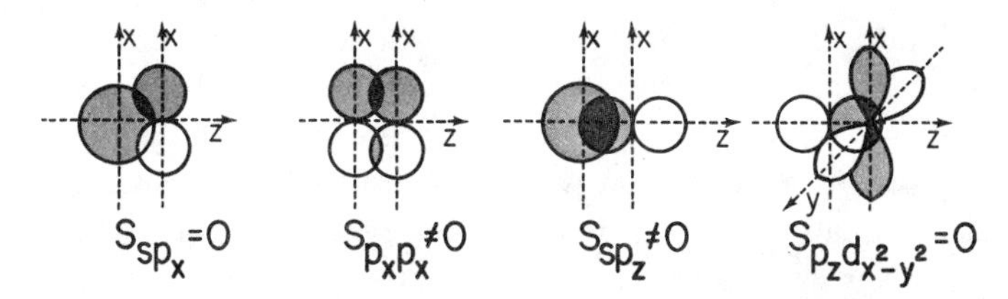

Fig. 4-11. Possible combinations of s, p, and d orbitals with zero or nonzero overlap.

4-5. Diatomic Molecules

Using the LCAO method, we are going to build, *qualitatively*, a set of molecular orbitals for various molecules. There are several principles which guide the construction of molecular orbitals. One is that in order for two orbitals to form a molecular orbital, they must have the same symmetry about the internuclear axis. In Fig. 4-11 are shown several possible combinations of orbitals. Those with overlap different from zero are allowable combinations. If the overlap integral, S_{ij}, is zero, we discover that the exchange integral, H_{ij}, is also zero, and there is no net interaction between the two orbitals.

A second fact which helps in constructing linear combinations is that combining orbitals of very different energies leads to only small interactions. Figure 4-9 provided one example of this. As the energy difference between ϕ_A and ϕ_B increases, the stabilization (and destabilization) energy of the molecular orbital decreases. In constructing molecular orbitals for N_2, for example, we find that trying to combine the 1s atomic orbitals with any of the valence orbitals (2s or 2p) is of little use because there is very little interaction between orbitals of different shells.

The major aid, however, is the fact that *each possible molecular orbital must form a basis for some irreducible representation of the point group of the molecule*. This statement, which means that electronic wave functions must form bases for irreducible representations, is entirely analagous to the statement that vibrational wave functions must form bases for irreducible representations (Section 3-5). The three principles just mentioned are best understood by seeing them in action.

Suppose we wish to construct molecular orbitals for the second row diatomic molecules. The atomic orbitals one would logically start with would be the ls, 2s, and 2p orbitals, since these are the orbitals occupied in the free atoms. As stated above, it is not very useful to try to combine the ls and the valence orbitals with each other. So let us consider these as two separate problems. We have already solved the ls problem because we know that the combinations $ls_A + ls_B$ and $ls_A - ls_B$ are the ones we are seeking. There is a systematic way to determine the symmetries of the molecular orbitals, however, and it will be useful to introduce it here. Starting with a given number of atomic orbitals, we *must* generate that same number of molecular orbitals. The molecular orbitals must have the same symmetries as the atomic orbitals of which they are composed. Let's see how this works for the two ls orbitals of a homonuclear diatomic molecule. The two orbitals and the symmetry elements of the point group $D_{\infty h}$ are shown in Fig. 4-12. Under the identity operation, each atomic orbital goes into itself, so the character under E is 2. C_∞^ϕ also takes each orbital into itself, as does σ_v; so the characters under these operations are also 2. The operations i, S_∞^ϕ, and C_2 cause each orbital to move to the other atom. The character under each of these operations is zero. Thus, the reducible representation for which these two atomic orbitals form a basis is

$D_{\infty h}$	E	$2C_\infty^\phi$	$\infty\sigma_v$	i	$2S_\infty^\phi$	∞C_2	
Γ (ls)	2	2	2	0	0	0	$= \sigma_g^+ + \sigma_u^+$

which reduces to $\sigma_g^+ + \sigma_u^+$. This means that we must make molecular orbitals having the symmetries $\sigma_g^+ + \sigma_u^+$. (The notation for orbitals will utilize small lettlers. Capital letters will be reserved for molecular states, to be introduced in Chapter Five.)

The orbitals ψ_b and ψ^* derived by the variation method do, in fact, satisfy this symmetry requirement. Using the pictures at the left in Fig. 4-10, satisfy yourself that the characters in Table 4-6 are correct. In deriving these characters, each molecular orbital must be considered as a single entity. For example, the character under i for ψ^* is -1 because inversion changes the molecular orbital into minus itself.

Let us determine the symmetries of the valence orbitals now. These eight functions are shown in Fig. 4-13. The character under E is, of course, 8. Upon rotation about the C_∞ axis, the s and p_z orbitals are unchanged and contribute 4 to the trace. Figure 4-14 shows the effect of this rotation on p_x and p_y. These functions transform just as x and y, for which the rotation matrix eq. 4-34 applies (c.f. eq. 1-36):

$$R_\phi \begin{bmatrix} x \\ y \end{bmatrix} = \begin{bmatrix} \cos\phi & \sin\phi \\ -\sin\phi & \cos\phi \end{bmatrix} \begin{bmatrix} x \\ y \end{bmatrix} \qquad (4\text{-}34)$$

Table 4-6. Characters of ψ_b and ψ^*

$D_{\infty h}$	E	$2C_\infty^\phi$	$\infty\sigma_v$	i	$2S_\infty^\phi$	∞C_2	
ψ_b	1	1	1	1	1	1	$=\sigma_g^+$
ψ^*	1	1	1	-1	-1	-1	$=\sigma_u^+$

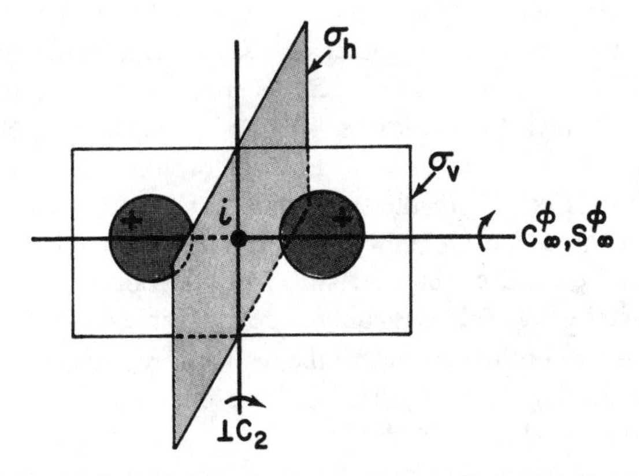

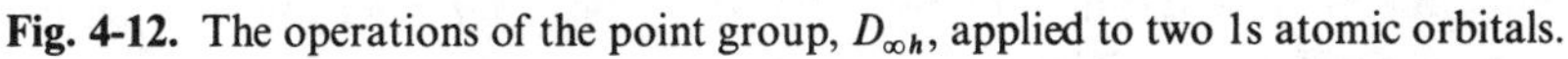
Fig. 4-12. The operations of the point group, $D_{\infty h}$, applied to two 1s atomic orbitals.

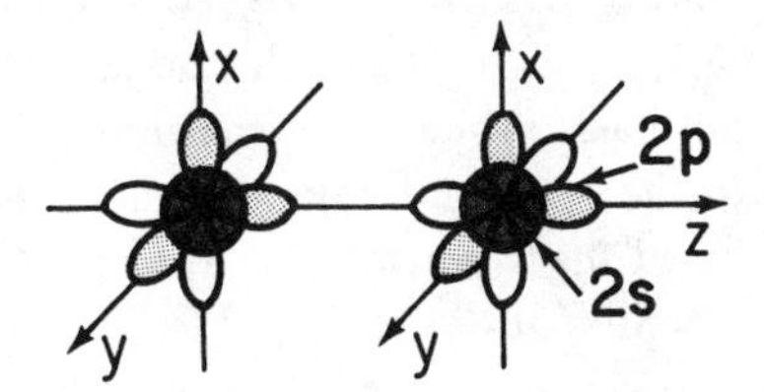

Fig. 4-13. 2s and 2p basis atomic orbitals used to construct the molecular orbitals of a diatomic molecule.

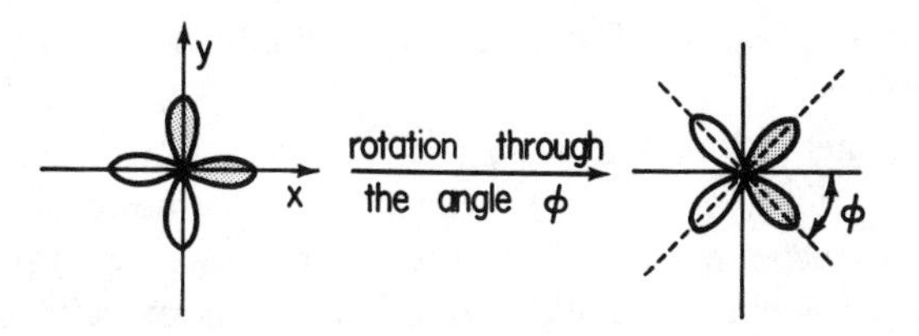

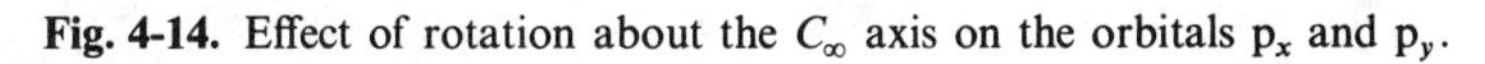
Fig. 4-14. Effect of rotation about the C_∞ axis on the orbitals p_x and p_y.

The character of the rotation is $2\cos\phi$ for each set of p_x and p_y orbitals. This makes sense because a 90° rotation takes p_y into p_x and p_x into $-p_y$. For $\phi = 90°$ the character should be zero and it is. At $\phi = 180°$ it is -2 because each orbital goes into minus itself. There are two sets of p_x and p_y orbitals in our basis set, so the net character under the C_∞^ϕ rotation is $4\cos\phi$. Hence the total character for all eight functions in our basis set is $4 + 4\cos\phi$. Now consider σ_v. We need to pick just one of the infinite number of mirror planes. Can we pick any σ_v plane we please? The answer is yes because we had a theorem in Chapter One saying that all operations in the same class have the same character. So let us choose the *xz* plane as our mirror plane. Applying $\sigma(xz)$ leaves s, p_x, and p_z unchanged. So the character from these six orbitals is $+6$. Each p_y orbital goes to minus itself, giving a character of -2 for the two orbitals. The total character for this operation performed on all eight basis orbitals is $+4$. The operations i, S_∞^ϕ, and C_2 all move the atomic orbitals from their original site and hence have a character of zero. This is summarized in Table 4-7 as $\Gamma(2s, 2p)$.

The next step will be to decompose the reducible representation $\Gamma(2s, 2p)$. Since we are dealing with an infinite point group, we will have to do the decomposition by eye. Let's divide $\Gamma(2s, 2p)$ into the two component reducible representations $\Gamma(2s, 2p_z)$ and $\Gamma(2p_x, 2p_y)$ in Table 4-7. After some reflection you should be able to decompose the former into $2\sigma_g^+ + 2\sigma_u^+$. Scanning the $D_{\infty h}$ character table, you think about $4\cos\phi$ a few times and discover that π_g and π_u add up to $\Gamma(2p_x, 2p_y)$. We have now found six irreducible representations, two of which are doubly degenerate. The symmetries of the eight basis orbitals are determined.

The construction of the eight molecular orbitals is next. Clearly, the 2s orbitals can be combined the same way the 1s orbitals were combined to give a σ_g^+ and a σ_u^+ combination. In fact, Fig. 4-15(a) shows that the p_z orbitals can be combined in this manner as well. All that is left are the p_x and p_y orbitals. The two combinations $(p_x + p_x)$ and $(p_y + p_y)$ in Fig. 4-15(b), *together*, form a basis for the doubly degenerate representation π_u. Similarly, the combinations $(p_x - p_x)$ and $(p_y - p_y)$ form a basis for π_g. We have now used up all the irreducible representations into which $\Gamma(2s, 2p)$ decomposes; so the job of finding linear combinations of appropriate symmetry is done.

Just as symmetry coordinates do not necessarily correspond to normal modes of vibration (Section 3-7), neither must the molecular orbitals we have just constructed correspond to the "best" (lowest energy) molecular orbitals. In particular, we could take the two sets of valence molecular orbitals which transform as σ_g^+ (*viz.*, $2s + 2s$ and $2p_z - 2p_z$) and recombine

Table 4-7. Characters of the Atomic Orbitals

$D_{\infty h}$	E	$2C_\infty^\phi$	$\infty\sigma_v$	i	$2S_\infty^\phi$	∞C_2	
$\Gamma(2s, 2p)$	8	$4+4c\phi$	4	0	0	0	
$\Gamma(2s, 2p_z)$	4	4	4	0	0	0	$= 2\sigma_g^+ + 2\sigma_u^+$
$\Gamma(2p_x, 2p_y)$	4	$4c\phi$	0	0	0	0	$= \pi_g + \pi_u$

them in some new linear combination and perhaps obtain a lower energy (and a higher energy) molecular orbital. Mixing the 2s and $2p_z$ orbitals is the same as hybridizing to form sp orbitals, and this happens to varying extents in all molecules. For our purposes we will use the unhybridized orbitals, but it should be remembered that even in the LCAO approximation these are not the best orbitals we could construct. Regardless of their energies, the symmetries of these orbitals are correct.

What about the relative energies of the molecular orbitals we have constructed? The σ_g^+(1s) bonding combination will be lower in energy than any other molecular orbital because it is the bonding combination of the lowest energy basis atomic orbitals. σ_u^+(1s) will be of practically the same energy as σ_g^+(1s) because there is nearly zero overlap between the two 1s atomic orbitals. The energies of both of these molecular orbitals are nearly equal to the energy of the isolated atomic orbitals. Just considering bonding

Fig. 4-15. (a) Bonding (σ_g^+) and antibonding (σ_u^+) combinations of p_z orbitals. (b) Bonding (π_u) and antibonding (π_g) combinations of p_x and p_y orbitals. There are *two sets* of degenerate orbitals in each.

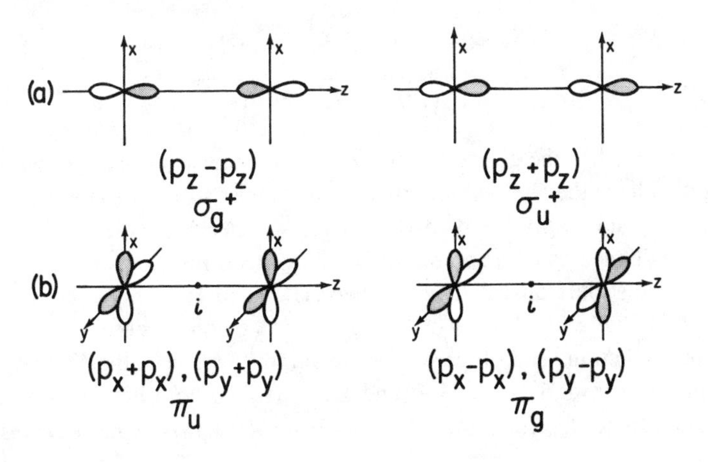

or antibonding character, we can say that $\sigma_g^+(2s)$ is lower than $\sigma_u^+(2s)$, and $\sigma_g^+(2p_z)$ is below $\sigma_u^+(2p_z)$. Further, $\pi_u(2p_x, 2p_y)$ ought to be of lower energy than $\pi_g(2p_x, 2p_y)$.

To gain further insight into the possible relationships of the energy levels, one can consider two extreme possibilities. In one case the two atoms are separated by an infinite distance. The 1s, 2s, and 2p orbitals form just three energy levels. As the atoms are brought closer together, molecular orbitals may be formed by the overlap of the atomic orbitals. When the two atoms come to zero separation we achieve the "united atom" limit. We will have formed a single atom of twice the nuclear charge of each of the original atoms. The path of transformation of each orbital is shown in Fig. 4-16. We find, for example, that the σ_u^+ combination of s orbitals goes into a p_z orbital of σ_u^+ symmetry in the united atom limit. All of these transformations are collected in the *correlation diagram*, Fig. 4-17. The correlation diagram shows the range of possible orderings of molecular orbital energies. There is no reason why the lines connecting the two extremes (united and separated atoms) must be linear. Figure 4-17 is strictly schematic. The molecular orbitals are numbered (*e.g.*, $1\sigma_g^+$, $2\sigma_g^+$, $3\sigma_g^+$) according to increasing energy for each symmetry species. The positions at which selected diatomic molecules are believed to occur are indicated in Fig. 4-17.

Let us see how the correlation diagram can be used to predict some of the properties of molecules. According to Fig. 4-17, the lowest energy orbitals for H_2 and He_2 are $1\sigma_g^+$ (ψ_b) and $1\sigma_u^+$ (ψ^*). H_2^+, H_2, and He_2^+ are all predicted to be stable species because each has more bonding than antibonding electrons:

H_2^+	H_2	He_2^+	He_2	
—	—	↑	↑↓	$1\sigma_u^+$
↑	↑↓	↑↓	↑↓	$1\sigma_g^+$

Further, H_2 has the greatest excess of bonding electrons and consequently ought to have the shortest internuclear distance and the strongest chemical bond. Considering a bond to consist of two electrons, the quantity *bond order* is defined as (the number of bonding electrons minus the number of antibonding electrons) divided by two. H_2^+ and He_2^+ have bond orders of 1/2. H_2 has a bond order of 1, and He_2 has a bond order of 0. Some pertinent predictions and observations are summarized in Table 4-8. The close similarity in r_e and D_0 of H_2^+ and He_2^+ (which both have a bond order of 1/2) is somewhat fortuitous. Tables 4-9 and 4-10 provide more representa-

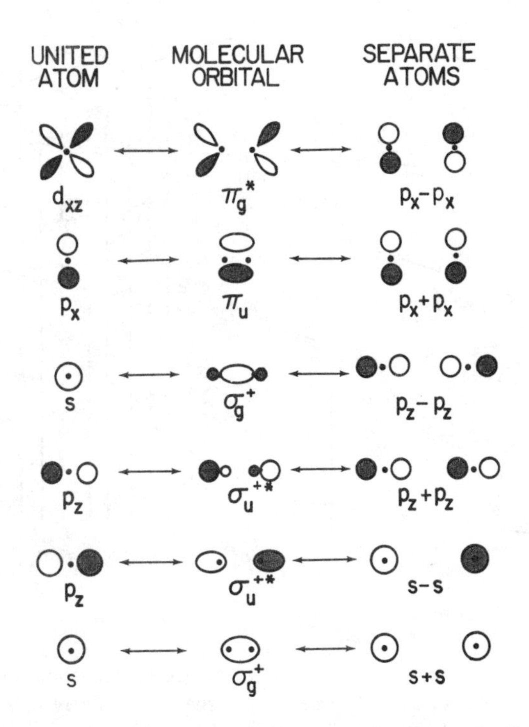

Fig. 4-16. Schematic changes in the orbitals on going from the limit of separate atomic orbitals, to molecular orbitals, to the united atom. The symmetry of each orbital is preserved throughout.

tive data on the dependence of r_e and D_0 on bond order for different molecules.

By feeding electrons into the orbitals of our correlation diagram, it is possible to build up the first row diatomic molecules. We must remember that the $1\pi_u$ and $3\sigma_g^+$ orbital energies *cross* between N_2 and O_2. Therefore, up to N_2 we use Scheme I of Fig. 4-18, and at O_2 we switch to Scheme II. Table 4-9 summarizes information about the first row diatomic molecules. Two illustrations emphasizing the relationship of bond energy and bond length to bond order are shown in Fig. 4-19.

There are several noteworthy points in Table 4-9. The bond lengths in 4-9 are all greater than those of Table 4-8. The reason is that the valence orbitals have the quantum number $n=2$ in 4-9 and $n=1$ in 4-8. The $n=2$ orbitals are much bigger than the $n=1$ orbitals. Another point is that O_2 has two unpaired electrons. Hund's first rule (Section 4-2-B) says that

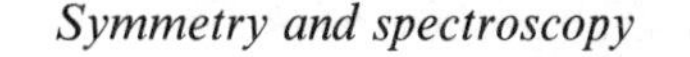

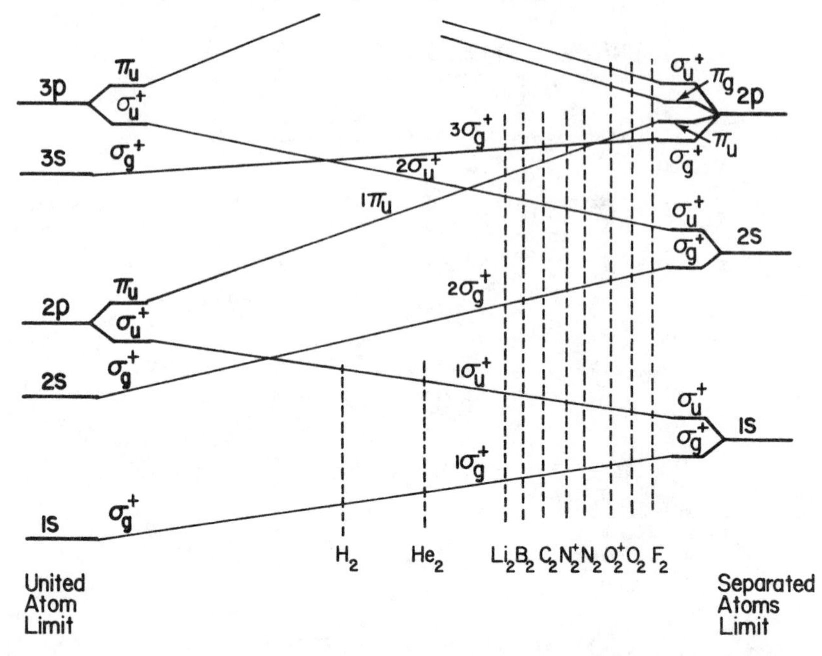

Fig. 4-17. Correlation diagram for diatomic molecules. The approximate positions of the diatomic molecules are indicated by dashed lines. These lines only indicate the order of the orbitals for each molecule and have no quantitative significance.

electrons are placed in each separate degenerate orbital before filling any one of them up. Further, the electron spins are aligned in the same direction. Thus, the various oxygen species would have the following configurations:

↑ —	↑ ↑	↑↓ ↑	↑↓ ↑↓	$1\pi_g{}^*$
$O_2{}^+$	O_2	$O_2{}^-$	$O_2{}^{2-}$	

The configuration of B_2 is $(1\sigma_g{}^+)^2\ (1\sigma_u{}^+)^2\ (2\sigma_g{}^+)^2\ (2\sigma_u{}^+)^1\ (1\pi_u)^2\ (3\sigma_g{}^+)^1$ instead of $(1\sigma_g{}^+)^2\ (1\sigma_u{}^+)^2\ (2\sigma_g{}^+)^2\ (2\sigma_u{}^+)^2\ (1\pi_u)^2$ which we would have expected. The reason is apparently that the $2\sigma_u{}^+$, $1\pi_u$, and $3\sigma_g{}^+$ orbitals are so close together for this molecule that the *spin pairing energy* is larger than the energy gained by putting another electron in the $2\sigma_u{}^+$ or $1\pi_u$ orbitals. The spin pairing energy is the energy necessary to hold two electrons in the same orbital. It is principally the Coulombic energy of repulsion of the two electrons.

Suppose you want to construct a molecular orbital diagram for the heteronuclear diatomic molecule *AB*. Figure 4-9 showed how orbitals from two different kinds of atoms could combine. In general, atomic orbitals of

Table 4-8. Properties of Hydrogen and Helium Species

Molecule	Predictions		Observations		
	Electronic Configuration	Bond Order	Bond Length (r_e) (Å)	Bond Enthalpy (D_0) (kcal/mol)	(kJ/mol)
H_2^+	$(1\sigma_g^+)^1$	$\frac{1}{2}$	1.06	61	260
H_2	$(1\sigma_g^+)^2$	1	0.74	103	431
He_2^+	$(1\sigma_g^+)^2(1\sigma_u^+)^1$	$\frac{1}{2}$	1.08	~60	~250
He_2	$(1\sigma_g^+)^2(1\sigma_u^+)^2$	0	---	--	---

the more electronegative atom will be lower in energy than the corresponding orbitals of the less electronegative atom. In Fig. 4-20 atom B is more electronegative than atom A. Note that the symmetry labels in this mo scheme are those appropriate for $C_{\infty v}$ instead of $D_{\infty h}$. Table 4-10 lists some pertinent properties of several heteronuclear diatomic molecules. The ordering of the 1π and $5\sigma^+$ orbitals is generally uncertain. The case of BN is interesting. The 1π and $5\sigma^+$ levels are so close in energy that the energy needed to promote a 1π electron to the $5\sigma^+$ orbital is less than the spin pairing energy. Therefore, the configuration is $(1\sigma^+)^2(2\sigma^+)^2(3\sigma^+)^2(4\sigma^+)^2(1\pi)^3(5\sigma^+)^1$.

Fig. 4-18. The two energy level orderings found for first row diatomic molecules.

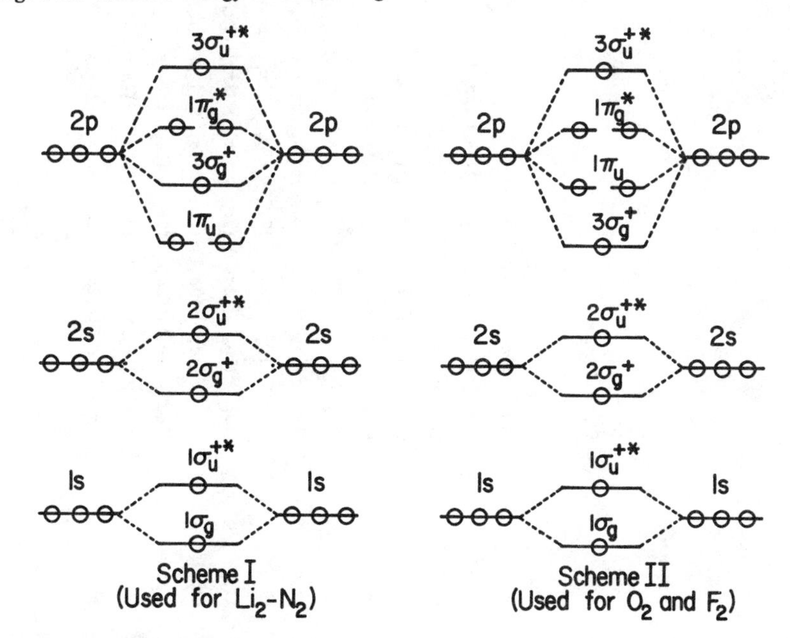

Table 4-9. Properties of Homonuclear Diatomic Molecules

Molecule	Predictions		Observations		
	Electronic Configuration	Bond Order	Bond Length (r_e) (Å)	Bond Enthalpy (D_0) (kcal/mol)	(kJ/mol)
Li_2	$(1\sigma_g^+)^2(1\sigma_u^+)^2(2\sigma_g^+)^2$	1	2.672	26	111
Be_2	$(1\sigma_g^+)^2(1\sigma_u^+)^2(2\sigma_g^+)^2(2\sigma_u^+)^2$	0	---	--	---
B_2	$(1\sigma_g^+)^2(1\sigma_u^+)^2(2\sigma_g^+)^2(2\sigma_u^+)^1(1\pi_u)^2(3\sigma_g)^1$ (see text)	2	1.589	69	290
C_2^+	$(1\sigma_g^+)^2(1\sigma_u^+)^2(2\sigma_g^+)^2(2\sigma_u^+)^2(1\pi_u)^3$	$1\frac{1}{2}$	-----	130	544
C_2	$(1\sigma_g^+)^2(1\sigma_u^+)^2(2\sigma_g^+)^2(2\sigma_u^+)^2(1\pi_u)^4$	2	1.3117	150	628
N_2^+	$(1\sigma_g^+)^2(1\sigma_u^+)^2(2\sigma_g^+)^2(2\sigma_u^+)^2(1\pi_u)^4(3\sigma_g^+)^1$	$2\frac{1}{2}$	1.116	204	854
N_2	$(1\sigma_g^+)^2(1\sigma_u^+)^2(2\sigma_g^+)^2(2\sigma_u^+)^2(1\pi_u)^4(3\sigma_g^+)^2$	3	1.0976	225	941
O_2^+	$(1\sigma_g^+)^2(1\sigma_u^+)^2(2\sigma_g^+)^2(2\sigma_u^+)^2(3\sigma_g^+)^2(1\pi_u)^4(1\pi_g)^1$	$2\frac{1}{2}$	1.1227	156	653
O_2	$(1\sigma_g^+)^2(1\sigma_u^+)^2(2\sigma_g^+)^2(2\sigma_u^+)^2(3\sigma_g^+)^2(1\pi_u)^4(1\pi_g)^2$	2	1.20741	118	494
O_2^-	$(1\sigma_g^+)^2(1\sigma_u^+)^2(2\sigma_g^+)^2(2\sigma_u^+)^2(3\sigma_g^+)^2(1\pi_u)^4(1\pi_g)^3$	$1\frac{1}{2}$	1.26	90	380
O_2^{2-}	$(1\sigma_g^+)^2(1\sigma_u^+)^2(2\sigma_g^+)^2(2\sigma_u^+)^2(3\sigma_g^+)^2(1\pi_u)^4(1\pi_g)^4$	1	1.49	---	---
F_2^+	$(1\sigma_g^+)^2(1\sigma_u^+)^2(2\sigma_g^+)^2(2\sigma_u^+)^2(3\sigma_g^+)^2(1\pi_u)^4(1\pi_g)^3$	$1\frac{1}{2}$	-----	65	270
F_2	$(1\sigma_g^+)^2(1\sigma_u^+)^2(2\sigma_g^+)^2(2\sigma_u^+)^2(3\sigma_g^+)^2(1\pi_u)^4(1\pi_g)^4$	1	1.418	36	150
Ne_2	$(1\sigma_g^+)^2(1\sigma_u^+)^2(2\sigma_g^+)^2(2\sigma_u^+)^2(3\sigma_g^+)^2(1\pi_u)^4(1\pi_g)^4(3\sigma_u^+)^2$	0	---	--	---

Table 4-10. Properties of Heteronuclear Diatomic Molecules

Molecule	Predictions		Observations		
	Electronic Configuration	Bond Order	Bond Length (r_e) (Å)	Bond Enthalpy (D_0) (kcal/mol)	(kJ/mol)
BeO	$(1\sigma^+)^2 (2\sigma^+)^2 (3\sigma^+)^2 (4\sigma^+)^2 (1\pi)^4$	2	1.3308	124	519
BeF	$(1\sigma^+)^2 (2\sigma^+)^2 (3\sigma^+)^2 (4\sigma^+)^2 (1\pi)^4 (5\sigma^+)^1$	$2\frac{1}{2}$	1.3614	92	380
BN	$(1\sigma^+)^2 (2\sigma^+)^2 (3\sigma^+)^2 (4\sigma^+)^2 (1\pi)^3 (5\sigma^+)^1$ (see text)	2	1.281	92	380
BO	$(1\sigma^+)^2 (2\sigma^+)^2 (3\sigma^+)^2 (4\sigma^+)^2 (1\pi)^4 (5\sigma^+)^1$	$2\frac{1}{2}$	1.2049	185	774
BF	$(1\sigma^+)^2 (2\sigma^+)^2 (3\sigma^+)^2 (4\sigma^+)^2 (1\pi)^4 (5\sigma^+)^2$	3	1.262	195	816
CN^+	$(1\sigma^+)^2 (2\sigma^+)^2 (3\sigma^+)^2 (4\sigma^+)^2 (1\pi)^4$	2	1.1727	130	544
CN	$(1\sigma^+)^2 (2\sigma^+)^2 (3\sigma^+)^2 (4\sigma^+)^2 (1\pi)^4 (5\sigma^+)^1$	$2\frac{1}{2}$	1.1718	188	787
CN^-	$(1\sigma^+)^2 (2\sigma^+)^2 (3\sigma^+)^2 (4\sigma^+)^2 (1\pi)^4 (5\sigma^+)^2$	3	1.14	---	---
CO^+	$(1\sigma^+)^2 (2\sigma^+)^2 (3\sigma^+)^2 (4\sigma^+)^2 (1\pi)^4 (5\sigma^+)^1$	$2\frac{1}{2}$	1.1151	195	816
CO	$(1\sigma^+)^2 (2\sigma^+)^2 (3\sigma^+)^2 (4\sigma^+)^2 (1\pi)^4 (5\sigma^+)^2$	3	1.1282	256	1070
CF	$(1\sigma^+)^2 (2\sigma^+)^2 (3\sigma^+)^2 (4\sigma^+)^2 (1\pi)^4 (5\sigma^+)^2 (2\pi)^1$	$2\frac{1}{2}$	1.270	106	444
NO^+	$(1\sigma^+)^2 (2\sigma^+)^2 (3\sigma^+)^2 (4\sigma^+)^2 (1\pi)^4 (5\sigma^+)^2$	3	1.0619	254	1060
NO	$(1\sigma^+)^2 (2\sigma^+)^2 (3\sigma^+)^2 (4\sigma^+)^2 (1\pi)^4 (5\sigma^+)^2 (2\pi)^1$	$2\frac{1}{2}$	1.150	162	678

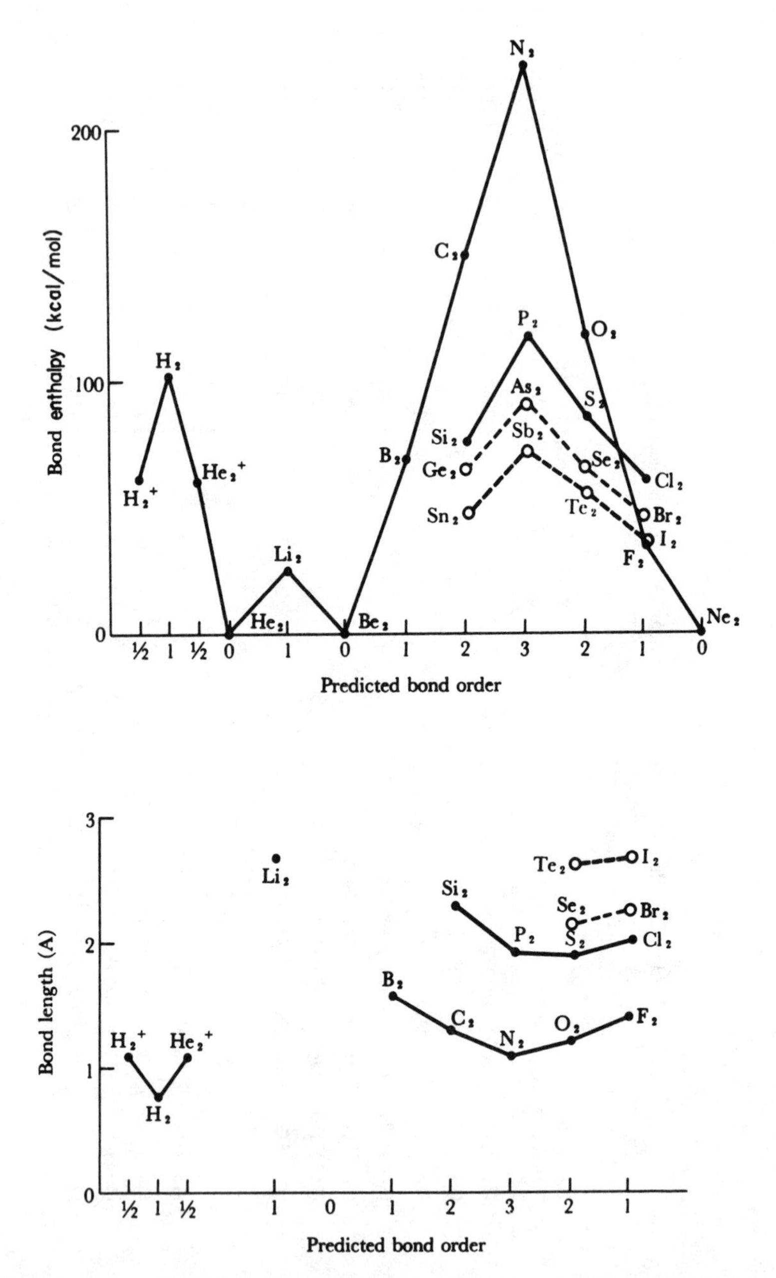

Fig. 4-19. Relation of bond energy and bond length to bond order. Reproduced from R.E. Dickerson, H.B. Gray, and G.P. Haight, Jr., *Chemical Principles*, W.A. Benjamin, N.Y., 1970.

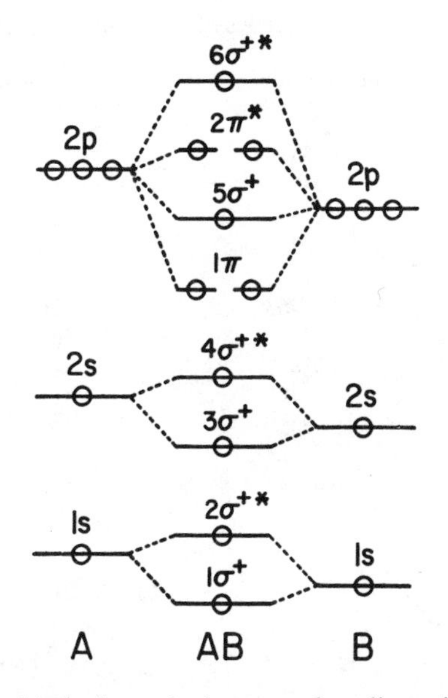

Fig. 4-20. Molecular orbital scheme for heteronuclear diatomic molecules analagous to Scheme I, Fig. 4-18. Some heteronuclear diatomic molecules will probably have the ordering of orbitals analagous to Scheme II, Fig. 4-18 as well.

The final class of diatomic molecules we will consider will be the HA molecules, of which HF is an example. The qualitative placement of the basis atomic orbitals in Fig. 4-21 is based on the known orbital energies of H and F. The H(1s) orbital is too high in energy to interact significantly with any but the 2p orbitals of F. Of these, only the F($2p_z$) orbital has the right symmetry to interact with the H(1s) orbital. The F($2p_x$) and F($2p_y$) orbitals are rigorously *nonbonding*. They cannot interact with the H(1s) orbital because there is zero overlap between them and the 1s orbital. The F(1s) and F(2s) orbitals are essentially nonbonding because their energies are so low compared to the H(1s) energy. For less electronegative elements than F, the $2\sigma^+$ level in Fig. 4-21 may be bonding since the 2s orbital will be closer in energy to the H(1s) orbital. Some properties of HA molecules are given in Table 4-11. Assignment of bond orders would be ambiguous since the $2\sigma^+$ and $3\sigma^+$ orbitals are of uncertain character.

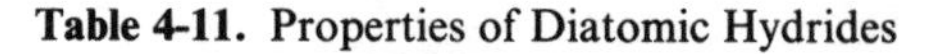

Table 4-11. Properties of Diatomic Hydrides

Molecule	Bond Length (r_e) (Å)	Bond Enthalpy (D_0) (kcal/mol)	(kJ/mol)
HLi	1.5953	58	240
HBe	1.3431	53	220
HB^+	1.215	50	210
HB	1.2325	70	290
HC^+	1.131	98	410
HC	1.1198	80	330
HN^+	1.084	103	431
HN	1.038	85	360
HO^+	1.0289	111	464
HO	0.9706	107	448
HF^+	-----	89	370
HF	0.9175	134	561

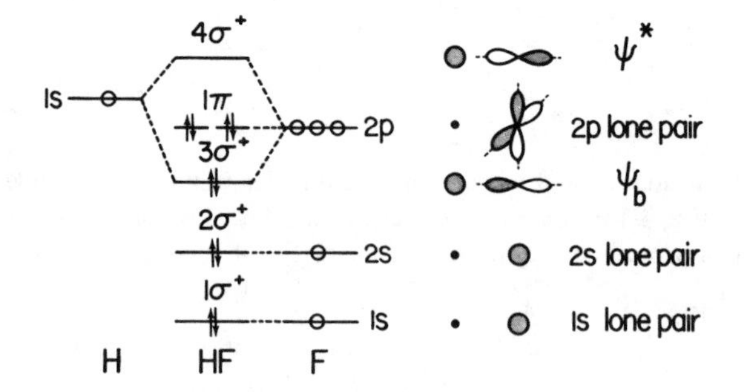

Fig. 4-21. Molecular orbital scheme for HF.

Problems

4-13. Fill in the following table.

molecule	$\bar{\nu}$ (cm^{-1})	force constant (N m^{-1})	bond order
NH	3133		
NF	1115		
O_2	1555		
N_2	2331		
NO	1876		

4-14. For each set of molecules, which will have the highest stretching frequency: (a) NO, NO^+, NO^-; (b) CP, CP^+, CP^-; (c) SiS, SiS^-, SiS^{2-}?

4-15. Order the following reactions according to increasing enthalpy: (a) $O_2^- \rightarrow O^- + O$; (b) $O_2 \rightarrow 2O$; (c) $O_2^+ \rightarrow O^+ + O$.

4-16. Which is most stable: N_2^+, NO^+, O_2^+, Li_2^+, Be_2^+? Which has the longest bond: CN^+, CN, CN^-, NO^+?

4-6. Polyatomic Molecules

We will now examine several examples of molecular orbital schemes for small molecules. We will especially emphasize the symmetries of the orbitals and will try to construct mo diagrams using information contained in the Lewis structure of each molecule.

A. *Carbon Dioxide.* The procedure for determining the symmetries of the molecular orbitals is the same as for diatomic molecules. It will be necessary to determine how the basis atomic orbitals transform and to construct linear combinations of appropriate symmetries. We will ignore the 1s orbitals because they do not interact significantly with the valence orbitals or with each other. We therefore have 2s and 2p orbitals on three atoms to consider (Fig. 4-22). Since we are starting with twelve atomic orbitals, we must generate twelve molecular orbitals in our final scheme. It is convenient to determine the symmetries of the basis orbitals in a few separate groups, as indicated in Table 4-12. It is necessary to consider equivalent atomic orbitals (such as the two O(2s) orbitals) together because some of the symmetry operations interchange these orbitals.

For guidance in constructing appropriate linear combinations of atomic orbitals, we will make use of the Lewis structure of CO_2:

$$\ddot{\underset{..}{O}}=C=\ddot{\underset{..}{O}}$$

Implicit in this structure is a good deal of information. For example, we know that each double bond implies one sigma bond and one pi bond.† When we construct the mo scheme, we expect to find two occupied sigma bonding orbitals and two occupied pi bonding orbitals. The organic chemists in the audience will immediately realize that if each oxygen atom uses a $2p_z$ orbital directed at carbon to form a sigma bond and another 2p orbital (say $2p_x$) to form a pi bond, the lone pair electrons must reside in the remaining O(2s) and O($2p_y$) orbitals. In constructing suitable LCAO's, then, we will try to produce lone pair orbitals from a set of oxygen 2s and 2p orbitals.

In constructing LCAO's of σ_g^+ symmetry, Table 4-12 tells us that the C(2s), O(2s), and O($2p_z$) orbitals may all contribute. Further, these atomic

†In this chapter we will write the word "sigma" when we describe an orbital which is axially symmetric and "pi" when we refer to an orbital which has one node colinear with the internuclear axis. The greek letters "σ" and "π" will be used only when the orbitals actually possess σ or π symmetry in the point groups $D_{\infty h}$ or $C_{\infty v}$. Thus, the pi bonding orbital of ethylene (Section 4-6-C) has b_{1u} symmetry in the point group D_{2h} and is a "pi" orbital but not a "π" orbital.

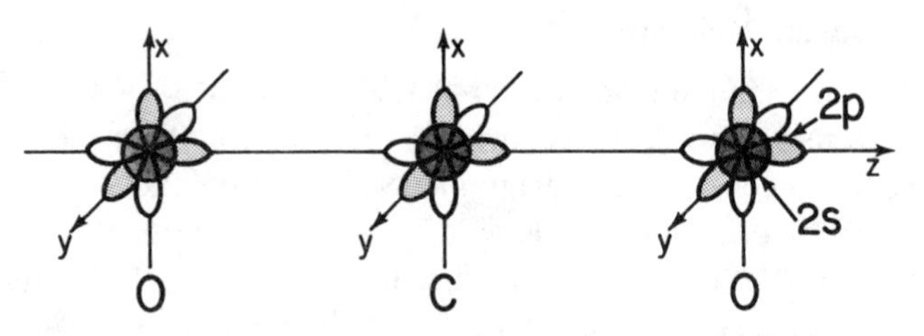

Fig. 4-22. Valence atomic orbitals of CO_2.

orbitals are the only ones which may contribute because they are the only ones whose representations include σ_g^+. The lowest energy valence atomic orbitals of CO_2 are the O(2s) orbitals since oxygen has a greater nuclear charge than carbon. Since we would like to construct a lone pair orbital that contains essentially only O(2s) character, we can write the combination $O(2s)_A + O(2s)_B$ shown at the bottom left of Fig. 4-23. We also expect from the Lewis structure that it should be possible to draw a sigma bonding orbital using the C(2s) and O($2p_z$) orbitals. The combination shown at the middle left of Fig. 4-23 is a C–O sigma bonding orbital of σ_g^+ symmetry. We have now constructed two linear combinations of this symmetry, but Table 4-12 tells us we need a third. From every bonding combination of orbitals, it is always possible to construct an antibonding combination of the same symmetry by reversing a few of the signs. By this means, we construct the σ_g^+ antibonding orbital shown at the top left of Fig. 4-23. In the mo scheme we will designate these three mo's $1\sigma_g^+$, $2\sigma_g^+$, and $3\sigma_g^+$ in order of increasing energy.

Table 4-12 tells us that we need to construct three mo's of σ_u^+ symmetry, using the same basis orbitals required for the σ_g^+ orbitals. At the right of

Table 4-12. Symmetries of the Valence Atomic Orbitals of CO_2 Shown in Fig. 4-22

$D_{\infty h}$	E	$2C_\infty^\phi$	$\infty\sigma_v$	i	$2S_\infty^\phi$	∞C_2	
C(2s)	1	1	1	1	1	1	$= \sigma_g^+$
C($2p_z$)	1	1	1	−1	−1	−1	$= \sigma_u^+$
C($2p_x$,$2p_y$)	2	$2c\phi$	0	−2	$2c\phi$	0	$= \pi_u$
2O(2s)	2	2	2	0	0	0	$= \sigma_g^+ + \sigma_u^+$
2O($2p_z$)	2	2	2	0	0	0	$= \sigma_g^+ + \sigma_u^+$
2O($2p_x$,$2p_y$)	4	$4c\phi$	0	0	0	0	$= \pi_u + \pi_g$

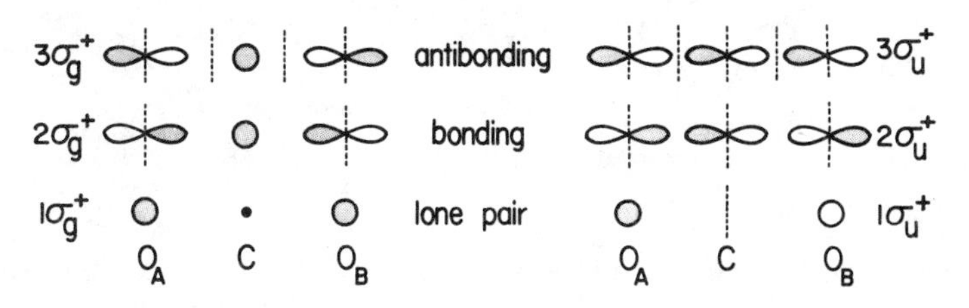

Fig. 4-23. $\sigma_g{}^+$ and $\sigma_u{}^+$ molecular orbitals of CO_2. Nodes (planes at which the wave functions change sign) are shown by dotted lines. The $1\sigma_g{}^+$ and $1\sigma_u{}^+$ orbitals are called lone pairs because the two O(2s) orbitals are too far apart for significant bonding or antibonding interaction. Lone pair orbitals are essentially nonbonding.

Fig. 4-23 we see how the same three kinds of orbitals at the left of this figure can be constructed, but with $\sigma_u{}^+$ symmetry. The six mo's in Fig. 4-23 account for both sigma bonds of CO_2, as well as two of the lone pairs. According to Table 4-12, we need to construct *four* pi orbitals (two degenerate pairs) of π_u symmetry using $C(2p_x, 2p_y)$ and $O(2p_x, 2p_y)$ orbitals. Since the Lewis structure of CO_2 leads us to expect two pi bonding orbitals, we construct the fully bonding LCAO's labelled $1\pi_u$ in Fig. 4-24. A corresponding antibonding pair of orbitals is labelled $2\pi_u$ in the same figure. By a process of elimination, we expect that the remaining mo of π_g symmetry must be a pair of $O(2p_x, 2p_y)$ nonbonding orbitals. The only possible way to draw a molecular orbital of π_g symmetry using any of the valence orbitals of CO_2 is shown at the right of Fig. 4-24. There can be no contribution from any carbon orbitals because the C(2p) orbitals all have *u*

Fig. 4-24. π_u and π_g molecular orbitals of CO_2.

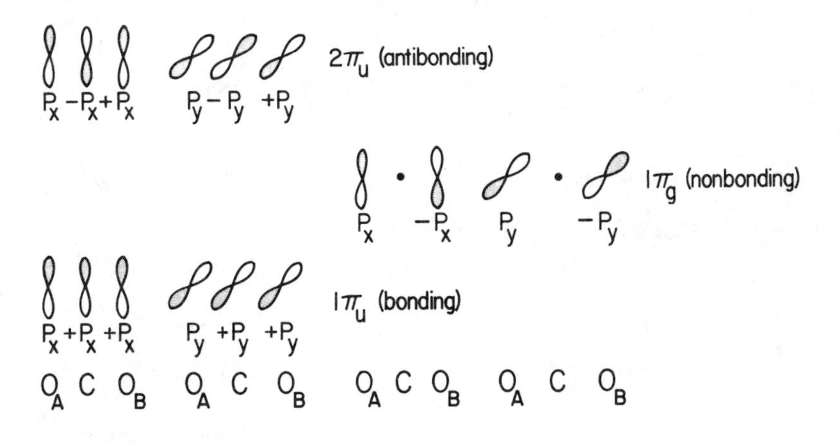

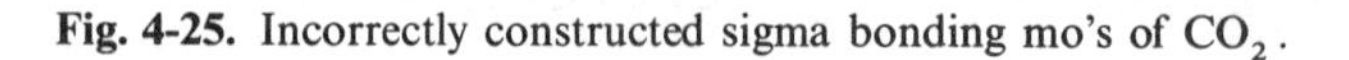

Fig. 4-25. Incorrectly constructed sigma bonding mo's of CO_2.

symmetry, and the required mo has *g* symmetry. The C(2s) orbital, which does have *g* symmetry, has zero net overlap with the $O(2p_x)$ and $O(2p_y)$ orbitals and therefore cannot interact with them.

When trying to construct sigma bonding linear combinations of atomic orbitals for the first time, the incorrectly drawn orbitals I and II in Fig. 4-25 often result. Why are these pictures incorrect while the ones in Fig. 4-23 labelled $2\sigma_g^+$ and $2\sigma_u^+$ are correct? The reason is that I and II do not form bases for irreducible representations of the point group $D_{\infty h}$. Under the operations E, C_∞^ϕ, and σ_v, I and II each go into themselves, which is fine, but orbital I goes into orbital II upon inversion. The σ_g^+ and σ_u^+ orbitals must go into plus and minus themselves, respectively, under the operation *i*. Similarly, S_∞^ϕ and C_2 will not work properly for I and II. When in doubt, you should check that the LCAO's you construct do indeed form bases for irreducible representations.

Fig. 4-26. Qualitative mo scheme for CO_2. Compare these linear combinations of atomic orbitals to the more accurately calculated mo's in Fig. 4-27.

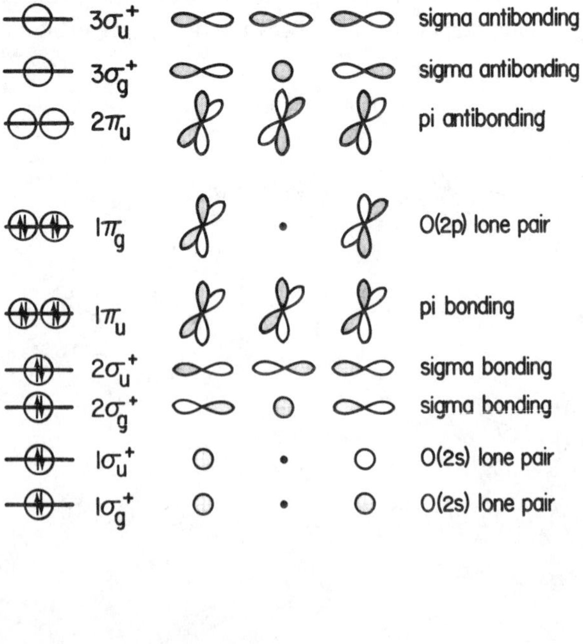

$3\sigma_u^+$ sigma antibonding
$3\sigma_g^+$ sigma antibonding
$2\pi_u$ pi antibonding
$1\pi_g$ O(2p) lone pair
$1\pi_u$ pi bonding
$2\sigma_u^+$ sigma bonding
$2\sigma_g^+$ sigma bonding
$1\sigma_u^+$ O(2s) lone pair
$1\sigma_g^+$ O(2s) lone pair

We will now assemble the linear combinations of atomic orbitals in Figs. 4-23 and 4-24 to create a complete mo diagram, shown in Fig. 4-26. We expect that the O(2s) lone pair orbitals will be lowest in energy because these two molecular orbitals are composed of the two lowest energy basis atomic orbitals. Among the remaining orbitals, we generally expect the order sigma bonding < pi bonding < lone pairs < pi antibonding < sigma antibonding. If two orbitals are of the same type, the one with fewer nodes will be of lower energy. Orbitals of the same type are, for example, the $2\sigma_g^+$ and $2\sigma_u^+$ orbitals which are both sigma bonding. The $2\sigma_g^+$ orbital has two nodes and the $2\sigma_u^+$ orbital has three nodes (Fig. 4-23). When the orbitals are ordered in this manner, and the 16 valence electrons of CO_2 are added to the diagram, we see that the mo scheme conforms closely to the predictions of the Lewis structure. Namely, the occupied orbitals account for two sigma bonds, two pi bonds, and four lone pairs. The chief difference between the molecular orbital treatment and the valence bond treatment (essentially summarized by the Lewis structure itself) is that in molecular orbital theory the electrons are delocalized throughout the molecule, and not confined between adjacent atoms.

An energy level diagram indicating correct relative spacings for the molecular orbitals is shown in Fig. 4-27. In this figure are also shown more accurately drawn molecular orbitals based on a quantitative calculation of the electronic structure of CO_2, and not just the qualitative considerations we have discussed. As an example of the differences between the qualitative scheme in Fig. 4-26 and the more accurate calculations, in the $1\sigma_g^+$ [O(2s) lone pair] orbital in Fig. 4-26, there is no contribution from the C(2s) orbital. A calculation indicates, however, that the C(2s) orbital does mix somewhat with the O(2s) orbitals. Therefore, the more accurately constructed $1\sigma_g^+$ orbital in Fig. 4-27 should be described as [O(2s) lone pair] + [C$-$O sigma bond]. The relation between the qualitative linear combinations of Fig. 4-26 and the calculated linear combinations of Fig. 4-27 is entirely analagous to the relation of symmetry coordinates and normal modes of vibration (Section 3-7).

Problems

4-17. Construct a molecular orbital scheme for the linear molecule BeH_2, using only the valence orbitals. Draw a picture of each mo, order them by increasing energy, and place the proper number of electrons in the diagram. How many sigma bonding orbitals are there? How many pi bonding orbitals? How many lone pairs? How many unpaired electrons?

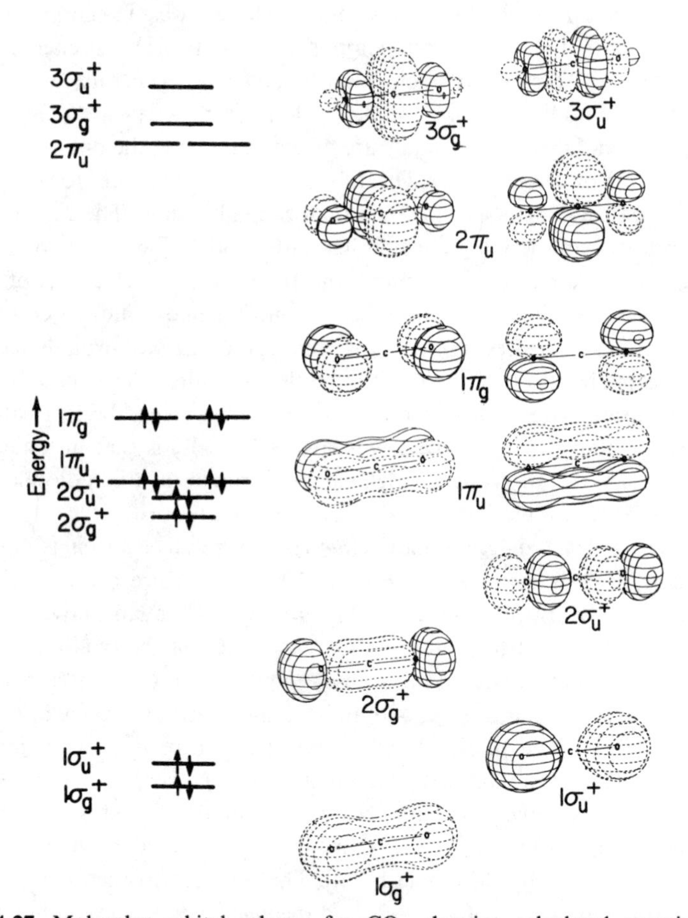

Fig. 4-27. Molecular orbital scheme for CO_2 showing calculated energies and orbitals. Solid and dashed lines correspond to regions of the wave function of opposite sign. The contours indicated correspond to electron densities of 0.0675 electrons/Å^3 for one-electron wave functions and were chosen merely for satisfactory visual display of the orbitals. Contour diagrams reproduced from W. L. Jorgensen and L. Salem, *The Organic Chemist's Book of Orbitals*, Academic Press, N.Y., 1973.

4-18. *A priori*, one might expect methylene, CH_2, to be either bent or linear. The mo scheme for the linear species will be the same as for BeH_2 (Problem 4.17) with the appropriate number of electrons. Construct a molecular orbital scheme for bent CH_2 using the *hybrid* orbitals drawn below as basis functions. How many unpaired electrons do you expect for the linear and bent species? State what factor could lead to there being two unpaired electrons in *both* cases. Experimental evidence indicates that in its ground state CH_2 does have unpaired electrons.

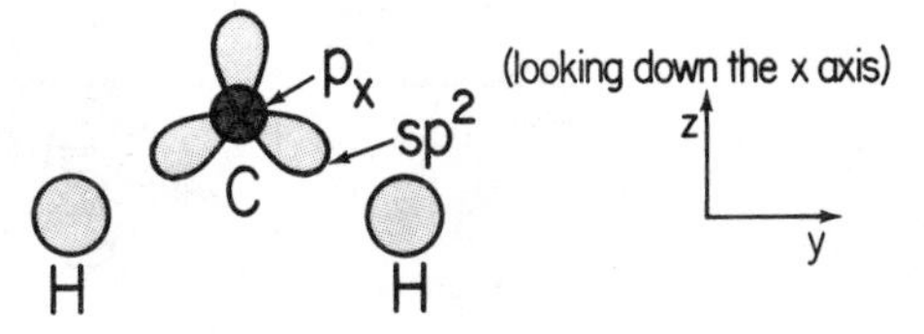

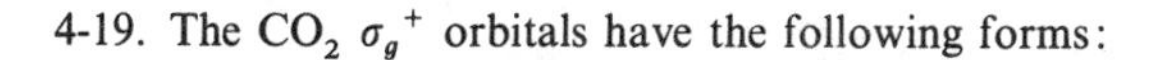

4-19. The CO_2 σ_g^+ orbitals have the following forms:

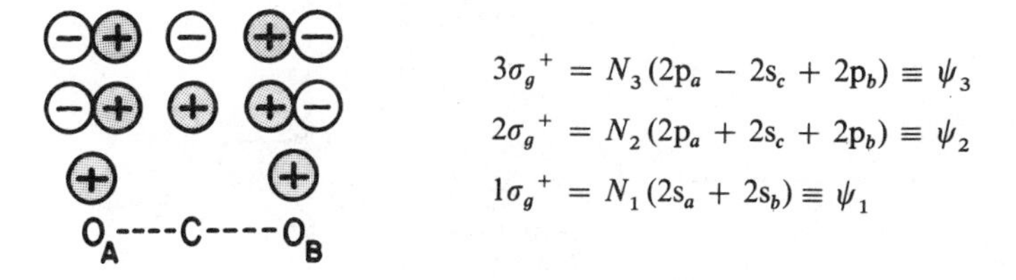

$$3\sigma_g^+ = N_3(2p_a - 2s_c + 2p_b) \equiv \psi_3$$

$$2\sigma_g^+ = N_2(2p_a + 2s_c + 2p_b) \equiv \psi_2$$

$$1\sigma_g^+ = N_1(2s_a + 2s_b) \equiv \psi_1$$

The letters p and s refer to p and s orbitals; and the subscripts *a*, *b*, and *c* refer to O_A, O_B, and carbon, respectively. To simplify matters, assume that the overlap integrals $\int(2s_a)(2s_b)d\tau$ and $\int(2p_a)(2p_b)d\tau$ are zero. (a) Defining $S = \int(2p_a)(2s_c)d\tau$, determine the normalization coefficients N_1, N_2, and N_3. (b) Using eq. 4-11, determine the energies of the three molecular orbitals in terms of the following integrals:

$$H_{aa} = \int (2p_a)\,\mathscr{H}\,(2p_a)d\tau$$

$$H_{ab} = \int (2p_a)\,\mathscr{H}\,(2p_b)d\tau$$

$$H_{ac} = \int (2p_a)\,\mathscr{H}\,(2s_c)d\tau$$

$$H_{cc} = \int (2s_c)\,\mathscr{H}\,(2s_c)d\tau$$

$$H'_{aa} = \int (2s_a)\,\mathscr{H}\,(2s_a)d\tau$$

$$H'_{ab} = \int (2s_a)\,\mathscr{H}\,(2s_b)d\tau$$

(c) Using the variation method (Section 4-4), derive the form of the secular determinant one gets by using the linear combinations ψ_1, ψ_2, and ψ_3 as bases for a more accurate determination of the correct forms of the σ_g^+ molecular orbitals. That is, allow ψ_1, ψ_2, ψ_3 to mix to form "better" mo's.

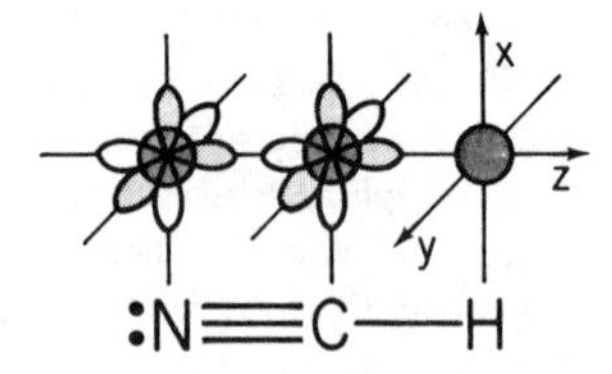

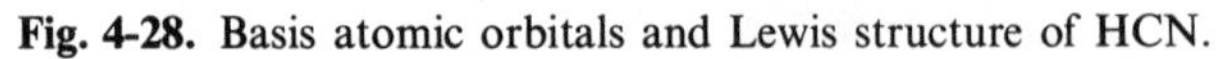

Fig. 4-28. Basis atomic orbitals and Lewis structure of HCN.

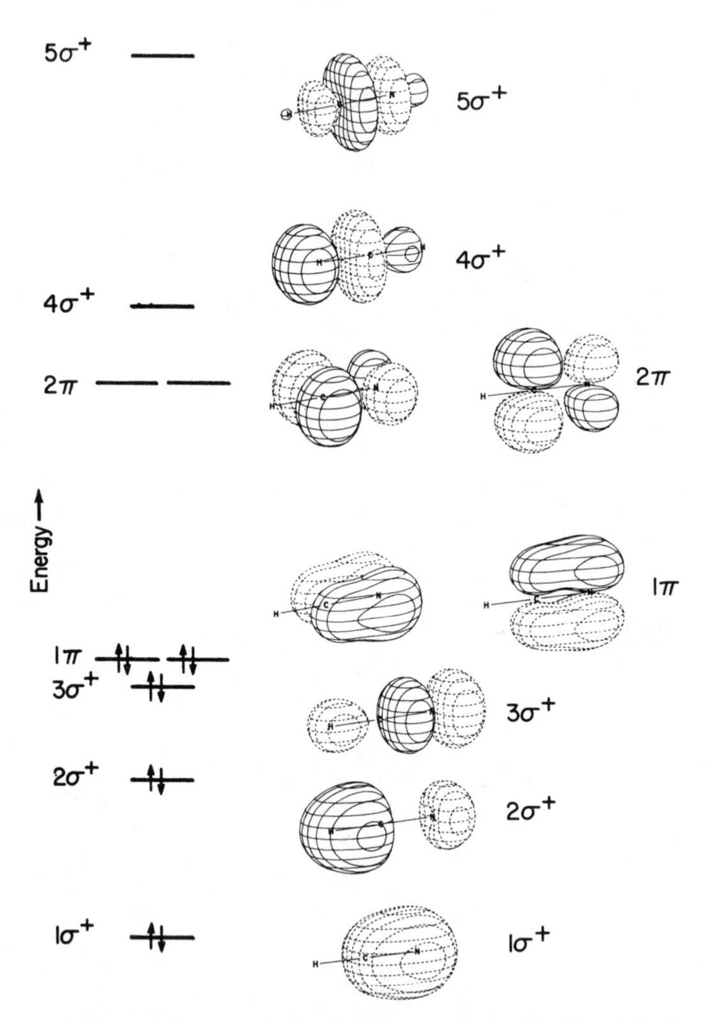

Fig. 4-29. Molecular orbital scheme for HCN, as described in Fig. 4-27. Contour diagrams reproduced from W.L. Jorgensen and L. Salem, *The Organic Chemist's Book of Orbitals*, Academic Press, N.Y., 1973.

B. *Hydrogen Cyanide.* Using the basis atomic orbitals in Fig. 4-28, and calling the molecular axis the z axis, we derive the symmetries in Table 4-13. The calculated mo scheme appears in Fig. 4-29. Now you examine Fig. 4-29 and wonder why the $1\sigma^+$ orbital is drawn as a localized orbital between just two atoms when we just finished saying that you could not draw such localized orbitals for CO_2. The answer is in two parts. One is that the orbital drawn in Fig. 4-29 *does* transform properly under the operations of the point group $C_{\infty v}$, but so would a more delocalized orbital extending over the whole molecule. The second part of the answer is that the only way to choose between a localized and a delocalized orbital is by a calculation. Just such a calculation went into the construction of Fig. 4-29.

Another kind of representation of the HCN orbitals is given in Fig. 4-30. Plotted on the vertical axis is the electron density at each point in a

Table 4-13. Symmetries of the Valence Atomic Orbitals of HCN

$C_{\infty v}$	E	$2C_\infty^\phi$	$\infty\sigma_v$	
H(1s) + C(2s) + N(2s)	3	3	3	$= 3\sigma^+$
$C(2p_z) + N(2p_z)$	2	2	2	$= 2\sigma^+$
$C(2p_x, 2p_y) + N(2p_x, 2p_y)$	4	$4c\phi$	0	$= 2\pi$

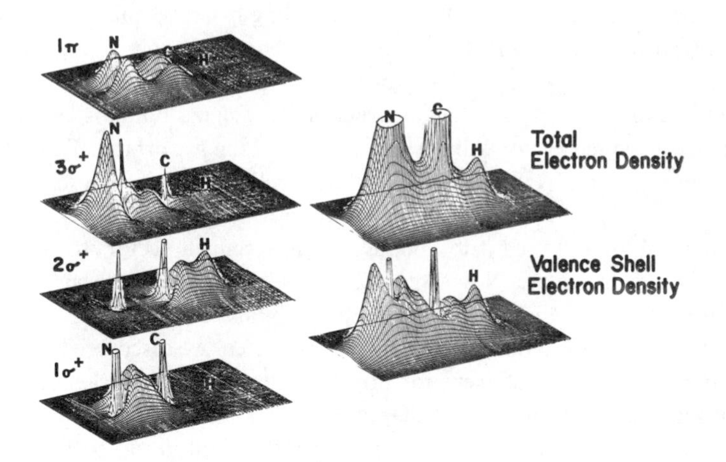

Fig. 4-30. Electron density plots of HCN reproduced from J.B. Robert, H. Marsmann, I Absara, and J.R. Van Wazer, *J. Amer. Chem. Soc.*, *93*, 3320 (1971).

plane cut through the molecule. The pictures show density in each of the occupied orbitals as well as a composite of all the valence orbitals. There is also one picture showing *total* electron density, including the inner N(1s) and C(1s) orbitals. These marvelous drawings even show that the valence electrons in the C–N bond are drawn toward the more electronegative nitrogen atom.

Problems

4-20. Consider only the pi orbitals of a linear C_4 species, which is found in the gas phase above molten carbon. Draw a picture of each pi orbital and order these orbitals by increasing energy. Which species, C_4, C_4^+, or C_4^-, would you expect to be the most stable?

C. *Ethylene.* Using the coordinate system in Fig. 4-31, the valence atomic orbitals transform as in Table 4-14. We ordinarily think of using sp^2 hybrid orbitals on carbon to make the sigma bonds and p_z orbitals for the pi bonds. The sp^2 orbitals utilize the carbon s, p_x, and p_y orbitals which transform as $2a_g + b_{1g} + b_{2u} + 2b_{3u}$. The four H(1s) orbitals transform as $a_g + b_{1g} + b_{2u} + b_{3u}$. From these ten orbitals we need to form four C–H bonding orbitals, four C–H antibonding orbitals, one C–C sigma bonding orbital, and one C–C sigma antibonding orbital. We further expect that the C–C and C–H orbitals of the same symmetry will be mixed. As one example of the construction of a sigma bonding orbital of the correct symmetry, we will derive the form of the b_{2u} orbital. Table 4-14 indicates that the H(1s) and C(2p$_y$) orbitals will contribute to a molecular orbital of b_{2u} symmetry. In Section 3-7 we introduced the projection operator (eq. 3-120) and used it to operate on displacement vectors to generate symmetry coordinates. Now we will allow the appropriate projection operator to act on the H(1s) and C(2p$_y$) orbitals. In Fig. 3-67, in which the ν(CH)(b_{3u}) vibration of C_2H_4 was constructed, we discovered that the D_2 character table is sufficient for problems dealing with ethylene. Use of all of the operations of the D_{2h} character table is unnecessary. In Fig. 4-32 we apply the b_2 projection operator of the D_2 point group to the H(1s) orbitals to produce a *group orbital* (a combination of atomic orbitals) of symmetry b_{2u}. (You should satisfy yourself that the group orbital so constructed does have b_{2u} symmetry using all of the operations of D_{2h}.)

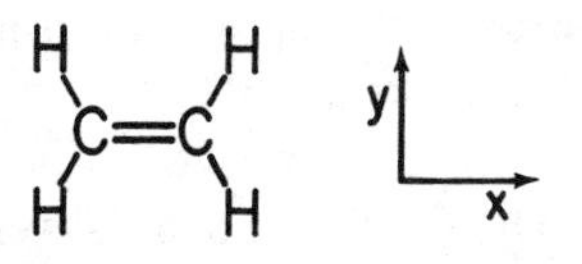

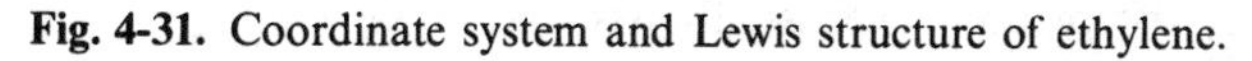

Fig. 4-31. Coordinate system and Lewis structure of ethylene.

Table 4-14. Symmetries of the Valence Atomic Orbitals of Ethylene

D_{2h}	E	$C_2(z)$	$C_2(y)$	$C_2(x)$	i	$\sigma(xy)$	$\sigma(xz)$	$\sigma(yz)$	
4H(1s)	4	0	0	0	0	4	0	0	$= a_g + b_{1g} + b_{2u} + b_{3u}$
2C(2s)	2	0	0	2	0	2	2	0	$= a_g + b_{3u}$
2C(2p$_x$)	2	0	0	2	0	2	2	0	$= a_g + b_{3u}$
2C(2p$_y$)	2	0	0	−2	0	2	−2	0	$= b_{1g} + b_{2u}$
2C(2p$_z$)	2	0	0	−2	0	−2	2	0	$= b_{2g} + b_{1u}$

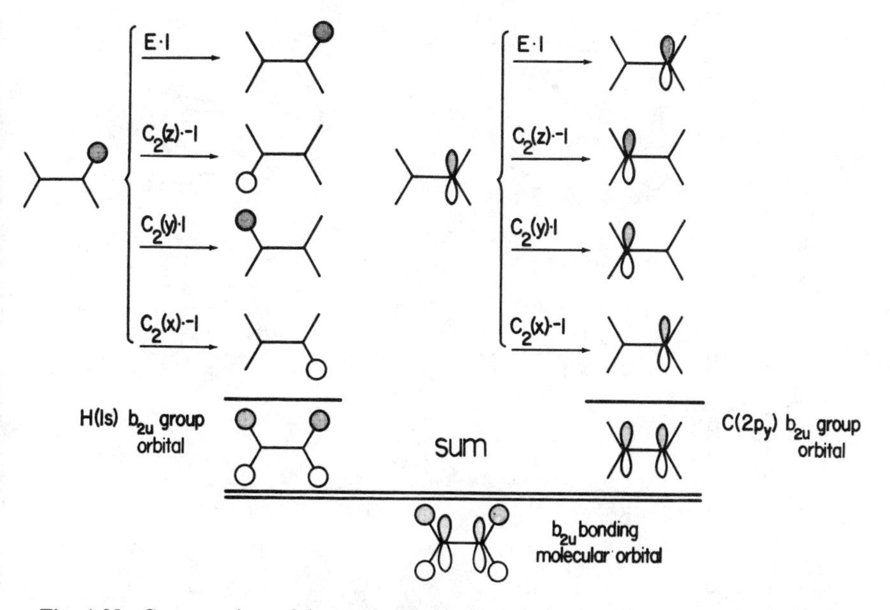

Fig. 4-32. Construction of b_{2u} group orbitals and the bonding molecular orbital. Compare this mo to the one labelled $1b_{2u}$ in Fig. 4-33.

Similarly, we can construct a C(2p$_y$) group orbital, and combine it with the H(1s) group orbital to make a molecular orbital of symmetry b_{2u} (Fig. 4-32). The orbital so constructed can be described as C–H sigma bonding and C–C pi bonding. The C–C pi bond, however, is in the plane of the molecule and is not the pi bond we usually associate with ethylene. The latter is perpendicular to the plane of the molecule and has no C–H character.

The calculated mo diagram for ethylene is shown in Fig. 4-33. Since the C–H and C–C bond energies are roughly 330–420 kJ/mol and the pi

Fig. 4-33. Valence molecular orbitals of ethylene. Contour diagrams are as described in Fig. 4-27 and reproduced from W.L. Jorgensen and L. Salem, *The Organic Chemist's Book of Orbitals*, Academic Press, N.Y., 1973.

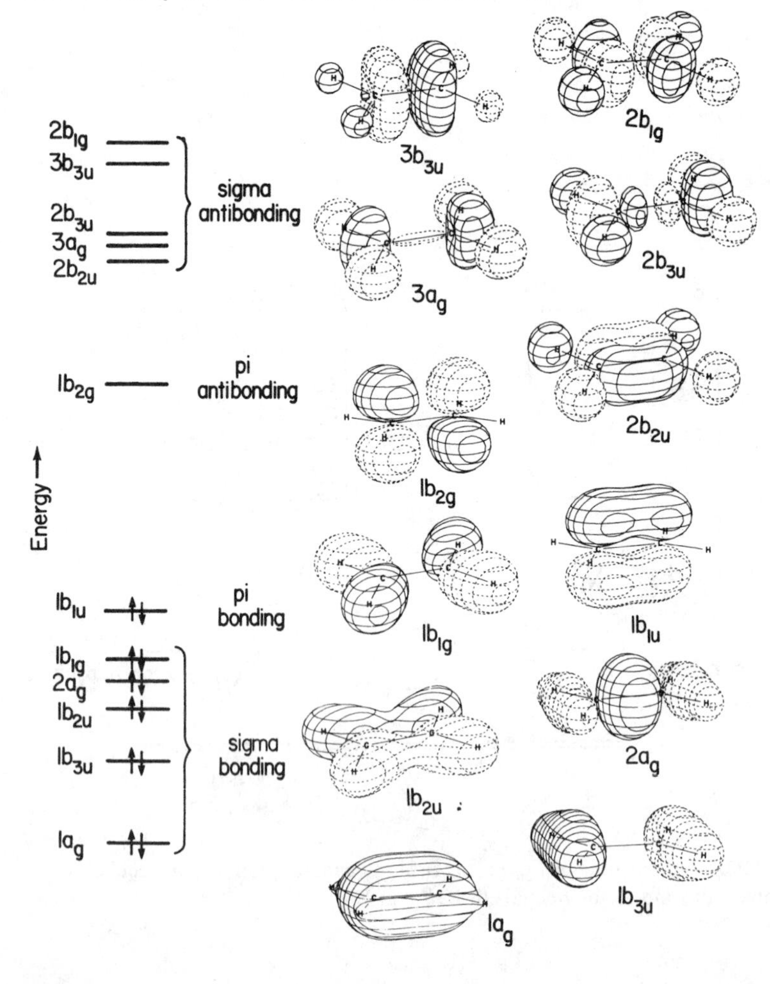

bond energy is about 250 kJ/mol, it is reasonable to find the sigma bonding orbitals below the pi bonding orbitals, and the pi antibonding orbitals below the sigma antibonding orbitals. Notice that the C–C pi bond is not a π bond, but a b_{1u} bond. The name "pi" remains from the symmetry of this type of bond in $D_{\infty h}$ symmetry. The photoelectron spectrum (Fig. 4-34) of ethylene shows the expected six ionization potentials in the valence orbital energy region. Notice that the calculated orbital energies are in good qualitative agreement with the observed photoelectron peaks, but not in very good quantitative agreement. This is an accurate reflection of the state of the art of mo calculations.

Fig. 4-34. Experimental He II (30.4 nm, 40.8 eV) photoelectron spectrum of ethylene, and calculated energies of the occupied valence orbitals. Adapted from C.R. Brundle, M.B. Robin, H. Basch, M. Pinsky, and A. Bond, *J. Amer. Chem. Soc.*, *92*, 3863 (1970).

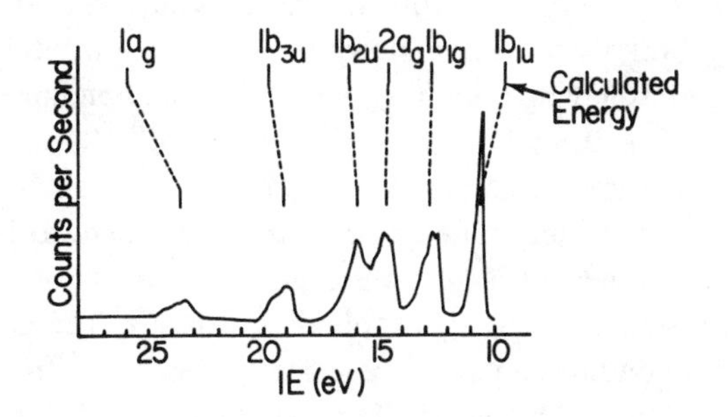

Problems

4-21. How many sigma bonding orbitals will there be in cyclopropene? How many sigma antibonding orbitals? How many pi bonding orbitals? How many pi antibonding orbitals? How many lone pairs? Make a diagram showing the relative energies of these orbitals.

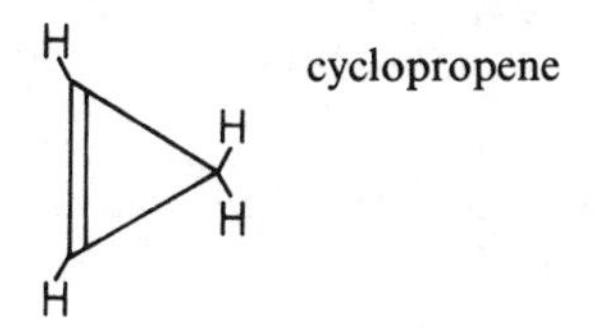

4-22. Propose a molecular orbital scheme for the vinyl cation with the geometry below. By "propose a molecular orbital scheme" we mean: (a) draw the basis atomic orbitals and determine how they transform; (b) construct appropriate linear combinations of the basis orbitals to make reasonable molecular orbitals; (c) order the mo's by increasing energy as best you can; and (d) place the proper number of electrons in the diagram.

$$\mathrm{H_2C{=}\overset{+}{C}{-}H} \qquad \text{vinyl cation}$$

4-23. Propose a molecular orbital scheme for acetylene, $HC{\equiv}CH$.

D. *Formaldehyde.* Using the coordinate system in Fig. 4-35, the basis atomic orbitals transform as in Table 4-15. The Lewis structure of formaldehyde leads us to expect three sigma bonding orbitals, three sigma antibonding orbitals, one pi bonding orbital, one pi antibonding orbital, and two oxygen lone pairs. These predictions are borne out by the calculated forms of the orbitals in Fig. 4-36 with the exception that the O(2s) lone pair of a_1 symmetry is thoroughly mixed with the C–O and C–H sigma bonding orbitals of a_1 symmetry to yield three orbitals which can only be described as having both lone pair and sigma bonding character.

Let us rationalize the relative energies of the orbitals in Fig. 4-36 as follows: There are three sigma bonding orbitals which ought to be of lowest energy and three sigma antibonding orbitals which ought to be of highest energy. The O(2s) lone pair is at lower energy than the pi orbitals, and, as just stated, is mixed with the sigma bonding orbitals. The O($2p_x$) orbital is essentially nonbonding so its energy is nearly unchanged from the energy of a pure O(2p) orbital. The two C–O pi bonding and antibonding orbitals will necessarily fall above and below the position of the O($2p_x$) orbital because they are stabilized and destabilized with respect to the energy of a pure O(2p) orbital. The essential new feature of the formaldehyde mo scheme, compared to that of ethylene (Fig. 4-33), is that the highest filled orbital is a nonbonding (lone pair) orbital found between the filled pi bonding orbital and empty pi antibonding orbital.

E. *Methyl Cation, CH_3^+.* The planar methyl cation is treated here because it illustrates the procedure to use in constructing degenerate

Fig. 4-35. Coordinate system and Lewis structure of formaldehyde.

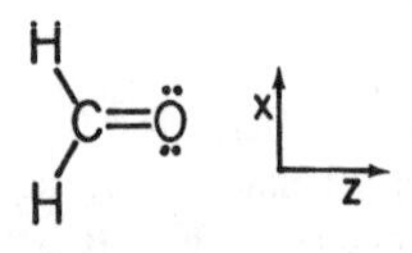

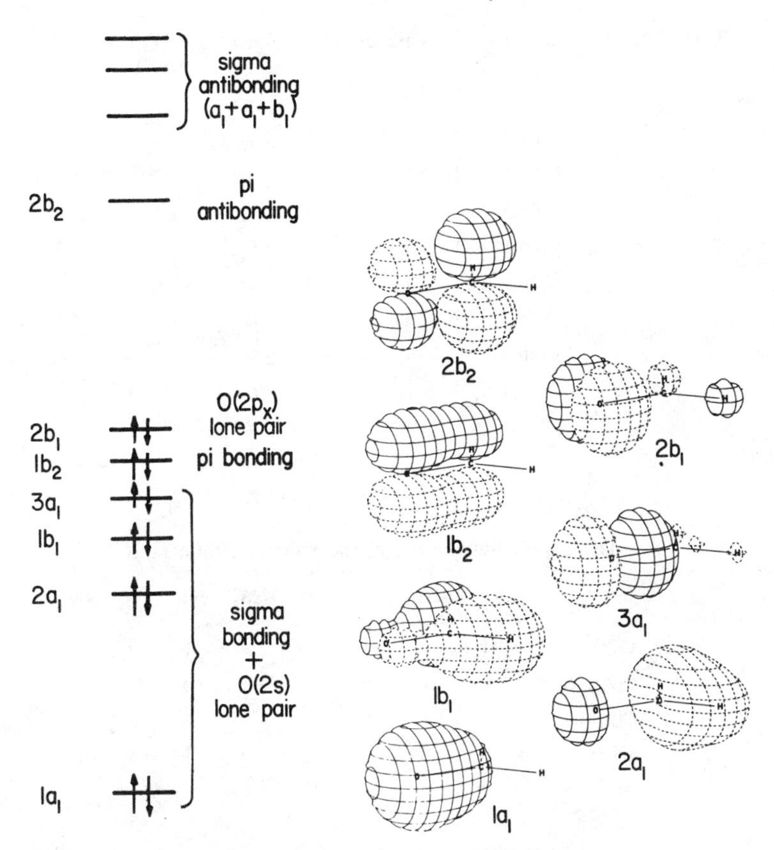

Fig. 4-36. Valence molecular orbitals of formaldehyde. Contour diagrams are as described in Fig. 4-27 and reproduced from W.L. Jorgensen and L. Salem, *The Organic Chemist's Book of Orbitals*, Academic Press, N.Y., 1973.

orbitals. Using the coordinate system in Fig. 4-37, the basis functions transform as shown in Table 4-16. This table, plus the structure of CH_3^+, tells us that we have to construct three sigma bonding orbitals of symmetry $a_1' + e'$ and three sigma antibonding orbitals of symmetry $a_1' + e'$, since the s, p_x, and p_y orbitals, which constitute the sigma framework, possess these symmetries. The C($2p_z$) orbital does not have the proper symmetry to combine with any other valence orbital; so it must be a nonbonding molecular orbital by itself.

The a_1' bonding and antibonding orbitals are totally symmetric and are easily constructed from the C(2s) and H(1s) orbitals as shown in Fig.

Table 4-15. Symmetries of the Valence Atomic Orbitals of Formaldehyde

C_{2v}	E	C_2	$\sigma(xz)$	$\sigma(yz)$	
2H(1s)	2	0	2	0	$= a_1 + b_1$
C(2s)	1	1	1	1	$= a_1$
C($2p_x$)	1	−1	1	−1	$= b_1$
C($2p_y$)	1	−1	−1	1	$= b_2$
C($2p_z$)	1	1	1	1	$= a_1$
O(2s)	1	1	1	1	$= a_1$
O($2p_x$)	1	−1	1	−1	$= b_1$
O($2p_y$)	1	−1	−1	1	$= b_2$
O($2p_z$)	1	1	1	1	$= a_1$

Table 4-16. Symmetries of the Valence Orbitals of CH_3^+

D_{3h}	E	$2C_3$	$3C_2$	σ_h	$2S_3$	$3\sigma_v$	
C(2s)	1	1	1	1	1	1	$= a_1'$
C($2p_x$,$2p_y$)	2	−1	0	2	−1	0	$= e'$
C(p_z)	1	1	−1	−1	−1	1	$= a_2''$
3H(1s)	3	0	1	3	0	1	$= a_1' + e'$

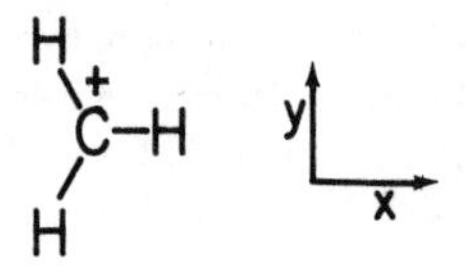

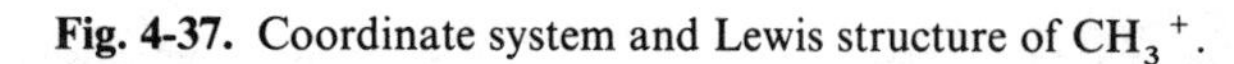
Fig. 4-37. Coordinate system and Lewis structure of CH_3^+.

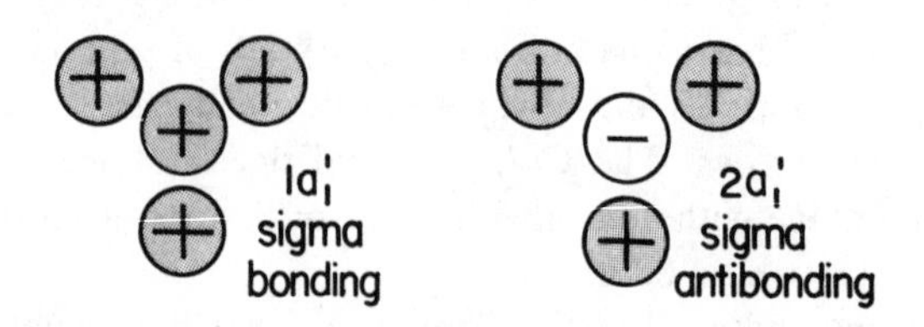

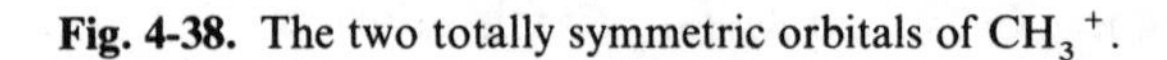
Fig. 4-38. The two totally symmetric orbitals of CH_3^+.

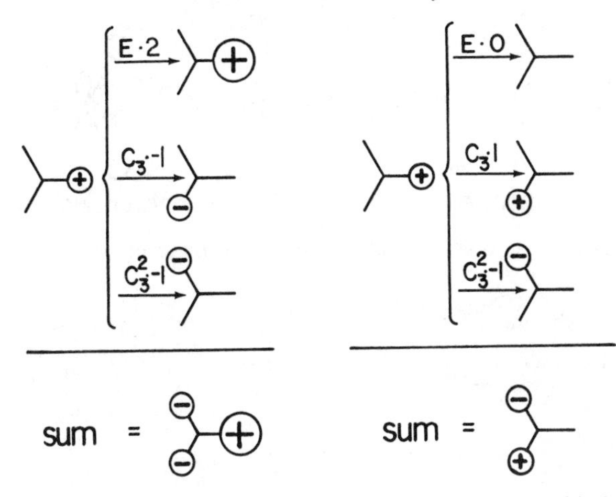

Fig. 4-39. Use of the projection operator to make H(1s) group orbitals of e' symmetry.

4-38. To construct degenerate e' orbitals from the H(1s) and C($2p_x$,$2p_y$) orbitals, we apply the projection operator using the real characters of the E representation of the point group C_3, as we have done previously for the stretching vibrations of BCl_3 in Section 3-7 (Fig. 3-66). The construction of linear combinations of H(1s) orbitals (group orbitals) with e' symmetry is shown in Fig. 4-39. The C($2p_x$) and C($2p_y$) orbitals together already possess e' symmetry and need no further transformation. The group orbital at the left of Fig. 4-39 can be combined with the C($2p_x$) orbital to give an orbital with C–H bonding character, and the group orbital at the right of Fig. 4-39 can be combined with the C($2p_y$) orbital (Fig. 4-40). By reversing the signs of the C(2p) orbitals, one can construct the corresponding e' antibonding orbitals also shown in Fig. 4-40.

In assembling the orbitals into a molecular orbital diagram (Fig. 4-41), we place the bonding orbitals lowest, followed by the nonbonding orbital, followed by the antibonding orbital. The six valence electrons of CH_3^+ just fill the bonding orbitals. In the nearly planar neutral methyl radical, $CH_3\cdot$, with seven valence electrons, the highest occupied orbital will be the C($2p_z$) nonbonding orbital. The photoelectron spectrum of $CH_3\cdot$ exhibits the first IE at 9.82 eV. This may be compared to the first IE of methane, 12.6 eV, which is representative of the energy of a C–H sigma bonding orbital and is probably close to the IE of the $1e'$ orbitals of CH_3^+.[†]

[†]The photoelectron spectra of $CH_3\cdot$ and CH_4 appear in T. Koenig, T. Balle, and W. Snell, *J. Amer. Chem. Soc.*, *97*, 662 (1975); and A.W. Potts and W.C. Price, *Proc. Roy. Soc.* (London) *A*, *326*, 165 (1972), respectively.

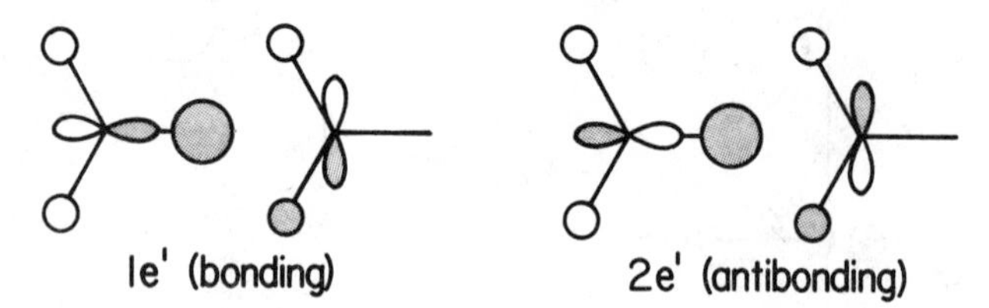

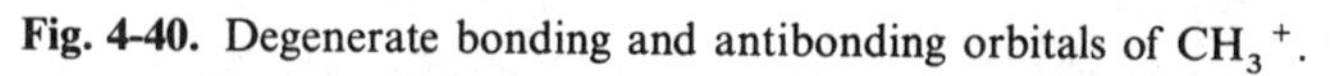

Fig. 4-40. Degenerate bonding and antibonding orbitals of $CH_3{}^+$.

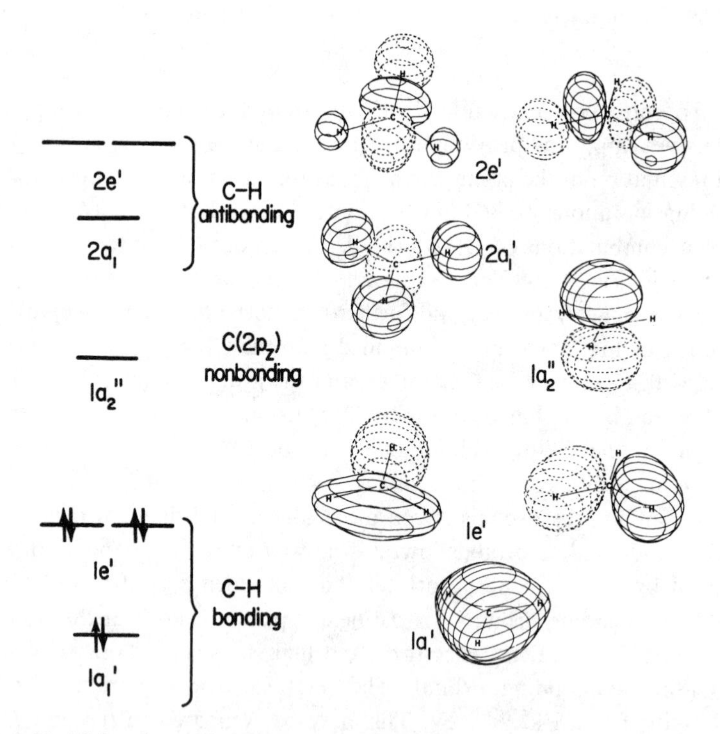

Fig. 4-41. Molecular orbital scheme for $CH_3{}^+$. Orbital contours are as described in Fig. 4-27 and reproduced from W.L. Jorgensen and L. Salem, *The Organic Chemist's Book of Orbitals*, Academic Press, N.Y. 1973.

Problems

4-24. The three pi mo's of the cyclopropenium cation, $C_3H_3^+$, are drawn below. In this picture, only the top half of each pi orbital is shown. The half below the plane of the page would have the opposite sign at each carbon atom. In the wave functions given below, $S = \int\phi_1\phi_2 d\tau = \int\phi_1\phi_3 d\tau = \int\phi_2\phi_3 d\tau$ and the functions ϕ_1, ϕ_2, and ϕ_3 are C(2p) atomic orbitals centered on each atom. Show that the energies of the two e' orbitals are identical. Your expression for the energy of each orbital will be in terms of overlap, exchange, and Coulomb integrals. Note that $H_{11} = H_{22} = H_{33}$ and $H_{12} = H_{13} = H_{23}$.

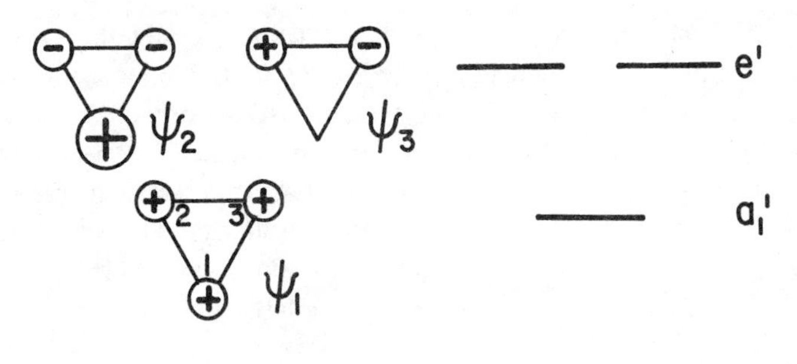

$$\psi_1 = [3(1 + 2S)]^{-\frac{1}{2}}(\phi_1 + \phi_2 + \phi_3)$$
$$\psi_2 = [6(1 - S)]^{-\frac{1}{2}}(2\phi_1 - \phi_2 - \phi_3)$$
$$\psi_3 = [2(1 - S)]^{-\frac{1}{2}}(\phi_2 - \phi_3)$$

4-25. Propose a molecular orbital scheme (see Problem 4-22) for the pi electron network of trimethylenemethane. How many unpaired electrons will the D_{3h} species have?

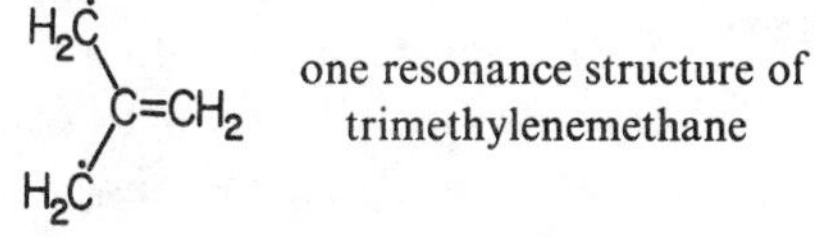

one resonance structure of trimethylenemethane

4-26. Propose a molecular orbital scheme (see Problem 4-22) for the pi electron network of cyclobutadiene in D_{4h} symmetry. Show what would be the effect on the final energy level diagram of elongation of two opposite sides of the molecule.

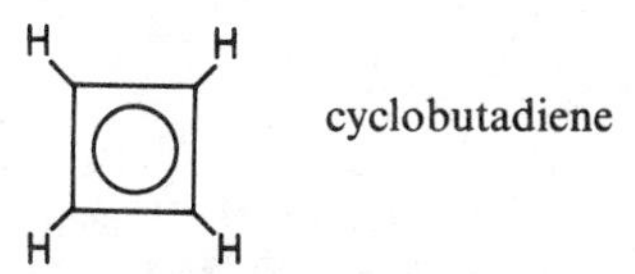

cyclobutadiene

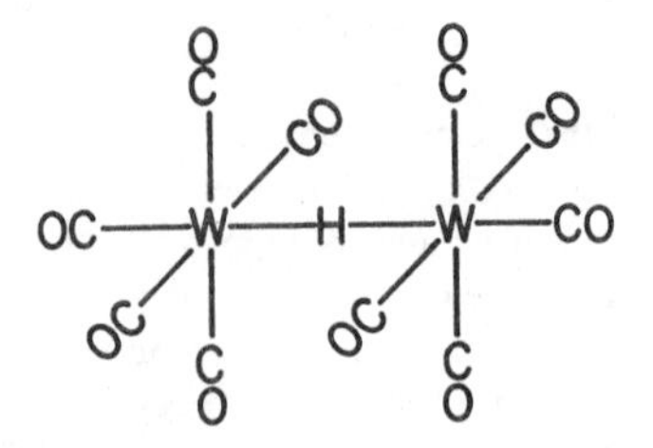

Fig. 4-42. The geometry of $HW_2(CO)_{10}^-$. For leading references, see D.C. Harris and H.B. Gray, *J. Amer. Chem. Soc.*, *97*, 3073 (1975).

F. *Electron Deficient Bonding.* This type of bonding, first recognized in diborane, B_2H_6, is found in a wide variety of stable inorganic species, as well as an ever increasing list of exotic, unstable organic species, such as CH_5^+. An example of a stable electron deficient species is the anion $HW_2(CO)_{10}^-$, with the structure in Fig. 4-42. In crystals of the tetraethylammonium salt, the W−H−W linkage is linear. The structure of this species suggests two W−H sigma bonds, but a count of electrons indicates that only two electrons are formally available for the entire W−H−W linkage.†

If we just consider some kind of hybrid orbital on each W atom directed at the H atom, the problem is one of constructing a molecular orbital scheme from the three basis orbitals shown in Fig. 4-43. Considering just the linear HW_2 unit, the symmetry is $D_{\infty h}$. Table 4-17 tells us that the H(1s) orbital can participate in molecular orbitals of symmetry σ_g^+, while the two W hybrid orbitals will participate in molecular orbitals of symmetry σ_g^+ and σ_u^+. We can draw a totally symmetric bonding orbital labelled $1\sigma_g^+$ in Fig. 4-44. This can be made into a totally symmetric antibonding orbital labelled $2\sigma_g^+$ by changing the sign of the H(1s) orbital. The only orbital of σ_u^+ symmetry which can be drawn is the one shown at the center of Fig. 4-44. Because this contains no H(1s) character and because the two W hybrid orbitals have little overlap, this orbital is essentially nonbonding (*i.e.*, only slightly antibonding). The two electrons occupy the only bonding orbital and a stable three-center two-electron unit results. Such bonding is typically not as strong as the much more common two-center two-electron bonding.

†Each W atom contributes six valence electrons, H contributes one, the lone pairs from CO contribute (10 × 2 =) 20 electrons, and the negative charge adds another electron, for a total of 34 valence electrons about the W−H−W unit. Ten pairs are formally localized in the W−C bonds, and six electrons on each W atom are directed between CO ligands and occupy the t_{2g} orbitals (Section 4-8). This leaves just two electrons formally associated with the W−H−W linkage.

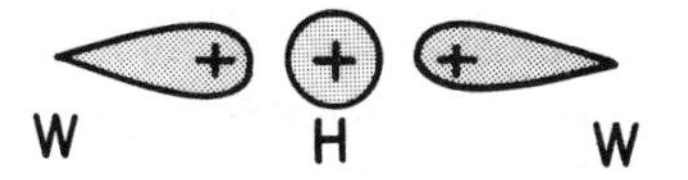

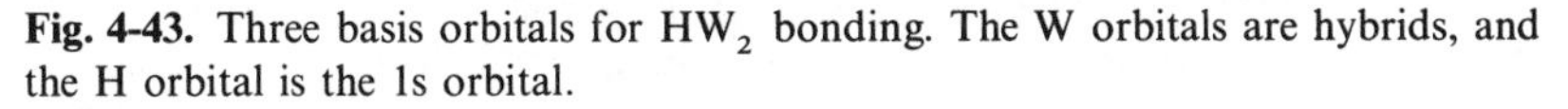

Fig. 4-43. Three basis orbitals for HW_2 bonding. The W orbitals are hybrids, and the H orbital is the 1s orbital.

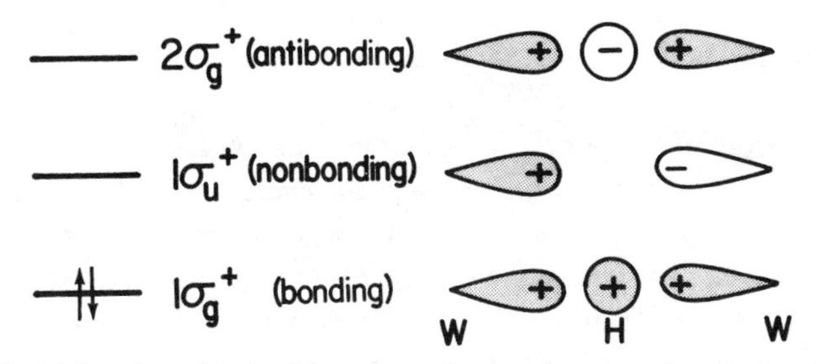

Fig. 4-44. Molecular orbital scheme for a three-center two-electron bond.

Table 4-17. Properties of the HW_2 Basis Orbitals in $D_{\infty h}$ Symmetry

$D_{\infty h}$	E	$2C_\infty^\phi$	$\infty\sigma_v$	i	$2S_\infty^\phi$	∞C_2	
H(1s)	1	1	1	1	1	1	$= \sigma_g^+$
2W(hybrids)	2	2	2	0	0	0	$= \sigma_g^+ + \sigma_u^+$

Problems

4-27. Derive the symmetries of the three basis orbitals in Fig. 4-43 in the true point group of $HW_2(CO)_{10}^-$, D_{4h}. Show that the three molecular orbitals in Fig. 4-44 transform as required in D_{4h} as well as in $D_{\infty h}$.

4-28. Using the coordinate system below, derive the symmetries of the four bridging B hybrid orbitals and two bridging H(1s) orbitals of diborane. By analogy to Fig. 4-44, construct the six bridging molecular orbitals of B_2H_6 and label each with the correct symmetry. Indicate which are bonding, nonbonding, and antibonding. Which will be occupied in diborane?

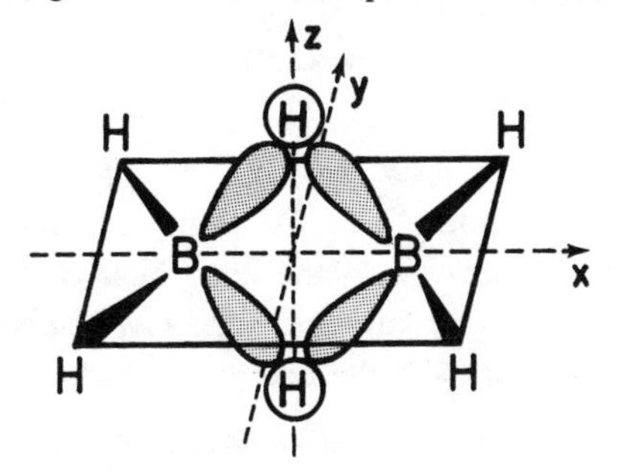

4-7. The Hückel Method

With an arsenal of drastic approximations it is possible to estimate the energies of molecular orbitals in a fairly simple manner. As crude as the method to be presented may seem, it gives answers with remarkable qualitative and semi-quantitative significance. The Hückel method is generally used for systems of pi electrons. In particular, we will limit ourselves to molecules containing only hydrogen and carbon, though the method can be extended to include sigma bonds and other kinds of atoms. Recall the case of ethylene (Section 4-6-C). We found a set of sigma bonding orbitals below the energy of the pi orbital. At higher energy than the pi orbital was the pi antibonding orbital, and at highest energy were the sigma antibonding orbitals. The highest filled and lowest unoccupied orbitals are pi orbitals. These are the orbitals most important to reactivity and spectroscopy. In the Hückel treatment outlined below we will completely ignore the sigma frameworks of hydrocarbons and only treat the pi orbitals.

The LCAO method involves construction of wave functions with undetermined coefficients. Our basis atomic orbitals in the Hückel method will be one 2p orbital from each carbon atom participating in the pi network. For butadiene, for example, we would have a wave function of the form in eq. 4-35:

$$\psi = a_1 2\mathrm{p}_1 + a_2 2\mathrm{p}_2 + a_3 2\mathrm{p}_3 + a_4 2\mathrm{p}_4 \qquad (4\text{-}35)$$

in which the subscripts 1–4 correspond to the four carbon atoms, and the a_i are undetermined coefficients. Using eq. 4-11 to calculate the energy of the wave function, and applying the variation treatment outlined in Section 4-4, we are led to a secular determinant of the form in eq. 4-36, analogous to eq. 4-29:

$$\begin{vmatrix} H_{11} - E & H_{12} - ES_{12} & H_{13} - ES_{13} & H_{14} - ES_{14} \\ H_{21} - ES_{21} & H_{22} - E & H_{23} - ES_{23} & H_{24} - ES_{24} \\ H_{31} - ES_{31} & H_{32} - ES_{32} & H_{33} - E & H_{34} - ES_{34} \\ H_{41} - ES_{41} & H_{42} - ES_{42} & H_{43} - ES_{43} & H_{44} - E \end{vmatrix} = 0 \qquad (4\text{-}36)$$

We use the notation $H_{ij} = \int (2\mathrm{p}_i)\mathscr{H}(2\mathrm{p}_j)\,\mathrm{d}\tau$.

Now for the approximations. The Coulomb integrals, H_{ii}, represent roughly the energy of an electron in the 2p orbital of carbon number i. We will assume all such integrals are equal and give them the value α. The exchange integrals, H_{ij}, represent the energy of interaction of p orbitals

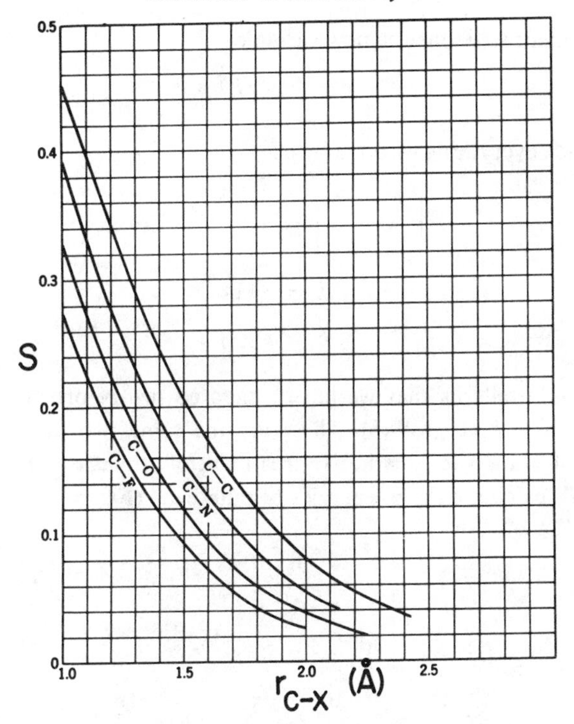

Fig. 4-45. Pi (2p,2p) overlap integrals. Reproduced from A. Streitwieser, Jr., *Molecular Orbital Theory for Organic Chemists*, John Wiley & Sons, N.Y., 1961.

located on atoms i and j. We will assume that if atoms i and j are adjacent, H_{ij} has the value β. If atoms i and j are not nearest neighbors, we set H_{ij} equal to zero. Note that relative to the energy of an electron at infinity, the values of α and β are *negative.* They represent stabilizing interactions. The overlap integrals, S_{ij}, are the easiest. We will set them all equal to zero, except S_{ii} which we have already correctly set equal to unity in eq. 4-36. To get some idea of the grossness of this approximation, a plot of the pi overlap integral is given in Fig. 4-45. At a C–C distance of 1.4 Å, typical of a pi system, S_{ij} has a value of about 0.25. Nonetheless, we are going to call it zero.

To see what these approximations do in the case of ethylene, the secular equation reduces to the form in eq. 4-37:

$$\begin{vmatrix} \alpha - E & \beta \\ \beta & \alpha - E \end{vmatrix} = 0 \tag{4-37}$$

It is convenient to define the new variable

$$x = (\alpha - E)/\beta \tag{4-38}$$

and to divide everything through by β:

$$\begin{vmatrix} x & 1 \\ 1 & x \end{vmatrix} = x^2 - 1 = 0$$

$$x = \pm 1$$

$$E = \alpha \pm \beta \tag{4-39}$$

Equation 4-39 tells us that we have generated *two* pi orbitals of energies $\alpha + \beta$ and $\alpha - \beta$ (Fig. 4-46). These are the same two pi orbitals which came out of our earlier qualitative treatment of ethylene.

Now let's calculate the pi energies of butadiene (Fig. 4-47):

$$\begin{vmatrix} x & 1 & 0 & 0 \\ 1 & x & 1 & 0 \\ 0 & 1 & x & 1 \\ 0 & 0 & 1 & x \end{vmatrix} = 0 \tag{4-40}$$

$$x(x^3 - 2x) - (x^2 - 1) = x^4 - 3x^2 + 1 = 0$$

$$x = \pm \left(\frac{3 \pm \sqrt{5}}{2}\right)^{\frac{1}{2}}$$

$$\begin{aligned} E_1 &= \alpha + \beta \left(\frac{3 + \sqrt{5}}{2}\right)^{\frac{1}{2}} = \alpha + 1.62\beta \\ E_2 &= \alpha + \beta \left(\frac{3 - \sqrt{5}}{2}\right)^{\frac{1}{2}} = \alpha + 0.62\beta \\ E_3 &= \alpha - \beta \left(\frac{3 - \sqrt{5}}{2}\right)^{\frac{1}{2}} = \alpha - 0.62\beta \\ E_4 &= \alpha - \beta \left(\frac{3 + \sqrt{5}}{2}\right)^{\frac{1}{2}} = \alpha - 1.62\beta \end{aligned} \tag{4-41}$$

The four energy levels which result from this calculation are shown in Fig. 4-48. The total energy of the four pi electrons of butadiene is $4\alpha + 4.48\beta$. The energy of two separate ethylene molecules is $4\alpha + 4\beta$. The butadiene molecule is therefore 0.48β more stable than two isolated ethylene double

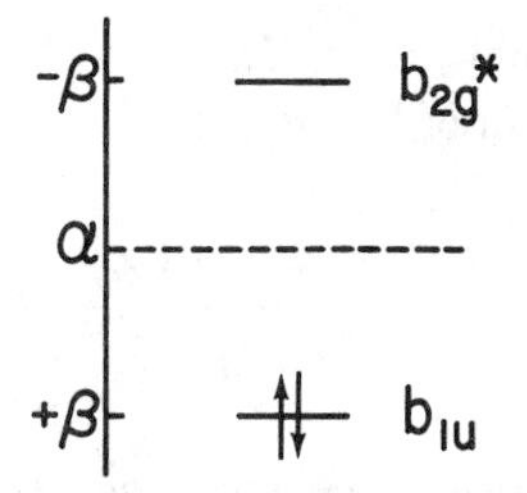

Fig. 4-46. Hückel pi energy levels of ethylene. These are the same pi orbitals of Fig. 4-33. Since β is a negative number, energies with positive values of β are lower than those with negative values of β.

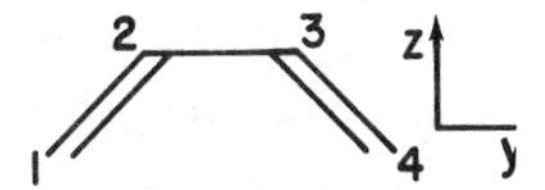

Fig. 4-47. Coordinate system and numbering system for butadiene.

bonds. Experimentally, this amounts to some 12 kJ/mol. This extra stabilization energy is called the *resonance* or *delocalization energy* and is common to all conjugated pi systems. (A conjugated system has a series of alternating single and double bonds.) It is due to the delocalization of the pi electrons throughout the entire pi network.

Now let's look at the wave functions for the butadiene pi orbitals. Each molecular orbital is a linear combination of four C(2p) orbitals (eq. 4-35). The variation treatment we have just outlined gives the numerical values of the coefficients of each orbital by using the energies from eqs. 4-41 in the equations for butadiene analogous to eqs. 4-27 and 4-28. That is, we must solve the four simultaneous equations 4-42 for each value of E in eqs. 4-41.

$$\begin{aligned}
c_1(H_{11}-E) + c_2H_{12} + c_3H_{13} + c_4H_{14} &= 0 \\
c_1H_{21} + c_2(H_{22}-E) + c_3H_{23} + c_4H_{24} &= 0 \\
c_1H_{31} + c_2H_{32} + c_3(H_{33}-E) + c_4H_{34} &= 0 \\
c_1H_{41} + c_2H_{42} + c_3H_{43} + c_4(H_{44}-E) &= 0
\end{aligned} \qquad (4\text{-}42)$$

Now we will see what group theory says about the symmetries of these four orbitals and compare the predictions of group theory to the numerical

Hückel results. In the point group C_{2v}, the orbitals transform as follows, using the coordinate system of Fig. 4-47:

C_{2v}	E	C_2	$\sigma(xz)$	$\sigma(yz)$	
$4C(2p_x)$	4	0	0	−4	$= 2a_2 + 2b_1$

(You could have just as well used the *trans* conformation of butadiene, instead of the *cis* form shown in Fig. 4-47, but the forms of the resulting orbitals would be identical. Only the point group and symmetry labels would be different.) Two linear combinations of p orbitals of symmetry a_2 and two of symmetry b_1 must be constructed. The functions in eqs. 4-43 are the most general ones which satisfy these requirements. The b_1 functions, for example, are symmetric to $\sigma(xz)$ and antisymmetric to C_2 and $\sigma(yz)$. By comparison, solution of eqs. 4-42 with the four values of E in eqs. 4-41 gives the wave function eqs. 4-44, which are drawn in Fig. 4-49.

$$\begin{aligned}
\psi(a_2) &= ap_1 + bp_2 - bp_3 - ap_4 \\
\psi(a_2) &= cp_1 - dp_2 + dp_3 - cp_4 \\
\psi(b_1) &= ep_1 + fp_2 + fp_3 + ep_4 \\
\psi(b_1) &= gp_1 - hp_2 - hp_3 + gp_4
\end{aligned} \tag{4-43}$$

$$\begin{aligned}
\psi_1(b_1) &= 0.371p_1 + 0.600p_2 + 0.600p_3 + 0.371p_4 \\
\psi_2(a_2) &= 0.600p_1 + 0.371p_2 - 0.371p_3 - 0.600p_4 \\
\psi_3(b_1) &= 0.600p_1 - 0.371p_2 - 0.371p_3 + 0.600p_4 \\
\psi_4(a_2) &= 0.371p_1 - 0.600p_2 + 0.600p_3 - 0.371p_4
\end{aligned} \tag{4-44}$$

The numerical solution therefore gives us orbitals of the same form predicted by group theory. An important point to note about the orbitals in Fig. 4-49 is that the orbital energy increases with the number of nodes. This is a general phenomenon.

Figures 4-46 and 4-48 lead us to expect that one occupied pi orbital of butadiene should be lower in energy than that of ethylene and one should be higher. This is nicely confirmed by the photoelectron spectra of these compounds in Fig. 4-50. The first *IE* of ethylene is observed near 10.5 eV, and the first two *IE*'s of butadiene are observed near 9.2 and 11.5 *eV*. That the peaks at still higher energy are due to sigma electrons is confirmed by the shift of the third peak to higher energy in the spectrum of 1,1,4,4-tetrafluorobutadiene. The fluorine substituents commonly stabilize sigma orbitals while having little effect on hydrocarbon pi orbitals.

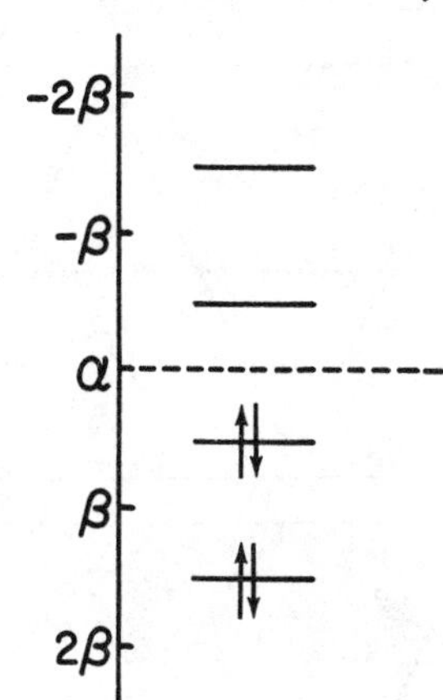

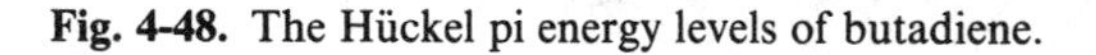

Fig. 4-48. The Hückel pi energy levels of butadiene.

As another example, we shall work the case of the allyl system (Fig. 4-51).

C_{2v}	E	C_2	$\sigma(xz)$	$\sigma(yz)$	
$3C(2p_x)$	3	-1	1	-3	$= a_2 + 2b_1$

$$\begin{vmatrix} x & 1 & 0 \\ 1 & x & 1 \\ 0 & 1 & x \end{vmatrix} = x^3 - 2x = 0 \tag{4-45}$$

$$x = 0, \pm\sqrt{2} \tag{4-46}$$

$$\begin{aligned} E_1 &= \alpha + \sqrt{2}\beta \\ E_2 &= \alpha \\ E_3 &= \alpha - \sqrt{2}\beta \end{aligned} \tag{4-47}$$

$$\begin{aligned} \psi_1(b_1) &= 0.500p_1 + 0.707p_2 + 0.500p_3 \\ \psi_2(a_2) &= 0.707p_1 \qquad\qquad\quad - 0.707p_3 \\ \psi_3(b_1) &= 0.500p_1 - 0.707p_2 + 0.500p_3 \end{aligned} \tag{4-48}$$

The three pi orbitals in eqs. (4-48) are analogous to the three sigma orbitals found in three-center two-electron bonding (Section 4-6-F). ψ_1 is bonding, ψ_2 is nonbonding, and ψ_3 is antibonding. The allyl cation has a total energy $2\alpha + 2\sqrt{2}\beta$ (Fig. 4-52). The allyl radical has energy $3\alpha + 2\sqrt{2}\beta$ and the anion has energy $4\alpha + 2\sqrt{2}\beta$. The third and fourth electrons of the radical and anion go into the nonbonding orbital.

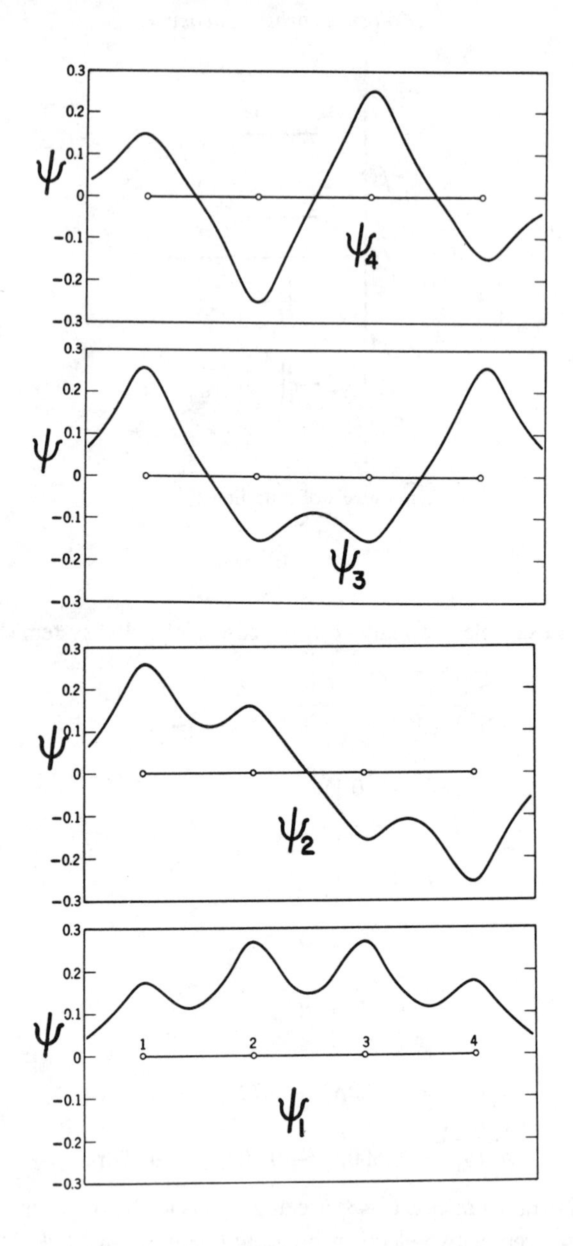

Fig. 4-49. Hückel pi wave functions for butadiene. From A. Streitwieser, Jr., *Molecular Orbital Theory for Organic Chemists*, John Wiley & Sons, N.Y., 1961.

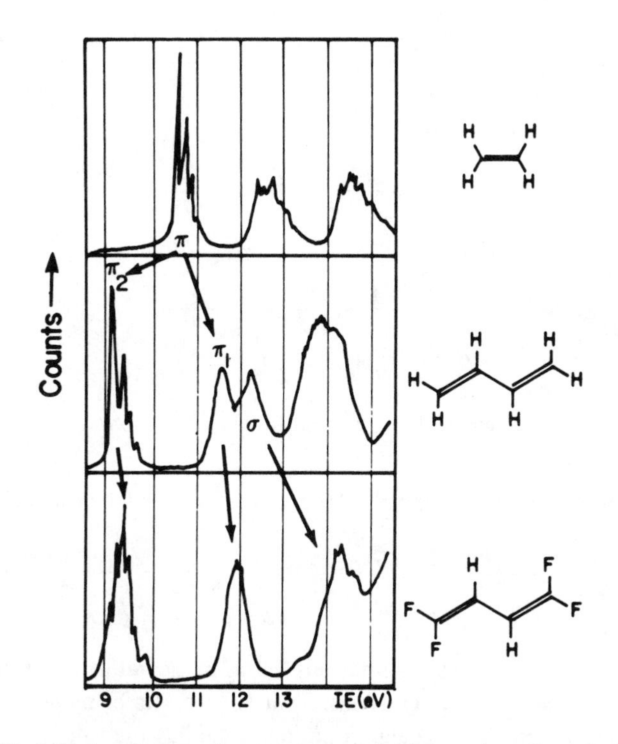

Fig. 4-50. He I (58.4 nm) photoelectron spectra of ethylene, butadiene, and 1,1,4,4-tetrafluorobutadiene, showing that one pi orbital energy of the dienes is greater than that of ethylene and one is less. Additional structure of the bands will be discussed in Chapter Five, Adapted from H. Bock and P.D. Mollère, *J. Chem. Ed.*, *51*, 506 (1974).

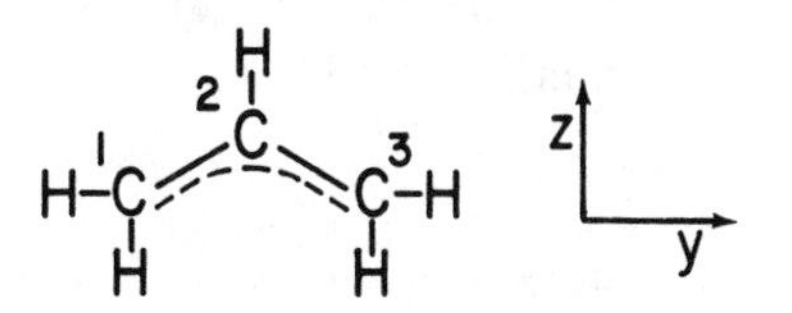

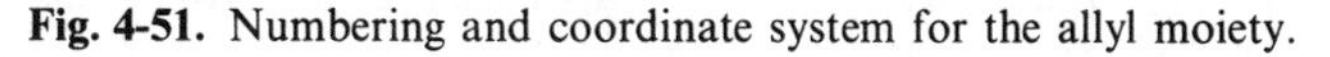

Fig. 4-51. Numbering and coordinate system for the allyl moiety.

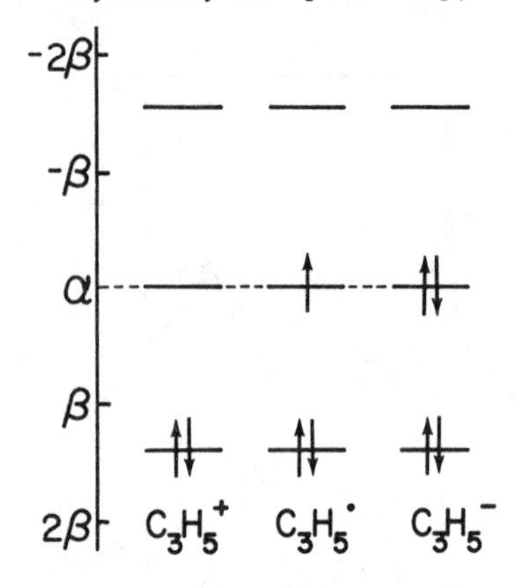

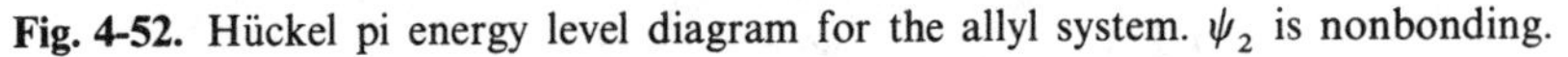

Fig. 4-52. Hückel pi energy level diagram for the allyl system. ψ_2 is nonbonding.

Finally, we shall examine the Hückel treatment of benzene (Fig. 4-53).

D_{6h}	E	$2C_6$	$2C_3$	C_2	$3C_2'$	$3C_2''$	i	$2S_3$	$2S_6$	σ_h	$3\sigma_d$	$3\sigma_v$
$6C(2p_z)$	6	0	0	0	-2	0	0	0	0	-6	0	2

$$= b_{2g} + e_{1g} + a_{2u} + e_{2u}$$

These four irreducible representations tell us immediately that there will be only *four* different pi energy levels. That is, two sets of molecular orbitals will be degenerate. The secular equation, 4-49, has the solutions $x = \pm 2$,

$$\begin{vmatrix} x & 1 & 0 & 0 & 0 & 1 \\ 1 & x & 1 & 0 & 0 & 0 \\ 0 & 1 & x & 1 & 0 & 0 \\ 0 & 0 & 1 & x & 1 & 0 \\ 0 & 0 & 0 & 1 & x & 1 \\ 1 & 0 & 0 & 0 & 1 & x \end{vmatrix} = 0 \qquad (4\text{-}49)$$

± 1, ± 1. These give the energy level diagram, Fig. 4-54, which tells us that the delocalization energy of benzene, relative to three isolated double bonds, is 2β. This comes to some 150 kJ/mol experimentally.† The forms of the pi orbitals are given in Fig. 4-55.

†The alert reader may recall that $0.48\beta = 12$ kJ/mol for butadiene and $2\beta = 150$ kJ/mol for benzene. Obviously the Hückel β is not constant from one molecule to another and is larger for benzene than for butadiene. It is reasonably constant within a given class of compounds such as straight chain conjugated polyenes.

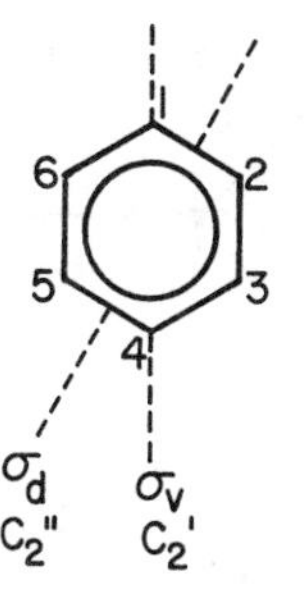

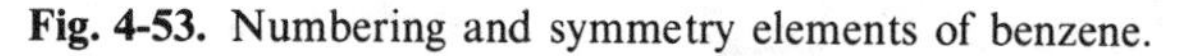

Fig. 4-53. Numbering and symmetry elements of benzene.

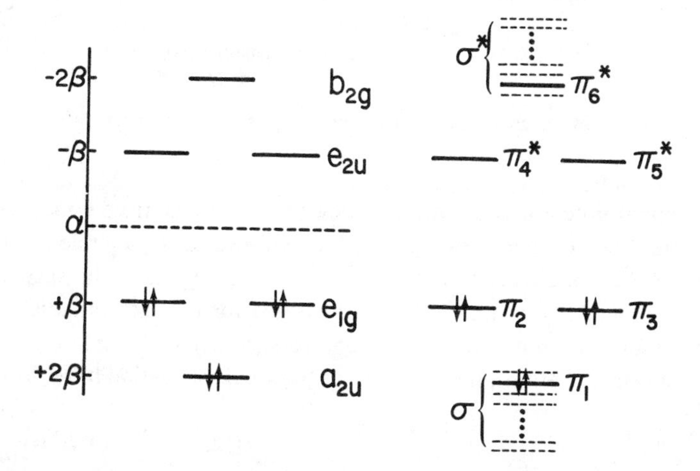

Fig. 4-54. Hückel pi energy level diagram for benzene (left) and the more complete mo diagram (right) showing that the sigma bonding orbitals overlap the lowest pi bonding orbital in energy. The degeneracies of the sigma orbitals are not indicated. The symbols "σ" and "π" refer to orbital type and not to orbital symmetry, as the point group is D_{6h}..

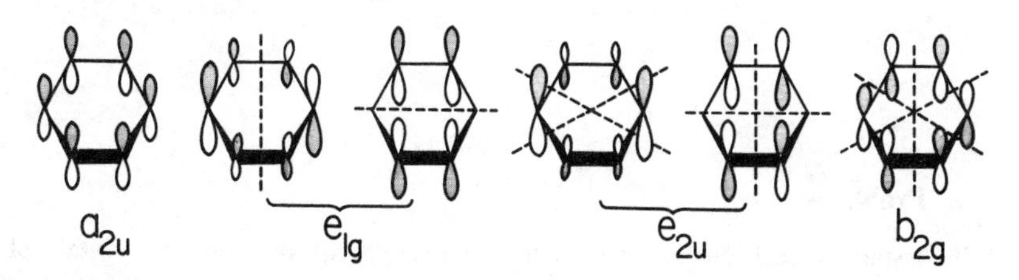

Fig. 4-55. Pi orbitals of benzene. Dotted lines are nodes. In each figure, we are looking at the $C(2p_z)$ orbitals perpendicular to the plane of the molecule, which is going into and out of the page.

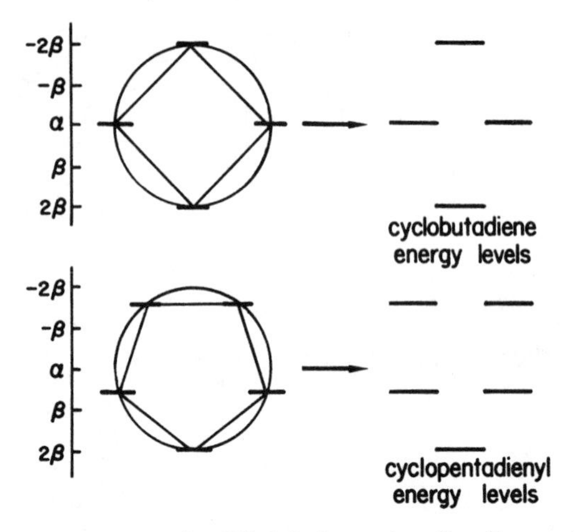

Fig. 4-56. Shortcut for generating Hückel pi energies of cyclic unsaturated systems.

You might note a useful pattern for cyclic pi systems. If you inscribe the geometric ring with one vertex pointing downward inside a circle of radius 2β, the Hückel pi energy levels come at the energies where the vertices of the figure intersect the circle. This is illustrated for the cyclobutadiene and cyclopentadienyl structures in Fig. 4-56. The geometry of Fig. 4-56 tells us that for a ring of n atoms, the solutions of the secular equation have the form

$$x_k = -2\cos\frac{2k\pi}{n} \quad \left(k = 0, \pm 1, \pm 2, \ldots \begin{cases} \pm\dfrac{n-1}{2} & \text{for odd } n \\ \pm\dfrac{n}{2} & \text{for even } n \end{cases} \right) \tag{4-50}$$

Just to be complete, for a straight chain of n atoms, the solutions of the Hückel secular determinant are of the form 4-51.

$$x_k = -2\cos\frac{k\pi}{n+1} \qquad (k = 1, 2, 3, \ldots, n) \tag{4-51}$$

Problems

4-29. Using Hückel theory, determine the energies of the four pi orbitals of trimethylenemethane (Problem 4-25).

4-30. (a) Determine the symmetries of the four pi orbitals of methylenecyclopropene and write the most general pi wave functions which possess these symmetries.

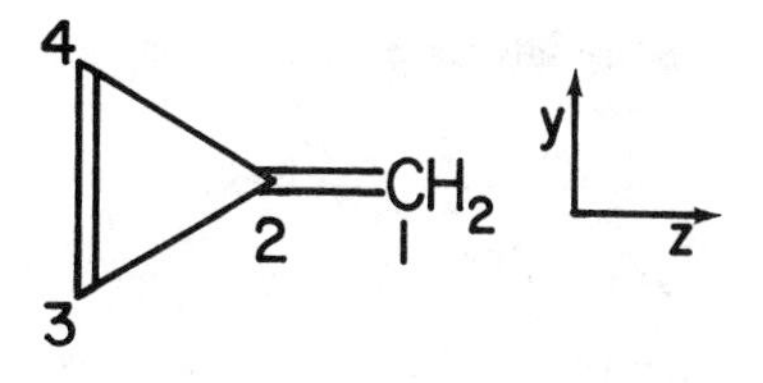

methylenecyclopropene

(b) Set up the secular determinant derived from Hückel theory for this molecule. (c) The roots of the secular determinant are $x = -2.170, -0.311, +1.000$, and $+1.481$. Draw an energy level diagram for this molecule and determine the delocalization energy of the neutral species. (d) The wave function for the lowest pi mo is $\psi = 0.278p_1 + 0.612p_2 + 0.524p_3 + 0.524p_4$. Using eq. 4-11, solve for the energy of this orbital in terms of α and β using the approximations of Hückel theory on pages 288–289. Compare your answer to that given by the appropriate root of the secular determinant.

4-31. By setting up and solving the appropriate simultaneous equations, calculate the coefficients of the pi orbitals of the allyl system. First solve for c_2 and c_3 in terms of c_1 and then normalize each wave function to obtain the coefficients in eq. 4-48. When normalizing the wave functions, remember the Hückel approximation that $S_{ij} = 0$ unless $i = j$.

4-8. Transition Metal Complexes

In this section we will consider only the most common idealized structure of transition metal complexes in which the metal is surrounded by six ligands at the vertices of an octahedron (Fig. 4-57). We suppose that each ligand has some kind of sigma orbital pointing at the transition metal, and that there is no pi bonding between the metal and ligands. The characters of the ligand sigma orbitals and the transition metal s, p, and d valence shell orbitals are given in Table 4-18. As one example of how these characters are obtained, the complete C_3 matrix for the metal d orbitals is given in eq. 4-52.

$$C_3 \cdot \begin{bmatrix} d_{xz} \\ d_{yz} \\ d_{xy} \\ d_{2z^2-x^2-y^2} \\ d_{x^2-y^2} \end{bmatrix} = \begin{bmatrix} 0 & 0 & 1 & 0 & 0 \\ 1 & 0 & 0 & 0 & 0 \\ 0 & 1 & 0 & 0 & 0 \\ 0 & 0 & 0 & -\frac{1}{2} & \frac{\sqrt{3}}{2} \\ 0 & 0 & 0 & -\frac{\sqrt{3}}{2} & -\frac{1}{2} \end{bmatrix} \begin{bmatrix} d_{xz} \\ d_{yz} \\ d_{xy} \\ d_{2z^2-x^2-y^2} \\ d_{x^2-y^2} \end{bmatrix} \tag{4-52}$$

Table 4-18. Symmetries of the Valence Orbitals Used in ML_6 Bonding

O_h	E	$8C_3$	$6C_2$	$6C_4$	$3C_2(=C_4^2)$	i	$6S_4$	$8S_6$	$3\sigma_h$	$6\sigma_d$	
6 ligand σ	6	0	0	2	2	0	0	0	4	2	$= a_{1g} + t_{1u} + e_g$
1 metal s	1	1	1	1	1	1	1	1	1	1	$= a_{1g}$
3 metal p	3	0	−1	1	−1	−3	−1	0	1	1	$= t_{1u}$
5 metal d	5	−1	1	−1	1	5	−1	−1	1	1	$= e_g + t_{2g}$

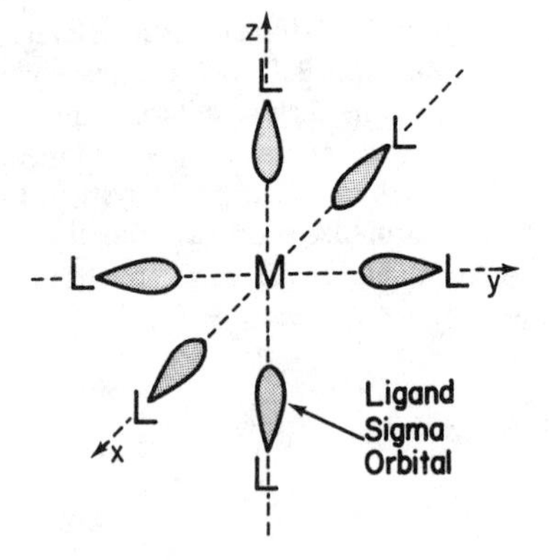

Fig. 4-57. Octahedral transition metal complex.

The d_{xz}, d_{yz}, and d_{xy} orbitals form one set of t_{2g} symmetry while the d_{z^2} and $d_{x^2-y^2}$ orbitals form another set of symmetry e_g.† This is the origin of the labels e_g and t_{2g} which you may have seen in crystal field theory.

The basis orbitals in Table 4-18 may be combined by the LCAO method and a molecular orbital scheme constructed (Fig. 4-58). In this scheme the $1a_{1g}$, $1e_g$, and $1t_{1u}$ sigma bonding orbitals are formally filled by the twelve ligand electrons. The $1t_{2g}$ and $2e_g$* orbitals would be occupied by whatever electrons the metal contributes. In the case of six metal electrons, such as in $Co(H_2O)_6{}^{3+}$, the $1t_{2g}$ orbitals are completely filled and the lowest energy excitation is $1t_{2g} \rightarrow 2e_g$*. The energy separation between the $1t_{2g}$ and $2e_g$* orbitals is commonly called Δ_o, the crystal field splitting energy. The forms of the ML_6 molecular orbitals are shown in Fig. 4-59.

† d_{z^2} is short for $d_{2z^2-x^2-y^2}$.

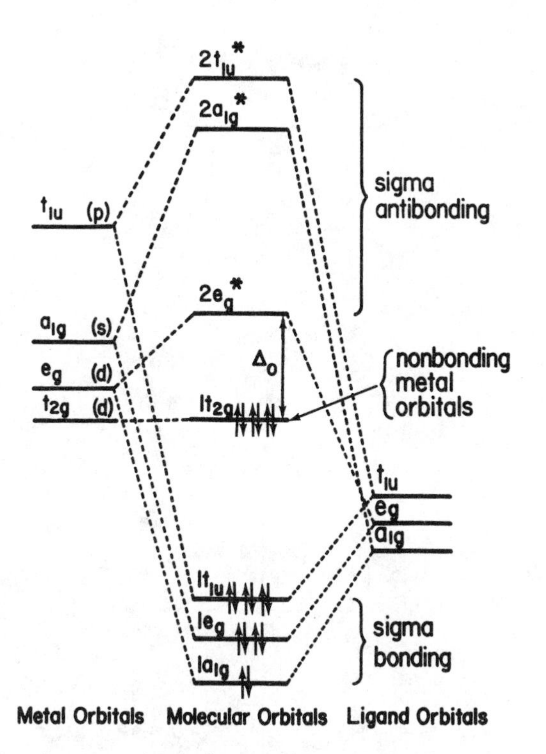

Fig. 4-58. Schematic energy level diagram for an octahedral d^6 transition metal complex.

Problems

4-32. Propose a molecular orbital scheme for a tetrahedral d^8 ML_4 transition metal complex including only sigma orbitals of the ligands. Draw a picture of each mo and make an energy level diagram with the proper number of electrons in each orbital.

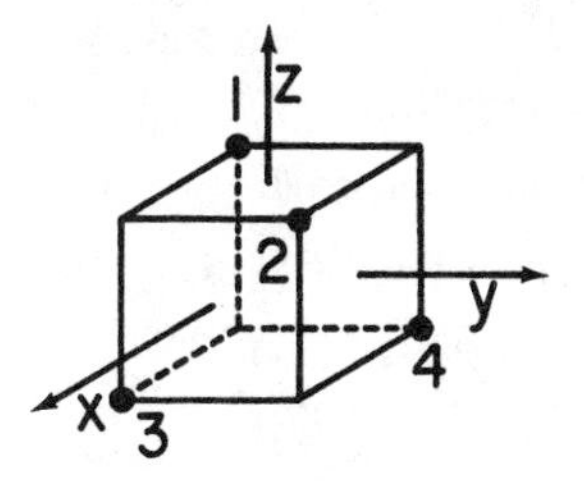

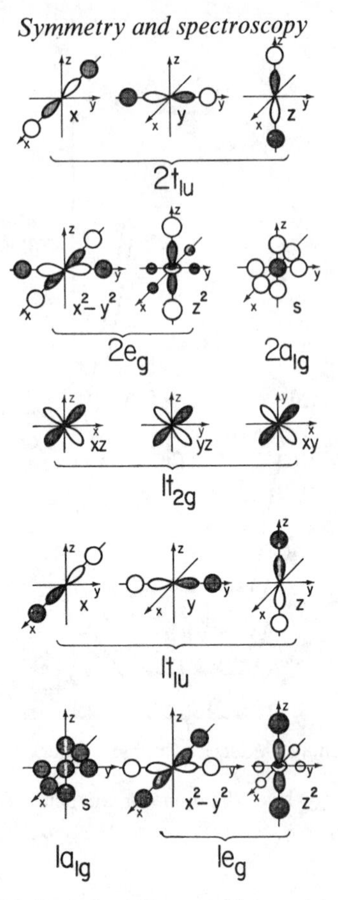

Fig. 4-59. Molecular orbitals of the ML_6 transition metal complex. Light and dark lobes are of different signs.

Additional Problems

4-33. For a one-electron atom the quantum mechanical average value of $1/r$, the inverse of the electron-nuclear distance, calculated with eq. 2-43, is $\int_{r=0}^{\infty}\int_{\theta=0}^{\pi}\int_{\phi=0}^{2\pi} \psi^* \frac{1}{r} \psi\, r^2 \sin\theta\, dr\, d\theta\, d\phi = z/(a_0 n^2)$. Use this result to calculate the average potential energy, $\langle V \rangle$, of an electron in the field of the nucleus, which will be given by $\langle V \rangle = \int_{r=0}^{\infty}\int_{\theta=0}^{\pi}\int_{\phi=0}^{2\pi} \psi^* \left(-\frac{Ze^2}{4\pi\epsilon_0 r}\right) \psi\, r^2 \sin\theta\, dr\, d\theta\, d\phi$. Using the relation $\langle K \rangle = T - \langle V \rangle$, where T is the total energy and $\langle K \rangle$ is the average kinetic energy, calculate the average kinetic energy of the electron.

Since the average kinetic energy is related to the average of the square of the speed by $\langle K \rangle = (1/2)m_e \langle v^2 \rangle$, find the average speed of an electron in the 1s orbital of hydrogen and of C^{5+}. Compare these speeds to the speed of light.

4-34. Use eq. 4-11 to calculate the energy of a harmonic oscillator if ψ_0 is approximated as a half cosine wave extending from $-x_0$ to $+x_0$. Show that this is a greater energy than the true energy, E_0, of the harmonic oscillator.

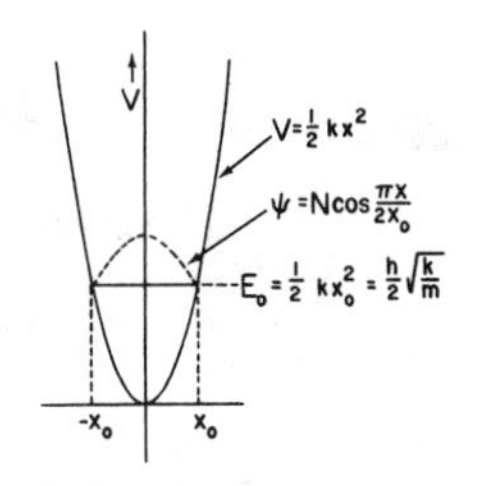

4-35. Using a table of microstates, derive the Russell-Saunders terms which come from the configuration p^3.

4-36. Propose a molecular orbital scheme (see Problem 4-22) for the pi electrons of the following molecules:

a. carbon suboxide, O=C=C=C=O b. allene, $H_2C{=}C{=}CH_2$

c. tetramethylenecyclobutane d. trimethylenecyclopropane

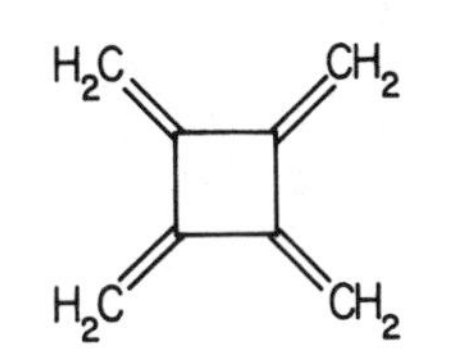

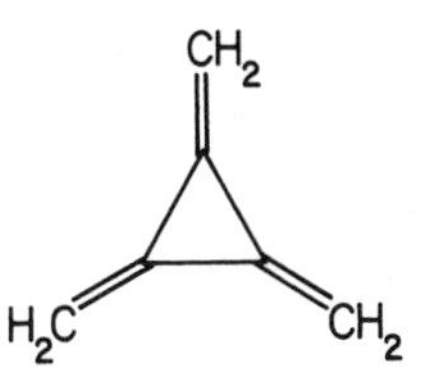

4-37. Propose a molecular orbital scheme for methane using the coordinate system in Problem 4-32. Draw each mo and label its symmetry. The photoelectron spectrum of CH_4 exhibits its first two bands with maxima near 14 and 23 eV, with relative areas 3:1, respectively. To what ionizations are these peaks due?

4-38. Using only your character tables and a table of trigonometric functions, write the explicit forms of the pi electron wave functions of the planar cyclopentadienyl anion.

4-39. Calculate the Hückel pi energy levels and delocalization energy of 3-methylene-1,4-pentadiene.

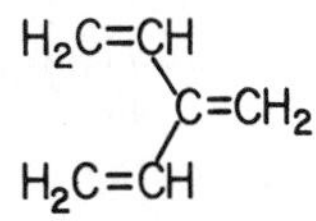

4-40. The energies (from photoelectron spectra) of the highest occupied orbitals of the aromatic compounds below are correlated with the orbital energies of benzene as follows:

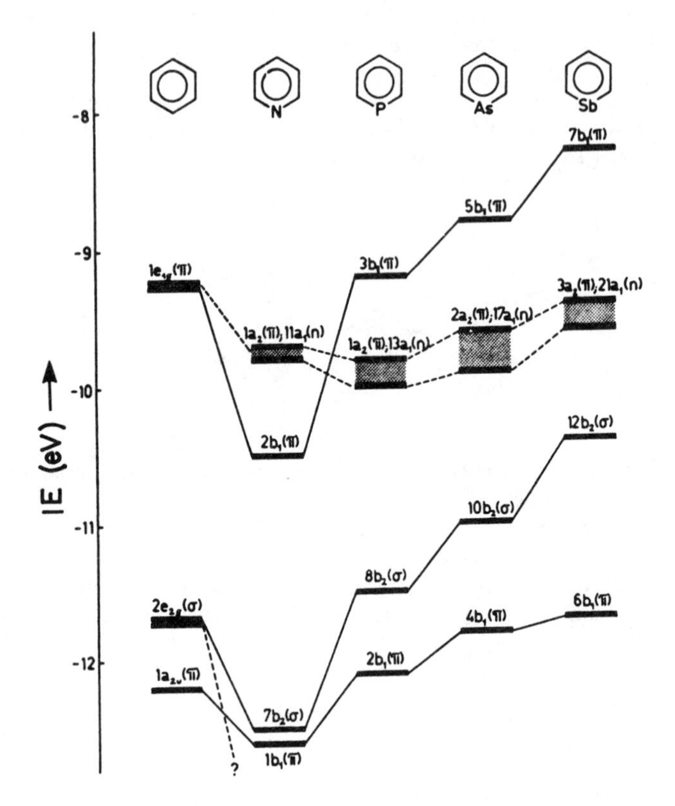

The correlation is done by quantitatively calculating the effect that substitution of one CH group of benzene by a group V atom will have on the energy of each benzene orbital. In the correlation diagram, sets of energy levels connected by a shaded region are of uncertain order. (a) Using Fig. 4-55 as a reference, and given that the $1a_{2u}$ orbital of benzene correlates with the $1b_1(\pi)$ orbitals of the C_5H_5X compounds, draw the coordinate system used for the C_5H_5X molecules. That is, in which plane is the molecule? (b) Draw the $2b_1(\pi)$, $1a_2(\pi)$, $13a_1(n)$, and $3b_1(\pi)$ orbitals of phosphabenzene. (c) Which orbital, $2b_1$ or $3b_1$, of phosphabenzene has a greater amount of phosphorus p orbital character? For which of these orbitals will the energy be more dependent on the nature of the atom X of C_5H_5X? Do the correlation assignments bear out this prediction? (Reference: C. Batich *et al.*, *J. Amer. Chem. Soc.*, *95*, 928 [1973].)

4-41. For each molecule whose photoelectron spectrum is shown below, construct a qualitative pi orbital mo scheme. Draw a picture of each occupied orbital, label each with the proper symmetry, and order the occupied orbitals by energy. When drawing the orbitals of the tetraene, for example, use the Newman projection below in which each pi orbital is seen end on. In the occupied orbitals, the pi interactions between atoms 1 and 2, 3 and 4, 6 and 7, and

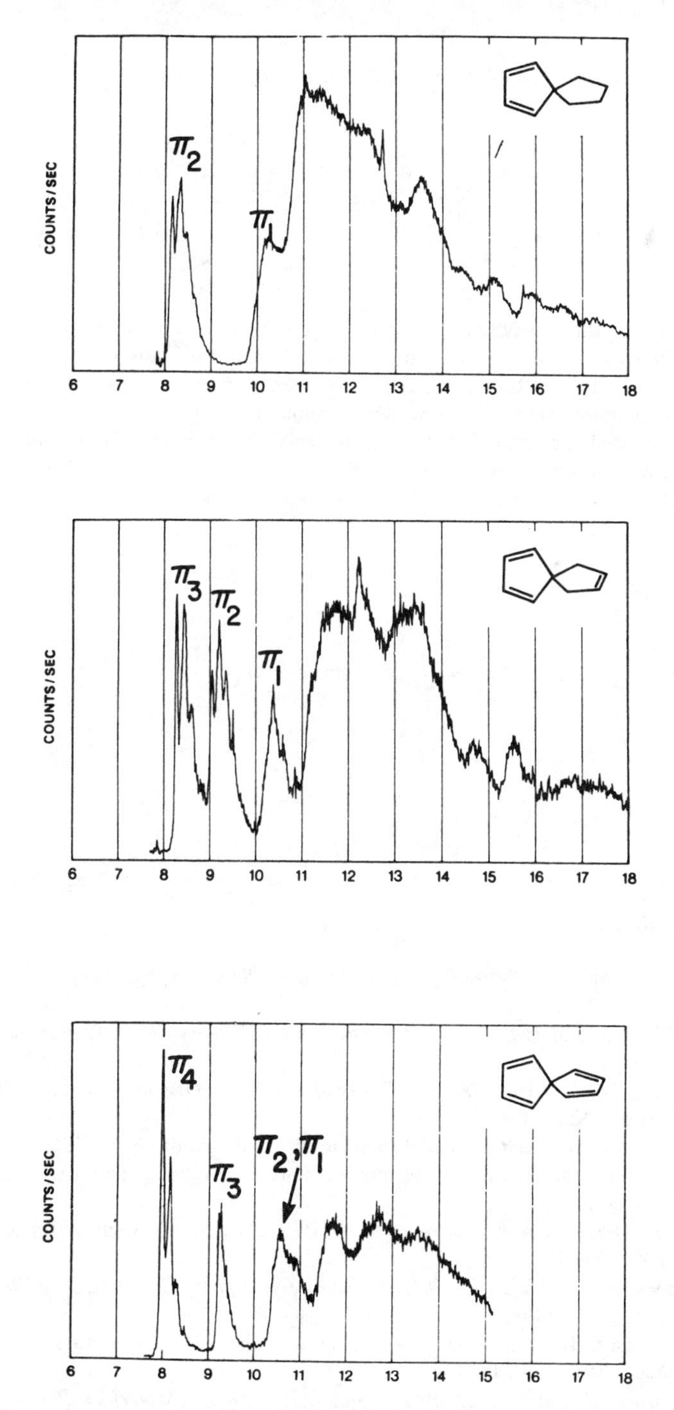

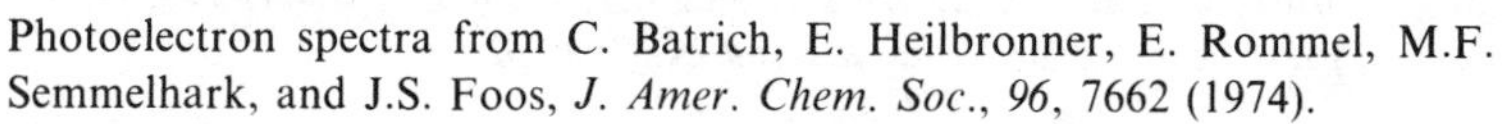
Photoelectron spectra from C. Batrich, E. Heilbronner, E. Rommel, M.F. Semmelhark, and J.S. Foos, *J. Amer. Chem. Soc.*, *96*, 7662 (1974).

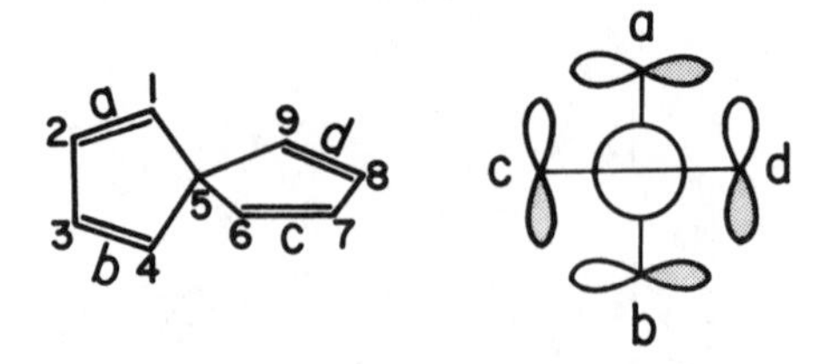

8 and 9 will always be bonding. Therefore the Newman projection of the orbitals shown above gives the signs of all eight p orbitals, even though only four are shown. Rationalize the energies and number of bands present in the pi orbital region ($\lesssim$ 11 eV) of each compound.

4-42. What are the symmetry species of the ten $C(2p_z)$ orbitals of ferrocene? Draw the ten carbon pi group orbitals. Which Fe valence orbitals (3d, 4s, and 4p) can interact with each cyclopentadienyl group orbital?

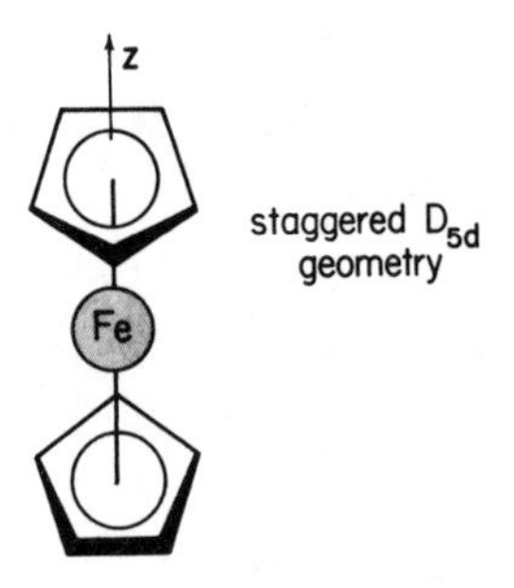

Related Reading

A.D. Baker and D. Betteridge, *Photoelectron Spectroscopy*, Pergamon Press, Oxford, 1972.

C.J. Ballhausen and H.B. Gray, *Molecular Orbital Theory*, W.A. Benjamin, N.Y., 1965.

R.L. Flurry, Jr., *Molecular Orbital Theory of Bonding in Organic Molecules*, Marcel Dekker, N.Y., 1968.

H.B. Gray, *Electrons and Chemical Bonding*, W.A. Benjamin, N.Y., 1965.

W.L. Jorgensen and L. Salem, *The Organic Chemist's Book of Orbitals*, Academic Press, N.Y., 1973.

M. Orchin and H.H. Jaffé, *Symmetry, Orbitals and Spectra*, John Wiley & Sons, N.Y., 1971.

A. Streitwieser, Jr., *Molecular Orbital Theory for Organic Chemists*, John Wiley & Sons, N.Y., 1961.

A. Streitwieser, Jr., and P.H. Owens, *Orbital and Electron Density Diagrams*, Macmillan, N.Y., 1973.

D.W. Turner, C. Baker, A.D. Baker, and C.R. Brundle, *Molecular Photoelectron Spectroscopy*, John Wiley & Sons, N.Y., 1970.

5 · *Electronic spectroscopy*

5-1. Introduction

Light of energy sufficient to cause electronic transitions simultaneously promotes vibrational and rotational transitions. It is unfortunate that the absorption bands of most solution phase spectra are so broad that the vibrational and rotational information is totally obscured. However, at low temperature and in the gas phase, the presence of vibrational structure in an electronic absorption spectrum is the rule, not the exception, and provides the key to the interpretation of the spectrum. Since rotational fine structure is not generally resolved in the electronic spectra of polyatomic molecules, we are going to ignore rotational energy levels in this chapter. However, since an electronic absorption spectrum may contain a great deal of vibrational information, one needs to know about both electronic and vibrational states in order to interpret the spectrum. We will use all of the tools of the first four chapters for our study of electronic spectroscopy.†

5-2. Another Look at Molecular Vibrations

In Chapter Three we found it reasonable to represent the vibrational energy levels of the electronic ground state of a diatomic molecule as horizontal

†A much more detailed and complete treatment of molecular spectroscopy than is presented in this book has been written in three fine volumes by G. Herzberg: *Molecular Spectra and Molecular Structure, I: Spectra of Diatomic Molecules* (2nd ed., 1950); *II: Infrared and Raman Spectra of Polyatomic Molecules* (1945); *III: Electronic Spectra and Electronic Structure of Polyatomic Molecules* (1966), Van Nostrand Reinhold, N.Y. These books are recommended to all serious students of molecular spectroscopy.

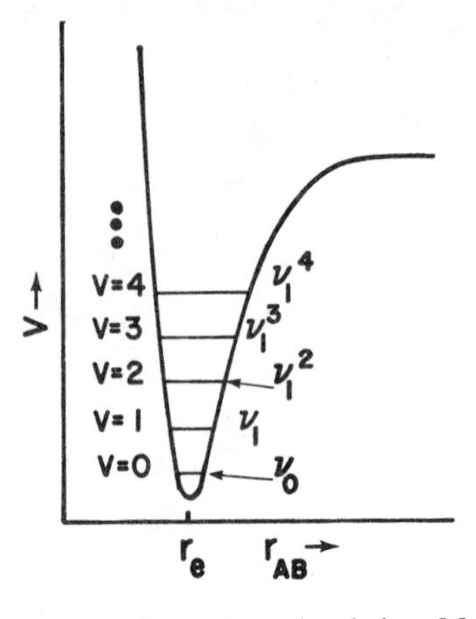

Fig. 5-1. Vibrational levels of a diatomic molecule in a Morse potential well.

lines in a Morse potential well (Fig. 5-1). The lowest energy level, in which v, the vibrational quantum number, is zero is the vibrational ground state, designated ν_0. The $v = 1$ state, designated ν_1, is attained by the fundamental transition. Succeeding vibrational energy levels are designated ν_1^2, ν_1^3, ν_1^4, . . . in Fig. 5-1. The designation ν_1^4, for example, means that the vibration ν_1 is in the $v = 4$ state. For a diatomic molecule, the subscript 1 is superfluous because the molecule has only one kind of vibration. For polyatomic molecules, however, the subscript indicates which of the many possible vibrations is being considered.

Polyatomic molecules with n atoms possess $3n-6$ (or $3n-5$ for linear molecules) vibrational degrees of freedom, and each vibration will generally require a different potential well for its description. For example, Fig. 5-2 shows three different wells needed to describe the three vibrations of the bent triatomic molecule, SO_2. Symmetric stretching of the $S-O$ bonds ought to be represented by a Morse-like potential because the potential energy will rise rapidly as the $S-O$ distances decrease, and level off as the $S-O$ distances increase very much from their equilibrium value. Asymmetric stretching, however, ought to give rise to a symmetric potential well because as one $S-O$ bond shortens, the other lengthens. Hence the potential energy rises symmetrically on either side of the equilibrium bond length. Bending is altogether different because the molecule has *two* equally stable equilibrium geometries, shown in Fig. 5-2. The energy barrier between the two equilibrium positions is considerable so the molecule is effectively locked into one side of the well or the other.

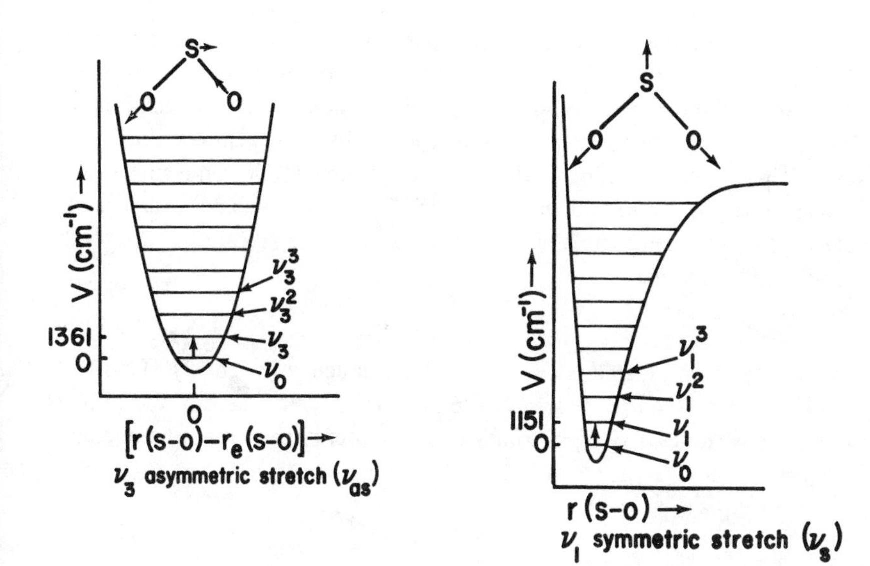

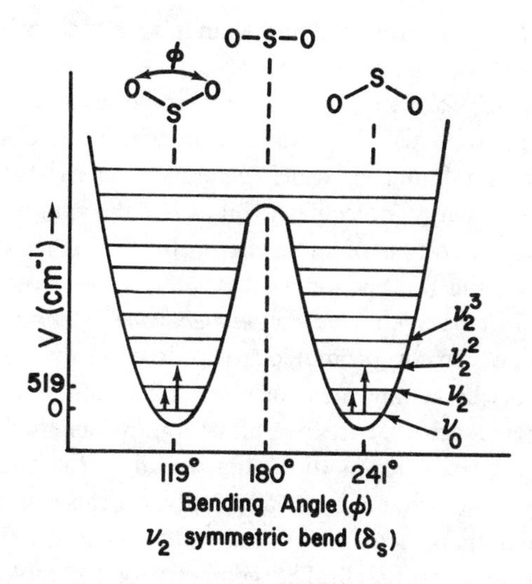

Fig. 5-2. Three potential wells needed to describe the three vibrations of SO_2. Vertical arrows in each well represent infrared energy transitions between vibrational states. Inversion doubling (Section 3-6-F) is negligible for this molecule and is not shown.

The fundamental vibrational frequencies of SO_2 are: ν_1, 1151; ν_2, 519; ν_3, 1361 cm^{-1}. In order to describe the three normal modes of vibration on a single diagram, we would need four dimensions—one for each normal coordinate and one for the potential energy. Since we cannot do this in any simple way, we will generally describe all three modes on a single two-dimensional Morse-like diagram. In Fig. 5-3 the vibrational ground state is called ν_0 and represents the state in which all three vibrations occur with their zero-point energies:

$$\bar{\nu}_0 = \sum_{i=1}^{3n-6} (1/2)\overline{\omega}_i \tag{5-1}$$

In this equation $(1/2)\overline{\omega}_i$ is the zero point frequency (in cm^{-1}) of normal mode number i and there are $3n - 6$ such modes. In Fig. 5-3 we see that the first overtone of ν_2 ($\nu_2{}^2$) comes at lower energy than the fundamentals of ν_1 and ν_3.

5-3. Basic Notions

A. *Origin of Spectra.* Infrared and Raman spectroscopy deal with transitions between vibrational energy levels *within a single electronic potential well.* Four such transitions are shown by vertical arrows in Fig. 5-2. The electronic ground state will be that state in which the electronic energy of the molecule is minimized, which generally means that the electrons occupy the lowest energy molecular orbitals. (If the spin pairing energy is greater than the separation of molecular orbitals, higher energy orbitals will be occupied, as in B_2, Section 4-5.) Electronic spectroscopy deals with transitions of molecules *between different electronic potential wells.* In terms of mo theory, an electron is promoted from a low energy orbital to a higher energy orbital, or the arrangement of electrons within a set of degenerate orbitals is changed such that the overall energy is increased. The result of such an electronic transition is disruption of all of the established forces within the molecule. Often, but by no means always, this results in a lowering of the vibrational energies of the excited state potential well compared to those of the ground state well. The geometry of the molecule may also change in the excited state.

The solid arrow in Fig. 5-4 represents a transition between the vibrational ground states of two different electronic states. Since the molecule goes from vibrational state $v = 0$ in the electronic ground state to the vibrational state $v' = 0$ in the electronic excited state, the transition is called

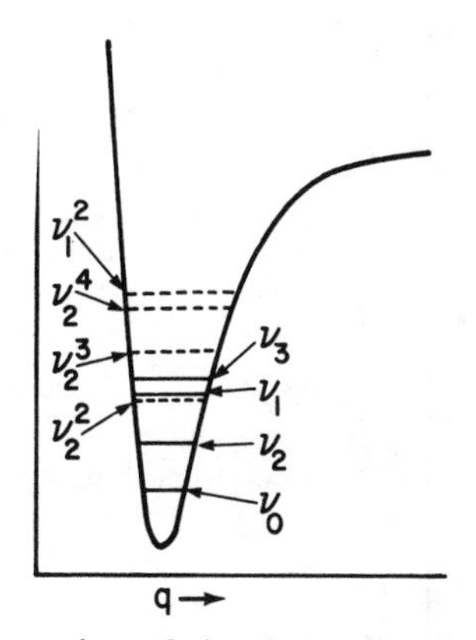

Fig. 5-3. Schematic representation of the three vibrational modes of SO_2 on a single diagram.

the 0-0 transition (read "zero-zero"). We will use primed quantum numbers for the excited electronic state and unprimed quantum numbers for the ground electronic state throughout this chapter. The 0-0 transition is a "pure" electronic transition because it involves no quanta of vibrational energy. The dashed arrow in Fig. 5-4 is a *vibronic transition* involving one quantum of electronic energy and one quantum of vibrational energy. (But this is still a one-photon process.) In this case, ν_1 is excited from the $v = 0$ level in the electronic ground state to the $v' = 1$ level in the electronic excited state. This vibronic transition will be designated 1_0^1:

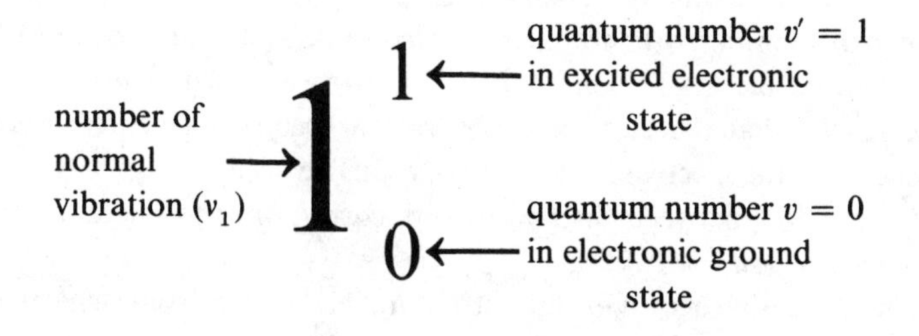

The notation will be developed more fully in Sections 5-4-D and 5-8.

In general then, the electronic absorption spectrum of a molecule may be a complex series of vibronic transitions built upon the pure electronic 0-0 transition. For example, the absorption spectrum of carbon monoxide

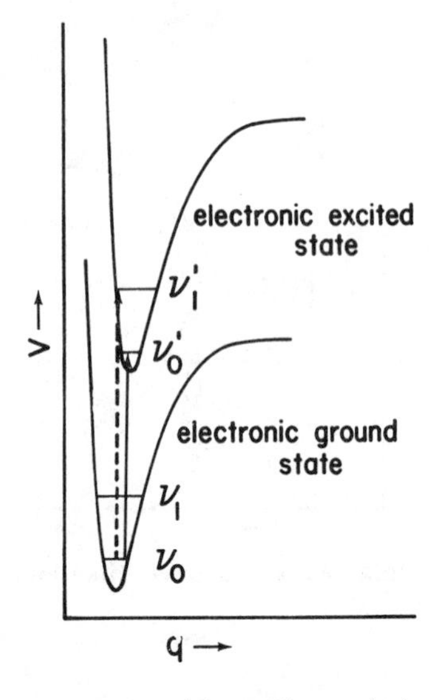

Fig. 5-4. Examples of electronic transitions. The vertical arrow is the 0-0 transition and the dashed arrow is a transition designated 1_0^1.

(Fig. 5-5) shows a series of vibronic absorptions built upon the 0-0 transition which comes at 64,703 cm^{-1} (155 nm). The vibronic transitions are to the high energy side of the 0-0 transition because they require enough energy to stimulate both the electronic and vibrational transitions. The spacings between the vibronic transitions allow us to say that the vibrational frequency of the molecule *in its excited electronic state* is $\bar{\omega}_e = 1516$ cm^{-1}, and the anharmonicity constant, $\bar{\omega}_e x_e$, is 17.25 cm^{-1}. This value of $\bar{\omega}_e$ should be compared to the ground electronic state value of 2170.2 cm^{-1}. We see that the C—O bond is considerably weakened in the excited state. The reason that the electronic absorption spectrum gives vibrational frequencies of the electronic excited state can be seen in Fig. 5-4. The difference in energy between the two arrows corresponds to one quantum of ν_1 *in the upper well.*

The appearance of vibronic structure in the electronic spectrum of a polyatomic molecule is very dependent on the physical state of the sample. Figures 5-6 and 5-7 show how vibrational structure tends to sharpen in non-polar solvents and at low temperature. The most vibrational structure is generally seen in the vapor phase. These generalizations apply as well to the line widths of pure vibrational spectra.

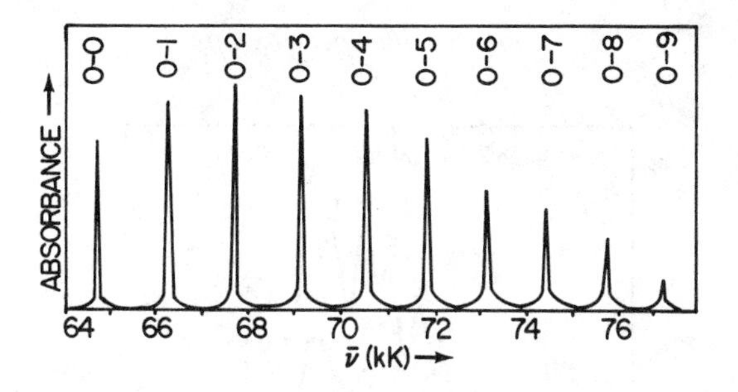

Fig. 5-5. Schematic $^1\Sigma^+ \rightarrow {}^1\Pi$ electronic absorption spectrum of CO showing vibronic transitions to higher energy than the pure electronic 0-0 transition. The notation for electronic states is developed is Section 5-3-D.

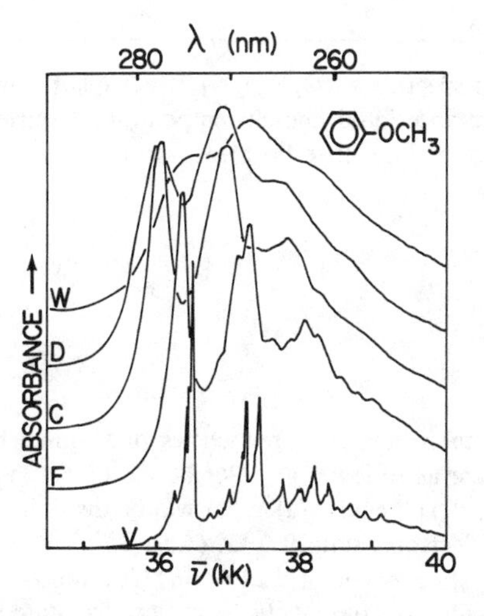

Fig. 5-6. Absorption spectra of anisole: (V) vapor, (F) in perfluorooctane, (C) in cyclohexane, (D) in dioxane, and (W) in water. Spectra are displaced vertically for clarity. From G.L. Tilley, Ph. D. Thesis, Purdue Univ., 1967.

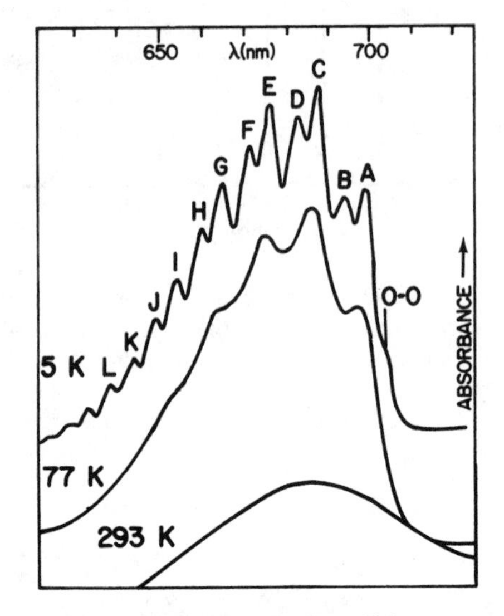

Fig. 5-7. Absorption spectra of $[(C_4H_9)_4N]_2[Re_2Cl_8]$ at room temperature, liquid nitrogen temperature and liquid helium temperature. Courtesy C. Cowman, California Institute of Technology. (See Problem 5-56.)

Problems

5-1. The fundamental vibrational frequencies of formaldehyde in the electronic ground state are as follows: ν_1, 2766.4; ν_2, 1746.1; ν_3, 1500.6; ν_4, 1167.3; ν_5, 2843.4; ν_6, 1251.2 cm^{-1}. (a) What will be the total energy of the molecule in the ground electronic state at 0 K? (Answer in cm^{-1} and call the minimum of the ground state potential well 0 cm^{-1}.) Suppose the energy of the 0-0 electronic transition to the $^1A_2$ $(n \rightarrow \pi^*)$ excited state (the notation is unimportant for now) is designated E_{00}. The vibrational frequencies in this excited state are: ν_1', 2847; ν_2', 1173; ν_3', 887; ν_4', 124.6; ν_5', 2968; ν_6', 904 cm^{-1}. (V.A. Job, V. Sethuraman and K.K. Innes, *J. Mol. Spec.*, *30*, 365 [1969].) (b) The transition from the ground state (lowest vibrational state of the lowest electronic state) to the excited electronic state, $^1A_2$, in which ν_3' is also excited by one quantum is designated 3_0^1. What will be the energy of this transition? (c) Draw a diagram like the one in Fig. 5-4 showing the transition from the first excited state of ν_4 in the electronic ground state to the ground vibrational state of the $^1A_2$ electronic excited state. This transition is designated 4_1^0. What is the energy of this transition?

B. *Electronic Configurations.* The electronic configuration of a molecule is a statement of how many electrons are in each molecular orbital. We know from Chapter Four that the ground state configuration of, for example, N_2 is:

$$N_2 \ (\textit{ground state}): (1\sigma_g^+)^2 (1\sigma_u^+)^2 (2\sigma_g^+)^2 (2\sigma_u^+)^2 (1\pi_u)^4 (3\sigma_g^+)^2 (1\pi_g)^0$$

It is not customary to include the unoccupied orbital, $1\pi_g$, but we did so because this orbital is occupied in the first excited state configuration, which is written as follows:

$$N_2 \ (\textit{first excited state}): (1\sigma_g^+)^2 (1\sigma_u^+)^2 (2\sigma_g^+)^2 (2\sigma_u^+)^2 (1\pi_u)^4 (3\sigma_g^+)^1 (1\pi_g)^1$$

Specifying the electronic configuration of a molecule is *not* sufficient to determine its energy. Not only do we not know the rotational and vibrational states of the molecule, but we do not even know how the electrons are distributed in the occupied orbitals. For example, in the first excited state configuration of N_2, we have not specified whether the electron spins are parallel (unpaired) or antiparallel (paired) and this makes quite a difference in the energy of the molecule. A more complete description of the state of a molecule lies at the heart of electronic spectroscopy.

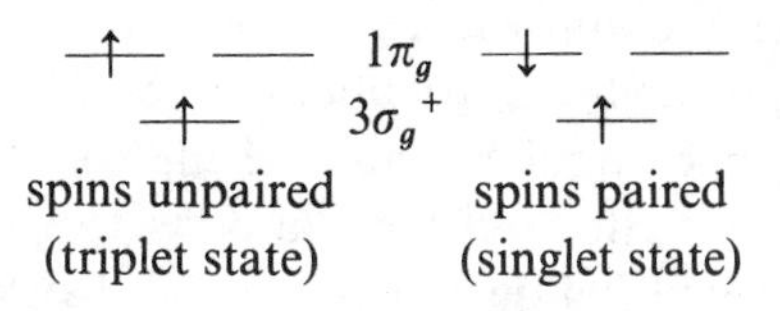

Problems

5-2. Write the electronic configurations of the following species: (a) BN (ground state); (b) Si_2^+ (ground state); (c) HCN (first excited state); (d) CH_3^+ (ground state and first excited state); (e) $CH_3\cdot$ (neutral molecule, ground state assuming planar geometry); (f) butadiene (first excited state); (g) cyclopentadienyl anion (first excited state); (h) cyclopentadienyl radical (ground state); (i) $Co(NH_3)_6^{3+}$ (ground and first excited states of the ML_6 framework); (j) CO_2 (first excited state of the linear molecule).

C. *Wave Functions, the Pauli Principle and Spin Degeneracy.* It is a reasonable approximation to factor the total wave function for a molecule into a product of wave functions as follows:

$$\psi = \psi_v \psi_{es} = \psi_v \psi_e \psi_s \tag{5-2}$$

ψ_v is the nuclear (vibrational) wave function; ψ_{es} is the total electronic wave function; ψ_e is the electronic orbital wave function, and ψ_s is the electron spin wave function. The separation of nuclear and electronic wave functions is the Born-Oppenheimer approximation discussed in Section 4-4. The separation of electronic orbital and spin wave functions (*i.e.*, $\psi_{es} = \psi_e \psi_s$) is generally valid for the lighter elements of the periodic table, and therefore for most organic compounds. When the separation of spin and orbital wave functions is not a good approximation, there is said to be spin-orbit coupling.

In Chapter Four the Pauli Principle was stated in the form "no two electrons can have the same set of quantum numbers." A somewhat more restrictive statement, from which the above statement can be derived, is the following: "The total many-electron wave function, Ψ, must be antisymmetric with respect to exchange of two electrons." What does this mean? If a total wave function $\Psi(1,2\ldots n)$ involves terms for n electrons and we interchange electrons i and j wherever they appear, we have not changed the total probability distribution Ψ^2 (or $\Psi^*\Psi$) of all n electrons. Since Ψ^2 must be unchanged by this exchange of electrons, Ψ must go into either $+\Psi$ or $-\Psi$ after exchange. In order to agree with all observations of measurable physical quantities, we find that Ψ must go into only $-\Psi$ upon exchange of two electrons and not to $+\Psi$.†

Let's apply this to an atomic electronic configuration to get a better feeling for the meaning of the Pauli Principle. Consider the Li atom with the configuration $(1s)^2(2s)^1$. Each electron can have a spin quantum

†The Pauli Principle must apply as well to interchange of any two equivalent elementary particles in an atom or molecule. Particles with *half integral spin*, such as electrons ($s = 1/2$), protons ($s = 1/2$), or ^{65}Cu nuclei ($s = 3/2$) are called *fermions* and obey the Pauli Principle as stated in the text; *viz.*, Ψ must be antisymmetric with respect to the exchange of two fermions. Particles with *integral spin* such as deuterons ($s = 1$), alpha particles (^4He, $s = 0$), or photons ($s = 1$) are called *bosons*. Ψ must be *symmetric* with respect to the exchange of two bosons. Probably the most outstanding example of the importance of nuclear spin is the case of ortho and para hydrogen (H_2) which have very different physical properties. Yet these two molecules differ only in that their nuclear spins are paired (para) or unpaired (ortho).

number of $+1/2$ or $-1/2$. We will call the corresponding spin "wave functions" α and β, respectively. We can try writing a wave function which is the product of three one-electron wave functions:

$$\psi_{es}(\text{Li}) = (100\alpha|1) \cdot (100\beta|2) \cdot (200\alpha|3) \tag{5-3}$$

$n\ \ell\ m_\ell\ m_s$ electron number

where the function $(100\alpha|1)$, for example, means that electron number 1 has quantum number $n = 1$, $\ell = 0$, $m_\ell = 0$, and $m_s = +1/2$. But this wave function is neither symmetric nor antisymmetric with respect to the exchange of two electrons, say 2 and 3:

$$\psi_{es}(\text{Li}) \xrightarrow{\text{2,3 exchange}} (100\alpha|1)\,(100\beta|3)\,(200\alpha|2) \neq \pm\,(100\alpha|1)\,(100\beta|2)\,(200\alpha|3) \tag{5-4}$$

In order to write a wave function which is antisymmetric to the exchange of any two electrons, we need six terms:

$$\begin{aligned}\psi_{es}(\text{Li}) = (1/\sqrt{6})\,[&(100\alpha|1)\,(100\beta|2)\,(200\alpha|3) + (100\beta|1)\,(200\alpha|2)\,(100\alpha|3)\\ &+ (200\alpha|1)\,(100\alpha|2)\,(100\beta|3) - (200\alpha|1)\,(100\beta|2)\,(100\alpha|3)\\ &- (100\alpha|1)\,(200\alpha|2)\,(100\beta|3) - (100\beta|1)\,(100\alpha|2)\,(200\alpha|3)]\end{aligned} \tag{5-5}$$

These six terms thoroughly confuse matters by giving each electron an equal probability of being in each orbital. Further, no two electrons have the same quantum numbers in any one term, which is the way the Pauli Principle is applied to such a wave function. The signs of the individual terms were chosen to make the total function antisymmetric to the exchange of any two electrons. For example, interchanging the labels of electrons 2 and 3 gives

$$\begin{aligned}\psi'_{es}(\text{Li}) = (1/\sqrt{6})\,[&(100\alpha|1)\,(100\beta|3)\,(200\alpha|2) + (100\beta|1)\,(200\alpha|3)\,(100\alpha|2)\\ &+ (200\alpha|1)\,(100\alpha|3)\,(100\beta|2) - (200\alpha|1)\,(100\beta|3)\,(100\alpha|2)\\ &- (100\alpha|1)\,(200\alpha|3)\,(100\beta|2) - (100\beta|1)\,(100\alpha|3)\,(200\alpha|2)\end{aligned} \tag{5-6}$$

which is just the negative of eq. 5-5. You should verify this term for term.

The factor $1/\sqrt{6}$ in these equations is a normalization constant. A more convenient way to write the function in eq. 5-5 is with a determinant:

$$\psi_{es}(\mathrm{Li}) = (1/\sqrt{3!}) \begin{vmatrix} (100\alpha|1) & (100\alpha|2) & (100\alpha|3) \\ (100\beta|1) & (100\beta|2) & (100\beta|3) \\ (200\alpha|1) & (200\alpha|2) & (200\alpha|3) \end{vmatrix} \tag{5-7}$$

which is generally abbreviated to the form in eq. 5-8 in which only the

$$\psi_{es}(\mathrm{Li}) \equiv (1/\sqrt{3!}) \left| (100\alpha|1)\,(100\beta|2)\,(200\alpha|3) \right| \tag{5-8}$$

diagonal elements of the full determinant are written. The normalization coefficient for an $N \times N$ wave function of the form 5-7 will always be $1/\sqrt{N!}$. A determinant wave function such as eq. 5-7 satisfies both statements of the Pauli Principle. If two electrons have the same quantum numbers, two rows of the determinant are identical and the value of the determinant is zero. If we interchange the labels of any two electrons, we interchange two rows of the determinant, which just changes the sign of the determinant.

For a molecule the total wave function will depend on ψ_v as well as ψ_{es}. It will always be necessary to make ψ_{es} antisymmetric to the exchange of electrons because ψ_v does not depend on the electrons and so must be symmetric to electron exchange. The condition that ψ_{es} be antisymmetric allows us to make either ψ_e or ψ_s antisymmetric if the other is symmetric. It precludes choosing both to be symmetric or both to be antisymmetric since the product would then be symmetric.

We know that in the absence of an external magnetic field, the direction of electron spin (α or β) is meaningless because there is no reference direction to compare to the direction of the electron spin. However, addition of a second electron immediately establishes a reference direction (each electron is a reference for the other) and it is then meaningful to speak of spin functions α and β. First we note that the spin functions are orthonormal (orthogonal and normalized—eqs. 2-37 and 2-39).

$$\begin{aligned} \int_{e_1} \alpha(1)\alpha(1)d\tau_1 &= 1 \\ \int_{e_1} \beta(1)\beta(1)d\tau_1 &= 1 \\ \int_{e_1} \alpha(1)\beta(1)d\tau_1 &= 0 \end{aligned} \tag{5-9}$$

The symbols $\int_{e_1} (\ldots)d\tau_1$ say that the integral applies to electron number 1 only. Two-electron spin wave functions must also be orthonormal and we need to specify which electron, 1 or 2, has spin α or spin β. For example, the wave functions $\alpha(1)\alpha(2)$ and $\beta(1)\beta(2)$ are orthogonal by the following reasoning:

$$\begin{aligned}\int_{e_1}\int_{e_2}[\alpha(1)\alpha(2)][\beta(1)\beta(2)]d\tau_1\,d\tau_2 &= \left[\int_{e_1}\alpha(1)\beta(1)d\tau_1\right]\left[\int_{e_2}\alpha(2)\beta(2)d\tau_2\right]\\ &= \quad 0 \quad \cdot \quad 0\\ &= 0\end{aligned} \tag{5-10}$$

In eq. 5-10 we have split the double integral involving two electrons into two integrals each involving only one electron. The single integrals are then evaluated with eq. 5-9.

We can now construct the spin wave functions for a system of two electrons. These functions can be either symmetric or antisymmetric with respect to exchange of electrons. The function $\alpha(1)\alpha(2)$ is symmetric because $\alpha(1)\alpha(2) = +\alpha(2)\alpha(1)$. The function $\beta(1)\beta(2)$ is similarly symmetric. The function $\alpha(1)\beta(2)$ is neither symmetric nor antisymmetric: $\alpha(1)\beta(2) \neq \pm\alpha(2)\beta(1)$. But look at the linear combinations in eq. 5-11:

$$\begin{aligned}\psi &= (1/\sqrt{2})[\alpha(1)\beta(2) + \beta(1)\alpha(2)]\\ \psi' &= (1/\sqrt{2})[\alpha(1)\beta(2) - \beta(1)\alpha(2)]\end{aligned} \tag{5-11}$$

ψ', for example, is antisymmetric:

$$\begin{aligned}\psi' \xrightarrow{\text{1,2 exchange}} & (1/\sqrt{2})[\alpha(2)\beta(1) - \beta(2)\alpha(1)]\\ &= -(1/\sqrt{2})[\alpha(1)\beta(2) - \beta(1)\alpha(2)] = -\psi'\end{aligned} \tag{5-12}$$

By the same procedure we would find that ψ is symmetric. The factor $1/\sqrt{2}$ is needed for normalization and can be obtained as follows:

$$\begin{aligned}1 &\equiv \int \psi'\psi'\,d\tau\\ &= \int_{e_1}\int_{e_2} N[\alpha(1)\beta(2) - \beta(1)\alpha(2)]N[\alpha(1)\beta(2) - \beta(1)\alpha(2)]d\tau_1\,d\tau_2 \qquad (5\text{-}13)\\ &= N^2\int_{e_1}\int_{e_2}[\alpha(1)\beta(2)\alpha(1)\beta(2) - \alpha(1)\beta(2)\beta(1)\alpha(2)\\ &\quad -\beta(1)\alpha(2)\alpha(1)\beta(2) + \beta(1)\alpha(2)\beta(1)\alpha(2)]d\tau_1\,d\tau_2\\ &= N^2\left[\int_{e_1}\alpha(1)\alpha(1)d\tau_1\int_{e_2}\beta(2)\beta(2)d\tau_2 - \int_{e_1}\alpha(1)\beta(1)d\tau_1\int_{e_2}\beta(2)\alpha(2)d\tau_2\right.\\ &\quad \left.-\int_{e_1}\beta(1)\alpha(1)d\tau_1\int_{e_2}\alpha(2)\beta(2)d\tau_2 + \int_{e_1}\beta(1)\beta(1)d\tau_1\int_{e_2}\alpha(2)\alpha(2)d\tau_2\right]\end{aligned}$$

$$1 = N^2[1 - 0 - 0 + 1]$$

$$N = \pm 1/\sqrt{2}$$

We arbitrarily choose the + sign for N.

We have just constructed the only four valid two-electron spin wave functions. It is customary to abbreviate the function $\alpha(1)\beta(2)$ as $\alpha\beta$, in which the order of the electron numbers is implicit in the order of the spin functions. We summarize our four two-electron spin wave functions as follows:

$$\left.\begin{aligned} \psi_1 &= \alpha\alpha \\ \psi_2 &= \beta\beta \\ \psi_3 &= (1/\sqrt{2})[\alpha\beta + \beta\alpha] \end{aligned}\right\} \text{symmetric}$$
$$\psi_4 = (1/\sqrt{2})[\alpha\beta - \beta\alpha] \quad \text{antisymmetric} \tag{5-14}$$

Finally we come to molecular electronic wave functions. Consider the ground state of a hydrogen molecule with the configuration $(1\sigma_g{}^+)^2(1\sigma_u{}^+)^0$ which we will abbreviate as $(\sigma_g{}^+)^2$. The orbital wave function must be

$$\psi_e = \sigma_g{}^+(1)\,\sigma_g{}^+(2) \tag{5-15}$$

which is symmetric to exchange of electrons. There is no other orbital wave function we could have written since both electrons occupy the $\sigma_g{}^+$ orbital. Since ψ_e is symmetric, ψ_s must be antisymmetric. The one antisymmetric spin function of eqs. 5-14 gives us the following total electronic wave function:

$$\begin{aligned} \psi_{es}(16) = \psi_e\psi_s &= \sigma_g{}^+(1)\sigma_g{}^+(2)\,(1/\sqrt{2})[\alpha\beta - \beta\alpha] \\ &\equiv (1/\sqrt{2})\,\sigma_g{}^+\,\sigma_g{}^+\,[\alpha\beta - \beta\alpha] \end{aligned} \tag{5-16}$$

where the order of electrons 1 and 2 is implicit in the order of the terms.

Now consider the first excited state of H_2 with the configuration $(\sigma_g{}^+)^1$ $(\sigma_u{}^+)^1$. We can write symmetric and antisymmetric orbital wave functions for this configuration:

$$\begin{aligned} \psi_e(\text{sym}) &= (1/\sqrt{2})[\sigma_g{}^+\sigma_u{}^+ + \sigma_u{}^+\sigma_g{}^+] \\ \psi_e(\text{antisym}) &= (1/\sqrt{2})[\sigma_g{}^+\sigma_u{}^+ - \sigma_u{}^+\sigma_g{}^+] \end{aligned} \tag{5-17}$$

Combining the symmetric ψ_e with the antisymmetric ψ_s, and combining the antisymmetric ψ_e with the symmetric ψ_s, we get

$$\psi_{es}(18) = (1/2)[\sigma_g{}^+\sigma_u{}^+ + \sigma_u{}^+\sigma_g{}^+][\alpha\beta - \beta\alpha] \tag{5-18}$$

$$\psi_{es}(19a) = (1/\sqrt{2})[\sigma_g{}^+\sigma_u{}^+ - \sigma_u{}^+\sigma_g{}^+][\alpha\alpha] \tag{5-19a}$$

$$\psi_{es}(19b) = (1/\sqrt{2})[\sigma_g{}^+\sigma_u{}^+ - \sigma_u{}^+\sigma_g{}^+][\beta\beta] \tag{5-19b}$$

$$\psi_{es}(19c) = (1/2)[\sigma_g{}^+\sigma_u{}^+ - \sigma_u{}^+\sigma_g{}^+][\alpha\beta + \beta\alpha] \tag{5-19c}$$

All four of these functions are antisymmetric. $\psi_{es}(18)$ represents a single nondegenerate state of the molecule with the energy $E(18)$. $\psi_{es}(19a, 19b, 19c)$ represent three degenerate states of the molecule in the absence of an external magnetic field. The energy of these three states, $E(19)$, will be somewhat lower than $E(18)$. We call the energy level $E(19)$ a *triplet* state because it is triply degenerate. $E(18)$ is a *singlet* state because it is nondegenerate. These are diagrammed as follows:

——	—↓—	—↑—	$1\sigma_u^+$
—↑↓—	—↑—	—↑—	$1\sigma_g^+$
$\psi_{es}(16)$	$\psi_{es}(18)$	$\psi_{es}(19)$	
ground state (singlet)	excited singlet state	excited triplet state	
(spins paired)	(spins paired)	(spins unpaired)	

An energy level diagram of these three states would be as shown in Fig. 5-8. There are two possible electronic transitions represented by the vertical arrows in this figure. The nomenclature ($^3\Sigma_u^+$, *etc.*) will be explained in Section 5-3-D.

We have just seen two examples of *spin multiplicity*. Zero unpaired electrons lead to a single spin wave function which defines a *singlet* state and is diagrammed in the following ways:

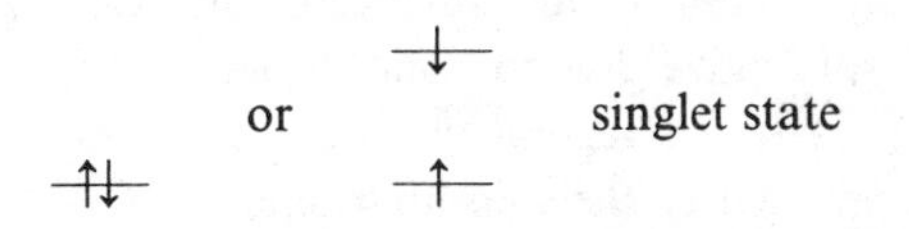

One unpaired electron has two physically distinguishable states in the presence of an external magnetic field:

direction of external field ⇑ —↑— or —↓— doublet state

The state in which the spin angular momentum is parallel to the external field is of lower energy than the state in which it is antiparallel. In the absence of an external field, the two states are degenerate and define a *doublet* state. (Transitions between the parallel and antiparallel states in the presence of an external field are the subject of another kind of spectroscopy—electron spin resonance.) Two unpaired electrons were seen in

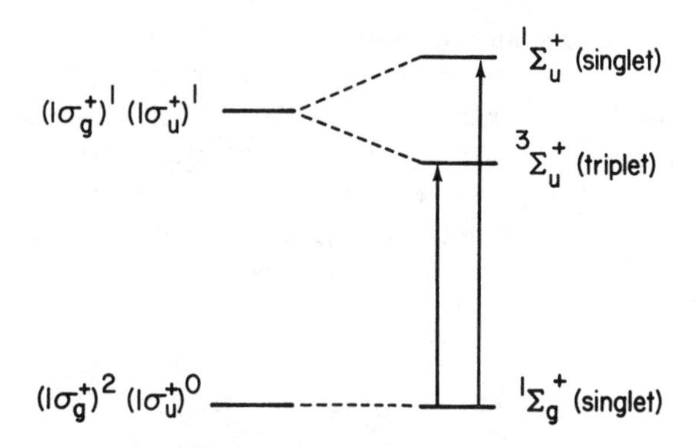

Fig. 5-8. The three lowest energy electronic states of H_2 and two possible electronic transitions.

our previous discussion to define three degenerate spin wave functions forming a *triplet* state. We will represent the triplet state by diagrams such as

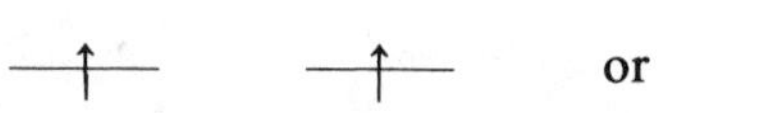

The diagram on the left applies to two unpaired electrons in degenerate orbitals, and the diagram on the right applies to two unpaired electrons in nondegenerate orbitals. In either case, one can write three spin wave functions which are degenerate in the absence of an external magnetic field.[†] Three unpaired spins give rise to four degenerate spin wave functions defining a *quartet* state. In general, *n unpaired electrons will define n* + 1 *spin wave functions*. That is, the spin multiplicity is $n + 1$ or $2s + 1$, where s is the total spin of the n unpaired electrons.

[†] The three wave functions are rigorously degenerate for atoms, which have spherical symmetry, but are not rigorously degenerate for most molecules. The two unpaired electrons difine a net spin angular momentum which may be along, for example, the x, y, or z axes of a molecule such as naphthalene:

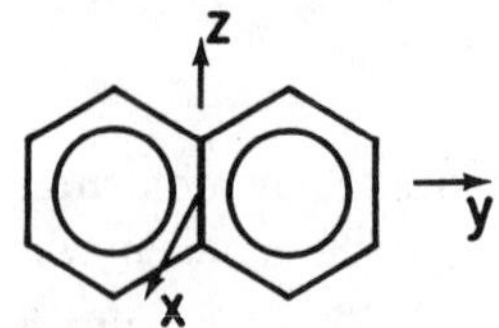

The state in which the angular momentum is in the x direction does *not* have the same energy as the states in which the angular momentum is in the y or z directions. The differences in energy are exceedingly small—microwave energies—so we do not normally detect them. This lack of degeneracy in the absence of a magnetic field is called *zero field splitting*.

Problems

5-3. Show that the following wave functions violate the Pauli Principle: (a) $\psi_{es}(A) = (\sigma_g^+ \sigma_g^+)(\alpha\alpha)$; (b) $\psi_{es}(B) = (1/\sqrt{2})(\sigma_g^+ \sigma_u^+ + \sigma_u^+ \sigma_g^+)(\beta\beta)$.

5-4. By performing an integration over spin coordinates, as in eq. 5-13, show that the wave functions ψ_2 and ψ_3 of eq. 5-14 are orthogonal. Do the same for ψ_3 and ψ_4.

5-5. What are the possible spin degeneracies (multiplicities) of the electronic states which arise from the following configurations: (a) $(\sigma_g^+)^2$; (b) $(\sigma_g^+)^1 (\sigma_u^+)^1$; (c) $(\sigma_g^+)^1 (\pi_u)^1$; (d) $(\sigma_g^+)^1(\sigma_u^+)^1(\sigma_g^+)^1$; (e) $(\sigma_g^+)^1(\sigma_u^+)^1(\sigma_g^+)^1(\pi_u)^2$?

D. *Stationary States.* The stationary states of a molecule are the discrete energy levels which are solutions of the Schrödinger equation for that molecule:

$$\mathscr{H}_0 \Psi = i\hbar \frac{\partial \Psi}{\partial t} \tag{5-20}$$

where $\mathscr{H}_0$ contains no time-dependent terms. Equation 5-20 is just the full time-dependent Schrödinger equation introduced in Section 2-3-B. The eigenvalues (energies) are independent of time because $\mathscr{H}_0$ is independent of time. (In Section 3-4 we found that transitions between stationary states can be induced by adding a time-dependent term, $\mathscr{H}(t)$, to the hamiltonian: $\mathscr{H} = \mathscr{H}_0 + \mathscr{H}(t)$. $\mathscr{H}(t)$ might be, for example, the oscillating electric field of a light wave.) In this section we will delineate the rules for determining the orbital symmetries and spin multiplicities of the stationary states which arise from particular configurations of molecules.

We start by noting that all completely filled molecular orbitals are totally symmetric and will not contribute to the net symmetry of the electronic state. This is analogous to the case of atoms, for which any filled subshell has the quantum numbers $L = S = J = 0$, and there is no contribution to the Russell-Saunders term (Section 4-2-C). For molecules, a filled molecular orbital has the full symmetry of the molecule. So we will confine our attention strictly to incompletely filled molecular orbitals. We divide our treatment into six cases:

Case 1. All occupied orbitals are fully occupied. As just stated, the electron distribution is totally symmetric. Since there are no unpaired electrons, the spin multiplicity is unity and the state is a singlet state. For a molecule of $D_{\infty h}$ symmetry, such as H_2, the state would be called $^1\Sigma_g^+$ (read "singlet-sigma-g-plus"). The ground state of ethylene (D_{2h} symmetry) would be labelled ^1A_g.

Case 2. The only unoccupied orbital is singly occupied. Since there is one unpaired electron, the spin multiplicity must be two. If the symmetry of the singly occupied orbital is χ, the electronic state is labelled ^{2}X. (X is the capital of the Greek letter χ.) For example, the ground state of N_2^+ is $(1\sigma_g^+)^2 \ldots (1\pi_g)^4 (3\sigma_g^+)^1$; so the electronic state is $^2\Sigma_g^+$ (read "doublet-sigma-g-plus"). The dots in the configuration represent filled orbitals between those indicated. The ground state of O_2^+ is $(1\sigma_g^+)^2 \ldots (1\pi_u)^4 (1\pi_g)^1$ so the electronic state is $^2\Pi_g$. The labels $^2\Pi_g$, $^1\Sigma_g^+$, *etc.*, are sometimes called *terms* or *term symbols*.

Case 3. There are two singly occupied orbitals. We exclude the case of two electrons in a degenerate orbital (*e.g.*, $(\pi)^2$), which is Case 5. The two electrons can be either paired or unpaired, so the spin multiplicity can be either 1 (singlet) or 3 (triplet). The symmetries of the possible states are given by the direct product of the symmetries of the partially occupied orbitals. For example, the first excited configuration of ethylene is $\ldots (b_{1u})^1 (b_{2g})^1$; so the resultant states are singlets and triplets of symmetry $b_{1u} \times b_{2g} = B_{3u}$. Hence this configuration gives rise to the states $^1B_{3u}$ and $^3B_{3u}$. *In general, the triplet state will be of lower energy than the singlet state from the same configuration.*† The first excited configuration of N_2 is $\ldots (3\sigma_g^+)^1 (1\pi_g)^1$, which gives rise to $^3\Pi_g$ and $^1\Pi_g$ states. The configuration $(\pi_u)^1 (\pi_g)^1$ would yield singlet and triplet states of symmetry $\pi_u \times \pi_g = \Sigma_u^+ + \Sigma_u^- + \Delta_u$. There would therefore be *six* states from this configuration: $^1\Sigma_u^+$, $^1\Sigma_u^-$, $^1\Delta_u$, $^3\Sigma_u^+$, $^3\Sigma_u^-$, and $^3\Delta_u$.

Case 4. A degenerate orbital is lacking a single electron. For example, C_2^+ has the configuration $\ldots (2\sigma_u^+)^2 (1\pi_u)^3$. A degenerate orbital lacking one electron is formally equivalent to the same degenerate orbital occupied by only a single electron. The C_2^+ configuration, therefore, defines a $^2\Pi_u$ state.

To help see that it is reasonable that a configuration such as $(\pi_g)^3$ is equivalent to $(\pi_g)^1$, we can look at what happens when the symmetry of the molecule is lowered. The $1\pi_g$ molecular orbital of CO_2 splits into two nondegenerate orbitals of symmetry a_2 and b_2 when the molecule is bent from $D_{\infty h}$ to C_{2v} geometry (Fig. 5-9). If the $1\pi_g$ orbital were incompletely filled, as in CO_2^+, and the molecule were bent, the $(1\pi_g)^3$ configuration would give rise to either $(b_2)^2(a_2)^1$ or $(a_2)^2(b_2)^1$ configurations, both of which are doublet states ($^2A_2$ and $^2B_2$, respectively). It is therefore reasonable that the state of the linear molecule is an orbitally degenerate doublet

†The reason this is so is not simply stated and has been misstated in several texts. For a discussion, see R.L. Snow and J.L. Bills, *J. Chem. Ed.*, *51*, 585 (1974).

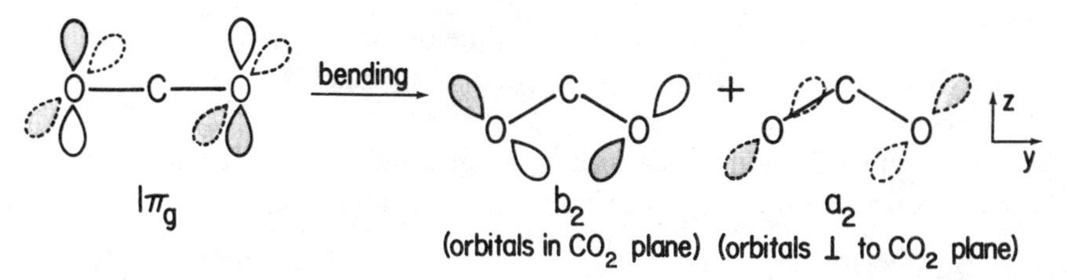

Fig. 5-9. Degenerate π_g orbitals of linear CO_2 split into nondegenerate orbitals of symmetry b_2 and a_2 when the molecule is bent.

state ($^2\Pi_g$) since it gives rise to two orbitally nondegenerate doublet states when the molecule is distorted. Note also that the configuration $(1\pi_g)^1$ would have the same behavior on bending:

$$\underset{^2\Pi_g}{(1\pi_g)^1} \longrightarrow \underset{^2B_2}{(b_2)^1(a_2)^0} \quad \text{or} \quad \underset{^2A_2}{(a_2)^1(b_2)^0}$$

It would be very *unreasonable* if the configuration $(1\pi_g)^3$ gave rise to four degenerate states given by the triple direct product $\pi_g \times \pi_g \times \pi_g = \Pi_g + \Pi_g + \Pi_g + \Phi_g$.

Case 5. Two electrons in a degenerate orbital. We expect singlet and triplet states since the electrons can be paired or unpaired. If χ is the symmetry of the orbital, we *do not* generate a singlet and triplet state for each representation of the product $\chi \times \chi$. Consider the ground state of O_2 which has the configuration ... $(1\pi_g)^2$. How many wave functions, degenerate or not, are possible? We have two electrons and two orbitals so the problem boils down to the question, "How many different ways are there to put two electrons into the four boxes below?"

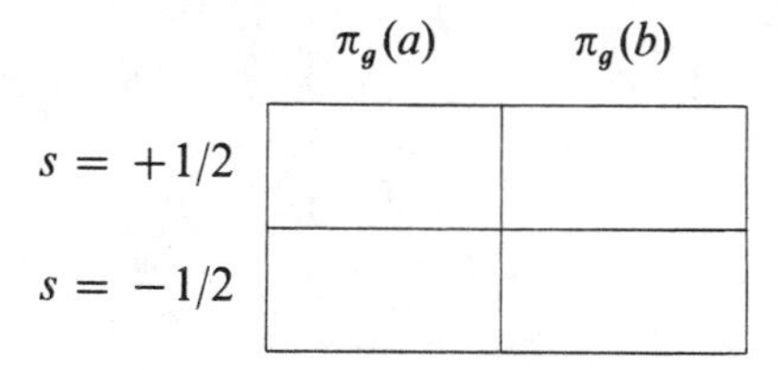

The answer is that there are six different ways, which are drawn below.

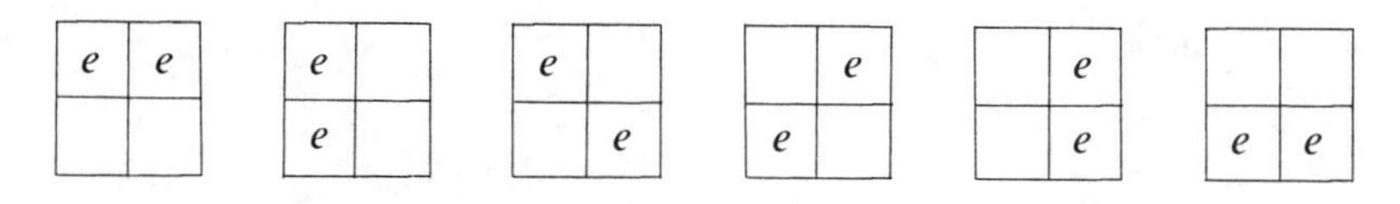

Since the electrons can be distributed in the orbitals in six ways, we expect that there will be a total of six different wave functions, some of which may be degenerate.

Now suppose we just tried forming singlet and triplet states for each of the terms of the direct product $\pi_g \times \pi_g$. We would get

$^1\Sigma_g^+$	+ $^1\Sigma_g^-$	+ $^1\Delta_g$	+ $^3\Sigma_g^+$	+ $^3\Sigma_g^-$	+ $^3\Delta_g$
1×1	1×1	1×2	3×1	3×1	3×2
= one state	= one state	= two degenerate states	= three degenerate states	= three degenerate states	= six degenerate states

= sixteen wave functions

The degeneracy of each term is equal to (spin multiplicity × orbital degeneracy). So, for example, $^3\Delta_g$ represents $3 \times 2 = 6$ degenerate wave functions. The total degeneracy of the states we have written is sixteen. But we just finished saying that we should only have six wave functions for the configuration $(\pi_g)^2$. How do we know which of the sixteen terms to keep and which to throw out? The answer is that some of the terms violate the Pauli Principle. We know from eq. 5-14 that the triplet spin wave functions are symmetric and the singlet spin wave functions are antisymmetric to the exchange of two electrons. Therefore, the orbital

Table 5-1. Symmetric and Antisymmetric Components of the Direct Product $e' \times e'$ in the Point Group D_{3h}

D_{3h}	E	$2C_3$	$3C_2$	σ_h	$2S_3$	$3\sigma_v$	
a_1'	1	1	1	1	1	1	
a_2'	1	1	−1	1	1	−1	
→ e'	2	−1	0	2	−1	0	
a_1''	1	1	1	−1	−1	−1	
a_2''	1	1	−1	−1	−1	1	
e''	2	−1	0	−2	1	0	
R^2	E	C_3^2	E	E	C_3^2	E	
$\chi(R^2)$	2	−1	2	2	−1	2	
$[\chi(R)]^2$	4	1	0	4	1	0	
χ^+	3	0	1	3	0	1	$= a_1' + e'$
χ^-	1	1	−1	1	1	−1	$= a_2'$

part of a triplet state term must be antisymmetric and the orbital part of a singlet wave function must be symmetric to the exchange of electrons.

The characters of the symmetric (χ^+) and antisymmetric (χ^-) components of the direct product are given by the following formulas:

$$\begin{aligned} \text{characters } \chi^+ &= \tfrac{1}{2}\{[\chi(R)]^2 + \chi(R^2)\} \\ \text{characters } \chi^- &= \tfrac{1}{2}\{[\chi(R)]^2 - \chi(R^2)\} \end{aligned} \tag{5-21}$$

$\chi(R)$ is the character for the operation R and $\chi(R^2)$ is the character of the square of the operation R. As an example, in Table 5-1 we find that the configuration $(e')^2$ for the point group D_{3h} gives rise to orbitally symmetric wave functions of symmetry $a_1' + e'$ and an orbitally antisymmetric wave function of symmetry a_2'. We can find this more quickly with the aid of Appendix B. For the point group D_{3h} we find the following table of direct products in Appendix B:

$C_{3v}, D_3, D_{3d}, D_{3h}$	A_1	A_2	E
A_1	A_1	A_2	E
A_2		A_1	E
E			$A_1 + [A_2] + E$

For the entry $E \times E$ (under E and across from E) we see $A_1 + [A_2] + E$. The term in brackets is the antisymmetric component of the direct product. (To this result we must add the prime-double-prime product rule, *viz.*, $' \times ' = 1$.)

Getting back to the configuration $(\pi_g)^2$, we find in Appendix B that the direct product $\Pi \times \Pi$ gives three terms: Σ^+, $[\Sigma^-]$, Δ. Noting that the subscript multiplication is $g \times g = g$, we find orbital symmetries $\Sigma_g{}^+$ and Δ_g for the symmetric wave functions and $\Sigma_g{}^-$ for the antisymmetric wave function. Since the symmetric orbital functions must go with the antisymmetric (singlet) spin wave function, and the antisymmetric orbital function goes with the symmetric (triplet) spin wave functions, we can write the states of the configuration $(\pi_g)^2$ as follows:

$$\begin{array}{lccccc} \text{states of } (\pi_g)^2 = & {}^1\Sigma_g{}^+ & + & {}^1\Delta_g & + & {}^3\Sigma_g{}^- \\ & 1 \times 1 & & 1 \times 2 & & 3 \times 1 \\ & = 1 \text{ wave} & & = 2 \text{ wave} & & = 3 \text{ wave} \\ & \text{function} & & \text{functions} & & \text{functions} \\ & = \text{six wave functions} & & & & \end{array} \tag{5-22}$$

What this means is that the ground state configuration of O_2 gives rise to six different wave functions which have three different energies. Only the $^3\Sigma_g^-$ function is the ground state of the molecule. The other two terms describe excited states of the O_2 molecule:

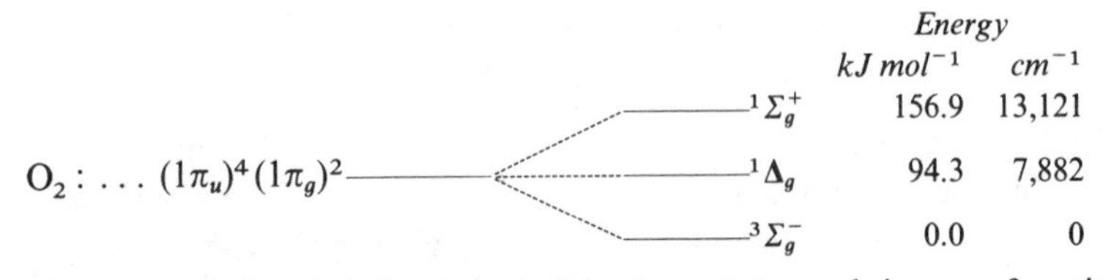

We defer a more detailed discussion of the three states and six wave functions until Section 5-5-C.

Case 6. *n* electrons in an *n*-fold degenerate orbital. In Case 5 we treated the case of two electrons in any degenerate orbital. The case of two electrons in a twofold degenerate orbital is a special case of Case 6. We now treat the examples of three electrons in a threefold degenerate orbital and four electrons in a fourfold degenerate orbital.

With three electrons, the possible multiplicities are quartet (↑↑↑) and doublet (↑↑↓). The character, χ, under each operation, R, for each spin multiplicity is:

$$\chi(\text{doublet}) = (1/3)\,\{[\chi(R)]^3 - \chi(R^3)\}$$
$$\chi(\text{quartet}) = (1/6)\,\{[\chi(R)]^3 - 3\chi(R)\chi(R^2) + 2\chi(R^3)\} \quad (5\text{-}23)$$

These equations are applied in the same manner as eqs. 5-21, which were illustrated in Table 5-1. For four electrons in a fourfold degenerate orbital, the possible spin multiplicities are quintet (↑↑↑↑), triplet (↑↑↑↓), and singlet (↑↓↑↓). The characters for each are given by:

$$\chi(\text{singlet}) = (1/12)\,\{[\chi(R)]^4 - 4\chi(R)\chi(R^3) + 3[\chi(R^2)]^2\} \quad (5\text{-}24)$$
$$\chi(\text{triplet}) = (1/8)\,\{[\chi(R)]^4 - 2[\chi(R)]^2\chi(R^2) + 2\chi(R^4) - [\chi(R^2)]^2\}$$
$$\chi(\text{quintet}) = (1/24)\,\{[\chi(R)]^4 - 6[\chi(R)]^2\chi(R^2) + 8\chi(R)\chi(R^3) - 6\chi(R^4) + 3[\chi(R^2)]^2\}$$

The configuration $(t_{2g})^3$, which is encountered in many transition metal problems, would be treated with eqs. 5-23. Only for molecules with icosahedral symmetry will you ever encounter any orbital degeneracy higher than three.† These six cases, and combinations thereof, will allow you to treat virtually any configuration you are likely to encounter.

A Hairy Example. The rules for Cases 1–6 can be extended to more complicated configurations which can be thought of as the products of

† To handle still more complicated configurations, the reader is referred to D.I. Ford, *J. Chem. Ed.*, *49*, 336 (1972), in which eqs. 5-21, 5-23, and 5-24 are derived.

individual cases. For example, the ground state configuration of B_2 is $(1\sigma_g^+)^2\ (1\sigma_u^+)^2\ (2\sigma_g^+)^2\ (2\sigma_u^+)^1\ (1\pi_u)^2\ (3\sigma_g^+)^1$. To what electronic states does this configuration give rise? The states which each incompletely occupied orbital would produce, *by itself*, are as follows:

configuration	states	Case
$(2\sigma_u^+)^1$	$^2\Sigma_u^+$	2
$(1\pi_u)^2$	$^3\Sigma_g^-, {}^1\Sigma_g^+, {}^1\Delta_g$	5
$(3\sigma_g^+)^1$	$^2\Sigma_g^+$	2

The possible states of the molecule are obtained from the direct product of all of these terms, with due consideration of the multiplicities. The orbital symmetries of the various states will therefore be as follows:

$$(^2\Sigma_u^+)\begin{pmatrix} ^3\Sigma_g^- \\ ^1\Sigma_g^+ \\ ^1\Delta_g \end{pmatrix}(^2\Sigma_g^+) = \begin{pmatrix} \Sigma_u^- \\ \Sigma_u^+ \\ \Delta_u \end{pmatrix} \tag{5-25}$$

The spin multiplicities can be obtained from Table 5-2. As one example of the reasoning behind this table, consider the term doublet × quartet. The two spin vectors, $\vec{S}_i$ and $\vec{S}_j$, can be added only to give the resultant magnitudes $S_i + S_j$, $S_i + S_j - 1, \ldots S_i - S_j$. For doublet × quartet, $S_i = 3/2$ and $S_j = 1/2$. The resultant total spins are $3/2 + 1/2 = 2$ (quintet state) and $3/2 - 1/2 = 1$ (triplet state). Using these multiplicity rules, we can work out the top multiplication of eq. 5-25 as follows:

$$\begin{aligned} \underbrace{(^2\Sigma_u^+)(^3\Sigma_g^-)}(^2\Sigma_g^+) &= (^2\Sigma_u^- + {}^4\Sigma_u^-)(^2\Sigma_g^+) \\ &= {}^1\Sigma_u^- + {}^3\Sigma_u^- + {}^3\Sigma_u^- + {}^5\Sigma_u^- \end{aligned} \tag{5-26}$$

Table 5-2. Multiplicities Arising from the Combination of Terms

Separate Terms	Product
singlet × singlet	singlet
singlet × doublet	doublet
singlet × triplet	triplet
singlet × quartet	quartet
doublet × doublet	singlet + triplet
doublet × triplet	doublet + quartet
doublet × quartet	triplet + quintet
triplet × triplet	singlet + triplet + quintet
triplet × quartet	doublet + quartet + sextet
quartet × quartet	singlet + triplet + quintet + septet

Table 5-3. Rules for Common Electronic Configurations

Case	Configuration	Resulting Terms
1	all orbitals filled	singlet—totally symmetric representation
2[a]	$\ldots(\chi)^1$	$^2\mathrm{X}$
3[b]	$\ldots(\chi_i)^1(\chi_j)^1$	$^1(\chi_i \times \chi_j) + {}^3(\chi_i \times \chi_j)$
4[c]	$\ldots(\chi_d)^{2d-n} = (\chi_d)^n$	$^2\mathrm{X}$ (if $n = 1$)
5	$\ldots(\chi_d)^2$	^{1}(symmetric product) + ^{3}(antisymmetric product); Use eqs. 5-21 or Appendix B
6	$\ldots(\chi_d)^d$	Use eqs. 5-23 or 5-24.

[a] χ is the symmetry of an orbital. X, the capital of χ, is used to designate states.
[b] $\chi_i \times \chi_j$ is the direct product of χ_i and χ_j. It can contain one or more terms.
[c] χ_d is a d-fold degenerate representation.

When we complete the other two rows of multiplication of eq. 5-25, we find a total of twenty-four wave functions which are divided into eight energy levels (terms) given explicitly as $^1\Sigma_u^- + (2)^3\Sigma_u^- + {}^1\Sigma_u^+ + {}^3\Sigma_u^+ + {}^5\Sigma_u^- + {}^1\Delta_u + {}^3\Delta_u$. The coefficients in parentheses denote more than one of that kind of term. To repeat, *the ground state configuration of B_2 defines twenty-four different wave functions and eight different energy levels*! There would be seven possible electronic transitions just within this ground state manifold of terms.

The rules of Cases 1–6 are summarized in Table 5-3. These five cases take care of the vast majority of configurations you will encounter. It is possible to derive the Russell-Saunders terms for atoms using the rules in this section—see Problem 5-54.

Problems

5-6. What terms are generated by each configuration of Problem 5-2?

5-7. What terms are generated by the configuration $(e_{1g})^2(e_{2u})^2$ in D_{6h} symmetry? If you can handle this problem, you understand Section 5-3-D.

5-8. Using eqs. 5-23, derive the terms which arise from the configuration $(t_{2g})^3$ in the point group O_h.

5-4. Selection Rules

A. *Transition Moments, Band Intensities, and the Franck-Condon Principle.* Using a simplified one-dimensional model, it was shown in Section 3-4 that the probability that a transition between two states will be induced by the oscillating electric field of a light wave is proportional to the square of the transition moment integral. For an electronic transition, the transition moment integral has the form in eq. 5-27:

$$M = \int \psi'^* \, \hat{\mu} \, \psi \, d\tau \tag{5-27}$$

where the prime denotes an excited state and the dipole moment operator, $\hat{\mu}$, can now be divided into two components. One depends on nuclear coordinates ($\hat{\mu}_n$) and the other depends on electron coordinates ($\hat{\mu}_e$). Writing $\psi = \psi_{es}\psi_v$, we get

$$\begin{aligned} M &= \int \psi^*_{e's'} \psi^*_{v'} (\hat{\mu}_n + \hat{\mu}_e) \, \psi_{es} \psi_v \, d\tau \\ &= \int \psi^*_{e's'} \, \psi^*_{v'} \, \hat{\mu}_n \psi_{es} \psi_v \, d\tau + \int \psi^*_{e's'} \, \psi^*_{v'} \, \hat{\mu}_e \psi_{es} \psi_v \, d\tau \\ &= \underbrace{\int \psi^*_{e's'} \, \psi_{es} d\tau_{es}}_{=\,0} \int \psi^*_{v'} \, \hat{\mu}_n \psi_v d\tau_n + \underbrace{\int \psi^*_{v'} \, \psi_v d\tau_n}_{\text{Franck-Condon factor}} \int \psi^*_{e's'} \, \hat{\mu}_e \psi_{es} d\tau_{es} \end{aligned} \tag{5-28}$$

We were able to break each term into a product of two integrals because $\hat{\mu}_n$ depends only on nuclear coordinates, and $\hat{\mu}_e$ depends only on electron coordinates. Further, the integral on the bottom line on the left is zero because the two electronic wave functions are orthogonal. The vibrational integral on the right, labelled the *Franck-Condon factor*, is not necessarily zero because the two vibrational wave functions do not belong to the same electronic state and therefore need not be orthogonal.[†]

Equation 5-28 can be written out more completely as follows:

$$\boxed{M = \int \psi^*_{v'} \psi_v d\tau_n \underbrace{\int \psi^*_{e'} \, \hat{\mu}_e \psi_e d\tau_e}_{\text{basis of orbital selection rules}} \underbrace{\int \psi^*_{s'} \psi_s d\tau_s}_{\text{basis of spin selection rules}}} \tag{5-29}$$

since $\hat{\mu}_e$ does not operate on the spin coordinate. Equation 5-29 will be the basis of the electronic selection rules. If any of the integrals is zero, the

[†] The Franck-Condon factor is actually usually taken as the square of the integral we have so labelled. This is because the transition probability is proportional to the square of M.

transition is formally forbidden. We say, therefore, that there are vibrational, orbital, and spin selection rules for electronic transitions. It is found that some selection rules are obeyed more rigorously than others. Since eq. 5-29 is the result of a series of approximations, not the least of which is the factoring of the total wave function into nuclear, orbital, and spin functions, it is reasonable that the selection rules are not rigorous.†

The individual selection rules will be illustrated and discussed at length in the sections which follow, but at this time it is appropriate to discuss the relative "allowedness" of transitions allowed by one or more of the selection rules. If a transition is forbidden by the spin selection rule ($\int \psi_{s'}^* \psi_s d\tau = 0$), the extinction coefficient, ϵ (Section 2-2-C), is generally in the range 10^{-5}–10^0 M^{-1} cm^{-1}.†† This is true regardless of the vibrational or orbital allowedness. If the transition is spin-allowed but orbitally-forbidden (regardless of the vibrational allowedness) ϵ is usually in the range 10^0–10^3. If both spin and orbital integrals are nonzero, the transition is said to be "fully allowed" and ϵ is observed to be 10^3–10^5. These values of ϵ are necessarily approximate and some illustrations of each case are given later. The vibrational selection rules were not considered explicitly in these absorption intensity rules, but, rather, serve to modulate the intensity of a transition within the above stated limits. The ranges of ϵ, which are important to remember, are summarized in Table 5-4.

Although ϵ values are by far the most common expression of absorption intensity, two other terms are occasionally encountered. The *integrated absorption coefficient*, I, is defined by the relationships

$$I = \int_0^\infty \epsilon(\nu)d\nu \quad \text{or} \quad \bar{I} = \int_0^\infty \epsilon(\bar{\nu})d\bar{\nu} = I/c \tag{5-30}$$

These represent the area under an absorption band when ϵ is plotted against ν or $\bar{\nu}$. I is proportional to the square of the transition moment integral. The *oscillator strength*, f, is a dimensionless quantity given by

$$f = 4.33 \times 10^{-9} \int_0^\infty \epsilon(\bar{\nu})d\bar{\nu} \tag{5-31}$$

It represents the ratio of the observed, integrated, absorption coefficient to that calculated for a single electron in a three-dimensional harmonic potential

†Another point to realize is that we are only considering electric dipole transitions. Transitions induced by the oscillating magnetic field of light are about five orders of magnitude weaker than electric dipole transitions. But since electronic transitions vary in intensity by about ten orders of magnitude, magnetic dipole transitions (and even weaker electric quadrupole transitions) cannot be entirely ignored.

††For compounds containing heavy elements, such as those of the third transition metal series, ϵ can be as high as $\sim 10^2$ for "spin-forbidden transitions because of extensive spin-orbit coupling.

Table 5-4. Intensities of Electronic Transitions

Governing Condition	Approximate Range of ϵ Values (M^{-1} cm^{-1})
spin-forbidden	10^{-5}–10^{0}
spin-allowed, orbitally-forbidden	10^{0}–10^{3}
spin- and orbitally-allowed	10^{3}–10^{5}

well. The maximum value of f for a fully allowed transition is of the order of unity.

In eq. 5-28 the quantity $\int \psi_{v'}^{*} \psi_{v} d\tau_{n}$ was identified as the Franck-Condon factor. This integral represents the overlap of vibrational wave functions of the ground and excited electronic states. The magnitude of this integral acts to modulate the intensity of absorption bands whose general range of magnitude is determined primarily by the spin and orbital selection rules. Since most ground state molecules are in the $v = 0$ vibrational state, we will only consider the $v = 0$ wave function for the ground state in the discussion which follows. In Fig. 5-10, a case is illustrated in which maximum overlap occurs between the $v = 0$ and $v' = 2$ vibrational wave functions. The resulting absorption spectrum might have the qualitative appearance

Fig. 5-10. Illustration of the Franck-Condon Principle. The potential curves are drawn to give maximal overlap between the $v = 0$ and $v' = 2$ vibrational wave functions.

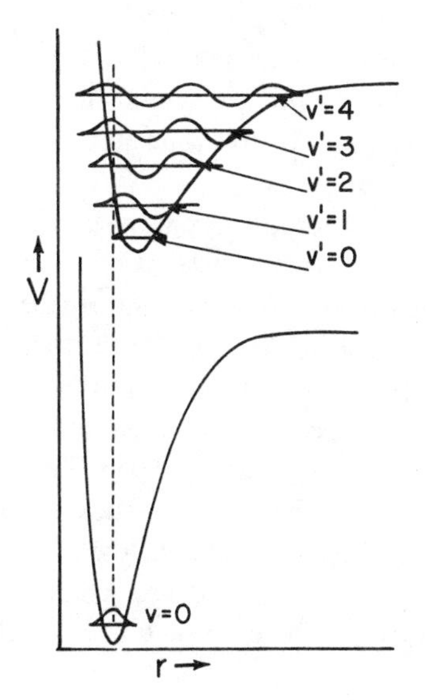

of the CO spectrum, Fig. 5-5, in which the 0-2 transition is the most intense. Clearly, relative left-right displacement of the two potential wells in Fig. 5-10 would serve to increase some overlaps and decrease others, and could produce a variety of intensity distributions of the resultant absorption band.

A classical formulation of the Franck-Condon Principle goes as follows: "An electronic transition is so fast, compared to nuclear motion, that the nuclei still have nearly the same position and momentum immediately after the transition as before." Consider Fig. 5-11 (a) in which the upper well has a slightly greater internuclear separation than the lower well at the energy minimum. A transition from the ground state to the vibronic level labelled A (with the internuclear separation still nearly r_e^0) involves little change in the position or momentum of the nuclei. The momentum, which is close to zero in the ground state, is close to zero in the excited state at the separation r_e^0 because we have drawn the level labelled A such that at $r = r_e^0$ there is a turning point in the vibration and the momentum must go through zero as it changes sign (when the nuclei change their direction of motion). A transition to the level B is less likely because at the internuclear separation $r = r_e^0$ there is considerable momentum in the level B. A transition to the level C is very unlikely because r cannot be equal to r_e^0 according to classical mechanics. This figure illustrates the practical consequence of the Franck-Condon Principle; *viz.*, the vibrational state whose turning point is vertically above the equilibrium internuclear separation of the ground state will give rise to the most intense absorption. The predictions of this classical mechanical statement and the quantum mechanical Franck-Condon factor are not always in precise agreement, but will usually be close. The quantum mechanical prediction should be used in case of disagreement.

We are now in a position to consider an important detail in photoelectron spectroscopy. In Fig. 5-12 is the photoelectron spectrum of formaldehyde showing rich vibronic structure. The spectrum can be interpreted in terms of a diagram such as Fig. 5-4 in which the excited state well applies to an *ionized* state of the molecule. As indicated by the vertical arrows in Fig. 5-4, vibronic spacings in the spectrum correspond to *vibrational energies of the ionic state*. In the lowest energy band ($2b_1$) in Fig. 5-12, the most intense transition is the 0-0 transition. The energy of the 0-0 transition is called the *adiabatic ionization energy*. In the $1b_2$ band, the most intense transition is called the *vertical ionization energy* because it corresponds to a vertical transition on a Franck-Condon diagram such as Fig. 5-11. In many spectra,

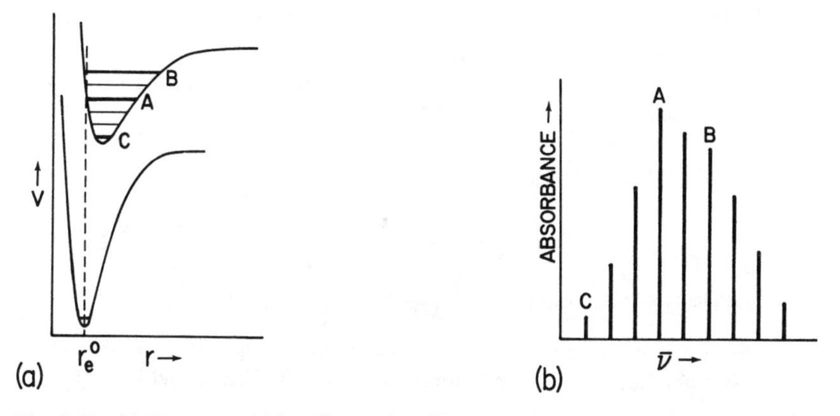

Fig. 5-11. (a) Two potential wells used to illustrate the Franck-Condon Principle. (b) The absorption spectrum which might result from the two wells in (a).

vibrational structure is obscured and only broad bands are observed. The band maximum is the vertical ionization energy, and the adiabatic ionization energy can only be estimated as the onset of the band. When using Koopmans' theorem (Section 4-3) to determine orbital energies, it is the adiabatic *IE* that is desired, not the vertical *IE*. Since we cannot always determine the adiabatic *IE*, we often settle for the vertical *IE*. The same is true of an electronic absorption spectrum where the 0-0 band is the desired measure of separation between electronic states, but the band maximum is usually all we can measure.

Fig. 5-12. He I photoelectron spectrum of formaldehyde. The spectrum above 21 eV is a He II spectrum. From C.R. Brundle, M.B. Robin, N.A. Kuebler, and H. Basch, *J. Amer. Chem. Soc.*, *94*, 1451 (1972).

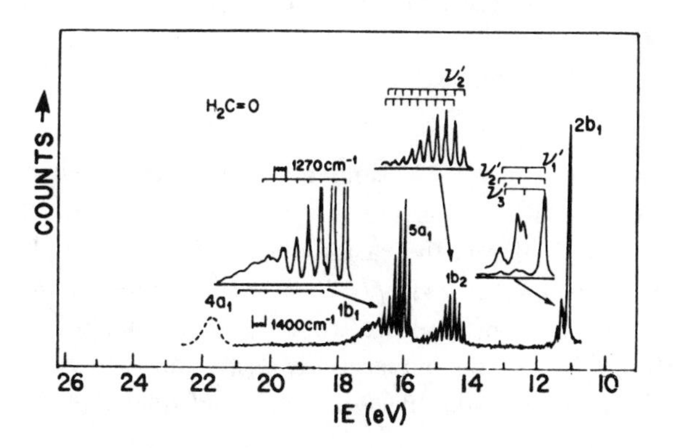

Problems

5-9. Draw a schematic vibronic absorption spectrum of a diatomic molecule for a case in which the minima of the potential wells which describe the ground and excited states occur at the same internuclear separation.

5-10. Two common idealized band shapes are called Gaussian and Lorentzian. The equations for these curves, when plotting ϵ *vs.* $\bar{\nu}$, are

$$\text{Lorentzian: } \epsilon = \epsilon_{max} \frac{\Delta\bar{\nu}_{1/2}^2/4}{(\Delta\bar{\nu}_{1/2}^2/4) + (\bar{\nu} - \bar{\nu}_{max})^2} = \epsilon_{max} \frac{\Gamma^2}{\Gamma^2 + (\bar{\nu} - \bar{\nu}_{max})^2}$$

$$\text{Gaussian: } \epsilon = \epsilon_{max} \exp\left[-\frac{4(\ln 2)(\bar{\nu} - \bar{\nu}_{max})^2}{\Delta\bar{\nu}_{1/2}^2}\right] = \epsilon_{max} \exp\left[-\frac{(\ln 2)(\bar{\nu} - \bar{\nu}_{max})^2}{\Gamma^2}\right]$$

where ϵ_{max} is the value of the molar extinction coefficient at the absorption maximum, $\bar{\nu}_{max}$. $\Delta\bar{\nu}_{1/2}$ is the width of the peak at half height and $\Gamma = \Delta\bar{\nu}_{1/2}/2$.

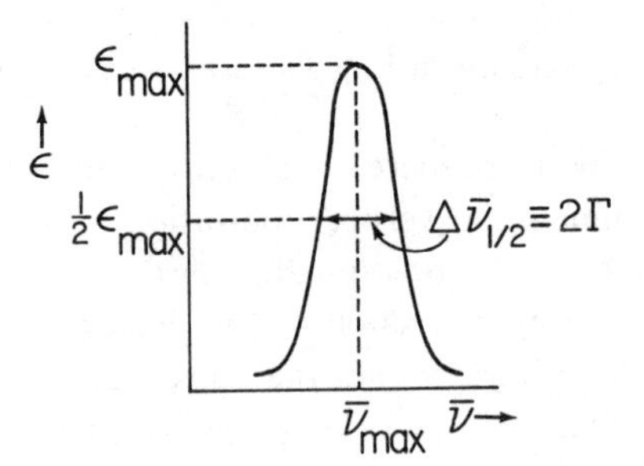

(a) Derive a general formula for the integrated absorption coefficient, $\bar{I}$, of each type of absorption band. You may wish to use the definite integrals below:

$$\int_0^\infty \frac{a\,dx}{a^2 + x^2} = \frac{\pi}{2} \qquad \int_0^\infty e^{-ax^2} dx = \frac{1}{2}\sqrt{\frac{\pi}{a}} \qquad (a > 0)$$

(b) The 0-0 absorption band of the ${}^1A_g \rightarrow {}^1B_{3u}$ transition of pyrazine (see Problem 5-34, Section 5-8-B) exhibits $\epsilon_{max} = 3100\ M^{-1}cm^{-1}$, $\Delta\bar{\nu}_{1/2} = 46\ cm^{-1}$; and $\bar{\nu}_{max} = 30{,}840\ cm^{-1}$. Assuming a Gaussian line shape, calculate $\bar{I}$ for this absorption. (c) Calculate the oscillator strength of this transition.

B. *Spin Selection Rules.* The spin selection rules, based on eq. 5-29, state that the integral $\int \psi_{s'}^* \psi_s d\tau_s$ must be nonzero if the transition is to be allowed. In practice, this is the easiest selection rule to apply because *a transition is spin-allowed if and only if the multiplicities of the two states involved are identical.* This follows from the orthogonality of spin wave functions. If $\psi_{s'} \neq \psi_s$, the integral $\int \psi_{s'}^* \psi_s d\tau_s$ must be zero. Thus, for example, singlet→singlet, doublet→doublet, and quartet→quartet transitions are allowed, but singlet→triplet and quartet→doublet transitions are forbidden. As indicated in Table 5-4, the spin selection rule is the strictest of the electronic selection rules.

As an example, we will show that the triplet→singlet transition involving the spin functions $(1/\sqrt{2})\ (\alpha\beta - \beta\alpha)$ (singlet) and $\alpha\alpha$ (triplet) is forbidden:

$$\psi_s{}'\psi_s d\tau = \frac{1}{\sqrt{2}}\int_{e_{1,2}} [\alpha(1)\beta(2) - \beta(1)\alpha(2)][\alpha(1)\alpha(2)]\,d\tau_{s_{1,2}}$$

$$= \frac{1}{\sqrt{2}}\left[\int_{e_{1,2}} \alpha(1)\beta(2)\alpha(1)\alpha(2)\,d\tau_{s_{1,2}} - \int_{e_{1,2}} \beta(1)\alpha(2)\alpha(1)\alpha(2)\,d\tau_{s_{1,2}}\right]$$

$$= \frac{1}{\sqrt{2}}\left[\int_{e_1} \alpha(1)\alpha(1)\,d\tau_{s_1}\int_{e_2} \beta(2)\alpha(2)\,d\tau_{s_2} - \int_{e_1} \beta(1)\alpha(1)\,d\tau_{s_1}\int_{e_2} \alpha(2)\alpha(2)\,d\tau_{s_2}\right]$$

$$= \frac{1}{\sqrt{2}}\left[(1)(0) - (0)(1)\right] = 0 \quad \text{spin-forbidden} \qquad (5\text{-}32)$$

Problems

5-11. (a) Show that the singlet → singlet transition from the state whose spin wave function is $(1/\sqrt{2})(\alpha\beta - \beta\alpha)$ to the state whose wave function is also $(1/\sqrt{2})$ $(\alpha\beta - \beta\alpha)$ is spin-allowed. (b) Show that the transition $(1/\sqrt{2})(\alpha\beta - \beta\alpha) \rightarrow$ $(1/\sqrt{2})(\alpha\beta + \beta\alpha)$ is forbidden. What are the multiplicities of the initial and final states?

C. *Orbital Selection Rules.* The orbital selection rules, also derived from eq. 5-29, state that the integral $\int\psi_{e'}^{*}\hat{\mu}_e\psi_e d\tau_e$ must be nonzero if the transition is to be orbitally allowed. This is completely analogous to the vibrational selection rule of Chapter Three. *An electronic transition is orbitally-allowed if and only if the triple direct product $\Gamma(\psi_{e'}) \times \Gamma(\hat{\mu}_e) \times \Gamma(\psi_e)$ contains the totally symmetric irreducible representation of the point group of the molecule.* The dipole moment operator once again has three components which transform as x, y, and z.

As an example, the spin-allowed transition ${}^1\Sigma_g{}^+ \rightarrow {}^1\Sigma_u{}^+$ is orbitally-allowed because the triple direct product contains $\sigma_g{}^+$:

$$\int\psi_{e'}^{*}\hat{\mu}_e\psi_e d\tau_e \sim \Sigma_u{}^+\begin{pmatrix}\sigma_u{}^+\\ \pi_u\end{pmatrix}\Sigma_g{}^+ = \begin{pmatrix}\sigma_g{}^+\\ \pi_g\end{pmatrix}\longleftarrow\text{allowed} \qquad (5\text{-}33)$$

A transition which is both spin- and orbitally-allowed is called *fully allowed* and can be very intense. It should be immediately obvious that *for a totally symmetric ground state, only transitions to excited states which have the same symmetry as at least one of the components of the dipole moment operator will be orbitally-allowed.* The absorption due to the spin-forbidden transition $^1\Sigma_g^+ \rightarrow {}^3\Sigma_u^+$, though orbitally-allowed, will be very weak because the spin selection rule is the strictest of all.

The spin-allowed electronic transition to the first excited state of benzene, $^1A_{1g} \rightarrow {}^1B_{2u}$ is orbitally-forbidden because the dipole moment operator transforms as $a_{2u}(z) + e_{1u}(x,y)$ in the point group D_{6h}:

$$A_{1g}\begin{pmatrix} a_{2u} \\ e_{1u} \end{pmatrix} B_{2u} = \begin{pmatrix} b_{1g} \\ e_{2g} \end{pmatrix} \quad \text{forbidden} \tag{5-34}$$

Such spin-allowed but orbitally-forbidden transitions exhibit intensities between those of fully allowed and spin-forbidden transitions.

Problems

5-12. The ground state electron configuration of ethylene is ... $(b_{1u})^2\,(b_{2g})^0$. What electronic states are generated by the ground and first excited state configurations? In what range would ϵ be expected for each possible transition from the ground state?

5-13. What is the ground state of a planar methyl radical whose configuration is ... $(a_1')^2\,(e')^4\,(a_2'')^1$? What excited state(s) is generated by the transition $e' \rightarrow a_2''$? What will be the range of ϵ for such a transition(s)? If $CH_3\cdot$ were pyramidal (as NH_3, C_{3v} symmetry) the ground state configuration would be ... $(a_1)^2\,(e)^4\,(a_1)^1$. Answer the same questions for pyramidal $CH_3\cdot$.

D. *Vibronic Selection Rules.* Equation 4-29 can be rearranged as follows:

$$M = \int \psi_{v'}^* \psi_v \, d\tau_n \int \psi_{e'}^* \hat{\mu}_e \psi_e \, d\tau_e \int \psi_{s'}^* \psi_s \, d\tau_s$$

$$M = \underbrace{\int \psi_{e'}^* \psi_{v'}^* \hat{\mu}_e \psi_e \psi_v \, d\tau_{en}}_{\text{basis of vibronic selection rules}} \int \psi_{s'}^* \psi_s \, d\tau_s \tag{5-35}$$

In the preceding discussion of orbital selection rules we have implicitly considered only 0-0 ("pure electronic") transitions for which both $\psi_{v'}$

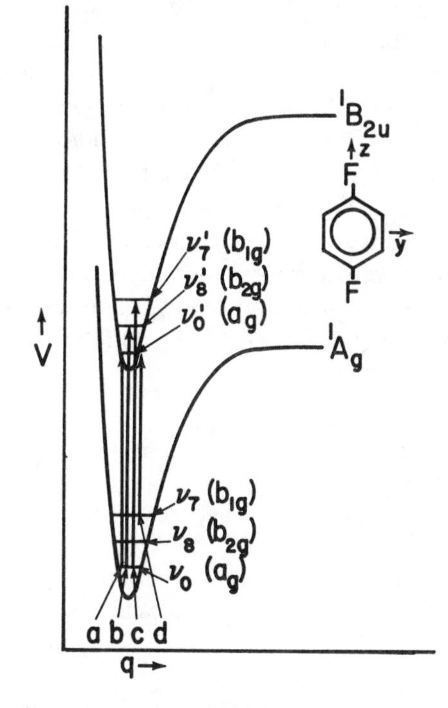

Fig. 5-13. Schematic representation of the $^1B_{2u}$ and ^1A_g electronic potential wells of p-difluorobenzene showing the vibrational ground states and the vibrational levels ν_7 and ν_8. The transitions are designated as follows: a, 0-0; b, 8_0^1; c, 7_0^1; d, 7_1^0.

($= \psi^*_{v'}$) and ψ_v are totally symmetric. In eq. 5-35 we have combined the vibrational and electronic integrals to allow for the case in which $\psi_{v'}$ or ψ_v is not totally symmetric. *The condition for a vibronic transition to be allowed is that the integral* $\int \psi^*_{e'} \psi^*_{v'} \hat{\mu}_e \psi_e \psi_v \, d\tau_{en}$ *be nonzero.*

Let's consider some selected vibronic transitions of the molecule p-difluorobenzene which has D_{2h} symmetry. In Fig. 5-13 we show two of the vibrational energy levels of the ground and lowest excited electronic state of this molecule. Each of the thirty normal modes of vibration of p-difluorobenzene would require a different well to describe its potential. So putting more than one mode in a single well is just a convenient way to talk about vibronic transitions, but not a realistic way. The 0-0 transition is labelled a in this figure. The transitions labelled b and c are called 8_0^1 and 7_0^1, respectively, because, for example, in transition b we are exciting ν_8 from the $v = 0$ to the $v' = 1$ level. Transition d is designated 7_1^0 because ν_7 goes from $v = 1$ to $v' = 0$. Such a transition is called a *hot band* because the molecule is initially in an excited vibrational state in the ground electronic state. The intensity of such a transition will generally increase with temperature; since the molecule is more likely to be vibrationally excited at higher temperature. Experimentally, one determines if an observed absorption is a hot band by heating or cooling the sample and looking at the increased or decreased intensity of the absorption.

Is the spin-allowed pure electronic transition ${}^1A_g \rightarrow {}^1B_{2u}$ of p-difluorobenzene orbitally-allowed? The answer is "yes" because the dipole moment operator transforms as $b_{3u}(x)$, $b_{2u}(y)$, and $b_{1u}(z)$ in the point group D_{2h}:

$$B_{2u}\begin{pmatrix} b_{3u} \\ b_{2u} \\ b_{1u} \end{pmatrix} A_g = \begin{pmatrix} b_{1g} \\ a_g \\ b_{3g} \end{pmatrix} \longleftarrow \text{allowed} \tag{5-36}$$

Evaluating the vibronic allowedness of this transition with the integral $\int \psi_{e'}^* \psi_{v'}^* \hat{\mu}_e \psi_e \psi_v d\tau_{en}$ must, of course, give the same result since $\psi_{v'}$ and ψ_v both transform as a_g:

$$\underset{\substack{\uparrow \\ \psi_{v'}}}{B_{2u}a_g}\begin{pmatrix} b_{3u} \\ b_{2u} \\ b_{1u} \end{pmatrix} \underset{\substack{\uparrow \\ \psi_v}}{A_g a_g} = \begin{pmatrix} b_{1g} \\ a_g \\ b_{3g} \end{pmatrix} \longleftarrow \text{allowed} \tag{5-37}$$

This transition moment integral also tells us that the transition is y-polarized since it is the y component of the direct product which is totally symmetric. What about the vibronic transition 8_0^1 if ν_8 has b_{2g} symmetry? The appropriate direct product is

$$B_{2u}b_{2g}\begin{pmatrix} b_{3u} \\ b_{2u} \\ b_{1u} \end{pmatrix} A_g a_g = \begin{pmatrix} b_{3g} \\ b_{2g} \\ b_{1g} \end{pmatrix} \qquad \text{forbidden} \tag{5-38}$$

which tells us that the transition is forbidden and will be orders of magnitude less intense than the 0-0 transition. A similar set of direct products tells us that the transition 7_0^1 is fully allowed and x-polarized. The hot band transition 7_1^0 is also fully allowed and x-polarized:

$$\underset{\substack{\uparrow \\ \psi_{v'}}}{B_{2u}\,a_g} \begin{pmatrix} b_{3u} \\ b_{2u} \\ b_{1u} \end{pmatrix} \underset{\substack{\uparrow \\ \psi_v}}{A_g\,b_{1g}} = \begin{pmatrix} a_g \\ b_{1g} \\ b_{2g} \end{pmatrix} \begin{matrix} \longleftarrow \text{ allowed} \\ \\ \\ \end{matrix} \tag{5-39}$$

This does not mean that 7_1^0 will be an intense transition because ν_7 may hardly be populated in the electronic ground state. For that matter, the fact that 7_0^1 is fully allowed does not mean that it will be a strong absorption either. It only means that the transition 7_0^1 is likely to be much more intense than the forbidden transition 8_0^1. Both could, however, be very weak.

The most important aspect of vibronic coupling is illustrated by molecules for which the 0-0 transition is orbitally-forbidden. Such is the case of the lowest spin-allowed electronic transition of benzene, ${}^1A_{1g} \rightarrow {}^1B_{2u}$. Since $\hat{\mu}_e$ transforms as $a_{2u}(z)$ and $e_{1u}(x,y)$, the 0-0 transition is forbidden:

$$B_{2u} \begin{pmatrix} a_{2u} \\ e_{1u} \end{pmatrix} A_{1g} = \begin{pmatrix} b_{1g} \\ e_{2g} \end{pmatrix} \tag{5-40}$$

But this integrand tells us that any vibration of the molecule with b_{1g} or e_{2g} symmetry might serve as a *vibronic origin*, making the transition vibronically-allowed. For example, ν_{10} (e_{2g}) gives rise to the (x, y)-polarized 10_0^1 vibronically-allowed transition:

$$\underset{\substack{\uparrow \\ \psi_{v'}}}{B_{2u}\,e_{2g}} \begin{pmatrix} a_{2u} \\ e_{1u} \end{pmatrix} \underset{\substack{\uparrow \\ \psi_v}}{A_{1g}\,a_{1g}} = e_{1u} \begin{pmatrix} a_{2u} \\ e_{1u} \end{pmatrix} a_{1g} = \begin{pmatrix} e_{1g} \\ \underline{\underline{a_{1g}}} + a_{2g} + e_{2g} \end{pmatrix} \tag{5-41}$$

The hot band transition 10_1^0 would also be vibronically-allowed since this only reverses the positions of $\psi_{v'}$ and ψ_v in the transition moment integrand.

We see, therefore, that by coupling an orbitally-forbidden electronic transition with a vibrational transition, it is possible that the selection rules can be satisfied and some intensity can be expected. Such vibronically-allowed, orbitally-forbidden transitions generally fall in the range $\epsilon \simeq 10^0 - 10^3$. If there were no vibronic coupling, the transitions might be even weaker.

Problems

5-14. Show that the hot band transition $^1A_g(\nu_8) \rightarrow {}^1B_{2u}(\nu_7')$ of p-difluorobenzene shown below is vibronically-allowed and z-polarized. The notation means that the molecule is excited from the $\nu_8\,(b_{2g})$ vibrationally excited electronic ground state to an electronic excited state in which $\nu_7'(b_{1g})$ is vibrationally excited. The notation for such a transition would be $8^0_1\,7^1_0$.

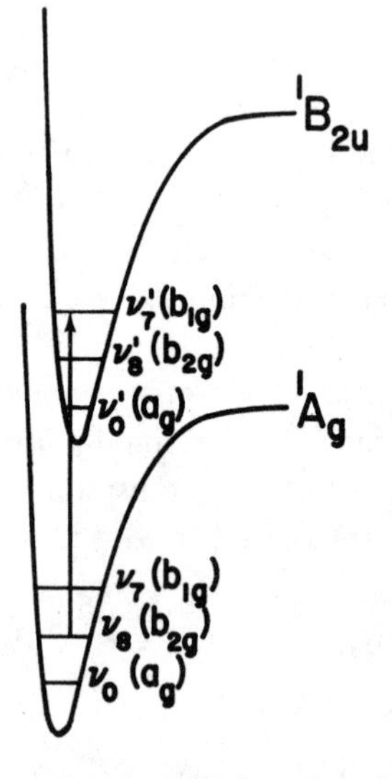

Transition $A_g\ (\nu_8) \rightarrow B_{2u}\ (\nu_7')$

5-15. The vibration ν_{10} of benzene has symmetry e_{2g} and energy $\nu_{10} = 608\ \text{cm}^{-1}$ in the ground electronic state ($^1A_{1g}$). In the $^1B_{2u}$ excited state, the vibrational energy is $\nu_{10}' = 520\ \text{cm}^{-1}$. What is the energy of the hot band transition 10^0_1 if the energy of the transition 10^1_0 is 38,611 cm^{-1}? Illustrate this transition in the manner of Fig. 5-13.

5-16. (a) If the Co atom of *trans*-dichlorobis(ethylenediamine)cobalt(III) only "senses" its nearest neighbors, the effective local symmetry about Co will be D_{4h}. The ground state of this d^6 species is $^1A_{1g}$. The first three singlet excited states, in no particular order, are $^1A_{2g}$, $^1B_{2g}$, and ^1E_g.

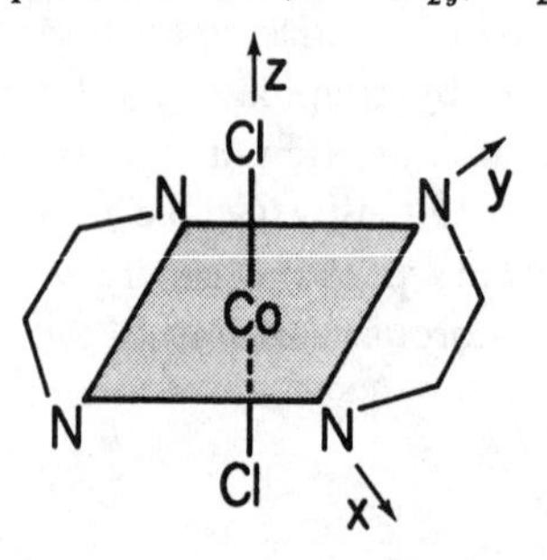

Fill in the table below showing the symmetries of the integrands of the various possible transition moment integrals. Are any of the transitions allowed in either z- or (x,y)-polarization?

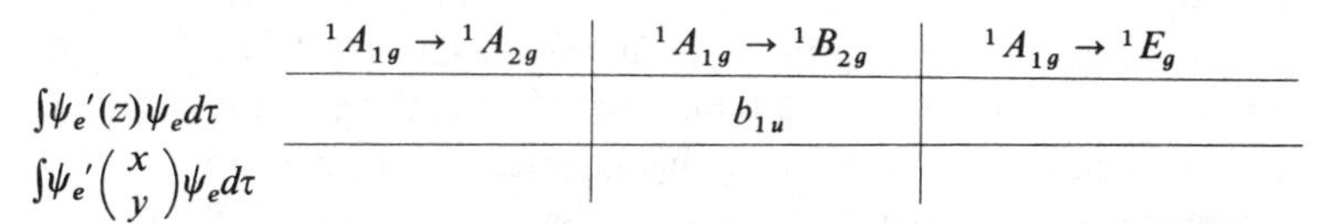

	${}^1A_{1g} \rightarrow {}^1A_{2g}$	${}^1A_{1g} \rightarrow {}^1B_{2g}$	${}^1A_{1g} \rightarrow {}^1E_g$
$\int \psi_e'(z)\psi_e d\tau$		b_{1u}	
$\int \psi_e' \begin{pmatrix} x \\ y \end{pmatrix} \psi_e d\tau$			

(b) The normal modes of vibration of the CoN_4Cl_2 skeleton transform as $2a_{1g} + b_{1g} + b_{2g} + e_g + 2a_{2u} + b_{1u} + 3e_u$. Will any of the electronic transitions in part (a) be vibronically allowed, using any of these fifteen vibrations as a vibronic origin? That is, will any of the vibronic transitions i_0^1 be allowed, where i is the number of one of the normal modes? You should be able to fill in the table below by inspection.

	Polarization with Vibronic Coupling	
	z	x,y
${}^1A_{1g} \rightarrow {}^1A_{2g}$		
${}^1A_{1g} \rightarrow {}^1B_{2g}$	allowed	
${}^1A_{1g} \rightarrow {}^1E_g$		

(c) The spectrum of this species is shown below. Can you assign any of the bands on the basis of your analysis?

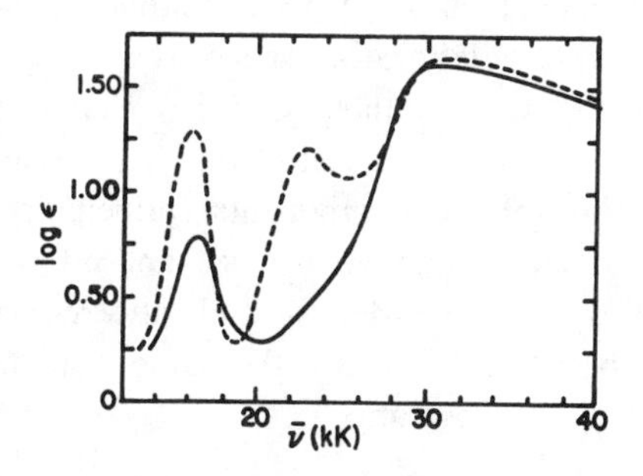

Single-crystal absorption spectrum of *trans*-$[Co(en)_2Cl_2]Cl \cdot HCl \cdot 2H_2O$. The solid line shows the spectrum with z-polarized light and the dashed line is with (x,y)-polarized light. From S. Yamada and R. Tsuchida, *Bull. Chem. Soc. Jap.*, *25*, 127 (1952).

5-5. The Electronic Spectra of Some Diatomic Molecules

A. *Hydrogen.* The ground state configuration of H_2 (Table 4-8) is $(1\sigma_g{}^+)^2$ which defines a ${}^1\Sigma_g{}^+$ state. The first excited configuration is $(1\sigma_g{}^+)^1(1\sigma_u{}^+)^1$ which gives rise to ${}^3\Sigma_u{}^+$ and ${}^1\Sigma_u{}^+$ states. We can therefore draw the energy level diagram shown in Fig. 5-8 and expect ${}^1\Sigma_g{}^+ \rightarrow {}^3\Sigma_u{}^+$ and ${}^1\Sigma_g{}^+ \rightarrow {}^1\Sigma_u{}^+$ transitions. The singlet-singlet transition is both spin- and orbitally-allowed and gives rise to an intense absorption with rich vibrational and rotational structure and a 0-0 band at 90,196 cm^{-1} (110 nm, 1100 Å). The unit 1000 cm^{-1} is called one *kilokayser* (kK) and we will generally use kK units through the remainder of this chapter. As an indication of the anti-bonding nature of the $1\sigma_u{}^+$ orbital, the ground state (${}^1\Sigma_g{}^+$) vibrational frequency of H_2 is $\varpi_e = 4395$ cm^{-1}. But in the ${}^1\Sigma_u{}^+$ state the vibrational frequency is $\varpi_e = 1357$ cm^{-1}. As can be seen by the shaded area in Fig. 5-14, the Franck-Condon factor will be largest for fairly highly excited states of ${}^1\Sigma_u{}^+$. The Franck-Condon factor for the 0-0 transition is very small, but the Franck-Condon factor for the 0-0 transition of the ${}^1\Sigma_g{}^+ \rightarrow {}^1\Pi_u$ transition is much larger. The ${}^1\Sigma_g{}^+ \rightarrow {}^3\Sigma_u{}^+$ transition is expected at even lower energy than the ${}^1\Sigma_g{}^+ \rightarrow {}^1\Sigma_u{}^+$ transition, but the singlet-triplet transition is spin-forbidden and is expected to be very weak. In fact, this triplet state of H_2 is unstable and gives rise not to discrete absorptions but to *continuous* absorption. A potential curve for this unstable state is illustrated in Fig. 5-14. Above the asymptote given by the energy of the two completely dissociated (${}^2S$) hydrogen atoms, any energy is allowed and there is no quantization of states.

The complexity of electronic spectra is illustrated by the fact that no fewer than 39 electronic states have been identified for H_2. Fortunately (or unfortunately), ordinary laboratory spectroscopy is generally restricted to the region < 50 kK so only the lowest few electronic states of most polyatomic molecules can be readily observed. The reason for this energy restriction is that atmospheric components (principally O_2, but N_2 at higher energy) absorb energies above ~ 50 kK and interfere with measurements of other substances. In order to study the region > 50 kK, one must evacuate the spectrometer; hence the name "vacuum ultraviolet" is given to this region of the spectrum.

Problems

5-17. The adiabatic (Section 5.4-A) *IE* of H_2 is 15.43 eV and the experimental vibrational constants of the resulting ionic ${}^2\Sigma_g{}^+$ state are $\varpi_e = 2319$ cm^{-1} and $\varpi_e x_e = 58$ cm^{-1}. Compute the positions (in eV) of the first five vibrational bands in the photoelectron spectrum of H_2. Given the observed spectrum below, draw qualitative potential energy curves for the neutral ground state and first ionic state of H_2, showing the relation of the potential wells neces-

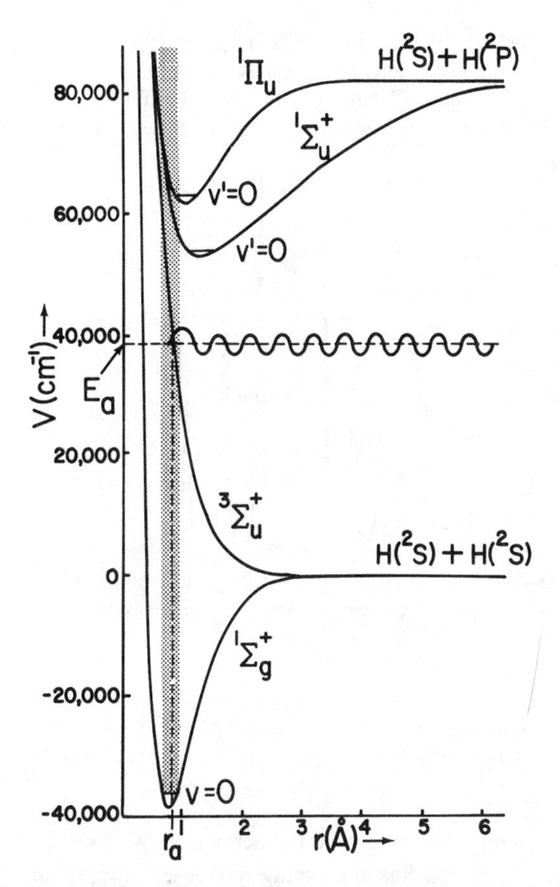

Fig. 5-14. Potential curves for some lower states of H_2. The shaded region is that in which Franck-Condon factors for transitions from $v = 0$ in the $^1\Sigma_g{}^+$ state will be largest. The unstable $^3\Sigma_u{}^+$ state has no minimum in its potential. The particular wave function shown tells us that for some arbitrary energy, E_a, the atoms have only a tiny probability of being separated by a distance less than r_a. The atoms could be at any distance apart greater than r_a with nearly uniform probability. A similar kind of wave function could be drawn at any arbitrary energy above the asymptote of a Morse potential curve (cf. Fig. 2-8). The data for these curves come from W. Kolos and L. Wolniewicz, *J. Chem. Phys.*, *43*, 2429 (1965); *45*, 509 (1966).

sary to produce the intensity pattern in the spectrum. Given that D_2 will have essentially the same potential wells as H_2, but with smaller vibrational spacings, will the Franck-Condon maximum in the photoelectron spectrum of D_2 occur at a higher or lower vibrational quantum number than that of H_2?

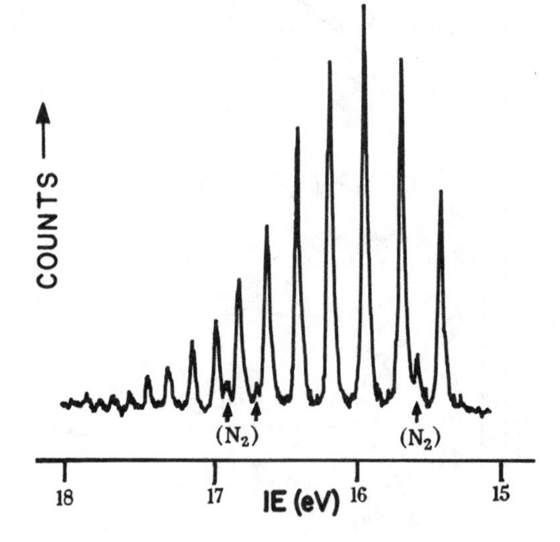

He I photoelectron spectrum of H_2 from D.W. Turner, *Proc. Roy. Soc. A*, *307*, 55 (1968). The photoelectron spectrum of D_2 appears in A.B. Conford, D.C. Frost, C.A. McDowell, J.L. Ragle, and I.A. Stenhouse, *Chem. Phys. Lett.*, *5*, 486 (1970).

B. *Nitrogen.* According to Fig. 4-18, Scheme I, we anticipate a ${}^1\Sigma_g{}^+$ ground state for N_2 which has the configuration ... $(1\pi_u)^4 (3\sigma_g{}^+)^2 (1\pi_g)^0 (3\sigma_u{}^+)^0$. The first excited state configuration is ... $(1\pi_u)^4 (3\sigma_g{}^+)^1 (1\pi_g)^1 (3\sigma_u{}^+)^0$ which yields the states ${}^1\Pi_g$ and ${}^3\Pi_g$. The state diagram and transition energies are shown in Fig. 5-15. The ${}^1\Sigma_g{}^+ \leftrightarrow {}^1\Pi_g$ spin-allowed, orbitally-forbidden transition is observed in both absorption and emission spectra at 145 nm. A state identified not as ${}^3\Pi_g$ but as ${}^3\Sigma_u{}^+$ has been observed 49.8 kK above the ground state. The vibrational frequency of this ${}^3\Sigma_u{}^+$ state is $\bar{\omega}_e = 1460\ \text{cm}^{-1}$. This state must arise from an even more highly excited configuration of N_2, as indicated in Fig. 5-15. Triplet-triplet emission from a state about 10 kK higher in energy than ${}^3\Sigma_u{}^+$ is attributed to the ${}^3\Pi_g \rightarrow {}^3\Sigma_u{}^+$ transition which is both spin- and orbitally-allowed. The vibrational frequency of the ${}^3\Pi_g$ state is $\bar{\omega}_e = 1734\ \text{cm}^{-1}$ and the vibrational frequency of the ${}^1\Pi_g$ state derived from the same configuration is $\bar{\omega}_e = 1692\ \text{cm}^{-1}$. Perhaps the most important feature of the spectrum of N_2 is the lack of absorption in the normally accessible region < 50 kK.

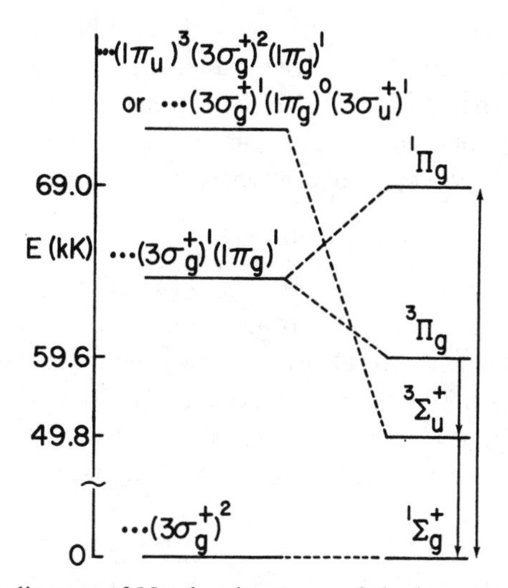

Fig. 5-15. State diagram of N_2 showing some of the lowest energy observed transitions. Upward arrows represent transitions observed in absorption and downward arrows transitions observed in emission.

Problems

5-18. Show that the $^1\Sigma_g^+ \rightarrow {}^1\Pi_g$ transition of N_2 is orbitally forbidden. How could you determine this by inspection? Show that the $^3\Pi_g \rightarrow {}^3\Sigma_u^+$ emission is orbitally allowed. What other states are produced by the two possible configurations of Fig. 5-15 which might give rise to the $^3\Sigma_u^+$ state?

C. *Oxygen.* The ground state configuration of oxygen is ... $(1\pi_u)^4$ $(1\pi_g)^2\,(3\sigma_u^+)^0$ (Fig. 4-18, Scheme II). This configuration gives rise to the *three* states $^3\Sigma_g^-$, $^1\Delta_g$, and $^1\Sigma_g^+$. The ground state is determined by Hund's first rule (Section 4-2-B) which says that the state of highest spin multiplicity, $^3\Sigma_g^-$, is the ground state. The order of $^1\Delta_g$ and $^1\Sigma_g^+$ is determined by applying Hund's second rule (Section 4-2-C) which says that among states of the same multiplicity, the one with the greatest orbital angular momentum is usually the most stable.

What has orbital angular momentum got to do with the term symbols? To answer this we must look at the properties of molecular orbitals once again. The discussion which follows applies to linear molecules only. For such molecules, electrons in σ orbitals have no net angular momentum.

Electrons in π orbitals have one unit (one unit $= \hbar$) of angular momentum about the internuclear axis. Electrons in δ and ϕ orbitals have two and three units of angular momentum, respectively. Calling the angular momentum of each electron $\vec{\lambda}$, the total angular momentum of the molecule, $\vec{\Lambda}$, is the vector sum of the individual angular momenta:

$$\Lambda \equiv |\vec{\Lambda}| = |\Sigma \vec{\lambda}_i| \tag{5-42}$$

The arrows designate vectors and Λ is the magnitude of $\vec{\Lambda}$. Electrons in fully occupied orbitals contribute no net angular momentum to the molecule since their momenta are all paired and cancel out. So we will only consider the electrons in incompletely occupied orbitals.

O_2 has two electrons in π orbitals so each electron has one unit of angular momentum. These can be added vectorially as follows:

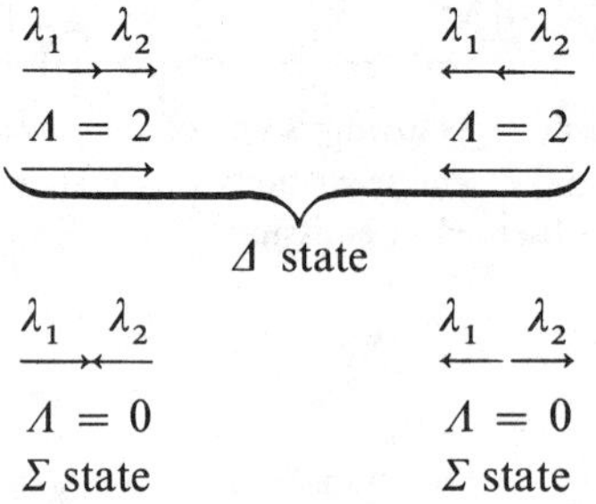

$\lambda_1 \; \lambda_2$ ($\rightarrow \rightarrow$) $\quad \Lambda = 2$ ($\longrightarrow$) $\qquad$ $\lambda_1 \; \lambda_2$ ($\leftarrow \leftarrow$) $\quad \Lambda = 2$ ($\longleftarrow$)

Δ state

$\lambda_1 \; \lambda_2$ ($\rightarrow \leftarrow$) $\qquad$ $\lambda_1 \; \lambda_2$ ($\leftarrow \rightarrow$)

$\Lambda = 0$ $\qquad$ $\Lambda = 0$

Σ state $\qquad$ Σ state

The state is designated by a capital Greek letter: $\Sigma, \Pi, \Delta, \Phi \ldots$ correspond to $\Lambda = 0, 1, 2, 3 \ldots$ respectively. The $(\pi)^2$ configuration of O_2 therefore produces one Δ state and two Σ states. The meanings of the subscripts and superscripts on the term symbols will be discussed at the end of this section. According to Hund's second rule, then, the ${}^1\Delta_g$ state will be lower in energy than the ${}^1\Sigma_g{}^+$ state.

The ground state configuration of O_2 gives the electronic states ${}^3\Sigma_g{}^-$ $< {}^1\Delta_g < {}^1\Sigma_g{}^+$. The first excited configuration, $\ldots (1\pi_u)^3(1\pi_g)^3$, gives rise to the states ${}^3\Delta_u$, ${}^3\Sigma_u{}^+$, ${}^3\Sigma_u{}^-$, ${}^1\Delta_u$, ${}^1\Sigma_u{}^+$, and ${}^1\Sigma_u{}^-$. We can say from Hund's rules that the triplets will be below the singlets and that the Δ state should be below the Σ states of the same multiplicity. Any further ordering of the Σ states requires a good experiment or a very good calculation, or both. The resulting state diagram and the four lowest observed transitions of O_2 are shown in Fig. 5-16. A transition to the ${}^3\Delta_u$ state has not been observed. Since the ground state is a triplet, the triplet-triplet transitions are spin-allowed and the triplet-singlet transitions are forbidden.

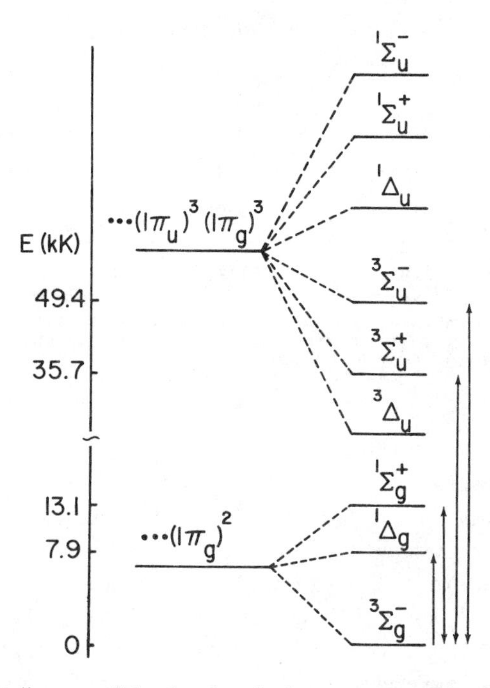

Fig. 5-16. State diagram of O_2 showing the lowest energy observed transitions.

Information about each of these states is collected in Table 5-5. Note that the energies of the three states which arise from the ground configuration are relatively close, compared to the energy needed to promote an electron from the π_u to the π_g* orbital. Notice also that since the bond order is formally 2 in the three lowest states, and 1 in the next two states, the bond lengths and vibrational frequencies take a discontinuous jump between these two sets of states. Oxygen in its $^1\Delta_g$ state is intrinsically much more long-lived than in its $^1\Sigma_g^+$ state. O_2 ($^1\Delta_g$) decays to O_2 ($^3\Sigma_g^-$), but O_2 ($^1\Sigma_g^+$) goes to O_2 ($^1\Delta_g$). The lifetimes of the $^3\Sigma_u^+$ and $^3\Sigma_u^-$ states could be as short as 10^{-8} s since transition to the ground state is spin-allowed.

To add a touch of reality to this discussion, the absorption spectrum of O_2 gas at 1.5×10^7 N m^{-2} (150 atm) in a 0.065 m cell is shown in Fig. 5-17. At this pressure and temperature the concentration of O_2 is 6.2 M. The absorbance of the $^3\Sigma_g^- \rightarrow {}^1\Delta_g$ 0-0 transition at 1269 nm is about 0.1; so $\epsilon = A/c\ell = 0.002$ M^{-1} cm^{-1}. But not only do we see the 0-0 and 0–1 components of the $^1\Delta_g$ and $^1\Sigma_g^+$ states, we see a series of weak

Table 5-5. Low Lying States of O_2[a]

State	Configuration	Energy of 0-0 Absorption cm^{-1}	nm	$\bar{\omega}_e$ (cm^{-1})	r_e (Å)	Lifetime (s)[b]
$^3\Sigma_g^-$	$(\pi_u)^4(\pi_g)^2$	0	0	1580	1.2074	—
$^1\Delta_g$	$(\pi_u)^4(\pi_g)^2$	7882.39	1268.65	1509	1.2155	2700[c]
$^1\Sigma_g^+$	$(\pi_u)^4(\pi_g)^2$	13120.9	762.143	1433	1.2268	7.1[d]
$^3\Sigma_u^+$	$(\pi_u)^3(\pi_g)^3$	35713	280.01	819	1.42	—
$^3\Sigma_u^-$	$(\pi_u)^3(\pi_g)^3$	49363	202.58	700	1.60	—

[a] Data from G. Herzberg, *Spectra of Diatomic Molecules*, Van Nostrand Reinhold Co., N.Y., 1950.
[b] Lifetimes are extrapolated to zero pressure. The lifetime will be shortened in the presence of other matter.
[c] W.H.J. Childs and R. Mecke, *Z. Physik*, *68*, 344 (1931).
[d] R.M. Badger, A.C. Wright, and R.F. Whitlock, *J. Chem. Phys.*, *43*, 4345 (1965).

absorptions between the $^1\Sigma_g^+$ and $^3\Sigma_u^+$ bands. These are assigned to simultaneous transitions of two molecules of O_2 interacting with a single photon. Thus, for example, the label $[^1\Sigma_g^+ + {}^1\Delta_g]$ implies that two molecules interact with one photon to produce one molecule in the $^1\Sigma_g^+$ state and one in the $^1\Delta_g$ state.† It is not unreasonable to invoke simultaneous transitions, since the concentration of O_2 is so high that the average separation of O_2 molecules is only about 6.4 Å and the wavelength of the light is thousands of Ångstroms. Conceivably, one could study the pressure dependence of the bands designated as simultaneous transitions to support this assignment.

We now address the question of the relation between wave functions and the symmetries of molecular states. How are the labels $^3\Sigma_g^-$, $^1\Delta_g$, and $^1\Sigma_g^+$ related to wave functions? The two $1\pi_g$ orbitals of O_2 will be called X and Y, defined as follows:

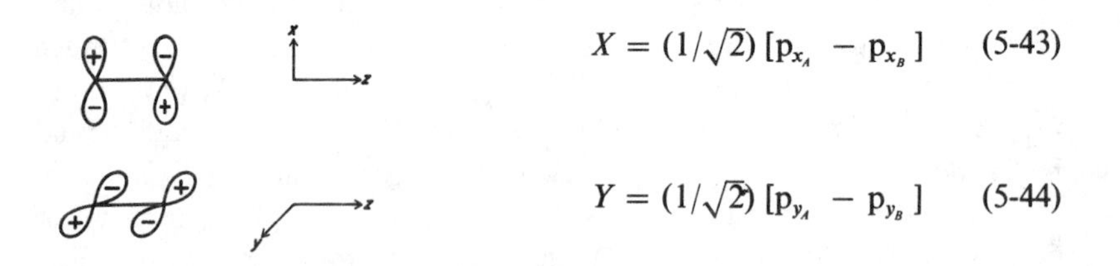

$$X = (1/\sqrt{2})\,[p_{x_A} - p_{x_B}] \qquad (5\text{-}43)$$

$$Y = (1/\sqrt{2})\,[p_{y_A} - p_{y_B}] \qquad (5\text{-}44)$$

† Vibronic spacings in this spectrum are about 1500 cm^{-1}; so, for example, the designation $2[^1\Delta_g]$ 1,0 must mean that two molecules have been promoted to the $^1\Delta_g$ state but only one is vibrationally excited. The designations 2,0 and 3,0 are ambiguous for simultaneous transitions as they do not specify how the vibrational quanta are distributed between excited molecules.

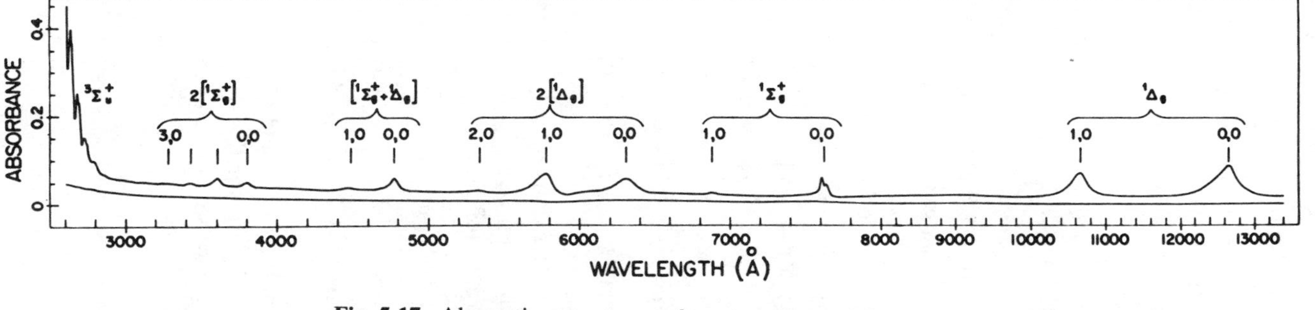

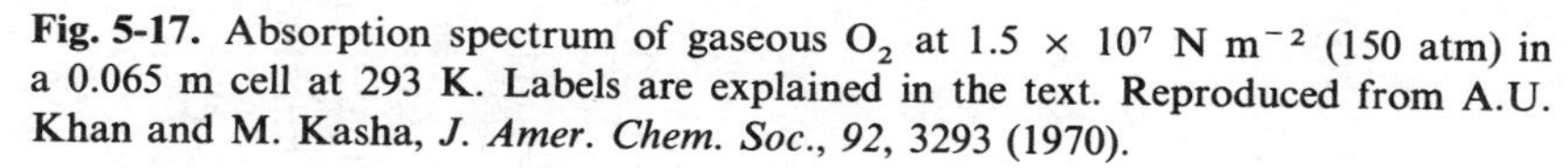

Fig. 5-17. Absorption spectrum of gaseous O_2 at 1.5×10^7 N m^{-2} (150 atm) in a 0.065 m cell at 293 K. Labels are explained in the text. Reproduced from A.U. Khan and M. Kasha, *J. Amer. Chem. Soc.*, *92*, 3293 (1970).

The notation p_{x_A} stands for a p_x atomic orbital on atom A. The possible spin wave functions for singlet and triplet states were given in eq. 5-14. The possible orbital wave functions are as follows:

$$\left.\begin{array}{l}(1/\sqrt{2})\,[X(1)X(2) + Y(1)Y(2)] \\ (1/\sqrt{2})\,[X(1)X(2) - Y(1)Y(2)] \\ (1/\sqrt{2})\,[X(1)Y(2) + Y(1)X(2)]\end{array}\right\} \text{symmetric to electron exchange} \tag{5-45}$$

$$(1/\sqrt{2})\,[X(1)Y(2) - Y(1)X(2)] \quad \text{antisymmetric to electron exchange} \tag{5-46}$$

Dropping the electron numbers 1 and 2, and combining the symmetric orbital wave functions with the antisymmetric spin function, and combining the antisymmetric orbital wave function with the symmetric spin functions, we get the complete electronic wave functions:

$$\left.\begin{array}{l}\psi_1 = (1/\sqrt{2})\,[XY - YX]\,(\alpha\alpha) \\ \psi_2 = (1/\sqrt{2})\,[XY - YX]\,(\beta\beta) \\ \psi_3 = (1/2)\,[XY - YX]\,(\alpha\beta + \beta\alpha)\end{array}\right\} \quad {}^3\Sigma_g^- \tag{5-47}$$

$$\left.\begin{array}{l}\psi_4 = (1/2)\,[XY + YX]\,(\alpha\beta - \beta\alpha) \\ \psi_5 = (1/2)\,[XX - YY]\,(\alpha\beta - \beta\alpha)\end{array}\right\} \quad {}^1\Delta_g \tag{5-48}$$

$$\psi_6 = (1/2)\,[XX + YY]\,(\alpha\beta - \beta\alpha) \quad {}^1\Sigma_g^+ \tag{5-49}$$

Let's find out if these wave functions actually transform as the symmetry labels purport. In the point group $D_{\infty h}$, the functions X and Y defined in eqs. 5-43 and 5-44 transform as follows:†

	E	C_∞^ϕ	σ_v	i	S_∞^ϕ	C_2
X	X	$X\cos\phi - Y\sin\phi$	X	X	$-X\cos\phi + Y\sin\phi$	X
Y	Y	$X\sin\phi + Y\cos\phi$	$-Y$	Y	$-X\sin\phi - Y\cos\phi$	$-Y$

Since there are an infinite number of σ_v and C_2 operations, we have taken just one of each ($\sigma(xz)$ and $C_2(y)$) for this demonstration. The choice of operations is arbitrary, but once we have chosen a set of operations it is

† In this example, we have chosen rotation about the C_∞ axis to be counterclockwise when looking down the z axis from the $+z$ direction.

necessary to use that set throughout. Cosϕ and sinϕ will be abbreviated C and S, respectively. We will work through the transformational properties of ψ_4 and ψ_5 to show that they transform *together* as Δ_g. To do this we note that the product of a symmetry operation, R, and a function such as $XY + YX$ is

$$R[XY + YX] = [(RX)(RY) + (RY)(RX)] \tag{5-50}$$

The order of X and Y must be maintained in each multiplication because the electron numbers are implicit in that order. That is, $X(1)\,Y(2) \neq Y(1)\,X(2)$. So let us operate on ψ_4 and ψ_5 with each operation of the point group:

$$E\psi_4 = XY + YX = \psi_4 \tag{5-51a}$$
$$E\psi_5 = XX - YY = \psi_5 \tag{5-51b}$$

$$\begin{aligned} C_\infty^\phi\,\psi_4 &= (\mathrm{C}X - \mathrm{S}Y)(\mathrm{S}X + \mathrm{C}Y) + (\mathrm{S}X + \mathrm{C}Y)(\mathrm{C}X - \mathrm{S}Y) \\ &= \mathrm{CS}XX + \mathrm{C}^2XY - \mathrm{S}^2YX - \mathrm{SC}YY + \mathrm{CS}XX - \mathrm{S}^2XY + \\ &\quad\ \mathrm{C}^2YX - \mathrm{CS}YY \\ &= 2\mathrm{CS}XX + (\mathrm{C}^2 - \mathrm{S}^2)XY + (\mathrm{C}^2 - \mathrm{S}^2)YX - 2\mathrm{CS}YY \\ &= (XY + YX)(\mathrm{C}^2 - \mathrm{S}^2) + (XX - YY)(2\mathrm{CS}) \\ &= \psi_4(\mathrm{C}^2 - \mathrm{S}^2) + \psi_5(2\mathrm{CS}) \end{aligned} \tag{5-52a}$$

$$\begin{aligned} C_\infty^\phi\,\psi_5 &= (XY + YX)(-2\mathrm{CS}) + (XX - YY)(\mathrm{C}^2 - \mathrm{S}^2) \\ &= \psi_4(-2\mathrm{CS}) + \psi_5(\mathrm{C}^2 - \mathrm{S}^2) \end{aligned} \tag{5-52b}$$

$$\sigma_v\psi_4 = -(XY + YX) = -\psi_4 \tag{5-53a}$$
$$\sigma_v\psi_5 = XX + YY = \psi_5 \tag{5-53b}$$

$$i\psi_4 = XY + YX = \psi_4 \tag{5-54a}$$
$$i\psi_5 = XX - YY = \psi_5 \tag{5-54b}$$

$$\begin{aligned} S_\infty^\phi\,\psi_4 &= (-\mathrm{C}X + \mathrm{S}Y)(-\mathrm{S}X - \mathrm{C}Y) + (-\mathrm{S}X - \mathrm{C}Y)(-\mathrm{C}X + \mathrm{S}Y) \\ &= \psi_4(\mathrm{C}^2 - \mathrm{S}^2) + \psi_5(2\mathrm{CS}) \end{aligned} \tag{5-55a}$$
$$S_\infty^\phi\,\psi_5 = \psi_4(-2\mathrm{CS}) + \psi_5(\mathrm{C}^2 - \mathrm{S}^2) \tag{5-55b}$$

$$C_2\psi_4 = -\,(XY + YX) = -\psi_4 \tag{5-56a}$$
$$C_2\psi_5 = XX + YY = \psi_5 \tag{5-56b}$$

Equations 5-52 and 5-55 show that ψ_4 and ψ_5 are not individually bases for any irreducible representation of $D_{\infty h}$ because they are mixed by the operations C_∞^ϕ and S_∞^ϕ. But *together* they transform as follows:

$$E\begin{pmatrix}\psi_4\\ \psi_5\end{pmatrix}=\begin{pmatrix}1 & 0\\ 0 & 1\end{pmatrix}\begin{pmatrix}\psi_4\\ \psi_5\end{pmatrix}\qquad \text{trace}=2 \tag{5-57}$$

$$C_\infty^\phi\begin{pmatrix}\psi_4\\ \psi_5\end{pmatrix}=\begin{pmatrix}C^2-S^2 & 2CS\\ -2CS & C^2-S^2\end{pmatrix}\begin{pmatrix}\psi_4\\ \psi_5\end{pmatrix}\qquad \begin{aligned}\text{trace}&=2(C^2-S^2)\\ &=2\cos 2\phi\end{aligned} \tag{5-58}$$

$$\sigma_v\begin{pmatrix}\psi_4\\ \psi_5\end{pmatrix}=\begin{pmatrix}-1 & 0\\ 0 & 1\end{pmatrix}\begin{pmatrix}\psi_4\\ \psi_5\end{pmatrix}\qquad \text{trace}=0 \tag{5-59}$$

$$i\begin{pmatrix}\psi_4\\ \psi_5\end{pmatrix}=\begin{pmatrix}1 & 0\\ 0 & 1\end{pmatrix}\begin{pmatrix}\psi_4\\ \psi_5\end{pmatrix}\qquad \text{trace}=2 \tag{5-60}$$

$$S_\infty^\phi\begin{pmatrix}\psi_4\\ \psi_5\end{pmatrix}=\begin{pmatrix}C^2-S^2 & 2CS\\ -2CS & C^2-S^2\end{pmatrix}\begin{pmatrix}\psi_4\\ \psi_5\end{pmatrix}\qquad \text{trace}=2\cos 2\phi \tag{5-61}$$

$$C_2\begin{pmatrix}\psi_4\\ \psi_5\end{pmatrix}=\begin{pmatrix}-1 & 0\\ 0 & 1\end{pmatrix}\begin{pmatrix}\psi_4\\ \psi_5\end{pmatrix}\qquad \text{trace}=0 \tag{5-62}$$

You should now be very impressed because these are precisely the characters of the Δ_g representation of $D_{\infty h}$!

We have just taken great pains to show that ψ_4 and ψ_5 transform as Δ_g and we have already established that the spin multiplicity of each function is unity. Therefore, these two functions define a ${}^1\Delta_g$ electronic state. In the same manner, you can show that ψ_1 has all of the proper characters for the Σ_g^- irreducible representation and that ψ_6 transforms as Σ_g^+. We went through all this work to show explicitly that a term such as ${}^1\Delta_g$ is not an abstract device of group theory but actually describes the symmetry of two molecular wave functions. So when you come across terms like ${}^1B_{2u}$ and ${}^3E_{1u}$, you will realize that there are actually wave functions with these symmetries.

Problems

5-19. Show that ψ_6 (eq. 5-49) and ψ_1 (eqs. 5-47) transform as Σ_g^+ and Σ_g^-, respectively.

5-20. Show how the angular momenta of two electrons, one in a π orbital and one in a δ orbital, can add to give Π and Φ states. Using angular momentum, show what states are produced by the configuration $(\pi)^3$.

5-21. What molecular state of O_2^+ is produced by the lowest energy ionization of O_2 ($^3\Sigma_g^-$)? Of O_2 ($^1\Delta_g$)? The first *IE* of O_2 ($^3\Sigma_g^-$) is observed at 12.08 eV (D.W. Turner and D.P. May, *J. Chem. Phys.*, *45*, 471 [1966]). Using the data in Table 5-5, at what energy should the first ionization of O_2 ($^1\Delta_g$) occur? It is observed at 11.09 eV (N. Jonathan, D.J. Smith, and K.J. Ross, *J. Chem. Phys.*, *53*, 3758 [1970]). Will the states of O_2^+ produced by the second ionization of O_2 ($^3\Sigma_g^-$) and O_2 ($^1\Delta_g$) be the same?

D. *Spin-Orbit Splitting in the Photoelectron Spectra of the Dihalogens.* In Section 5-5-C we found that Σ, Π, Δ, Φ ... states of a diatomic molecule possess 0, 1, 2, 3... units ($= \hbar$) of angular momentum about the internuclear axis. In Section 4-2-C we found that the energy of an atom depends on its total (orbital plus spin) angular momentum. The same is true of molecules, and for linear molecules the determination of total angular momentum is particularly easy. For a $^2\Pi$ state, for example, $\Lambda = 1$ and $S = 1/2$. The total angular momentum can be $\Lambda + S, \Lambda + S - 1, \Lambda + S - 2 \ldots \Lambda - S$, which is just 3/2 or 1/2 in this case. A $^2\Pi$ state therefore splits into two levels, $^2\Pi_{3/2}$ and $^2\Pi_{1/2}$, when spin-orbit coupling is significant.

The photoelectron spectra of the diatomic halogen molecules nicely illustrate spin-orbit splitting. The configuration of F_2 (Fig. 4-18, Scheme II) is ... $(\pi_u)^4 (\pi_g)^4$. The lowest ionic state has the configuration ... $(\pi_u)^4 (\pi_g)^3$, which defines $^2\Pi_{g,3/2}$ and $^2\Pi_{g,1/2}$ levels. At an adiabatic *IE* of 15.70 eV we observe the spectrum of F_2^+ in which each vibrational transition is split into two components (Fig. 5-18[a]). The vibrational splitting is 1050 ± 40 cm^{-1} and the spin-orbit splitting is 337 ± 40 cm^{-1}. In the corresponding spectrum of Br_2^+ at 10.51 eV (Fig. 5-18[b]), the vibrational splitting is 360 cm^{-1} and the spin-orbit splitting is 2820 cm^{-1}, giving rise to two separate regions corresponding to the $^2\Pi_{g,3/2}$ and $^2\Pi_{g,1/2}$ ionic states. The data for the halogens are summarized in Table 5-6.

Table 5-6. Vibrational and Spin-Orbit Constants for Dihalogen Molecules[a]

	F_2	F_2^+	Cl_2	Cl_2^+	Br_2	Br_2^+	I_2	I_2^+
r_e (Å)	1.435	1.326	1.988	1.892	2.283	~2.2	2.666	—
$\bar{\omega}_e$ (cm^{-1})	892.1	1054.5	564.9	645.6	323.2	~360	214.6	~220
spin-orbit splitting (cm)$^{-1}$	—	337	—	645	—	2820	—	5125

[a] Data refer to the ground electronic states: $^1\Sigma_g^+$ for X_2 and $^2\Pi_g$ for X_2^+. Taken from A.B. Conford, D.C. Frost, C.A. McDowell, J.L. Ragle, and I.A. Stenhouse, *J. Chem. Phys.*, *54*, 2651 (1971).

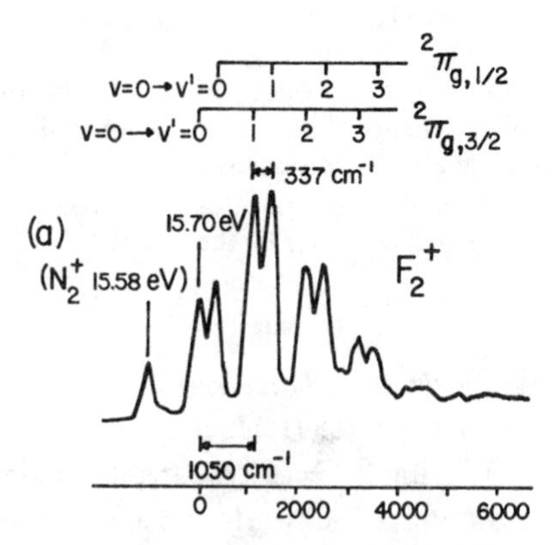

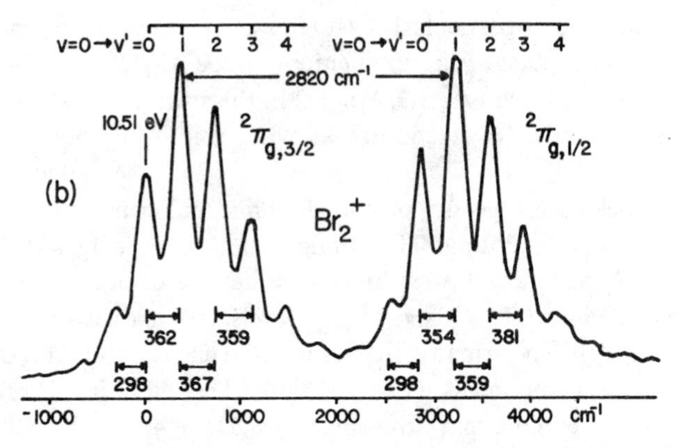

Fig. 5-18. He I photoelectron spectra of (a) F_2 and (b) Br_2, showing fine structure of the lowest ionic states. From A.B. Conford, D.C. Frost, C.A. McDowell, J.L. Ragle, and I.A. Stenhouse, *J. Chem. Phys.*, *54*, 2651 (1971).

Problems

5-22. Explain why r_e is smaller, and $\overline{\omega}_e$ greater, for $X_2{}^+$ than for X_2 in Table 5-6.

5-23. Based on Fig. 3-17, what is the ratio of Br_2 in the $v = 1$ state to Br_2 in the $v = 0$ state in the electronic ground state? To what transition would you assign the band 298 cm^{-1} below the 0-0 transition in Fig. 5-18(b)?

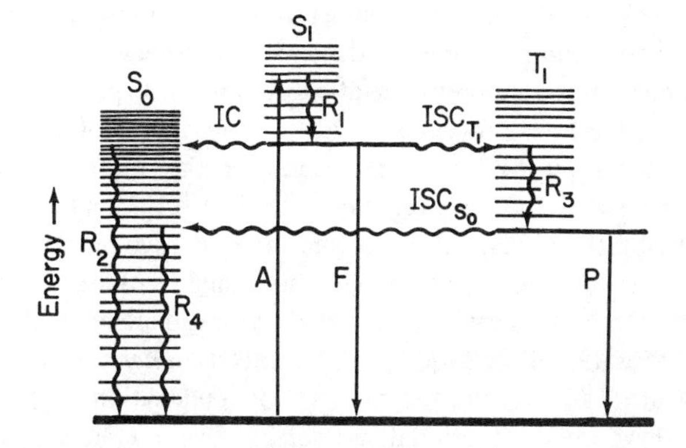

Fig. 5-19. Jablonski diagram illustrating possible electronic processes following absorption of a photon. Key: *A*-absorption, *F*-fluorescence, *P*-phosphorescence, *IC*-internal conversion, *ISC*-intersystem crossing, *R*-vibrational relaxation. S_0 is the singlet ground state and S_1 and T_1 are the lowest singlet and triplet excited states. Straight arrows represent processes involving photons and wiggly arrows represent processes that do not involve photons.

5-6. The Fate of Absorbed Energy

Once a molecule has absorbed a photon and finds itself in an excited electronic state, the excess energy can be eliminated, and the molecule returned to the ground state, by both radiative (emission of light) and nonradiative processes. Radiative processes include *fluorescence* and *phosphorescence* which will be defined shortly. Nonradiative processes include (a) relaxation to the ground state with dissipation of heat through molecular collisions; (b) thermal relaxation to states of intermediate energy from which photons may be emitted to complete the journey back to the ground state, and (c) the use of the absorbed energy to promote chemical reactions. Of these possibilities, the promotion of a chemical reaction (photochemistry) is an enormous field in itself and is not properly the subject of this book.

The processes of radiative and nonradiative decay are conveniently illustrated on the Jablonski diagram, Fig. 5-19. In this figure straight arrows represent processes involving photons and wiggly arrows represent thermal processes. The heavy horizontal line at the bottom of the diagram represents the energy of the ground vibrational state of the ground electronic state, S_0, taken as a singlet. S_1 and T_1 are the lowest excited singlet and triplet states. Since S_1 and T_1 are from the same configuration, T_1 is lower than S_1 in energy. This diagram treats the single most common case in which the ground state is a singlet and absorption (A) produces some singlet excited state. In general, the excited molecule will be both electronically and vibrationally excited after photon absorption. The process labelled R_1, the first to occur after absorption, is loss of vibrational energy through collisions with other molecules, heating the medium as a whole. Such *vibrational relaxation* requires only 10^{-11}–10^{-9} s to occur. From the ground vibrational state of S_1, several nonradiative processes are possible. Isoenergetic *internal conversion* (IC) to a highly excited vibrational state of S_0 would be followed by very rapid vibrational relaxation (R_2) to the ground vibrational state of S_0. Alternatively, *intersystem crossing* to the triplet state (ISC_{T_1}), followed by vibrational relaxation (R_3), leaves a molecule in the ground vibrational state of T_1. From T_1 it is possible to have intersystem crossing to the singlet state S_0 (ISC_{S_0}) followed by vibrational relaxation (R_4) to the ground vibrational state of S_0.

Although Fig. 5-19 only illustrates the case in which absorption populates S_1, some higher state, say S_2, could just as well have been populated. In nearly all cases, the most rapid process following excitation to any state higher than S_1 is nonradiative internal conversion down to S_1.† We can visualize such a process with the aid of Fig. 5-20 which shows some selected vibrational wave functions of the states S_0, S_1, and S_2. The vibrational wave function of a molecule in the $v = 0$ level of S_2 might overlap the $v = 3$ wave function of S_1 as drawn in Fig. 5-20. The overlap of the two wave functions ($\int \psi_{v'} \psi_v d\tau_n$) is just the Franck-Condon factor which governs the probability of changing from one state to the other by both radiative and nonradiative processes. Further vibrational relaxation to the $v = 0$ level of S_1 might then be followed by internal conversion from S_1 to S_0. As drawn in Fig. 5-20, there is not much overlap between the vibrational wave functions of S_1 and S_0; so this internal conversion process might be relatively slow (improbable). The slower the rate of internal conversion, the greater the likelihood that a competing process, such as intersystem crossing (ISC_{T_1}), can occur. One can add potential wells for the triplet states to Fig. 5-20 to explain intersystem crossing in the same manner.

We will continue our discussion of the Jablonski diagram but must first establish a vocabulary of chemical kinetics. For a unimolecular process

†The most notable exception is azulene, whose emission is derived entirely from S_2 (M. Beer and H.C. Longuet-Higgins, *J. Chem. Phys.*, *23*, 1390 [1955]; G. Viswanath and M. Kasha, *J. Chem. Phys.*, *24*, 574 [1956]). Another interesting case is that of pyrene in which emission is observed from both S_1 and S_2 (H. Baba, A. Nakajima, M. Aoi, and K. Chihara, *J. Chem. Phys.*, *55*, 2433 [1971]).

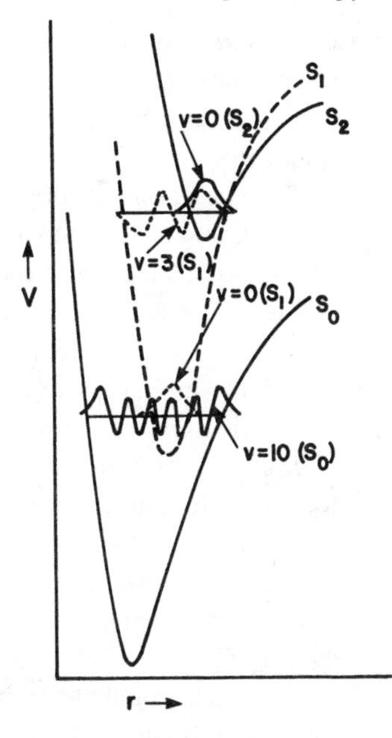

Fig. 5-20. Potential wells for the states S_0, S_1, and S_2 showing selected vibrational wave functions. There is a large overlap between the $v = 0$ (S_2) and $v = 3$ (S_1) wave functions so the probability of internal conversion $(S_2 \rightarrow S_1)$ is great. The $v = 0$ (S_1) and $v = 10$ (S_0) wave functions have much less overlap, so the probability of internal conversion from S_1 to S_0 is much less than from S_2 to S_1.

whose rate depends only on the concentration of some molecule, say A, the chemical reaction

$$A \xrightarrow{k} \text{products} \tag{5-63}$$

in which k is a rate constant, will have the rate law

$$-\frac{d[A]}{dt} = k[A] \tag{5-64}$$

whose solution has the form

$$[A]_t = [A]_0 e^{-kt} = [A]_0 e^{-t/\tau} \tag{5-65}$$

where $[A]_t$ is the concentration of A at time t, $[A]_0$ is the concentration of A at time $t = 0$, and τ, the *lifetime* for the reaction, is defined as $1/k$. Physically, τ is the time needed for the concentration of A to decrease to $1/e$ times its value at time $t = 0$. The processes of internal conversion, intersystem crossing, fluorescence and phosphorescence are all unimolecular; so we can speak of rate constants and lifetimes for these processes with the meanings defined in this paragraph.†

Getting back to the Jablonski diagram, Fig. 5-19, *fluorescence* (F) is radiative emission involving no change of spin multiplicity ($S_1 \rightarrow S_0$), and *phosphorescence* (P) is radiative emission accompanied by a change of spin multiplicity ($T_1 \rightarrow S_0$). It should be clear from the Jablonski diagram that the relative importance of fluorescence and phosphorescence will be determined by the relative rates of F and P compared to the rates of IC, ISC_{T_1}, and ISC_{S_0}. Fluorescence lifetimes (τ_f) are typically very short (10^{-4}–10^{-8} s). The lifetime for IC is quite variable and may be longer or shorter than τ_f. The rate of ISC_{T_1} is also variable, but can be greatly enhanced by the introduction of paramagnetic or heavy atom species which facilitate "spin-forbidden" processes. Fluorescence is usually decreased by the presence of O_2 which enhances ISC_{T_1}.

Since phosphorescence is a spin-forbidden process, its rate is relatively slow and its lifetime relatively long (10^{-4}–10^{2} s). In order for a molecule which has arrived in the T_1 manifold to phosphoresce, ISC_{S_0} and other completing processes must be relatively slow. Some of these competing processes include (a) collisions of two triplets to give two singlets in a spin-allowed process and (b) quenching of the triplet by impurities. At low temperature, in an environment where molecular diffusion is greatly reduced, many organic compounds do exhibit phosphorescence. Such an environment is called a matrix and some common matrix isolation procedures include the following: (a) Freezing a dilute solution in a solvent which forms a glass at low temperature, typically 77 K (liquid nitrogen temperature). (b) Isolating the molecule (called the guest) in a crystal of its completely deuterated analog (known as the host). Such a system is called an ideal mixed crystal because the guest is in its own crystalline environment but is energetically independent of that environment. The S_1 and T_1 states of the guest are lower than those of the host so these states are called energy traps. At low temperature, all emission is from the guest. (c) Freezing a dilute solution of the molecule of interest in a heavy noble gas (Ar-Xe). This creates a known crystalline environment which enhances phosphorescence.

Although we commonly distinguish fluorescence from phosphorescence in organic compounds, which are composed of relatively light atoms, we

† To see how the absorption spectrum may be used to calculate the emission lifetime, see J.B. Birks and D.J. Dyson, *Proc. Roy. Soc.* (London) *A 275*, 135 (1963); and S.J. Strickler and R.A. Berg, *J. Chem. Phys.*, *37*, 814 (1962).

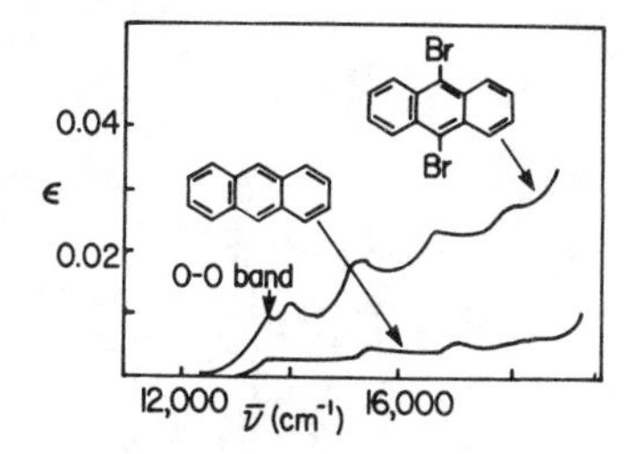

Fig. 5-21. Absorption spectra in the region of the lowest singlet-triplet transitions of anthracene and 9, 10-dibromoanthracene in CS_2. From S.P. McGlynn, T. Azumi, and M. Kasha, *J. Chem. Phys.*, *40*, 507 (1964).

often refer to emission from molecules with heavier atoms, such as transition metal complexes, as "luminescence." This purposefully ambiguous term is used because the validity of spin as a separate wave function begins to break down and hence the classification of processes as "spin-allowed" or "spin-forbidden" begins to lose meaning. As one illustration of the effect of a heavy atom on a spin-forbidden process, in Fig. 5-21 we see that substituting two Br atoms onto anthracene increases the intensity of the lowest singlet-triplet absorption band. Similar effects can be obtained by merely dissolving a compound with light atoms in a solvent with heavy atoms.

In Fig. 5-22 we see that the $S_0 \rightarrow S_1$ absorption and $S_1 \rightarrow S_0$ emission spectra of N-methylcarbazole bear an approximate mirror image relationship. We also notice that the 0-0 bands of absorption and emission do not precisely coincide. We shall now offer explanations of both of these observations. The mirror image relationship can be understood using Fig. 5-23 in which we see that emission from the $v = 0$ level of S_1 to various vibrational levels of S_0 must have less energy than absorptions from the $v = 0$ level of S_0 to various vibrational levels of S_1. Furthermore, since the spacings between vibrational levels in S_0 and S_1 are roughly the same, the absorption and emission spectra will have similar spacings between vibronic transitions. The reason that the 0-0 bands of absorption and emission do not precisely coincide is that immediately after absorption an excited state is produced with the same arrangement of atoms in the absorbing molecule and the same arrangement of solvent molecules about the absorbing molecule found in the ground state. After some short time, but before emission occurs, the excited molecule and nearby solvent molecules rearrange to accommodate the excited state electron distribution. Such rearrangement can include a change in the geometry of the excited molecule and consider-

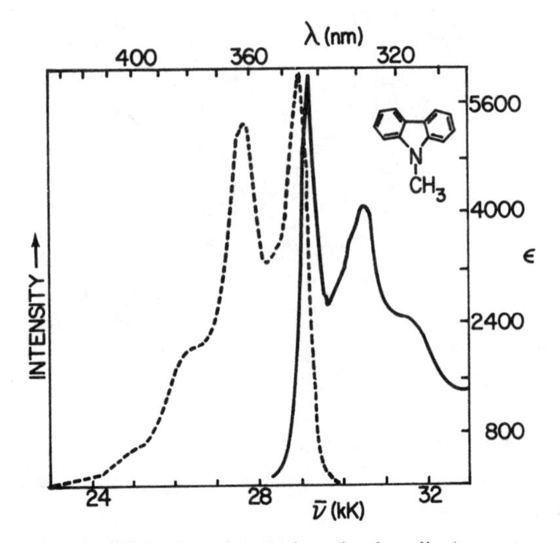

Fig. 5-22. Absorption (solid line) and emission (broken line) spectra of N-methylcarbazole in cyclohexane solution illustrating the approximate mirror image relationship between absorption and emission. From. I.B. Berlman, *Handbook of Fluorescence Spectra of Aromatic Molecules*, Academic Press, N.Y., 1971.

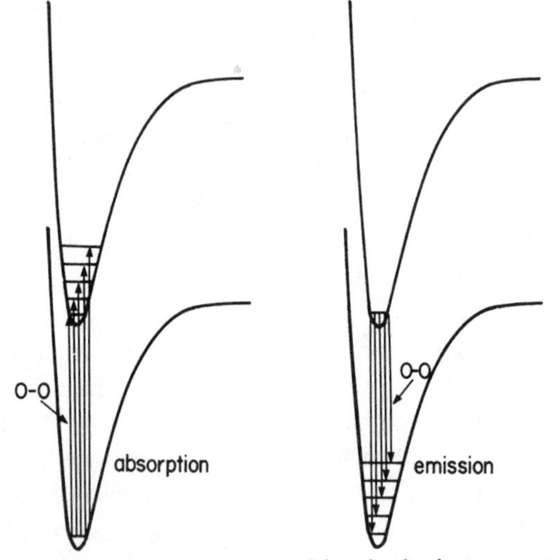

Fig. 5-23. Diagram showing that the 0-0 transition is the lowest energy vibronic absorption and the highest energy vibronic emission. The absorption and emission spectra will bear a rough mirror image relationship if the spacings between vibrational states in each electronic state are similar and if the transition probabilities are similar.

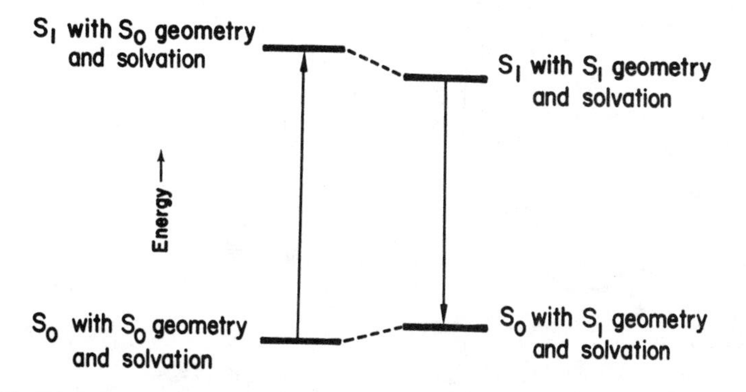

Fig. 5-24. Diagram showing why the 0-0 band of emission is at slightly lower energy than the 0-0 band of absorption.

able reorientation of the solvent molecules. This rearrangement lowers the energy of the excited molecule slightly and raises the energy of the ground state to which that molecule can decay by emission. The ground state energy is increased because immediately after emission there will be a ground state molecule with an excited state geometry and solvent sphere. Hence the 0-0 band of absorption will generally be at slightly higher energy than the 0-0 band of emission (Fig. 5-24). That the properties of different electronic states can be very different is illustrated by the molecule 4-amino-4′-nitrobiphenyl. The dipole moment of the S_0 state is 6.4 D and the dipole moment of the S_1 state is 18.0 D.† Such a substantial difference is bound to alter the solvent sphere markedly.

H_2N–C₆H₄–C₆H₄–NO_2 4-amino-4′-nitrobiphenyl
$\mu(S_0) = 6.4$ D $\quad \mu(S_1) = 18.0$ D

A fairly complete illustration of the phenomena described in this section, the absorption and emission spectra of 1-chloronaphthalene, are shown in Fig. 5-25. A state diagram with rate constants for ISC_{T_1} , ISC_{S_0} IC, F, and P is shown in Fig. 5-26. In this instance 94% of the molecules decay

†The dipole moment of a molecule in an excited state may be measured by the *Stark effect*. In such an experiment the sample is placed in an external electric field. The variation in the spectrum as a function of the applied field depends on the dipole moment of the molecule. See, for example, D.E. Freeman, W. Klemperer, and J.R. Lombardi, *J. Chem. Phys.*, *40*, 604 (1964).

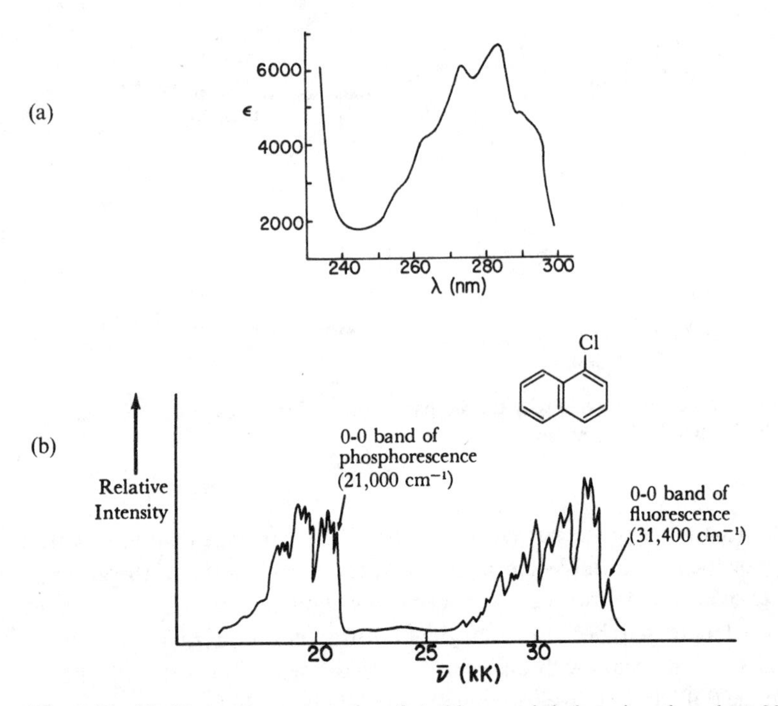

Fig. 5-25. (a) Absorption spectrum of 1-chloronaphthalene in ethanol at 298 K. (From J. Ferguson, *J. Chem. Soc.*, *304* (1954).) (b) Emission spectrum of 1-chloronaphthalene at 77 K in the mixed solvent "EPA" (5:5:2 ether: isopentane: ethanol by volume). EPA is used for many low temperature studies because it forms a glass instead of a fractured crystalline solid which is opaque. (From N.J. Turro, *Molecular Photochemistry*, W.A. Benjamin, N.Y., 1967.)

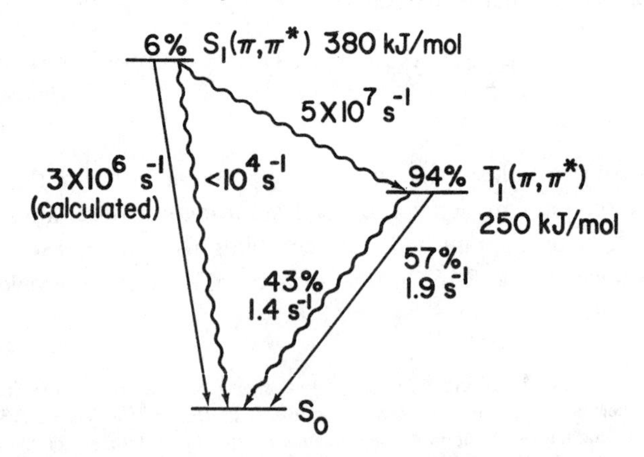

Fig. 5-26. State diagram for 1-chloronaphthalene at 77 K. Data from N.J. Turro, *Molecular Photochemistry*, W.A. Benjamin, N.Y., 1967.

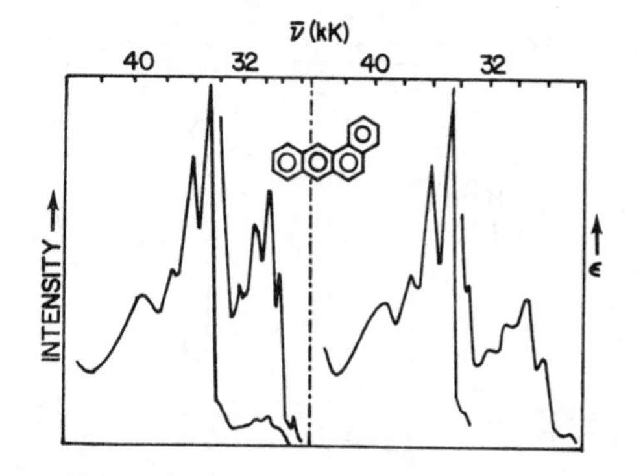

Fig. 5-27. Fluorescence excitation spectrum (left) and absorption spectrum (right) of 1,2-benzanthracene in ethanol. Fluorescence was monitored at ~ 24 kK. From C.A. Parker, *Nature*, *182*, 1002 (1958).

through T_1† before reaching S_0 and phosphorescence greatly predominates over fluorescence. Intersystem crossing from T_1 to S_0, however, competes nearly equally with phosphorescence.

It is possible to determine the absorption spectrum of a compound by monitoring its emission at a fixed wavelength while irradiating it at variable wavelengths. Such a spectrum is called an *excitation spectrum*. For example, in Fig. 5-27 we see the excitation and absorption spectrum of 1,2-benzanthracene. The excitation spectrum is a plot of emission intensity at ~426 nm *vs.* irradiation frequency. The excitation spectrum is very similar to the absorption spectrum.

Although we are generally avoiding the topic of photochemistry in this book, one particularly important and interesting aspect of photochemistry will be mentioned. This is the use of laser initiated photochemistry for isotopic separation. In Fig. 5-28 is seen part of the visible absorption spectrum of ICl. The electronic transition is from the ground state ($^1\Sigma^+$) to the lowest triplet state ($^3\Pi_1$). As an example of the notation in the figure, the numbers 19,0 marking the band head at 6016.402 Å (16,621.23 cm^{-1}) indicate that $I^{37}Cl$ is excited from $^1\Sigma^+$ ($v = 0$) to $^3\Pi_1$ ($v = 19$). The unlabelled structure in the spectrum (which is not all noise) is due to *individual rotational transitions*. In the experiment we will discuss,†† the $I^{37}Cl$ 18,0 transition was preferentially excited with a tunable dye laser

† It has been possible to measure the *absorption spectrum* of T_1 of several molecules by irradiating a sample at high laser power and then measuring the absorbance of the sample at a particular wavelength ~10^{-8} s after irradiation. See, for example, D.V. Bent and E. Hayon, *J. Amer. Chem. Soc.*, *97*, 2599, 2606, 2612 (1975).

†† D.D.-S. Liu, S. Datta, and R.N. Zare, *J. Amer. Chem. Soc.*, *97*, 2557 (1975).

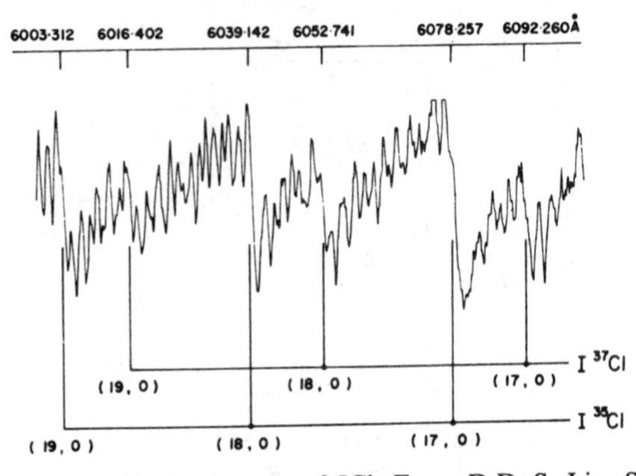

Fig. 5-28. Visible absorption spectrum of ICl. From D.D.-S. Liu, S. Datta, and R.N. Zare, *J. Amer. Chem. Soc.*, *97*, 2557 (1975).

(λ = 6053 Å, bandwidth = 3 Å, power = 10mW). The excited molecule then reacted as follows:

$$\text{I*Cl} + \begin{matrix} \text{Cl} & & \text{H} \\ & \text{C=C} & \\ \text{H} & & \text{Cl} \end{matrix} \xrightarrow{h\nu} \text{ICl} + \begin{matrix} \text{Cl} & & \text{*Cl} \\ & \text{C=C} & \\ \text{H} & & \text{H} \end{matrix} \qquad (5\text{-}66)$$

In eq. 5-66 *Cl denotes the Cl atom initially in ICl. Reaction 5-66 does not occur without photoactivation and since $I^{37}Cl$ is preferentially excited by the laser, the product *cis*-dichloroethylene is *enriched in* ^{37}Cl. In one reaction the dichloroethylene $^{37}Cl/^{35}Cl$ ratio changed from its natural value of 0.332 to a final value of 0.366. Such isotopic enrichment by laser initiated photochemistry promises to produce lower cost isotopically enriched compounds, as well as cheaper ^{235}U for nuclear reactors. In the latter case, preferential laser ionization of ^{235}U atoms, followed by removal of the ions by an electric field, appears promising.

To close this section we mention the culturally interesting optical brighteners, such as derivatives of bistriazyldiaminostilbenedisulphonic acid:

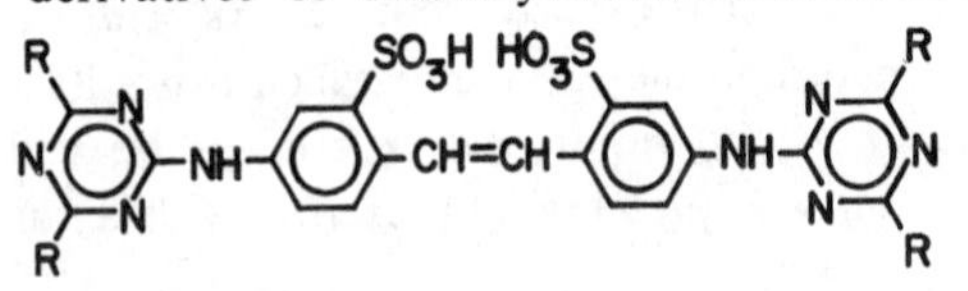

These compounds produce white fluorescence and are added to white fabrics to give them a bright white appearance. Most new, men's T-shirts glow in the dark under near uv irradiation because of these additives. The whiteness of such fabrics has little to do with how clean the material is.

Problem

5-24. The amount of energy carried by one mole of photons of a particular wavelength is an *einstein* of energy. The *quantum yield* for any photochemical process is defined as

$$\Phi = \frac{\text{the number of molecules undergoing the process}}{\text{the number of photons absorbed by the system}}$$

Let's consider the possibility of an electronically excited molecule losing its extra energy in three ways:

a. Emission of a photon.
b. Energy transfer to an acceptor, A, to produce an excited state of the acceptor, A^*. This is called *quenching* by the acceptor.
c. Collisional deactivation to produce heat.

These processes and their rate laws are shown in the table below:

Process	Equation	Rate Law
absorption	$D + h\nu \rightarrow D^*$	$-d[D]/dt = I_a$(einsteins liter^{-1}s^{-1})
emission	$D^* \rightarrow D + h\nu$	$-d[D^*]/dt = k_1[D^*]$
quenching	$D^* + A \rightarrow A^* + D$	$-d[D^*]/dt = k_2[D^*][A]$
deactivation	$D^* \rightarrow D +$ heat	$-d[D^*]/dt = k_3[D^*]$

I_a is the rate of light absorption (= moles of D per liter per second absorbing light). Under steady illumination, the system will reach a steady state in which the rate of formation of D^* will equal the rate of disappearance of D^*.

(a) Write an expression for the net rate of production of D^*, $d[D^*]/dt$, and set this equal to zero for the steady state. You will be able to solve this equation for I_a in terms of the rate constants and the concentrations of A and D^*. (b) The quantum yield for emission from D^* *in the absence of A* is just

$$\Phi_0 = k_1[D^*]/I_a$$

Using your expression for I_a from part (a), set $[A] = 0$ and obtain an expression for Φ_0 in terms of the various rate constants. (c) The quantum yield for emission with A present is

$$\Phi_A = k_1[D^*]/I_a$$

where $[A] \neq 0$ in the expression for I_a. Write an expression for Φ_A. (d) Φ_0 and Φ_A can be determined experimentally. Write an expression for the ratio Φ_0/Φ_A and reduce this to the form $\Phi_0/\Phi_A = 1 + K[A]$, where K is a composite of several constants. This is called the Stern-Volmer expression. A plot of Φ_0/Φ_A will be linear with a slope of K if bimolecular quenching occurs.

5-25. A state diagram for benzophenone is shown below. Essentially 100% of emission is from T_1 under the conditions of this problem. Call the T_1 state D^* of Problem 5-24. (a) What are the values of k_1 and k_3 defined in Problem 5-24? (b) What is the value of Φ_0? (c) A Stern-Volmer plot for some quencher, A, has a slope of 5×10^2 M^{-1}. What is the value of k_2?

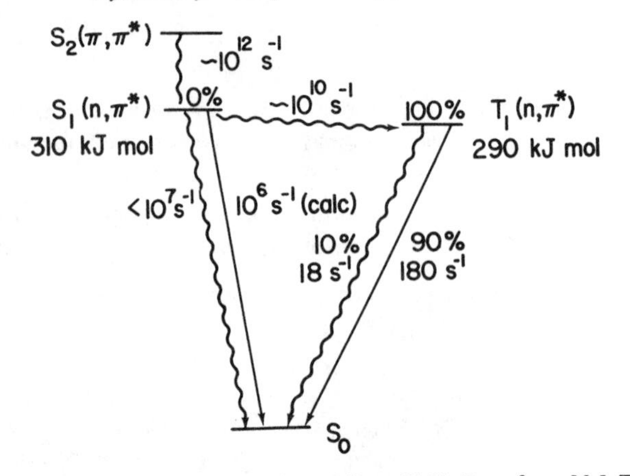

State diagram for benzophenone, $(C_6H_5)_2CO$, at 77 K. Data from N.J. Turro, *Molecular Photochemistry*, W.A. Benjamin, N.Y., 1967.

5-7. Single Bonds, Double Bonds, and Lone Pairs

A. *Saturated Compounds.* Compounds possessing only single bonds, with no mutiple bonds or lone pairs, exhibit $\sigma \rightarrow \sigma^*$ electronic transitions at energies much higher than we can ordinarily observe. At such high energies (> 50 kK) we encounter another class of transitions called *Rydberg transitions*. Consider the qualitative mo scheme for methane in Fig. 5-29. The energies of the σ^* orbitals are calculated to be very close to each other. But at energies comparable to those of the σ^* orbitals are the higher-than-valence-shell atomic orbitals. Transitions to these energy levels, called Rydberg transitions, are possible and make the assignment of the spectrum of methane, Fig. 5-30, ambiguous. It is not known whether the spectrum arises from $\sigma \rightarrow \sigma^*$ transitions, Rydberg transitions, or both.† One would think that a calculation could answer this question for so simple a molecule as CH_4, but different kinds of calculations are, thus far, in disagreement.

In saturated compounds possessing lone pairs, the lone pair (nonbonding) energy level generally falls between those of the sigma bonding and sigma antibonding orbitals. The spectrum of CH_3Cl (Fig. 5-31) exhibits a weak transition near 58 kK ($\epsilon < 1000$) assigned as $n(Cl) \rightarrow \sigma^*$, where the nonbonding electrons originate on the Cl atom. The next transition, with a maximum near 64 kK, is assigned to the Rydberg transition, $n(Cl) \rightarrow 4s(Cl)$.

† For a discussion see M.B. Robin, *Higher Excited States of Polyatomic Molecules*, Vol. I., Academic Press, N.Y. 1974.

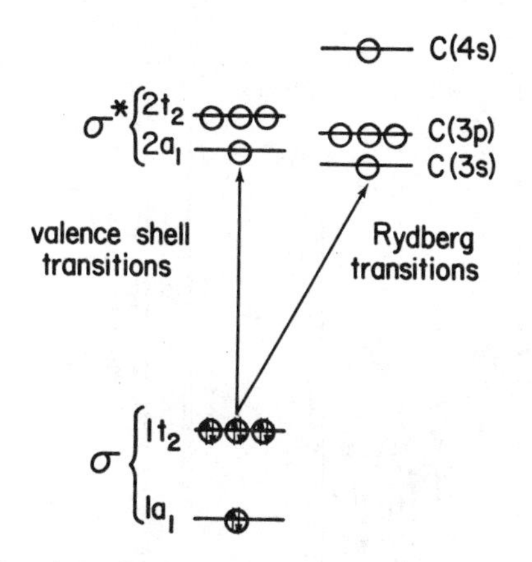

Fig. 5-29. Valence shell and Rydberg transitions of CH_4.

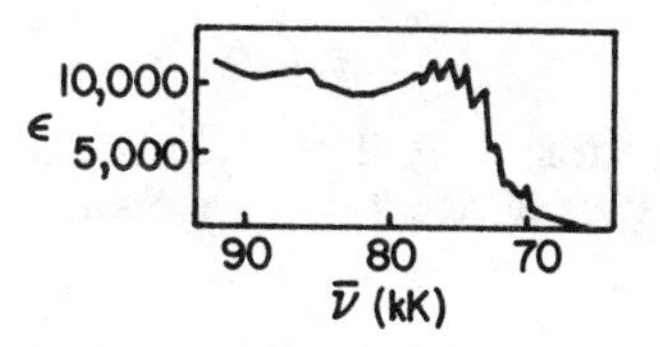

Fig. 5-30. Vacuum uv spectrum of methane. From J.W. Raymonda and W.T. Simpson, *J. Chem. Phys.*, *47*, 430 (1967).

A molecular orbital scheme for H_2O, showing all valence orbitals, is given in Fig. 5-32. The first $n \rightarrow \sigma^*$ transition is observed at about 60 kK (170 nm, $\epsilon \approx 1000$).† Water, methanol, and other simple alcohols and hydrocarbons are frequently used solvents for ultraviolet spectroscopy because they do not absorb strongly at wavelengths above about 210 nm.

B. *Double Bonds.* The pi orbitals of hydrocarbons generally fall between the energies of the σ and σ^* orbitals, so $\pi \rightarrow \pi^*$ transitions will be at lower energy than $\sigma \rightarrow \sigma^*$ transitions. The ground state configuration of ethylene†† (Section 4-6-C) is ... $(b_{1u})^2$ and the first excited configura-

†The spectrum can be found in M.B. Robin, *Higher Excited States of Polyatomic Molecules*, Vol. I., Academic Press, N.Y., 1974.

††The coordinate system used in this book for ethylene was given in Fig. 4-31. In most literature the coordinate system used is one in which the x and z axes in Fig. 4-31 are interchanged. In such a coordinate system the lowest excited state of ethylene is B_{1u} instead of B_{3u}. Our departure from the established coordinate system was accidental and was not changed to avoid introducing errors into the text. For references to the spectrum of ethylene, see G. Herzberg, *Electronic Spectra of Polyatomic Molecules*, Van Nostrand Reinhold, N.Y., 1966.

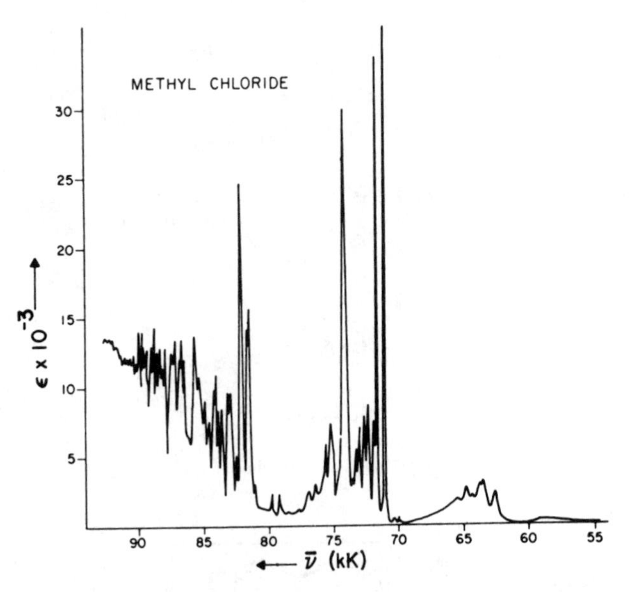

Fig. 5-31. Vacuum uv spectrum of CH_3Cl. Reproduced from J.W. Raymonda, L.O. Edwards, and B.R. Russell, *J. Amer. Chem. Soc.*, *96*, 1708 (1974).

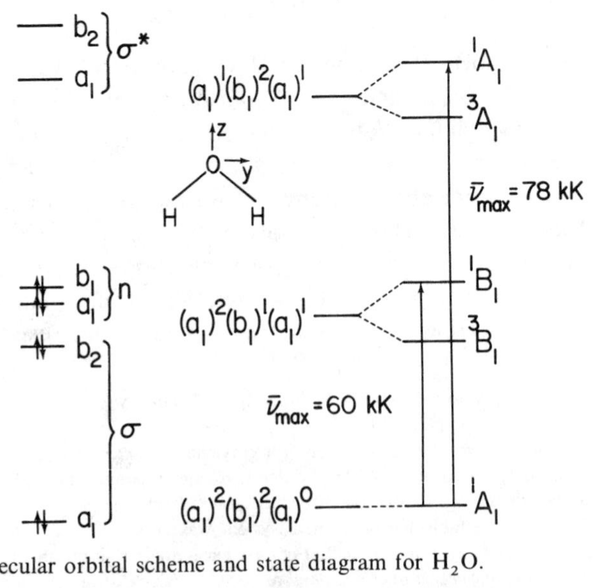

Fig. 5-32. Molecular orbital scheme and state diagram for H_2O.

tion is . . . $(b_{1u})^1 (b_{2g})^1$. This $\pi \rightarrow \pi^*$ electron promotion gives rise to $^1A_{1g} \rightarrow {}^3B_{3u}$ and $^1A_{1g} \rightarrow {}^1B_{3u}$ transitions. The former is manifested by a series of exceedingly weak vibronic transitions in the region 340-260 nm with the 0-0 band estimated to come at less than 29 kK. The spin-allowed transition gives rise to a strong band which rises to a maximum near 162 nm (68 kK). The 0-0 transition has been assigned in a region of very weak absorption at 40 kK. The spin pairing energy for this B_{3u} state (difference between singlet and triplet 0-0 energies) is therefore > 11 kK. Both the singlet and triplet excited states of ethylene feature a 90° rotation about the C–C bond, resulting in D_{2d} geometry. Occupation of the π^* orbital lowers the C=C stretching frequency from 1623 cm^{-1} in the ground state to 850 cm^{-1} in the $^1B_{3u}$ state.

Conjugation of double bonds increases the number of pi orbitals and decreases the spacing between orbitals, lowering the energy of the first absorption band. The pi energies of ethylene, butadiene, and hexatriene are compared in Fig. 5-33 and the energy of the first spin-allowed transition maximum is seen to decrease in a regular manner as the number of conjugated double bonds increases. The plant pigment β-carotene, with eleven conjugated double bonds, exhibits its lowest energy $\pi \rightarrow \pi^*$ transition near 460 nm (22 kK) with an enormous extinction coefficient of more than 10^5 M^{-1} cm^{-1} (Fig. 5-34).

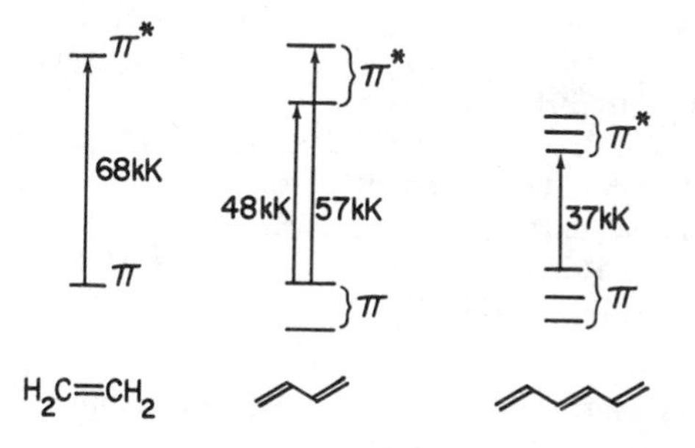

Fig. 5-33. Comparison of the pi energy levels of some simple alkenes showing that the energy of the first $\pi \rightarrow \pi^*$ transition decreases as the number of conjugated bonds increase.

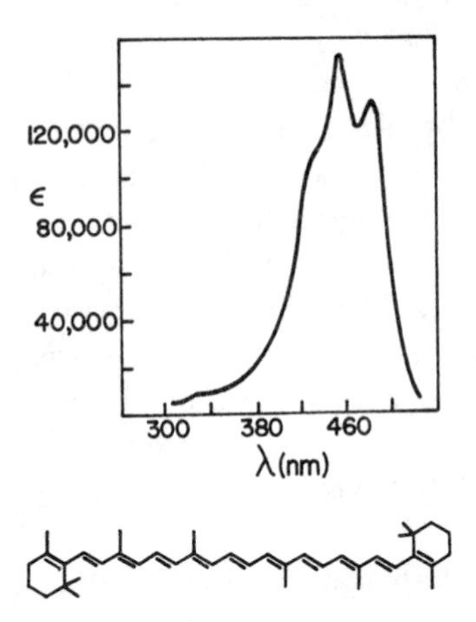

Fig. 5-34. Spectrum of all-*trans*-β-carotene. From H.H. Jaffé and M. Orchin, *Theory and Applications of Ultraviolet Spectroscopy*, John Wiley & Sons, N.Y., 1962.

Problems

5-26. To what excited states do the transitions $1t_2 \rightarrow 2a_1$ and $1t_2 \rightarrow 2t_2$ of CH_4 (Fig. 5-29) give rise? How many fully allowed $\sigma \rightarrow \sigma^*$ transitions should there be?

5-27. What are the ground and first excited electronic configurations of acetylene? To what electronic transitions do these configurations give rise? The lowest singlet-singlet transition of acetylene gives an absorption band in the region 240–210 nm with the 0-0 transition assigned at 42 kK. The singlet excited state is bent, with C_{2h} symmetry. Draw a picture showing the geometry of this excited state.

5-28. The spectra of conjugated polyenes, $H(CH{=}CH)_nH$, with $n = 2$–10 (excluding $n = 9$) exhibit the following lowest energy spin-allowed absorption maxima in hydrocarbon solvents: $n = 2$, 217 nm; $n = 3$, 268 nm; $n = 4$, 304 nm; $n = 5$, 334 nm; $n = 6$, 364 nm; $n = 7$, 390 nm; $n = 8$, 410 nm; $n = 10$, 447 nm. Make a graph of the transition energy (in kK) *vs.* n. (Sources: F. Sondheimer, D.A. Ben-Efriam, and R. Wolovsky, *J. Amer. Chem. Soc.*, *83*, 1675 (1961); G.F. Woods and L.H. Schwartzman, *J. Amer. Chem. Soc.*, *71*, 1396 (1949).

C. *Lone Pairs and Double Bonds.* An approximate mo diagram for formaldehyde† is shown in Fig. 5-35. The lowest energy transition, labelled a, is $n_p \rightarrow \pi^*$. The designation n_p refers to a lone pair localized approximately in the $O(2p_x)$ orbital. This electron promotion gives rise to $^1A_1 \rightarrow$

† For a review of the electronic structure and spectra of formaldehyde, see D.C. Moule and A.D. Walsh, *Chem. Rev.*, *75*, 67 (1975).

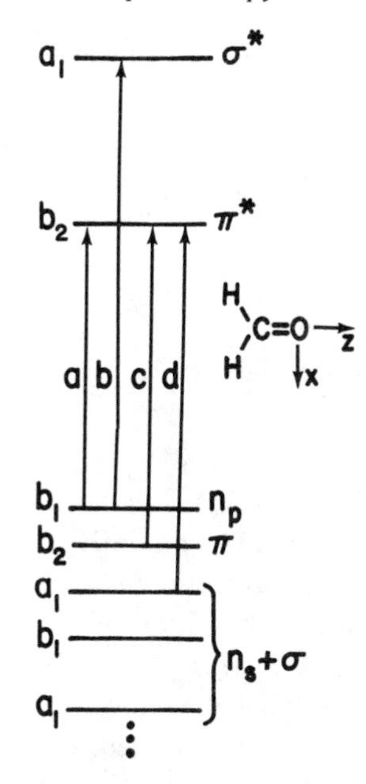

Fig. 5-35. Approximate energy level diagram of formaldehyde using the calculated energies of S. Aung, R.M. Pitzer, and S.I. Chan, *J. Chem. Phys.*, *45*, 3457 (1966).

$^3A_2$ and $^1A_1 \rightarrow {}^1A_2$ absorptions with 0-0 bands at 25.2 and 28.2 kK. respectively. Hence the approximate spin pairing energy of the A_2 excited state is only 3 kK. Both the $^1A_2$ and $^3A_2$ states of formaldehyde deviate from planarity by about 30°, with pyramidal C_s structures. The pyramidal structure leads to a double minimum potential well and considerable inversion doubling of the vibronic absorption bands (Section 3-6-F). On the basis of Fig. 5-35 one would expect that the second lowest transition would be $\pi \rightarrow \pi^*$. However, none of the transitions b, c, or d has been identified in the spectrum of formaldehyde. The next observed transition, at 57 kK, is assigned as the Rydberg transition, $n_{\text{p}} \rightarrow$ O(3s). In fact, a whole series of Rydberg transitions identified as $n_{\text{p}} \rightarrow$ 3s, 4s, 5s, . . . , 3p, 4p, 5p, . . . , 3d, 4d, *etc.*, has been identified for formaldehyde.

A weak ($\epsilon \approx 20$) absorption in the vicinity of 280 nm is common to aliphatic ketones. This is the spin-allowed, orbitally-forbidden $n \rightarrow \pi^*$ transition. Table 5-7 shows that the position of the $n \rightarrow \pi^*$ transition is quite dependent on the immediate environment of the carbonyl group. The $n \rightarrow \sigma^*$ and $\pi \rightarrow \pi^*$ transitions are at higher energy than we can ordinarily observe. Not only does λ_{max} vary with the structure of the carbonyl com-

Table 5-7. $n \rightarrow \pi^*$ Transitions of Simple Carbonyl Compounds[a]

Compound	λ_{max} (nm)	ϵ_{max}	Solvent
CH_3CHO	293.4	11.8	Hexane
CH_3CO_2H	204	41	Ethanol
$CH_3CO_2CH_2CH_3$	204	60	Water
CH_3CONH_2	214	--	Water
CH_3COCl	235	53	Hexane
CH_3COCH_3	279	14.8	Hexane

[a] Data from H.H. Jaffé and M. Orchin, *Theory and Applications of Ultraviolet Spectroscopy*, John Wiley & Sons, N.Y., 1962.

pound, but also with the solvent. For example, λ_{max} of the $n \rightarrow \pi^*$ transition of acetone is found at 264.5 nm in H_2O, 270 nm in methanol, 272 nm in ethanol, 277 nm in chloroform, and 279 nm in hexane. This variation is explained in terms of stabilization of the lone pair in the ground state by hydrogen bonding. The stronger the H–bond, the more stable is the "n" orbital and the greater the energy needed to promote an electron to the π^* orbital. Different kinds of electronic transitions are affected in characteristic ways by variations in solvent and pressure. *Solvent is a most important variable in any kind of spectroscopy.* To introduce some terminology, the shift of a band to higher energy is called a *hypsochromic* or *blue shift*. A shift to lower energy is a *bathochromic* or *red shift*. An increase in intensity is a *hyperchromic effect*, and a decrease is a *hypochromic effect*.

The extension of the ideas of the last few pages to more complex *chromophores* is not difficult if we are content with qualitative analysis of the low energy region of the spectrum. (The term chromophore refers to a molecule or part of a molecule principally responsible for a particular absorption band. An *auxochrome* is a part of the molecule which does not give rise to a spectrum of its own but which modifies the spectrum of the chromophore). Suppose we make a carbonyl group part of a conjugated pi system as follows:

$$\rangle C=\underset{|}{C}-\overset{|}{C}=\ddot{\underset{..}{O}}$$

The qualitative mo scheme is obtained by allowing the appropriate orbitals of the C=C and C=$\ddot{\underset{..}{O}}$ groups to interact, as shown in Fig. 5-36. The four pi orbitals will be qualitatively similar to those of butadiene ($H_2C=CH-CH=CH_2$) and the only new feature of the mo scheme is the addition of two oxygen lone pairs. We expect from Fig. 5-36 a low energy $n_p \rightarrow \pi_3{}^*$ transition and a higher energy $\pi_2 \rightarrow \pi_3{}^*$ transition. The spectrum of mesityl

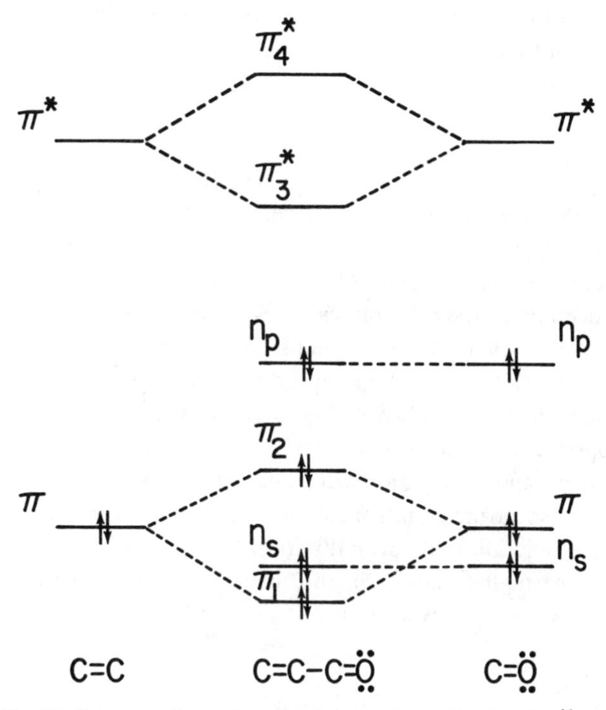

Fig. 5-36. Qualitative mo scheme for the chromophore C=C–C=Ö obtained by the interaction of the C=C and C=Ö groups.

oxide in hexane solvent, for example, exhibits the $n_p \rightarrow \pi_3{}^*$ absorption maximum at 327 nm (31 kK, ϵ = 98) and the $\pi_2 \rightarrow \pi_3{}^*$ maximum at 230 nm (44 kK, ϵ = 12,600).

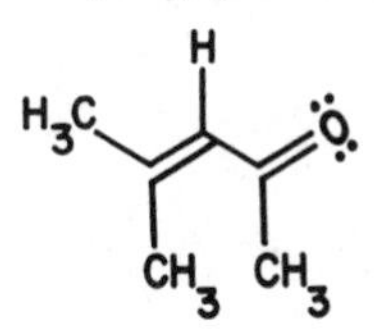

mesityl oxide
4-methyl-3-pentene-2-one

$n_p \rightarrow \pi_3{}^*$ 31 kK
$\pi_2 \rightarrow \pi_3{}^*$ 44 kK

As an instructive example, let's see if we can't "understand" the spectrum of nitromethane, CH_3NO_2, which exhibits two maxima at 270 nm (37 kK, $\epsilon \approx 20$) and 210 nm (48 kK, $\epsilon \approx 16{,}000$). The two resonance structures of a nitro group are as follows:

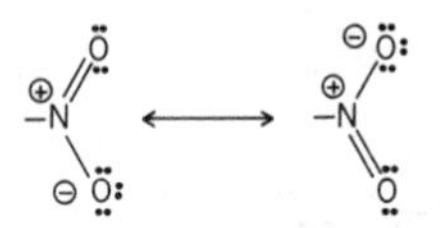

We expect that the highest energy occupied orbitals will involve the four pi electrons and two O(2p) lone pairs. (The two O(2s) lone pairs will be mixed with the sigma orbitals.) The pi orbitals will be qualitatively similar to the pi orbitals of the allyl radical in Section 4-7 and are shown in Fig. 5-37. The two in-plane O(2p) orbitals combine to form symmetric (n_+) and antisymmetric (n_-) nonbonding orbitals. The qualitative energy level scheme ought to be as shown in Fig. 5-38, in which the symmetry labels are appropriate to the local C_{2v} symmetry of the nitro group. The placement of the n_+ and n_- orbitals with respect to π_2 is not at all certain. But in view of the spectrum of nitromethane, we interpret the weak low energy band as the spin-allowed, orbitally-forbidden $n_- \rightarrow \pi_3{}^*$ ($^1A_1 \rightarrow {}^1A_2$) transition. The higher energy, intense band is probably the fully allowed $\pi_2 \rightarrow \pi_3{}^*$ ($^1A_1 \rightarrow {}^1A_1$) transition.

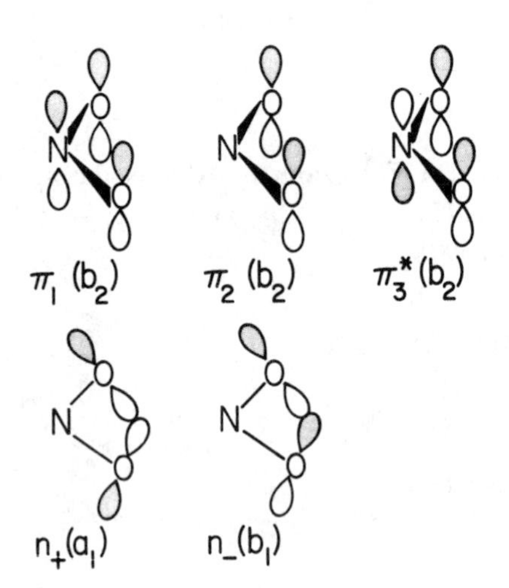

Fig. 5-37. Pi and O(2p) nonbonding orbitals of a nitro group. The atoms are in the *xz* plane.

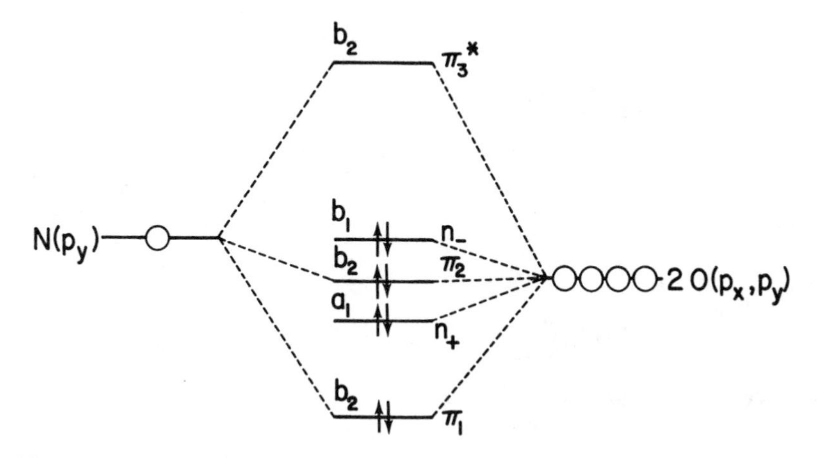

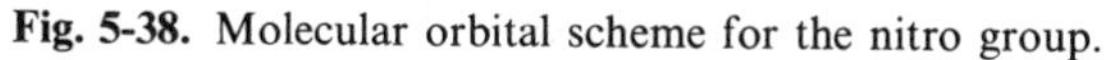
Fig. 5-38. Molecular orbital scheme for the nitro group.

Problems

5-29. Show how a polarization study or a vibronic analysis of the 210 nm band of nitromethane might be used to establish whether the transition is $\pi_2 \rightarrow \pi_3{}^*$ or $n_+ \rightarrow \pi_3{}^*$. (See Figs. 5-37 and 5-38.)

5-30. Propose a molecular orbital scheme for the pi and nonbonding orbitals of azomethane (*trans*-$CH_3\ddot{N}{=}\ddot{N}CH_3$) which exhibits a weak band at 347 nm (29 kK) and an intense band in the vacuum uv. How would you interpret this spectrum?

5-31. To what transitions would the arrow labelled "d" in Fig. 5-35 give rise? What is the polarization of the spin-allowed transition?

D. *Aromatic Compounds.* Your intuition should tell you, by now, that the lowest energy transitions of a benzene ring ought to be of the $\pi \rightarrow \pi^*$ variety. The high symmetry of a benzene ring adds some complexity to the possible low energy transitions. The mo scheme of benzene (Fig. 4-54) tells us that the highest occupied orbital is degenerate (e_{1g}) and so is the lowest unoccupied orbital (e_{2u}). The ground state of benzene is ${}^1A_{1g}$, and the lowest excited configuration defines the six states in Fig. 5-39. Since there are three singlet excited states, we expect three spin-allowed transitions to result from the lowest energy $\pi \rightarrow \pi^*$ electron promotion.

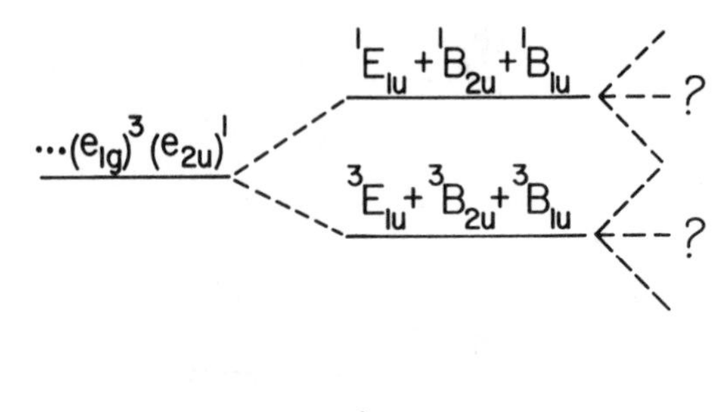

Fig. 5-39. Partial state diagram of benzene.

The orbital selection rules tell us that only the ${}^1A_{1g} \rightarrow {}^1E_{1u}$ transition is fully allowed:

$$A_{1g} \rightarrow E_{1u}: \quad A_{1g}(a_{1g}) \begin{pmatrix} a_{2u} \\ e_{1u} \end{pmatrix} E_{1u}(a_{1g}) = \begin{pmatrix} e_{1g} \\ \underline{\underline{a_{1g}}} + a_{2g} + e_{2g} \end{pmatrix} \tag{5-67}$$

$$A_{1g} \rightarrow B_{1u}: \quad A_{1g}(a_{1g}) \begin{pmatrix} a_{2u} \\ e_{1u} \end{pmatrix} B_{1u}(a_{1g}) = \begin{pmatrix} b_{2g} \\ e_{2g} \end{pmatrix} \tag{5-68}$$

$$A_{1g} \rightarrow B_{2u}: \quad A_{1g}(a_{1g}) \begin{pmatrix} a_{2u} \\ e_{1u} \end{pmatrix} B_{2u}(a_{1g}) = \begin{pmatrix} b_{1g} \\ e_{2g} \end{pmatrix} \tag{5-69}$$

A low resolution spectrum of benzene is shown in Fig. 5-40. It seems reasonable to assign the most intense absorption band near 180 nm to the fully allowed ${}^1A_{1g} \rightarrow {}^1E_{1u}$ transition. Without additional analysis, we can say that the two less intense bands near 260 and 200 nm are probably the two spin-allowed, orbitally-forbidden transitions ${}^1A_{1g} \rightarrow {}^1B_{1u}$ and ${}^1A_{1g} \rightarrow {}^1B_{2u}$. A basis for distinguishing which band belongs to which transition will be discussed in Section 5-8-B when we analyze the vibronic structure of the 260 nm band. The lowest energy absorption near 340 nm, with an ϵ value of about 10^{-3}, must be a singlet-triplet transition. We will analyze the ${}^3B_{1u}$ assignment of this band in detail in Section 5-8-C when we discuss the phosphorescence of benzene.

The 260 nm band of benzene is often called the benzenoid band. Lowering the symmetry of the molecule with substituents does not perturb the pi orbitals so much as to change the gross features of the $\pi \rightarrow \pi^*$ absorption spectrum. Bathochromic substituents, such as CN, CO_2H, and NH_2,

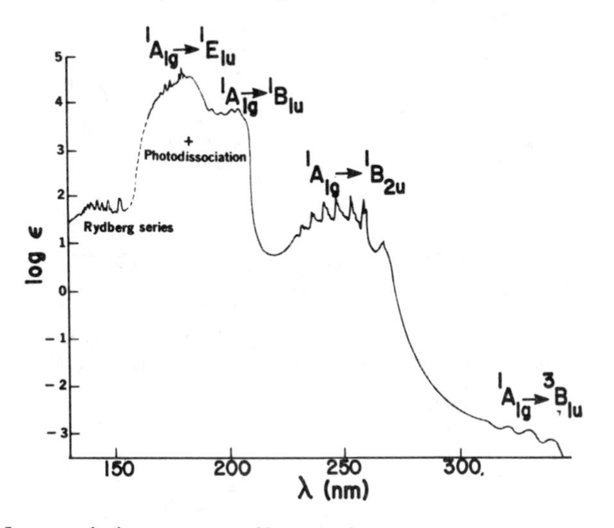

Fig. 5-40. Low resolution spectrum of benzene from K.S. Pitzer, *Quantum Chemistry*, Prentice-Hall, Englewood Cliffs, New Jersey, 1953.

can shift the second absorption maximum ($^1B_{1u}$) into the accessible region >210 nm and can also shift the benzenoid band to lower energy. Polynuclear aromatic compounds, with fused benzene rings, exhibit their lowest energy $\pi \rightarrow \pi^*$ transitions at lower energy as the size of the pi system increases. The spectra of most aromatic compounds, especially in nonpolar solvents, retain a characteristic wealth of vibronic structure, as in Fig. 5-41.

Now what happens if we introduce a lone pair, as in pyridine? We expect that the lone pair orbital energy will be near that of the upper occupied

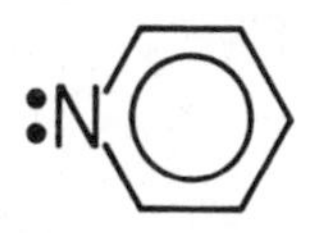

pi orbitals. We further expect that the introduction of a nitrogen atom into the benzene ring will remove the pi orbital degeneracy and lower the energies of most of the orbitals (because of the greater nuclear charge of nitrogen). All of this is summarized in Fig. 5-42. After this discussion you might anticipate that the spectrum of pyridine is drastically different from that of benzene. In fact, the two are remarkably similar (Fig. 5-43) in the low energy region. We rationalize this by saying that the expected low intensity $n \rightarrow \pi_4^*$ transition of pyridine comes near or under the intense $\pi_3 \rightarrow \pi_4^*$ absorption band.

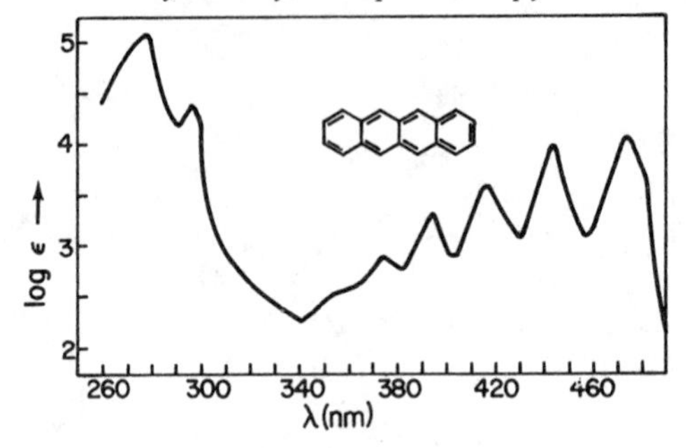

Fig. 5-41. Spectrum of tetracene in benzene solution. From R.A. Friedel and M. Orchin, *Ultraviolet Spectra of Aromatic Compounds*, John Wiley & Sons, N.Y., 1951.

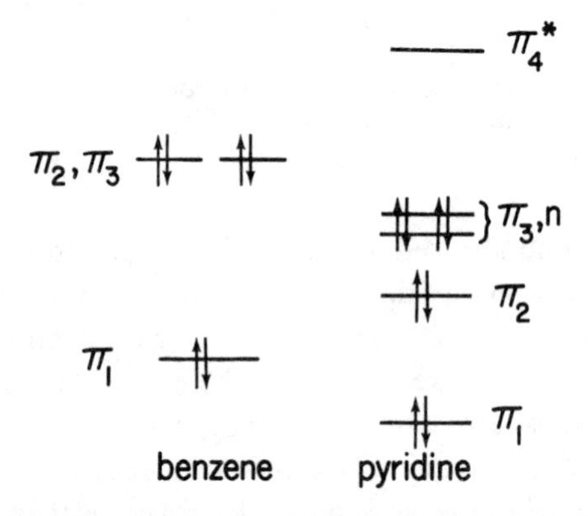

Fig. 5-42. Comparison of benzene and pyridine pi orbital energy regions of the molecular orbital diagram. The pyridine π_3 and n (lone pair) orbitals are nearly degenerate, with the order uncertain.

Problems

5-32. Based on the benzene pi orbitals (Fig. 4-55) draw an approximate picture of each of the six pi orbitals of pyridine and label the symmetry of each. Choose the yz plane as the molecular plane. Of the two sets of degenerate pi bonding orbitals of benzene, we expect that the ones involving more N(2p) character in pyridine will be lower in energy. On this basis, order the six nondegenerate pyridine pi orbitals by energy. What are the states involved in the lowest energy singlet-singlet $\pi \rightarrow \pi^*$ transition? Is it an orbitally-allowed transition?

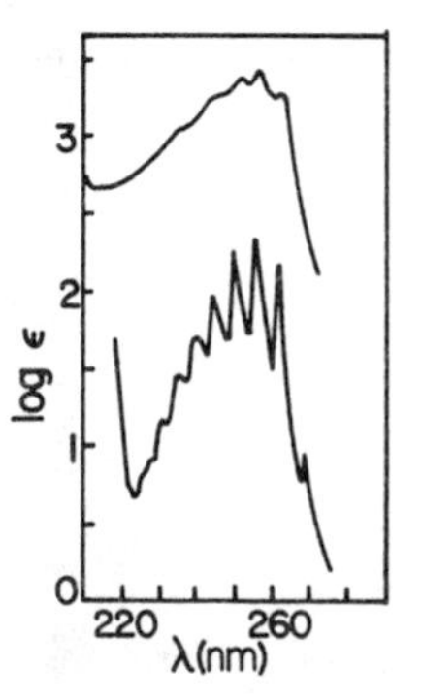

Fig. 5-43. Comparison of the spectra of pyridine (in 95% ethanol, upper trace) and benzene (in cyclohexane, lower trace). From R.A. Friedel and M. Orchin, *Ultraviolet Spectra of Aromatic Compounds*, John Wiley & Sons, N.Y., 1951.

5-8. Vibronic Analysis

Thus far we have stressed the use of energies and intensities of absorption bands in the assignment of electronic spectra. In suitable cases, the analysis of vibronic structure can be the most important (and most challenging) aspect of the assignment of a spectrum. In this section, we will develop more fully the notation for vibronic transitions and perform sample analyses involving orbitally-allowed and orbitally-forbidden transitions. We will also analyze the vibronic structure of the emission spectrum of benzene.

A. *p-Difluorobenzene.* This molecule has 30 nondegenerate modes of vibration listed in Table 5-8. Approximate drawings of the highest occupied and lowest unoccupied pi orbitals are given in Fig. 5-44. The b_{1g} and b_{2g}

Fig. 5-44. Approximate drawings of the highest filled and lowest unfilled pi orbitals of p-difluorobenzene. Shown are the upper lobes of the $C(2p_x)$ orbitals which lie perpendicular to the plane of the molecule. Shaded and unshaded orbitals are of opposite sign.

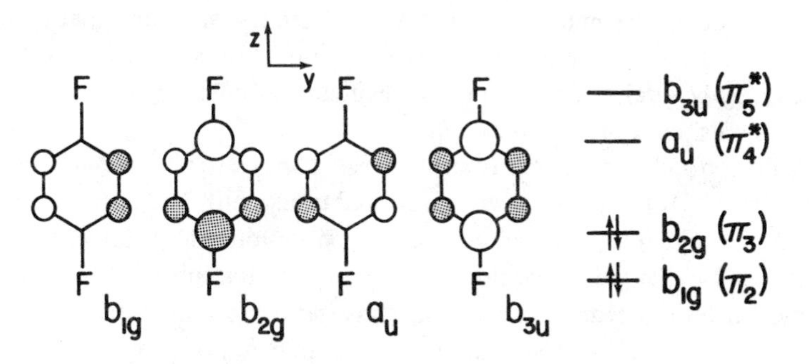

Table 5-8. Vibrations of p-Difluorobenzene[a]

Number	$\bar{\nu}$ (cm^{-1})	Symmetry	Number	$\bar{\nu}$ (cm^{-1})	Symmetry
1	3084	a_g	16	943	a_u
2	1617	a_g	17	406	a_u
3	1245	a_g	18	3050	b_{1u}
4	1142	a_g	19	1511	b_{1u}
5	858	a_g	20	1212	b_{1u}
6	451	a_g	21	1012	b_{1u}
7	800	b_{1g}	22	737	b_{1u}
8	928	b_{2g}	23	3080	b_{2u}
9	692	b_{2g}	24	1437	b_{2u}
10	375	b_{2g}	25	1285	b_{2u}
11	3084	b_{3g}	26	1085	b_{2u}
12	1617	b_{3g}	27	350	b_{2u}
13	1285	b_{3g}	28	833	b_{3u}
14	635	b_{3g}	29	509	b_{3u}
15	427	b_{3g}	30	163	b_{3u}

[a] From J.H.S. Green, W. Kynaston, and M.H. Paisley, *J. Chem. Soc.*, 473 (1963).

orbitals are derived from the e_{1g} orbitals of benzene, and the a_u and b_{3u} orbitals are derived from the e_{2u} orbitals of benzene. The lowest energy electron promotion is $\ldots (b_{2g})^2 (a_u)^0 \rightarrow \ldots (b_{2g})^1 (a_u)^1$, which defines ${}^1A_g \rightarrow {}^3B_{2u}$ and ${}^1A_g \rightarrow {}^1B_{2u}$ transitions. We will analyze the spin-allowed transition whose spectrum appears in Fig. 5-45.

It was shown in eq. 5-36 (Section 5-4-D) that the 0-0 transition is fully allowed. When this is the case, the "strongest band at lowest energy" is usually assigned as the 0-0 absorption. This is indicated in Fig. 5-45. We have intentionally used an ambiguous phrase ("strongest band at lowest energy") because the procedure is not well defined and any assignment must be regarded as tentative until the remainder of the spectrum can be assigned in a consistent manner. Even then, there is no law that says that a completely self-consistent assignment of the entire spectrum has any particular relation to reality. Very few analyses of vibronic absorption spectra are truly definitive. In general, they only provide evidence to favor or disfavor certain assignments, but not to prove them.

Which vibrations of p-difluorobenzene can serve as *vibronic origins* (Section 5-4-D) for an $A_g \rightarrow B_{2u}$ electronic transition? The possibilities are tested in Table 5-9. We have not tried any vibrations of u symmetry in this table because they will be forbidden by virtue of their u subscript. Transitions forbidden by g-u symmetry are said to be *parity* forbidden. According to Table 5-9, vibrations of symmetry a_g, b_{1g}, and b_{3g} can serve as vibronic

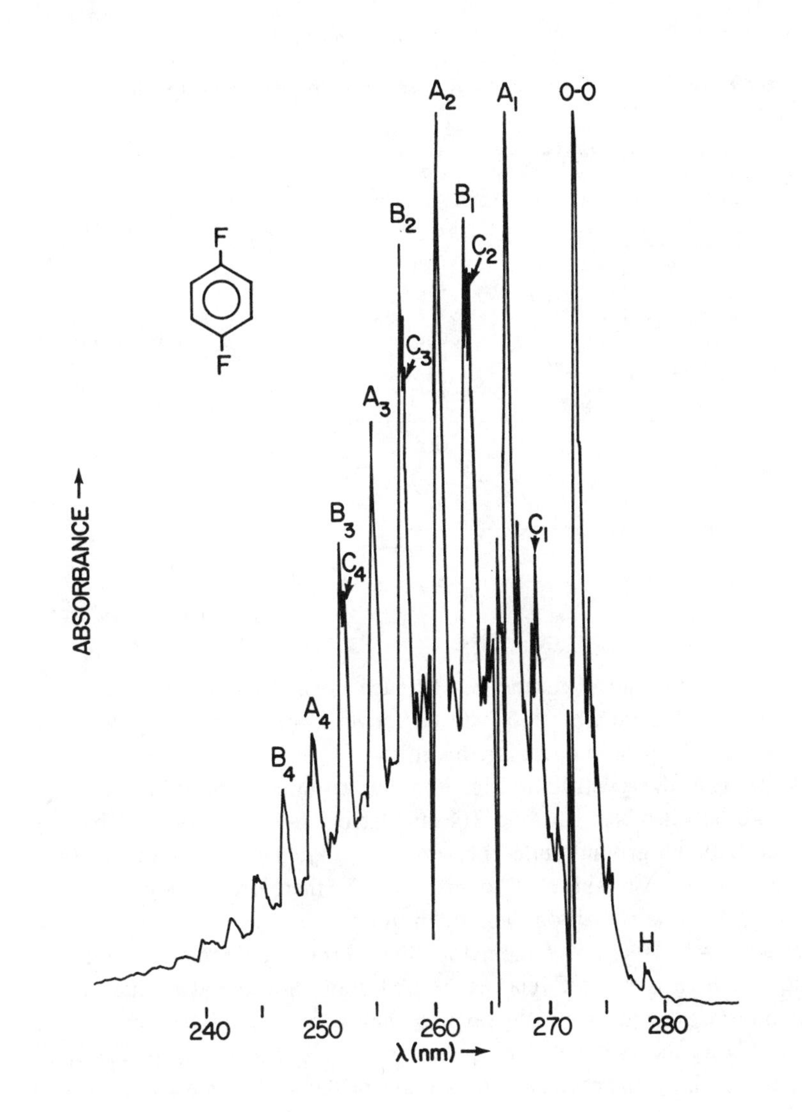

Fig. 5-45. Absorption spectrum of p-difluorobenzene.

Table 5-9. Testing the Allowedness of Possible Vibronic Origins for the $A_g \rightarrow B_{2u}$ Transition

$\psi_{v'}$	$\psi_{e'}$	$\psi_{e'}\psi_{v'}$	$\hat{\mu}$	$\psi_e \psi_v$	M	
			b_{3u}		b_{1g}	
a_g	B_{2u}	b_{2u}	b_{2u}	$a_g =$	a_g	allowed, y polarized
			b_{1u}		b_{3g}	
			b_{3u}		a_g	allowed, x polarized
b_{1g}	B_{2u}	b_{3u}	b_{2u}	$a_g =$	b_{1g}	
			b_{1u}		b_{2g}	
			b_{3u}		b_{3g}	
b_{2g}	B_{2u}	a_u	b_{2u}	$a_g =$	b_{2g}	
			b_{1u}		b_{1g}	
			b_{3u}		b_{2g}	
b_{3g}	B_{2u}	b_{1u}	b_{2u}	$a_g =$	b_{3g}	
			b_{1u}		a_g	allowed, z polarized

origins and p-difluorobenzene has 12 such normal modes. This does not say that all of these vibrations will be active in the electronic spectrum. It only makes them candidates for this role.

Looking at the spectrum in Fig. 5-45, the strongest band at higher energy than 0-0, labelled A_1, is found 814 cm^{-1} from 0-0. Turning to Table 5-8, we see that the ground state energies of ν_5 (a_g, 858 cm^{-1}) and ν_7 (b_{1g}, 800 cm^{-1}) are reasonably close to this value. We therefore try the assignments 5_0^1 or 7_0^1 for this transition. The definition of these symbols was given in Section 5-3-A. (We stress again that the vibronic assignments cannot be unique. There is no law relating excited state vibrational frequencies to ground state frequencies. We expect that excited state frequencies will usually be in the vicinity of the ground state frequencies, but this is by no means certain. For an actual comparison of the ground state and $n \rightarrow \pi^*$ excited state vibrational frequencies of formaldehyde, see Problem 5-1.) The next strongest absorption at higher energy, A_2, is found 1628 cm^{-1} from 0-0. This is *twice* the energy separation of A_1 and 0-0, and should alert you to the possibility that A_2 is an overtone of A_1. In fact, at least four more members of this series, all separated by about 814 cm^{-1}, can be identified in Fig. 5-45. The first four are labelled A_1–A_4. We still cannot tell if the vibration involved is ν_5 (a_g) or ν_7 (b_{1g}) because the overtones of each are allowed. As an example, we work the cases 7_0^3 and 7_0^4. The notation 7_0^m means that ν_7 is excited from $v = 0$ in the ground electronic state to $v' = m$ in the excited electronic state. The symmetry of an overtone of a nondegenerate vibration is just given by the symmetry of the mode itself if v' is odd and by the totally symmetric representation if v' is even. That is $\psi_{v'}$ $(7_0^3) \sim b_{1g}$ and $\psi_{v'}$ $(7_0^4) \sim a_g$. (See Section 3-6-E if you need to review this.) But the transition moment integral tells us that any $\psi_{v'}$ of a_g or b_{1g} symmetry will be allowed (Table 5-9); so all overtones of ν_7 are allowed. We leave the assignment of the A_i series of vibronic transitions in limbo for now.

The strongest unassigned band in the spectrum, B_1, is found 1268 cm^{-1} from 0-0. This leads us to suspect that $\nu_3(a_g$, 1245 cm^{-1}) or $\nu_{13}(b_{3g}$, 1285 cm^{-1}) might be involved and that the transition might be 3_0^1 or 13_0^1. Now an interesting feature of the spectrum appears. At intervals of about 814 cm^{-1} to higher energy than B_1, we discover a series of transitions seemingly built upon B_1. These are shown schematically in Fig. 5-46. Such a series of equally spaced vibronic absorptions is called a *vibronic progression.* The vibration responsible for the first spacing (1268 cm^{-1}) is called the *vibronic origin.* The vibration responsible for all subsequent spacings (814 cm^{-1}) is the *progression forming mode.*

The observations of the last two paragraphs give us courage to make some assignments. Since progression forming modes are almost always totally symmetric, the 814 cm^{-1} spacing will be assigned to ν_5'. The A_i progression is therefore assigned 5_0^1, 5_0^2, 5_0^3, . . . and not 7_0^1, 7_0^2, 7_0^3, . . . This is also consistent with the expectation that the vibrational frequency is lower in the excited state than in the ground state. The excited state frequency (814 cm^{-1}) is lower than ν_5 (858 cm^{-1}) but higher than ν_7 (800 cm^{-1}). The B_i progression must be of the form X_0^1, $5_0^1 X_0^1$, $5_0^2 X_0^1$, $5_0^3 X_0^1$, . . . The notation means that the vibronic origin (ν_x, still unidentified) is excited from $v = 0$ to $v' = 1$ in each case and that ν_5 adds in units of 814 cm^{-1} to each transition. In $5_0^2 X_0^1$, for example, ν_x goes from $v = 0$ to $v' = 1$, and ν_5 goes from $v = 0$ to $v' = 2$. These two vibrational transitions, plus the ${}^1A_g \rightarrow {}^1B_{2u}$ electronic transition, are all stimulated simultaneously by the same photon. We suggest that ν_x is ν_{13} and not ν_3 since the excited state

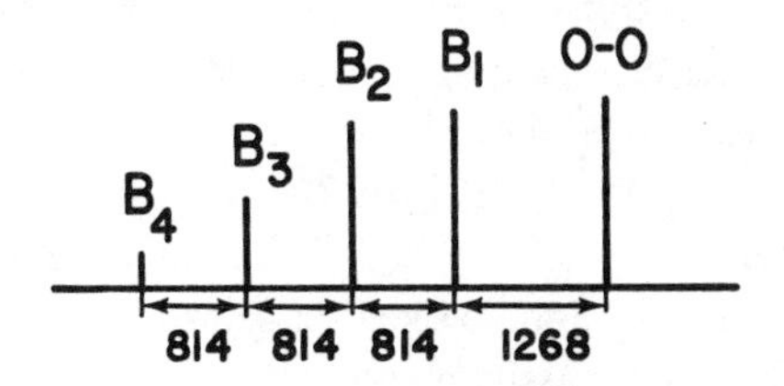

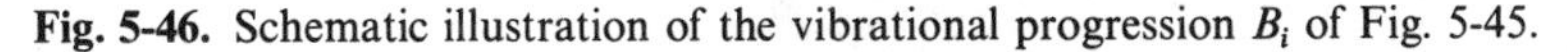

Fig. 5-46. Schematic illustration of the vibrational progression B_i of Fig. 5-45.

vibrational frequency (1268 cm^{-1}) is higher than ν_3 (1245 cm^{-1}) but lower than ν_{13} (1285 cm^{-1}). This assignment cannot be certain, however.

Vibronic transitions labelled C_i in Fig. 5-45 form another progression of the form Y_0^1, $5_0^1 Y_0^1$, $5_0^2 Y_0^1$, $5_0^3 Y_0^1$, ... The separation of C_1 and 0-0 is 402 cm^{-1} and subsequent separations are 814 cm^{-1}. For C_1, 6_0^1 (a_g, 451 cm^{-1}) and 15_0^1 (b_{3g}, 427 cm^{-1}) are excellent candidates.

Since ν_5 accounts for the strongest vibronic origin and is also the strongest progression forming mode, we might expect that ν_5 will give rise to a relatively strong *hot band*. Indeed, 858 cm^{-1} *to lower energy* than 0-0 is a band labelled H. This energy separation is the *ground state energy* of ν_5 and the hot band is assigned 5_1^0. The notation means that ν_5 goes from $v = 1$ to $v' = 0$. The band is called a hot band because the population of $v = 1$ in the electronic ground state will increase with temperature, and so will the intensity of band H. This can be confirmed by heating or cooling the sample. The appearance of 5_1^0 is simultaneously the most convincing evidence that 0-0 and 5_0^1 are correctly assigned.

We will not analyze the spectrum of p-difluorobenzene any further except for one small point. Why are there so many bands very close to the 0-0 band when the lowest vibrational frequency of this molecule is 163 cm^{-1} and the lowest energy allowed vibronic origin is about 427 cm^{-1}? We offer one example of an allowed *combination band*, $10_0^1 15_1^0$, whose energy will be near that of the 0-0 band. The symmetry of the combined vibrational wave function is $b_{2g}(\nu_{10}) \times b_{3g}(\nu_{15}) = b_{1g}$, so the vibronic transition is allowed. The energy will be the 0-0 energy *plus* ν_{10}' *minus* ν_{15}. ν_{15} is 427 cm^{-1} and ν_{10}' will be somewhere near 375 cm^{-1}, the value of ν_{10}. The energy of the combination band will therefore be approximately $E_{0\text{-}0} + 375 - 427 = E_{0\text{-}0} - 52$ cm^{-1}. Thus, a band about 52 cm^{-1} to the right of 0-0 in Fig. 5-45 could be readily explained.

To summarize this section, the assignments of the major vibronic transitions are listed in Table 5-10. These vibrational assignments by no means *prove* that the electronic transition is $A_g \rightarrow B_{2u}$, but they are consistent with the electronic transition. In principal, one should try to fit the vibronic absorption spectrum to different possible electronic transition assignments and see if any other electronic assignments are acceptable.

B. *The 260 nm Absorption Band of Benzene.* In Section 5-7-D we decided, on the basis of band intensities, that the 260 nm band of benzene (Fig. 5-40) is either ${}^1A_{1g} \rightarrow {}^1B_{2u}$ or ${}^1A_{1g} \rightarrow {}^1B_{1u}$. We will analyze part of the medium resolution spectrum in Fig. 5-47 to see if the vibronic structure lends support to either possible electronic assignment.

Table 5-10. Assignment of the Vibronic Transitions of p-Difluorobenzene, Fig. 5-45

Band	Assignment	Band	Assignment	Band	Assignment
A_1	5_0^1	B_1	13_0^1	C_1	6_0^1 or 15_0^1
A_2	5_0^2	B_2	$5_0^1 13_0^1$	C_2	$5_0^1(6_0^1$ or $15_0^1)$
A_3	5_0^3	B_3	$5_0^2 13_0^1$	C_3	$5_0^2(6_0^1$ or $15_0^1)$
A_4	5_0^4	B_4	$5_0^3 13_0^1$	C_4	$5_0^3(6_0^1$ or $15_0^1)$
H	5_1^0				

The orbital selection rules (eqs. 5-68 and 5-69, Section 5-7-D) tell us that neither electronic transition is orbitally allowed, but that vibrations of b_{2g} or e_{2g} symmetry may serve as vibronic origins for the ${}^1A_{1g} \rightarrow {}^1B_{1u}$ transition. Vibrations of symmetry b_{1g} or e_{2g} may serve as vibronic origins for the ${}^1A_{1g} \rightarrow {}^1B_{2u}$ transition. The thirty normal modes of benzene include many degeneracies (Table 5-11). The vibrations of benzene include some of e_{2g} and b_{2g} symmetry, but none of b_{1g} symmetry. It will therefore not be possible to confirm the ${}^1A_{1g} \rightarrow {}^1B_{2u}$ assignment by the presence of a b_{1g} vibronic origin. We could confirm a ${}^1A_{1g} \rightarrow {}^1B_{1u}$ assignment by noting the presence of a b_{2g} vibronic origin. The absence of a b_{2g} vibronic origin would not confirm either electronic assignment, but would lend weak support to the ${}^1B_{2u}$ transition.

To analyze the spectrum of benzene, let's first think back a moment to Section 5-8-A. How might we have attacked the spectrum of p-difluorobenzene? Since the 0-0 transition was orbitally-allowed, we assigned it to the "strongest band at lowest energy." Next, we could have seen which

Table 5-11. Vibrational Frequencies of Benzene[a]

Number	ν (cm^{-1}) Gas	Liquid	Symmetry	Number	ν (cm^{-1}) Gas	Liquid	Symmetry
1	(3073)	(3062)	a_{1g}	11	674	675	a_{2u}
2	995	(993)	a_{1g}	12	(3057)	(3048)	b_{1u}
3	(1350)	1346	a_{2g}	13	(1010)	1010	b_{1u}
4	(990)	(991)	b_{2g}	14	(1309)	1309	b_{2u}
5	(707)	(707)	b_{2g}	15	(1146)	1146	b_{2u}
6	(846)	850	e_{1g}	16	3047	3036	e_{1u}
7	(3056)	(3048)	e_{2g}	17	1482	1479	e_{1u}
8	(1590)	1586	e_{2g}	18	1037	1035	e_{1u}
9	(1178)	1177	e_{2g}	19	967	969	e_{2u}
10	608	(606)	e_{2g}	20	399	404	e_{2u}

[a] Values in parentheses are calculated. Data from E.R. Bernstein, S.D. Colson, D.S. Tinti, and G.W. Robinson, *J. Chem. Phys.*, *48*, 4632 (1968).

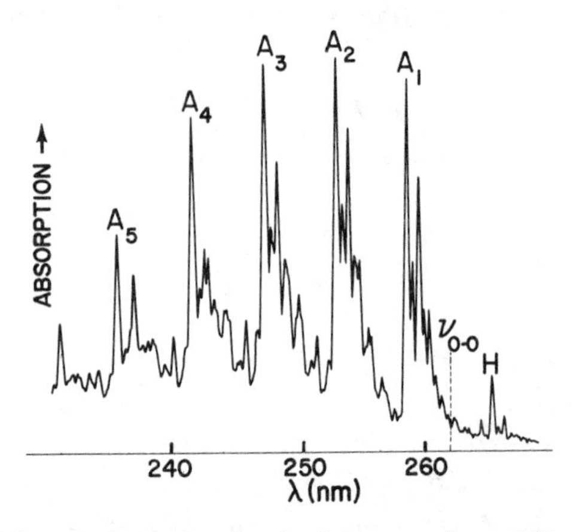

Fig. 5-47. The 260 nm absorption band of benzene from J.H. Callomon, T.M. Dunn, and I.M. Mills, *Phil. Trans. Roy. Soc.* (London), *259A*, 499 (1966).

vibrations are the strongest in the hot band region of the spectrum. This is not difficult because the energy spacings of hot bands from the 0-0 band correspond to *ground state* vibrational frequencies, which are known. Having done this, we could search the spectrum at higher energy for vibronic origins which correspond to the active hot band vibrations. For example, in the spectrum of p-difluorobenzene, we find that ν_5 gives rise to a strong hot band and is also the strongest vibronic origin.

The analysis of the benzene spectrum in Fig. 5-47 is similar, *except that the 0-0 band will be weak or absent* since it is orbitally-forbidden. Further, no totally symmetric modes may serve as vibronic origins. The "strongest band at lowest energy" will therefore correspond to an allowed vibronic origin. Our analysis of the 260 nm band relies on the fact that the small peak at 37,483 cm^{-1} (labelled H in Fig. 5-47) has a large temperature dependence and is certainly a hot band. It corresponds to a transition involving the vibration ν_x which is unknown at this time. The band labelled A_1, at 38,611 cm^{-1}, is *assumed* to represent a vibronic origin involving ν_x. In other words, we are calling band H, X_1^0, and band A_1, X_0^1. As shown in Fig. 5-48, the energy of X_0^1 is $E_{0\text{-}0} + \nu_x'$. The energy of X_1^0 is $E_{0\text{-}0} - \nu_x$. Therefore,

$$(E_{0\text{-}0} + \nu_x') - (E_{0\text{-}0} - \nu_x) = 38{,}611 - 37{,}483 = 1128\ \text{cm}^{-1} = \nu_x' + \nu_x \tag{5-70}$$

Assuming that ν_x' is probably close to ν_x, inspection of the b_{2g} and e_{2g} vibrations of benzene (Table 5-11) indicates that ν_{10} is the most likely candidate. Under this assumption, ν_{10}' has the energy

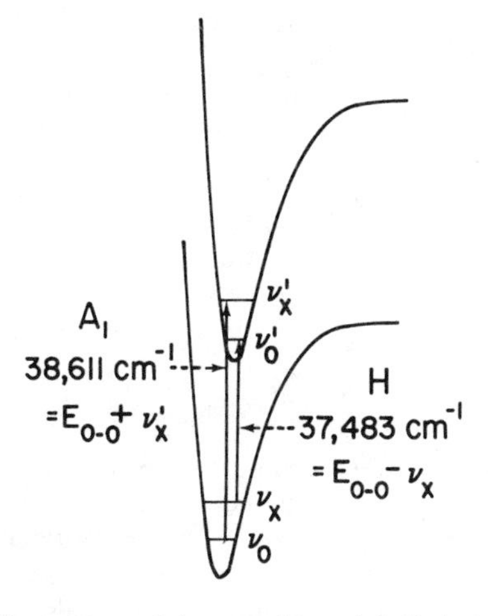

Fig. 5-48. Schematic illustration of the transitions labelled A_1 and H in Fig. 5-47.

$$(\nu_{10}' + \nu_{10}) - \nu_{10} = 1128 - 608 = 520 \text{ cm}^{-1} = \nu_{10}' \qquad (5\text{-}71)$$

The spectral origin (the position of the 0-0 band) must lie at 38,091 cm^{-1}:

$$(E_{0\text{-}0} + \nu_{10}') - \nu_{10}' = 38{,}611 - 520 = 38{,}091 \text{ cm}^{-1} = E_{0\text{-}0} \quad (5\text{-}72)$$

The remaining gross features of the spectrum involve a totally symmetric mode of energy 920 cm^{-1} adding on in multiples to the ν_{10} vibronic origin to give the vibronic progression A_i. This totally symmetric mode most reasonably corresponds to ν_2 whose ground state frequency is 995 cm^{-1}. The various members of the progression $2_0^m 10_0^1$ have an intensity pattern similar to what might be expected from the variation in magnitude of the Franck-Condon factor associated with a symmetrically enlarged molecule in the excited state. No vibronic transition associated with either b_{2g} mode (ν_4 or ν_5) could be found. Similarly, in the fluorescence spectrum of benzene, no transitions to ν_4 or ν_5 could be found. This negative evidence does not prove either electronic assignment, but it lends weak support to the $^1B_{2u}$ assignment. That is the extent of the power of a typical vibronic analysis.

Problems

5-33. In the electronic spectrum of p-difluorobenzene (Fig. 5-45), the totally symmetric mode ν_5 served as a strong vibronic origin. Why are no totally symmetric modes observed as vibronic origins in the spectrum of benzene (Fig. 5-47)?

5-34. Part of the mo scheme of pyrazine is shown below.

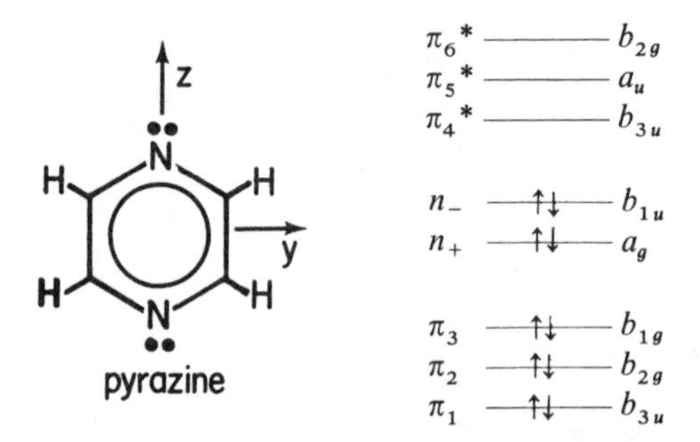

(a) Draw pictures of the n_+ and n_- lone pair orbitals. (b) Write the ground state term of pyrazine. (c) Assuming the mo scheme is qualitatively correct with respect to the relative spacings between orbitals, write the first two excited state configurations and the terms which arise from them. (d) Evaluate the symmetries of the transition moment integrals for the 0-0 bands of the two lowest energy spin-allowed transitions. Is either orbitally-allowed? For each of these transitions, what are the symmetries of the vibrations which can serve as vibronic origins? (e) In the spectrum of pyrazine on the next page, these two 0-0 bands are labelled A and B. Assign A and B to the proper electronic transitions. (f) The fundamental frequencies for pyrazine in its ground electronic state are listed below:

Symmetry	ν_i	cm^{-1}	Symmetry	ν_i	cm^{-1}
a_g	1	3054	b_{2g}	13	919
a_g	2	1578	b_{2g}	14	703
a_g	3	1230	b_{2u}	15	3066
a_g	4	1015	b_{2u}	16	1418
a_g	5	596	b_{2u}	17	1346
a_u	6	?	b_{2u}	18	1063
a_u	7	363	b_{3g}	19	3041
b_{1g}	8	757	b_{3g}	20	1524
b_{1u}	9	3066	b_{3g}	21	1118
b_{1u}	10	1484	b_{3g}	22	641?
b_{1u}	11	1135	b_{3u}	23	804
b_{1u}	12	1021	b_{3u}	24	416

The bands labelled X_1, X_2, X_3, and X_4 in the spectrum form a progression associated with the 0-0 transition A. Label these four transitions i_n^m, $i_n^m j_{n'}^{m'}$, *etc.* Your assignment should be reasonable in light of your answer to part (d). (g) Do the same for the Y and Z progressions associated with B. Suggest an assignment for the band at 30,242 cm^{-1}. (References for this problem are K.K. Innes, J.P. Byrne, and I.G. Ross, *J. Mol. Spec.*, *22*, 125 (1967); W.R. Moomaw and J.F. Skinner, *J. Chem. Ed.*, *48*, 304 (1971). The spectrum has not been calibrated, so the absolute accuracy of the band positions is not as high as stated. However, the differences in energy between bands should be accurate. The positions listed for the 0-0 bands in the first reference above are 30,875.8 and ~30,425 cm^{-1}. The mo scheme for pyrazine given in this problem is not well established. For an alternative scheme, see P. Bischof, R. Gleiter, and P. Hofmann, *Chem. Commun.*, 767 (1974) and references cited therein.)

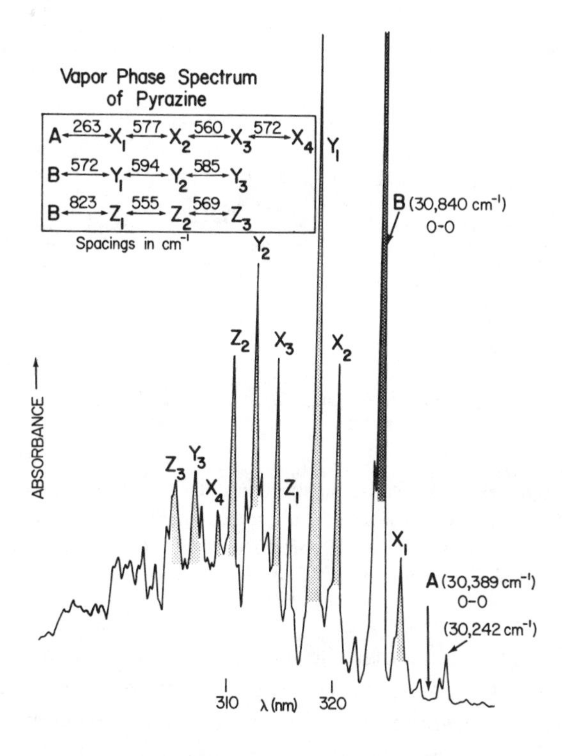

C. *The Phosphorescence of Benzene.* Very few other, if any, vibronic analyses are as satisfying in their clarity and decisiveness as the one to be presented now. In Fig. 5-49 is shown a portion of the high resolution emission spectrum of benzene doped into a crystal of C_6D_6 and cooled to 4 K. The emission lifetime is 8.7 s, indicating that the emission is phosphorescence. The 0-0 band, at *highest* energy in the emission spectrum, comes near 337 nm (29,658.2 cm^{-1}). This is very nearly the same position as the 0-0 band of singlet-triplet absorption (Fig. 5-40); so the same triplet state accounts for the absorption and emission spectra. This is very reasonable as one expects that phosphorescence will originate from the lowest triplet state.

When analyzing the spectrum of a crystalline sample, one must take into account the *site symmetry*, which is the symmetry of the environment of a molecule in the crystal. The crystal structure of benzene (and presum-

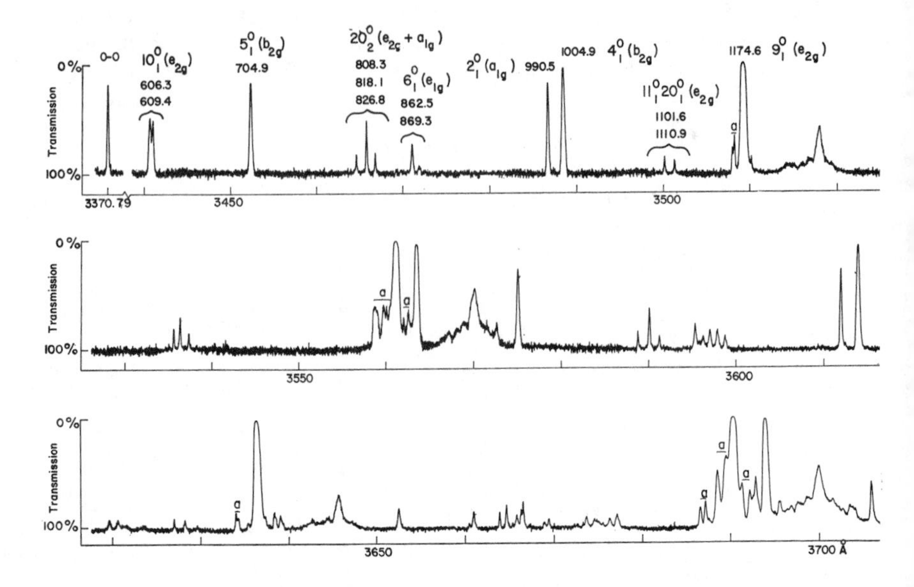

Fig. 5-49. High resolution emission spectrum of C_6H_6 doped in a crystal of C_6D_6 and cooled to 4 K. Bands labelled "*a*" are assigned to the natural abundance of $^{13}C^{12}C_5H_6$ present in sample. Reproduced from E.R. Bernstein S.D. Colson, D.S. Tinti, and G.W. Robinson, *J. Chem. Phys.*, *48*, 4632 (1968).

ably of C_6D_6) is such that the environment of each benzene molecule possesses only C_i symmetry. This means that the benzene electronic and vibrational states cannot rigorously be classified according to any symmetry except inversion. All of the g representations of D_{6h} transform as a_g in C_i symmetry, and all of the u representations transform as a_u. All degeneracy in D_{6h} is removed in C_i. Since all of the possible triplet excited states have u symmetry, and the ground state has g symmetry, all of the triplet → singlet transitions are orbitally-allowed in the crystalline environment and we may expect a moderate intensity for the 0-0 band.

But if all of the transitions are orbitally-allowed, and if all vibrations of g symmetry may serve as vibronic origins, what good is an analysis of the spectrum of this crystal? The answer is that placing the highly symmetric benzene molecule in the crystalline environment of lower symmetry does not destroy the symmetry of the individual molecule. If the forces exerted by neighboring molecules are small compared to the forces holding one molecule together, we expect that intensities of vibronic transitions will still reflect the D_{6h} selection rules. That is, transitions allowed by both D_{6h} and C_i selection rules will be stronger than those allowed only

Table 5-12. Allowed Vibronic Origins for the Possible Triplet→Singlet Transitions of Benzene Using the Selection Rules for D_{6h} Symmetry

Triplet State	Vibronic Origins
E_{1u}	$a_{1g}, a_{2g}, e_{1g}, e_{2g}$
B_{1u}	b_{2g}, e_{2g}
B_{2u}	b_{1g}, e_{2g}

by the less stringent C_i selection rules. If this is the case, we can make a table of the D_{6h} benzene vibrations expected to serve as vibronic origins for each of the possible triplet → singlet transitions. From the transition moment integrals, eqs. 5-67–5-69, we construct Table 5-12.

We will analyze a portion of the spectrum shown in Fig. 5-49. The beauty of an emission spectrum is that the vibrational energies should correspond to ground state (known) energies. The ground state vibrational frequencies of benzene were given in Table 5-11. Now the first peak to lower energy than 0-0 is a doublet at 606.3 and 609.4 cm^{-1}. Surveying Table 5-11, we see that $\nu_{10}(e_{2g})$ has an energy of 606 cm^{-1}. The splitting in the emission spectrum is a weak perturbation due to the C_i site symmetry. The degenerate e_{2g} vibrations of the D_{6h} molecule are no longer degenerate in C_i symmetry. You can see, however, that the splitting due to the crystalline environment (3 cm^{-1}) is only a small fraction of the energy of the vibrations. The next band at 704.9 cm^{-1} must be the nondegenerate $\nu_5(b_{2g})$ with a ground state energy of 707 cm^{-1}. Three bands near 818 cm^{-1} do not correspond to any ground state fundamental frequency. But the first overtone of ν_{20} ($2 \times 404 = 808$ cm^{-1}) provides a good fit. The symmetry of $\nu_{20}{}^2$ is the symmetric product of $e_{2u} \times e_{2u} = e_{2g} + a_{1g}$ (Appendix B). The e_{2g} component is further split in C_i symmetry; so $\nu_{20}{}^2$ yields *three* vibronic transitions in the emission spectrum. In a similar manner, the remaining bands of Fig. 5-49 are assigned as indicated on the figure. Only combinations of g symmetry were selected for this assignment since we expect the g-u selection rule to be rigorously obeyed in the crystalline environment. The notation for these emission bands is of the form X_1^0 since ν_x goes from the zero vibrational level of the electronic excited state to an excited vibrational level of the electronic ground state.

A perusal of the assignments in Fig. 5-49 indicates that the e_{2g} and b_{2g} vibrations serve as the strongest vibronic origins. Looking at Table 5-12, we have a rather strong case for assigning the phosphorescent transition $^3B_{1u} \rightarrow {}^1A_{1g}$. The remainder of the spectrum in Fig. 5-49, as well as the spectrum of 1,3,5-$C_6H_3D_3$ (Fig. 5-50, Problem 5.35), is fully consistent with the portion we have analyzed and you will rarely see a more conclusive vibronic analysis.

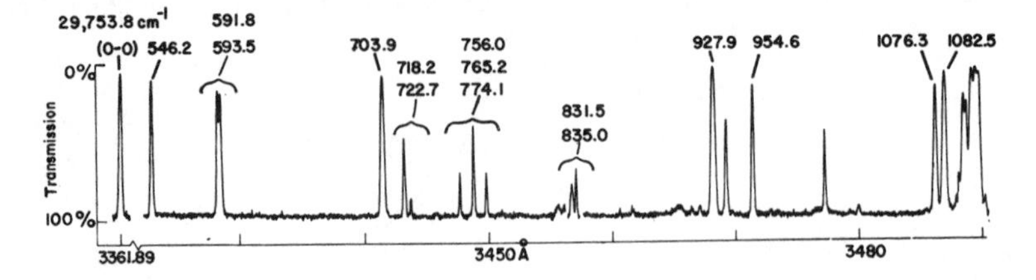

Fig. 5-50. 1,3,5-$C_6H_3D_3$ emission spectrum for Problem 5-35. Reproduced from E.R. Bernstein, S.D. Colson, D.S. Tinti, and G.W. Robinson, *J. Chem. Phys.*, *48*, 4632 (1968).

Problems

5-35. The vibrational frequencies of 1,3,5-$C_6H_3D_3$ are listed below.

1,3,5-$C_6H_3D_3$

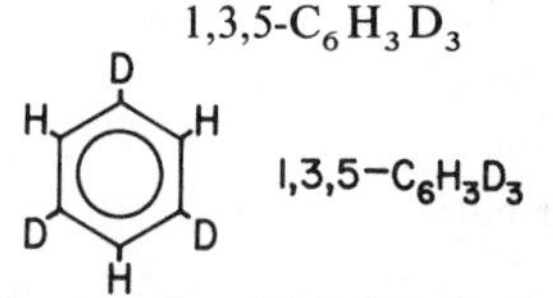

Vibrational Frequencies of 1,3,5-$C_6H_3D_3$ (D_{3h} Symmetry)[a]

Number	$\bar{\nu}$(cm^{-1}) Gas	Liquid	Symmetry	Number	$\bar{\nu}$(cm^{-1}) Gas	Liquid	Symmetry
1	(3074)	(3062)	a_1'	11	3063	3553	e'
2	(2294)	2282	a_1'	12	2282	2274	e'
3	(1004)	1003	a_1'	13	1580	1575	e'
4	(956)	955	a_1'	14	1414	1412	e'
5	(1321)	1322	a_2'	15	1101	1101	e'
6	(1253)	(1252)	a_2'	16	833	833	e'
7	(912)	(911)	a_2'	17	594	594	e'
8	917	918	a_2''	18	(924)	(926)	e''
9	697	697	a_2''	19	(707)	711	e''
10	531	533	a_2''	20	(370)	375	e''

[a] Values in parentheses are calculated. Data from E.R. Bernstein, S.D. Colson, D.S. Tinti, and G.W. Robinson, *J. Chem. Phys.*, *48*, 4632 (1968).

The electron configuration of this molecule is $(a_2'')^2(e'')^4(e'')^0(a_2'')^0$. In working this problem, you may wish to refer to the benzene emission spectrum analysis of Section 5-8-C and note the correlation of the irreducible representations of D_{3h} with those of D_{6h} given below:

D_{6h}	D_{3h}
a_{1g}	a_1'
b_{1u}	a_1'

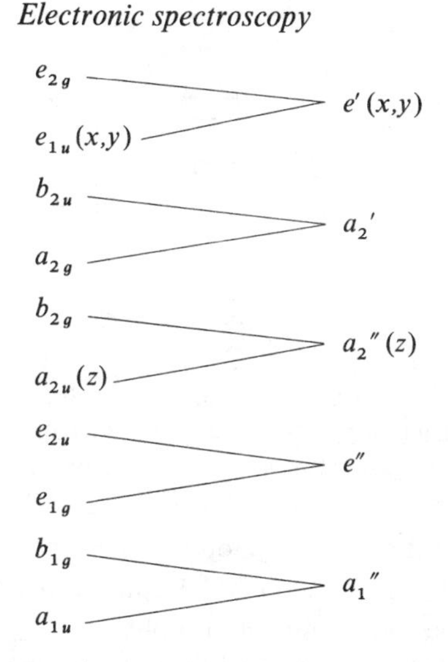

(a) What is the symmetry and spin multiplicity of the ground state? (b) What terms are generated by the lowest excited configuration? (c) Write the three possible transitions which would lead to phosphorescence from the lowest excited configuration. (d) By evaluating the symmetries of the transition moment integrals, determine the modes which may serve as vibronic origins for each of the three electronic transitions. (e) The phosphorescence spectrum of 1,3,5-$C_6H_3D_3$ doped into a crystal of C_6D_6 (τ_p = 8.7 s, 4 K) is shown in Fig. 5-50. Suggest an assignment for the transitions whose energies are given. Label each one X_n^m, $X_n^m Y_r^s$, *etc.* (f) From which electronic state is the phosphorescence originating?

5-9. Transition Metal Complexes

A. *Energy Levels and States.* In Section 4.8 we introduced a molecular orbital scheme for a sigma bonded octahedral ML_6 transition metal complex. The essential features of the mo scheme are reproduced in Fig. 5-51. The sigma bonding orbitals are formally filled by electrons donated by the six ligands. The metal d electrons will occupy the nonbonding t_{2g} orbitals and, if necessary, the e_g antibonding orbitals which are also predominantly of metal d character. Transitions of the type $t_{2g} \rightarrow e_g$ will generally be of lowest energy and are called *d-d transitions*. Transitions from filled sigma orbitals (or filled ligand pi orbitals, if there are any) to the metal d orbitals might occur at higher energy and are called ligand-to-metal ($L \rightarrow M$) *charge transfer transitions*. It is also possible to have $M \rightarrow L$ charge transfer, such as $e_g \rightarrow$ (ligand π^*) with suitable ligands. d-d transitions often fall in the visible region of the spectrum and account for the characteristic colors of transition metal complexes. In this section we treat some of the most

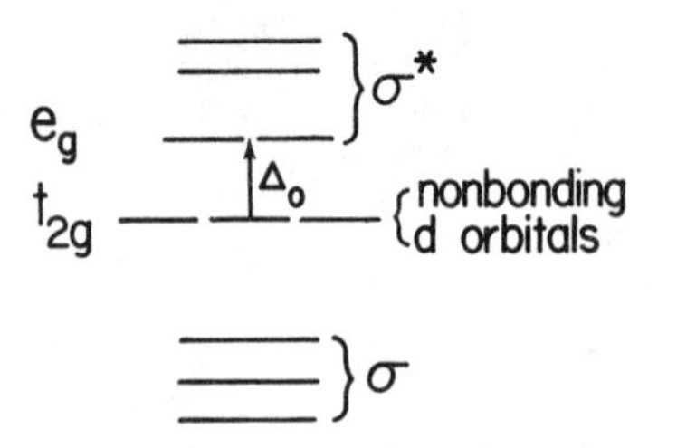

Fig. 5-51. Important features of the mo scheme for an octahedral ML_6 transition metal complex. Only sigma bonding is considered. Δ_o, the crystal field splitting energy, is the energy gap between the metal t_{2g} and e_g orbitals.

basic aspects of d-d spectra. For the most part, we treat only first row transition metal complexes for which spin-orbit coupling is relatively small. The consequences of spin-orbit coupling are very significant for heavier metals, but this subject is beyond the scope of the present text.[†]

The t_{2g} and e_g orbitals are composed principally of the metal (d_{xz}, d_{yz}, d_{xy}) and (d_{z^2}, $d_{x^2-y^2}$) orbitals, respectively. Complexes containing 1–3 d electrons have the configurations drawn below:

e_g / t_{2g}

d^1 $\quad$ d^2 $\quad$ d^3

The spins are parallel according to Hund's rule. The position of a fourth d electron is not unique. If the energy difference between the t_{2g} and e_g orbitals is large, the fourth electron is paired with one of the t_{2g} electrons to give a *low spin configuration.* This costs energy, however, as two electrons in the same orbital repel each other more than electrons in different orbitals. For first row transition metal ions, this *spin pairing energy, P,* is in the approximate range 12–25 kK (140–300 kJ mole^{-1}). If *P* is greater than Δ_0, the electron will occupy the e_g orbital to give a *high spin configuration.*

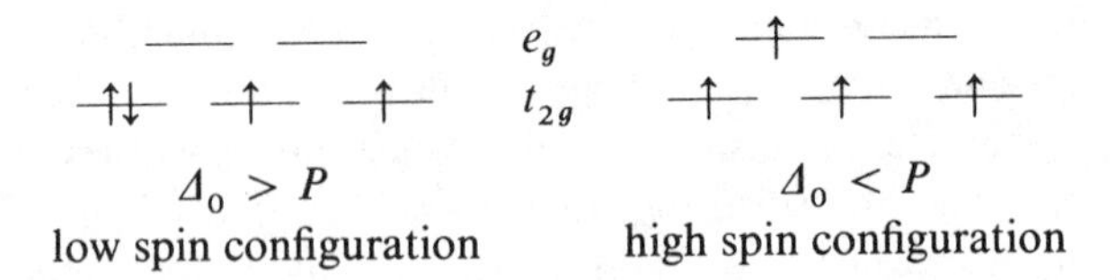

[†] For more detailed treatment of transition metal spectra, the interested reader is referred to B.N. Figgis, *Introduction to Ligand Fields*, John Wiley & Sons, N.Y., 1966 for a readable discussion. Another detailed but less readable text is C.J. Ballhausen, *Ligand Field Theory*, McGraw-Hill, N.Y., 1962.

Table 5-13. Values of Δ_0 for $M(H_2O)_6{}^{n+}$ Complexes[a]

Ion	d electrons	Δ_0 (kK)	Ion	d electrons	Δ_0 (kK)
Ti^{3+}	1	20.3	Mn^{2+}	5	7.5
V^{3+}	2	18.0	Fe^{3+}	5	14.0
V^{2+}	3	11.8	Fe^{2+}	6	10.0
Cr^{3+}	3	17.6	Co^{2+}	7	10.0
Cr^{2+}	4	14.0	Ni^{2+}	8	8.6
Mn^{3+}	4	21.0	Cu^{2+}	9	13.0

[a] All complexes are high spin. Data from T.M. Dunn, D.S. McClure, and R.G. Pearson, *Some Aspects of Crystal Field Theory*, Harper & Row, N.Y., 1965.

The relative magnitudes of Δ_0 and P will also govern the configurations of d^5, d^6, and d^7 complexes:

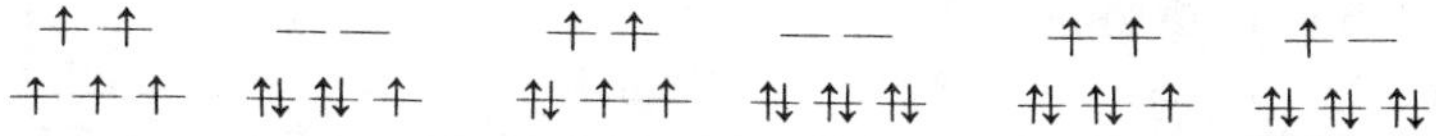

↑ ↑	— —	↑ ↑	— —	↑ ↑	↑ —
↑ ↑ ↑	↑↓ ↑↓ ↑	↑↓ ↑ ↑	↑↓ ↑↓ ↑↓	↑↓ ↑↓ ↑	↑↓ ↑↓ ↑↓
high spin	low spin	high spin	low spin	high spin	low spin
$\Delta_0 < P$	$\Delta_0 > P$	$\Delta_0 < P$	$\Delta_0 > P$	$\Delta_0 < P$	$\Delta_0 > P$
d^5		d^6		d^7	

There will be no ambiguity in the configurations d^8, d^9, and d^{10}, since the t_{2g} orbitals must be filled in all cases.

Several trends in the values of Δ_0 are noteworthy. (As usual, there are exceptions to these rules.) Δ_0 increases by about 50% on going from the first row to the second row of transition metals. A further increase of about 25% is observed on going from the second to the third row. For example, the values of Δ_0 for $Co(NH_3)_6{}^{3+}$, $Rh(NH_3)_6{}^{3+}$, and $Ir(NH_3)_6{}^{3+}$ are 22.9, 34.1, and 41.0 kK, respectively. This increase of Δ_0 accounts for the almost exclusive low spin configurations observed for second and third row complexes. The value of Δ_0 also increases as the oxidation state of the metal increases (Table 5-13). The order of increasing Δ_0 as the ligand is varied is generally the same for different metals. This rather remarkable statement says, for example, that the value of Δ_0 is greater for $M(NH_3)_6{}^{3+}$ than for $M(H_2O)_6{}^{3+}$, regardless of the nature of M. This order of increasing "field strength" of ligands is called the spectrochemical series. The order for some representative ligands is shown below.

$$I^- < Br^- < S^{2-} < NCS^- < Cl^- < NO_3{}^- < F^- < OH^- < C_2O_4{}^{2-} < H_2O < SCN^- < CH_3CN < NH_3 < H_2NCH_2CH_2NH_2 < \text{(bipy)} < \text{(phen)} < NO_2{}^- < \text{(phosphine)} < CN^- < CO$$

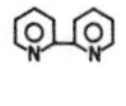

This order is not without exceptions, especially for metals in unusual oxidation states. Table 5-14 gives some data to illustrate the spectrochemical series and provides representative values of Δ_0.

To interpret the electronic spectra of transition metal complexes, we need to know how the energies of the various states which arise from the d^n configurations vary with the crystal field splitting, Δ_0. We will not attempt any quantitative treatment of this subject, but the qualitative treatment of the d^2 configuration provides some insight into transition metal energy level diagrams. We consider two extreme cases. Suppose first that $\Delta_0 = 0$. That is, the transition metal ion feels no influence of its ligands and is effectively free in the gas phase. All five d orbitals are degenerate and the states of the system will be given by the states of a free atom with the d^2 configuration: ^3F, ^3P, ^1G, ^1D, and ^1S (Table 4-4, Section 4-2-C). What happens to each energy level as Δ_0 increases from zero? What we want to know is how each state of the spherical ion splits as the symmetry is lowered from K_h to O_h. It is not necessary to use the full symmetries of K_h and O_h, but just the rotational symmetries of K and O. To find $\chi(\alpha)$, the character of the state whose orbital angular momentum is L under the rotation operation C_α, we use the following formula:†

$$\chi(\alpha) = \frac{\sin(L + 1/2)\alpha}{\sin(\alpha/2)} \tag{5-73}$$

Let's see what this means for the term ^3F. The character of the F state ($L = 3$) under the operation C_3 ($\alpha = 2\pi/3$), for example, is

$$\chi(C_3) = \frac{\sin(7/2)(2\pi/3)}{\sin(2\pi/3)(1/2)} = \frac{\sin(7\pi/3)}{\sin(\pi/3)} = \frac{\sin(\pi/3)}{\sin(\pi/3)} = 1 \tag{5-74}$$

In a similar way, we can find all of the characters of Table 5-15.†† Table 5-15 tells us that the F state will split into the states $A_2 + T_1 + T_2$ as an octahedral ligand field is applied to the free d^2 ion.

†Equation 5-73 is partially derived in F.A. Cotton, *Chemical Applications of Group Theory*, John Wiley & Sons, N.Y., 1971. A very readable discussion of other methods of generating Fig. 5-52 is given in this reference.

††For $\alpha = 0$, eq. 5-73 is indeterminate. But L' Hôpital's rule (Problem 3-22) gives:

$$\lim_{\alpha \to 0} \chi(\alpha) = \frac{(L + 1/2)\cos(L + 1/2)\alpha}{(1/2) \quad \cos(\alpha/2)} = 2L + 1$$

Table 5-14. Values of Δ_0 (kK) for Some Co^{3+}, Rh^{3+}, and Ir^{3+} Complexes[a]

Species	Δ_0 (M=Co)	Δ_0 (M=Rh)	Δ_0 (M=Ir)
MCl_6^{3-}	—	20.3	25.0
$M(dtp)_3$	14.2	22.0	26.6
$M(H_2O)^{3+}$	18.2	27.0	—
$M(NH_3)^{3+}$	22.9	34.1	41.0
$M(en)_3^{3+}$	23.2	34.6	41.4
$M(CN)_6^{3-}$	33.5	45.5	—

[a] dtp = $(CH_3CH_2O)_2PS_2^-$; en = $H_2NCH_2CH_2NH_2$. Data from J.E. Huheey, *Inorganic Chemistry: Principles of Structure and Reactivity*, Harper and Row, N.Y., 1972.

We can begin to construct a *correlation diagram*, somewhat analogous to Fig. 4-17, in which one extreme is the free ion with zero ligand field and the other extreme is the d^2 ion in a strong octahedral ligand field (Fig. 5-52). In the strong field limit are the configurations $(t_{2g})^2$, $(t_{2g})^1(e_g)^1$, and $(e_g)^2$, in order of increasing energy. You should be able to determine the states which arise from each of these configurations by application of the rules of Section 5-3-D. The results are given in Table 5-16. We summarize this information for the d^2 configuration at the right of Fig. 5-52. It is found, for example, that the $(t_{2g})^2$ configuration gives rise to ${}^1A_{1g}$, ${}^1E_g$, ${}^1T_{2g}$, and ${}^3T_{1g}$ states.

Now all that remains in the correlation diagram is to connect the proper states on the left with those on the right. In doing this we must conserve the symmetry and multiplicity of each state. That is, a ${}^3T_{1g}$ state in the weak field limit must be connected to a ${}^3T_{1g}$ state in the strong field limit. Further, states of the same symmetry may not cross. To use these rules in practice, we start with the lowest state at the left and connect it to the lowest state of the same symmetry and multiplicity at the right. Proceeding in this

Table 5-15. Symmetry Species of Atomic States in Octahedral Symmetry

O	E	$6C_4$	$3C_2(=C_4^2)$	$8C_3$	$6C_2$	
S (L = 0)	1	1	1	1	1	$= A_1$
P (L = 1)	3	1	−1	0	−1	$= T_1$
D (L = 2)	5	−1	1	−1	1	$= E + T_2$
F (L = 3)	7	−1	−1	1	−1	$= A_2 + T_1 + T_2$
G (L = 4)	9	1	1	0	1	$= A_1 + E + T_1 + T_2$
H (L = 5)	11	1	−1	−1	−1	$= E + 2T_1 + T_2$
I (L = 6)	13	−1	1	1	1	$= A_1 + A_2 + E + T_1 + 2T_2$

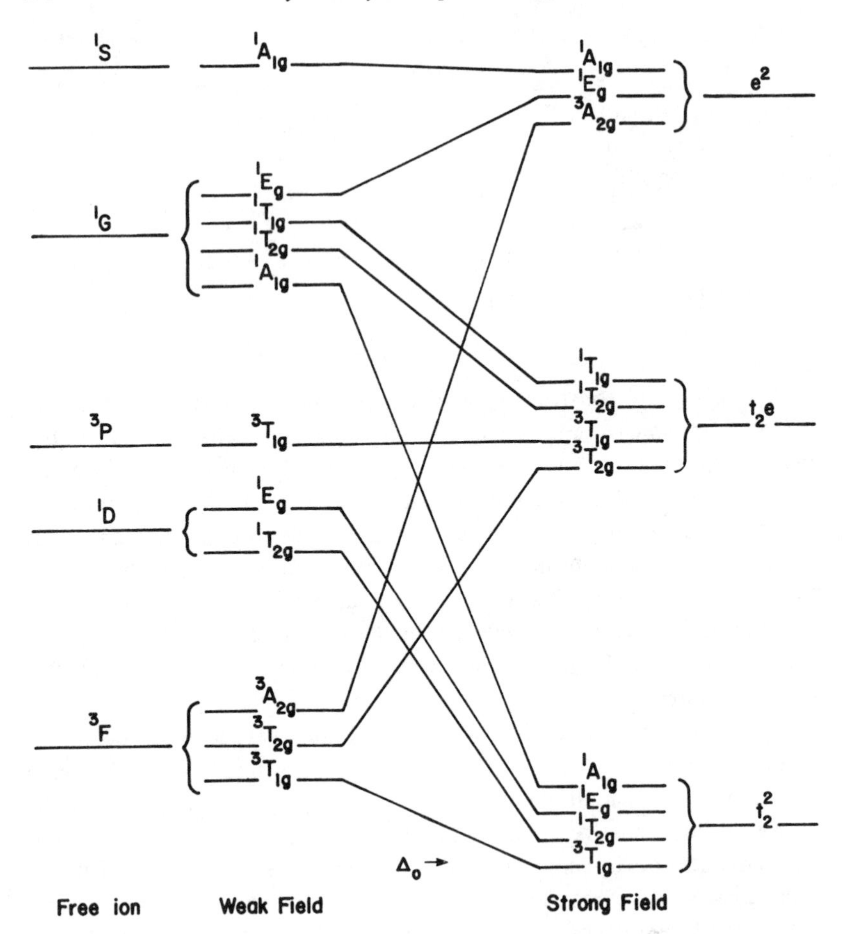

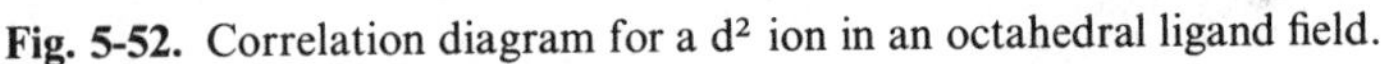

Fig. 5-52. Correlation diagram for a d² ion in an octahedral ligand field.

manner up the left side of the correlation diagram we can connect each state in the weak field limit with a unique state in the strong field limit.

The qualitative correlation diagram, Fig. 5-52, gives us some information about how the order of the states of the d² complex will vary with Δ_0. As in the correlation diagram for diatomic molecules, Fig. 4-17, the connections between the two extreme cases need not be linear, so Fig. 5-52 is strictly schematic. It does tell us, for example, that at large values of Δ_0 we expect the order of the triplet states to be ${}^3T_{1g} < {}^3T_{2g} < {}^3T_{1g} < {}^3A_{2g}$. But at small values of Δ_0 the order will be ${}^3T_{1g} < {}^3T_{2g} < {}^3A_{2g} < {}^3T_{1g}$. This qualitative diagram should be compared to the calculated variation in energy levels given in Fig. 5-53.

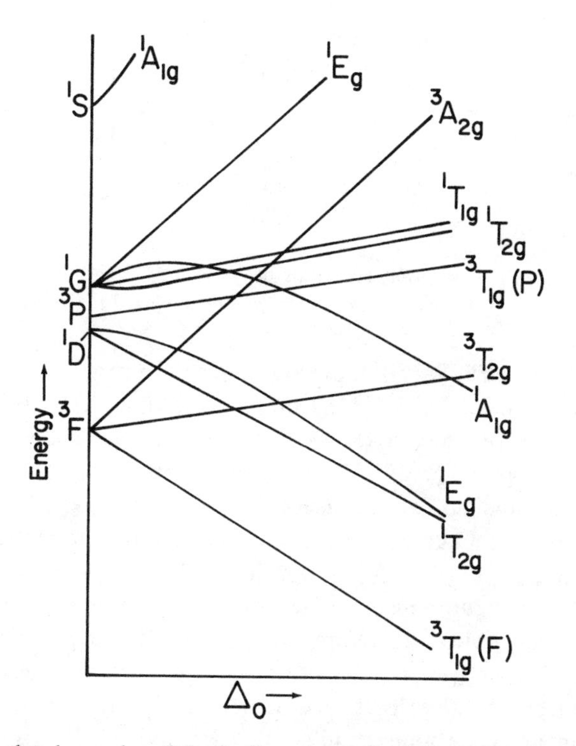

Fig. 5-53. Calculated energies of the various states of a d² transition metal species as a function of Δ_0. From F.A. Cotton and G. Wilkinson, *Advanced Inorganic Chemistry*, John Wiley & Sons, N.Y., 1972.

Problems

5-36. Using Table 5-16, determine the symmetry and spin multiplicity of the ground state of the following octahedral complexes: $Fe(CN)_6{}^{3-}$ (low spin), $Cr(CO)_6$ (low spin), $Cr(H_2O)_6{}^{3+}$ (high spin).

5-37. Why does thiocyanate appear twice in the spectrochemical series? Draw a picture of the metal-thiocyanate linkage for each entry, based on nearby ligands in the spectrochemical series.

5-38. Draw a correlation diagram similar to Fig. 5-52 for a *tetrahedral* d² complex. Remember that the order of metal orbitals is $e < t_2$ in a tetrahedral environment.

5-39. Draw a correlation diagram similar to Fig. 5-52 for an octahedral d³ complex, given the order of atomic states: ${}^4F < {}^4P < {}^2G < {}^2H < {}^2P < {}^2D < {}^2D$. What is the ground state term for all values of Δ_0?

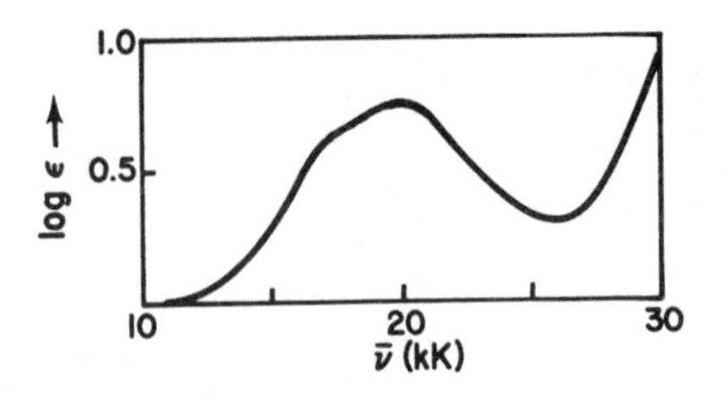

Fig. 5-54. Visible absorption spectrum of $Ti(H_2O)_6{}^{3+}$ From H. Hartman, H.L. Schläfer, and K.H. Hansen, *Z. Anorg. Allg. Chem.*, *284*, 153 (1956).

B. *d-d Spectra.* We will now examine some energy level diagrams and spectra for a few different d electron configurations. The simplest is the d^1 configuration, of which $Ti(H_2O)_6{}^{3+}$ is an example. The visible spectrum of this violet species (Fig. 5-54) shows an absorption band with a maximum near 20 kK and an extinction coefficient which suggests that it is spin-allowed but orbitally-forbidden. This is readily interpreted as a ${}^2T_{2g}$ $[(t_{2g})^1(e_g)^0] \rightarrow {}^2E_g\,[(t_{2g})^0(e_g)^1]$ transition. At this point we should note that d^{10-n} configurations give rise to the same states as d^n configurations. The d^9 configuration, therefore, is also expected to give rise to a single d-d transition, ${}^2E_g\,[(t_{2g})^6(e_g)^3] \rightarrow {}^2T_{2g}\,[(t_{2g})^5(e_g)^4]$, in which the states are reversed from the d^1 case. In most Cu(II) (d^9) complexes such a transition results in an absorption band at the red end of the visible spectrum, accounting for the blue or green colors of Cu(II) compounds.

Even the spectrum which results from so simple a configuration as d^1 does not exhibit a symmetric absorption band (Fig. 5-54). An interpretation of this is provided by the *Jahn-Teller theorem* which states that any nonlinear molecule in an orbitally degenerate state will undergo a distortion which lowers the energy of the molecule and removes the degeneracy of the state. By orbital degeneracy, we mean any state whose orbital degeneracy is $\geqslant$ 2, *i.e.*, E and T states. The ground state of $Ti(H_2O)_6{}^{3+}$, ${}^2T_{2g}$, is orbitally degenerate and a distortion in which two opposite ligands move closer to the Ti atom occurs, as shown in Fig. 5-55.† This reduces the symmetry of the complex to D_{4h} and gives a ground state term of ${}^2B_{2g}$. The single $t_{2g} \rightarrow e_g$ transition of the O_h complex is replaced by two transitions of similar energy in the D_{4h} complex: ${}^2B_{2g} \rightarrow {}^2B_{1g}$ and ${}^2B_{2g} \rightarrow {}^2A_{1g}$. The two transitions must overlap extensively to produce the asymmetric

†The Jahn-Teller theorem states that distortion will occur to lower the energy of the molecule, but it does not say how much or what type of distortion will occur. In some cases the distortion may be so small that we cannot detect it.

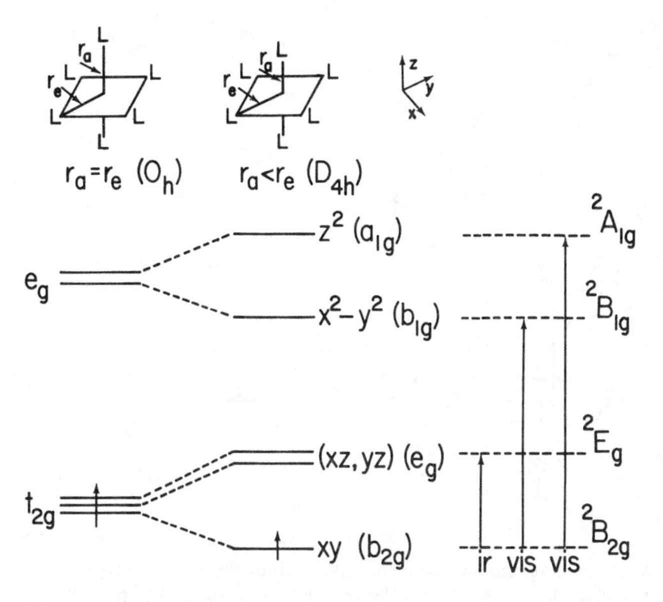

Fig. 5-55. Distortion of a d[1] octahedral complex by compression of bonds on the *z* axis and the resulting energy level diagram and expected transitions.

absorption band in Fig. 5-54. A very low energy $^2B_{2g} \rightarrow {}^2E_g$ transition is expected in the infrared region of the spectrum as a result of this distortion.

The d^2 configuration $(t_{2g})^2(e_g)^0$ yields a $^3T_{1g}$ ground state. Table 5-16 tells us that higher energy configurations of the d^2 ion give three more triplet states ($^3T_{1g}$, $^3T_{2g}$, and $^3A_{2g}$) to which spin-allowed d-d transitions may occur. This is shown in the state diagram, Fig. 5-56 (a), and in the calculated energy level diagram, Fig. 5-53 (sometimes called an Orgel diagram). Based on Fig. 5-53 we would expect three low energy d-d transitions for a d^2 species, and these are observed for $V(H_2O)_6{}^{3+}$ at the following energies:

$$^3T_{1g} \rightarrow {}^3T_{2g} \qquad 17.4 \text{ kK}$$

$$^3T_{1g} \rightarrow {}^3T_{1g} \qquad 25.2 \text{ kK}$$

$$^3T_{1g} \rightarrow {}^3A_{2g} \qquad 34.5 \text{ kK}$$

By fitting these observed energies to the spacings between the triplet states in Fig. 5-53, one can determine the value of Δ_0 for this particular complex.

The $d^{10-2} = d^8$ complexes should give rise to the same groups of states as the d^2 complexes, but in the reverse order. This is illustrated in

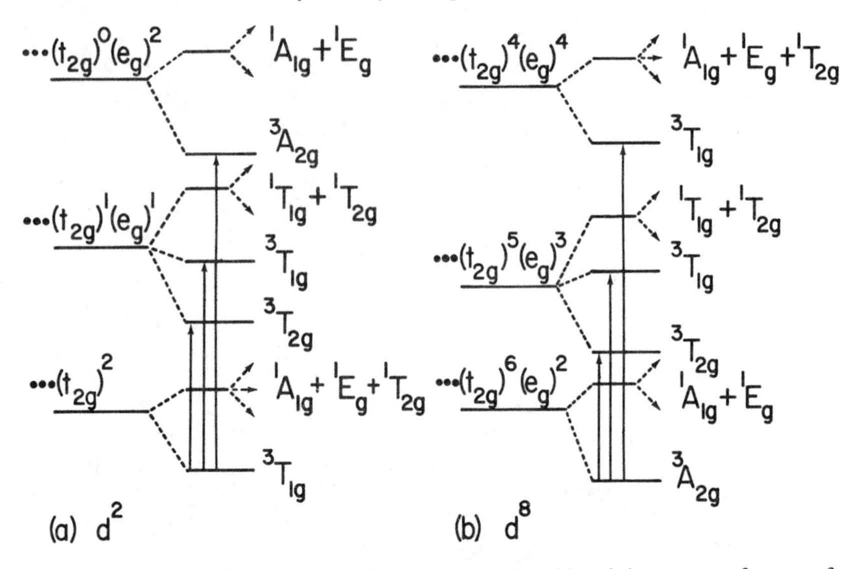

Fig. 5-56. State diagram showing the inverse relationship of the groups of states of d^2 and d^8 complexes. A similar inverse relationship holds for all d^n and d^{10-n} configurations.

Fig. 5-56 (b) and the spectra of two Ni(II) (d^8) complexes indeed show three bands due to the spin-allowed d-d transitions (Fig. 5-57).

Pink $Co(H_2O)_6^{2+}$ is a high spin d^7 ion of configuration $(t_{2g})^5(e_g)^2$, yielding a $^4T_{1g}$ ground state. This can be seen from Table 5-16 by replacing $(t_{2g})^5(e_g)^2$ by the equivalent configuration $(t_{2g})^1(e_g)^2$. The other configurations of d^7 give rise to three quartet states to which we might expect d-d transitions. A partial energy level diagram, Fig. 5-58, leads us to expect that the lowest energy transition is $^4T_{1g} \rightarrow {}^4T_{2g}$. This is seen in the near ir at 8350 cm^{-1} (1200 nm). The $^4T_{1g}(F) \rightarrow {}^4T_{1g}(P)$ transition is the main feature of the visible spectrum, Fig. 5-59. The $^4T_{1g} \rightarrow {}^4A_{2g}$ transition is expected at slightly lower energy (~560 nm) and is apparently buried under the stronger $^4T_{1g} \rightarrow {}^4T_{1g}$ absorption band.

The spectrum of the pale pink complex $Mn(H_2O)_6^{2+}$ appears in Fig. 5-60 and features a myriad of weak peaks. The high spin d^5 configuration gives rise to a $^6A_{1g}$ ground state and you should be able to reason that any d-d transition of this species *must* be spin-forbidden. Table 5-16 shows that there are a multitude of states for a d^5 species; so the large number of bands in the spectrum is not unexpected.

High spin Fe^{3+} is also a d^5 ion and gives rise to several very weak absorptions in the visible region of the spectrum (and often some strong

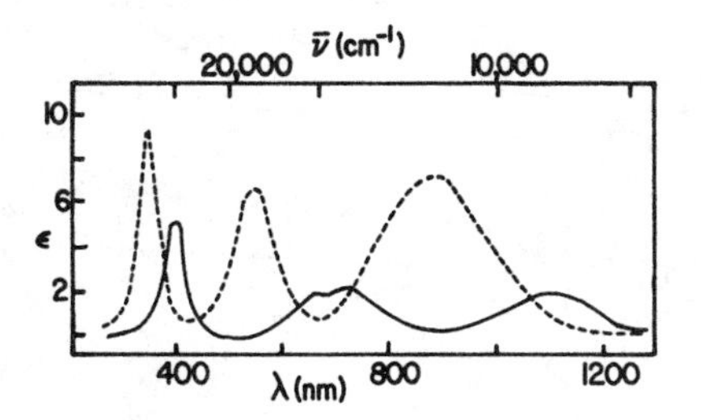

Fig. 5-57. Absorption spectrum of $Ni(H_2O)_6^{2+}$ (——) and $Ni(H_2NCH_2CH_2NH_2)_3^{2+}$ (----). From F.A. Cotton and G. Wilkinson, *Advanced Inorganic Chemistry*, John Wiley & Sons, N.Y., 1972.

Table 5-16. States of $(t_{2g})^x(e_g)^y$ Configurations

Electronic configuration		Electronic states
Free ion	Ion subject to octahedral symmetry	
d^1, d^9	$(e_g)^1$	^2E_g
	$(t_{2g})^1$	$^2T_{2g}$
d^2, d^8	$(e_g)^2$	$^3A_{2g}$, $^1A_{1g}$, ^1E_g
	$(t_{2g})^1(e_g)^1$	$^3T_{1g}$, $^3T_{2g}$, $^1T_{1g}$, $^1T_{2g}$
	$(t_{2g})^2$	$^3T_{1g}$, $^1A_{1g}$, ^1E_g, $^1T_{2g}$
d^3, d^7	$(e_g)^3$	^2E_g
	$(t_{2g})^1(e_g)^2$	$^4T_{1g}$, 2^2T_{1g}, 2^2T_{2g}
	$(t_{2g})^2(e_g)^1$	$^4T_{1g}$, $^4T_{2g}$, $^2A_{2g}$, 2^2T_{1g}, 2^2T_{2g}, 2^2E_g, $^2A_{1g}$
	$(t_{2g})^3$	$^4A_{2g}$, ^2E_g, $^2T_{1g}$, $^2T_{2g}$
d^4, d^6	$(e_g)^4$	$^1A_{1g}$
	$(t_{2g})^1(e_g)^3$	$^3T_{1g}$, $^3T_{2g}$, $^1T_{1g}$, $^1T_{2g}$
	$(t_{2g})^2(e_g)^2$	$^5T_{2g}$, ^3E_g, 3^3T_{1g}, 2^3T_{2g}, 2^1A_{1g}, $^1A_{2g}$, 3^1E_g, $^1T_{1g}$, 3^1T_{2g}, $^3A_{2g}$
	$(t_{2g})^3(e_g)^1$	^5E_g, $^3A_{1g}$, $^3A_{2g}$, 2^3E_g, 2^3T_{1g}, 2^3T_{2g}, $^1A_{1g}$, $^1A_{2g}$, ^1E_g, 2^1T_{1g}, 2^1T_{2g}
	$(t_{2g})^4$	$^3T_{1g}$, $^1A_{1g}$, ^1E_g, $^1T_{2g}$
d^5	$(t_{2g})^1(e_g)^4$	$^2T_{2g}$
	$(t_{2g})^2(e_g)^3$	$^4T_{1g}$, $^4T_{2g}$, $^2A_{1g}$, $^2A_{2g}$, 2^2E_g, 2^2T_{1g}, 2^2T_{2g}
	$(t_{2g})^3(e_g)^2$	$^6A_{1g}$, $^4T_{1g}$, $^4A_{2g}$, 2^4E_g, $^4A_{1g}$, $^4T_{2g}$, 2^2A_{1g}, $^2A_{2g}$, 3^2E_g, 4^2T_{1g}, 4^2T_{2g}
	$(t_{2g})^4(e_g)^1$	$^4T_{1g}$, $^4T_{2g}$, $^2A_{1g}$, $^2A_{2g}$, 2^2E_g, 2^2T_{1g}, 2^2T_{2g}
	$(t_{2g})^5$	$^2T_{2g}$

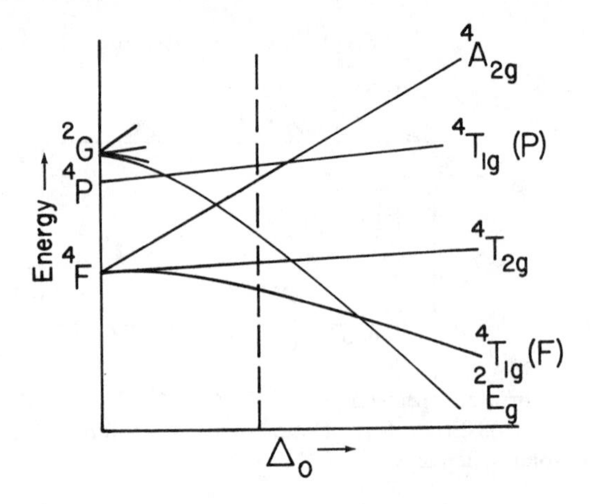

Fig. 5-58. Partial energy level diagram for a d^7 ion in an octahedral ligand field. The dashed line is the position of Δ_0 for $Co(H_2O)_6^{2+}$. From F.A. Cotton and G. Wilkinson, *Advanced Inorganic Chemistry*, John Wiley & Sons, N.Y., 1972.

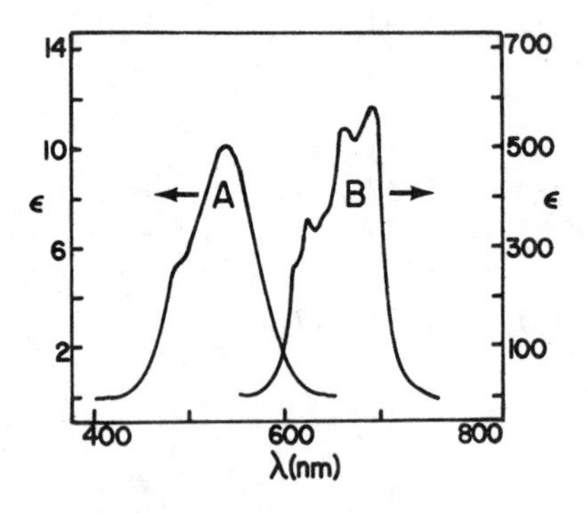

Fig. 5-59. Visible spectrum of $Co(H_2O)_6^{2+}$ (A) and $CoCl_4^{2-}$ (B). From F.A. Cotton and G. Wilkinson, *Advanced Inorganic Chemistry*, John Wiley & Sons, N.Y., 1972.

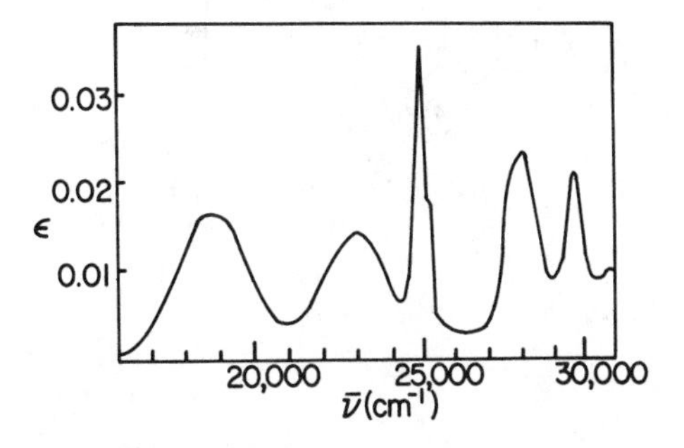

Fig. 5-60. Visible spectrum of $Mn(H_2O)_6^{2+}$. From C.K. Jorgensen, *Acta Chem. Scand.*, *8*, 1502 (1954).

charge transfer absorptions). In the dimeric oxo-bridged d^5 species

$$[(HEDTA)Fe-O-Fe(HEDTA)]^{2-}$$

$$\text{HEDTA} = \begin{matrix} HOCH_2CH_2 \diagdown & & \diagup CH_2CO_2^- \\ & NCH_2CH_2N & \\ ^-O_2C\,CH_2 \diagup & & \diagdown CH_2CO_2^- \end{matrix}$$

$[(HEDTA)Fe]_2O^{2-}$, weak sextet-quartet transitions are observed at low energy, as expected for each individual high spin Fe(III) in a roughly octahedral environment. But at higher energy are observed several intense ($\epsilon \approx 10^3$) bands which cannot be interpreted in terms of isolated Fe(III) ions or charge transfer bands (Fig. 5-61). As shown in Table 5-17, the energies of these bands are equal to sums of the energies of the spin-forbidden transitions. Band *e*, for example, is interpreted in terms of the simultaneous transition *a* + *b*. Although both *a* and *b* are spin-forbidden and weak, *e* is spin-allowed and strong:

$$\underset{^6A_1}{\uparrow\uparrow\uparrow\uparrow\uparrow} \xrightarrow{a} \underset{^4T_1}{\uparrow\uparrow\uparrow\uparrow\downarrow} \qquad \text{spin-forbidden}$$

$$\underset{^6A_1}{\downarrow\downarrow\downarrow\downarrow\downarrow} \xrightarrow{b} \underset{^4T_2}{\downarrow\downarrow\downarrow\downarrow\uparrow} \qquad \text{spin-forbidden}$$

$$\underset{s\,=\,0}{\uparrow\uparrow\uparrow\uparrow\uparrow + \downarrow\downarrow\downarrow\downarrow\downarrow} \xrightarrow{e} \underset{s\,=\,0}{\uparrow\uparrow\uparrow\uparrow\downarrow + \downarrow\downarrow\downarrow\downarrow\uparrow} \qquad \text{spin-allowed}$$

Transitions such as *e* have been named simultaneous pair excitations. Such pair excitations are reminiscent of the pair excitations invoked to explain the spectrum of gaseous O_2 under high pressure (Fig. 5-17).

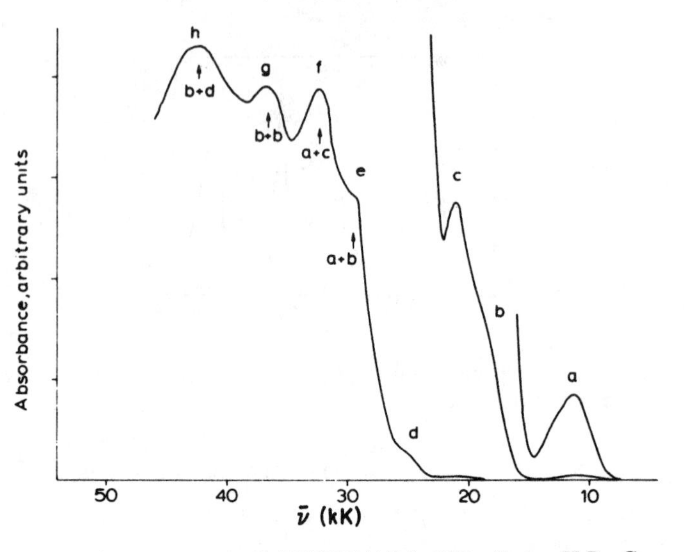

Fig. 5-61. Absorption spectrum of $[(HEDTA)Fe]_2O^{2-}$. From H.B. Gray and H.J. Schugar, "Electronic Structures of Iron Complexes," in *Inorganic Biochemistry* (G.L. Eichhorn, ed.), Vol. I., Elsevier, Amsterdam, 1973.

Table 5-17. Spectral Data for $[(HEDTA)Fe]_2O^{2-}$ [a]

Band	$\bar{\nu}_{max}$(kK)	Assignment
a	11.2	$^6A_1 \rightarrow {}^4T_1$
b	18.2	$^6A_1 \rightarrow {}^4T_2$
c	21.0	$^6A_1 \rightarrow ({}^4A_1, {}^4E)$
d	24.4	$^6A_1 \rightarrow {}^4T_2$
e	29.2	$a + b = 29.4$
f	32.5	$a + c = 32.2$
g	36.8	$b + b = 36.4$
h	42.6	$b + d = 42.6$

[a] From H.J. Schugar, G.R. Rossman, C.G. Barraclough, and H.B. Gray, *J. Amer. Chem. Soc.*, *94*, 2683 (1972).

Problems

5-40. Use a drawing similar to the ones on page 397 to show that any d-d transition of a high spin d^5 complex must be spin-forbidden.

5-41. Is it sensible that the absorption bands of $Ni(H_2O)_6{}^{2+}$ are at lower energy than the corresponding bands of $Ni(H_2NCH_2CH_2NH_2)_3{}^{2+}$ in Fig. 5-57?

5-42. The visible spectrum of the deep blue tetrahedral $CoCl_4{}^{2-}$ ion is shown in Fig. 5-59. As is often observed, the intensity of the absorption is much greater than the absorption intensity of octahedral complexes of the same metal. Tetrahedral Co(II) species also exhibit two, lower energy, spin-allowed d-d transitions in the near infrared between 3000 and 8000 cm^{-1}. Given that the order of the e and t_2 orbitals is reversed on going from octahedral to tetrahedral symmetry, draw a state diagram similar to Fig. 5-56 showing these three transitions. Give an example of an octahedral complex which will have the same qualitative state diagram that you just drew.

5-43. A Tanabe-Sugano energy level diagram for a d^6 octahedral complex is shown below. For our purposes, the ordinate and abscissa are given in arbitrary energy units. Arrows representing d-d transitions are drawn at appropriate values of Δ_0 for CoF_6^{3-} and $Co(H_2NCH_2CH_2NH_2)_3^{3+}$. Knowing what you do about the spectrochemical series, which arrows apply to which complex? Why is there a discontinuity in the slopes of the lines at one particular value of Δ_0?

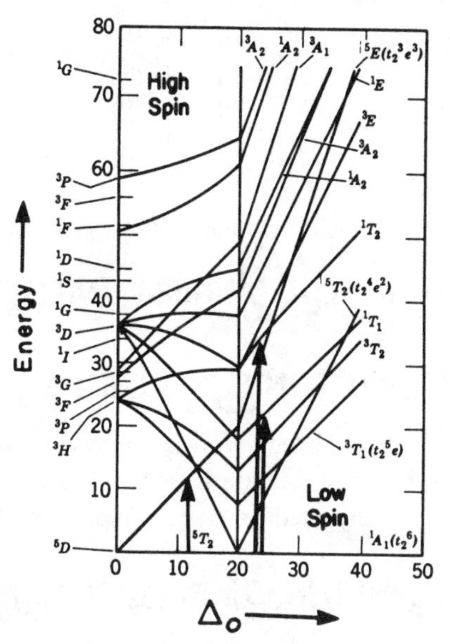

Tanabe-Sugano diagram for d^6 octahedral species showing transitions at energies appropriate to CoF_6^{3-} and $Co(H_2NCH_2CH_2NH_2)_3^{3+}$. The arrows for the low spin complex should actually be superimposed at the same value of Δ_0. Adapted from Y. Tanabe and S. Sugano, *J. Phys. Soc.* (Japan), *9*, 766 (1954).

C. *Charge Transfer.*† In the high energy region of most transition metal spectra are intense bands attributable to charge transfer transitions. For example, in the spectrum of $CoCl_4^{2-}$ (whose d-d spectrum is shown in Fig 5-59) we find a strong band at 43 kK ($\epsilon \approx 4000$). In tetrahedral complexes the order of the t_2 and e orbitals is reversed from that of the t_{2g} and e_g

† For a discussion of charge transfer spectra, see A.B.P. Lever, *J. Chem. Ed.*, *51*, 612 (1974).

orbitals in O_h symmetry. So the configuration of this d^8 species is $(e)^4(t_2)^4$. In the charge transfer transition, an electron localized in the Cl(3p) orbitals (called ligand pi or $L\pi$) is promoted to the predominantly metal-in-character t_2 orbitals.

It is also possible to have intermolecular charge transfer, as in the ion pair below. Solutions of this salt exhibit a band near 14 kK ($\epsilon \approx 180$) absent in the separate spectra of cation and anion in the presence of other kinds of counterions. The transition has been interpreted in terms of anion → cation charge transfer.[†]

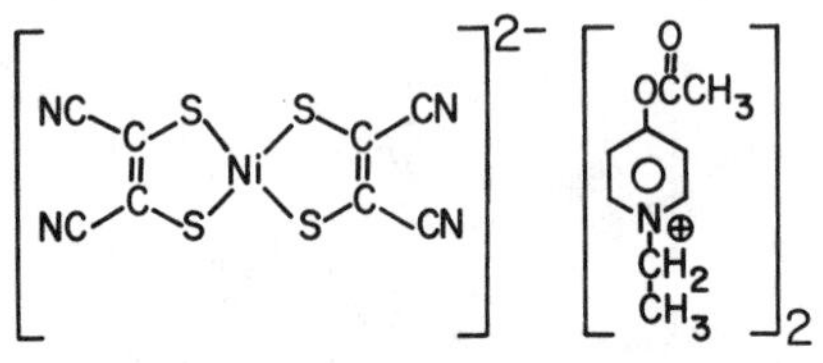

The Ru(II) and Ru(III) species, *A* and *B*, do not give rise to any absorption

$$(H_3N)_5Ru(II)N\bigcirc NRu(II)(NH_3)_5{}^{4+}$$

$$(H_3N)_5Ru(III)N\bigcirc NRu(III)(NH_3)_5{}^{6+}$$

bands in the near ir region of the spectrum. However, the mixed Ru(II)-Ru(III) species, *C*, exhibits an intense band at 6370 cm^{-1} (1570 nm, $\epsilon \approx$ 6000) which has been interpreted as an intramolecular Ru(II)Ru(III) →

$$(H_3N)_5Ru(II)N\bigcirc NRu(III)(NH_3)_5{}^{5+}$$

C

Ru(III)Ru(II) charge transfer transition.[††] The energy of the transition represents that needed to place a Ru(III) ion in an arrangement of ligands suitable for Ru(II), and *vice versa.* This interpretation is strongly supported by photoelectron (ESCA) studies,[†††] and similar near ir transitions have been observed in a variety of similar "mixed valence" compounds.

[†] I.G. Dance and P.J. Solstad, *J. Amer. Chem. Soc.*, *95*, 7256 (1973).
[††] C. Creutz and H. Taube, *J. Amer. Chem. Soc.*, *91*, 3988 (1969).
[†††] P.H. Citrin, *J. Amer. Chem. Soc.*, *95*, 6472 (1973).

Problems

5-44. Rationalize the order of energies of the following pairs of transitions:

	Complex	Energy(kK)	Assignment
(a)	$FeCl_4^{2-}$	45.5	$L\pi \rightarrow M(e)$
	$FeBr_4^{2-}$	40.9	$L\pi \rightarrow M(e)$
(b)	$OsCl_6^{3-}$	35.4	$L\pi \rightarrow M(t_{2g})$
	$OsCl_6^{2-}$	27.0	$L\pi \rightarrow M(t_{2g})$

5-10. Concluding Remarks

It appears to this writer that while we seem to be able to adequately, or usually adequately, rationalize the appearance of electronic and vibrational spectra, assignments are often less than satisfying. When we lack the number of bands we expect, we can (reasonably) say that some are too weak to see or are buried beneath stronger bands. When we find too many bands, we can invoke any of a dozen or so reasons (Jahn-Teller distortion, spin-orbit coupling, combination bands, Fermi-resonance) why the bands appear. Richard Feynman has stated in his marvelous physics texts that, "the basis of a science is its ability to *predict*. To predict means to tell what will happen in an experiment that has never been done before." Clearly, we are nowhere near the stage of being able to predict the appearance of molecular spectra from first principles. As we approach that capability (which could take a long time), some, if not many, or even most, of the ideas expressed in this volume will fall by the wayside. It should always be remembered that what is *real* is some inexplicable wiggly line in a spectrum, and what is a figment of our imaginations is a $^2T_2 \rightarrow {}^2A_1$ transition. If a "full" understanding of molecular spectra is at the peak of a small mountain, we are going to shed many more layers of theory before we beat a path to the pinnacle of that mountain.

On that note, I bid you good night.

Additional Problems

5-45. Determine the symmetries of the first two singlet excited states of Li_2 to which orbitally-allowed transitions are observed at $E_{0-0} = 14{,}021$ and $20{,}398\ cm^{-1}$. Draw a state diagram showing these transitions.

5-46. The CO^+ cation can be seen in oxyhydrogen flames, in which CO is ionized and promoted to electronically excited states. Three prominent emission lines from such a flame are observed at 20,408; 25,226; and 45,634 cm^{-1}. (In fact, the first of these emissions is also observed in the tails of comets.) The ground state configuration of CO^+ is...$(4\sigma^+)^2(1\pi)^4(5\sigma^+)^1$ and the first two excited configurations are ... $(4\sigma^+)^2(1\pi)^3(5\sigma^+)^2$ and ... $(4\sigma^+)^1(1\pi)^4(5\sigma^+)^2$. How can you explain these three emission bands in terms of the configurations just mentioned?

5-47. What are the multiplicities (*just* the multiplicities) of the states which can arise from the electronic configuration $(e)^3(a_1)^1(t_2)^1$?

5-48. The configuration of a diatomic molecule is $(\sigma^+)^2(\pi)^2(\sigma^+)^0$. What is the orbital symmetry and spin multiplicity of the ground state? (Remember that only *one* state can be the ground state.) The first excited state is $(\sigma^+)^2(\pi)^1(\sigma^+)^1$. What excited state terms are generated? The electronic spectrum exhibits the following absorption bands: 10 kK ($\epsilon = 10^{-3}$), 22 kK ($\epsilon = 10^{-2}$), 30 kK ($\epsilon = 10^4$), and 31 kK ($\epsilon = 10^{-3}$). Suggest an assignment of each band.

5-49. To what states does the configuration $(g_u)^4$ in the point group I_h give rise?

5-50. The lowest energy fully allowed transition of CO is observed at 64.7 kK. It is assigned ${}^1\Sigma^+ \rightarrow {}^1\Pi$. The next two fully-allowed transitions are both assigned ${}^1\Sigma^+ \rightarrow {}^1\Sigma^+$ and are observed at 86.9 and 91.9 kK. Make a state diagram for CO and show these three transitions. Can you assign the second and third excited configurations?

5-51. Using eq. 5-73, show that the f orbitals transform as $t_2 + g$ in the point group I, which includes only the identity and proper rotation operations of I_h. For f orbitals, $L = 3$. How will the f orbitals transform in the point group I_h?

5-52. Photoionization of CO, in which an electron is removed from the 1π orbital (Fig. 4-20), produces peaks at *IE*'s of 16.53, 16.73, 16.92, 17.11, 17.29, 17.47, 17.64, 17.81, 17.98, 18.15, 18.31, 18.47, ... eV. From these data, calculate $\bar{\omega}_e$ and $\bar{\omega}_e x_e$ for the resulting ${}^2\Pi$ state. (See Problem 3-39, Section 3-10.)

5-53. The lowest energy electronic transition of HCN (Section 4-6-B) is expected to be $\pi \rightarrow \pi^*$ and the second ought to be $n \rightarrow \pi^*$. If the molecule is in the $n \rightarrow \pi^*$ excited state, will it be a stronger or weaker Lewis base than the ground state molecule? The ground state dipole moment is indicated schematically below:

$$\underset{\longleftarrow\!\!\!+}{N \equiv C - H}$$

How will the dipole moment change in the $n \rightarrow \pi^*$ excited state? Protonated nitrogen, NNH^+, is isoelectronic with HCN and ought to have a similar mo scheme. Do you expect the $n \rightarrow \pi^*$ transition of NNH^+ to occur at higher or lower energy than the same transition of HCN? Why?

5-54. The point group for atoms, which have spherical symmetry, is called K_h (*Kugel Gruppe*). This group includes E, i, an infinite number of mirror planes, and an infinite number of C_n axes of every order. The somewhat simpler group, K, has the infinite number of proper rotation axes and the element E, but no center of inversion or mirror planes. The character table for this simpler group is given in Appendix A and a partial direct product table is given in Appendix B. The names of the irreducible representations are S, P, D, F, . . . *etc*., and the orbitals of the same name form bases for each of these representations. Using the rules developed in Section 5-3-D and using the character and direct product tables for K, derive the Russell-Saunders terms for the following atomic configurations: (a) $1s^2 2s^2 2p^2$; (b) $1s^2 2s^2 2p^1 3p^1$; (c) $4d^1 5s^1$; (d) $4d^1 5p^1$; (e) $3d^2$; (f) $2p^3$ (a challenging application of eqs. 5-23—see text in Appendix B).

5-55. The selection rules for atomic spectra are as follows:

$$\Delta L = 0 \text{ or } \pm 1 \text{ (but } L = 0 \rightarrow L = 0 \text{ is forbidden)}$$
$$\Delta S = 0$$
$$\Delta J = 0 \text{ or } \pm 1 \text{ (but } J = 0 \rightarrow J = 0 \text{ is forbidden)}$$

Using these rules, which of the following transitions are allowed: (a) ${}^2S_{1/2} \rightarrow {}^2P_{3/2}$; (b) ${}^2S_{1/2} \rightarrow {}^2P_{1/2}$; (c) ${}^2S_{1/2} \rightarrow {}^2D_{3/2}$; (d) ${}^1P_1 \rightarrow {}^1D_2$; (e) ${}^3P_1 \rightarrow {}^3F_2$; (f) ${}^3S_1 \rightarrow {}^3S_1$? Using the transition moment integral in eq. 5-29 and the character and direct product tables for the group K, show that the predictions of group theory on the allowedness of these transitions agree with those of the ΔL selection rule just stated.

5-56. The $Re_2Cl_8^{2-}$ anion has been shown by x-ray crystallography to possess the eclipsed D_{4h} geometry below. The anion is believed to have a *quadruple*

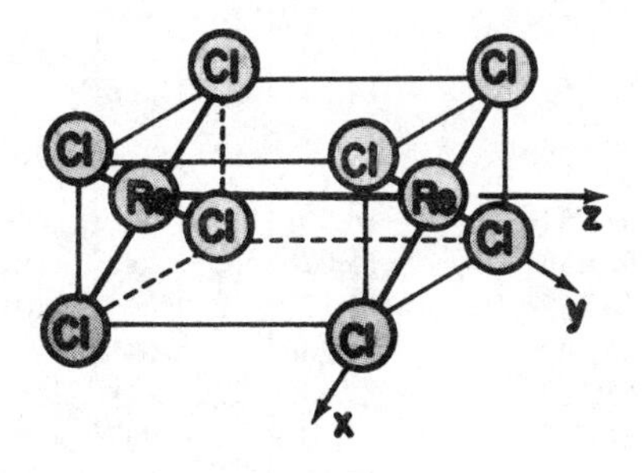

bond between the two metal atoms formed by the overlap of d orbitals on these two atoms. The bonding consists of a σ bond, two π bonds, and a δ bond. These have zero, one, and two nodes, respectively, colinear with the bond axis:

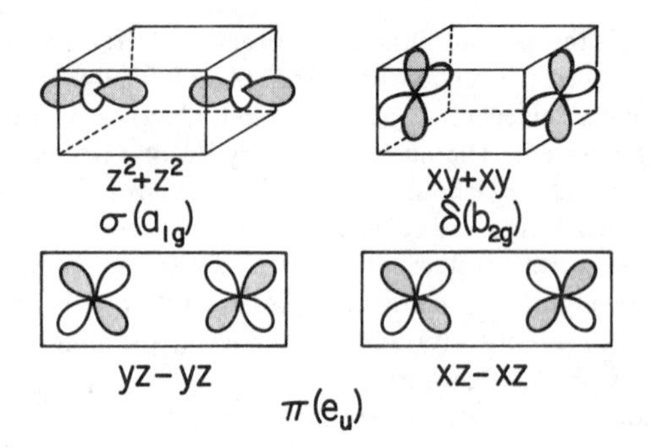

The mo scheme for this anion has been the subject of considerable controversy. (For leading references, see J.G. Norman, Jr. and H.J. Kolari, *J. Amer. Chem. Soc.*, *97*, 33 (1975); R.J.H. Clark and M.L. Franks, *Ibid.*, 2691 [1975].) A partial scheme, which we adopt for this problem, is shown below:

b_{1u}	——	$\delta^*(xy - xy)$
b_{2g}	$\downarrow\uparrow$	$\delta(xy + xy)$
a_{1g}	$\downarrow\uparrow$	$\sigma(z^2 + z^2)$
e_u	$\downarrow\uparrow$ $\downarrow\uparrow$	$\pi(yz - yz, xz - xz)$

The z-polarized spectrum of the first spin-allowed transition of $[(n\text{-}C_4H_9)_4N]_2[Re_2Cl_8]$ at three temperatures was shown in Fig. 5-7. At liquid nitrogen temperature, a single vibronic progression appears. At liquid helium temperature, this splits into two progressions. The entire absorption band is z-polarized.

Band	Position (nm)	Position (cm^{-1})	Spacings (cm^{-1})					
0-0	705.50	14,174						
			126					
A	699.29	14,300		225				
			99			367		
B	694.50	14,399			241			
			142				345	
C	687.69	14,541		246				
			104					383
D	682.84	14,645			241			
			137			350		
E	676.51	14,782		246				
			109				381	
F	671.54	14,891			244			
			135					355
G	665.50	15,026		246				
			111			375		
H	660.65	15,137			240			
			129				359	
I	655.03	15,266		248				
			119					370
J	650.00	15,385			241			
			122			376		
K	644.85	15,507		247				
			125					
L	639.70	15,632						

(a) Is z-polarization consistent with the proposed mo scheme? (b) The vibrations of $Re_2Cl_8{}^{2-}$ transform as $3a_{1g} + 2b_{1g} + b_{2g} + 3e_g + a_{1u} + 2a_{2u} + b_{1u} + 2b_{2u} + 3e_u$. Which modes may serve as vibronic origins for this electronic transition in z polarization? (c) The positions and spacings between the peaks of the spectrum are listed above. The position of the 0-0 band is not well-defined and is somewhat arbitrarily taken at 705.5 nm. For each peak, *A–L*, we have listed one more significant figure than is really justified by the data so that we can estimate spacings between peaks to the nearest wave number. Why can we assume that the first shoulder at 705.5 nm is 0-0 and not a hot band? (d) A tentative assignment of the vibrational spectrum of $Re_2Cl_8{}^{2-}$ is given below. Suggest an assignment of the electronic spectrum (peaks *A–L*), labelling each vibronic transition as a fundamental (i_0^1), overtone (i_0^n), or combination band ($i_0^n j_0^m$). Give the average value for the wave number of any excited state vibration you use in assigning the spectrum. Make the assignment as simple as you can.

Observed and Calculated Frequencies of $Re_2Cl_8{}^{2-}$[a]

Symmetry	Number	Frequency (cm^{-1}) Observed	Calculated	Symmetry	Number	Frequency (cm^{-1}) Observed	Calculated
a_{1g}	1	—	359	a_{1u}	10	—	—
a_{1g}	2	274	274	a_{2u}	11	347	347
a_{1g}	3	(115)	117	a_{2u}	12	164	168
b_{1g}	4	—	356	b_{1u}	13	—	189
b_{1g}	5	(115)	122	b_{2u}	14	—	356
b_{2g}	6	—	189	b_{2u}	15	—	122
e_g	7	—	351	e_u	16	335	334
e_g	8	—	185	e_u	17	154	155
e_g	9	(140)	145	e_u	18	124	119

[a] From W.K. Bratton, F.A. Cotton, M. Debeau, and R.A. Walton, *J. Coord. Chem.*, *1*, 121 (1971). Frequencies in parentheses are of questionable validity.

5-57. Comparing the observed vibrational spacings in the photoelectron spectrum below to the N_2 ground state value of $\bar{\nu} = 2331\ cm^{-1}$, what can you say about the bonding or antibonding nature of the three highest filled orbitals of N_2?

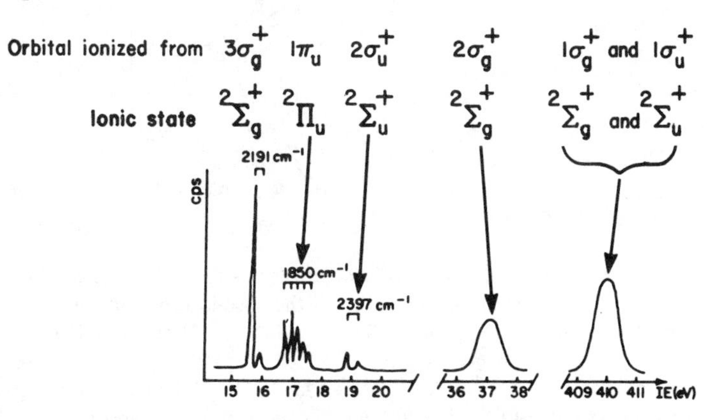

He I photoelectron spectrum of N_2 reproduced from H. Bock and P.D. Mollère, *J. Chem. Ed. 51*, 506 (1974).

5-58. Derive correlation diagrams similar to Fig. 5-52 for octahedral and tetrahedral d^8 complexes, using the same order of atomic states as in Fig. 5-52.

5-59. The figure below shows the time dependent spectrum of a thiocyanate complex which rearranges spontaneously to an isothiocyanate complex. At $t = 0$ only thiocyanate is present. At $t = \infty$, only isothiocyanate is present. The equilibrium constant is large and the reverse reaction is negligible.

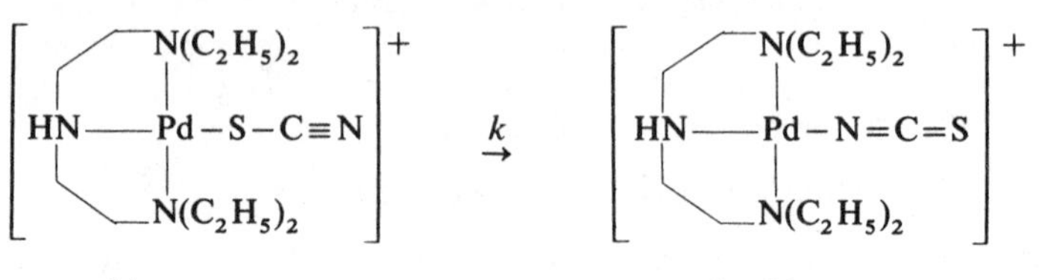

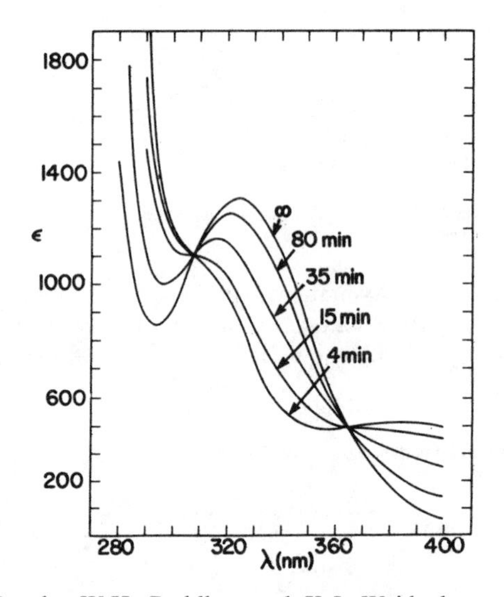

From F. Basolo, W.H. Baddley, and K.J. Weidenbaum, *J. Amer. Chem. Soc.*, *88*, 1576 (1966).

When you look at a series of spectra such as these, the first very remarkable point is that at some wavelengths all the spectra cross. In our spectra these *isosbestic points* are found at 307 and 364 nm. Why should this be? Let's consider a hypothetical sample in a cell of pathlength 1.00 cm so we can ignore the term ℓ in eq. 2-15 (Beer's law). The absorbance at any particular wavelength due to the two species of concentrations c_i and c_j is

$$A = \epsilon_i c_i + \epsilon_j c_j$$

where the extinction coefficients apply at the wavelength in question. In general, the spectra of the two components will cross at some points. At these points, $\epsilon_i = \epsilon_j$. Now when we consider pure I going to pure J,

$$c_i + c_j \equiv c_0$$
$$A = \epsilon_i c_i + \epsilon_j c_j = \epsilon_i c_i + \epsilon_j(c_0 - c_i)$$
$$A = (\epsilon_i - \epsilon_j)c_i + \epsilon_j c_0$$

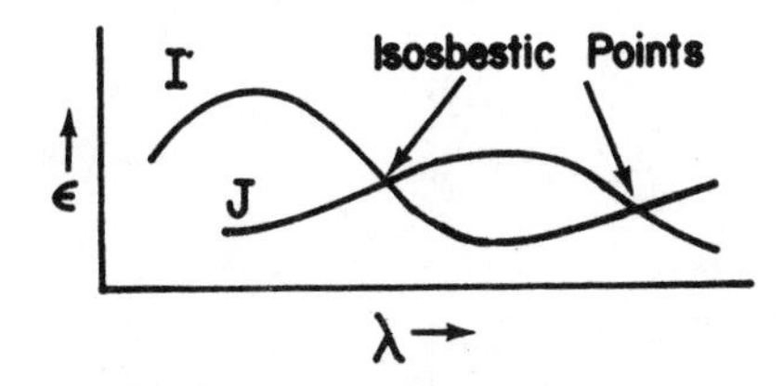

At the points where the spectra of I and J cross, $\epsilon_i = \epsilon_j$. Therefore, A = $\epsilon_j c_0$. But $\epsilon_j c_0$ is a constant. Hence if the two spectra cross at a particular point when the total concentration of I and J is c_0, the spectrum of *any* mixture of I and J with total concentration c_0 will pass through that point. The existence of isosbestic points is strong evidence that only two species are present in solution. (But see Problem 5-60.)

Now back to our isomerization:

$$I \xrightarrow{k} J$$

$$\text{Rate} = -dc_i/dt = dc_j/dt = kc_i$$
$$\int_{c_0}^{c_i} dc_i/c_i = -\int_0^t kdt$$
$$\ln c_i/c_0 = -kt$$
$$\log_{10} c_i/c_0 = -kt/2.303 \qquad \text{(a)}$$

At any particular wavelength at a time t,

$$A = \epsilon_i c_i + \epsilon_j c_j = \epsilon_i c_i + \epsilon_j(c_0 - c_i)$$

At time $t = \infty$, $c_i = 0$ so $A_\infty = \epsilon_j c_0$. At *any* time, t:

$$A_\infty - A = (\epsilon_j - \epsilon_i)c_i$$
$$c_i = (A_\infty - A)/(\epsilon_j - \epsilon_i) \qquad \text{(b)}$$

Equation (b) gives us the concentration of species I or J at any time t. This allows us to measure the rate of the reaction. Substituting (b) into (a):

$$\log_{10}[(A_\infty - A)/c_0(\epsilon_j - \epsilon_i)] = -kt/2.303$$
$$\log_{10}(A_\infty - A) = -kt/2.303 + \log_{10} c_0(\epsilon_j - \epsilon_i)$$

Make a plot of $\log_{10}(A_\infty - A)$ *vs.* t, the slope of which is $-k/2.303$ and whose intercept is $\log_{10}c_0(\epsilon_j - \epsilon_i)$. These equations should be useful at any wavelength. Try them at 330 and 340 nm using the absorbance data below. $c_0 = 5.00 \times 10^{-4}$ M. ϵ_j can be determined from the $t = \infty$ spectrum. Determine ϵ_i at both 330 and 340 nm. Express the rate constant, k, determined at these two wavelengths in units of s^{-1}.

Time (min)	ϵ	Absorbance	
4	733	0.366	330 nm
15	844	0.422	
35	1012	0.506	
80	1188	0.594	
∞	1277	0.638	
4	560	0.280	340 nm
15	679	0.340	
35	836	0.418	
80	1017	0.508	
∞	1119	0.560	

5-60. Consider the following equilibria in which XY is an ambidentate ligand capable of binding through atoms X or Y:

$$M + XY \overset{K_1}{\rightleftharpoons} M{-}XY$$

$$M + XY \overset{K_2}{\rightleftharpoons} M{-}YX$$

Show that the region of the spectrum where XY absorbs may exhibit an isosbestic point as the concentration of XY is varied, despite the fact that there are more than two chromophores present in solution. This illustrates the important point that a solution with $2 + n$ species will exhibit isosbestic points if there are n relations between the concentrations of the species. In this case there are 3 species ($n = 1$) and the relation between them is $[M{-}XY]/[M{-}YX] = K_1/K_2$. Observation of isosbestic points does not necessarily imply, therefore, that there are only two chromophores in solution. (Reference: D.V. Stynes, *Inorg. Chem.*, *14*, 453 (1975).)

Related Reading

G.M. Barrow, *Introduction to Molecular Spectroscopy*, McGraw-Hill, N.Y., 1962.

F.A. Cotton, *Chemical Applications of Group Theory,* John Wiley & Sons, N.Y., 1971.

B.N. Figgis, *An Introduction to Ligand Fields,* John Wiley & Sons, N.Y., 1966.

F. Grum, "Visible and Ultraviolet Spectroscopy" in *Techniques of Chemistry* (A. Weissberger and B.W. Rossiter, eds.), John Wiley and Sons, N.Y., 1972.

G. Herzberg, *Electronic Spectra of Polyatomic Molecules,* Van Nostrand Reinhold, N.Y., 1966.

G. Herzberg, *Spectra of Diatomic Molecules,* Van Nostrand Reinhold, N.Y., 1950.

G. Herzberg, *The Spectra and Structure of Simple Free Radicals,* Cornell University Press, Ithaca, N.Y., 1971.
H.H. Jaffé and M. Orchin, *Theory and Applications of Ultraviolet Spectroscopy,* John Wiley & Sons, N.Y., 1962.
C.A. Parker, *Photoluminescence of Solutions,* Elsevier, Amsterdam, 1968.
M.B. Robin, *Higher Excited States of Polyatomic Molecules,* Academic Press, N.Y., 1974.
J.I. Steinfeld, *Molecules and Radiation,* Harper and Row, N.Y., 1974.
N.J. Turro, *Molecular Photochemistry,* W.A. Benjamin, N.Y., 1967.

Besides the landmark works of Herzberg mentioned in a footnote at the beginning of this chapter, the interested reader is especially referred to the book by Steinfeld which provides a logical next step after reading this chapter.

Appendix A
Character tables

Abbreviations: c, cosine; s, sine.

Nonaxial groups

C_1	E
A	1

C_s	E	σ_h			
A'	1	1	x, y, R_z	x^2, y^2, z^2, xy	$xz^2, yz^2, x(x^2-3y^2), y(3x^2-y^2)$
A''	1	−1	z, R_x, R_y	yz, xz	$z^3, xyz, z(x^2-y^2)$
$\Gamma_{x,y,z}$	3	1			

C_i	E	i			
A_g	1	1	R_x, R_y, R_z	$x^2, y^2, z^2, xy, xz, yz$	
A_u	1	−1	x, y, z		$z^3, xyz, z(x^2-y^2), xz^2, yz^2, x(x^2-3y^2), y(3x^2-y^2)$
$\Gamma_{x,y,z}$	3	−3			

C_n Groups

C_2	E	C_2			
A	1	1	z, R_z	x^2, y^2, z^2, xy	$z^3, xyz, z(x^2-y^2)$
B	1	−1	x, y, R_x, R_y	yz, xz	$xz^2, yz^2, x(x^2-3y^2), y(3x^2-y^2)$
$\Gamma_{x,y,z}$	3	−1			

$\epsilon = e^{2\pi i/3}$

C_3	E	C_3	C_3^2			
A	1	1	1	z, R_z	x^2+y^2, z^2	$z^3, x(x^2-3y^2), y(3x^2-y^2)$
E	$\begin{Bmatrix} 1 \\ 1 \end{Bmatrix}$	$\begin{matrix} \epsilon \\ \epsilon^* \end{matrix}$	$\begin{matrix} \epsilon^* \\ \epsilon \end{matrix}$	$(x, y), (R_x, R_y)$	$(x^2-y^2, xy), (yz, xz)$	$(xz^2, yz^2), [xyz, z(x^2-y^2)]$
$\Gamma_{x,y,z}$	3	0	0			

C_4	E	C_4	C_2	C_4^3			
A	1	1	1	1	z, R_z	x^2+y^2, z^2	z^3
B	1	−1	1	−1		x^2-y^2, xy	$xyz, z(x^2-y^2)$
E	$\begin{Bmatrix} 1 \\ 1 \end{Bmatrix}$	$\begin{matrix} i \\ -i \end{matrix}$	$\begin{matrix} -1 \\ -1 \end{matrix}$	$\begin{matrix} -i \\ i \end{matrix}$	$(x, y), (R_x, R_y)$	(yz, xz)	$(xz^2, yz^2)[x(x^2-3y^2), y(3x^2-y^2)]$
$\Gamma_{x,y,z}$	3	1	−1	1			

C_5	E	C_5	C_5^2	C_5^3	C_5^4		$\epsilon = e^{2\pi i/5}$	
A	1	1	1	1	1	$z,\ R_z$	$x^2+y^2,\ z^2$	z^3
E_1	$\begin{cases}1\\1\end{cases}$	$\begin{matrix}\epsilon\\\epsilon^*\end{matrix}$	$\begin{matrix}\epsilon^2\\\epsilon^{2*}\end{matrix}$	$\begin{matrix}\epsilon^{2*}\\\epsilon^2\end{matrix}$	$\begin{matrix}\epsilon^*\\\epsilon\end{matrix}$	$(x,\ y),\ (R_x,\ R_y)$	$(yz,\ xz)$	$(xz^2,\ yz^2)$
E_2	$\begin{cases}1\\1\end{cases}$	$\begin{matrix}\epsilon^2\\\epsilon^{2*}\end{matrix}$	$\begin{matrix}\epsilon^*\\\epsilon\end{matrix}$	$\begin{matrix}\epsilon\\\epsilon^*\end{matrix}$	$\begin{matrix}\epsilon^{2*}\\\epsilon^2\end{matrix}$		$(x^2-y^2,\ xy)$	$[xyz,\ z(x^2-y^2)],\ [x(x^2-3y^2),\ y(3x^2-y^2)]$
$\Gamma_{x,y,z}$	3	$1+2c\frac{2\pi}{5}$	$1+2c\frac{4\pi}{5}$	$1+2c\frac{4\pi}{5}$	$1+2c\frac{2\pi}{5}$			

C_6	E	C_6	C_3	C_2	C_3^2	C_6^5		$\epsilon = e^{2\pi i/6}$	
A	1	1	1	1	1	1	$z,\ R_z$	$x^2+y^2,\ z^2$	z^3
B	1	−1	1	−1	1	−1			$x(x^2-3y^2),\ y(3x^2-y^2)$
E_1	$\begin{cases}1\\1\end{cases}$	$\begin{matrix}\epsilon\\\epsilon^*\end{matrix}$	$\begin{matrix}-\epsilon^*\\-\epsilon\end{matrix}$	$\begin{matrix}-1\\-1\end{matrix}$	$\begin{matrix}-\epsilon\\-\epsilon^*\end{matrix}$	$\begin{matrix}\epsilon^*\\\epsilon\end{matrix}$	$(x,\ y)$ $(R_x,\ R_y)$	$(xz,\ yz)$	$(xz^2,\ yz^2)$
E_2	$\begin{cases}1\\1\end{cases}$	$\begin{matrix}-\epsilon^*\\-\epsilon\end{matrix}$	$\begin{matrix}-\epsilon\\-\epsilon^*\end{matrix}$	$\begin{matrix}1\\1\end{matrix}$	$\begin{matrix}-\epsilon^*\\-\epsilon\end{matrix}$	$\begin{matrix}-\epsilon\\-\epsilon^*\end{matrix}$		$(x^2-y^2,\ xy)$	$[xyz,\ z(x^2-y^2)]$
$\Gamma_{x,y,z}$	3	2	0	−1	0	2			

C_7	E	C_7	C_7^2	C_7^3	C_7^4	C_7^5	c_7^6	$\epsilon = e^{2\pi i/7}$		
A	1	1	1	1	1	1	1	z, R_z	x^2+y^2, z^2	z^3
E_1	1	ϵ	ϵ^2	ϵ^3	ϵ^{3*}	ϵ^{2*}	ϵ^*	(x, y)	(xz, yz)	(xz^2, yz^2)
	1	ϵ^*	ϵ^{2*}	ϵ^{3*}	ϵ^3	ϵ^2	ϵ	(R_x, R_y)		
E_2	1	ϵ^2	ϵ^{3*}	ϵ^*	ϵ	ϵ^3	ϵ^{2*}		(x^2-y^2, xy)	$[xyz, z(x^2-y^2)]$
	1	ϵ^{2*}	ϵ^3	ϵ	ϵ^*	ϵ^{3*}	ϵ^2			
E_3	1	ϵ^3	ϵ^*	ϵ^2	ϵ^{2*}	ϵ	ϵ^{3*}			$[x(x^2-3y^2), y(3x^2-y^2)]$
	1	ϵ^{3*}	ϵ	ϵ^{2*}	ϵ^2	ϵ^*	ϵ^3			
$\Gamma_{x,y,z}$	3	$1+2c\frac{2\pi}{7}$	$1+2c\frac{4\pi}{7}$	$1+2c\frac{6\pi}{7}$	$1+2c\frac{6\pi}{7}$	$1+2c\frac{4\pi}{7}$	$1+2c\frac{2\pi}{7}$			

C_8	E	C_8	C_4	C_2	C_4^3	C_8^3	C_8^5	C_8^7	$\epsilon = e^{2\pi i/8}$		
A	1	1	1	1	1	1	1	1	z, R_z	x^2+y^2, z^2	z^3
B	1	-1	1	1	1	-1	-1	-1			
E_1	1	ϵ	i	-1	$-i$	$-\epsilon^*$	$-\epsilon$	ϵ^*	(x, y)	(xz, yz)	(xz^2, xy^2)
	1	ϵ^*	$-i$	-1	i	$-\epsilon$	$-\epsilon^*$	ϵ	(R_x, R_y)		
E_2	1	i	-1	1	-1	$-i$	i	$-i$		(x^2-y^2, xy)	$[xyz, z(x^2-y^2)]$
	1	$-i$	-1	1	-1	i	$-i$	i			
E_3	1	$-\epsilon$	i	-1	$-i$	ϵ^*	ϵ	$-\epsilon^*$			$[x(x^2-3y^2), y(3x^2-y^2)]$
	1	$-\epsilon^*$	$-i$	-1	i	ϵ	ϵ^*	$-\epsilon$			
$\Gamma_{x,y,z}$	3	$1+\sqrt{2}$	1	-1	1	$1-\sqrt{2}$	$1-\sqrt{2}$	$1+\sqrt{2}$			

C_9	E	C_9	C_9^2	C_3	C_9^4	C_9^5	C_3^2	C_9^7	C_9^8	$\epsilon = e^{2\pi i/9}$		
A	1	1	1	1	1	1	1	1	1	z, R_z	$x^2 + y^2, z^2$	z^3
E_1	1	ϵ	ϵ^2	ϵ^3	ϵ^4	ϵ^{4*}	ϵ^{3*}	ϵ^{2*}	ϵ^*	(x, y) (R_x, R_y)	(xz, yz)	(xz^2, yz^2)
	1	ϵ^*	ϵ^{2*}	ϵ^{3*}	ϵ^{4*}	ϵ^4	ϵ^3	ϵ^2	ϵ			
E_2	1	ϵ^2	ϵ^4	ϵ^{3*}	ϵ^*	ϵ	ϵ^3	ϵ^{4*}	ϵ^{2*}		$(x^2 - y^2, xy)$	$[xyz, z(x^2 - y^2)]$
	1	ϵ^{2*}	ϵ^{4*}	ϵ^3	ϵ	ϵ^*	ϵ^{3*}	ϵ^4	ϵ^2			
E_3	1	ϵ^3	ϵ^{3*}	1	ϵ^3	ϵ^{3*}	1	ϵ^3	ϵ^{3*}			$[x(x^2 - 3y^2), y(3x^2 - y^2)]$
	1	ϵ^{3*}	ϵ^3	1	ϵ^{3*}	ϵ^3	1	ϵ^{3*}	ϵ^3			
E_4	1	ϵ^4	ϵ^*	ϵ^3	ϵ^{2*}	ϵ^2	ϵ^{3*}	ϵ	ϵ^{4*}			
	1	ϵ^{4*}	ϵ	ϵ^{3*}	ϵ^2	ϵ^{2*}	ϵ^3	ϵ^*	ϵ^4			
$\Gamma_{x,y,z}$	3	$1 + 2c\frac{2\pi}{9}$	$1 + 2c\frac{4\pi}{9}$	0	$1 + 2c\frac{8\pi}{9}$	$1 + 2c\frac{8\pi}{9}$	0	$1 + 2c\frac{4\pi}{9}$	$1 + 2c\frac{2\pi}{9}$			

C_n Groups (cont.)

C_{10}	E	C_{10}	C_5	C_{10}^3	C_5^2	C_2	C_5^3	C_{10}^7	C_5^4	C_{10}^9
A	1	1	1	1	1	1	1	1	1	1
B	1	-1	1	-1	1	-1	1	-1	1	-1
E_1	1	ϵ	ϵ^2	$-\epsilon^{2*}$	$-\epsilon^*$	-1	$-\epsilon$	$-\epsilon^2$	ϵ^{2*}	ϵ^*
	1	ϵ^*	ϵ^{2*}	$-\epsilon^2$	$-\epsilon$	-1	$-\epsilon^*$	$-\epsilon^{2*}$	ϵ^2	ϵ
E_2	1	ϵ^2	$-\epsilon^*$	$-\epsilon$	ϵ^{2*}	1	ϵ^2	$-\epsilon^*$	$-\epsilon$	ϵ^{2*}
	1	ϵ^{2*}	$-\epsilon$	$-\epsilon^*$	ϵ^2	1	ϵ^{2*}	$-\epsilon$	$-\epsilon^*$	ϵ^2
E_3	1	$-\epsilon^{2*}$	$-\epsilon$	ϵ^*	ϵ^2	-1	ϵ^{2*}	ϵ	$-\epsilon^*$	$-\epsilon^2$
	1	$-\epsilon^2$	$-\epsilon^*$	ϵ	ϵ^{2*}	-1	ϵ^2	ϵ^*	$-\epsilon$	$-\epsilon^{2*}$
E_4	1	$-\epsilon^*$	ϵ^{2*}	ϵ^2	$-\epsilon$	1	$-\epsilon^*$	ϵ^{2*}	ϵ^2	$-\epsilon$
	1	$-\epsilon$	ϵ^2	ϵ^{2*}	$-\epsilon^*$	1	$-\epsilon$	ϵ^2	ϵ^{2*}	$-\epsilon^*$
$\Gamma_{x,y,z}$	3	$1+2c\frac{\pi}{5}$	$1+2c\frac{2\pi}{5}$	$1-2c\frac{2\pi}{5}$	$1-2c\frac{\pi}{5}$	-1	$1-2c\frac{\pi}{5}$	$1-2c\frac{2\pi}{5}$	$1+2c\frac{2\pi}{5}$	$1+2c\frac{\pi}{5}$

C_{10} (cont.)		$\epsilon = e^{2\pi i/10}$	
A	z, R_z	x^2+y^2, z^2	z^3
B			
E_1	(x, y) (R_x, R_y)	(xz, yz)	(xz^2, yz^2)
E_2		(x^2-y^2, xy)	$[xyz, z(x^2-y^2)]$
E_3			$[x(x^2-3y^2), y(3x^2-y^2)]$
E_4			

C_{nv} Groups

C_{2v}	E	C_2	$\sigma_v(xz)$	$\sigma_v'(yz)$			
A_1	1	1	1	1	z	x^2, y^2, z^2	$z^3, z(x^2-y^2)$
A_2	1	1	−1	−1	R_z	xy	xyz
B_1	1	−1	1	−1	x, R_y	xz	$xz^2, x(x^2-3y^2)$
B_2	1	−1	−1	1	y, R_x	yz	$yz^2, y(3x^2-y^2)$
$\Gamma_{x,y,z}$	3	−1	1	1			

C_{3v}	E	$2C_3$	$3\sigma_v$			
A_1	1	1	1	z	x^2+y^2, z^2	$z^3, x(x^2-3y^2)$
A_2	1	1	−1	R_z		$y(3x^2-y^2)$
E	2	−1	0	$(x, y), (R_x, R_y)$	$(x^2-y^2, xy)(xz, yz)$	$(xz^2, yz^2), [xyz, z(x^2-y^2)]$
$\Gamma_{x,y,z}$	3	0	1			

C_{4v}	E	$2C_4$	C_2	$2\sigma_v$	$2\sigma_d$			
A_1	1	1	1	1	1	z	x^2+y^2, z^2	z^3
A_2	1	1	1	−1	−1	R_z		
B_1	1	−1	1	1	−1		x^2-y^2	$z(x^2-y^2)$
B_2	1	−1	1	−1	1		xy	xyz
E	2	0	−2	0	0	$(x, y), (R_x, R_y)$	(xz, yz)	$(xz^2, yz^2), [x(x^2-3y^2), y(3x^2-y^2)]$
$\Gamma_{x,y,z}$	3	1	−1	1	1			

C_{nv} Groups (cont.)

C_{5v}	E	$2C_5$	$2C_5^2$	$5\sigma_v$			
A_1	1	1	1	1	z	x^2+y^2, z^2	z^3
A_2	1	1	1	−1	R_z		
E_1	2	$2c\frac{2\pi}{5}$	$2c\frac{4\pi}{5}$	0	$(x, y), (R_x, R_y)$	(xz, yz)	(xz^2, yz^2)
E_2	2	$2c\frac{4\pi}{5}$	$2c\frac{2\pi}{5}$	0		(x^2-y^2, xy)	$[xyz, z(x^2-y^2)], [x(x^2-3y^2), y(3x^2-y^2)]$
$\Gamma_{x,y,z}$	3	$1+2c\frac{2\pi}{5}$	$1+2c\frac{4\pi}{5}$	1			

C_{6v}	E	$2C_6$	$2C_3$	C_2	$3\sigma_v$	$3\sigma_d$			
A_1	1	1	1	1	1	1	z	x^2+y^2, z^2	z^3
A_2	1	1	1	1	−1	−1	R_z		
B_1	1	−1	1	−1	1	−1			$x(x^2-3y^2)$
B_2	1	−1	1	−1	−1	1			$y(3x^2-y^2)$
E_1	2	1	−1	−2	0	0	$(x, y), (R_x, R_y)$	(xz, yz)	(xz^2, yz^2)
E_2	2	−1	−1	2	0	0		(x^2-y^2, xy)	$[xyz, z(x^2-y^2)]$
$\Gamma_{x,y,z}$	3	2	0	−1	1	1			

C_{7v}	E	$2C_7$	$2C_7^2$	$2C_7^3$	$7\sigma_v$			
A_1	1	1	1	1	1	z	x^2+y^2, z^2	z^3
A_2	1	1	1	1	-1	R_z		
E_1	2	$2c\frac{2\pi}{7}$	$2c\frac{4\pi}{7}$	$2c\frac{6\pi}{7}$	0	$(x, y), (R_x, R_y)$	(xz, yz)	(xz^2, yz^2)
E_2	2	$2c\frac{4\pi}{7}$	$2c\frac{6\pi}{7}$	$2c\frac{2\pi}{7}$	0		(x^2-y^2, xy)	$[xyz, z(x^2-y^2)]$
E_3	2	$2c\frac{6\pi}{7}$	$2c\frac{2\pi}{7}$	$2c\frac{4\pi}{7}$	0			$[x(x^2-3y^2), y(3x^2-y^2)]$
$\Gamma_{x,y,z}$	3	$1+2c\frac{2\pi}{7}$	$1+2c\frac{4\pi}{7}$	$1+2c\frac{6\pi}{7}$	1			

C_{8v}	E	$2C_8$	$2C_4$	$2C_8^3$	C_2	$4\sigma_v$	$4\sigma_d$			
A_1	1	1	1	1	1	1	1	z	x^2+y^2, z^2	z^3
A_2	1	1	1	1	1	-1	-1	R_z		
B_1	1	-1	1	-1	1	1	-1			
B_2	1	-1	1	-1	1	-1	1			
E_1	2	$\sqrt{2}$	0	$-\sqrt{2}$	-2	0	0	$(x, y), (R_x, R_y)$	(xz, yz)	(xz^2, yz^2)
E_2	2	0	-2	0	2	0	0		(x^2-y^2, xy)	$[xyz, z(x^2-y^2)]$
E_3	2	$-\sqrt{2}$	0	$\sqrt{2}$	-2	0	0			$[x(x^2-3y^2), y(3x^2-y^2)]$
$\Gamma_{x,y,z}$	3	$1+\sqrt{2}$	1	$1-\sqrt{2}$	-1	1	1			

C_{nv} Groups (cont.)

C_{9v}	E	$2C_9$	$2C_9^2$	$2C_3$	$2C_9^4$	$9\sigma_v$			
A_1	1	1	1	1	1	1	z	$x^2+y^2,\ z^2$	z^3
A_2	1	1	1	1	1	-1	R_z		
E_1	2	$2c\,\frac{2\pi}{9}$	$2c\,\frac{4\pi}{9}$	-1	$2c\,\frac{8\pi}{9}$	0	$(x, y), (R_x, R_y)$	(xz, yz)	(xz^2, yz^2)
E_2	2	$2c\,\frac{4\pi}{9}$	$2c\,\frac{8\pi}{9}$	-1	$2c\,\frac{2\pi}{9}$	0		(x^2-y^2, xy)	$[xyz, z(x^2-y^2)]$
E_3	2	-1	-1	2	-1	0			$[x(x^2-3y^2), y(3x^2-y^2)]$
E_4	2	$2c\,\frac{8\pi}{9}$	$2c\,\frac{2\pi}{9}$	-1	$2c\,\frac{4\pi}{9}$	0			
$\Gamma_{x,y,z}$	3	$1+2c\,\frac{2\pi}{9}$	$1+2c\,\frac{4\pi}{9}$	0	$1+2c\,\frac{8\pi}{9}$	1			

C_{10v}	E	$2C_{10}$	$2C_5$	$2C_{10}^3$	$2C_5^2$	C_2	$5\sigma_v$	$5\sigma_d$			
A_1	1	1	1	1	1	1	1	1	z	$x^2+y^2,\ z^2$	z^3
A_2	1	1	1	1	1	1	-1	-1	R_z		
B_1	1	-1	1	-1	1	-1	1	-1			
B_2	1	-1	1	-1	1	-1	-1	1			
E_1	2	$2c\,\frac{\pi}{5}$	$2c\,\frac{2\pi}{5}$	$-2c\,\frac{2\pi}{5}$	$-2c\,\frac{\pi}{5}$	-2	0	0	$(x, y), (R_x, R_y)$	(xz, yz)	(xz^2, yz^2)
E_2	2	$2c\,\frac{2\pi}{5}$	$-2c\,\frac{\pi}{5}$	$-2c\,\frac{\pi}{5}$	$2c\,\frac{2\pi}{5}$	2	0	0		(x^2-y^2, xy)	$[xyz, z(x^2-y^2)]$
E_3	2	$-2c\,\frac{2\pi}{5}$	$-2c\,\frac{\pi}{5}$	$2c\,\frac{\pi}{5}$	$2c\,\frac{2\pi}{5}$	-2	0	0			$[x(x^2-3y^2), y(3x^2-y^2)]$
E_4	2	$-2c\,\frac{\pi}{5}$	$2c\,\frac{2\pi}{5}$	$2c\,\frac{2\pi}{5}$	$-2c\,\frac{\pi}{5}$	2	0	0			
$\Gamma_{x,y,z}$	3	$1+2c\,\frac{\pi}{5}$	$1+2c\,\frac{2\pi}{5}$	$1-2c\,\frac{2\pi}{5}$	$1-2c\,\frac{\pi}{5}$	-1	1	1			

C_{nh} Groups

C_{2h}	E	C_2	i	σ_h			
A_g	1	1	1	1	R_z	x^2, y^2, z^2, xy	
B_g	1	−1	1	−1	R_x, R_y	xz, yz	
A_u	1	1	−1	−1	z		$z^3, xyz, z(x^2-y^2)$
B_u	1	−1	−1	1	x, y		$xz^2, yz^2, x(x^2-3y^2), y(3x^2-y^2)$
$\Gamma_{x,y,z}$	3	−1	−3	1			

C_{3h}	E	C_3	C_3^2	σ_h	S_3	S_3^5		$\epsilon = e^{2\pi i/3}$	
A'	1	1	1	1	1	1	R_z	x^2+y^2, z^2	$x(x^2-3y^2), y(3x^2-y^2)$
E'	$\begin{cases}1\\1\end{cases}$	ϵ ϵ^*	ϵ^* ϵ	1 1	ϵ ϵ^*	ϵ^* ϵ	(x, y)	(x^2-y^2, xy)	(xz^2, yz^2)
A''	1	1	1	−1	−1	−1	z		z^3
E''	$\begin{cases}1\\1\end{cases}$	ϵ ϵ^*	ϵ^* ϵ	−1 −1	$-\epsilon$ $-\epsilon^*$	$-\epsilon^*$ $-\epsilon$	(R_x, R_y)	(xz, yz)	$[xyz, z(x^2-y^2)]$
$\Gamma_{x,y,z}$	3	0	0	1	−2	−2			

C_{4h}	E	C_4	C_2	C_4^3	i	S_4^3	σ_h	S_4			
A_g	1	1	1	1	1	1	1	1	R_z	x^2+y^2, z^2	
B_g	1	−1	1	−1	1	−1	1	−1		x^2-y^2, xy	
E_g	$\begin{cases}1\\1\end{cases}$	i $-i$	−1 −1	$-i$ i	1 1	i $-i$	−1 −1	$-i$ i	(R_x, R_y)	(xz, yz)	
A_u	1	1	1	1	−1	−1	−1	−1	z		z^3
B_u	1	−1	1	−1	−1	1	−1	1			$xyz, z(x^2-y^2)$
E_u	$\begin{cases}1\\1\end{cases}$	i $-i$	−1 −1	$-i$ i	−1 −1	$-i$ i	1 1	i $-i$	(x, y)		$(xz^2, yz^2), [x(x^2-3y^2), y(3x^2-y^2)]$
$\Gamma_{x,y,z}$	3	1	−1	1	−3	−1	1	−1			

C_{nh} Groups (cont.)

C_{5h}	E	C_5	C_5^2	C_5^3	C_5^4	σ_h	S_5	S_5^7	S_5^3	S_5^9
A'	1	1	1	1	1	1	1	1	1	1
E_1'	1	ϵ	ϵ^2	ϵ^{2*}	ϵ^*	1	ϵ	ϵ^2	ϵ^{2*}	ϵ^*
	1	ϵ^*	ϵ^{2*}	ϵ^2	ϵ	1	ϵ^*	ϵ^{2*}	ϵ^2	ϵ
E_2'	1	ϵ^2	ϵ^*	ϵ	ϵ^{2*}	1	ϵ^2	ϵ^*	ϵ	ϵ^{2*}
	1	ϵ^{2*}	ϵ	ϵ^*	ϵ^2	1	ϵ^{2*}	ϵ	ϵ^*	ϵ^2
A''	1	1	1	1	1	−1	−1	−1	−1	−1
E_1''	1	ϵ	ϵ^2	ϵ^{2*}	ϵ^*	−1	$-\epsilon$	$-\epsilon^2$	$-\epsilon^{2*}$	$-\epsilon^*$
	1	ϵ^*	ϵ^{2*}	ϵ^2	ϵ	−1	$-\epsilon^*$	$-\epsilon^{2*}$	$-\epsilon^2$	$-\epsilon$
E_2''	1	ϵ^2	ϵ^*	ϵ	ϵ^{2*}	−1	$-\epsilon^2$	$-\epsilon^*$	$-\epsilon$	$-\epsilon^{2*}$
	1	ϵ^{2*}	ϵ	ϵ^*	ϵ^2	−1	$-\epsilon^{2*}$	$-\epsilon$	$-\epsilon^*$	$-\epsilon^2$
$\Gamma_{x,y,z}$	3	$1 + 2c\,\frac{2\pi}{5}$	$1 + 2c\,\frac{4\pi}{5}$	$1 + 2c\,\frac{4\pi}{5}$	$1 + 2c\,\frac{2\pi}{5}$	1	$-1 + 2c\,\frac{2\pi}{5}$	$-1 + 2c\,\frac{4\pi}{5}$	$-1 + 2c\,\frac{4\pi}{5}$	$-1 + 2c\,\frac{2\pi}{5}$

$\epsilon = e^{2\pi i/5}$

C_{5h} (cont.)			
A'	R_z	$x^2 + y^2,\ z^2$	
E_1'	(x, y)		$(xz^2,\ yz^2)$
E_2'		$(x^2 - y^2,\ xy)$	$[x(x^2 - 3y^2),\ y(3x^2 - y^2)]$
A''	z		z^3
E_1''	$(R_x,\ R_y)$	$(xz,\ yz)$	
E_2''			$[xyz,\ z(x^2 - y^2)]$

C_{6h}	E	C_6	C_3	C_2	C_3^2	C_6^5	i	S_3^5	S_6^5	σ_h	S_6	S_3	$\epsilon = e^{2\pi i/6}$		
A_g	1	1	1	1	1	1	1	1	1	1	1	1	R_z	$x^2+y^2,\ z^2$	
B_g	1	−1	1	−1	1	−1	1	−1	1	−1	1	−1			
E_{1g}	1	ϵ	$-\epsilon^*$	−1	$-\epsilon$	ϵ^*	1	ϵ	$-\epsilon^*$	−1	$-\epsilon$	ϵ^*	$(R_x,\ R_y)$	$(xz,\ yz)$	
	1	ϵ^*	$-\epsilon$	−1	$-\epsilon^*$	ϵ	1	ϵ^*	$-\epsilon$	−1	$-\epsilon^*$	ϵ			
E_{2g}	1	$-\epsilon^*$	$-\epsilon$	1	$-\epsilon^*$	$-\epsilon$	1	$-\epsilon^*$	$-\epsilon$	1	$-\epsilon^*$	$-\epsilon$		$(x^2-y^2,\ xy)$	
	1	$-\epsilon$	$-\epsilon^*$	1	$-\epsilon$	$-\epsilon^*$	1	$-\epsilon$	$-\epsilon^*$	1	$-\epsilon$	$-\epsilon^*$			
A_u	1	1	1	1	1	1	−1	−1	−1	−1	−1	−1	z		z^3
B_u	1	−1	1	−1	1	−1	−1	1	−1	1	−1	1			$x(x^2-3y^2),\ y(3x^2-y^2)$
E_{1u}	1	ϵ	$-\epsilon^*$	−1	$-\epsilon$	ϵ^*	−1	$-\epsilon$	ϵ^*	1	ϵ	$-\epsilon^*$	$(x,\ y)$		$(xz^2,\ yz^2)$
	1	ϵ^*	$-\epsilon$	−1	$-\epsilon^*$	ϵ	−1	$-\epsilon^*$	ϵ	1	ϵ^*	$-\epsilon$			
E_{2u}	1	$-\epsilon^*$	$-\epsilon$	1	$-\epsilon^*$	$-\epsilon$	−1	ϵ^*	ϵ	−1	ϵ^*	ϵ			$[xyz,\ z(x^2-y^2)]$
	1	$-\epsilon$	$-\epsilon^*$	1	$-\epsilon$	$-\epsilon^*$	−1	ϵ	ϵ^*	−1	ϵ	ϵ^*			
$\Gamma_{x,y,z}$	3	2	0	−1	0	2	−3	−2	0	1	0	−2			

C_{nh} Groups (cont.)

C_{7h}	E	C_7	C_7^2	C_7^3	C_7^4	C_7^5	C_7^6	σ_h	S_7	S_7^9	S_7^3
A'	1	1	1	1	1	1	1	1	1	1	1
E_1'	1	ϵ	ϵ^2	ϵ^3	ϵ^{3*}	ϵ^{2*}	ϵ^*	1	ϵ	ϵ^2	ϵ^3
	1	ϵ^*	ϵ^{2*}	ϵ^{3*}	ϵ^3	ϵ^2	ϵ	1	ϵ^*	ϵ^{2*}	ϵ^{3*}
E_2'	1	ϵ^2	ϵ^{3*}	ϵ^*	ϵ	ϵ^3	ϵ^{2*}	1	ϵ^2	ϵ^{3*}	ϵ^*
	1	ϵ^{2*}	ϵ^3	ϵ	ϵ^*	ϵ^{3*}	ϵ^2	1	ϵ^{2*}	ϵ^3	ϵ
E_3'	1	ϵ^3	ϵ^*	ϵ^2	ϵ^{2*}	ϵ	ϵ^{3*}	1	ϵ^3	ϵ^*	ϵ^2
	1	ϵ^{3*}	ϵ	ϵ^{2*}	ϵ^2	ϵ^*	ϵ^3	1	ϵ^{3*}	ϵ	ϵ^{2*}
A''	1	1	1	1	1	1	1	-1	-1	-1	-1
E_1''	1	ϵ	ϵ^2	ϵ^3	ϵ^{3*}	ϵ^{2*}	ϵ^*	-1	$-\epsilon$	$-\epsilon^2$	$-\epsilon^3$
	1	ϵ^*	ϵ^{2*}	ϵ^{3*}	ϵ^3	ϵ^2	ϵ	-1	$-\epsilon^*$	$-\epsilon^{2*}$	$-\epsilon^{3*}$
E_2''	1	ϵ^2	ϵ^{3*}	ϵ^*	ϵ	ϵ^3	ϵ^{2*}	-1	$-\epsilon^2$	$-\epsilon^{3*}$	$-\epsilon^*$
	1	ϵ^{2*}	ϵ^3	ϵ	ϵ^*	ϵ^{3*}	ϵ^2	-1	$-\epsilon^{2*}$	$-\epsilon^3$	$-\epsilon$
E_3''	1	ϵ^3	ϵ^*	ϵ^2	ϵ^{2*}	ϵ	ϵ^{3*}	-1	$-\epsilon^3$	$-\epsilon^*$	$-\epsilon^2$
	1	ϵ^{3*}	ϵ	ϵ^{2*}	ϵ^2	ϵ^*	ϵ^3	-1	$-\epsilon^{3*}$	$-\epsilon$	$-\epsilon^{2*}$
$\Gamma_{x,y,z}$	3	$1 + 2c\,\frac{2\pi}{7}$	$1 + 2c\,\frac{4\pi}{7}$	$1 + 2c\,\frac{6\pi}{7}$	$1 + 2c\,\frac{6\pi}{7}$	$1 + 2c\,\frac{4\pi}{7}$	$1 + 2c\,\frac{2\pi}{7}$	-3	$-1 + 2c\,\frac{2\pi}{7}$	$-1 + 2c\,\frac{4\pi}{7}$	$-1 + 2c\,\frac{6\pi}{7}$

C_{7h} (cont.)	S_7^{11}	S_7^{5}	S_7^{13}		$\epsilon = e^{2\pi i/7}$	
A'	1	1	1	R_z	$x^2 + y^2, z^2$	
E_1'	ϵ^{3*}	ϵ^{2*}	ϵ^{*}	(x, y)		(xz^2, yz^2)
	ϵ^{3}	ϵ^{2}	ϵ			
E_2'	ϵ	ϵ^{3}	ϵ^{2*}		$(x^2 - y^2, xy)$	
	ϵ^{*}	ϵ^{3*}	ϵ^{2}			
E_3'	ϵ^{2*}	ϵ	ϵ^{3*}			$[x(x^2 - 3y^2), y(3x^2 - y^2)]$
	ϵ^{2}	ϵ^{*}	ϵ^{3}			
A''	-1	-1	-1	z		z^3
E_1''	$-\epsilon^{3*}$	$-\epsilon^{2*}$	$-\epsilon^{*}$	(R_x, R_y)	(xz, yz)	
	$-\epsilon^{3}$	$-\epsilon^{2}$	$-\epsilon$			
E_2''	$-\epsilon$	$-\epsilon^{3}$	$-\epsilon^{2*}$			$[xyz, z(x^2 - y^2)]$
	$-\epsilon^{*}$	$-\epsilon^{3*}$	$-\epsilon^{2}$			
E_3''	$-\epsilon^{2*}$	$-\epsilon$	$-\epsilon^{3*}$			
	$-\epsilon^{2}$	$-\epsilon^{*}$	$-\epsilon^{3}$			
$\Gamma_{x,y,z}$	$-1 + 2c\,\frac{6\pi}{7}$	$-1 + 2c\,\frac{4\pi}{7}$	$-1 + 2c\,\frac{2\pi}{7}$			

C_{nh} Groups (cont.)

C_{8h}	E	C_8	C_4	C_2	C_4^3	C_8^3	C_8^5	C_8^7	i	S_8^5	S_4^3	σ_h	S_4	S_8^7	S_8	S_8^3
A_g	1	1	1	1	1	1	1	1	1	1	1	1	1	1	1	1
B_g	1	-1	1	1	1	-1	-1	-1	1	-1	1	1	1	-1	-1	-1
E_{1g}	1	ϵ	i	-1	$-i$	$-\epsilon^*$	$-\epsilon$	ϵ^*	1	ϵ	i	-1	$-i$	$-\epsilon^*$	$-\epsilon$	ϵ^*
	1	ϵ^*	$-i$	-1	i	$-\epsilon$	$-\epsilon^*$	ϵ	1	ϵ^*	$-i$	-1	i	$-\epsilon$	$-\epsilon^*$	ϵ
E_{2g}	1	i	-1	1	-1	$-i$	i	$-i$	1	i	-1	1	-1	$-i$	i	$-i$
	1	$-i$	-1	1	-1	i	$-i$	i	1	$-i$	-1	1	-1	i	$-i$	i
E_{3g}	1	$-\epsilon$	i	-1	$-i$	ϵ^*	ϵ	$-\epsilon^*$	1	$-\epsilon$	i	-1	$-i$	ϵ^*	ϵ	$-\epsilon^*$
	1	$-\epsilon^*$	$-i$	-1	i	ϵ	ϵ^*	$-\epsilon$	1	$-\epsilon^*$	$-i$	-1	i	ϵ	ϵ^*	$-\epsilon$
A_u	1	1	1	1	1	1	1	1	-1	-1	-1	-1	-1	-1	-1	-1
B_u	1	-1	1	1	1	-1	-1	-1	-1	1	-1	-1	-1	1	1	1
E_{1u}	1	ϵ	i	-1	$-i$	$-\epsilon^*$	$-\epsilon$	ϵ^*	-1	$-\epsilon$	$-i$	1	i	ϵ^*	ϵ	$-\epsilon^*$
	1	ϵ^*	$-i$	-1	i	$-\epsilon$	$-\epsilon^*$	ϵ	-1	$-\epsilon^*$	i	1	$-i$	ϵ	ϵ^*	$-\epsilon$
E_{2u}	1	i	-1	1	-1	$-i$	i	$-i$	-1	$-i$	1	-1	1	i	$-i$	i
	1	$-i$	-1	1	-1	i	$-i$	i	-1	i	1	-1	1	$-i$	i	$-i$
E_{3u}	1	$-\epsilon$	i	-1	$-i$	ϵ^*	ϵ	$-\epsilon^*$	-1	ϵ	$-i$	1	i	$-\epsilon^*$	$-\epsilon$	ϵ^*
	1	$-\epsilon^*$	$-i$	-1	i	ϵ	ϵ^*	$-\epsilon$	-1	ϵ^*	i	1	$-i$	$-\epsilon$	$-\epsilon^*$	ϵ
$\Gamma_{x,y,z}$	3	$1+\sqrt{2}$	1	-1	1	$1-\sqrt{2}$	$1-\sqrt{2}$	$1+\sqrt{2}$	-3	$-1-\sqrt{2}$	-1	1	-1	$-1+\sqrt{2}$	$-1+\sqrt{2}$	$-1-\sqrt{2}$

C_{8h} (cont.)		$\epsilon = e^{2\pi i/8}$	
A_g	R_z	$x^2 + y^2, z^2$	
B_g			
E_{1g}	(R_x, R_y)	(xz, yz)	
E_{2g}		$(x^2 - y^2, xy)$	
E_{3g}			
A_u	z		z^3
B_u			
E_{1u}	(x, y)		(xz^2, yz^2)
E_{2u}			$[xyz, z(x^2 - y^2)]$
E_{3u}			$[x(x^2 - 3y^2), y(3x^2 - y^2)]$

C_{nh} Groups (cont.)

C_{9h}	E	C_9	C_9^2	C_3	C_9^4	C_9^5	C_3^2	C_9^7	C_9^8	σ_h	S_9
A'	1	1	1	1	1	1	1	1	1	1	1
E_1'	1	ϵ	ϵ^2	ϵ^3	ϵ^4	ϵ^{4*}	ϵ^{3*}	ϵ^{2*}	ϵ^*	1	ϵ
	1	ϵ^*	ϵ^{2*}	ϵ^{3*}	ϵ^{4*}	ϵ^4	ϵ^3	ϵ^2	ϵ	1	ϵ^*
E_2'	1	ϵ^2	ϵ^4	ϵ^{3*}	ϵ^*	ϵ	ϵ^3	ϵ^{4*}	ϵ^{2*}	1	ϵ^2
	1	ϵ^{2*}	ϵ^{4*}	ϵ^3	ϵ	ϵ^*	ϵ^{3*}	ϵ^4	ϵ^2	1	ϵ^{2*}
E_3'	1	ϵ^3	ϵ^{3*}	1	ϵ^3	ϵ^{3*}	1	ϵ^3	ϵ^{3*}	1	ϵ^3
	1	ϵ^{3*}	ϵ^3	1	ϵ^{3*}	ϵ^3	1	ϵ^{3*}	ϵ^3	1	ϵ^{3*}
E_4'	1	ϵ^4	ϵ^*	ϵ^3	ϵ^{2*}	ϵ^2	ϵ^{3*}	ϵ	ϵ^{4*}	1	ϵ^4
	1	ϵ^{4*}	ϵ	ϵ^{3*}	ϵ^2	ϵ^{2*}	ϵ^3	ϵ^*	ϵ^4	1	ϵ^{4*}
A''	1	1	1	1	1	1	1	1	1	-1	-1
E_1''	1	ϵ	ϵ^2	ϵ^3	ϵ^4	ϵ^{4*}	ϵ^{3*}	ϵ^{2*}	ϵ^*	-1	$-\epsilon$
	1	ϵ^*	ϵ^{2*}	ϵ^{3*}	ϵ^{4*}	ϵ^4	ϵ^3	ϵ^2	ϵ	-1	$-\epsilon^*$
E_2''	1	ϵ^2	ϵ^4	ϵ^{3*}	ϵ^*	ϵ	ϵ^3	ϵ^{4*}	ϵ^{2*}	-1	$-\epsilon^2$
	1	ϵ^{2*}	ϵ^{4*}	ϵ^3	ϵ	ϵ^*	ϵ^{3*}	ϵ^4	ϵ^2	-1	$-\epsilon^{2*}$
E_3''	1	ϵ^3	ϵ^{3*}	1	ϵ^3	ϵ^{3*}	1	ϵ^3	ϵ^{3*}	-1	$-\epsilon^3$
	1	ϵ^{3*}	ϵ^3	1	ϵ^{3*}	ϵ^3	1	ϵ^{3*}	ϵ^3	-1	$-\epsilon^{3*}$
E_4''	1	ϵ^4	ϵ^*	ϵ^3	ϵ^{2*}	ϵ^2	ϵ^{3*}	ϵ	ϵ^{4*}	-1	$-\epsilon^4$
	1	ϵ^{4*}	ϵ	ϵ^{3*}	ϵ^2	ϵ^{2*}	ϵ^3	ϵ^*	ϵ^4	-1	$-\epsilon^{4*}$
$\Gamma_{x,y,z}$	3	$1 + 2c\,\frac{2\pi}{9}$	$1 + 2c\,\frac{4\pi}{9}$	0	$1 + 2c\,\frac{8\pi}{9}$	$1 + 2c\,\frac{8\pi}{9}$	0	$1 + 2c\,\frac{4\pi}{9}$	$1 + 2c\,\frac{2\pi}{9}$	-3	$-1 + 2c\,\frac{2\pi}{9}$

C_{9h} (cont.)	S_9^{11}	S_9^{3}	S_9^{13}	S_9^{5}	S_9^{15}	S_9^{7}	S_9^{17}	$\epsilon = e^{2\pi i/9}$		
A'	1	1	1	1	1	1	1	R_z	$x^2 + y^2, z^2$	
E_1'	ϵ^2	ϵ^3	ϵ^4	ϵ^{4*}	ϵ^{3*}	ϵ^{2*}	ϵ^*	(x, y)		(xz^2, yz^2)
	ϵ^{2*}	ϵ^{3*}	ϵ^{4*}	ϵ^4	ϵ^3	ϵ^2	ϵ			
E_2'	ϵ^4	ϵ^{3*}	ϵ^*	ϵ	ϵ^3	ϵ^{4*}	ϵ^{2*}		$(x^2 - y^2, xy)$	
	ϵ^{4*}	ϵ^3	ϵ	ϵ^*	ϵ^{3*}	ϵ^4	ϵ^2			
E_3'	ϵ^{3*}	1	ϵ^3	ϵ^{3*}	1	ϵ^3	ϵ^{3*}			$[x(x^2 - 3y^2), y(3x^2 - y^2)]$
	ϵ^3	1	ϵ^{3*}	ϵ^3	1	ϵ^{3*}	ϵ^3			
E_4'	ϵ^*	ϵ^3	ϵ^{2*}	ϵ^2	ϵ^{3*}	ϵ	ϵ^{4*}			
	ϵ	ϵ^{3*}	ϵ^2	ϵ^{2*}	ϵ^3	ϵ^*	ϵ^4			
A''	−1	−1	−1	−1	−1	−1	−1	z		z^3
E_1''	$-\epsilon^2$	$-\epsilon^3$	$-\epsilon^4$	$-\epsilon^{4*}$	$-\epsilon^{3*}$	$-\epsilon^{2*}$	$-\epsilon^*$	(R_x, R_y)	(xz, yz)	
	$-\epsilon^{2*}$	$-\epsilon^{3*}$	$-\epsilon^{4*}$	$-\epsilon^4$	$-\epsilon^3$	$-\epsilon^2$	$-\epsilon$			
E_2''	$-\epsilon^4$	$-\epsilon^{3*}$	$-\epsilon^*$	$-\epsilon$	$-\epsilon^3$	$-\epsilon^{4*}$	$-\epsilon^{2*}$			$[xyz, z(x^2 - y^2)]$
	$-\epsilon^{4*}$	$-\epsilon^3$	$-\epsilon$	$-\epsilon^*$	$-\epsilon^{3*}$	$-\epsilon^4$	$-\epsilon^2$			
E_3''	$-\epsilon^{3*}$	−1	$-\epsilon^3$	$-\epsilon^{3*}$	−1	$-\epsilon^3$	$-\epsilon^{3*}$			
	$-\epsilon^3$	−1	$-\epsilon^{3*}$	$-\epsilon^3$	−1	$-\epsilon^{3*}$	$-\epsilon^3$			
E_4''	$-\epsilon^*$	$-\epsilon^3$	$-\epsilon^{2*}$	$-\epsilon^2$	$-\epsilon^{3*}$	$-\epsilon$	$-\epsilon^{4*}$			
	$-\epsilon$	$-\epsilon^{3*}$	$-\epsilon^2$	$-\epsilon^{2*}$	$-\epsilon^3$	$-\epsilon^*$	$-\epsilon^4$			
$\Gamma_{x,y,z}$	$-1 + 2c\frac{4\pi}{9}$	0	$-1 + 2c\frac{8\pi}{9}$	$-1 + 2c\frac{8\pi}{9}$	0	$-1 + 2c\frac{4\pi}{9}$	$-1 + 2c\frac{2\pi}{9}$			

C_{10h}	E	C_{10}	C_5	C_{10}^3	C_5^2	C_2	C_5^3	C_{10}^7	C_5^8	C_{10}^9	i	S_5^3	S_{10}^7
A_g	1	1	1	1	1	1	1	1	1	1	1	1	1
B_g	1	-1	1	-1	1	-1	1	-1	1	-1	1	-1	1
E_{1g}	1	ϵ	ϵ^2	$-\epsilon^{2*}$	$-\epsilon^*$	-1	$-\epsilon$	$-\epsilon^2$	ϵ^{2*}	ϵ^*	1	ϵ	ϵ^2
	1	ϵ^*	ϵ^{2*}	$-\epsilon^2$	$-\epsilon$	-1	$-\epsilon^*$	$-\epsilon^{2*}$	ϵ^2	ϵ	1	ϵ^*	ϵ^{2*}
E_{2g}	1	ϵ^2	$-\epsilon^*$	$-\epsilon$	ϵ^{2*}	1	ϵ^2	$-\epsilon^*$	$-\epsilon$	ϵ^{2*}	1	ϵ^2	$-\epsilon^*$
	1	ϵ^{2*}	$-\epsilon$	$-\epsilon^*$	ϵ^2	1	ϵ^{2*}	$-\epsilon$	$-\epsilon^*$	ϵ^2	1	ϵ^{2*}	$-\epsilon$
E_{3g}	1	$-\epsilon^{2*}$	$-\epsilon$	ϵ^*	ϵ^2	-1	ϵ^{2*}	ϵ	$-\epsilon^*$	$-\epsilon^2$	1	$-\epsilon^{2*}$	$-\epsilon$
	1	$-\epsilon^2$	$-\epsilon^*$	ϵ	ϵ^{2*}	-1	ϵ^2	ϵ^*	$-\epsilon$	$-\epsilon^{2*}$	1	$-\epsilon^2$	$-\epsilon^*$
E_{4g}	1	$-\epsilon^*$	ϵ^{2*}	ϵ^2	$-\epsilon$	1	$-\epsilon^*$	ϵ^{2*}	ϵ^2	$-\epsilon$	1	$-\epsilon^*$	ϵ^{2*}
	1	$-\epsilon$	ϵ^2	ϵ^{2*}	$-\epsilon^*$	1	$-\epsilon$	ϵ^2	ϵ^{2*}	$-\epsilon^*$	1	$-\epsilon$	ϵ^2
A_u	1	1	1	1	1	1	1	1	1	1	-1	-1	-1
B_u	1	-1	1	-1	1	-1	1	-1	1	-1	-1	1	-1
E_{1u}	1	ϵ	ϵ^2	$-\epsilon^{2*}$	$-\epsilon^*$	-1	$-\epsilon$	$-\epsilon^2$	ϵ^{2*}	ϵ^*	-1	$-\epsilon$	$-\epsilon^2$
	1	ϵ^*	ϵ^{2*}	$-\epsilon^2$	$-\epsilon$	-1	$-\epsilon^*$	$-\epsilon^{2*}$	ϵ^2	ϵ	-1	$-\epsilon^*$	$-\epsilon^{2*}$
E_{2u}	1	ϵ^2	$-\epsilon^*$	$-\epsilon$	ϵ^{2*}	1	ϵ^2	$-\epsilon^*$	$-\epsilon$	ϵ^{2*}	-1	$-\epsilon^2$	ϵ^*
	1	ϵ^{2*}	$-\epsilon$	$-\epsilon^*$	ϵ^2	1	ϵ^{2*}	$-\epsilon$	$-\epsilon^*$	ϵ^2	-1	$-\epsilon^{2*}$	ϵ
E_{3u}	1	$-\epsilon^{2*}$	$-\epsilon$	ϵ^*	ϵ^2	-1	ϵ^{2*}	ϵ	$-\epsilon^*$	$-\epsilon^2$	-1	ϵ^{2*}	ϵ
	1	$-\epsilon^2$	$-\epsilon^*$	ϵ	ϵ^{2*}	-1	ϵ^2	ϵ^*	$-\epsilon$	$-\epsilon^{2*}$	-1	ϵ^2	ϵ^*
E_{4u}	1	$-\epsilon^*$	ϵ^{2*}	ϵ^2	$-\epsilon$	1	$-\epsilon^*$	ϵ^{2*}	ϵ^2	$-\epsilon$	-1	ϵ^*	$-\epsilon^{2*}$
	1	$-\epsilon$	ϵ^2	ϵ^{2*}	$-\epsilon^*$	1	$-\epsilon$	ϵ^2	ϵ^{2*}	$-\epsilon^*$	-1	ϵ	$-\epsilon^2$
$\Gamma_{x,y,z}$	3	$1 + 2c\,\frac{\pi}{5}$	$1 + 2c\,\frac{2\pi}{5}$	$1 - 2c\,\frac{2\pi}{5}$	$1 - 2c\,\frac{\pi}{5}$	-1	$1 - 2c\,\frac{\pi}{5}$	$1 - 2c\,\frac{2\pi}{5}$	$1 + 2c\,\frac{2\pi}{5}$	$1 + 2c\,\frac{\pi}{5}$	-3	$-1 - 2c\,\frac{\pi}{5}$	$-1 - 2c\,\frac{2\pi}{5}$

C_{10h} (cont.)	S_5^9	S_{10}^9	σ_h	S_{10}	S_5	S_{10}^3	S_5^7		$\epsilon = e^{2\pi i/10}$	
A_g	1	1	1	1	1	1	1	R_z	$x^2 + y^2, z^2$	
B_g	−1	1	−1	1	−1	1	−1			
E_{1g}	$-\epsilon^{2*}$	$-\epsilon^*$	−1	$-\epsilon$	$-\epsilon^2$	ϵ^{2*}	ϵ^*	(R_x, R_y)	(xz, yz)	
	$-\epsilon^2$	$-\epsilon$	−1	$-\epsilon^*$	$-\epsilon^{2*}$	ϵ^2	ϵ			
E_{2g}	$-\epsilon$	ϵ^{2*}	1	ϵ^2	$-\epsilon^*$	$-\epsilon$	ϵ^{2*}		$(x^2 - y^2, xy)$	
	$-\epsilon^*$	ϵ^2	1	ϵ^{2*}	$-\epsilon$	$-\epsilon^*$	ϵ^2			
E_{3g}	ϵ^*	ϵ^2	−1	ϵ^{2*}	ϵ	$-\epsilon^*$	$-\epsilon^2$			
	ϵ	ϵ^{2*}	−1	ϵ^2	ϵ^*	$-\epsilon$	$-\epsilon^{2*}$			
E_{4g}	ϵ^2	$-\epsilon$	1	$-\epsilon^*$	ϵ^{2*}	ϵ^2	$-\epsilon$			
	ϵ^{2*}	$-\epsilon^*$	1	$-\epsilon$	ϵ^2	ϵ^{2*}	$-\epsilon^*$			
A_u	−1	−1	−1	−1	−1	−1	−1	z		z^3
B_u	1	−1	1	−1	1	−1	1			
E_{1u}	ϵ^{2*}	ϵ^*	1	ϵ	ϵ^2	$-\epsilon^{2*}$	$-\epsilon^*$	(x, y)		(xz^2, yz^2)
	ϵ^2	ϵ	1	ϵ^*	ϵ^{2*}	$-\epsilon^2$	$-\epsilon$			
E_{2u}	ϵ	$-\epsilon^{2*}$	−1	$-\epsilon^2$	ϵ^*	ϵ	$-\epsilon^{2*}$			$[xyz, z(x^2 - y^2)]$
	ϵ^*	$-\epsilon^2$	−1	$-\epsilon^{2*}$	ϵ	ϵ^*	$-\epsilon^2$			
E_{3u}	$-\epsilon^*$	$-\epsilon^2$	1	$-\epsilon^{2*}$	$-\epsilon$	ϵ^*	ϵ^2			$[x(x^2 - 3y^2), y(3x^2 - y^2)]$
	$-\epsilon$	$-\epsilon^{2*}$	1	$-\epsilon^2$	$-\epsilon^*$	ϵ	ϵ^{2*}			
E_{4u}	$-\epsilon^2$	ϵ	−1	ϵ^*	$-\epsilon^{2*}$	$-\epsilon^2$	ϵ			
	$-\epsilon^{2*}$	ϵ^*	−1	ϵ	$-\epsilon^2$	$-\epsilon^{2*}$	ϵ^*			
$\Gamma_{x,y,z}$	$-1 + 2c\frac{2\pi}{5}$	$-1 + 2c\frac{\pi}{5}$	1	$-1 + 2c\frac{\pi}{5}$	$-1 + 2c\frac{2\pi}{5}$	$-1 - 2c\frac{2\pi}{5}$	$-1 - 2c\frac{\pi}{5}$			

D_n Groups

D_2	E	$C_2(z)$	$C_2(y)$	$C_2(x)$			
A	1	1	1	1		x^2, y^2, z^2	xyz
B_1	1	1	−1	−1	z, R_z	xy	$z^3, z(x^2 - y^2)$
B_2	1	−1	1	−1	y, R_y	xz	$yz^2, y(3x^2 - y^2)$
B_3	1	−1	−1	1	x, R_x	yz	$xz^2, x(x^2 - 3y^2)$
$\Gamma_{x,y,z}$	3	−1	−1	−1			

D_3	E	$2C_3$	$3C_2$			
A_1	1	1	1		$x^2 + y^2, z^2$	$x(x^2 - 3y^2)$
A_2	1	1	−1	z, R_z		$z^3, y(3x^2 - y^2)$
E	2	−1	0	(x, y) (R_x, R_y)	$(x^2 - y^2, xy), (xz, yz)$	$(xz^2, yz^2), [xyz, z(x^2 - y^2)]$
$\Gamma_{x,y,z}$	3	0	−1			

D_4	E	$2C_4$	$C_2(= C_4^2)$	$2C_2'$	$2C_2''$			
A_1	1	1	1	1	1		$x^2 + y^2, z^2$	
A_2	1	1	1	−1	−1	z, R_z		z^3
B_1	1	−1	1	1	−1		$x^2 - y^2$	xyz
B_2	1	−1	1	−1	1		xy	$z(x^2 - y^2)$
E	2	0	−2	0	0	$(x, y), (R_x, R_y)$	(xz, yz)	$(xz^2, yz^2), [x(x^2 - 3y^2), y(3x^2 - y^2)]$
$\Gamma_{x,y,z}$	3	1	−1	−1	−1			

D_5	E	$2C_5$	$2C_5^2$	$5C_2$			
A_1	1	1	1	1		x^2+y^2, z^2	
A_2	1	1	1	−1	z, R_z		z^3
E_1	2	$2c\frac{2\pi}{5}$	$2c\frac{4\pi}{5}$	0	$(x, y), (R_x, R_y)$	(xz, yz)	(xz^2, yz^2)
E_2	2	$2c\frac{4\pi}{5}$	$2c\frac{2\pi}{5}$	0		(x^2-y^2, xy)	$[xyz, z(x^2-y^2)], [x(x^2-3y^2), y(3x^2-y^2)]$
$\Gamma_{x,y,z}$	3	$1+2c\frac{2\pi}{5}$	$1+2c\frac{4\pi}{5}$	−1			

D_6	E	$2C_6$	$2C_3$	C_2	$3C_2'$	$3C_2''$			
A_1	1	1	1	1	1	1		x^2+y^2, z^2	
A_2	1	1	1	1	−1	−1	z, R_z		z^3
B_1	1	−1	1	−1	1	−1			$x(x^2-3y^2)$
B_2	1	−1	1	−1	−1	1			$y(3x^2-y^2)$
E_1	2	1	−1	−2	0	0	$(x, y), (R_x, R_y)$	(xz, yz)	(xz^2, yz^2)
E_2	2	−1	−1	2	0	0		(x^2-y^2, xy)	$[xyz, z(x^2-y^2)]$
$\Gamma_{x,y,z}$	3	2	0	−1	−1	−1			

D_n Groups (cont.)

D_7	E	$2C_7$	$2C_7^2$	$2C_7^3$	$7C_2$			
A_1	1	1	1	1	1		x^2+y^2, z^2	
A_2	1	1	1	1	−1	z, R_z		z^3
E_1	2	$2c\frac{2\pi}{7}$	$2c\frac{4\pi}{7}$	$2c\frac{6\pi}{7}$	0	$(x, y), (R_x, R_y)$	(xz, yz)	(xz^2, yz^2)
E_2	2	$2c\frac{4\pi}{7}$	$2c\frac{6\pi}{7}$	$2c\frac{2\pi}{7}$	0		(x^2-y^2, xy)	$[xyz, z(x^2-y^2)]$
E_3	2	$2c\frac{6\pi}{7}$	$2c\frac{2\pi}{7}$	$2c\frac{4\pi}{7}$	0			$[x(x^2-3y^2), y(3x^2-y^2)]$
$\Gamma_{x,y,z}$	3	$1+2c\frac{2\pi}{7}$	$1+2c\frac{4\pi}{7}$	$1+2c\frac{6\pi}{7}$	−1			

D_8	E	$2C_8$	$2C_4$	$2C_8^3$	C_2	$4C_2'$	$4C_2''$			
A_1	1	1	1	1	1	1	1		x^2+y^2, z^2	
A_2	1	1	1	1	1	−1	−1	z, R_z		z^3
B_1	1	−1	1	−1	1	1	−1			
B_2	1	−1	1	−1	1	−1	1			
E_1	2	$\sqrt{2}$	0	$-\sqrt{2}$	−2	0	0	$(x, y), (R_x, R_y)$	(xz, yz)	(xz^2, yz^2)
E_2	2	0	−2	0	2	0	0		(x^2-y^2, xy)	$[xyz, z(x^2-y^2)]$
E_3	2	$-\sqrt{2}$	0	$\sqrt{2}$	−2	0	0			$[x(x^2-3y^2), y(3x^2-y^2)]$
$\Gamma_{x,y,z}$	3	$1+\sqrt{2}$	1	$1-\sqrt{2}$	−1	−1	−1			

D_9	E	$2C_9$	$2C_9^2$	$2C_3$	$2C_9^4$	$9C_2$			
A_1	1	1	1	1	1	1		x^2+y^2, z^2	
A_2	1	1	1	1	1	−1	z, R_z		z^3
E_1	2	$2c\frac{2\pi}{9}$	$2c\frac{4\pi}{9}$	−1	$2c\frac{8\pi}{9}$	0	$(x, y), (R_x, R_y)$	(xz, yz)	(xz^2, yz^2)
E_2	2	$2c\frac{4\pi}{9}$	$2c\frac{8\pi}{9}$	−1	$2c\frac{2\pi}{9}$	0		(x^2-y^2, xy)	$[xyz, z(x^2-y^2)]$
E_3	2	−1	−1	2	−1	0			$[x(x^2-3y^2), y(3x^2-y^2)]$
E_4	2	$2c\frac{8\pi}{9}$	$2c\frac{2\pi}{9}$	−1	$2c\frac{4\pi}{9}$	0			
$\Gamma_{x,y,z}$	3	$1+2c\frac{2\pi}{9}$	$1+2c\frac{4\pi}{9}$	0	$1+2c\frac{8\pi}{9}$	−1			

D_{10}	E	$2C_{10}$	$2C_5$	$2C_{10}^3$	$2C_5^2$	C_2	$5C_2'$	$5C_2''$			
A_1	1	1	1	1	1	1	1	1		x^2+y^2, z^2	
A_2	1	1	1	1	1	1	−1	−1	z, R_z		z^3
B_1	1	−1	1	−1	1	−1	1	−1			
B_2	1	−1	1	−1	1	−1	−1	1			
E_1	2	$2c\frac{\pi}{5}$	$2c\frac{2\pi}{5}$	$-2c\frac{2\pi}{5}$	$-2c\frac{\pi}{5}$	−2	0	0	$(x, y), (R_x, R_y)$	(xz, yz)	(xz^2, yz^2)
E_2	2	$2c\frac{2\pi}{5}$	$-2c\frac{\pi}{5}$	$-2c\frac{\pi}{5}$	$2c\frac{2\pi}{5}$	2	0	0		(x^2-y^2, xy)	$[xyz, z(x^2-y^2)]$
E_3	2	$-2c\frac{2\pi}{5}$	$-2c\frac{\pi}{5}$	$2c\frac{\pi}{5}$	$2c\frac{2\pi}{5}$	−2	0	0			$[x(x^2-3y^2), y(3x^2-y^2)]$
E_4	2	$-2c\frac{\pi}{5}$	$2c\frac{2\pi}{5}$	$2c\frac{2\pi}{5}$	$-2c\frac{\pi}{5}$	2	0	0			
$\Gamma_{x,y,z}$	3	$1+2c\frac{\pi}{5}$	$1+2c\frac{2\pi}{5}$	$1-2c\frac{2\pi}{5}$	$1-2c\frac{\pi}{5}$	−1	−1	−1			

D_{nd} Groups

D_{2d}	E	$2S_4$	C_2	$2C_2'$	$2\sigma_d$			
A_1	1	1	1	1	1		x^2+y^2, z^2	xyz
A_2	1	1	1	−1	−1	R_z		$z(x^2-y^2)$
B_1	1	−1	1	1	−1		x^2-y^2	
B_2	1	−1	1	−1	1	z	xy	z^3
E	2	0	−2	0	0	$(x, y), (R_x, R_y)$	(xz, yz)	$(xz^2, yz^2), [x(x^2-3y^2), y(3x^2-y^2)]$
$\Gamma_{x,y,z}$	3	−1	−1	−1	1			

D_{3d}	E	$2C_3$	$3C_2$	i	$2S_6$	$3\sigma_d$			
A_{1g}	1	1	1	1	1	1		x^2+y^2, z^2	
A_{2g}	1	1	−1	1	1	−1	R_z		
E_g	2	−1	0	2	−1	0	(R_x, R_y)	$(x^2-y^2, xy), (xz, yz)$	
A_{1u}	1	1	1	−1	−1	−1			$x(x^2-3y^2)$
A_{2u}	1	1	−1	−1	−1	1	z		$z^3, y(3x^2-y^2)$
E_u	2	−1	0	−2	1	0	(x, y)		$(xz^2, yz^2), [xyz, z(x^2-y^2)]$
$\Gamma_{x,y,z}$	3	0	−1	−3	0	1			

D_{4d}	E	$2S_8$	$2C_4$	$2S_8^3$	C_2	$4C_2'$	$4\sigma_d$			
A_1	1	1	1	1	1	1	1		x^2+y^2, z^2	
A_2	1	1	1	1	1	−1	−1	R_z		
B_1	1	−1	1	−1	1	1	−1			
B_2	1	−1	1	−1	1	−1	1	z		z^3
E_1	2	$\sqrt{2}$	0	$-\sqrt{2}$	−2	0	0	(x, y)		(xz^2, yz^2)
E_2	2	0	−2	0	2	0	0		(x^2-y^2, xy)	$[xyz, z(x^2-y^2)]$
E_3	2	$-\sqrt{2}$	0	$\sqrt{2}$	−2	0	0	(R_x, R_y)	(xz, yz)	$[x(x^2-3y^2), y(3x^2-y^2)]$
$\Gamma_{x,y,z}$	3	$-1+\sqrt{2}$	1	$-1-\sqrt{2}$	−1	−1	1			

D_{5d}	E	$2C_5$	$2C_5^2$	$5C_2$	i	$2S_{10}^3$	$2S_{10}$	$5\sigma_d$
A_{1g}	1	1	1	1	1	1	1	1
A_{2g}	1	1	1	−1	1	1	1	−1
E_{1g}	2	$2c\frac{2\pi}{5}$	$2c\frac{4\pi}{5}$	0	2	$2c\frac{2\pi}{5}$	$2c\frac{4\pi}{5}$	0
E_{2g}	2	$2c\frac{4\pi}{5}$	$2c\frac{2\pi}{5}$	0	2	$2c\frac{4\pi}{5}$	$2c\frac{2\pi}{5}$	0
A_{1u}	1	1	1	1	−1	−1	−1	−1
A_{2u}	1	1	1	−1	−1	−1	−1	1
E_{1u}	2	$2c\frac{2\pi}{5}$	$2c\frac{4\pi}{5}$	0	−2	$-2c\frac{2\pi}{5}$	$-2c\frac{4\pi}{5}$	0
E_{2u}	2	$2c\frac{4\pi}{5}$	$2c\frac{2\pi}{5}$	0	−2	$-2c\frac{4\pi}{5}$	$-2c\frac{2\pi}{5}$	0
$\Gamma_{x,y,z}$	3	$1+2c\frac{2\pi}{5}$	$1+2c\frac{4\pi}{5}$	−1	−3	$-1-2c\frac{2\pi}{5}$	$-1-2c\frac{4\pi}{5}$	1

D_{5d} (cont.)			
A_{1g}		$x^2+y^2,\ z^2$	
A_{2g}	R_z		
E_{1g}	$(R_x,\ R_y)$	$(xz,\ yz)$	
E_{2g}		$(x^2-y^2,\ xy)$	
A_{1u}			
A_{2u}	z		z^3
E_{1u}	$(x,\ y)$		$(xz^2,\ yz^2)$
E_{2u}			$[xyz,\ z(x^2-y^2)],\ [x(x^2-3y^2),\ y(3x^2-y^2)]$

D_{nd} Groups (cont.)

D_{6d}	E	$2S_{12}$	$2C_6$	$2S_4$	$2C_3$	$2S_{12}^5$	C_2	$6C_2'$	$6\sigma_d$			
A_1	1	1	1	1	1	1	1	1	1		x^2+y^2, z^2	
A_2	1	1	1	1	1	1	1	−1	−1	R_z		
B_1	1	−1	1	−1	1	−1	1	1	−1			
B_2	1	−1	1	−1	1	−1	1	−1	1	z		z^3
E_1	2	$\sqrt{3}$	1	0	−1	$-\sqrt{3}$	−2	0	0	(x, y)		(xz^2, yz^2)
E_2	2	1	−1	−2	−1	1	2	0	0		(x^2-y^2, xy)	
E_3	2	0	−2	0	2	0	−2	0	0			$[x(x^2-3y^2), y(3x^2-y^2)]$
E_4	2	−1	−1	2	−1	−1	2	0	0			$[xyz, z(x^2-y^2)]$
E_5	2	$-\sqrt{3}$	1	0	−1	$\sqrt{3}$	−2	0	0	(R_x, R_y)	(xz, yz)	
$\Gamma_{x,y,z}$	3	$-1+\sqrt{3}$	2	−1	0	$-1-\sqrt{3}$	−1	−1	1			

D_{7d}	E	$2C_7$	$2C_7^2$	$2C_7^3$	$7C_2$	i	$2S_{14}^5$	$2S_{14}^3$	$2S_{14}$	$7\sigma_d$
A_{1g}	1	1	1	1	1	1	1	1	1	1
A_{2g}	1	1	1	1	−1	1	1	1	1	−1
E_{1g}	2	$2c\frac{2\pi}{7}$	$2c\frac{4\pi}{7}$	$2c\frac{6\pi}{7}$	0	2	$2c\frac{2\pi}{7}$	$2c\frac{4\pi}{7}$	$2c\frac{6\pi}{7}$	0
E_{2g}	2	$2c\frac{4\pi}{7}$	$2c\frac{6\pi}{7}$	$2c\frac{2\pi}{7}$	0	2	$2c\frac{4\pi}{7}$	$2c\frac{6\pi}{7}$	$2c\frac{2\pi}{7}$	0
E_{3g}	2	$2c\frac{6\pi}{7}$	$2c\frac{2\pi}{7}$	$2c\frac{4\pi}{7}$	0	2	$2c\frac{6\pi}{7}$	$2c\frac{2\pi}{7}$	$2c\frac{4\pi}{7}$	0
A_{1u}	1	1	1	1	1	−1	−1	−1	−1	−1
A_{2u}	1	1	1	1	−1	−1	−1	−1	−1	1
E_{1u}	2	$2c\frac{2\pi}{7}$	$2c\frac{4\pi}{7}$	$2c\frac{6\pi}{7}$	0	−2	$-2c\frac{2\pi}{7}$	$-2c\frac{4\pi}{7}$	$-2c\frac{6\pi}{7}$	0
E_{2u}	2	$2c\frac{4\pi}{7}$	$2c\frac{6\pi}{7}$	$2c\frac{2\pi}{7}$	0	−2	$-2c\frac{4\pi}{7}$	$-2c\frac{6\pi}{7}$	$-2c\frac{2\pi}{7}$	0
E_{3u}	2	$2c\frac{6\pi}{7}$	$2c\frac{2\pi}{7}$	$2c\frac{4\pi}{7}$	0	−2	$-2c\frac{6\pi}{7}$	$-2c\frac{2\pi}{7}$	$-2c\frac{4\pi}{7}$	0
$\Gamma_{x,y,z}$	3	$1+2c\frac{2\pi}{7}$	$1+2c\frac{4\pi}{7}$	$1+2c\frac{6\pi}{7}$	−1	−3	$-1-2c\frac{2\pi}{7}$	$-1-2c\frac{4\pi}{7}$	$-1-2c\frac{6\pi}{7}$	1

D_{7d} (cont.)			
A_{1g}		$x^2 + y^2, z^2$	
A_{2g}	R_z		
E_{1g}	(R_x, R_y)	(xz, yz)	
E_{2g}		$(x^2 - y^2, xy)$	
E_{3g}			
A_{1u}			
A_{2u}	z		z^3
E_{1u}	(x, y)		(xz^2, yz^2)
E_{2u}			$[xyz, z(x^2 - y^2)]$
E_{3u}			$[x(x^2 - 3y^2, y(3x^2 - y^2)]$

D_{nd} Groups (cont.)

D_{8d}	E	$2S_{16}$	$2C_8$	$2S_{16}^3$	$2C_4$	$2S_{16}^5$	$2C_8^3$	$2S_{16}^7$	C_2	$8C_2'$	$8\sigma_d$
A_1	1	1	1	1	1	1	1	1	1	1	1
A_2	1	1	1	1	1	1	1	1	1	−1	−1
B_1	1	−1	1	−1	1	−1	1	−1	1	1	−1
B_2	1	−1	1	−1	1	−1	1	−1	1	−1	1
E_1	2	$2c\frac{\pi}{8}$	$\sqrt{2}$	$2c\frac{3\pi}{8}$	0	$-2c\frac{3\pi}{8}$	$-\sqrt{2}$	$-2c\frac{\pi}{8}$	−2	0	0
E_2	2	$\sqrt{2}$	0	$-\sqrt{2}$	−2	$-\sqrt{2}$	0	$\sqrt{2}$	2	0	0
E_3	2	$2c\frac{3\pi}{8}$	$-\sqrt{2}$	$-2c\frac{\pi}{8}$	0	$2c\frac{\pi}{8}$	$\sqrt{2}$	$-2c\frac{3\pi}{8}$	−2	0	0
E_4	2	0	−2	0	2	0	−2	0	2	0	0
E_5	2	$-2c\frac{3\pi}{8}$	$-\sqrt{2}$	$2c\frac{\pi}{8}$	0	$-2c\frac{\pi}{8}$	$\sqrt{2}$	$2c\frac{3\pi}{8}$	−2	0	0
E_6	2	$-\sqrt{2}$	0	$\sqrt{2}$	−2	$\sqrt{2}$	0	$-\sqrt{2}$	2	0	0
E_7	2	$-2c\frac{\pi}{8}$	$\sqrt{2}$	$-2c\frac{3\pi}{8}$	0	$2c\frac{3\pi}{8}$	$-\sqrt{2}$	$2c\frac{\pi}{8}$	−2	0	0
$\Gamma_{x,y,z}$	3	$-1+2c\frac{\pi}{8}$	$1+\sqrt{2}$	$-1+2c\frac{3\pi}{8}$	1	$-1-2c\frac{3\pi}{8}$	$1-\sqrt{2}$	$-1-2c\frac{\pi}{8}$	−1	−1	1

D_{8d} (cont.)			
A_1		$x^2 + y^2, z^2$	
A_2	R_z		
B_1			
B_2	z		z^3
E_1	(x, y)		(xz^2, yz^2)
E_2		$(x^2 - y^2, xy)$	
E_3			$[x(x^2 - 3y^2), y(3x^2 - y^2)]$
E_4			
E_5			
E_6			$[xyz, z(x^2 - y^2)]$
E_7	(R_x, R_y)	(xz, yz)	

D_{nd} Groups (cont.)

D_{9d}	E	$2C_9$	$2C_9^2$	$2C_3$	$2C_9^4$	$9C_2$	i	$2S_{18}^7$	$2S_{18}^5$	$2S_{18}^3$	$2S_{18}$	$9\sigma_d$
A_{1g}	1	1	1	1	1	1	1	1	1	1	1	1
A_{2g}	1	1	1	1	1	−1	1	1	1	1	1	−1
E_{1g}	2	$2c\frac{2\pi}{9}$	$2c\frac{4\pi}{9}$	−1	$2c\frac{8\pi}{9}$	0	2	$2c\frac{2\pi}{9}$	$2c\frac{4\pi}{9}$	−1	$2c\frac{8\pi}{9}$	0
E_{2g}	2	$2c\frac{4\pi}{9}$	$2c\frac{8\pi}{9}$	−1	$2c\frac{2\pi}{9}$	0	2	$2c\frac{4\pi}{9}$	$2c\frac{8\pi}{9}$	−1	$2c\frac{2\pi}{9}$	0
E_{3g}	2	−1	−1	2	−1	0	2	−1	−1	2	−1	0
E_{4g}	2	$2c\frac{8\pi}{9}$	$2c\frac{2\pi}{9}$	−1	$2c\frac{4\pi}{9}$	0	2	$2c\frac{8\pi}{9}$	$2c\frac{2\pi}{9}$	−1	$2c\frac{4\pi}{9}$	0
A_{1u}	1	1	1	1	1	1	−1	−1	−1	−1	−1	−1
A_{2u}	1	1	1	1	1	−1	−1	−1	−1	−1	−1	1
E_{1u}	2	$2c\frac{2\pi}{9}$	$2c\frac{4\pi}{9}$	−1	$2c\frac{8\pi}{9}$	0	−2	$-2c\frac{2\pi}{9}$	$-2c\frac{4\pi}{9}$	1	$-2c\frac{8\pi}{9}$	0
E_{2u}	2	$2c\frac{4\pi}{9}$	$2c\frac{8\pi}{9}$	−1	$2c\frac{2\pi}{9}$	0	−2	$-2c\frac{4\pi}{9}$	$-2c\frac{8\pi}{9}$	1	$-2c\frac{2\pi}{9}$	0
E_{3u}	2	−1	−1	2	−1	0	−2	1	1	−2	1	0
E_{4u}	2	$2c\frac{8\pi}{9}$	$2c\frac{2\pi}{9}$	−1	$2c\frac{4\pi}{9}$	0	−2	$-2c\frac{8\pi}{9}$	$-2c\frac{2\pi}{9}$	1	$-2c\frac{4\pi}{9}$	0
$\Gamma_{x,y,z}$	3	$1+2c\frac{2\pi}{9}$	$1+2c\frac{4\pi}{9}$	0	$1+2c\frac{8\pi}{9}$	−1	−3	$-1-2c\frac{2\pi}{9}$	$-1-2c\frac{4\pi}{9}$	0	$-1-2c\frac{8\pi}{9}$	1

D_{9d} (cont.)			
A_{1g}		$x^2 + y^2, z^2$	
A_{2g}	R_z		
E_{1g}	(R_x, R_y)	(xz, yz)	
E_{2g}		$(x^2 - y^2, xy)$	
E_{3g}			
E_{4g}			
A_{1u}			
A_{2u}	z		z^3
E_{1u}	(x, y)		(xz^2, yz^2)
E_{2u}			$[xyz, z(x^2 - y^2)]$
E_{3u}			$[x(x^2 - 3y^2, y(3x^2 - y^2)]$
E_{4u}			

D_{nd} Groups (cont.)

D_{10d}	E	$2S_{20}$	$2C_{10}$	$2S_{20}^3$	$2C_5$	$2S_{20}^5$	$2S_{10}^3$	$2S_{20}^7$	$2C_5^2$	$2S_{20}^9$	C_2	$10C_2'$	$10\sigma_d$
A_1	1	1	1	1	1	1	1	1	1	1	1	1	1
A_2	1	1	1	1	1	1	1	1	1	1	1	−1	−1
B_1	1	−1	1	−1	1	−1	1	−1	1	−1	1	1	−1
B_2	1	−1	1	−1	1	−1	1	−1	1	−1	1	−1	1
E_1	2	$2c\frac{\pi}{10}$	$2c\frac{\pi}{5}$	$2c\frac{3\pi}{10}$	$2c\frac{2\pi}{5}$	0	$-2c\frac{2\pi}{5}$	$-2c\frac{3\pi}{10}$	$-2c\frac{\pi}{5}$	$-2c\frac{\pi}{10}$	−2	0	0
E_2	2	$2c\frac{\pi}{5}$	$2c\frac{2\pi}{5}$	$-2c\frac{2\pi}{5}$	$-2c\frac{\pi}{5}$	−2	$-2c\frac{\pi}{5}$	$-2c\frac{2\pi}{5}$	$2c\frac{2\pi}{5}$	$2c\frac{\pi}{5}$	2	0	0
E_3	2	$2c\frac{3\pi}{10}$	$-2c\frac{2\pi}{5}$	$-2c\frac{\pi}{10}$	$-2c\frac{\pi}{5}$	0	$2c\frac{\pi}{5}$	$2c\frac{\pi}{10}$	$2c\frac{2\pi}{5}$	$-2c\frac{3\pi}{10}$	−2	0	0
E_4	2	$2c\frac{2\pi}{5}$	$-2c\frac{\pi}{5}$	$-2c\frac{\pi}{5}$	$2c\frac{2\pi}{5}$	2	$2c\frac{2\pi}{5}$	$-2c\frac{\pi}{5}$	$-2c\frac{\pi}{5}$	$2c\frac{2\pi}{5}$	2	0	0
E_5	2	0	−2	0	2	0	−2	0	2	0	−2	0	0
E_6	2	$-2c\frac{2\pi}{5}$	$-2c\frac{\pi}{5}$	$2c\frac{\pi}{5}$	$2c\frac{2\pi}{5}$	−2	$2c\frac{2\pi}{5}$	$2c\frac{\pi}{5}$	$-2c\frac{\pi}{5}$	$-2c\frac{2\pi}{5}$	2	0	0
E_7	2	$-2c\frac{3\pi}{10}$	$-2c\frac{2\pi}{5}$	$2c\frac{\pi}{10}$	$-2c\frac{\pi}{5}$	0	$2c\frac{\pi}{5}$	$-2c\frac{\pi}{10}$	$2c\frac{2\pi}{5}$	$2c\frac{3\pi}{10}$	−2	0	0
E_8	2	$-2c\frac{\pi}{5}$	$2c\frac{2\pi}{5}$	$2c\frac{2\pi}{5}$	$-2c\frac{\pi}{5}$	2	$-2c\frac{\pi}{5}$	$2c\frac{2\pi}{5}$	$2c\frac{2\pi}{5}$	$-2c\frac{\pi}{5}$	2	0	0
E_9	2	$-2c\frac{\pi}{10}$	$2c\frac{\pi}{5}$	$-2c\frac{3\pi}{10}$	$2c\frac{2\pi}{5}$	0	$-2c\frac{2\pi}{5}$	$2c\frac{3\pi}{10}$	$-2c\frac{\pi}{5}$	$2c\frac{\pi}{10}$	−2	0	0
$\Gamma_{x,y,z}$	3	$-1+2c\frac{\pi}{10}$	$1+2c\frac{\pi}{5}$	$-1+2c\frac{3\pi}{10}$	$1+2c\frac{2\pi}{5}$	−1	$1-2c\frac{2\pi}{5}$	$-1-2c\frac{3\pi}{10}$	$1-2c\frac{\pi}{5}$	$-1-2c\frac{\pi}{10}$	−1	−1	1

D_{10d} (cont.)			
A_1		$x^2 + y^2, z^2$	
A_2	R_z		
B_1			
B_2	z		z^3
E_1	(x, y)		(xz^2, yz^2)
E_2		$(x^2 - y^2, xy)$	
E_3			$[x(x^2 - 3y^2), y(3x^2 - y^2)]$
E_4			
E_5			
E_6			
E_7			
E_8			$[xyz, z(x^2 - y^2)]$
E_9	(R_x, R_y)	(xz, yz)	

D_{nh} Groups

D_{2h}	E	$C_2(z)$	$C_2(y)$	$C_2(x)$	i	$\sigma(xy)$	$\sigma(xz)$	$\sigma(yz)$			
A_g	1	1	1	1	1	1	1	1		x^2, y^2, z^2	
B_{1g}	1	1	−1	−1	1	1	−1	−1	R_z	xy	
B_{2g}	1	−1	1	−1	1	−1	1	−1	R_y	xz	
B_{3g}	1	−1	−1	1	1	−1	−1	1	R_x	yz	
A_u	1	1	1	1	−1	−1	−1	−1			xyz
B_{1u}	1	1	−1	−1	−1	−1	1	1	z		$z^3, z(x^2-y^2)$
B_{2u}	1	−1	1	−1	−1	1	−1	1	y		$yz^2, y(3x^2-y^2)$
B_{3u}	1	−1	−1	1	−1	1	1	−1	x		$xz^2, x(x^2-3y^2)$
$\Gamma_{x,y,z}$	3	−1	−1	−1	−3	1	1	1			

D_{3h}	E	$2C_3$	$3C_2$	σ_h	$2S_3$	$3\sigma_v$			
A_1'	1	1	1	1	1	1		x^2+y^2, z^2	$x(x^2-3y^2)$
A_2'	1	1	−1	1	1	−1	R_z		$y(3x^2-y^2)$
E'	2	−1	0	2	−1	0	(x, y)	(x^2-y^2, xy)	(xz^2, yz^2)
A_1''	1	1	1	−1	−1	−1			
A_2''	1	1	−1	−1	−1	1	z		z^3
E''	2	−1	0	−2	1	0	(R_x, R_y)	(xz, yz)	$[xyz, z(x^2-y^2)]$
$\Gamma_{x,y,z}$	3	0	−1	1	−2	1			

D_{4h}	E	$2C_4$	C_2	$2C_2'$	$2C_2''$	i	$2S_4$	σ_h	$2\sigma_v$	$2\sigma_d$			
A_{1g}	1	1	1	1	1	1	1	1	1	1		$x^2 + y^2, z^2$	
A_{2g}	1	1	1	−1	−1	1	1	1	−1	−1	R_z		
B_{1g}	1	−1	1	1	−1	1	−1	1	1	−1		$x^2 - y^2$	
B_{2g}	1	−1	1	−1	1	1	−1	1	−1	1		xy	
E_g	2	0	−2	0	0	2	0	−2	0	0	(R_x, R_y)	(xz, yz)	
A_{1u}	1	1	1	1	1	−1	−1	−1	−1	−1			
A_{2u}	1	1	1	−1	−1	−1	−1	−1	1	1	z		z^3
B_{1u}	1	−1	1	1	−1	−1	1	−1	−1	1			xyz
B_{2u}	1	−1	1	−1	1	−1	1	−1	1	−1			$z(x^2 - y^2)$
E_u	2	0	−2	0	0	−2	0	2	0	0	(x, y)		$(xz^2, yz^2), [x(x^2 - 3y^2), y(3x^2 - y^2)]$
$\Gamma_{x,y,z}$	3	1	−1	−1	−1	−3	−1	1	1	1			

D_{nh} Groups (cont.)

D_{5h}	E	$2C_5$	$2C_5^2$	$5C_2$	σ_h	$2S_5$	$2S_5^3$	$5\sigma_v$			
A_1'	1	1	1	1	1	1	1	1		x^2+y^2, z^2	
A_2'	1	1	1	−1	1	1	1	−1	R_z		
E_1'	2	$2c\frac{2\pi}{5}$	$2c\frac{4\pi}{5}$	0	2	$2c\frac{2\pi}{5}$	$2c\frac{4\pi}{5}$	0	(x, y)		(xz^2, yz^2)
E_2'	2	$2c\frac{4\pi}{5}$	$2c\frac{2\pi}{5}$	0	2	$2c\frac{4\pi}{5}$	$2c\frac{2\pi}{5}$	0		(x^2-y^2, xy)	$[x(x^2-3y^2), y(3x^2-y^2)]$
A_1''	1	1	1	1	−1	−1	−1	−1			
A_2''	1	1	1	−1	−1	−1	−1	1	z		z^3
E_1''	2	$2c\frac{2\pi}{5}$	$2c\frac{4\pi}{5}$	0	−2	$-2c\frac{2\pi}{5}$	$-2c\frac{4\pi}{5}$	0	$(R_x\ R_y)$	(xz, yz)	
E_2''	2	$2c\frac{4\pi}{5}$	$2c\frac{2\pi}{5}$	0	−2	$-2c\frac{4\pi}{5}$	$-2c\frac{2\pi}{5}$	0			$[xyz, z(x^2-y^2)]$
$\Gamma_{x,y,z}$	3	$1+2c\frac{2\pi}{5}$	$1+2c\frac{4\pi}{5}$	−1	1	$-1+2c\frac{2\pi}{5}$	$-1+2c\frac{4\pi}{5}$	1			

D_{6h}	E	$2C_6$	$2C_3$	C_2	$3C_2'$	$3C_2''$	i	$2S_3$	$2S_6$	σ_h	$3\sigma_d$	$3\sigma_v$			
A_{1g}	1	1	1	1	1	1	1	1	1	1	1	1		x^2+y^2, z^2	
A_{2g}	1	1	1	1	−1	−1	1	1	1	1	−1	−1	R_z		
B_{1g}	1	−1	1	−1	1	−1	1	−1	1	−1	1	−1			
B_{2g}	1	−1	1	−1	−1	1	1	−1	1	−1	−1	1			
E_{1g}	2	1	−1	−2	0	0	2	1	−1	−2	0	0	(R_x, R_y)	(xz, yz)	
E_{2g}	2	−1	−1	2	0	0	2	−1	−1	2	0	0		(x^2-y^2, xy)	
A_{1u}	1	1	1	1	1	1	−1	−1	−1	−1	−1	−1			
A_{2u}	1	1	1	1	−1	−1	−1	−1	−1	−1	1	1	z		z^3
B_{1u}	1	−1	1	−1	1	−1	−1	1	−1	1	−1	1			$x(x^2-3y^2)$
B_{2u}	1	−1	1	−1	−1	1	−1	1	−1	1	1	−1			$y(3x^2-y^2)$
E_{1u}	2	1	−1	−2	0	0	−2	−1	1	2	0	0	(x, y)		(xz^2, yz^2)
E_{2u}	2	−1	−1	2	0	0	−2	1	1	−2	0	0			$[xyz, z(x^2-y^2)]$
$\Gamma_{x,y,z}$	3	2	0	−1	−1	−1	−3	−2	0	1	1	1			

D_{nh} Groups (cont.)

D_{7h}	E	$2C_7$	$2C_7^2$	$2C_7^3$	$7C_2$	σ_h	$2S_7$	$2S_7^5$	$2S_7^3$	$7\sigma_v$
A_1'	1	1	1	1	1	1	1	1	1	1
A_2'	1	1	1	1	−1	1	1	1	1	−1
E_1'	2	$2c\frac{2\pi}{7}$	$2c\frac{4\pi}{7}$	$2c\frac{6\pi}{7}$	0	2	$2c\frac{2\pi}{7}$	$2c\frac{4\pi}{7}$	$2c\frac{6\pi}{7}$	0
E_2'	2	$2c\frac{4\pi}{7}$	$2c\frac{6\pi}{7}$	$2c\frac{2\pi}{7}$	0	2	$2c\frac{4\pi}{7}$	$2c\frac{6\pi}{7}$	$2c\frac{2\pi}{7}$	0
E_3'	2	$2c\frac{6\pi}{7}$	$2c\frac{2\pi}{7}$	$2c\frac{4\pi}{7}$	0	2	$2c\frac{6\pi}{7}$	$2c\frac{2\pi}{7}$	$2c\frac{4\pi}{7}$	0
A_1''	1	1	1	1	1	−1	−1	−1	−1	−1
A_2''	1	1	1	1	−1	−1	−1	−1	−1	1
E_1''	2	$2c\frac{2\pi}{7}$	$2c\frac{4\pi}{7}$	$2c\frac{6\pi}{7}$	0	−2	$-2c\frac{2\pi}{7}$	$-2c\frac{4\pi}{7}$	$-2c\frac{6\pi}{7}$	0
E_2''	2	$2c\frac{4\pi}{7}$	$2c\frac{6\pi}{7}$	$2c\frac{2\pi}{7}$	0	−2	$-2c\frac{4\pi}{7}$	$-2c\frac{6\pi}{7}$	$-2c\frac{2\pi}{7}$	0
E_3''	2	$2c\frac{6\pi}{7}$	$2c\frac{2\pi}{7}$	$2c\frac{4\pi}{7}$	0	−2	$-2c\frac{6\pi}{7}$	$-2c\frac{2\pi}{7}$	$-2c\frac{4\pi}{7}$	0
$\Gamma_{x,y,z}$	3	$1+2c\frac{2\pi}{7}$	$1+2c\frac{4\pi}{7}$	$1+2c\frac{6\pi}{7}$	−1	1	$-1+2c\frac{2\pi}{7}$	$-1+2c\frac{4\pi}{7}$	$-1+2c\frac{6\pi}{7}$	1

D_{7h} (cont.)			
A_1'		$x^2 + y^2, z^2$	
A_2'	R_z		
E_1'	(x, y)		(xz^2, yz^2)
E_2'		$(x^2 - y^2, xy)$	
E_3'			$[x(x^2 - 3y^2), y(3x^2 - y^2)]$
A_1''			
A_2''	z		z^3
E_1''	(R_x, R_y)	(xz, yz)	
E_2''			$[xyz, z(x^2 - y^2)]$
E_3''			

D_{nh} Groups (cont.)

D_{8h}	E	$2C_8$	$2C_4$	$2C_8^3$	C_2	$4C_2'$	$4C_2''$	i	$2S_8^3$	$2S_4$	$2S_8$	σ_h	$4\sigma_v$	$4\sigma_d$
A_{1g}	1	1	1	1	1	1	1	1	1	1	1	1	1	1
A_{2g}	1	1	1	1	1	−1	−1	1	1	1	1	1	−1	−1
B_{1g}	1	−1	1	−1	1	1	−1	1	−1	1	−1	1	1	−1
B_{2g}	1	−1	1	−1	1	−1	1	1	−1	1	−1	1	−1	1
E_{1g}	2	$\sqrt{2}$	0	$-\sqrt{2}$	−2	0	0	2	$\sqrt{2}$	0	$-\sqrt{2}$	−2	0	0
E_{2g}	2	0	−2	0	2	0	0	2	0	−2	0	2	0	0
E_{3g}	2	$-\sqrt{2}$	0	$\sqrt{2}$	−2	0	0	2	$-\sqrt{2}$	0	$\sqrt{2}$	−2	0	0
A_{1u}	1	1	1	1	1	1	1	−1	−1	−1	−1	−1	−1	−1
A_{2u}	1	1	1	1	1	−1	−1	−1	−1	−1	−1	−1	1	1
B_{1u}	1	−1	1	−1	1	1	−1	−1	1	−1	1	−1	−1	1
B_{2u}	1	−1	1	−1	1	−1	1	−1	1	−1	1	−1	1	−1
E_{1u}	2	$\sqrt{2}$	0	$-\sqrt{2}$	−2	0	0	−2	$-\sqrt{2}$	0	$\sqrt{2}$	2	0	0
E_{2u}	2	0	−2	0	2	0	0	−2	0	2	0	−2	0	0
E_{3u}	2	$-\sqrt{2}$	0	$\sqrt{2}$	−2	0	0	−2	$\sqrt{2}$	0	$-\sqrt{2}$	2	0	0
$\Gamma_{x,y,z}$	3	$1+\sqrt{2}$	1	$1-\sqrt{2}$	−1	−1	−1	−3	$-1-\sqrt{2}$	−1	$-1+\sqrt{2}$	1	1	1

D_{8h} (cont.)			
A_{1g}		$x^2 + y^2, z^2$	
A_{2g}	R_z		
B_{1g}			
B_{2g}			
E_{1g}	(R_x, R_y)	(xz, yz)	
E_{2g}		$(x^2 - y^2, xy)$	
E_{3g}			
A_{1u}			
A_{2u}	z		z^3
B_{1u}			
B_{2u}			
E_{1u}	(x, y)		(xz^2, yz^2)
E_{2u}			$[xyz, z(x^2 - y^2)]$
E_{3u}			$[x(x^2 - 3y^2), y(3x^2 - y^2)]$

D_{nh} Groups (cont.)

D_{9h}	E	$2C_9$	$2C_9^2$	$2C_3$	$2C_9^4$	$9C_2$	σ_h	$2S_9$	$2S_9^7$	$2S_3$	$2S_9^5$	$9\sigma_v$
A_1'	1	1	1	1	1	1	1	1	1	1	1	1
A_2'	1	1	1	1	1	−1	1	1	1	1	1	−1
E_1'	2	$2c\frac{2\pi}{9}$	$2c\frac{4\pi}{9}$	−1	$2c\frac{8\pi}{9}$	0	2	$2c\frac{2\pi}{9}$	$2c\frac{4\pi}{9}$	−1	$2c\frac{8\pi}{9}$	0
E_2'	2	$2c\frac{4\pi}{9}$	$2c\frac{8\pi}{9}$	−1	$2c\frac{2\pi}{9}$	0	2	$2c\frac{4\pi}{9}$	$2c\frac{8\pi}{9}$	−1	$2c\frac{2\pi}{9}$	0
E_3'	2	−1	−1	2	−1	0	2	−1	−1	2	−1	0
E_4'	2	$2c\frac{8\pi}{9}$	$2c\frac{2\pi}{9}$	−1	$2c\frac{4\pi}{9}$	0	2	$2c\frac{8\pi}{9}$	$2c\frac{2\pi}{9}$	−1	$2c\frac{4\pi}{9}$	0
A_1''	1	1	1	1	1	1	−1	−1	−1	−1	−1	−1
A_2''	1	1	1	1	1	−1	−1	−1	−1	−1	−1	1
E_1''	2	$2c\frac{2\pi}{9}$	$2c\frac{4\pi}{9}$	−1	$2c\frac{8\pi}{9}$	0	−2	$-2c\frac{2\pi}{9}$	$-2c\frac{4\pi}{9}$	1	$-2c\frac{8\pi}{9}$	0
E_2''	2	$2c\frac{4\pi}{9}$	$2c\frac{8\pi}{9}$	−1	$2c\frac{2\pi}{9}$	0	−2	$-2c\frac{4\pi}{9}$	$-2c\frac{8\pi}{9}$	1	$-2c\frac{2\pi}{9}$	0
E_3''	2	−1	−1	2	−1	0	−2	1	1	−2	1	0
E_4''	2	$2c\frac{8\pi}{9}$	$2c\frac{2\pi}{9}$	−1	$2c\frac{4\pi}{9}$	0	−2	$-2c\frac{8\pi}{9}$	$-2c\frac{2\pi}{9}$	1	$-2c\frac{4\pi}{9}$	0
$\Gamma_{x,y,z}$	3	$1+2c\frac{2\pi}{9}$	$1+2c\frac{4\pi}{9}$	0	$1+2c\frac{8\pi}{9}$	−1	1	$-1+2c\frac{2\pi}{9}$	$-1+2c\frac{4\pi}{9}$	−2	$-1+2c\frac{8\pi}{9}$	1

D_{9h} (cont.)			
A_1'		$x^2 + y^2, z^2$	
A_2'	R_z		
E_1'	(x, y)		(xz^2, yz^2)
E_2'		$(x^2 - y^2, xy)$	
E_3'			$[x(x^2 - 3y^2), y(3x^2 - y^2)]$
E_4'			
A_1''			
A_2''	z		z^3
E_1''	(R_x, R_y)	(xz, yz)	
E_2''			$[xyz, z(x^2 - y^2)]$
E_3''			
E_4''			

D_{nh} Groups (cont.)

D_{10h}	E	$2C_{10}$	$2C_5$	$2C_{10}^3$	$2C_5^2$	C_2	$5C_2'$	$5C_2''$	i	$2S_5^3$	$2S_{10}^3$	$2S_5$	$2S_{10}$
A_{1g}	1	1	1	1	1	1	1	1	1	1	1	1	1
A_{2g}	1	1	1	1	1	1	−1	−1	1	1	1	1	1
B_{1g}	1	−1	1	−1	1	−1	1	−1	1	−1	1	−1	1
B_{2g}	1	−1	1	−1	1	−1	−1	1	1	−1	1	−1	1
E_{1g}	2	$2c\frac{\pi}{5}$	$2c\frac{2\pi}{5}$	$-2c\frac{2\pi}{5}$	$-2c\frac{\pi}{5}$	−2	0	0	2	$2c\frac{\pi}{5}$	$2c\frac{2\pi}{5}$	$-2c\frac{2\pi}{5}$	$-2c\frac{\pi}{5}$
E_{2g}	2	$2c\frac{2\pi}{5}$	$-2c\frac{\pi}{5}$	$-2c\frac{\pi}{5}$	$2c\frac{2\pi}{5}$	2	0	0	2	$2c\frac{2\pi}{5}$	$-2c\frac{\pi}{5}$	$-2c\frac{\pi}{5}$	$2c\frac{2\pi}{5}$
E_{3g}	2	$-2c\frac{2\pi}{5}$	$-2c\frac{\pi}{5}$	$2c\frac{\pi}{5}$	$2c\frac{2\pi}{5}$	−2	0	0	2	$-2c\frac{2\pi}{5}$	$-2c\frac{\pi}{5}$	$2c\frac{\pi}{5}$	$2c\frac{2\pi}{5}$
E_{4g}	2	$-2c\frac{\pi}{5}$	$2c\frac{2\pi}{5}$	$2c\frac{2\pi}{5}$	$-2c\frac{\pi}{5}$	2	0	0	2	$-2c\frac{\pi}{5}$	$2c\frac{2\pi}{5}$	$2c\frac{2\pi}{5}$	$-2c\frac{\pi}{5}$
A_{1u}	1	1	1	1	1	1	1	1	−1	−1	−1	−1	−1
A_{2u}	1	1	1	1	1	1	−1	−1	−1	−1	−1	−1	−1
B_{1u}	1	−1	1	−1	1	−1	1	−1	−1	1	−1	1	−1
B_{2u}	1	−1	1	−1	1	−1	−1	1	−1	1	−1	1	−1
E_{1u}	2	$2c\frac{\pi}{5}$	$2c\frac{2\pi}{5}$	$-2c\frac{2\pi}{5}$	$-2c\frac{\pi}{5}$	−2	0	0	−2	$-2c\frac{\pi}{5}$	$-2c\frac{2\pi}{5}$	$2c\frac{2\pi}{5}$	$2c\frac{\pi}{5}$
E_{2u}	2	$2c\frac{2\pi}{5}$	$-2c\frac{\pi}{5}$	$-2c\frac{\pi}{5}$	$2c\frac{2\pi}{5}$	2	0	0	−2	$-2c\frac{2\pi}{5}$	$2c\frac{\pi}{5}$	$2c\frac{\pi}{5}$	$-2c\frac{2\pi}{5}$
E_{3u}	2	$-2c\frac{2\pi}{5}$	$-2c\frac{\pi}{5}$	$2c\frac{\pi}{5}$	$2c\frac{2\pi}{5}$	−2	0	0	−2	$2c\frac{2\pi}{5}$	$2c\frac{\pi}{5}$	$-2c\frac{\pi}{5}$	$-2c\frac{2\pi}{5}$
E_{4u}	2	$-2c\frac{\pi}{5}$	$2c\frac{2\pi}{5}$	$2c\frac{2\pi}{5}$	$-2c\frac{\pi}{5}$	2	0	0	−2	$2c\frac{\pi}{5}$	$-2c\frac{2\pi}{5}$	$-2c\frac{2\pi}{5}$	$2c\frac{\pi}{5}$
$\Gamma_{x,y,z}$	3	$1+2c\frac{\pi}{5}$	$1+2c\frac{2\pi}{5}$	$1-2c\frac{2\pi}{5}$	$1-2c\frac{\pi}{5}$	−1	−1	−1	−3	$-1-2c\frac{\pi}{5}$	$-1-2c\frac{2\pi}{5}$	$-1+2c\frac{2\pi}{5}$	$1+2c\frac{\pi}{5}$

D_{10h} (cont.)	σ_h	$5\sigma_d$	$5\sigma_v$			
A_{1g}	1	1	1		$x^2 + y^2, z^2$	
A_{2g}	1	−1	−1	R_z		
B_{1g}	−1	1	−1			
B_{2g}	−1	−1	1			
E_{1g}	−2	0	0	(R_x, R_y)	(xz, yz)	
E_{2g}	2	0	0		$(x^2 - y^2, xy)$	
E_{3g}	−2	0	0			
E_{4g}	2	0	0			
A_{1u}	−1	−1	−1			
A_{2u}	−1	1	1	z		z^3
B_{1u}	1	−1	1			
B_{2u}	1	1	−1			
E_{1u}	2	0	0	(x, y)		(xz^2, yz^2)
E_{2u}	−2	0	0			$[xyz, z(x^2 - y^2)]$
E_{3u}	2	0	0			$[x(x^2 - 3y^2), y(3x^2 - y^2)]$
E_{4u}	−2	0	0			
$\Gamma_{x,y,z}$	1	1	1			

S_{2n} Groups

S_4	E	S_4	C_2	S_4^3			
A	1	1	1	1	R_z	x^2+y^2, z^2	$xyz, z(x^2-y^2)$
B	1	−1	1	−1	x	x^2-y^2, xy	z^3
E	$\begin{cases}1\\1\end{cases}$	$\begin{matrix}i\\-i\end{matrix}$	$\begin{matrix}-1\\-1\end{matrix}$	$\begin{rcases}-i\\i\end{rcases}$	$(x, y), (R_x, R_y)$	(xz, yz)	$(xz^2, yz^2), [x(x^2-3y^2), y(3x^2-y^2)]$
$\Gamma_{x,y,z}$	3	−1	−1	−1			

S_6	E	C_3	C_3^2	i	S_6^5	S_6		$\epsilon = e^{2\pi i/3}$	
A_g	1	1	1	1	1	1	R_z	x^2+y^2, z^2	
E_g	$\begin{cases}1\\1\end{cases}$	$\begin{matrix}\epsilon\\\epsilon^*\end{matrix}$	$\begin{matrix}\epsilon^*\\\epsilon\end{matrix}$	$\begin{matrix}1\\1\end{matrix}$	$\begin{matrix}\epsilon\\\epsilon^*\end{matrix}$	$\begin{rcases}\epsilon^*\\\epsilon\end{rcases}$	(R_x, R_y)	$(x^2-y^2, xy), (xz, yz)$	
A_u	1	1	1	−1	−1	−1	z		$z^3, x(x^2-3y^2), y(3x^2-y^2)$
E_u	$\begin{cases}1\\1\end{cases}$	$\begin{matrix}\epsilon\\\epsilon^*\end{matrix}$	$\begin{matrix}\epsilon^*\\\epsilon\end{matrix}$	$\begin{matrix}-1\\-1\end{matrix}$	$\begin{matrix}-\epsilon\\-\epsilon^*\end{matrix}$	$\begin{rcases}-\epsilon^*\\-\epsilon\end{rcases}$	(x, y)		$(xz^2, yz^2), [xyz, z(x^2-y^2)]$
$\Gamma_{x,y,z}$	3	0	0	−3	0	0			

S_8	E	S_8	C_4	S_8^3	C_2	S_8^5	C_4^3	S_8^7		$\epsilon = e^{2\pi i/8}$	
A	1	1	1	1	1	1	1	1	R_z	$x^2 + y^2,\ z^2$	
B	1	-1	1	-1	1	-1	1	-1	z		z^3
E_1	1	ϵ	i	$-\epsilon^*$	-1	$-\epsilon$	$-i$	ϵ^*	$(x,\ y),\ (R_x,\ R_y)$		$(xz^2,\ yz^2)$
	1	ϵ^*	$-i$	$-\epsilon$	-1	$-\epsilon^*$	i	ϵ			
E_2	1	i	-1	$-i$	1	i	-1	$-i$		$(x^2 - y^2,\ xy)$	$[xyz,\ z(x^2 - y^2)]$
	1	$-i$	-1	i	1	$-i$	-1	i			
E_3	1	$-\epsilon^*$	$-i$	ϵ	-1	ϵ^*	i	$-\epsilon$		$(xz,\ yz)$	$[x(x^2 - 3y^2),\ y(3x^2 - y^2)]$
	1	$-\epsilon$	i	ϵ^*	-1	ϵ	$-i$	$-\epsilon^*$			
$\Gamma_{x,y,z}$	3	$-1 + \sqrt{2}$	1	$-1 - \sqrt{2}$	-1	$-1 - \sqrt{2}$	1	$-1 + \sqrt{2}$			

S_{2n} Groups (cont.)

S_{10}	E	C_5	C_5^2	C_5^3	C_5^4	i	S_{10}^7	S_{10}^9	S_{10}	S_{10}^3
A_g	1	1	1	1	1	1	1	1	1	1
E_{1g}	1	ϵ	ϵ^2	ϵ^{2*}	ϵ^*	1	ϵ	ϵ^2	ϵ^{2*}	ϵ^*
	1	ϵ^*	ϵ^{2*}	ϵ^2	ϵ	1	ϵ^*	ϵ^{2*}	ϵ^2	ϵ
E_{2g}	1	ϵ^2	ϵ^*	ϵ	ϵ^{2*}	1	ϵ^2	ϵ^*	ϵ	ϵ^{2*}
	1	ϵ^{2*}	ϵ	ϵ^*	ϵ^2	1	ϵ^{2*}	ϵ	ϵ^*	ϵ^2
A_u	1	1	1	1	1	−1	−1	−1	−1	−1
E_{1u}	1	ϵ	ϵ^2	ϵ^{2*}	ϵ^*	−1	$-\epsilon$	$-\epsilon^2$	$-\epsilon^{2*}$	$-\epsilon^*$
	1	ϵ^*	ϵ^{2*}	ϵ^2	ϵ	−1	$-\epsilon^*$	$-\epsilon^{2*}$	$-\epsilon^2$	$-\epsilon$
E_{2u}	1	ϵ^2	ϵ^*	ϵ	ϵ^{2*}	−1	$-\epsilon^2$	$-\epsilon^*$	$-\epsilon$	$-\epsilon^{2*}$
	1	ϵ^{2*}	ϵ	ϵ^*	ϵ^2	−1	$-\epsilon^{2*}$	$-\epsilon$	$-\epsilon^*$	$-\epsilon^2$
$\Gamma_{x,y,z}$	3	$1 + 2c\,\frac{2\pi}{5}$	$1 + 2c\,\frac{4\pi}{5}$	$1 + 2c\,\frac{4\pi}{5}$	$1 + 2c\,\frac{2\pi}{5}$	−3	$-1 - 2c\,\frac{2\pi}{5}$	$-1 - 2c\,\frac{4\pi}{5}$	$-1 - 2c\,\frac{4\pi}{5}$	$-1 - 2c\,\frac{2\pi}{5}$

S_{10}(cont.)	$\epsilon = e^{2\pi i/10}$		
A_g	R_z	(x^2+y^2, z^2)	
E_{1g}	(R_x, R_y)	(xz, yz)	
E_{2g}		(x^2-y^2, xy)	
A_u	z		z^3
E_{1u}	(x, y)		(xz^2, yz^2)
E_{2u}			$[xyz, z(x^2-y^2)], [x(x^2-3y^2), y(3x^2-y^2)]$

Cubic Groups

T	E	$4C_3$	$4C_3^2$	$3C_2$	$\epsilon = e^{2\pi i/3}$		
A	1	1	1	1		$x^2+y^2+z^2$	xyz
E	$\begin{cases}1\\1\end{cases}$	$\begin{matrix}\epsilon\\\epsilon^*\end{matrix}$	$\begin{matrix}\epsilon^*\\\epsilon\end{matrix}$	$\begin{rcases}1\\1\end{rcases}$		$(2z^2-x^2-y^2, x^2-y^2)$	
T	3	0	0	−1	$(R_x, R_y, R_z), (x, y, z)$	(xy, xz, yz)	$(x^3, y^3, z^3), [x(z^2-y^2), y(z^2-x^2), z(x^2-y^2)]$
$\Gamma_{x,y,z}$	3	0	0	−1			

Cubic Groups (cont.)

T_h	E	$4C_3$	$4C_3^2$	$3C_2$	i	$4S_6$	$4S_6^5$	$3\sigma_h$
A_g	1	1	1	1	1	1	1	1
A_u	1	1	1	1	−1	−1	−1	−1
E_g	1	ϵ	ϵ^*	1	1	ϵ	ϵ^*	1
	1	ϵ^*	ϵ	1	1	ϵ^*	ϵ	1
E_u	1	ϵ	ϵ^*	1	−1	$-\epsilon$	$-\epsilon^*$	−1
	1	ϵ^*	ϵ	1	−1	$-\epsilon^*$	$-\epsilon$	−1
T_g	3	0	0	−1	3	0	0	−1
T_u	3	0	0	−1	−3	0	0	1
$\Gamma_{x,y,z,}$	3	0	0	−1	−3	0	0	1

T_h (cont.)		$\epsilon = e^{2\pi i/3}$	
A_g		$x^2 + y^2 + z^2$	
A_u			xyz
E_g		$(2z^2 - x^2 - y^2, x^2 - y^2)$	
E_u			
T_g	(R_x, R_y, R_z)	(xz, yz, xy)	
T_u	(x, y, z)		(x^3, y^3, z^3), $[x(z^2 - y^2), y(z^2 - x^2), z(x^2 - y^2)]$

T_d	E	$8C_3$	$3C_2$	$6S_4$	$6\sigma_d$			
A_1	1	1	1	1	1		$x^2 + y^2 + z^2$	xyz
A_2	1	1	1	−1	−1			
E	2	−1	2	0	0		$(2z^2 - x^2 - y^2, x^2 - y^2)$	
T_1	3	0	−1	1	−1	(R_x, R_y, R_z)		$[x(z^2 - y^2), y(z^2 - x^2), z(x^2 - y^2)]$
T_2	3	0	−1	−1	1	(x, y, z)	(xy, xz, yz)	(x^3, y^3, z^3)
$\Gamma_{x,y,z}$	3	0	−1	−1	1			

O	E	$6C_4$	$3C_2(= C_4^2)$	$8C_3$	$6C_2$			
A_1	1	1	1	1	1		$x^2 + y^2 + z^2$	
A_2	1	−1	1	1	−1			xyz
E	2	0	2	−1	0		$(2z^2 - x^2 - y^2, x^2 - y^2)$	
T_1	3	1	−1	0	−1	$(R_x, R_y, R_z), (x, y, z)$		(x^3, y^3, z^3)
T_2	3	−1	−1	0	1		(xy, xz, yz)	$[x(z^2 - y^2), y(z^2 - x^2), z(x^2 - y^2)]$
$\Gamma_{x,y,z}$	3	1	−1	0	−1			

Cubic Groups (cont.)

O_h	E	$8C_3$	$6C_2$	$6C_4$	$3C_2(=C_4^2)$	i	$6S_4$	$8S_6$	$3\sigma_h$	$6\sigma_d$		
A_{1g}	1	1	1	1	1	1	1	1	1	1		$x^2+y^2+z^2$
A_{2g}	1	1	−1	−1	1	1	−1	1	1	−1		
E_g	2	−1	0	0	2	2	0	−1	2	0		$(2z^2-x^2-y^2, x^2-y^2)$
T_{1g}	3	0	−1	1	−1	3	1	0	−1	−1	(R_x, R_y, R_z)	
T_{2g}	3	0	1	−1	−1	3	−1	0	−1	1		(xz, yz, xy)
A_{1u}	1	1	1	1	1	−1	−1	−1	−1	−1		
A_{2u}	1	1	−1	−1	1	−1	1	−1	−1	1		xyz
E_u	2	−1	0	0	2	−2	0	1	−2	0		
T_{1u}	3	0	−1	1	−1	−3	−1	0	1	1	(x, y, z)	(x^3, y^3, z^3)
T_{2u}	3	0	1	−1	−1	−3	1	0	1	−1		$[x(z^2-y^2), y(z^2-x^2), z(x^2-y^2)]$
$\Gamma_{x,y,z}$	3	0	−1	1	−1	−3	−1	0	1	1		

Icosahedral Group

I_h	E	$12C_5$	$12C_5^2$	$20C_3$	$15C_2$	i	$12S_{10}$	$12S_{10}^3$	$20S_6$	15σ		
A_g	1	1	1	1	1	1	1	1	1	1		$x^2+y^2+z^2$
T_{1g}	3	$2c\frac{\pi}{5}$	$2c\frac{3\pi}{5}$	0	−1	3	$2c\frac{3\pi}{5}$	$2c\frac{\pi}{5}$	0	−1	(R_x, R_y, R_z)	
T_{2g}	3	$2c\frac{3\pi}{5}$	$2c\frac{\pi}{5}$	0	−1	3	$2c\frac{\pi}{5}$	$2c\frac{3\pi}{5}$	0	−1		
G_g	4	−1	−1	1	0	4	−1	−1	1	0		
H_g	5	0	0	−1	1	5	0	0	−1	1		$(2z^2-x^2-y^2, x^2-y^2, xy, yz, xz)$
A_u	1	1	1	1	1	−1	−1	−1	−1	−1		
T_{1u}	3	$2c\frac{\pi}{5}$	$2c\frac{3\pi}{5}$	0	−1	−3	$-2c\frac{3\pi}{5}$	$-2c\frac{\pi}{5}$	0	1	(x, y, z)	
T_{2u}	3	$2c\frac{3\pi}{5}$	$2c\frac{\pi}{5}$	0	−1	−3	$-2c\frac{\pi}{5}$	$-2c\frac{3\pi}{5}$	0	1		
G_u	4	−1	−1	1	0	−4	1	1	−1	0		
H_u	5	0	0	−1	1	−5	0	0	1	−1		
$\Gamma_{x,y,z}$	3	$2c\frac{\pi}{5}$	$2c\frac{3\pi}{5}$	0	−1	−3	$-2c\frac{3\pi}{5}$	$-2c\frac{\pi}{5}$	0	1		

(In the point group I_h, the f orbitals transform as $T_{2u} + G_u$. We have not attempted to write analytical forms of the seven orbitals which possess just T_{2u} or G_u symmetry.)

Linear Groups

$C_{\infty v}$ †	E	$2C_\infty^\Phi$	$2C_\infty^{2\Phi}$	$2C_\infty^{3\Phi}$	...	$\infty\sigma_v$			
$A_1 \equiv \Sigma^+$	1	1	1	1	...	1	z	x^2+y^2, z^2	z^3
$A_2 \equiv \Sigma^-$	1	1	1	1	...	−1	R_z		
$E_1 \equiv \Pi$	2	$2c\phi$	$2c2\phi$	$2c3\phi$	...	0	$(x, y), (R_x, R_y)$	(xz, yz)	(xz^2, yz^2)
$E_2 \equiv \Delta$	2	$2c2\phi$	$2c4\phi$	$2c6\phi$	...	0		(x^2-y^2, xy)	$[xyz, z(x^2-y^2)]$
$E_3 \equiv \Phi$	2	$2c3\phi$	$2c6\phi$	$2c9\phi$	...	0			$[x(x^2-3y^2), y(3x^2-y^2)]$
$E_4 \equiv \Gamma$	2	$2c4\phi$	$2c8\phi$	$2c12\phi$	...	0			
⋮					⋮				
$\Gamma_{x,y,z}$	3	$1+2c\phi$	$1+2c2\phi$	$1+2c3\phi$	...	1			

† *N.B.* In any problem making use of this point group, the columns $2C_\infty^{2\Phi}$ and $2C_\infty^{3\Phi}$ may be ignored. The information they yield is redundant.

$D_{\infty h}$†	E	$2C_\infty^\phi$	$2C_\infty^{2\phi}$	$2C_\infty^{3\phi}$	$\cdots$	$\infty\sigma_v$	i	$2S_\infty^\phi$	$\cdots$	∞C_2		
Σ_g^+	1	1	1	1	$\cdots$	1	1	1	$\cdots$	1		x^2+y^2, z^2
Σ_g^-	1	1	1	1	$\cdots$	−1	1	1	$\cdots$	−1	R_z	
Π_g	2	$2c\phi$	$2c2\phi$	$2c3\phi$	$\cdots$	0	2	$-2c\phi$	$\cdots$	0	(R_x, R_y)	(xz, yz)
Δ_g	2	$2c2\phi$	$2c4\phi$	$2c6\phi$	$\cdots$	0	2	$2c2\phi$	$\cdots$	0		(x^2-y^2, xy)
Φ_g	2	$2c3\phi$	$2c6\phi$	$2c9\phi$	$\cdots$	0	2	$-2c3\phi$	$\cdots$	0		
Γ_g	2	$2c4\phi$	$2c8\phi$	$2c12\phi$	$\cdots$	0	2	$2c4\phi$	$\cdots$	0		
$\vdots$					$\vdots$				$\vdots$			
Σ_u^+	1	1	1	1	$\cdots$	1	−1	−1	$\cdots$	−1	z	z^3
Σ_u^-	1	1	1	1	$\cdots$	−1	−1	−1	$\cdots$	1		
Π_u	2	$2c\phi$	$2c2\phi$	$2c3\phi$	$\cdots$	0	−2	$2c\phi$	$\cdots$	0	(x, y)	(xz^2, yz^2)
Δ_u	2	$2c2\phi$	$2c4\phi$	$2c6\phi$	$\cdots$	0	−2	$-2c2\phi$	$\cdots$	0		$[xyz, z(x^2-y^2)]$
Φ_u	2	$2c3\phi$	$2c6\phi$	$2c9\phi$	$\cdots$	0	−2	$2c3\phi$	$\cdots$	0		$[x(x^2-3y^2), y(3x^2-y^2)]$
Γ_u	2	$2c4\phi$	$2c8\phi$	$2c12\phi$	$\cdots$	0	−2	$-2c4\phi$	$\cdots$	0		
$\vdots$					$\vdots$							
$\Gamma_{x,y,z}$	3	$1+2c\phi$	$1+2c2\phi$	$1+2c3\phi$	$\cdots$	1	−3	$-1+2c\phi$	$\cdots$	−1		

† *N.B.* In any problem making use of this point group, the columns $2C_\infty^{2\phi}$ and $2C_\infty^{3\phi}$ may be ignored. The information they yield is redundant.

Spherical Group

K†	E	∞C_∞^{ϕ}	$\infty C_\infty^{2\phi}$	...	
S	1	1	1		$x^2+y^2+z^2$
P	3	$1+2c\phi$	$1+2c2\phi$		$(x, y, z), (R_x, R_y, R_z)$
D	5	$1+2c\phi+2c2\phi$	$1+2c2\phi+2c4\phi$		$(2z^2-x^2-y^2, xz, yz, xy, x^2-y^2)$
F	7	$1+2c\phi+2c2\phi+2c3\phi$	$1+2c2\phi+2c4\phi+2c6\phi$		$[z^3, xz^2, yz^2, xyz, z(x^2-y^2), x(x^2-3y^2), y(3x^2-y^2)]$
G	9	$1+2c\phi+2c2\phi+2c3\phi+2c4\phi$	$1+2c2\phi+2c4\phi+2c6\phi+2c8\phi$		
.					
.					
.					

† For the irreducible representation for which the character under E is $2l+1$, the character under C_∞^{ϕ} will be $\left(1+\sum_{n=1}^{l} 2\cos n\phi\right) = [\sin(l+\frac{1}{2})\phi]/\sin\frac{1}{2}\phi$.

In any problem making use of this point group, the column $\infty C_\infty^{2\phi}$ may be ignored. The information it yields is redundant.

Appendix B
Direct products

Notes to Character Tables

1. Add the $g - u$ selection rules, *viz.*, $g \times g = g$; $g \times u = u$; $u \times u = g$.
2. Omit subscripts 1 and 2.
3. Add the prime-double prime selection rules, *viz.*, $' \times ' = '$; $' \times '' = ''$; $'' \times '' = '$.

Antisymmetric product (eq. 5-21) is in brackets.

C_s	A'	A''
A'	A'	A''
A''		A'

C_i	A_g	A_u
A_g	A_g	A_u
A_u		A_g

C_2, C_{2h}[1]	A	B
A	A	B
B		A

C_{2v}	A_1	A_2	B_1	B_2
A_1	A_1	A_2	B_1	B_2
A_2		A_1	B_2	B_1
B_1			A_1	A_2
B_2				A_1

D_2, D_{2h}[1]	A	B_1	B_2	B_3
A	A	B_1	B_2	B_3
B_1		A	B_3	B_2
B_2			A	B_1
B_3				A

C_3, C_{3h}[3], S_6[1]	A	E
A	A	E
E		$[A] + A + E$

$C_{3v}, D_3, D_{3d}^1, D_{3h}^3$	A_1	A_2	E
A_1	A_1	A_2	E
A_2		A_1	E
E			$A_1 + [A_2] + E$

C_4, C_{4h}^1, S_4	A	B	E
A	A	B	E
B		A	E
E			$[A] + A + 2B$

$C_{4v}, D_4, D_{2d}, D_{4h}^1$	A_1	A_2	B_1	B_2	E
A_1	A_1	A_2	B_1	B_2	E
A_2		A_1	B_2	B_1	E
B_1			A_1	A_2	E
B_2				A_1	E
E					$A_1 + [A_2] + B_1 + B_2$

C_5, C_{5h}^3, S_{10}^1	A	E_1	E_2
A	A	E_1	E_2
E_1		$[A] + A + E_2$	$E_1 + E_2$
E_2			$[A] + A + E_1$

$C_{5v}, D_5, D_{5d}^1, D_{5h}^3$	A_1	A_2	E_1	E_2
A_1	A_1	A_2	E_1	E_2
A_2		A_1	E_1	E_2
E_1			$A_1 + [A_2] + E_2$	$E_1 + E_2$
E_2				$A_1 + [A_2] + E_1$

C_6, C_{6h}^1	A	B	E_1	E_2
A	A	B	E_1	E_2
B		A	E_2	E_1
E_1			$[A] + A + E_2$	$2B + E_1$
E_2				$[A] + A + E_2$

C_{6v}, D_6, D_{6h}[1]	A_1	A_2	B_1	B_2	E_1	E_2
A_1	A_1	A_2	B_1	B_2	E_1	E_2
A_2		A_1	B_2	B_1	E_1	E_2
B_1			A_1	A_2	E_2	E_1
B_2				A_1	E_2	E_1
E_1					$A_1 + [A_2] + E_2$	$B_1 + B_2 + E_1$
E_2						$A_1 + [A_2] + E_2$

D_{6d}	A_1	A_2	B_1	B_2	E_1	E_2	E_3	E_4	E_5
A_1	A_1	A_2	B_1	B_2	E_1	E_2	E_3	E_4	E_5
A_2		A_1	B_2	B_1	E_1	E_2	E_3	E_4	E_5
B_1			A_1	A_2	E_5	E_4	E_3	E_2	E_1
B_2				A_1	E_5	E_4	E_3	E_2	E_1
E_1					$A_1 + [A_2] + E_2$	$E_1 + E_3$	$E_2 + E_4$	$E_3 + E_5$	$B_1 + B_2 + E_4$
E_2						$A_1 + [A_2] + E_4$	$E_1 + E_5$	$B_1 + B_2 + E_2$	$E_3 + E_5$
E_3							$A_1 + [A_2] + B_1 + B_2$	$E_1 + E_5$	$E_2 + E_4$
E_4								$A_1 + [A_2] + E_4$	$E_1 + E_3$
E_5									$A_1 + [A_2] + E_2$

$C_7, C_{7h}{}^3$	A	E_1	E_2	E_3
A	A	E_1	E_2	E_3
E_1		$[A] + A + E_2$	$E_1 + E_3$	$E_2 + E_3$
E_2			$[A] + A + E_3$	$E_1 + E_2$
E_3				$[A] + A + E_1$

$C_{7v}, D_7, D_{7d}{}^1, D_{7h}{}^3$	A_1	A_2	E_1	E_2	E_3
A_1	A_1	A_2	E_1	E_2	E_3
A_2		A_1	E_1	E_2	E_3
E_1			$A_1 + [A_2] + E_2$	$E_1 + E_3$	$E_2 + E_3$
E_2				$A_1 + [A_2] + E_3$	$E_1 + E_2$
E_3					$A_1 + [A_2] + E_1$

$C_8, C_{8h}{}^1, S_8$	A	B	E_1	E_2	E_3
A	A	B	E_1	E_2	E_3
B		A	E_3	E_2	E_1
E_1			$[A] + A + E_2$	$E_1 + E_3$	$2B + E_2$
E_2				$[A] + A + 2B$	$E_1 + E_3$
E_3					$[A] + A + E_2$

$D_{4d}, C_{8v}, D_8, D_{8h}{}^1$	A_1	A_2	B_1	B_2	E_1	E_2	E_3
A_1	A_1	A_2	B_1	B_2	E_1	E_2	E_3
A_2		A_1	B_2	B_1	E_1	E_2	E_3
B_1			A_1	A_2	E_3	E_2	E_1
B_2				A_1	E_3	E_2	E_1
E_1					$A_1 + [A_2] + E_2$	$E_1 + E_3$	$B_1 + B_2 + E_2$
E_2						$A_1 + [A_2] + B_1 + B_2$	$E_1 + E_3$
E_3							$A_1 + [A_2] + E_2$

D_{8d}	A_1	A_2	B_1	B_2	E_1	E_2	E_3	E_4	E_5	E_6	E_7
A_1	A_1	A_2	B_1	B_2	E_1	E_2	E_3	E_4	E_5	E_6	E_7
A_2		A_1	B_2	B_1	E_1	E_2	E_3	E_4	E_5	E_6	E_7
B_1			A_1	A_2	E_7	E_6	E_5	E_4	E_3	E_2	E_1
B_2				A_1	E_7	E_6	E_5	E_4	E_3	E_2	E_1
E_1					$A_1 + [A_2] + E_2$	$E_1 + E_3$	$E_2 + E_4$	$E_3 + E_5$	$E_4 + E_6$	$E_5 + E_7$	$B_1 + B_2 + E_6$
E_2						$A_1 + [A_2] + E_4$	$E_1 + E_5$	$E_2 + E_6$	$E_3 + E_7$	$B_1 + B_2 + E_4$	$E_5 + E_7$
E_3							$A_1 + [A_2] + E_6$	$E_1 + E_7$	$B_1 + B_2 + E_2$	$E_3 + E_7$	$E_4 + E_6$
E_4								$A_1 + [A_2] + B_1 + B_2$	$E_1 + E_7$	$E_2 + E_6$	$E_3 + E_5$
E_5									$A_1 + [A_2] + E_6$	$E_1 + E_5$	$E_2 + E_4$
E_6										$A_1 + [A_2] + E_4$	$E_1 + E_3$
E_7											$A_1 + [A_2] + E_2$

C_9, $C_{9h}{}^3$	A	E_1	E_2	E_3	E_4
A	A	E_1	E_2	E_3	E_4
E_1		$[A] + A + E_2$	$E_1 + E_3$	$E_2 + E_4$	$E_3 + E_4$
E_2			$[A] + A + E_4$	$E_1 + E_4$	$E_2 + E_3$
E_3				$[A] + A + E_3$	$E_1 + E_3$
E_4					$[A] + A + E_1$

$C_{9v}, D_9, D_{9d}^1, D_{9h}^3$	A_1	A_2	E_1	E_2	E_3	E_4
A_1	A_1	A_2	E_1	E_2	E_3	E_4
A_2		A_1	E_1	E_2	E_3	E_4
E_1			$A_1 + [A_2] + E_2$	$E_1 + E_3$	$E_2 + E_4$	$E_3 + E_4$
E_2				$A_1 + [A_2] + E_4$	$E_1 + E_4$	$E_2 + E_3$
E_3					$A_1 + [A_2] + E_3$	$E_1 + E_2$
E_4						$A_1 + [A_2] + E_1$

C_{10}, C_{10h}^1	A	B	E_1	E_2	E_3	E_4
A	A	B	E_1	E_2	E_3	E_4
B		A	E_4	E_3	E_2	E_1
E_1			$[A] + A + E_2$	$E_3 + E_4$	$E_2 + E_4$	$2B + E_2$
E_2				$[A] + A + E_4$	$2B + E_1$	$E_2 + E_4$
E_3					$[A] + A + E_4$	$E_1 + E_3$
E_4						$[A] + A + E_2$

C_{10v}, D_{10}, D_{10h}[1]	A_1	A_2	B_1	B_2	E_1	E_2	E_3	E_4
A_1	A_1	A_2	B_1	B_2	E_1	E_2	E_3	E_4
A_2		A_1	B_2	B_1	E_1	E_2	E_3	E_4
B_1			A_1	A_2	E_4	E_3	E_2	E_1
B_2				A_1	E_4	E_3	E_2	E_1
E_1					$A_1 + [A_2] + E_2$	$E_1 + E_3$	$E_2 + E_4$	$B_1 + B_2 + E_3$
E_2						$A_1 + [A_2] + E_4$	$B_1 + B_2 + E_1$	$E_2 + E_4$
E_3							$A_1 + [A_2] + E_4$	$E_1 + E_3$
E_4								$A_1 + [A_2] + E_2$

D_{10d}	A_1	A_2	B_1	B_2	E_1	E_2	E_3	E_4	E_5	E_6
A_1	A_1	A_2	B_1	B_2	E_1	E_2	E_3	E_4	E_5	E_6
A_2		A_1	B_2	B_1	E_1	E_2	E_3	E_4	E_5	E_6
B_1			A_1	A_2	E_9	E_8	E_7	E_6	E_5	E_4
B_2				A_1	E_9	E_8	E_7	E_6	E_5	E_4
E_1					$A_1 + [A_2] + E_2$	$E_1 + E_3$	$E_2 + E_4$	$E_3 + E_5$	$E_4 + E_6$	$E_5 + E_7$
E_2						$A_1 + [A_2] + E_4$	$E_1 + E_5$	$E_2 + E_6$	$E_3 + E_7$	$E_4 + E_8$
E_3							$A_1 + [A_2] + E_6$	$E_1 + E_7$	$E_2 + E_8$	$E_3 + E_9$
E_4								$A_1 + [A_2] + E_8$	$E_1 + E_9$	$B_1 + B_2 + E_2$
E_5									$A_1 + [A_2] + B_1 + B_2$	$E_1 + E_9$
E_6										$A_1 + [A_2] + E_8$
E_7										
E_8										
E_9										

D_{10d} (cont.)	E_7	E_8	E_9
A_1	E_7	E_8	E_9
A_2	E_7	E_8	E_9
B_1	E_3	E_2	E_1
B_2	E_3	E_2	E_1
E_1	$E_6 + E_8$	$E_7 + E_9$	$B_1 + B_2 + E_8$
E_2	$E_5 + E_9$	$B_1 + B_2 + E_6$	$E_7 + E_9$
E_3	$B_1 + B_2 + E_4$	$E_5 + E_9$	$E_6 + E_8$
E_4	$E_3 + E_9$	$E_4 + E_8$	$E_5 + E_7$
E_5	$E_2 + E_8$	$E_3 + E_7$	$E_4 + E_6$
E_6	$E_1 + E_7$	$E_2 + E_6$	$E_3 + E_5$
E_7	$A_1 + [A_2] + E_6$	$E_1 + E_5$	$E_2 + E_4$
E_8		$A_1 + [A_2] + E_4$	$E_1 + E_3$
E_9			$A_1 + [A_2] + E_2$

$O,\ O_h^1,\ T^2,\ T_d,\ T_h^{1,2}$	A_1	A_2	E	T_1	T_2
A_1	A_1	A_2	E	T_1	T_2
A_2		A_1	E	T_2	T_1
E			$A_1 + [A_2] + E$	$T_1 + T_2$	$T_1 + T_2$
T_1				$A_1 + E + [T_1] + T_2$	$A_2 + E + T_1 + T_2$
T_2					$A_1 + E + [T_1] + T_2$

D_∞, $C_{\infty v}$, $D_{\infty h}$[1]	Σ^+	Σ^-	Π	Δ	Φ	Γ	. . .
Σ^+	Σ^+	Σ^-	Π	Δ	Φ	Γ	
Σ^-		Σ^+	Π	Δ	Φ	Γ	
Π			$\Sigma^+ + [\Sigma^-] + \Delta$	$\Pi + \Phi$	$\Delta + \Gamma$	$\Phi + \mathrm{H}$	
Δ				$\Sigma^+ + [\Sigma^-] + \Gamma$	$\Pi + \mathrm{H}$	$\Delta + \mathrm{I}$	
Φ					$\Sigma^+ + [\Sigma^-] + \mathrm{I}$	$\Pi + \Theta$	
Γ						$\Sigma^+ + [\Sigma^-] + \mathrm{K}$	
.							
.							
.							

I, I_h[1]	A	T_1	T_2	G	H
A	A	T_1	T_2	G	H
T_1		$A + [T_1] + H$	$G + H$	$T_2 + G + H$	$T_1 + T_2 + G + H$
T_2			$A + [T_2] + H$	$T_1 + G + H$	$T_1 + T_2 + G + H$
G				$A + [T_1] + [T_2] + G + H$	$T_1 + T_2 + G + 2H$
H					$A + [T_1] + [T_2] + [G] + G + 2H$

K, K_h^1	S	P	D	F	. . .
S	S	P	D	F	
P		$S + [P] + D$	$P + D + F$	$D + F + G$	
D			$S + [P] + D + [F] + G$	$P + D + F + G + H$	
F				$S + [P] + D + [F] + G + [H] + I$	
.					
.					
.					

Direct products of imaginary characters. The characters for the E irreducible representation in the point group C_3 are imaginary:

C_3	E	C_3	C_3^2
A	1	1	1
E	$\left\{\begin{matrix} 1 \\ 1 \end{matrix}\right.$	$\begin{matrix} \epsilon \\ \epsilon^* \end{matrix}$	$\left.\begin{matrix} \epsilon^* \\ \epsilon \end{matrix}\right\}$

where $\epsilon = e^{2\pi i/3}$. To obtain the direct product $E \times E$ we multiply each line of E times both lines of E:

$$E \times E = \left\{\begin{matrix} (1 \quad \epsilon \quad \epsilon^*) \times \begin{pmatrix} 1 & \epsilon & \epsilon^* \\ 1 & \epsilon^* & \epsilon \end{pmatrix} = \begin{matrix} 1 & \epsilon^* & \epsilon \\ 1 & 1 & 1 \end{matrix} \\ \\ (1 \quad \epsilon^* \quad \epsilon\ ) \times \begin{pmatrix} 1 & \epsilon & \epsilon^* \\ 1 & \epsilon^* & \epsilon \end{pmatrix} = \begin{matrix} 1 & 1 & 1 \\ 1 & \epsilon & \epsilon^* \end{matrix} \end{matrix}\right\} = 2A + E$$

In performing these multiplications we used the following identities:

$$\begin{aligned} \epsilon \times \epsilon &= e^{4\pi i/3} = \epsilon^* \\ \epsilon \times \epsilon^* &= e^{6\pi i/3} = 1 \\ \epsilon^* \times \epsilon^* &= e^{8\pi i/3} = \epsilon \end{aligned}$$

All direct products of irreducible representations containing imaginary characters were obtained in a similar manner.

Direct products of trigonometric functions. In the point group C_{5v}, for example, some of the characters of degenerate representations contain trigonometric functions. Although direct products of these representations can be decomposed in analytical form (if you are good at trigonometry), it is easier to work direct products of such characters numerically. For example, the numerical form of the C_{5v} character table (to six decimal places) has the following form:

C_{5v}	E	$2C_5$	$2C_5^2$	$5\sigma_v$
A_1	1	1	1	-1
A_2	1	1	1	-1
E_1	2	0.618034	-1.618034	0
E_2	2	-1.618034	0.618034	0

The direct product $E_1 \times E_1$ is equal to

$$4 \qquad 0.381966 \qquad 2.618034 \qquad 0$$

which can be decomposed to the sum $A_1 + A_2 + E_2$ with eq. 1-64 (Section 1-8). All the direct products for finite point groups were evaluated in this numerical manner using a computer program.

Direct products in infinite point groups. Direct products in a point group such as K can be evaluated numerically by a judicious trial-and-error procedure. If we consider only the first two columns of operations and only the first four irreducible representations, the character table for the group K has the form on the left below:

K	E	∞C_∞^ϕ
S	1	1
P	3	$1 + 2c\phi$
D	5	$1 + 2c\phi + 2c2\phi$
F	7	$1 + 2c\phi + 2c2\phi + 2c3\phi$

$\Rightarrow$

K	E	$\infty C_\infty^{28.7°}$
S	1	1
P	3	2.754292
D	5	3.831834
F	7	3.967864

We obtained the form on the right by arbitrarily setting $\phi = 28.7°$. The direct product $P \times P$ has the characters

$$9 \qquad 7.586126$$

which can be decomposed by rather rapid trial and error to the sum $S + P + D$. Since the decomposition must be unique, we can evaluate direct products in infinite point groups in this manner.

Appendix C Overtones of degenerate vibrations

A recursion formula for the symmetry species of the vth wave function of a doubly degenerate vibration is

$$\chi_v(R) = \tfrac{1}{2}[\chi(R)\, \chi_{v-1}(R) + \chi(R^v)] \tag{C-1}$$

Here $\chi_v(R)$ is the character under the operation R for the vth energy level; $\chi(R)$ is the character under R for the degenerate irreducible representation; $\chi_{v-1}(R)$ is the character of the $(v-1)$th energy level; and $\chi(R^v)$ is the character of the operation R^v. As an example of the use of eq. C-1, Table C-1 shows how to derive the symmetry species of the $v = 1, 2, 3,$ and 4 wave functions of a vibration of symmetry e in the point group C_{4v}. Note that the symmetry species of the $v = 2$ wave function will always be the symmetric product of Appendix B.

Table C-1
Calculation of the Symmetry of the $v = 1, 2, 3$, and 4 Wave Functions of an e Mode in the C_{4v} Point Group

C_{4v}	E	$2C_4$	C_2	$2\sigma_v$	$2\sigma_d$	
A_1	1	1	1	1	1	
A_2	1	1	1	−1	−1	
B_1	1	−1	1	1	−1	
B_2	1	−1	1	−1	1	
E	2	0	−2	0	0	
	E	C_4	C_2	σ_v	σ_d	
R	E	C_4	C_2	σ_v	σ_d	
R^2	E	C_2	E	E	E	
R^3	E	C_4	C_2	σ_v	σ_d	
R^4	E	E	E	E	E	
$\chi(R)$	2	0	−2	0	0	
$\chi(R^2)$	2	−2	2	2	2	
$\chi(R^3)$	2	0	−2	0	0	
$\chi(R^4)$	2	2	2	2	2	
$\chi(R) \equiv \chi_1(R)$	2	0	−2	0	$0 = e$	$\sim \psi(1)$
$\tfrac{1}{2}[\chi(R)\chi_1(R) + \chi(R^2)] = \chi_2(R)$	3	−1	3	1	$1 = a_1 + b_1 + b_2$	$\sim \psi(2)$
$\tfrac{1}{2}[\chi(R)\chi_2(R) + \chi(R^3)] = \chi_3(R)$	4	0	−4	0	$0 = 2e$	$\sim \psi(3)$
$\tfrac{1}{2}[\chi(R)\chi_3(R) + \chi(R^4)] = \chi_4(R)$	5	1	5	1	$1 = 2a_1 + a_2 + b_1 + b_2$	$\sim \psi(4)$

The recursion formula for overtones of triply degenerate vibrations is

$$\chi_v(R) = \tfrac{1}{3}[2\chi(R)\chi_{v-1}(R) + \tfrac{1}{2}\{\chi(R^2) - [\chi(R)]^2\}\chi_{v-2}(R) + \chi(R^v)] \qquad \text{(C-2)}$$

where $\chi_1(R) \equiv \chi(R)$, $\chi_0(R) \equiv 1$, and $\chi_{-k}(R) \equiv 0$.

Some time-saving results from Wilson, Decius, and Cross are given below.

General rules:

$(g)^v = g$. $(u)^v = g$ if v is even, $= u$ if v is odd. $(')^v = (')$. $('')^v = (')$ if v is even, $= ('')$ if v is odd.

Doubly degenerate fundamentals:

For C_3, C_{3v}, C_{3h}, D_3, D_{3h}, T, T_d, T_h, O, O_h:

v even

Let $\dfrac{v-2}{2} = 3p + q$

where $p = 0, 1, 2, 3, \ldots$

$q = 0, 1, 2,$

$(E)^v = A_1 + E + p(A_1 + A_2 + 2E)$

$+ E$ if $q = 1$

or

$+ A_1 + A_2 + E$ if $q = 2$

v odd

Let $\dfrac{v+1}{2} = 3p + q$

where $p = 0, 1, 2, 3, \ldots$

$q = 0, 1, 2$

$(E)^v = p(A_1 + A_2 + 2E)$

$+ E$ if $q = 1$

or

$+ A_1 + A_2 + E$ if $q = 2$

For C_6, C_{6v}, C_{6h}, D_6, D_{3d}, D_{6h}, S_6: Use the same rules as for C_3, *etc.*, with the following modifications:

v even

$(E_1)^v$: Put subscript 2 on E

$(E_2)^v$: Put subscript 2 on E

v odd

$(E_1)^v$: Change A to B; put subscript 1 on E

$(E_2)^v$: Put subscript 2 on E

For C_4, C_{4v}, C_{4h}, D_4, D_{2d}, D_{4h}, S_4:

v even

Let $\dfrac{v}{2} = 2p + q$

where $p = 0, 1, 2, 3, \ldots$

$q = 0, 1$

$(E)^v = A_1 + p(A_1 + A_2 + B_1 + B_2)$

$+ q(B_1 + B_2)$

v odd

$(E)^v = \dfrac{v+1}{2} E$

For linear molecules:

v even	v odd
$(E_1)^v = A_1 + E_2 + E_4 + E_6 + \cdots + E_v$	$(E_1)^v = E_1 + E_3 + E_5 + \cdots + E_v$

where $A_1 = \Sigma^+$, $E_1 = \Pi$, $E_2 = \Delta$, $E_3 = \Phi$, $E_4 = \Gamma$, etc.

Triply degenerate fundamentals

For T_d, O, O_h:

v even	v odd
Let $\frac{v}{2} = 6p + q$	Let $\frac{v+1}{2} = 6p + q$
where $p = 0, 1, 2, 3, 4, \ldots$	where $p = 0, 1, 2, 3, \ldots$
$q = 0, 1, 2, 3, 4, 5$	$q = 0, 1, 2, 3, 4, 5$
$(T_1)^v = p\Gamma + p(3p + q - 3)\Gamma' + \Gamma_q$ if $q \neq 0$	$(T_1)^v = p\Gamma + p(3p + q - 3)\Gamma' + \Gamma_q$ if $q \neq 0$
$\Gamma = 7A_1 + 3A_2 + 9E + 9T_1 + 12T_2$	$\Gamma = A_1 + 4A_2 + 5E + 12T_1 + 9T_2$
$\Gamma' = A_1 + A_2 + 2E + 3T_1 + 3T_2$	$\Gamma' = A_1 + A_2 + 2E + 3T_1 + 3T_2$
$\Gamma_1 = A_1 + E + T_2$	$\Gamma_1 = T_1$
$\Gamma_2 = 2A_1 + 2E + T_1 + 2T_2$	$\Gamma_2 = A_2 + 2T_1 + T_2$
$\Gamma_3 = 3A_1 + A_2 + 3E + 2T_1 + 4T_2$	$\Gamma_3 = A_2 + E + 4T_1 + 2T_2$
$\Gamma_4 = 4A_1 + A_2 + 5E + 4T_1 + 6T_2$	$\Gamma_4 = 2A_2 + 2E + 6T_1 + 4T_2$
$\Gamma_5 = 5A_1 + 2A_2 + 7E + 6T_1 + 9T_2$	$\Gamma_5 = A_1 + 3A_2 + 3E + 9T_1 + 6T_2$
$(T_2)^v = (T_1)^v$	$(T_2)^v$: Same as $(T_1)^v$ but with subscripts 1 and 2 permuted

For T and T_h: Use the same rules but drop subscripts 1 and 2.

From E. B. Wilson, Jr., J. C. Decius, and P. C. Cross, *Molecular Vibrations*, McGraw-Hill, New York, 1955.

Appendix D The shapes of atomic orbitals

In this appendix are presented qualitative drawings showing the shapes of the s, p, d, and f atomic orbitals that are used frequently in this book. Lobes of dark and light shading are of opposite sign. Photographs of the models of the f orbitals are reproduced from C. Becker, *J. Chem. Ed.*, *41*, 358 (1964). For further discussion of the shapes of the f orbitals, the reader is referred to H. G. Friedman, Jr., G. R. Choppin, and D. G. Feuerbacher, *J. Chem. Ed.*, *41*, 354 (1964).

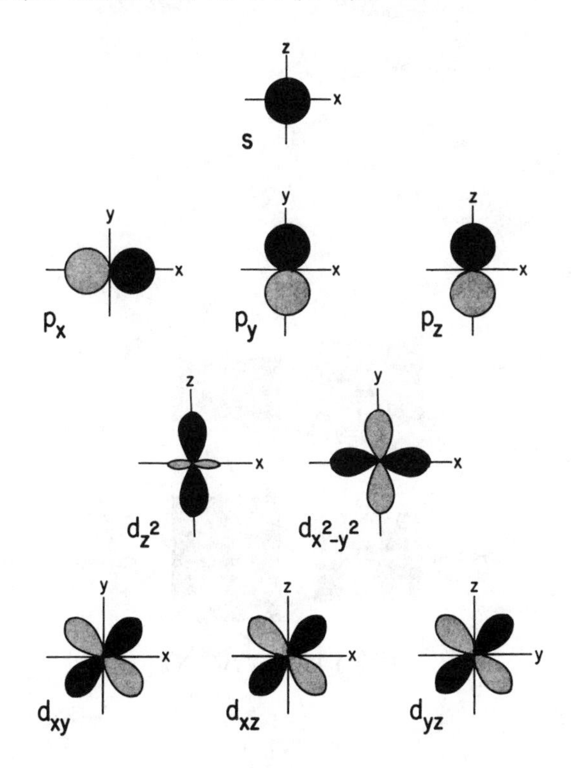

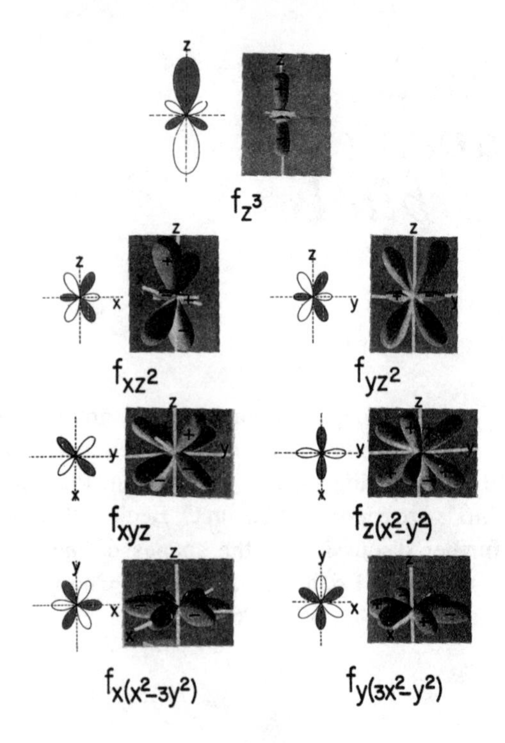
z
f_{z^3}
z
x
f_{xz^2}
z
y
f_{yz^2}
z
y
x
f_{xyz}
z
y
x
$f_{z(x^2-y^2)}$
y
z
x
$f_{x(x^2-3y^2)}$
y
x
$f_{y(3x^2-y^2)}$

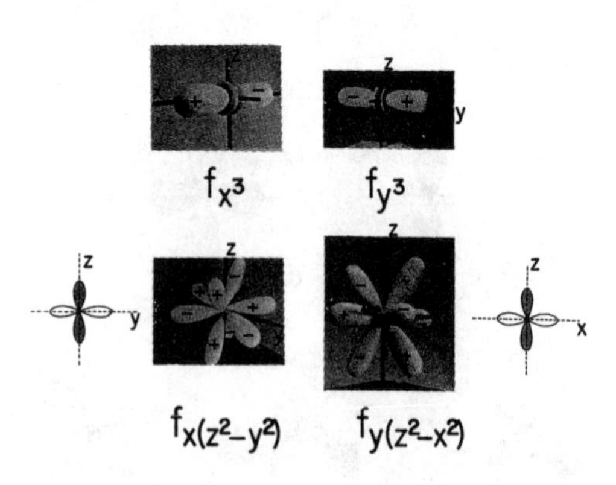
z
x
f_{x^3}
z
y
f_{y^3}
z
y
$f_{x(z^2-y^2)}$
z
x
$f_{y(z^2-x^2)}$

Appendix E *Physical constants*

e	elementary charge	1.6021892(46)	$\times 10^{-19}$ C
		4.803242(14)	$\times 10^{-10}$ esu
c	speed of light in vacuum	2.99792458(1.2)	$\times 10^{8}$ m s^{-1}
			$\times 10^{10}$ cm s^{-1}
h	Planck's constant	6.626176(36)	$\times 10^{-34}$ J s
			$\times 10^{-27}$ erg s
$\hbar$	$h/2\pi$	1.0545887(57)	$\times 10^{-34}$ J s
			$\times 10^{-27}$ erg s
N	Avogadro's number	6.022045(31)	$\times 10^{23}$ mol^{-1}
R	gas constant	8.31441(26)	J mol^{-1} K^{-1}
			$\times 10^{7}$ erg mol^{-1} K^{-1}
		8.20568(26)	$\times 10^{-5}$ m^3 atm mol^{-1} K^{-1}
		8.20568(26)	$\times 10^{-2}$ l atm mol^{-1} K^{-1}
		1.98719	cal mol^{-1} K^{-1}
k	Boltzmann's constant ($= R/N$)	1.380662(44)	$\times 10^{-23}$ J K^{-1}
			$\times 10^{-16}$ erg K^{-1}
m_e	electron rest mass	9.109534(47)	$\times 10^{-31}$ kg
			$\times 10^{-28}$ g
m_p	proton rest mass	1.6726485(86)	$\times 10^{-27}$ kg
			$\times 10^{-24}$ g
ϵ_0	dielectric constant (permittivity) of free space	8.854187818(71)	$\times 10^{-12}$ C^2 N^{-1} m^{-2}
G	gravitational constant	6.6720(41)	$\times 10^{-11}$ m^3 s^{-2} kg^{-1}
a_0	Bohr radius	5.2917706(44)	$\times 10^{-11}$ m
μ_B	Bohr magneton $\left(\frac{e\hbar}{2m_e c}\right)$	9.274078(36)	$\times 10^{-24}$ J T^{-1}
			$\times 10^{-21}$ erg G^{-1}
μ_e	electron magnetic moment	9.284832(36)	$\times 10^{-24}$ J T^{-1}
			$\times 10^{-21}$ erg G^{-1}

From E. R. Cohen and B. N. Taylor, *J. Phys. Chem. Ref. Data, 2,* 663 (1973). Numbers in parentheses are the one standard deviation uncertainties in the last digits.

Appendix F
Energy conversions

Unit =	J molecule^{-1}	J mol^{-1}	cal mol^{-1}	erg molecule^{-1}	eV molecule^{-1}	cm^{-1} molecule^{-1}
J molecule^{-1}	1	6.0220×10^{23}	1.4393×10^{23}	10^{7}	6.2415×10^{18}	5.0340×10^{22}
J mol^{-1}	1.6606×10^{-24}	1	2.3901×10^{-1}	1.6606×10^{-17}	1.0364×10^{-5}	8.3593×10^{-2}
cal mol^{-1}	6.9478×10^{-24}	4.184	1	6.9478×10^{-17}	4.3364×10^{-5}	3.4976×10^{-1}
erg molecule^{-1}	10^{-7}	6.0220×10^{16}	1.4393×10^{16}	1	6.2415×10^{11}	5.0340×10^{15}
eV molecule^{-1}	1.6022×10^{-19}	9.6485×10^{4}	2.3060×10^{4}	1.6022×10^{-12}	1	8.0655×10^{3}
cm^{-1} molecule^{-1}	1.9865×10^{-23}	1.1963×10^{1}	2.8591	1.9865×10^{-16}	1.2399×10^{-4}	1

Example of use: To convert from cal mol^{-1} to J mol^{-1}, multiply by 4.184 found under J mol^{-1} and across from cal mol^{-1}: 2.1000 cal mol^{-1} = (2.1000) × (4.184) = 8.7864 J mol^{-1}. Conversion factors were derived from the constants in Appendix E.

Appendix G
Answers to problems

Chapter 1

1-1. $PtCl_4^{2-}$: C_4, C_2, and S_4 axes through Pt ⊥ to the molecular plane. C_2 axes along Cl–Pt–Cl axes and bisecting these axes. Mirror planes ⊥ to the molecular plane colinear with the C_2 axes. The molecular plane is also a mirror plane, and the Pt atom is a center of inversion.
C_2H_4: Three ⊥ C_2 axes going through the center of the molecule and three ⊥ mirror planes that meet at the center of the molecule, which is a center of inversion.
SF_4: A C_2 axis passes through S. The plane of the page is a mirror plane, and there is another mirror plane ⊥ to the page and colinear with C_2.
C_3H_6: There are C_3 and S_3 axes ⊥ to the plane of the carbon ring, which is a mirror plane. There are three C_2 axes in the plane of the carbon ring, and three mirror planes ⊥ to the plane of the carbon ring.
$C_6H_4F_2$: The center of the molecule is an inversion center. There are three ⊥ C_2 axes that meet at the center of the molecule and three ⊥ mirror planes that also meet there.
ϕ_4C_4: There is a C_4 axis ⊥ to the center of the molecule, and C_2 axes pass through opposite phenyl rings. There are two C_2 axes passing through opposite edges of the C_4 ring.

1-2. (a) inactive—σ; (b) inactive—σ; (c) active; (d) inactive—i, σ; (e) inactive—σ; (f) inactive—i, σ; (g) active; (h) active; (i) active before randomized by rotations.

1-3. Identity $= G$. $(J\cdot L)\cdot M = P\cdot M = H$; $J\cdot(L\cdot M) = J\cdot K = H$. $(K\cdot H)\cdot P = J\cdot P = L$; $K\cdot(H\cdot P) = K\cdot M = L$. $O^{-1} = O$. $K^{-1} = L$. Possible orders of subgroups = 1, 2, 4. Order 2: (G, P), (G, H), (G, J), (G, O), (G, M). Order 4: (G, K, L, M), (G, H, M, P), (G, J, M, O). $H\cdot K\cdot H^{-1} = L$. $L\cdot O\cdot L^{-1} = J$. $P^{-1}\cdot P\cdot P = P$. All similarity transformations performed on H and P give only H and P as the products (e.g., $K\cdot H\cdot K^{-1} = P$).

1-4.

D_3	E	C_3	C_3^2	C_2	C_2'	C_2''
E	E	C_3	C_3^2	C_2	C_2'	C_2''
C_3	C_3	C_3^2	E	C_2'	C_2''	C_2
C_3^2	C_3^2	E	C_3	C_2''	C_2	C_2'
C_2	C_2	C_2''	C_2'	E	C_3^2	C_3
C_2'	C_2'	C_2	C_2''	C_3	E	C_3^2
C_2''	C_2''	C_2'	C_2	C_3^2	C_3	E

Subgroups: (E), (E, C_2), (E, C_2'), (E, C_2''), (E, C_3, C_3^2).
Classes: (E), (C_3, C_3^2), (C_2, C_2', C_2'').

1-5.

C_{3v}	E	C_3	C_3^2	σ_v	σ_v'	σ_v''
E	E	C_3	C_3^2	σ_v	σ_v'	σ_v''
C_3	C_3	C_3^2	E	σ_v'	σ_v''	σ_v
C_3^2	C_3^2	E	C_3	σ_v''	σ_v	σ_v'
σ_v	σ_v	σ_v''	σ_v'	E	C_3^2	C_3
σ_v'	σ_v'	σ_v	σ_v''	C_3	E	C_3^2
σ_v''	σ_v''	σ_v'	σ_v	C_3^2	C_3	E

1-6.

1-7.

D_{2h}	E	$C_2(z)$	$C_2(x)$	$C_2(y)$	i	$\sigma(xy)$	$\sigma(xz)$	$\sigma(yz)$
E	E	$C_2(z)$	$C_2(x)$	$C_2(y)$	i	$\sigma(xy)$	$\sigma(xz)$	$\sigma(yz)$
$C_2(z)$	$C_2(z)$	E	$C_2(y)$	$C_2(x)$	$\sigma(xy)$	i	$\sigma(yz)$	$\sigma(xz)$
$C_2(x)$	$C_2(x)$	$C_2(y)$	E	$C_2(z)$	$\sigma(yz)$	$\sigma(xz)$	$\sigma(xy)$	i
$C_2(y)$	$C_2(y)$	$C_2(x)$	$C_2(z)$	E	$\sigma(xz)$	$\sigma(yz)$	i	$\sigma(xy)$
i	i	$\sigma(xy)$	$\sigma(yz)$	$\sigma(xz)$	E	$C_2(z)$	$C_2(y)$	$C_2(x)$
$\sigma(xy)$	$\sigma(xy)$	i	$\sigma(xz)$	$\sigma(yz)$	$C_2(z)$	E	$C_2(x)$	$C_2(y)$
$\sigma(xz)$	$\sigma(xz)$	$\sigma(yz)$	$\sigma(xy)$	i	$C_2(y)$	$C_2(x)$	E	$C_2(z)$
$\sigma(yz)$	$\sigma(yz)$	$\sigma(xz)$	i	$\sigma(xy)$	$C_2(x)$	$C_2(y)$	$C_2(z)$	E

1-7 (*cont'd.*)

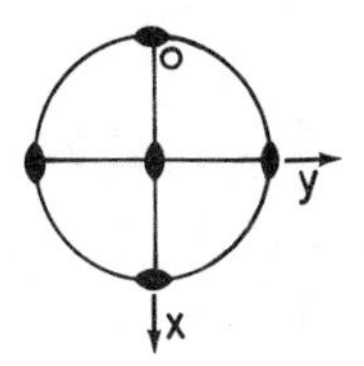

1-8.

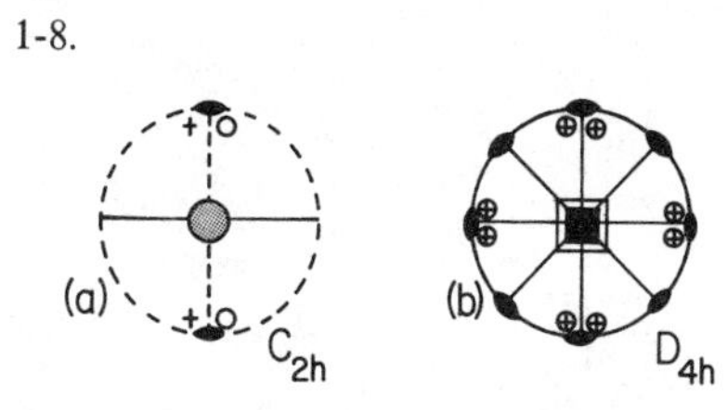

1-9.

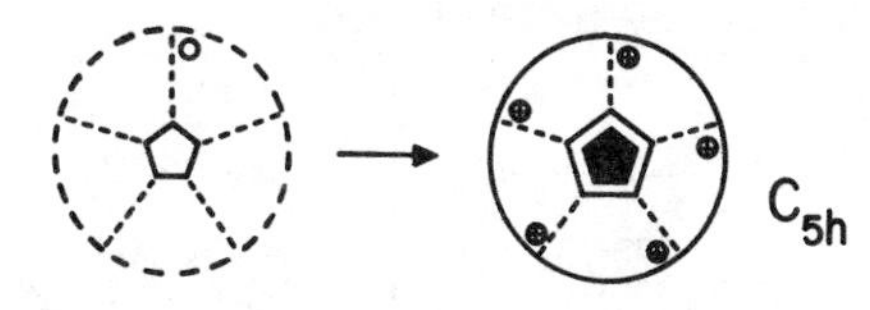

1-10. (a) $D_{\infty h}$; (b) $C_{\infty v}$; (c) D_{5h}; (d) C_{2v}; (e) C_{2v}; (f) C_s; (g) C_{2h}; (h) D_{3h}; (i) C_{2v}; (j) C_s; (k) D_{4d}; (l) D_{4h}; (m) D_{2h}; (n) T_d; (o) O_h; (p) T_d; (q) D_{2h}; (r) D_{3h}.

1-11. (a) $\begin{bmatrix} 11 \\ -1 \\ 7 \end{bmatrix}$ (b) $\begin{bmatrix} 11 & 2 & 4 \\ -1 & -4 & -1 \\ 7 & -2 & 2 \end{bmatrix}$

1-12.

$$\boldsymbol{E} = \begin{bmatrix} 1 & 0 & 0 \\ 0 & 1 & 0 \\ 0 & 0 & 1 \end{bmatrix} \quad \boldsymbol{C_2(z)} = \begin{bmatrix} -1 & 0 & 0 \\ 0 & -1 & 0 \\ 0 & 0 & 1 \end{bmatrix} \quad \boldsymbol{C_2(x)} = \begin{bmatrix} 1 & 0 & 0 \\ 0 & -1 & 0 \\ 0 & 0 & -1 \end{bmatrix}$$

$$\boldsymbol{C_2(y)} = \begin{bmatrix} -1 & 0 & 0 \\ 0 & 1 & 0 \\ 0 & 0 & -1 \end{bmatrix} \quad \boldsymbol{i} = \begin{bmatrix} -1 & 0 & 0 \\ 0 & -1 & 0 \\ 0 & 0 & -1 \end{bmatrix} \quad \boldsymbol{\sigma(xz)} = \begin{bmatrix} 1 & 0 & 0 \\ 0 & -1 & 0 \\ 0 & 0 & 1 \end{bmatrix}$$

$$\boldsymbol{\sigma(yz)} = \begin{bmatrix} -1 & 0 & 0 \\ 0 & 1 & 0 \\ 0 & 0 & 1 \end{bmatrix} \quad \boldsymbol{\sigma(xy)} = \begin{bmatrix} 1 & 0 & 0 \\ 0 & 1 & 0 \\ 0 & 0 & -1 \end{bmatrix}$$

1-13. $E = \begin{bmatrix} 1 & 0 & 0 \\ 0 & 1 & 0 \\ 0 & 0 & 1 \end{bmatrix}$ $C_3 = \begin{bmatrix} -1/2 & \sqrt{3}/2 & 0 \\ -\sqrt{3}/2 & -1/2 & 0 \\ 0 & 0 & 1 \end{bmatrix}$ $C_3^2 = \begin{bmatrix} -1/2 & -\sqrt{3}/2 & 0 \\ \sqrt{3}/2 & -1/2 & 0 \\ 0 & 0 & 1 \end{bmatrix}$

$$\sigma_v = \begin{bmatrix} 1 & 0 & 0 \\ 0 & -1 & 0 \\ 0 & 0 & 1 \end{bmatrix} \quad \sigma_v' = \begin{bmatrix} -1/2 & -\sqrt{3}/2 & 0 \\ -\sqrt{3}/2 & 1/2 & 0 \\ 0 & 0 & 1 \end{bmatrix} \quad \sigma_v'' = \begin{bmatrix} -1/2 & \sqrt{3}/2 & 0 \\ \sqrt{3}/2 & +1/2 & 0 \\ 0 & 0 & 1 \end{bmatrix}$$

$$C_3 \cdot C_3^2 = \begin{bmatrix} 1 & 0 & 0 \\ 0 & 1 & 0 \\ 0 & 0 & 1 \end{bmatrix}$$

1-14.

$$C_3^2 = \begin{bmatrix} 0 & 0 & 1 \\ 1 & 0 & 0 \\ 0 & 1 & 0 \end{bmatrix} \quad \sigma_v = \begin{bmatrix} 1 & 0 & 0 \\ 0 & 0 & 1 \\ 0 & 1 & 0 \end{bmatrix} \quad \sigma_v' = \begin{bmatrix} 0 & 1 & 0 \\ 1 & 0 & 0 \\ 0 & 0 & 1 \end{bmatrix}$$

$$\sigma_v'' = \begin{bmatrix} 0 & 0 & 1 \\ 0 & 1 & 0 \\ 1 & 0 & 0 \end{bmatrix} \quad Q^{-1} \cdot C_3 \cdot Q = \begin{bmatrix} 1 & 0 & 0 \\ 0 & -1/2 & 3/2\sqrt{3} \\ 0 & -3/2\sqrt{3} & -1/2 \end{bmatrix}$$

1-15. See Appendix A.
1-16. See Appendix A.
1-17. See Appendix A.
1-18. $6\,0\,0 = A_1 + A_2 + 2E$; $E \times E = A_1 + A_2 + E$.
1-19. $\Gamma_1 = B_1 + E$; $\Gamma_2 = A_1 + A_2$; $\Gamma_3 = A_1 + A_2 + B_1 + B_2 + 2E$; $\Gamma_4 = 2B_2 + E$.
1-20. A_2'; E'; $A_1'' + A_2'' + E''$; $A_1' + A_2' + E'$.
1-21. $2\Sigma^+ + 2\Pi$.
1-22. (a) C_{2v}; (b) C_{2v}; (c) D_{5h}; (d) C_{3v}; (e) C_{2h}; (f) C_s; (g) C_{2v}; (h) D_{3h}; (i) $C_{\infty v}$; (j) $D_{\infty h}$; (k) D_{3d}; (l) D_{3h}; (m) C_{2v} (since the ring is puckered. If you consider the planar carbon ring conformation, the symmetry is D_{4h}.) (n) D_{2h}.
1-23. C_{2h}.

1-24.

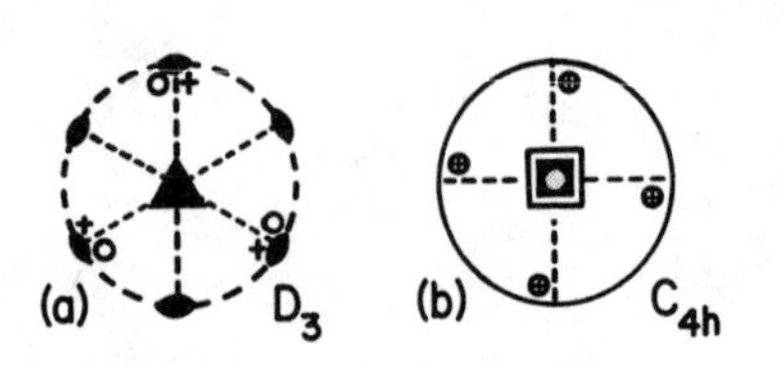

1-25.

$$U \cdot U^T = \begin{bmatrix} 1/\sqrt{2} & 1/\sqrt{2} & 0 \\ 0 & 0 & 1 \\ 1/\sqrt{2} & -1/\sqrt{2} & 0 \end{bmatrix} \cdot \begin{bmatrix} 1/\sqrt{2} & 0 & 1/\sqrt{2} \\ 1/\sqrt{2} & 0 & -1/\sqrt{2} \\ 0 & 1 & 0 \end{bmatrix}$$

$$= \begin{bmatrix} 1 & 0 & 0 \\ 0 & 1 & 0 \\ 0 & 0 & 1 \end{bmatrix} = U \cdot U^{-1}$$

$$U \cdot B \cdot U^{-1} = \begin{bmatrix} 3 & 3\sqrt{2} & 0 \\ 3\sqrt{2} & 9 & 0 \\ 0 & 0 & -1 \end{bmatrix}$$

1-26. See Appendix A.
1-27. See Appendix A.
1-28. See Appendix B.

Chapter 2

2-1.

λ(nm)	10	200	380	780
$\bar{\nu}$(cm^{-1})	1.0×10^6	5.00×10^4	2.63×10^4	1.28×10^4
ν(s^{-1})	3.0×10^{16}	1.50×10^{15}	7.89×10^{14}	3.84×10^{14}

λ(nm)	3000	3.00×10^4	3.00×10^5
$\bar{\nu}$(cm^{-1})	3333	333	33.3
ν(s^{-1})	1.0×10^{14}	1.0×10^{13}	1.0×10^{12}

2-2. 3.56×10^4.

2-3. 40% T means $A = \log(1/T) = 0.3979$. If a 0.0100 M solution has 40% T ($A = 0.3979$), a 0.0200 M solution will exhibit $A = 2 \times 0.3979 = 0.7958$, or $T = 10^{-A} = 0.160$ or % $T = 16.0$. Optical density is the same as absorbance (0.3979 and 0.7958).

2-4. The energy of one 366 nm photon is $hc/\lambda = 5.43 \times 10^{-19}$ J photon^{-1}. 150 J s^{-1}/5.43×10^{-19} J photon^{-1} = 2.76×10^{20} photon s^{-1}. The visible lamp will emit more photons since each photon carries less energy.

2-5. $\varepsilon = \varepsilon_0 \cos\left(\frac{2\pi x}{\lambda} + 2\pi\nu t\right)$ describes a wave moving to the left.

$$\text{Standing wave} = \varepsilon_0\left[\cos\left(\frac{2\pi x}{\lambda} - 2\pi\nu t\right) + \cos\left(\frac{2\pi x}{\lambda} + 2\pi\nu t\right)\right]$$

$$= 2\varepsilon_0 \cos\frac{2\pi x}{\lambda}\cos 2\pi\nu t.$$

The maxima and minima of this function occur at the same values of x, regardless of the value of t.

2-6. 10 eV electron: $\lambda = 3.88 \times 10^{-10}\ m$, both classically and relativistically.
1 MeV electron: $\lambda = 1.23 \times 10^{-12}\ m$, classically;
$8.72 \times 10^{-13}\ m$, relativistically.

2-7. 250 nm: $E = 4.79 \times 10^5\ \text{J mol}^{-1}$; $p = 1.60 \times 10^{-3}\ \text{kg m s}^{-1}\ \text{mol}^{-1}$.
500 nm: $E = 2.39 \times 10^5\ \text{J mol}^{-1}$; $p = 7.98 \times 10^{-4}\ \text{kg m s}^{-1}\ \text{mol}^{-1}$.
2000 cm^{-1}: $E = 2.39 \times 10^4\ \text{J mol}^{-1}$; $p = 7.98 \times 10^{-5}\ \text{kg m s}^{-1}\ \text{mol}^{-1}$.
Electron: $E = 3.97 \times 10^{-19}\ \text{J} \Rightarrow p = \sqrt{2mE} = 8.50 \times 10^{-25}\ \text{kg m s}^{-1}$.
$p(\text{electron})/p(\text{photon}) = 639$.

2-8. $\lambda = 1.67 \times 10^{-10}\ m \Rightarrow \theta = 50.9°$.

2-9. Maximum wavelength for ionization: Ni, 248 nm; Na, 538 nm. Kinetic energy if 230-nm light is used: Ni, 6.26×10^{-20} J; Na, 4.95×10^{-19} J.

2-10. For $\psi = e^{i\alpha x}$, $\frac{\partial^2\psi}{\partial x^2} = -\alpha^2\psi$ and eq. 2-35 becomes $-\frac{\hbar^2}{2m}(-\alpha^2\psi) + V\psi = E\psi$, yielding $\alpha = \sqrt{2m(E - V)}/\hbar$. Similarly, for $\psi = \cos\beta x$ or $\sin\beta x$, $\frac{\partial^2\psi}{\partial x^2} = -\beta^2\psi$, and $\beta = \sqrt{2m(E - V)}/\hbar$.

2-11. $c\psi$ is a solution of eq. 2-35 because $\frac{\partial^2(c\psi)}{\partial x^2} = c\frac{\partial^2\psi}{\partial x^2}$, and c will drop out on both sides of eq. 2-35. If $\psi = \psi_1 + c\psi_2$, eq. 2-35 becomes $-\frac{\hbar^2}{2m}\left[\frac{\partial^2\psi_1}{\partial x^2} + c\frac{\partial^2\psi_2}{\partial x^2}\right] + V(\psi_1 + c\psi_2) = E(\psi_1 + c\psi_2)$, which is satisfied since ψ_1 and ψ_2 are degenerate solutions of eq. 2-35.

2-12. $\frac{d^2}{dx^2}(\cos 4x) = -16\cos 4x$; eigenvalue $= -16$. $\left[-\frac{d^2}{dx^2} + x^2\right](xe^{-x^2/2}) = 3xe^{-x^2/2}$; eigenvalue $= 3$.

2-13. $E\psi(x)\phi(t) = i\hbar\frac{\partial}{\partial t}[\psi(x)\phi(t)] \Rightarrow E\phi(t) = i\hbar\frac{\partial\phi(t)}{\partial t}$. This is satisfied by $\phi(t) = e^{-iEt/\hbar}$, since $\frac{\partial}{\partial t}(e^{-iEt/\hbar}) = -iE\phi(t)/\hbar$, yielding $E\phi(t) = i\hbar(-iE)\phi(t)/\hbar = E\phi(t)$.

2-14. $\langle p\rangle = \int_{-\infty}^{\infty}\psi\left(-i\hbar\frac{\partial}{dx}\right)\psi\,dx = \int_{-\infty}^{\infty}\frac{i\hbar xe^{-x^2/2}}{2\sqrt{2\pi}}\,dx = 0$ (odd integrand).

$\langle p^2\rangle = \int_{-\infty}^{\infty}\psi\left(-\hbar^2\frac{\partial^2}{\partial x^2}\right)\psi\,dx = \hbar^2/4$. $\sigma_p = \sqrt{\langle p^2\rangle - \langle p\rangle^2} = \hbar/2$.

2-15. $A^2\int_0^{10}(1 + x)^{-2}\,dx = 1 \Rightarrow A = \sqrt{11/10}$.

$$\langle x\rangle = A^2\int_0^{10}\frac{x}{(1+x)^2}\,dx = A^2\left[\frac{1}{1+x} + \ln(1+x)\right]_0^{10} = 1.638.$$

$$\langle x^2\rangle = A^2\int_0^{10}\frac{x^2}{(1+x)^2}\,dx = A^2\left[(1+x) - 2\ln(1+x) - \frac{1}{1+x}\right]_0^{10} = 6.725.$$

$\langle x\rangle^2 = 1.638^2 = 2.682.$ $\sigma_x = 2.011.$

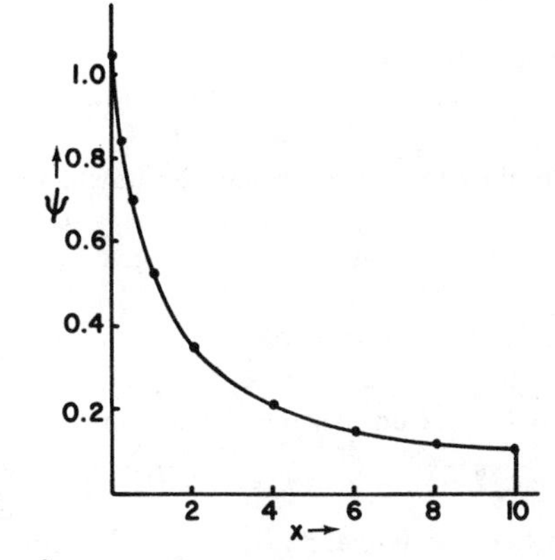

2-16. Linear: $x, \frac{d}{dx}, \frac{d^2}{dx^2}$. Operators in the table are linear.

2-17. $\frac{2}{a}\int_0^a \sin\frac{n\pi x}{a}\left(-\frac{\hbar^2}{2m}\frac{d^2}{dx^2}\right)\sin\frac{n\pi x}{a}\,dx = \frac{n^2\pi^2\hbar^2}{2ma^2}.$

2-18. $\frac{2}{a}\int_0^a \sin\frac{\pi x}{a}\sin\frac{2\pi x}{a}\,dx = 0\left(\text{using}\int \sin mx \sin nx\,dx = \frac{\sin(m-n)x}{2(m-n)} - \frac{\sin(m+n)x}{2(m+n)}\right).$

$\frac{2}{a}\int_0^a \sin\frac{\pi x}{a}\sin\frac{3\pi x}{a}\,dx = 0$ (using the same formula).

2-19. Probability $= \int_{.9a}^{a}\psi_n^2\,dx = \frac{2}{a}\int_{.9a}^{a}\sin^2\frac{n\pi x}{a}\,dx = \frac{2}{n\pi}\left[\frac{z}{2} - \frac{1}{4}\sin 2z\right]_{.9n\pi}^{n\pi}$ (using the substitution $z = n\pi x/a$). For $n = 1$, probability $= .00645$, $n = 8$, .119, and for $n = 100$, .100. Classical probability $= 1/10$, since the distribution should be uniform.

2-20. $\langle p\rangle = 0$ for any value of n. $\langle p^2\rangle = n^2\pi^2\hbar^2/a^2 \Rightarrow \hbar^2\pi^2/a^2$ for ψ_1 and $10^4\hbar^2\pi^2/a^2$ for ψ_{100}.

2-21. As one example, in eq. 2-64, if $b = 2a$, the energy levels ($n_x = 2$, $n_y = 2$), and ($n_x = 1$, $n_y = 4$) are degenerate.

2-22. $-\frac{\hbar^2}{2m}\left(\frac{\partial^2}{\partial x^2} + \frac{\partial^2}{\partial y^2}\right)\psi(x, y) = E\psi(x, y) \Rightarrow \psi = \sqrt{\frac{2}{a}}\sqrt{\frac{2}{b}}\sin\frac{n_x\pi x}{a}\sin\frac{n_y\pi y}{b}$ and

$E = \frac{\pi^2\hbar^2}{2m}\left[\frac{n_x^2}{a^2} + \frac{n_y^2}{b^2}\right]$. If b is not an integer multiple of a, the wave functions are not degenerate. If $b = a$, there will be some degeneracy.

2-23. For ψ_0: $\langle x \rangle = 0$, $\langle x^2 \rangle = 1/2\alpha$, $\sigma_x = \sqrt{1/2\alpha}$, $\langle p \rangle = 0$, $\langle p^2 \rangle = \hbar^2\alpha/2$, $\sigma_p = \hbar\sqrt{\alpha/2}$, and $\sigma_x\sigma_p = \hbar/2$. For ψ_1, $\langle x \rangle = \langle p \rangle = 0$, $\langle x^2 \rangle = 3/2\alpha$, $\langle p^2 \rangle = 3\hbar^2\alpha/2$, and $\sigma_x\sigma_p = 3\hbar/2$.

2-24. $$-\frac{\hbar^2}{2m}\frac{d^2}{dx^2}(e^{\pm\sqrt{2m(V-E)}x/\hbar}) + (V - E)e^{\pm\sqrt{2m(V-E)}x/\hbar}$$
$$= -\frac{\hbar^2}{2m}\left[\frac{2m}{\hbar^2}(V - E)e^{\pm\sqrt{2m(V-E)}x/\hbar} + (V - E)e^{\pm\sqrt{2m(V-E)}x/\hbar}\right] = 0.$$
(This holds if V is not a function of x.)

2-25. Left of barrier: $\psi = Ae^{i\alpha x} + Be^{-i\alpha x}$ ($\alpha = \sqrt{2mE}/\hbar$).
Right of barrier: $\psi = Ce^{i\beta x}$ ($\beta = \sqrt{2m(E + V)}/\hbar$).
(There is no D term since the particle cannot be moving to the left after passing through the barrier.) Continuity at the barrier ($x = 0$) requires $A + B = C$. Continuity of the first derivative at $x = 0$ gives $i\alpha A - i\alpha B = i\beta C$. The reflection coefficient is $\left|\frac{B}{A}\right|^2 = \frac{(\alpha - \beta)^2}{(\alpha + \beta)^2}$.

2-26. $V_0 = 0.01$ eV, $a = 100$ Å, $E/V_0 = 0.9 \Rightarrow T = 0.841$. $V_0 = 0.02$ eV, $a = 50$ Å, $E/V_0 = 0.9 \Rightarrow T = 0.915$. Consider the βa terms in eq. 2-93:
$$\beta a = \sqrt{2m(V_0 - E)}a/\hbar = \sqrt{2m(1 - E/V_0)V_0}a/\hbar.$$
βa is constant if E/V_0 is constant and $\sqrt{V_0}a$ (or $V_0 a^2$) is constant. Then consider the term $(\alpha^2 - \beta^2)^2/4\alpha^2\beta^2$, which can be rewritten
$$\frac{[2mE - 2m(V_0 - E)]^2}{4(2mE)[2m(V_0 - E)]} = \frac{(2E - V_0)^2}{4E(V_0 - E)} = \frac{[(2E/V_0) - 1]^2}{4[(E/V_0) - (E^2/V_0^2)]}.$$
This is also constant if E/V_0 is constant.

Chapter 3

3-1. The molecule goes from a vibrationally excited state to the vibrational ground state, emitting energy ($h\nu$) in the process.

3-2. Infrared active: a, c, d, f.

3-3. $B(N_2) = 3.997 \times 10^{-23}$ J $= 2.01$ cm^{-1} $\Rightarrow$ absorptions at 4.02, 8.04, and 12.06 cm^{-1}. $B(H_2) = 1.209 \times 10^{-21}$ J $= 60.85$ cm^{-1} $\Rightarrow$ absorptions at 121.7, 243.4, and 365.1 cm^{-1}.

3-4. $2B = 0.714$ cm^{-1} $= 1.418 \times 10^{-23}$ J $\Rightarrow r_e = 1.76 \times 10^{-10}$ m.

3-5. ΔE(in joules) $= h\omega = \hbar\sqrt{k/\mu} \Rightarrow k = \mu(\Delta E)^2/\hbar^2$. For H_2, $\Delta E = 4159.5$ cm^{-1} $= 8.2628 \times 10^{-20}$ J $\Rightarrow k = 514$ N m^{-1}. For D_2, $k = 530$ N m^{-1}.

3-6. $\frac{E(^{18}O_2)}{E(^{16}O_2)} = \frac{\omega(^{18}O_2)}{\omega(^{16}O_2)} = \sqrt{\frac{\mu(^{16}O_2)}{\mu(^{18}O_2)}} = \sqrt{\frac{8}{9}} \Rightarrow E(^{18}O_2)$ is predicted to be 775 cm^{-1}.

3-7.

N_2		harmonic	anharmonic
fundamental	$v=0 \rightarrow v=1$	2330.7	2330.7
first overtone	$v=0 \rightarrow v=2$	4661.4	4632.5
second overtone	$v=0 \rightarrow v=3$	6992.1	6905.4
third overtone	$v=0 \rightarrow v=4$	9322.8	9149.3

3-8. Harmonic potential: $\rho = 1/k$ (at $q = 0$)
Morse potential: $\rho = 1/2D_e\beta^2$ (at $q = 0$) $\Big\}\ k = 2D_e\beta^2$.

3-9. $\left(\frac{\alpha}{\pi}\right)^{\frac{1}{2}} \int_{-\infty}^{\infty} e^{-\alpha q^2}\, dq = \left(\frac{\alpha}{\pi}\right)^{\frac{1}{2}} \left(\frac{\pi}{\alpha}\right)^{\frac{1}{2}} = 1.$

3-10. $\left(\frac{\alpha}{\pi}\right)^{\frac{1}{4}} \left(\frac{4\alpha^3}{\pi}\right)^{\frac{1}{4}} \int_{-\infty}^{\infty} qe^{-\alpha q^2}\, dq = 0$ (odd integrand).

3-11. $\langle q\rangle = \langle p\rangle = 0; \langle q^2\rangle = 5/2\alpha; \langle p^2\rangle = 5\hbar^2\alpha/2; \sigma_q\sigma_p = 5\hbar/2.$

3-12. First note that the units of $\omega_e \left(= \frac{1}{2\pi}\sqrt{k_e/\mu}\right)$ are s^{-1}. The value of ω_e for HCl is $2988.9\ cm^{-1} = 100\left(\frac{cm}{m}\right)\cdot c\left(\frac{m}{s}\right)\cdot\bar{\nu}(cm^{-1}) = 8.960 \times 10^{13}\ s^{-1}$. Then using $\mu = 1.627 \times 10^{-27}$ kg, we find $D = 1.058 \times 10^{-26}\ J = 0.00053\ cm^{-1}$.

3-13. Heat the sample to increase the population of vibrationally excited molecules that give rise to anti-Stokes lines.

3-14. $^{37}Cl_2$.

3-15. $kT = 209\ cm^{-1}$. $n_j/n_i = 0.368$.

3-16. 500 K: $J = 8.8 \approx 9$; 100 K: $J = 3.7 \approx 4$.

3-17. (Graphs not shown.) Slope of eq. 3-58′ $\Rightarrow \bar{B}_2 = 1.886\ cm^{-1} \Rightarrow r_2 = 1.142$ Å
Slope of eq. 3-59′ $\Rightarrow \bar{B}_0 = 1.921\ cm^{-1} \Rightarrow r_0 = 1.131$ Å.
$\bar{B}_2 = \bar{B}_e - (5/2)\bar{\alpha}_e$, $\bar{B}_0 = \bar{B}_e - (1/2)\bar{\alpha}_e$ $\Rightarrow$ $\bar{B}_e = 1.929\ cm^{-1} \Rightarrow r_e = 1.129$Å, $\bar{\alpha}_e = 0.0175\ cm^{-1}$
The slopes of eqs. 3-59 and 3-59′ should be the same.

3-18. Following the procedure of Problem 3-12, the value of $\bar{\omega}_e = 2170.21\ cm^{-1} \Rightarrow \omega_e = 6.506 \times 10^{13}\ s^{-1}$. From eq. 3-24, $\bar{D}_e = \bar{D}_0 + (1/2)\,\bar{\omega}_e - (1/4)\bar{\omega}_e x_e$. Using $D_0 = 1076$ kJ/mol, we calculate $\bar{D}_0 = 8.995 \times 10^4\ cm^{-1}$ per molecule. Therefore $\bar{D}_e = 9.103 \times 10^4\ cm^{-1} = 1.808 \times 10^{-18}$ J, using the values of $\bar{\omega}_e$ and $\bar{\omega}_e x_e$ in Table 3-3. Equation 3-46 allows us to solve for $\beta = \pi\omega_e\sqrt{2\mu/D_e} = 2.294 \times 10^{10}\ m^{-1}$, since $\mu = 1.1387 \times 10^{-26}$ kg. Using these values of ω_e, μ, D_e, and β, and adding $r_e = 1.1282 \times 10^{-10}$ m (Table 3-8), we calculate $\alpha_e = 3.257 \times 10^{-25}\ J \Rightarrow \bar{\alpha}_e = 0.0164\ cm^{-1}$.

3-19. As in Problem 3-6, we find $\bar{\omega}(^{13}C^{16}O) \approx 2095\ cm^{-1}$. The P branch of $^{13}C^{16}O$ is most clearly seen around the $J = 20 \rightarrow J' = 19$ transition of $^{12}C^{16}O$. The intensity of the minor isotope is nearly 10% of the major isotope because the transmittance is related logarithmically to the concentration.

3-20. $v = 4.763 \times 10^2\ m\ s^{-1} \Rightarrow \Delta\lambda/\lambda = 1.589 \times 10^{-6}$. At $2000\ cm^{-1}$, $\lambda = 5 \times 10^{-6}$ m. A molecule moving toward the source sees light of wavelength $\lambda - \Delta\lambda = 4.99992055 \times 10^{-6}\ m = 2000.00318\ cm^{-1}$. A molecule moving away from the source sees light of wavelength $\lambda + \Delta\lambda = 5.000007945 \times$

10^{-6} m = 1999.99682 cm^{-1}. The Doppler broadening, therefore, is approximately 0.0064 cm^{-1}.

3-21. $e\int_{-\infty}^{\infty} \psi_1{}^* x \psi_0 \, dx = e\alpha\sqrt{\frac{2}{\pi}}\int_{-\infty}^{\infty} x^2 e^{-\alpha x^2} dx = \frac{e}{2}\sqrt{\frac{2}{\alpha}}.$

$e\int_{-\infty}^{\infty} \psi_2{}^* x \psi_0 \, dx = e\sqrt{\frac{\alpha}{2\pi}}\int_{-\infty}^{\infty} (2\alpha x^3 - x)e^{-\alpha x^2} dx = 0$ (odd integral).

$e\int_{-\infty}^{\infty} \psi_3{}^* x \psi_0 \, dx = \frac{e\alpha}{\sqrt{3\pi}}\int_{-\infty}^{\infty} (2\alpha x^4 - 3x^2)e^{-\alpha x^2} dx = \frac{e\alpha}{\sqrt{3\pi}}\left[\frac{3}{\alpha}\sqrt{\frac{\pi}{\alpha}} - \frac{3}{\alpha}\sqrt{\frac{\pi}{\alpha}}\right]$
$= 0.$

3-22. (a) $\frac{h'}{g'} = \frac{2\sin[(E_1 - E_0 - h\nu)t/2\hbar]\cos[(E_1 - E_0 - h\nu)t/2\hbar](-\pi t)}{2(E_1 - E_0 - h\nu)(-h)}.$

$\frac{h''}{g''} = \frac{\pi^2 t^2}{h^2}\{\cos^2[(E_1 - E_0 - h\nu)t/2\hbar] - \sin^2[(E_1 - E_0 - h\nu)t/2\hbar]\}$

$= \frac{\pi^2 t^2}{h^2}$ (when $E_1 - E_0 = h\nu$).

(b) At half height, $\frac{\pi^2 t^2}{2h^2} = \frac{\sin^2[(E_1 - E_0 - h\nu)t/2\hbar]}{(E_1 - E_0 - h\nu)^2}$,

or $\frac{\pi t}{h\sqrt{2}}(E_1 - E_0 - h\nu) = \sin[(E_1 - E_0 - h\nu)t\pi/h]$.

(c) $2(E_1 - E_0 - h\nu) = 2(1.39)h/\pi t = 0.01$ cm^{-1} $= 1.99 \times 10^{-25}$ J
$\Rightarrow t = 3 \times 10^{-9}$ s.

3-23. (a)

C_s	E	σ_h	
Γ_{tot}	9	3	$= 6a' + 3a''$

$\Gamma_{vib} = 3a'$

(b)

C_{3v}	E	$2C_3$	$3\sigma_v$	
Γ_{tot}	12	0	2	$= 3a_1 + a_2 + 4e$

$\Gamma_{vib} = 2a_1 + 2e$

(c)

T_d	E	$8C_3$	$3C_2$	$6S_4$	$6\sigma_d$	
Γ_{tot}	15	0	-1	-1	3	$= a_1 + e + t_1 + 3t_2$

$\Gamma_{vib} = a_1 + e + 2t_2$

(d)

$D_{\infty h}$	E	$2C_\infty{}^\phi$	$\infty\sigma_v$	i	$2S_\infty{}^\phi$	∞C_2
Γ_{tot}	12	$4 + 8c\phi$	4	0	0	0

$= 2\sigma_g{}^+ + 2\pi_g + 2\sigma_u{}^+ + 2\pi_u$

$\Gamma_{vib} = 2\sigma_g{}^+ + \pi_g + \sigma_u{}^+ + \pi_u$

(e)

D_{3h}	E	$2C_3$	$3C_2$	σ_h	$2S_3$	$3\sigma_v$	
Γ_{tot}	18	0	-2	6	0	2	$= 2a_1' + 2a_2' + 4e' + 2a_2'' + 2e''$

$\Gamma_{vib} = 2a_1' + a_2' + 3e' + a_2'' + e''$

(f)

C_{3v}	E	$2C_3$	$3\sigma_v$
Γ_{tot}	18	0	4

$= 5a_1 + a_2 + 6e$

$\Gamma_{vib} = 4a_1 + 4e$

(g)

O_h	E	$8C_3$	$6C_2$	$6C_4$	$3C_2$	i	$6S_4$	$8S_6$	$3\sigma_h$	$6\sigma_d$
Γ_{tot}	21	0	-1	3	-3	-3	-1	0	5	3

$= a_{1g} + e_g + t_{1g} + t_{2g} + 3t_{1u} + t_{2u}$

$\Gamma_{vib} = a_{1g} + e_g + t_{2g} + 2t_{1u} + t_{2u}$

(h)

D_{2d}	E	$2S_4$	$C_2(z)$	$2C_2'$	$2\sigma_d$
Γ_{tot}	21	-1	-3	-1	5

$= 3a_1 + a_2 + b_1 + 4b_2 + 6e$

$\Gamma_{vib} = 3a_1 + b_1 + 3b_2 + 4e$

3-24. $E = 1.9 \times 10^{-13}$ J per nucleus $= 9.6 \times 10^9$ cm^{-1} $= 2.9 \times 10^{20}$ s^{-1}. This is a γ-ray energy.

3-25. (a) All modes are infrared and Raman active.
(b) All modes are infrared and Raman active.
(c) All modes are Raman active. The two t_2 modes are infrared active.
(d) Raman: $2\sigma_g^+ + \pi_g$. Infrared: $\sigma_u^+ + \pi_u$.
(e) Raman: $2a_1' + 3e' + e''$. Infrared: $3e' + a_2''$. Inactive: a_2'.
(f) All modes are infrared and Raman active.
(g) Raman: $a_{1g} + e_g + t_{2g}$. Infrared: $2t_{1u}$. Inactive: t_{2u}.
(h) All modes are Raman active. $3b_2 + 4e$ are infrared active.

3-26. The structure must be linear, since a bent C_{2v} or C_s structure would not yield two infrared and one Raman active mode. The structure is most likely F-H-F, since the H-F-F structure would give three infrared and three Raman bands. The frequency that does not change upon deuteration must be ν_s, which involves no H or D motion.

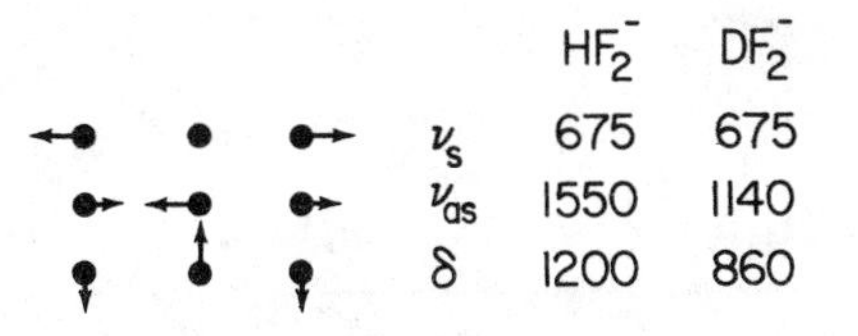

3-27. The lack of coincidental infrared and Raman frequencies suggests a center of inversion.

3-28. (a) a' – y and z polarized. [Note that the coordinate system in 3-23(a) is such that y and z transform as a', and x transforms as a''. The C_s character table in Appendix A assumes that the mirror plane is xy instead of yz.] Raman polarized: a'.
(b) $a_1 - z$; $e - (x, y)$. Raman polarized: a_1.
(c) $t_2 - (x, y, z)$. Raman polarized: a_1.
(d) $\sigma_u^+ - z$; $\pi_u - (x, y)$. Raman polarized: σ_g^+.

(e) $e' - (x, y); a_2'' - z$. Raman polarized: a_1'.
(f) $a_1 - z; e - (x, y)$. Raman polarized: a_1.
(g) $t_{1u} - (x, y, z)$. Raman polarized: a_{1g}.
(h) $b_2 - z; e - (x, y)$. Raman polarized: a_1.

3-29. Using formula C-1 in Appendix C, we get the following:

D_{3h}	E	$2C_3$	$3C_2$	σ_h	$2S_3$	$3\sigma_v$	
e'	2	-1	0	2	-1	0	
R	E	C_3	C_2	σ_h	S_3	σ_v	
R^2	E	$C_3{}^2$	E	E	$C_3{}^2$	E	
R^3	E	E	C_2	σ_h	σ_h	σ_v	
$\chi(R)$	2	-1	0	2	-1	0	
$\chi(R^2)$	2	-1	2	2	-1	2	
$\chi(R^3)$	2	2	0	2	2	0	
$\chi_1(R)$	2	-1	0	2	-1	0	$= e'$
$\chi_2(R)$	3	0	1	3	0	1	$= a_1' + e'$
$\chi_3(R)$	4	1	0	4	1	0	$= a_1' + a_2' + e'$

$v = 0 \sim a_1'; v = 1 \sim e'; v = 2 \sim a_1' + e'; v = 3 \sim a_1' + a_2' + e'$. The first overtone ($v = 2$) is allowed in both spectra.

3-30. NH_3: 2.1×10^{-11} s; PH_3: 3.5×10^{-8} s; AsH_3: 4.5×10^{6} s. AsR_3 compounds probably invert slowly enough to be optically active.

3-31.

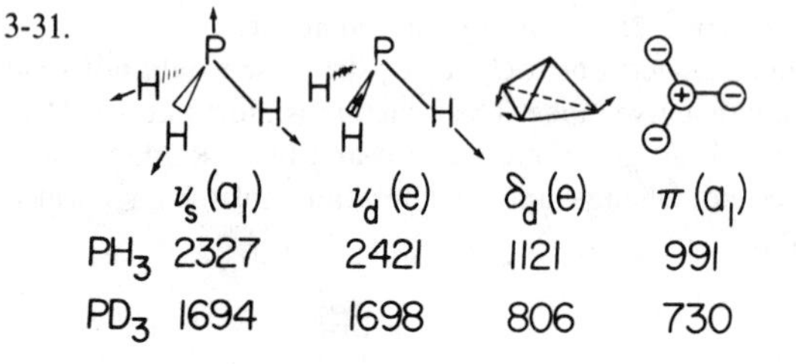

3-32.

D_{3h}	E	$2C_3$	$3C_2$	σ_h	$2S_3$	$3\sigma_v$	
in-plane	3	0	-1	3	0	-1	$= a_2' + e'$
out-of-plane	3	0	-1	-3	0	1	$= a_2'' + e''$

$$\delta_d(e') = \begin{cases} 2\Delta\alpha_1 - \Delta\alpha_2 - \Delta\alpha_3 \\ \Delta\alpha_2 - \Delta\alpha_3 \end{cases}$$

$$R_z(a_2') = \Delta\alpha_1 + \Delta\alpha_2 + \Delta\alpha_3$$

$$\pi(a_2'') = \Delta\beta_1 + \Delta\beta_2 + \Delta\beta_3$$

$$R_x, R_y(e'') = \begin{cases} 2\Delta\beta_1 - \Delta\beta_2 - \Delta\beta_3 \\ \Delta\beta_2 - \Delta\beta_3 \end{cases}$$

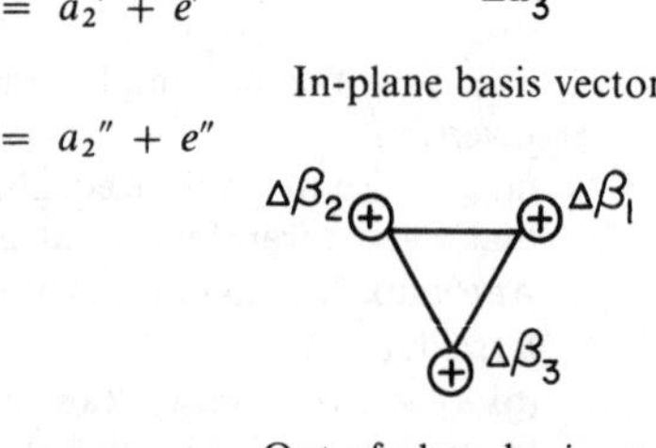

In-plane basis vectors

Out-of-plane basis vectors

3-33.

D_{5h}	E	$2C_5$	$2C_5^2$	$5C_2$	σ_h	$2S_5$	$2S_5^2$	$5\sigma_v$	
Γ_{stretch}	5	0	0	1	5	0	0	1	$= a_1' + e_1' + e_2'$

First, E_1 and E_2 of the C_5 character table are transformed to real form:

$$E_1 \begin{Bmatrix} 1 & \varepsilon & \varepsilon^2 & \varepsilon^{2*} & \varepsilon^* \\ 1 & \varepsilon^* & \varepsilon^{2*} & \varepsilon^2 & \varepsilon \end{Bmatrix} \Rightarrow \begin{Bmatrix} 1 & c\frac{2\pi}{5} & c\frac{4\pi}{5} & c\frac{4\pi}{5} & c\frac{2\pi}{5} \\ 0 & s\frac{2\pi}{5} & s\frac{4\pi}{5} & -s\frac{4\pi}{5} & -s\frac{2\pi}{5} \end{Bmatrix}$$

$$E_2 \begin{Bmatrix} 1 & \varepsilon^2 & \varepsilon^* & \varepsilon & \varepsilon^{2*} \\ 1 & \varepsilon^{2*} & \varepsilon & \varepsilon^* & \varepsilon^2 \end{Bmatrix} \Rightarrow \begin{Bmatrix} 1 & c\frac{4\pi}{5} & c\frac{2\pi}{5} & c\frac{2\pi}{5} & c\frac{4\pi}{5} \\ 0 & s\frac{4\pi}{5} & -s\frac{2\pi}{5} & s\frac{2\pi}{5} & -s\frac{4\pi}{5} \end{Bmatrix}$$

The stretching symmetry coordinates are then given by:

$$v_s(a_1') = \Delta_1 + \Delta_2 + \Delta_3 + \Delta_4 + \Delta_5$$

$$v_d(e_1') = \begin{cases} \Delta_1 + c\frac{2\pi}{5}\Delta_2 + c\frac{4\pi}{5}\Delta_3 + c\frac{4\pi}{5}\Delta_4 + c\frac{2\pi}{5}\Delta_5 \\ s\frac{2\pi}{5}\Delta_2 + s\frac{4\pi}{5}\Delta_3 - s\frac{4\pi}{5}\Delta_4 - s\frac{2\pi}{5}\Delta_5 \end{cases}$$

$$v_d(e_2') = \begin{cases} \Delta_1 + c\frac{4\pi}{5}\Delta_2 + c\frac{2\pi}{5}\Delta_3 + c\frac{2\pi}{5}\Delta_4 + c\frac{4\pi}{5}\Delta_5 \\ s\frac{4\pi}{5}\Delta_2 - s\frac{2\pi}{5}\Delta_3 + s\frac{2\pi}{5}\Delta_4 - s\frac{4\pi}{5}\Delta_5 \end{cases}$$

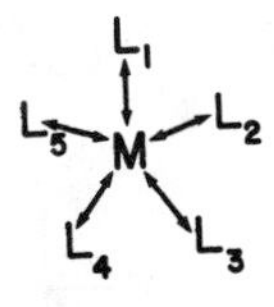

Stretching basis vectors

3-34.

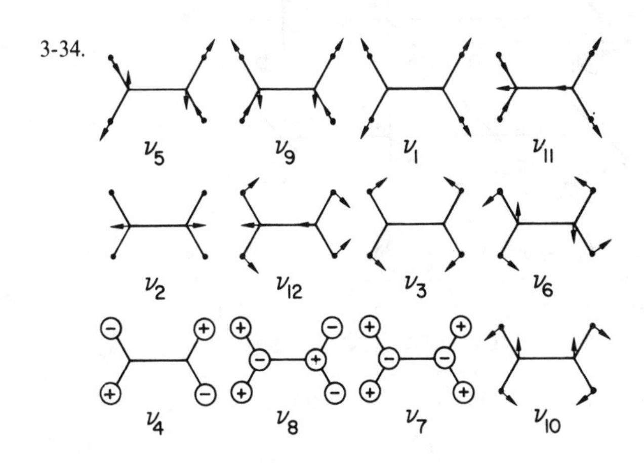

3-35. Monocapped trigonal prism:

C_{2v}	E	C_2	$\sigma(xz)$	$\sigma(yz)$	
Γ(CN)	7	1	1	3	$= 3a_1 + a_2 + b_1 + 2b_2$

infrared: $3a_1 + b_1 + 2b_2 = 6$ bands
Raman: $3a_1 + a_2 + b_1 + 2b_2 = 7$ bands
} 6 coincidences

Pentagonal bipyramid:

D_{5h}	E	$2C_5$	$2C_5^2$	$5C_2$	σ_h	$2S_5$	$2S_5^2$	$5\sigma_v$
Γ(CN)	7	2	2	1	5	0	0	3

$= 2a_1' + e_1' + e_2' + a_2''$

infrared: $e_1' + a_2'' = 2$ bands
Raman: $2a_1' + e_2' = 3$ bands
} no coincidences

Interpretation: solid—C_{2v}; solution—D_{5h}.

3-36. (a)

C_{2v}	E	C_2	$\sigma(xz)$	$\sigma(yz)$	
Γ(CO)	4	0	2	2	$= 2a_1 + b_1 + b_2$

(all infrared and Raman active)

D_{4h}	E	$2C_4$	C_2	$2C_2'$	$2C_2''$	i	$2S_4$	σ_h	$2\sigma_v$	$2\sigma_d$
Γ(CO)	4	0	0	2	0	0	0	4	2	0

$= a_{1g} + b_{1g} + e_u$

infrared: e_u Raman: $a_{1g} + b_{1g}$

(b) $Fe(CO)_4Cl_2$ probably has a *cis* structure with two overlapping bands.

(c)

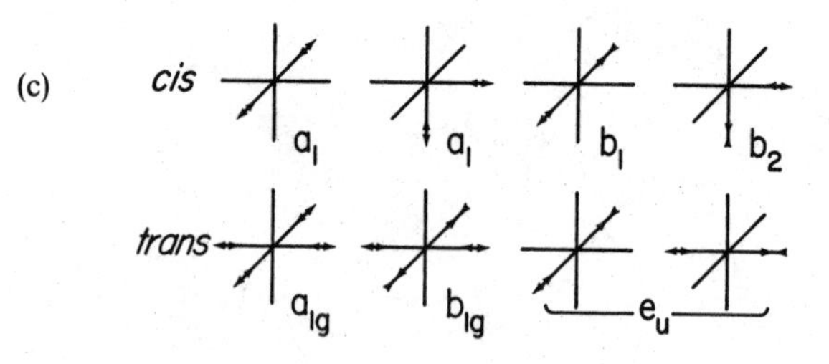

3-37. $C_3H_3^+$:

D_{3h}	E	$2C_3$	$3C_2$	σ_h	$2S_3$	$3\sigma_v$	
Γ(CH)	3	0	1	3	0	1	$= a_1' + e'$

C_4H_4:

D_{4h}	E	$2C_4$	C_2	$2C_2'$	$2C_2''$	i	$2S_4$	σ_h	$2\sigma_v$	$2\sigma_d$
Γ(CH)	4	0	0	2	0	0	0	4	2	0

$= a_{1g} + b_{1g} + e_u$

$C_5H_5^-$:	D_{5h}	E	$2C_5$	$2C_5^2$	$5C_2$	σ_h	$2S_5$	$2S_5^2$	$5\sigma_v$
	Γ(CH)	5	0	0	1	5	0	0	1

$$= a_1' + e_1' + e_2'$$

C_6H_6:	D_{6h}	E	$2C_6$	$2C_3$	C_2	$3C_2'$	$3C_2''$	i	$2S_3$	$2S_6$	σ_h	$3\sigma_d$	$3\sigma_v$
	Γ(CH)	6	0	0	0	2	0	0	0	0	6	0	2

$$= a_{1g} + e_{2g} + b_{1u} + e_{1u}$$

3-38. In Section 3.5 we found that Γ_{vib} for square planar XeF_4 transforms as $a_{1g} + b_{1g} + b_{2g} + a_{2u} + b_{2u} + 2e_u$. Subtracting Γ_{stretch} $(= a_{1g} + b_{1g} + e_u$, as for *trans* $M(CO)_4L_2$, Problem 3-36), we find $\Gamma_{\text{bend}} = \Gamma_{\text{vib}} - \Gamma_{\text{stretch}} = b_{2g} + a_{2u} + b_{2u} + e_u$. For the tetrahedral structure,

T_d	E	$8C_3$	$3C_2$	$6S_4$	$6\sigma_d$	
Γ_{tot}	15	0	-1	-1	3	$= a_1 + e + t_1 + 3t_2$
Γ_{stretch}	4	1	0	0	2	$= a_1 + t_2$

$$\Gamma_{\text{vib}} = a_1 + e + 2t_2$$
$$\Gamma_{\text{bend}} = e + t_2$$

The table in the text is derived from these symmetry species by inspection.

3-39. $\bar{\nu}_v = \bar{E}_v - \bar{E}_0 = [\bar{\omega}_e(v + \frac{1}{2}) - \bar{\omega}_e x_e(v + \frac{1}{2})^2] - [\bar{\omega}_e(\frac{1}{2}) - \bar{\omega}_e x_e(\frac{1}{2})^2]$ (from eq. 3-22)

$$\bar{\nu}_v/v = \underbrace{(\bar{\omega}_e - \bar{\omega}_e x_e)}_{\text{intercept}} - \underbrace{(\bar{\omega}_e x_e)}_{\text{slope}} v$$

These data are plotted below.

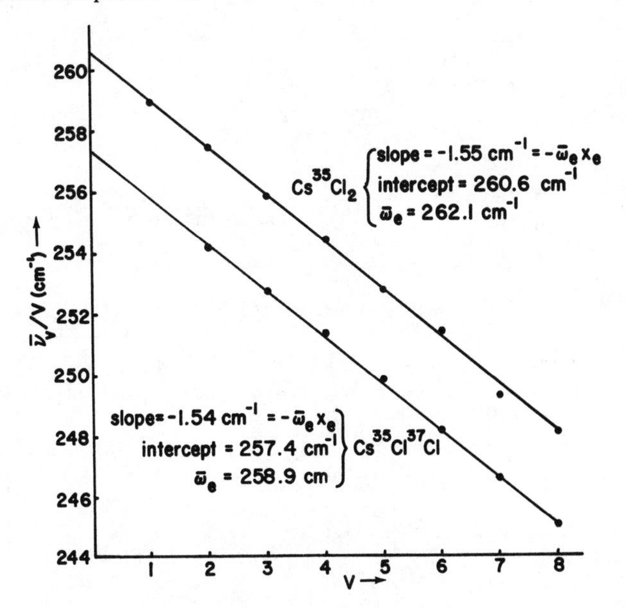

3-40. (a) CH_3CH_2OH (b) $H_3C\overset{O}{\overset{\|}{C}}CH_3$ (c) O(CH₂CH₂)₂O [ring: $O \langle {}^{CH_2CH_2}_{CH_2CH_2} \rangle O$] (d) F, F, F, F–C₆–CN [tetrafluorobenzonitrile: ring with F at four positions and CN]

(e) $CH_3CH_2CH_2NO_2$ (f) [benzene ring]–NH_2 with OCH_3 [Note that for spectra (e) and (f) other structures could have been drawn. Infrared spectra, alone, usually do not define a unique structure for all but the simplest compounds.]

3-41.

966	ν_s	a_1	p
933	ν_d	e	dp
620	δ_s	a_1	p
473	δ_d	e	dp

3-42.

$C_{\infty v}$	E	$2C_\infty^{\phi}$	$\infty\sigma_v$	
Γ_{tot}	9	$3+6c\phi$	3	$= 3\sigma^+ + 3\pi$

$\Gamma_{\text{vib}} = 2\sigma^+ + \pi$ (all infrared and Raman active)

X←→C—N $\quad$ X—C←→N

$\nu(CX)-\sigma^+$ $\quad$ $\nu(CN)-\sigma^+$ $\quad$ $\delta_d-\pi$

description:	$\nu(CX)$	$\nu(CN)$	δ_d
symmetry:	σ^+	σ^+	π
HCN	3311	2097	712
DCN	2630	1925	569
FCN	1077	2290	449
ClCN	714	2219	380
BrCN	574	2200	342
ICN	470	2158	321

3-43.

A	C_{2v}	4	4	4	2
B	C_{3v}	3	3	3	2
C	C_{4v}	2	3	2	1
D	C_s	4	4	4	3

The spectra are most consistent with structure D.

3-44.

D_{2h}	E	$C_2(z)$	$C_2(y)$	$C_2(x)$	i	$\sigma(xy)$	$\sigma(xz)$	$\sigma(yz)$
Γ_{tot}	24	0	−2	−2	0	8	2	2
Γ_{terminal}	4	0	0	0	0	4	0	0
Γ_{bridge}	4	0	0	0	0	4	0	0

$\Gamma_{\text{tot}} = 4a_g + 4b_{1g} + 2b_{2g} + 2b_{3g} + a_u + 3b_{1u} + 4b_{2u} + 4b_{3u}$

$\Gamma_{\text{terminal}} = a_g + b_{1g} + b_{2u} + b_{3u}$

$\Gamma_{\text{bridge}} = a_g + b_{1g} + b_{2u} + b_{3u}$

$$\Gamma_{\text{vib}} = 4a_g(\text{R, p}) + 3b_{1g}(\text{R, dp}) + b_{2g}(\text{R, dp}) + b_{3g}(\text{R, dp}) + a_u(\text{inactive}) + 2b_{1u}(\text{ir}, z) + 3b_{2u}(\text{ir}, y) + 3b_{3u}(\text{ir}, x)$$

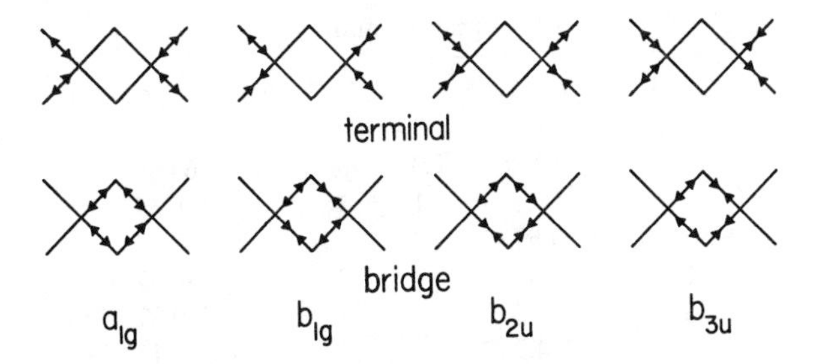

3-45. $Os(NH_3)_4(N_2)_2^{2+}$ appears to be the *cis* isomer because there are two $N \equiv N$ stretching frequencies. The *trans* isomer should only exhibit one infrared active mode. In I the formal oxidation state of Os is $+5/2$ if the two atoms are equal. If not, the one on the left is $+3$ and the one on the right is $+2$. In II both Os atoms are formally in the $+5/2$ state. In I each atom of the bridging N_2 has a different substituent ($X{-}N{\equiv}N{-}Y$), and one infrared active $N \equiv N$ stretching frequency is expected and observed. In II each atom of N_2 has the same substituent ($Y{-}N{\equiv}N{-}Y$), and no infrared active $N \equiv N$ stretching frequency is expected or observed.

3-46.

	$D_{\infty h}$	E	$2C_\infty^\phi$	$\infty\sigma_v$	i	$2S_\infty^\phi$	∞C_2
C–C–C–C	Γ_{tot}	12	$4+8\cos\phi$	4	0	0	0

$$= 2\sigma_g^+ + 2\sigma_u^+ + 2\pi_g + 2\pi_u$$

$$\Gamma_{vib} = 2\sigma_g^+ + \sigma_u^+ + \pi_g + \pi_u$$

$\nu_s(\sigma_g^+)$ $\quad$ $\nu_s(\sigma_g^+)$ $\quad$ $\nu_{as}(\sigma_u^+)$

$\delta_d(\pi_g)$ $\quad$ $\delta_d(\pi_u)$

3-47. (a) $CH_3CH_2OCH_2CH_3$ (b) $HC(=O)N(CH_3)_2$ (c) $CH_3C(=O)CH_2CH_3$

(d) o-xylene (benzene ring with two adjacent CH_3) (e) cyclohexanone (ring $= O$) (f) $CH_3CH{=}CHC{\equiv}N$

(g) $CH_3C(=O)CH{=}CH_2$

3-48. Using equations 3-58 and 3-59, we plot the following data and find the least-squares slopes indicated:

Fundamental

J	$2J+1$	$\bar{\nu}_R(J)-\bar{\nu}_P(J)$	J	$2J+3$	$\bar{\nu}_R(J)-\bar{\nu}_P(J+2)$
1	3	62.88	0	3	64.66
2	5	103.32	1	5	106.34
3	7	143.73	2	7	147.93
4	9	184.03	3	9	189.47
5	11	224.22	4	11	230.87
6	13	264.28	5	13	272.12
7	15	304.17	6	15	313.23
8	17	343.88	7	17	354.15
9	19	383.38	8	19	394.84

slope $= 20.04\ \text{cm}^{-1}$ $\Rightarrow \overline{B}_1 = 10.02\ \text{cm}^{-1}$

slope $= 20.65\ \text{cm}^{-1}$ $\Rightarrow \overline{B}_0 = 10.32\ \text{cm}^{-1}$

First Overtone

J	$2J+1$	$\bar{\nu}_R(J)-\bar{\nu}_P(J)$	J	$2J+3$	$\bar{\nu}_R(J)-\bar{\nu}_P(J+2)$
1	3	59.18	0	3	63.00
2	5	98.48	1	5	104.16
3	7	137.24	2	7	146.04

slope $= 19.52\ \text{cm}^{-1}$ $\Rightarrow \overline{B}_2 = 9.76\ \text{cm}^{-1}$

slope $= 20.76\ \text{cm}^{-1}$ $\Rightarrow \overline{B}_0 = 10.38\ \text{cm}^{-1}$

Second Overtone

J	$2J+1$	$\bar{\nu}_R(J)-\bar{\nu}_P(J)$	J	$2J+3$	$\bar{\nu}_R(J)-\bar{\nu}_P(J+2)$
1	3	57.19	0	3	62.63
2	5	95.31	1	5	104.30
3	7	133.26			

slope $= 19.02\ \text{cm}^{-1}$ $\Rightarrow \overline{B}_3 = 9.51\ \text{cm}^{-1}$

slope $= 20.84\ \text{cm}^{-1}$ $\Rightarrow \overline{B}_0 = 10.42\ \text{cm}^{-1}$

Using eq. 3-55, we calculate r_v (using $\mu = 1.627 \times 10^{-27}$ kg): $r_1 = 1.310$ Å; $r_2 = 1.327$ Å; $r_3 = 1.345$ Å; $r_0 = 1.291$ Å (from fundamental). Using eq. 3-60, we calculate B_e and α_e (using the $\overline{B}_0$ and $\overline{B}_1$ values from the fundamental data, which are the most accurate):

$$\left.\begin{aligned}\overline{B}_0 &= \overline{B}_e - (1/2)\bar{\alpha}_e\\ \overline{B}_1 &= \overline{B}_e - (3/2)\bar{\alpha}_e\end{aligned}\right\} \Rightarrow \begin{aligned}\overline{B}_e &= 10.47\ \text{cm}^{-1} \Rightarrow r_e = 1.282\ \text{Å}\\ \bar{\alpha}_e &= 0.30\ \text{cm}^{-1}\end{aligned}$$

3-49.

T_d	E	$8C_3$	$3C_2$	$6S_4$	$6\sigma_d$
Γ_{tot}	15	0	-1	-1	3 $= a_1 + e + t_1 + 3t_2$

$\Gamma_{vib} = a_1(\text{R}) + e(\text{R}) + 2t_2(\text{R, ir})$

Assignment: 3019—$\nu_d(t_2)$; 2917—$\nu_s(a_1)$; 1534—$\delta_d(e)$; 1306—$\delta_d(t_2)$.

3-50. Γ_{vib} was determined in Problem 3-23. Bands near 3000 cm^{-1} must be ν(CH). The band at 2249 must be ν(CN). Bands at 1440 and 1376 are characteristic

CH_3 bending frequencies. Assigning antisymmetric modes at higher frequencies than symmetric modes (which is not always correct), we get: 2999—$\nu_d(CH)(e)$; 2942—$\nu_s(CH)(a_1)$; 2249—$\nu(CN)(a_1)$; 1440—$\delta_d(CH_3)(e)$; 1376—$\delta_s(CH_3)(a_1)$. The remaining bands, which are not obvious without comparison to the spectra of other compounds, are assigned as follows: 1124—$\nu(CC)(a_1)$; 918—$\delta(HCC)(e)$; 380—$\delta(CCN)(e)$.

3-51. Labelling the structures A–F, from left to right, we derive the following for $\nu(CO)$ frequencies:

structure:	A	B	C	D	E	F
point group:	C_s	C_{2v}	C_s	D_{3h}	C_{2v}	C_s
infrared bands:	3	3	3	1	3	3
Raman bands:	3	3	3	2	3	3

Only structure D yields a single infrared band.

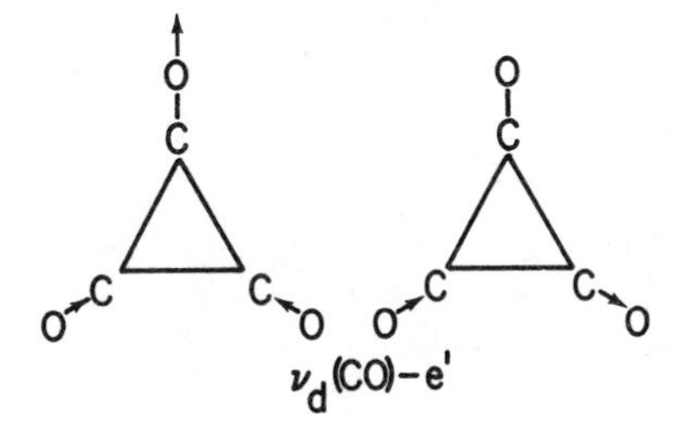

3-52.

D_{4d}	E	$2S_8$	$2C_4$	$2S_8^3$	C_2	$4C_2'$	$4\sigma_d$
$\Gamma(CO)$	10	0	2	0	2	0	4

$$= 2a_1(R) + 2b_2(ir) + e_1(ir) + e_2(R) + e_3(R)$$

The b_2 modes are z-polarized (parallel to the long axis of the molecule). In the observed spectrum, the bands at 2045 and 1980 cm^{-1} exhibit $A_\perp/A_\parallel < 1$ and are assigned symmetry b_2 (z-polarized). The band at 2009 exhibits $A_\perp/A_\parallel > 1$ and is assigned symmetry e_1 (x, y-polarized).

3-53.

C_{3v}	E	$2C_3$	$3\sigma_v$
Γ_{tot}	15	0	3

$= 4a_1 + a_2 + 5e$

$\Gamma_{vib} = 3a_1 + 3_e$

Assignment:			
2966	ν_1	a_1	$\nu_s(CH_3)$
1355	ν_2	a_1	$\delta_s(CH_3)$
732	ν_3	a_1	$\nu(CCl)$
3042	ν_4	e	$\nu_d(CH_3)$
1455	ν_5	e	$\delta_d(CH_3)$
1015	ν_6	e	$\delta_d(HCCl)$

The CH_3 characteristic frequencies were assigned as described for CH_3CN in Problem 3-50. The remaining two bands were assigned with the mode involving hydrogen motion at a higher frequency than the mode without hydrogen motion. The shape of the 2879-cm^{-1} band tells us that the symmetry is a_1. The first overtone of ν_5, with symmetry species $a_1 + e$, is the most likely candidate.

Chapter 4

4-1.

orbital	n	ℓ	m_ℓ	nodes
2s	2	0	0	1
3s	3	0	0	2
2p	2	1	$\pm 1, 0$	0
3d	3	2	$\pm 2, \pm 1, 0$	0
4f	4	3	$\pm 3, \pm 2, \pm 1, 0$	0
4d	4	2	$\pm 2, \pm 1, 0$	1

4-2.

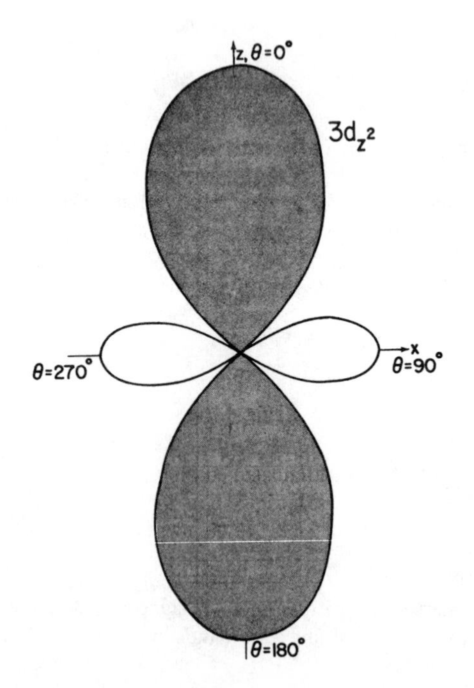

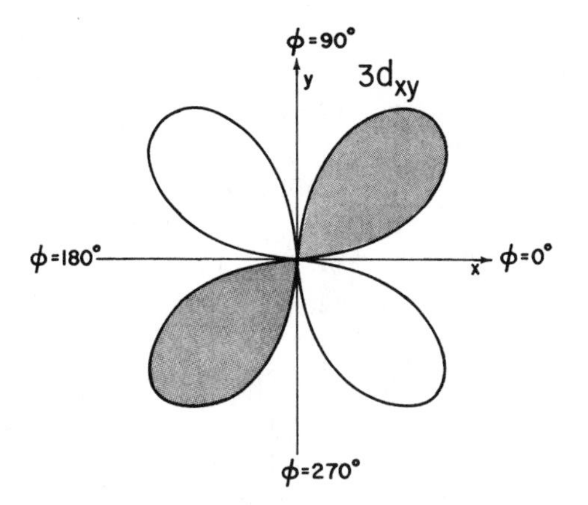

Angular part of $4f_{y(z^2-x^2)} = (r^2 \cos^2 \theta - r^2 \sin^2 \theta \cos^2 \phi)(r \sin \theta \sin \phi)$.

4-3. The transition energy for H going from the state with quantum number n_2 to the state with quantum number n_1 is given by:

$$\Delta E = E_2 - E_1 = \frac{-Z^2e^4\mu}{8\epsilon_0{}^2h^2n_2{}^2} - \left(\frac{-Z^2e^4\mu}{8\epsilon_0{}^2h^2n_1{}^2}\right) = \frac{Z^2e^4\mu}{8\epsilon_0{}^2h^2}\left(\frac{1}{n_1{}^2} - \frac{1}{n_2{}^2}\right)$$

Setting the constant term equal to $100hc\overline{R}_H$ in the Rydberg equation, we find $\overline{R}_H = Z^2e^4\mu/800\epsilon_0{}^2h^3c = 1.097 \times 10^5\ \text{cm}^{-1}$, where $Z = 1$ for hydrogen, $\mu = 9.1048 \times 10^{-31}$ kg, and e, ϵ_0, h, and c are expressed in SI units. The highest energy emission corresponds to the $n_2 = \infty \to n_1 = 1$ transition (i.e., capture of a free electron that lands in the $n = 1$ orbital). $E_{\infty \to 1} = 1.097 \times 10^5\ \text{cm}^{-1}$. This is equal to the ionization energy.

4-4. Cr would have the configuration $[\text{Ar}]4s^23d^4$ if the rules were not violated.

4-5. $1s^2$ means that the orbital angular momentum and the total spin are both zero $\Rightarrow$ ^1S term. d^1 means that the orbital angular momentum is 2 and the spin multiplicity is $2(\frac{1}{2}) + 1 = 2 \Rightarrow {}^2D$ term. The levels of ^2D are $^2D_{5/2}$ and $^2D_{3/2}$.

4-6. sp:

$M_L \backslash M_S$	1	0	-1
1	$(\overset{+}{0}_s, \overset{+}{1})$	$(\overset{+}{0}_s, \bar{1})$ $(\bar{0}_s, \overset{+}{1})$	$(\bar{0}_s, \bar{1})$
0	$(\overset{+}{0}_s, \overset{+}{0})$	$(\overset{+}{0}_s, \bar{0})$ $(\bar{0}_s, \overset{+}{0})$	$(\bar{0}_s, \bar{0})$
-1	$(\overset{+}{0}_s, -\overset{+}{1})$	$(\overset{+}{0}_s, -\bar{1})$ $(\bar{0}_s, -\overset{+}{1})$	$(\bar{0}_s, -\bar{1})$

$\Rightarrow$ ^3P, ^1P (Note that $(\overset{+}{0}, \overset{+}{0})$ does not violate the Pauli exclusion rule because the electrons are in different orbitals.)

d^2:

$M_L \backslash M_S$	1	0	-1
4		$(\overset{+}{2}, \bar{2})$	
3	$(\overset{+}{2}, \overset{+}{1})$	$(\overset{+}{2}, \bar{1})(\bar{2}, \overset{+}{1})$	$(\bar{2}, \bar{1})$
2	$(\overset{+}{2}, \overset{+}{0})$	$(\overset{+}{2}, \bar{0})(\bar{2}, \overset{+}{0})(\overset{+}{1}, \bar{1})$	$(\bar{2}, \bar{0})$
1	$(\overset{+}{1}, \overset{+}{0})(\overset{+}{2}, -\overset{+}{1})$	$(\overset{+}{1}, \bar{0})(\bar{1}, \overset{+}{0})$ $(\overset{+}{2}, -\bar{1})(\bar{2}, -\overset{+}{1})$	$(\bar{1}, \bar{0})(\bar{2}, -\bar{1})$
0	$(\overset{+}{2}, -\overset{+}{2})(\overset{+}{1}, -\overset{+}{1})$	$(\overset{+}{2}, -\bar{2})(\bar{2}, -\overset{+}{2})$ $(\overset{+}{1}, -\bar{1})$ $(\bar{1}, -\overset{+}{1})(\overset{+}{0}, \bar{0})$	$(\bar{2}, -\bar{2})(\bar{1}, -\bar{1})$
-1	$(-\overset{+}{1}, \overset{+}{0})(\overset{+}{1}, -\overset{+}{2})$	$(-\overset{+}{1}, \bar{0})(-\bar{1}, \overset{+}{0})$ $(-\overset{+}{2}, \bar{1})(-\bar{2}, \overset{+}{1})$	$(-\bar{1}, \bar{0})(\bar{1}, -\bar{2})$
-2	$(-\overset{+}{2}, \overset{+}{0})$	$(-\overset{+}{2}, \bar{0})(-\bar{2}, \overset{+}{0})(-\overset{+}{1}, -\bar{1})$	$(-\bar{2}, \bar{0})$
-3	$(-\overset{+}{2}, -\overset{+}{1})$	$(-\overset{+}{2}, -\bar{1})(-\bar{2}, -\overset{+}{1})$	$(-\bar{2}, -\bar{1})$
-4		$(-\overset{+}{2}, -\bar{2})$	

$\Rightarrow {}^3F, {}^3P, {}^1G, {}^1D, {}^1S$

d^3: (only a partial table is shown)

$M_L \backslash M_S$	3/2	1/2
6		
5		$(\overset{+}{2}, \bar{2}, \overset{+}{1})$
4		$(\overset{+}{2}, \bar{2}, \overset{+}{0})(\overset{+}{2}, \overset{+}{1}, \bar{1})$
3	$(\overset{+}{2}, \overset{+}{1}, \overset{+}{0})$	$(\overset{+}{2}, \overset{+}{1}, \bar{0})(\overset{+}{2}, \bar{1}, \overset{+}{0})(\bar{2}, \overset{+}{1}, \overset{+}{0})(\overset{+}{2}, \bar{2}, -\overset{+}{1})$
2	$(\overset{+}{2}, \overset{+}{1}, -\overset{+}{1})$	$(\overset{+}{2}, \overset{+}{1}, -\bar{1})(\overset{+}{2}, \bar{1}, -\overset{+}{1})(\bar{2}, \overset{+}{1}, -\overset{+}{1})(\overset{+}{2}, \overset{+}{0}, \bar{0})$ $(\overset{+}{2}, \bar{2}, -\overset{+}{2})(\overset{+}{1}, \bar{1}, \overset{+}{0})$
1	$(\overset{+}{2}, \overset{+}{1}, -\overset{+}{2})(\overset{+}{2}, -\overset{+}{1}, \overset{+}{0})$	$(\overset{+}{2}, \overset{+}{1}, -\bar{2})(\overset{+}{2}, \bar{1}, -\overset{+}{2})(\bar{2}, \overset{+}{1}, -\overset{+}{2})(\overset{+}{2}, -\overset{+}{1}, \bar{0})$ $(\overset{+}{2}, -\bar{1}, \overset{+}{0})(\bar{2}, -\overset{+}{1}, \overset{+}{0})(\overset{+}{1}, \overset{+}{0}, \bar{0})(\overset{+}{1}, \bar{1}, -\overset{+}{1})$
0	$(\overset{+}{2}, \overset{+}{0}, -\overset{+}{2})(\overset{+}{1}, \overset{+}{0}, -\overset{+}{1})$	$(\overset{+}{2}, \overset{+}{0}, -\bar{2})(\overset{+}{2}, \bar{0}, -\overset{+}{2})(\bar{2}, \overset{+}{0}, -\overset{+}{2})(\overset{+}{1}, \overset{+}{0}, -\bar{1})$ $(\overset{+}{1}, \bar{0}, -\overset{+}{1})(\bar{1}, \overset{+}{0}, -\overset{+}{1})(\overset{+}{2}, -\overset{+}{1}, -\bar{1})(-\overset{+}{2}, \overset{+}{1}, \bar{1})$

$\Rightarrow {}^4F, {}^4P, {}^2H, {}^2G, {}^2F, {}^2D, {}^2D, {}^2P$

4-7. Ground state: [Ne] $3s^1 \Rightarrow {}^2S_{1/2}$
First excited state: [Ne] $3p^1 \Rightarrow {}^2P_{3/2}, {}^2P_{1/2}$

$$\left.\begin{array}{l} 589.15788\text{ nm: } {}^2P_{3/2} \to {}^2S_{1/2} \\ 589.75537\text{ nm: } {}^2P_{1/2} \to {}^2S_{1/2} \end{array}\right\} \text{energy separation} = 17.196\text{ cm}^{-1}$$

4-8. N: $1s^22s^22p^3 \Rightarrow {}^4S_{3/2}$
(A) N^+: $1s^22s^22p^2 \Rightarrow {}^3P_2, {}^3P_1, {}^3P_0, {}^1D_2, {}^1S_0$
(B) N^+: $1s^22s^12p^3 \Rightarrow {}^5S_2, {}^3S_1, {}^3D_3, {}^3D_2, {}^3D_1, {}^1D_2, {}^3P_2, {}^3P_1, {}^3P_0, {}^1P_1$
(C) N^+: $1s^12s^22p^3 \Rightarrow$ same terms as (B)
By Hund's rules we expect the following order of transitions:

$$\begin{aligned} {}^4S_{3/2} &\to \text{(A) } {}^3P_0, {}^3P_1, {}^3P_2, {}^1D_2, {}^1S_0 \\ &\to \text{(B) } {}^5S_2, {}^3D_{3,2,1}, {}^3P_{2,1,0}, {}^3S_1, {}^1D_2, {}^1P_1 \\ &\to \text{(C) [same terms as (B)]} \end{aligned}$$

In fact, optical spectra indicate that the order of the terms for the ionic configuration (B) is ${}^5S < {}^3D < {}^3P < {}^1D < {}^3S < {}^1P$. The terms for the configuration sp^3 are derived from the following partial table of microstates:

$M_L \backslash M_S$	2	1	0
3			
2		$(\overset{+}{0}_s, \overset{+}{1}, \bar{1}, \overset{+}{0})$	$(\overset{+}{0}_s, \overset{+}{1}, \bar{1}, \bar{0})(\bar{0}_s, \overset{+}{1}, \bar{1}, \overset{+}{0})$
1		$(\overset{+}{0}_s, \overset{+}{1}, \bar{1}, -\overset{+}{1})(\overset{+}{0}_s, \overset{+}{1}, \overset{+}{0}, \bar{0})$	$(\overset{+}{0}_s, \overset{+}{1}, \bar{1}, -\bar{1})(\bar{0}_s, \overset{+}{1}, \bar{1}, -\overset{+}{1})$ $(\overset{+}{0}_s, \bar{1}, \overset{+}{0}, \bar{0})(\bar{0}_s, \overset{+}{1}, \overset{+}{0}, \bar{0})$
0	$(\overset{+}{0}_s, \overset{+}{1}, \overset{+}{0}, -\overset{+}{1})$	$(\overset{+}{0}_s, \overset{+}{1}, \overset{+}{0}, -\bar{1})(\overset{+}{0}_s, \overset{+}{1}, \bar{0}, -\overset{+}{1})$ $(\overset{+}{0}_s, \bar{1}, \overset{+}{0}, -\overset{+}{1})(\bar{0}_s, \overset{+}{1}, \overset{+}{0}, -\overset{+}{1})$	$(\bar{0}_s, \bar{1}, \overset{+}{0}, -\overset{+}{1})(\overset{+}{0}_s, \overset{+}{1}, \bar{0}, -\bar{1})$ $(\bar{0}_s, \overset{+}{1}, \bar{0}, -\overset{+}{1})(\overset{+}{0}_s, \bar{1}, \overset{+}{0}, -\bar{1})$ $(\bar{0}_s, \overset{+}{1}, \overset{+}{0}, -\bar{1})(\overset{+}{0}_s, \bar{1}, \bar{0}, -\overset{+}{1})$

4-9. $$E = \int_{-\infty}^{\infty} \psi_0^* \mathscr{H} \psi_0 \, dx = \left(\frac{\alpha}{\pi}\right)^{\frac{1}{2}} \int_{-\infty}^{\infty} e^{-\alpha x^2/2} \left(-\frac{\hbar^2}{2m}\frac{d^2}{dx^2} + \tfrac{1}{2}kx^2\right) e^{-\alpha x^2/2} \, dx$$
$$= \frac{\hbar}{2}\sqrt{\frac{k}{m}}$$

4-10. $$\mathscr{H} = -\frac{\hbar^2}{2m_e}(\nabla_1^2 + \nabla_2^2) \overbrace{- \frac{\hbar^2}{2m_A}\nabla_A^2 - \frac{\hbar^2}{2m_B}\nabla_B^2} - \frac{Z_Ae^2}{4\pi\epsilon_0 r_{A1}} - \frac{Z_Ae^2}{4\pi\epsilon_0 r_{A2}} - \frac{Z_Be^2}{4\pi\epsilon_0 r_{B1}} - \frac{Z_Be^2}{4\pi\epsilon_0 r_{B2}} + \frac{Z_AZ_Be^2}{4\pi\epsilon_0 R_{AB}} + \frac{e^2}{4\pi\epsilon_0 r_{12}}$$

The terms under the brace are eliminated by the Born-Oppenheimer approximation.

4-11. As R_{AB} increases, the system actually becomes two isolated hydrogen atoms.

4-12. $A^2\int(\phi_s + \phi_p)(\phi_s + \phi_p)d\tau = 1 = A^2[\int\phi_s\phi_s d\tau + 2\int\phi_s\phi_p d\tau + \int\phi_p\phi_p d\tau]$
$= A^2[1 + 0 + 1]$. Therefore $A = 1/\sqrt{2}$. Similarly, $B = 1/\sqrt{2}$.

$$E_a = \tfrac{1}{2}\int(\phi_s + \phi_p)\mathscr{H}(\phi_s + \phi_p)d\tau$$
$$= \tfrac{1}{2}\int(\phi_s\mathscr{H}\phi_s + \phi_p\mathscr{H}\phi_s + \phi_s\mathscr{H}\phi_p + \phi_p\mathscr{H}\phi_p)d\tau$$
$$= \tfrac{1}{2}\int(E_s\phi_s^2 + E_s\phi_p\phi_s + E_p\phi_s\phi_p + E_p\phi_p^2)d\tau = \tfrac{1}{2}(E_s + E_p).$$

Similarly, $E_b = \frac{1}{2}(E_s + E_p)$.

4-13.

NH	3133	539	1
NF	1115	590	2
O_2	1555	1140	2
N_2	2331	2240	3
NO	1876	1550	$2\frac{1}{2}$

4-14. (a) NO^+ (b) CP^- (c) SiS

4-15. (a) 380 kJ/mol (b) 494 kJ/mol (c) 653 kJ/mol

4-16. NO^+ (bond order = 3); CN^+ (bond order = 2)

4-17. 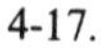

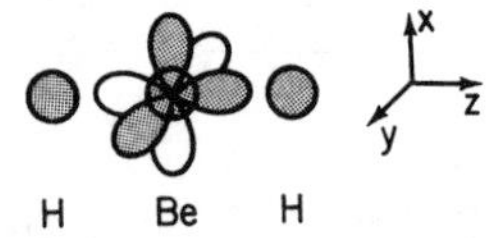

$D_{\infty h}$	E	$2C_\infty^\phi$	$\infty\sigma_v$	i	$2S_\infty^\phi$	∞C_2	
2H(1s)	2	2	2	0	0	0	$= \sigma_g^+ + \sigma_u^+$
Be(2s)	1	1	1	1	1	1	$= \sigma_g^+$
Be($2p_z$)	1	1	1	−1	−1	−1	$= \sigma_u^+$
Be($2p_x$, $2p_y$)	2	$2c\phi$	0	−2	$2c\phi$	0	$= \pi_u$

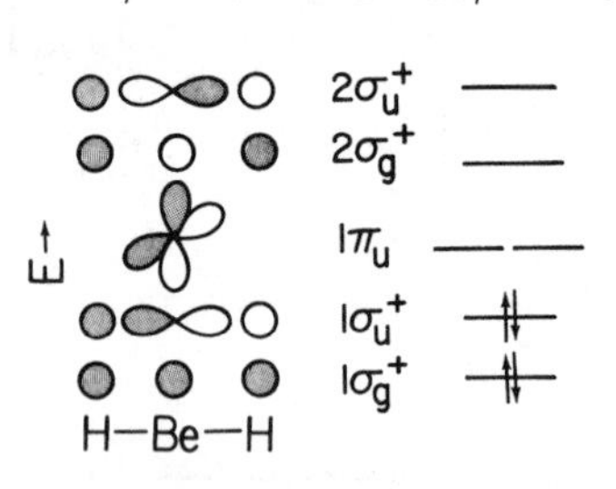

There are two sigma bonding orbitals, zero pi bonding orbitals, zero occupied lone pair orbitals, and zero unpaired electrons.

4-18.

C_{2v}	E	C_2	$\sigma(xz)$	$\sigma(yz)$	
2H(1s)	2	0	0	2	$= a_1 + b_2$
3C(sp^2)	3	1	1	3	$= 2a_1 + b_2$
C(p_x)	1	−1	1	−1	$= b_1$

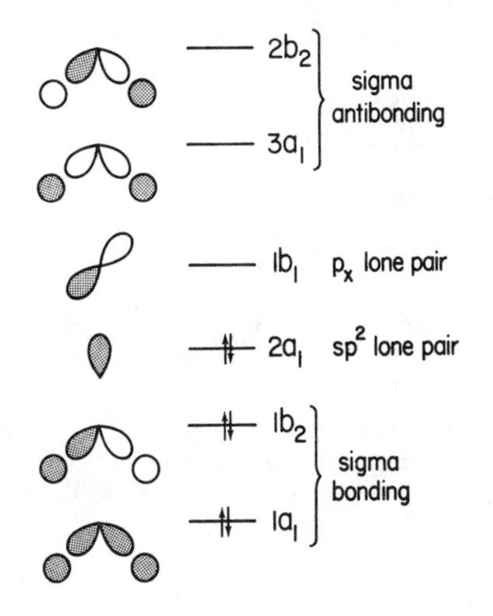

[Note that if the $1b_2$ and $2b_2$ sp^2 orbitals are added correctly, $1b_2$ and $2b_2$ are pure $C(2p_y)$ orbitals. The carbon $1a_1$ and $3a_1$ orbitals will be s^2p (not sp^2) hybrid orbitals directed between the H atoms. Linear CH_2 should have 2 unpaired electrons in the $1\pi_u$ orbitals of Problem 4-17. Bent CH_2 will have no unpaired electrons unless the $2a_1$ and $1b_1$ lone pair orbitals are closer in energy than the spin pairing energy (the energy of repulsion of two electrons in the $2a_1$ orbital].

4-19. (a) $1 = N_1^2\int(2s_a + 2s_b)^2\,d\tau = N_1^2\int[(2s_a2s_a) + 2(2s_a2s_b) + (2s_b2s_b)]\,d\tau$ $= N_1^2[1 + 0 + 1]$. Therefore, $N_1 = 1/\sqrt{2}$. Similarly, $N_2 = [1/(3 + 4S)]^{\frac{1}{2}}$ and $N_3 = [1/(3 - 4S)]^{\frac{1}{2}}$.

(b) $E_1 = N_1^2\int(2s_a + 2s_b)\mathscr{H}(2s_a + 2s_b)\,d\tau = N_1^2(2H'_{aa} + 2H'_{ab})$ (since $H'_{aa} = H'_{bb}$). Similarly, $E_2 = N_2^2(2H_{aa} + 4H_{ac} + 2H_{ab} + H_{cc})$ and $E_3 = N_3^2(2H_{aa} + 2H_{ab} - 4H_{ac} + H_{cc})$.

(c) The secular equation is:

$$\begin{vmatrix} H_{11}\text{-}E & H_{12}\text{-}ES_{12} & H_{13}\text{-}ES_{13} \\ H_{21}\text{-}ES_{21} & H_{22}\text{-}E & H_{23}\text{-}ES_{23} \\ H_{31}\text{-}ES_{31} & H_{32}\text{-}ES_{32} & H_{33}\text{-}E \end{vmatrix} = 0$$

where $H_{ij} = \int\psi_i^*\mathscr{H}\psi_j\,d\tau$ and $S_{ij} = \int\psi_i^*\psi_j\,d\tau$.

4-20.

$D_{\infty h}$	E	$2C_\infty^\phi$	$\infty\sigma_v$	i	$2S_\infty^\phi$	∞C_2	
$4(p_x,p_y)$	8	$8c\phi$	0	0	0	0	$= 2\pi_g + 2\pi_u$

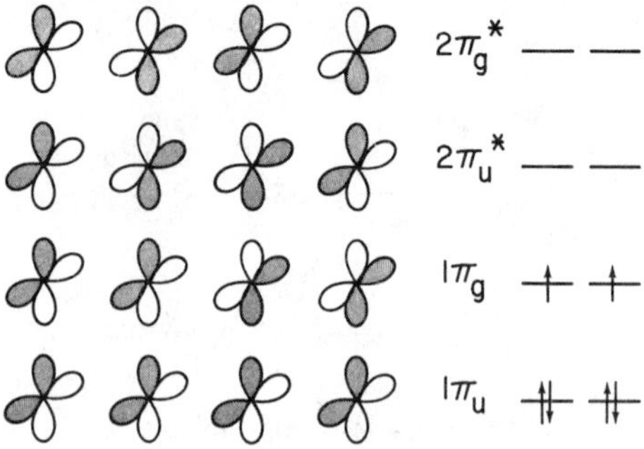

We expect C_4 to have six sigma bonding electrons in three sigma bonds and four lone pair electrons, with one lone pair on each terminal carbon atom. C_4 therefore has six bonding pi electrons. C_4^- will have seven bonding pi electrons and will be more stable.

4-21. 7, 7, 1, 1, 0. Energy: sigma < pi < pi* < sigma*.

4-22.

C_{2v}	E	C_2	$\sigma(xz)$	$\sigma(yz)$	
3H(1s)	3	1	3	1	$= 2a_1 + b_1$
2C(2s)	2	2	2	2	$= 2a_1$
2C($2p_x$)	2	−2	2	−2	$= 2b_1$
2C($2p_y$)	2	−2	−2	2	$= 2b_2$
2C($2p_z$)	2	2	2	2	$= 2a_1$

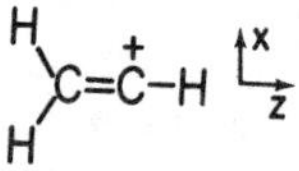

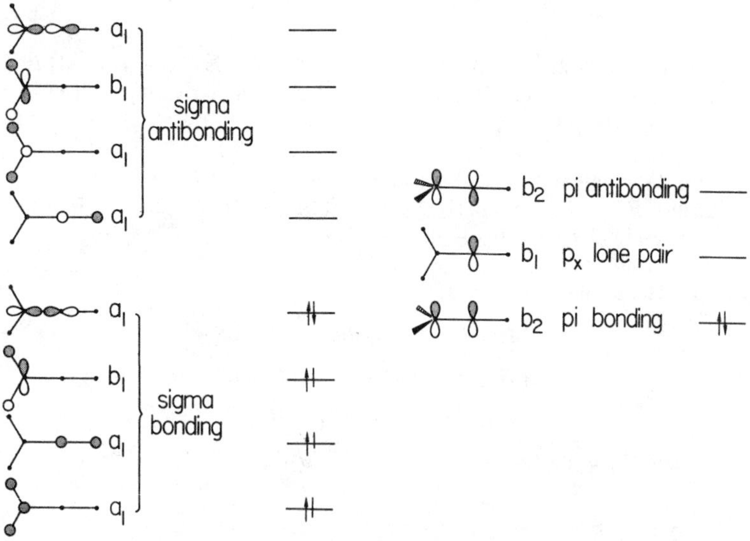

Orbitals of the same symmetry will be extensively mixed.

4-23.

$D_{\infty h}$	E	$2C_{\infty}^{\phi}$	$\infty\sigma_v$	i	$2S_{\infty}^{\phi}$	∞C_2	
2H(1s)	2	2	2	0	0	0	$= \sigma_g^{\ +} + \sigma_u^{\ +}$
2C(2s)	2	2	2	0	0	0	$= \sigma_g^{\ +} + \sigma_u^{\ +}$
$2C(2p_z)$	2	2	2	0	0	0	$= \sigma_g^{\ +} + \sigma_u^{\ +}$
$2C(2p_x, 2p_y)$	4	$4c\phi$	0	0	0	0	$= \pi_g + \pi_u$

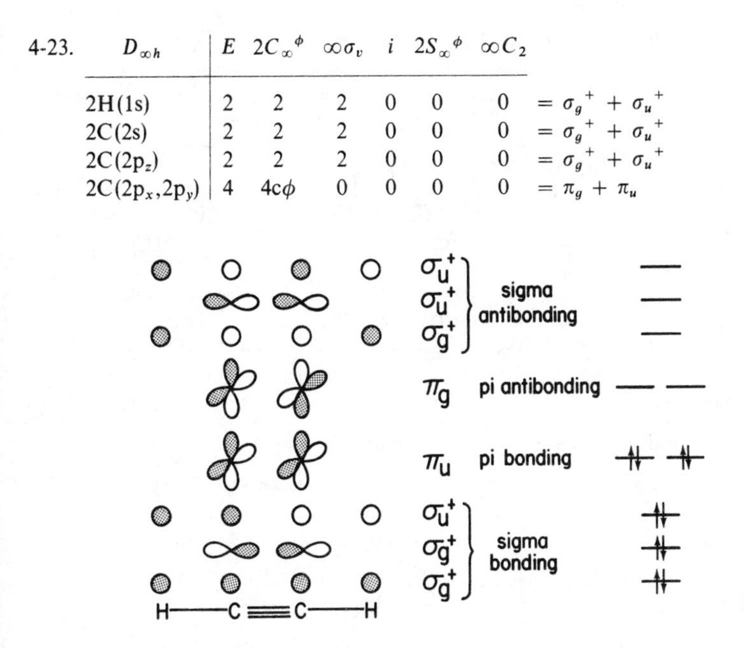

Orbitals of the same symmetry will be extensively mixed.

4-24.
$$E_2 = \frac{1}{6(1 - S)}\int(2\phi_1 - \phi_2 - \phi_3)\mathscr{H}(2\phi_1 - \phi_2 - \phi_3)\,d\tau$$
$$= \frac{1}{6(1 - S)}(4H_{11} - 2H_{12} - 2H_{13} - 2H_{21} + H_{22} + H_{23} - 2H_{31} + H_{32} + H_{33})$$
$$= \frac{1}{1 - S}(H_{ii} - H_{ij}).$$
$$E_3 = \frac{1}{2(1 - S)}\int(\phi_2 - \phi_3)\mathscr{H}(\phi_2 - \phi_3)\,d\tau = \frac{1}{1 - S}(H_{ii} - H_{ij}).$$

4-25.

D_{3h}	E	$2C_3$	$3C_2$	σ_h	$2S_3$	$3\sigma_v$	
$4C(2p_z)$	4	1	−2	−4	−1	2	$= 2a_2'' + e''$

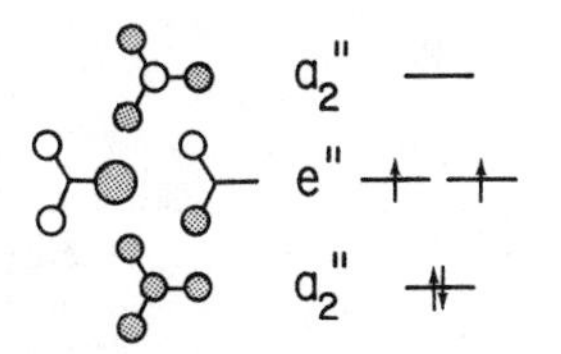

There will be two unpaired electrons.

4-26.

D_{4h}	E	$2C_4$	C_2	$2C_2'$	$2C_2''$	i	$2S_4$	σ_h	$2\sigma_v$	$2\sigma_d$	
$4C(p_z)$	4	0	0	−2	0	0	0	−4	2	0	$= e_g + a_{2u} + b_{2u}$

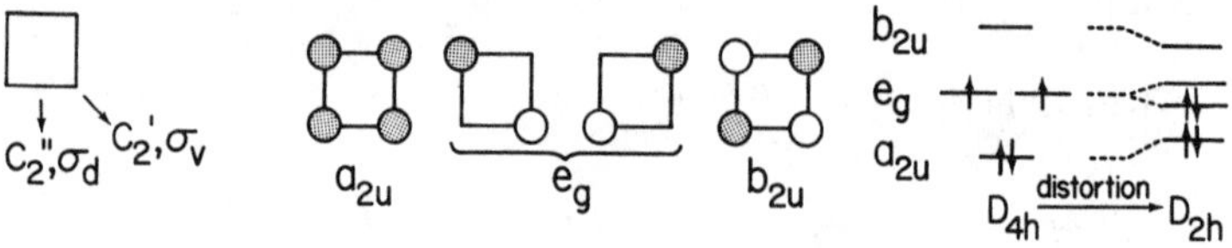

4-27.

D_{4h}	E	$2C_4$	C_2	$2C_2'$	$2C_2''$	i	$2S_4$	σ_h	$2\sigma_v$	$2\sigma_d$	
Fig. 4–43 orbitals	3	3	3	1	1	1	1	1	3	3	$= 2a_{1g} + a_{2u}$

Of the orbitals in Fig. 4–44, $\sigma_g^+ \sim a_{1g}$ and $\sigma_u^+ \sim a_{2u}$.

4-28.

D_{2h}	E	$C_2(z)$	$C_2(y)$	$C_2(x)$	i	$\sigma(xy)$	$\sigma(xz)$	$\sigma(yz)$	
4B(hybrids)	4	0	0	0	0	0	4	0	$= a_g + b_{2g} + b_{1u} + b_{3u}$
2H(1s)	2	2	0	0	0	0	2	2	$= a_g + b_{1u}$

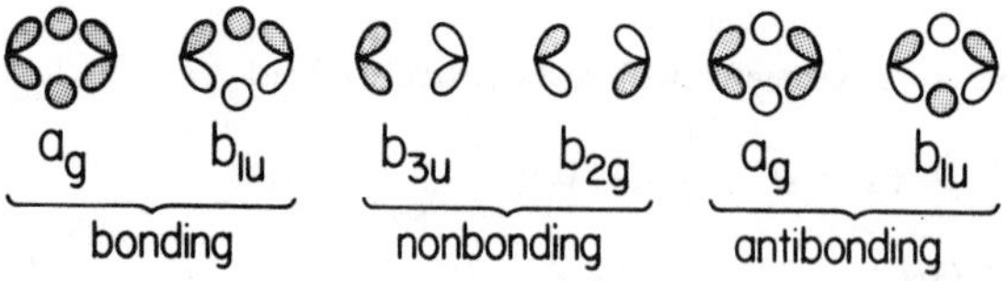

4-29.

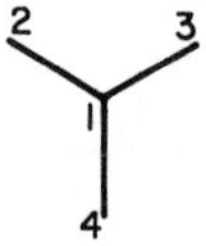

$$\begin{vmatrix} x & 1 & 1 & 1 \\ 1 & x & 0 & 0 \\ 1 & 0 & x & 0 \\ 1 & 0 & 0 & x \end{vmatrix} = 0 = x^4 - 3x^2 \Rightarrow x = 0, 0, \pm\sqrt{3}$$

$E = \alpha + \sqrt{3}\beta, \alpha, \alpha, \alpha - \sqrt{3}\beta$

4-30. (a)

C_{2v}	E	C_2	$\sigma(xz)$	$\sigma(yz)$	
$4C(2p_x)$	4	−2	2	−4	$= a_2 + 3b_1$

$\psi(b_1) = c_1p_1 + c_2p_2 + c_3(p_3 + p_4)$, $\psi(a_2) = p_3 - p_4$.

(b)

$$\begin{vmatrix} x & 1 & 0 & 0 \\ 1 & x & 1 & 1 \\ 0 & 1 & x & 1 \\ 0 & 1 & 1 & x \end{vmatrix} = 0$$

(c) -2β, $-\beta$, α, β, 2β

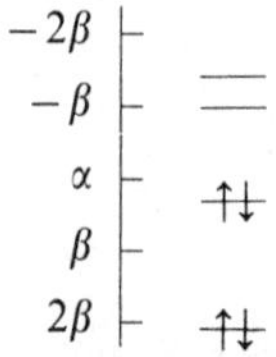

delocalization energy $= 0.962\ \beta$

(d) $E = \int(ap_1 + bp_2 + cp_3 + cp_4)\mathscr{H}(ap_1 + bp_2 + cp_3 + cp_4)d\tau$
$= a^2H_{11} + abH_{12} + acH_{13} + acH_{14} + baH_{21} + b^2H_{22} + bcH_{23} + bcH_{24} + caH_{31} + cbH_{32} + c^2H_{33} + c^2H_{34} + caH_{41} + cbH_{42} + c^2H_{43} + c^2H_{44}$
$= \alpha(a^2 + b^2 + 2c^2) + \beta(2ab + 4bc + 2c^2)$ (since $H_{ii} = \alpha$ and $H_{ij} = \beta$)
$= 1.00\alpha + 2.17\beta$

4-31. $c_1(H_{11} - E) + c_2H_{12} + c_3H_{13} = 0$
$c_1H_{21} + c_2(H_{22} - E) + c_3H_{23} = 0$
$c_1H_{31} + c_2H_{32} + c_3(H_{33} - E) = 0$

These equations must be solved simultaneously for each value of E ($= \alpha + \sqrt{2}\beta, \alpha, \alpha - \sqrt{2}\beta$). For ψ_1 ($E_1 = \alpha + \sqrt{2}\beta$), we get:
$c_1(-\sqrt{2}\beta) + c_2(\beta) + c_3(0) = 0 \Rightarrow c_2/c_1 = \sqrt{2}$
$c_1(\beta) + c_2(-\sqrt{2}\beta) + c_3(\beta) = 0$
$c_1(0) + c_2(\beta) + c_3(-\sqrt{2}\beta) = 0 \Rightarrow c_3/c_2 = 1/\sqrt{2} \Rightarrow c_3/c_1 = 1$

The form of ψ_1 is therefore $N_1(p_1 + \sqrt{2}p_2 + p_3)$. This is normalized as follows:
$1 = N_1^2\int(p_1 + \sqrt{2}p_2 + p_3)^2 d\tau$
$= N_1^2\int(p_1^2 + 2\sqrt{2}p_1p_2 + 2p_1p_3 + 2p_2^2 + 2\sqrt{2}p_2p_3 + p_3^2)d\tau$
$= N_1^2(1 + 0 + 0 + 2 + 0 + 1) \Rightarrow N_1 = \frac{1}{2}$
ψ_2 and ψ_3 are similarly derived.

4-32.

T_d	E	$8C_3$	$3C_2$	$6S_4$	$6\sigma_d$	
4 ligand sigma	4	1	0	0	2	$= a_1 + t_2$
1 metal s	1	1	1	1	1	$= a_1$
3 metal p	3	0	-1	-1	1	$= t_2$
5 metal d	5	-1	1	-1	1	$= t_2 + e$

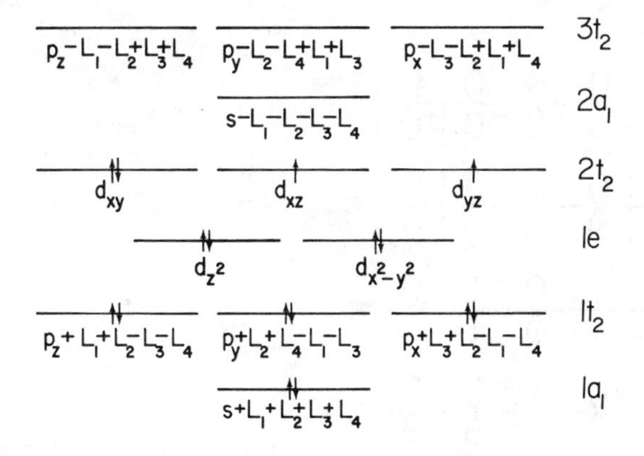

(The $2t_2$ orbitals will also have some antibonding ligand character: e.g., $d_{xy} - L_1 - L_2 + L_3 + L_4$.)

4-33. $\langle V \rangle = -\dfrac{Ze^2}{4\pi\epsilon_0}\displaystyle\iiint \psi^*\left(\frac{1}{r}\right)\psi r^2 \sin\theta \, dr\, d\theta\, d\phi = -\dfrac{Z^2e^2}{4\pi\epsilon_0 a_0 n^2} = -\dfrac{Z^2e^4\mu}{4\epsilon_0{}^2n^2h^2}$

$= 2T.$

$\langle K \rangle = T - \langle V \rangle = -T = \dfrac{Z^2e^4\mu}{8\epsilon_0{}^2n^2h^2}.$

$\therefore \sqrt{\langle v^2 \rangle} = \dfrac{Ze^2\sqrt{\mu}}{2\epsilon_0 nh\sqrt{m_e}} \approx \dfrac{Ze^2}{2\epsilon_0 nh}.$

For H(1s), $Z = 1$ and $\sqrt{\langle v^2 \rangle} = 2.19 \times 10^6 \text{ m s}^{-1} = 0.7\%\, c.$

For C^{5+}, $Z = 6$ and $\sqrt{\langle v^2 \rangle} = 2.63 \times 10^7 \text{ m s}^{-1} = 9\%\, c.$

4-34. $\psi = N \cos\dfrac{\pi x}{2x_0}.\quad N^2\displaystyle\int_{-x_0}^{x_0} \cos^2\frac{\pi x}{2x_0}\,dx = 1 \Rightarrow N = x_0{}^{-\frac{1}{2}}.$

$E = \dfrac{1}{x_0}\displaystyle\int_{-x_0}^{x_0} \cos\frac{\pi x}{2x_0}\left(-\frac{\hbar^2}{2m}\frac{d^2}{dx^2} + \tfrac{1}{2}kx^2\right)\cos\frac{\pi x}{2x_0}\,dx$

$= E_0\left(\dfrac{\pi^2}{4} + \dfrac{1}{3} - \dfrac{2}{\pi^2}\right) > E_0.$

4-35.

$M_L \backslash M_S$	3/2	1/2
3		
2		$(\overset{+}{1}, \bar{1}, \overset{+}{0})$
1		$(\overset{+}{1}, \bar{1}, -\overset{+}{1})(\overset{+}{1}, \overset{+}{0}, \bar{0})$
0	$(\overset{+}{1}, \overset{+}{0}, -\overset{+}{1})$	$(\overset{+}{1}, \overset{+}{0}, -\bar{1})(\overset{+}{1}, \bar{0}, -\overset{+}{1})(\bar{1}, \overset{+}{0}, -\overset{+}{1})$

$\Rightarrow {}^4S, {}^2D, {}^2P$

4-36. Carbon suboxide:

$D_{\infty h}$	E	$2C_\infty{}^\phi$	$\infty\sigma_v$	i	$2S_\infty{}^\phi$	∞C_2
$5(p_x, p_y)$	10	$10c\phi$	0	-2	$2c\phi$	0

$= 2\pi_g + 3\pi_u$

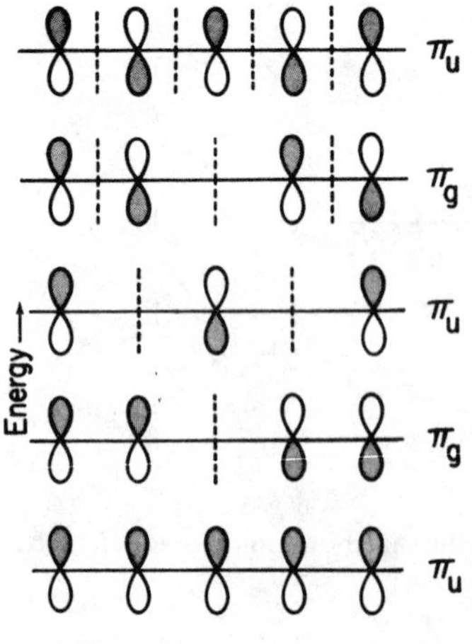

π_u

π_g

π_u

π_g

π_u

Energy

Only one component of each orbital is shown. The 12 pi electrons will occupy the three lowest energy levels. Nodes are indicated by dotted lines.

Allene:

D_{2d}	E	$2S_4$	C_2	$2C_2'$	$2\sigma_d$	
4C(2p)	4	0	−4	0	0	$= 2e$

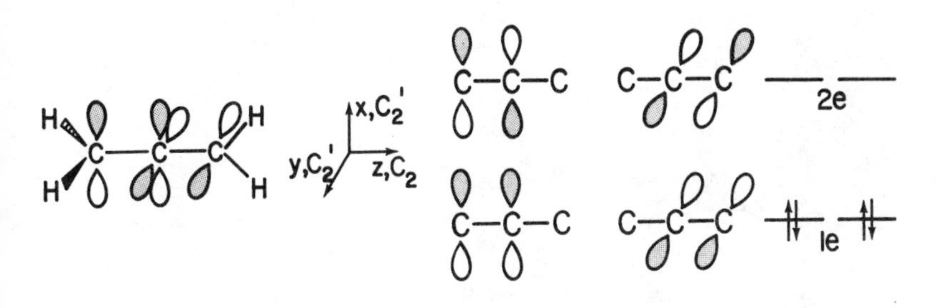

Tetramethylenecyclobutene:

D_{4h}	E	$2C_4$	C_2	$2C_2'$	$2C_2''$	i	$2S_4$	σ_h	$2\sigma_v$	$2\sigma_d$	
8C(2p)	8	0	0	−4	0	0	0	−8	4	0	$= 2e_g + 2a_{2u} + 2b_{2u}$

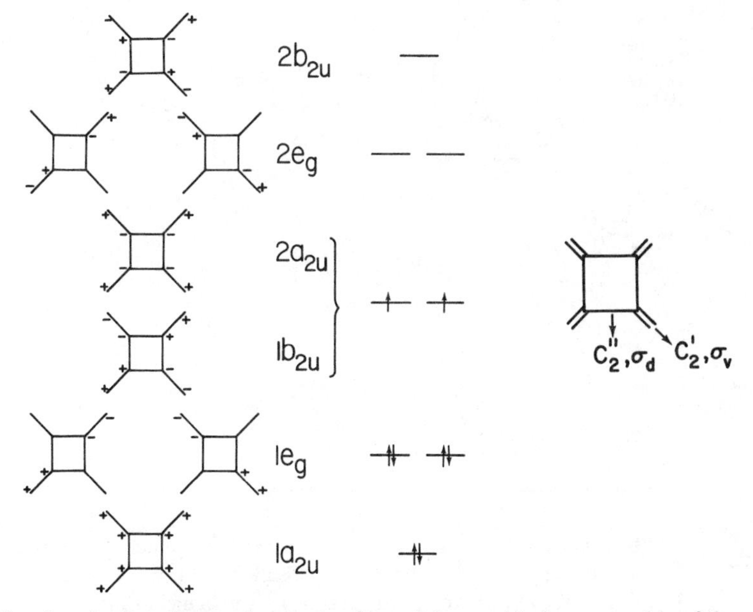

The signs in these figures denote the signs of the p orbitals on one side of the molecular plane. The $1b_{2u}$ and $2a_{2u}$ orbitals will be nearly degenerate, and, in fact, are degenerate within the Hückel approximation.

Trimethylenecyclopropane:

D_{3h}	E	$2C_3$	$3C_2$	σ_h	$2S_3$	$3\sigma_v$	
6C(2p)	6	0	−2	−6	0	2	$= 2a_2'' + 2e''$

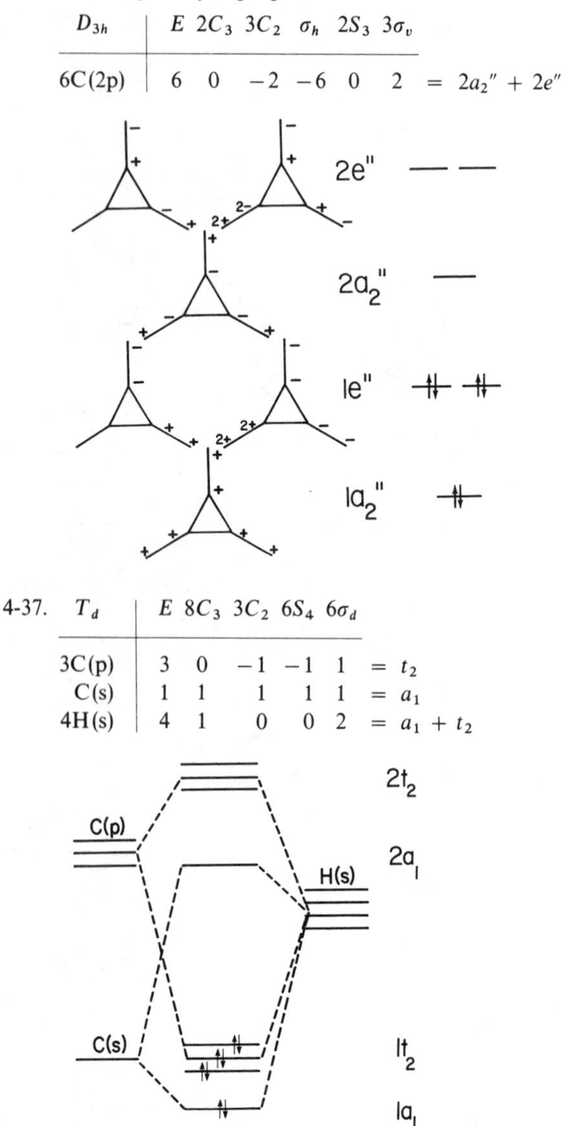

4-37.

T_d	E	$8C_3$	$3C_2$	$6S_4$	$6\sigma_d$	
3C(p)	3	0	−1	−1	1	$= t_2$
C(s)	1	1	1	1	1	$= a_1$
4H(s)	4	1	0	0	2	$= a_1 + t_2$

The forms of the $1a_1$, $1t_2$, $2a_1$, and $2t_2$ orbitals of methane are similar to the forms of the $1a_1$, $1t_2$, $2a_1$, and $3t_2$ orbitals of the tetrahedral ML_4 complex in Problem 4-32. The photoelectron spectrum is assigned as follows:

$$(1a_1)^2(1t_2)^6 \rightarrow (1a_1)^2(1t_2)^5\text{: 14 eV—area 3}$$
$$\rightarrow (1a_1)^1(1t_2)^6\text{: 23 eV—area 1}$$

4-38. (See Problem 3-33 for the transformation of imaginary characters into real form.)

D_{5h}	E	$2C_5$	$2C_5^2$	$5C_2$	σ_h	$2S_5$	$2S_5^3$	$5\sigma_v$	
5C(2p)	5	0	0	−1	−5	0	0	1	$= a_2'' + e_1'' + e_2''$

$\psi_1(a_2'') = N_1(p_1 + p_2 + p_3 + p_4 + p_5)$
$\psi_2(e_1'') = N_2(p_1 + .309p_2 - .809p_3 - .809p_4 + .309p_5)$
$\psi_3(e_1'') = N_3(\quad .951p_2 + .588p_3 - .588p_4 - .951p_5)$
$\psi_4(e_2'') = N_4(p_1 - .809p_2 + .309p_3 + .309p_4 - .809p_5)$
$\psi_5(e_2'') = N_5(\quad .588p_2 - .951p_3 + .951p_4 - .588p_5)$

4-39. $E = \alpha \pm \beta, \alpha \pm (2 + \sqrt{3})^{\frac{1}{2}}\beta, \alpha \pm (2 - \sqrt{3})^{\frac{1}{2}}\beta$; delocalization energy $= 0.899\ \beta$.

4-40. (*a*) The molecule is in the *yz* plane.

(*b*)

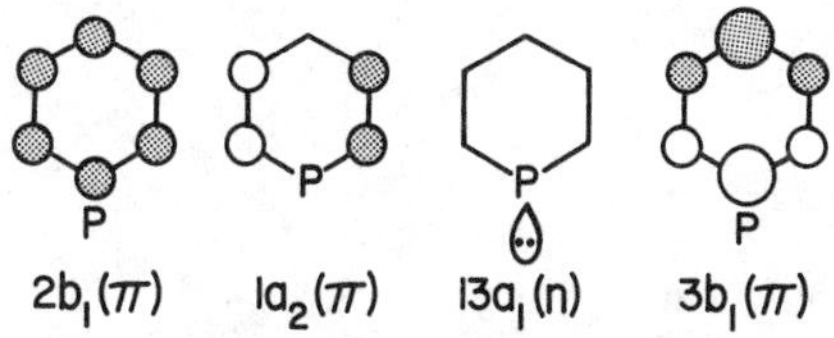

(*c*) $3b_1$ has more P(3p) character, and the energy of this orbital will vary more with the nature of X than the energy of $2b_1$. The correlation diagram agrees with this prediction.

4-41.

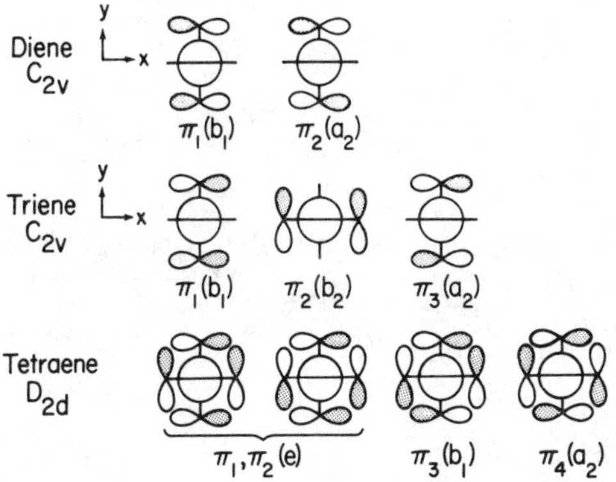

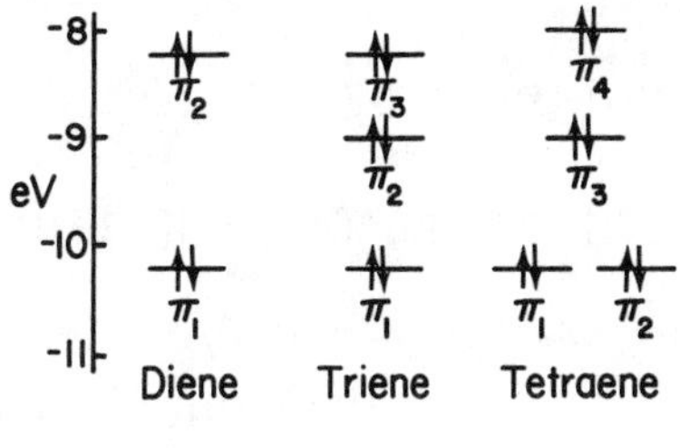

4-42.

D_{5d}	E	$2C_5$	$2C_5^2$	$5C_2$	i	$2S_{10}^3$	$2S_{10}$	$5\sigma_d$
10C(2p)	10	0	0	0	0	0	0	2

$$= a_{1g} + a_{2u} + e_{1g} + e_{2g} + e_{1u} + e_{2u}$$

The orbitals of Problem 4-38 can be used to make ferrocene carbon group orbitals out of the two cyclopentadienyl rings as follows: a_{1g}: $\psi_1 - \psi_1$; a_{2u}: $\psi_1 + \psi_1$; e_{1g}: $\psi_2 - \psi_2$, $\psi_3 - \psi_3$; e_{2g}: $\psi_4 - \psi_4$, $\psi_5 - \psi_5$; e_{1u}: $\psi_2 + \psi_2$, $\psi_3 + \psi_3$; e_{2u}: $\psi_4 + \psi_4$, $\psi_5 + \psi_5$. By inspection of the D_{5d} character table, the Fe orbitals interact with the cyclopentadienyl orbitals as follows: d_{z^2} – a_{1g}; d_{xz}, d_{yz} – e_{1g}; $d_{x^2-y^2}$, d_{xy} – e_{2g}; s – a_{1g}; p_z – a_{2u}; p_x, p_y – e_{1u}.

Chapter 5

5-1.

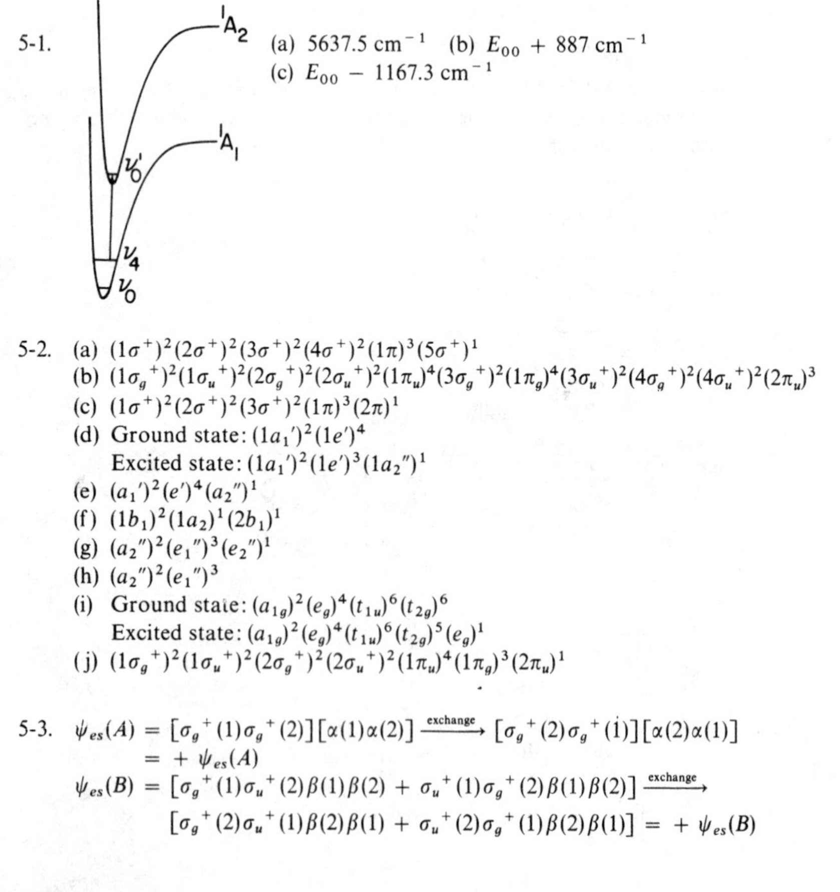

(a) 5637.5 cm^{-1} (b) E_{00} + 887 cm^{-1}
(c) E_{00} − 1167.3 cm^{-1}

5-2. (a) $(1\sigma^+)^2(2\sigma^+)^2(3\sigma^+)^2(4\sigma^+)^2(1\pi)^3(5\sigma^+)^1$
(b) $(1\sigma_g^+)^2(1\sigma_u^+)^2(2\sigma_g^+)^2(2\sigma_u^+)^2(1\pi_u)^4(3\sigma_g^+)^2(1\pi_g)^4(3\sigma_u^+)^2(4\sigma_g^+)^2(4\sigma_u^+)^2(2\pi_u)^3$
(c) $(1\sigma^+)^2(2\sigma^+)^2(3\sigma^+)^2(1\pi)^3(2\pi)^1$
(d) Ground state: $(1a_1')^2(1e')^4$
Excited state: $(1a_1')^2(1e')^3(1a_2'')^1$
(e) $(a_1')^2(e')^4(a_2'')^1$
(f) $(1b_1)^2(1a_2)^1(2b_1)^1$
(g) $(a_2'')^2(e_1'')^3(e_2'')^1$
(h) $(a_2'')^2(e_1'')^3$
(i) Ground state: $(a_{1g})^2(e_g)^4(t_{1u})^6(t_{2g})^6$
Excited state: $(a_{1g})^2(e_g)^4(t_{1u})^6(t_{2g})^5(e_g)^1$
(j) $(1\sigma_g^+)^2(1\sigma_u^+)^2(2\sigma_g^+)^2(2\sigma_u^+)^2(1\pi_u)^4(1\pi_g)^3(2\pi_u)^1$

5-3.

$$\psi_{es}(A) = [\sigma_g^+(1)\sigma_g^+(2)][\alpha(1)\alpha(2)] \xrightarrow{\text{exchange}} [\sigma_g^+(2)\sigma_g^+(1)][\alpha(2)\alpha(1)] = +\psi_{es}(A)$$

$$\psi_{es}(B) = [\sigma_g^+(1)\sigma_u^+(2)\beta(1)\beta(2) + \sigma_u^+(1)\sigma_g^+(2)\beta(1)\beta(2)] \xrightarrow{\text{exchange}} [\sigma_g^+(2)\sigma_u^+(1)\beta(2)\beta(1) + \sigma_u^+(2)\sigma_g^+(1)\beta(2)\beta(1)] = +\psi_{es}(B)$$

5-4. $\frac{1}{\sqrt{2}}\int_{e_1,e_2} \beta(1)\beta(2)[\alpha(1)\beta(2) + \beta(1)\alpha(2)]\,d\tau_{12} =$

$\frac{1}{\sqrt{2}}\left[\int_{e_1} \beta(1)\alpha(1)d\tau_1 \int_{e_2} \beta(2)\beta(2)\,d\tau_2 + \int_{e_1} \beta(1)\beta(1)d\tau_1 \int_{e_2} \beta(2)\alpha(2)\,d\tau_2\right] =$

$\frac{1}{\sqrt{2}}[0{\cdot}1 + 1{\cdot}0] = 0$. The orthogonality of ψ_3 and ψ_4 is shown in a similar manner.

5-5. (a) singlet (b) singlet, triplet (c) singlet, triplet (d) quartet, doublet (e) sextet, quartet, doublet

5-6. (a) $^3\Pi$, $^1\Pi$ (b) $^2\Pi_u$ (c) $^3\Sigma^+$, $^3\Sigma^-$, $^3\Delta$, $^1\Sigma^+$, $^1\Sigma^-$, $^1\Delta$ (d) $^1A_1'$ (ground state) and $^1E''$, $^3E''$ (excited state) (e) $^2A_2''$ (f) $^3B_2$, $^1B_2$ (g) $^3E_1'$, $^3E_2'$, $^1E_1'$, $^1E_2'$ (h) $^2E_1''$ (i) $^1A_{1g}$ (ground state) and $^3T_{1g}$, $^3T_{2g}$, $^1T_{1g}$, $^1T_{2g}$ (excited state) (j) $^3\Sigma_u^+$, $^3\Sigma_u^-$, $^3\Delta_u$, $^1\Sigma_u^+$, $^1\Sigma_u^-$, $^1\Delta_u$

5-7. $(e_{1g})^2 \Rightarrow {}^1A_{1g} + {}^1E_{2g} + {}^3A_{2g}$; $(e_{2u})^2 \Rightarrow {}^1A_{1g} + {}^1E_{2g} + {}^3A_{2g}$;
$(e_{1g})^2(e_{2g})^2 \Rightarrow (3)^1A_{1g} + {}^3A_{1g} + {}^5A_{1g} + {}^1A_{2g} + (2)^3A_{2g} + (3)^1E_{2g} + (2)^3E_{2g}$.

5-8. $^4A_{2g}$, ^2E_g, $^2T_{1g}$, $^2T_{2g}$

5-9.

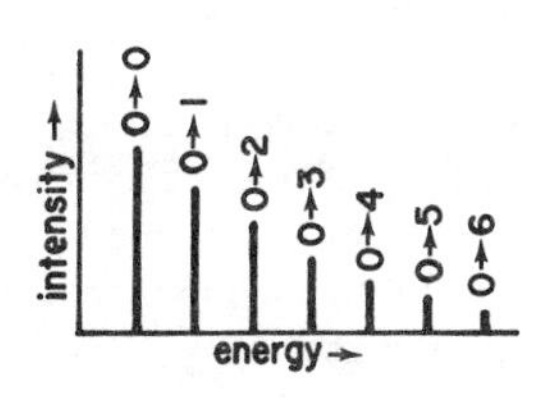

5-10. (a) Lorentzian: $\bar{I} = \int_0^\infty \epsilon(\bar{\nu})d\bar{\nu} = \frac{\pi\epsilon_{max}\Delta\bar{\nu}_{\frac{1}{2}}}{4}$; Gaussian: $\bar{I} = \frac{\epsilon_{max}\Delta\bar{\nu}_{\frac{1}{2}}}{4}\sqrt{\frac{\pi}{\ln 2}}$

(b) $\bar{I} = 7.59 \times 10^4\ M^{-1}\ cm^{-2}$ (c) $f = 3.29 \times 10^{-4}$

5-11. $\frac{1}{2}\int_{e_1,e_2} (\alpha\beta - \beta\alpha)^2\,d\tau_{12} = \frac{1}{2}\left[\int_{e_1} \alpha\alpha\,d\tau_1 \int_{e_2} \beta\beta\,d\tau_2 - \int_{e_1} \alpha\beta\,d\tau_1 \int_{e_1} \beta\alpha\,d\tau_2\right.$

$\left. - \int_{e_1} \beta\alpha\,d\tau_1 \int_{e_2} \alpha\beta\,d\tau_2 + \int_{e_1} \beta\beta\,d\tau_1 \int_{e_2} \alpha\alpha\,d\tau_2\right]$,

$= \frac{1}{2}[1 + 0 + 0 + 1] = 1$ (singlet → singlet, allowed).

Similarly, $\frac{1}{2}\int_{e_1,e_2} (\alpha\beta - \beta\alpha)(\alpha\beta + \beta\alpha)d\tau_{12} = 0$ (singlet → triplet, forbidden).

5-12. Ground state: ^1A_g; excited states: $^3B_{3u}$, $^1B_{3u}$; $^1A_g \to {}^3B_{3u}$: spin forbidden ⇒ $\epsilon \approx 10^{-5}$–1; $^1A_g \to {}^1B_{3u}$: fully allowed ⇒ $\epsilon \approx 10^3$–10^5.

5-13. Ground state: $^2A_2''$; excited state: $^2E'$; spin allowed but orbitally forbidden ⇒ $\epsilon \approx 1$–10^3. Pyramidal structure: $^2A_1 \to {}^2E$, fully allowed, $\epsilon \approx 10^3$–10^5.

5-14. $B_{2u}(b_{1g}) \begin{pmatrix} b_{3u}(x) \\ b_{2u}(y) \\ b_{1u}(z) \end{pmatrix} A_g(b_{2g}) \sim \begin{pmatrix} b_{2g} \\ b_{3g} \\ a_g \end{pmatrix}$ ← allowed, z-polarized

5-15. $E(10^0_1) = E(10^1_0) - \bar{\nu}_{10} - \bar{\nu}_{10}{}' = 37{,}483 \text{ cm}^{-1}$

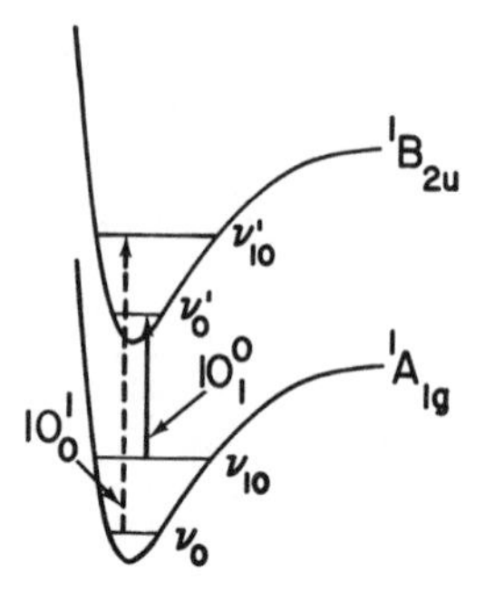

5-16. (*a*) a_{1u} b_{1u} e_u — e_u e_u $a_{1u} + a_{2u} + b_{1u} + b_{2u}$

(*b*) forbidden allowed — allowed allowed — allowed allowed

(*c*) The band near 22 kK is $^1A_{1g} \rightarrow {}^1A_{2g}$.

5-17. $E = IE + \text{vibrational energy} = IE + \bar{\omega}_e v - \bar{\omega}_e x_e v(v+1)$ (see eq. 3-23). Peaks are expected at 15.43, 15.70, 15.96, 16.21, 16.44, 16.65, ... eV. Since the vibrational levels of D_2 will be closer together than those of H_2, and since the value of r_0 will be very nearly the same for both, the Franck-Condon maximum in the photoelectron spectrum of D_2 will occur at a higher value of v' than in the spectrum of H_2.

5-18. $\Sigma_g^+ \begin{pmatrix} \pi_u \\ \sigma_u^+ \end{pmatrix} \Pi_g \sim \begin{pmatrix} \sigma_u^+ + \sigma_u^- + \Delta_u \\ \pi_u \end{pmatrix}$ (forbidden). This could be determined by inspection since the ground state is totally symmetric and the excited state does not transform as x, y, or z. $(1\pi_u)^3(3\sigma_g^+)^2(1\pi_g)^1 \Rightarrow {}^1\Sigma_u^+ + {}^1\Sigma_u^- + {}^1\Delta_u + {}^3\Sigma_u^+ + {}^3\Sigma_u^- + {}^3\Delta_u$. $(3\sigma_g^+)^1(3\sigma_u^+)^1 \Rightarrow {}^1\Sigma_u^+ + {}^3\Sigma_u^+$.

5-19. We will work the operations E, C_∞^ϕ, and σ_v only:

$E\psi_6 = \psi_6;$

$$C_\infty^\phi \psi_6 = \tfrac{1}{2}[(X\cos\phi - Y\sin\phi)^2 + (X\sin\phi + Y\cos\phi)^2]$$
$$= \tfrac{1}{2}[XX(\cos^2\phi + \sin^2\phi) + YY(\cos^2\phi + \sin^2\phi)] = \psi_6;$$

$\sigma_v\psi_6 = \tfrac{1}{2}[XX - (-Y)(-Y)] = \psi_6.$

$E\psi_1 = \psi_1;$

$$C_\infty^\phi \psi_1 = \frac{1}{\sqrt{2}}[(X\cos\phi - Y\sin\phi)(X\sin\phi + Y\cos\phi)$$
$$- (X\sin\phi + Y\cos\phi)(X\cos\phi - Y\sin\phi)]$$
$$= \frac{1}{\sqrt{2}}[XY(\cos^2\phi + \sin^2\phi) - YX(\cos^2\phi + \sin^2\phi)] = \psi_1;$$

$$\sigma_v\psi_1 = \frac{1}{\sqrt{2}}[X(-Y) - (-Y)X] = -\psi_1.$$

(Note that in Section 5.5-C we have taken rotation to be counterclockwise, which is opposite to the convention used earlier in this book. The direction of rotation should not affect any of our conclusions.)

5-20. Angular momentum: $\pi = 1$ unit; $\delta = 2$ units. $2 + 1 = 3 \Rightarrow \Phi$ state; $2 - 1 = 1 \Rightarrow \Pi$ state. The possible angular momentum microstates of $(\pi)^3$ are $1 + 1 + 1$, which is forbidden by the exclusion rule, and $1 - 1 + 1$, which gives a Π state.

5-21. $O_2(^3\Sigma_g^-)$ configuration is $\ldots(1\pi_g)^2$. The lowest energy ionization gives $\ldots(1\pi_g)^1 \Rightarrow {}^2\Pi_g$ state. $O_2(^1\Delta_g)$ has the same configuration as $O_2(^3\Sigma_g^-)$, and the lowest ionic state is the same $^2\Pi$ state produced by ionization of $O_2(^3\Sigma_g^-)$. Since $O_2(^1\Delta_g)$ lies 7882 cm^{-1} (= 0.98 eV) above $O_2(^3\Sigma_g^-)$, its IE should be $12.08 - 0.98 = 11.10$ eV. The states produced by the second ionization of $O_2(^1\Delta_g)$ will be different from those of $O_2(^3\Sigma_g^-)$ because the outer $(1\pi_g)^2$ shell will have a different arrangement of its electrons in each case:

$O_2(^3\Sigma_g^-)$ ionization gives: $\ldots(1\pi_u)^3(1\pi_g)^2$

$$^2\Pi_u \times {}^3\Sigma_g^- \Rightarrow {}^2\Pi_u + {}^4\Pi_u.$$

$O_2(^1\Delta_g)$ ionization gives: $\ldots(1\pi_u)^3(1\pi_g)^2$

$$^2\Pi_u \times {}^1\Delta_g = {}^2\Pi_u + {}^2\Phi_u.$$

5-22. The electron is removed from an antibonding orbital.

5-23. $Br_2(v = 1)/Br_2(v = 0) = 0.21$. The band 298 cm^{-1} below 0–0 is the hot band, $Br_2(v = 1) \rightarrow Br_2^+(v = 0)$.

5-24. (a) $\dfrac{d[D^*]}{dt} = 0 = I_a - k_1[D^*] - k_2[D^*][A] - k_3[D^*] \Rightarrow I_a = [D^*](k_1 + k_2[A] + k_3)$

(b) $\Phi_0 = \dfrac{k_1[D^*]}{[D^*](k_1 + k_3)} = \dfrac{k_1}{k_1 + k_3}$ (c) $\Phi_A = \dfrac{k_1}{k_1 + k_2[A] + k_3}$

(d) $\Phi_0/\Phi_A = \dfrac{k_1 + k_3 + k_2[A]}{k_1 + k_3} = 1 + \left(\dfrac{k_2}{k_1 + k_3}\right)[A]$

5-25. (a) $k_1 = 1.8 \times 10^2$ s^{-1}; $k_3 = 18$ s^{-1} (b) $\Phi_0 = 0.91$

(c) slope $= 5 \times 10^2$ $M^{-1} = \dfrac{k_2}{k_1 + k_3} \Rightarrow k_2 = 9.9 \times 10^4$ $M^{-1}s^{-1}$

5-26. $1t_2 \rightarrow 2a_1 \Rightarrow {}^1T_2, {}^3T_2$. $1t_2 \rightarrow 2t_2 \Rightarrow {}^1A_1, {}^1E, {}^1T_1, {}^1T_2, {}^3A_1, {}^3E, {}^3T_1, {}^3T_2$. There are two fully allowed $(^1A_1 \rightarrow {}^1T_2)$ transitions.

5-27. (See Problem 4-23.) ground state: $\ldots(\pi_u)^4 = {}^1\Sigma_g^+$; excited state: $\ldots(\pi_u)^3(\pi_g)^1 = {}^1\Sigma_u^+, {}^1\Sigma_u^-, {}^1\Delta_u, {}^3\Sigma_u^+, {}^3\Sigma_u^-, {}^3\Delta_u$. Excited state geometry: H–C–C–H (trans-bent: H on the right C pointing up, H on the left C pointing down)

5-28.

n	2	3	4	5	6	7	8	9	10
$\bar{E}$(kK):	46.1	37.3	32.9	29.9	27.5	25.6	24.4	—	22.4

5-29. $\pi_2 \rightarrow \pi_3^*$: $^1A_1 \rightarrow {}^1A_1$ (z-polarized). $n_+ \rightarrow \pi_3^*$: $^1A_1 \rightarrow {}^1B_2$ (y-polarized).

$\pi_2 \rightarrow \pi_3^*$: In z-polarization, only a_1 modes may serve as vibronic origins.
In x-polarization, only b_1 modes may serve as vibronic origins.
In y-polarization, only b_2 modes may serve as vibronic origins.

$n_+ \rightarrow \pi_3^*$: In z-polarization, only b_2 modes may serve as vibronic origins.
In x-polarization, only a_2 modes may serve as vibronic origins.
In y-polarization, only a_1 modes may serve as vibronic origins.

5-30.

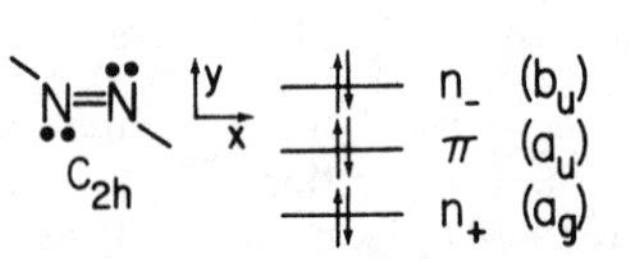

Both $n_- \rightarrow \pi^*$ and $\pi \rightarrow \pi^*$ are orbitally allowed. Since $n \rightarrow \pi^*$ transitions are generally much weaker than $\pi \rightarrow \pi^*$ transitions, and at lower energy, we assign the 347-nm band as $n \rightarrow \pi^*$ and the intense band $\pi \rightarrow \pi^*$.

5-31. ${}^1A_1 \rightarrow {}^1B_2$, ${}^3B_2$ (y-polarized)

5-32.

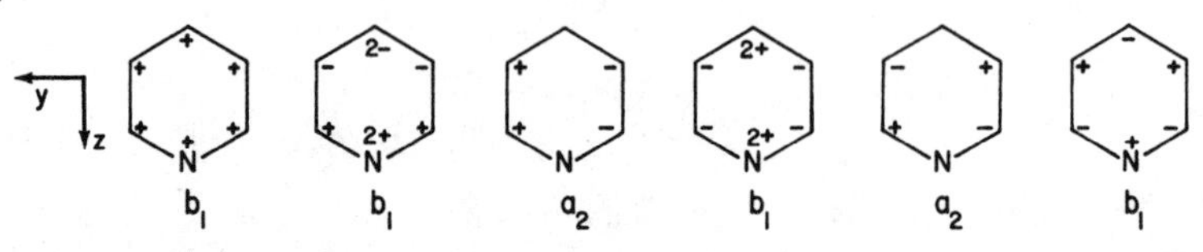

Lowest $\pi \rightarrow \pi^*$ is ${}^1A_1 \rightarrow {}^1B_2$ (allowed). (Even if the order of the pi orbitals is reversed, the transition is allowed.)

5-33. The electronic transition is orbitally forbidden, so coupling with a totally symmetric vibration cannot make the transition allowed.

5-34. (a)

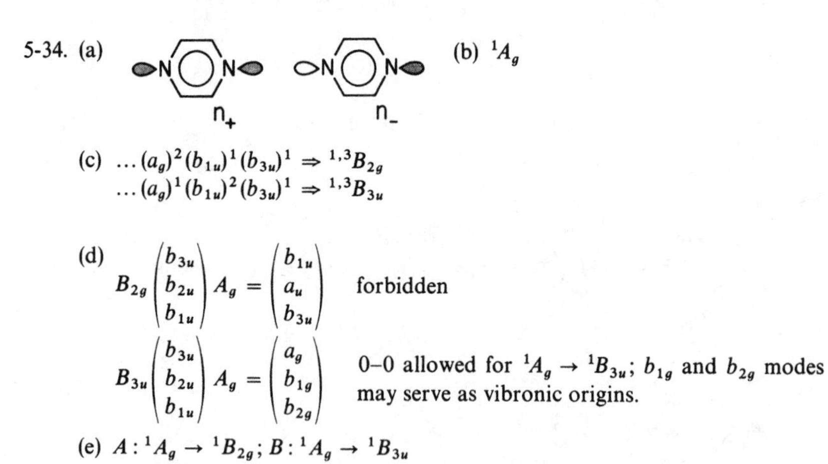

(b) ${}^1A_g$

(c) $\ldots (a_g)^2 (b_{1u})^1 (b_{3u})^1 \Rightarrow {}^{1,3}B_{2g}$
$\ldots (a_g)^1 (b_{1u})^2 (b_{3u})^1 \Rightarrow {}^{1,3}B_{3u}$

(d) $B_{2g} \begin{pmatrix} b_{3u} \\ b_{2u} \\ b_{1u} \end{pmatrix} A_g = \begin{pmatrix} b_{1u} \\ a_u \\ b_{3u} \end{pmatrix}$ forbidden

$B_{3u} \begin{pmatrix} b_{3u} \\ b_{2u} \\ b_{1u} \end{pmatrix} A_g = \begin{pmatrix} a_g \\ b_{1g} \\ b_{2g} \end{pmatrix}$ 0–0 allowed for ${}^1A_g \rightarrow {}^1B_{3u}$; b_{1g} and b_{2g} modes may serve as vibronic origins.

(e) $A: {}^1A_g \rightarrow {}^1B_{2g}$; $B: {}^1A_g \rightarrow {}^1B_{3u}$

Band	Assignment	Band	Assignment	Band	Assignment
X_1	7_0^1(possibly 24_0^1)	Y_1	5_0^1	Z_1	13_0^1
X_2	$7_0^1 5_0^1$	Y_2	5_0^2	Z_2	$13_0^1 5_0^1$
X_3	$7_0^1 5_0^2$	Y_3	5_0^3	Z_3	$13_0^1 5_0^2$
X_4	$7_0^1 5_0^3$				

The hot band at 30,242 cm^{-1} (598 cm^{-1} below band B) is 5_1^0.

5-35. (a) $^1A_1{}'$ (b) $(a_2{}'')^2(e'')^3(e'')^1 \Rightarrow {}^3E', {}^3A_1{}', {}^3A_2{}', {}^1E', {}^1A_1{}', {}^1A_2{}'$

(c) $^3E' \to {}^1A_1{}', {}^3A_1{}' \to {}^1A_1{}', {}^3A_2{}' \to {}^1A_1{}'$

(d)
$$E' \begin{pmatrix} a_2{}'' \\ e' \end{pmatrix} A_1{}' \sim \begin{pmatrix} e'' \\ a_1{}' + a_2{}' + e' \end{pmatrix}$$

$$A_1{}' \begin{pmatrix} a_2{}'' \\ e' \end{pmatrix} A_1{}' \sim \begin{pmatrix} a_2{}'' \\ e' \end{pmatrix}$$

$$A_2{}' \begin{pmatrix} a_2{}'' \\ e' \end{pmatrix} A_1{}' \sim \begin{pmatrix} a_1{}'' \\ e' \end{pmatrix}$$

symmetries of possible vibronic origins

(e) 546.2 cm^{-1}: $(a_2{}'')$ 10^0_1; 591.8, 593.5: (e') 17^0_1; 703.9: $(a_2{}'')$ 9^0_1; 718.2, 722.7: (e'') 19^0_1; 756.0, 765.2, 774.1: $(a_1{}' + e')$ 20^0_2; 831.5, 835.0: (e') 16^0_1; 927.9: $(a_2{}'')$ 8^0_1; 954.6: $(a_1{}')$ 4^0_1; 1076.3, 1082.5: $(a_2{}'' \times e'' = e')$ $9^0_1, 20^0_1$

(f) $^3A_1{}'$

5-36. $^2T_{2g}, {}^1A_{1g}, {}^4A_{2g}$.

5-37. First appearance: M–S–C≡N. Second appearance: M–N=C=S.

5-38.

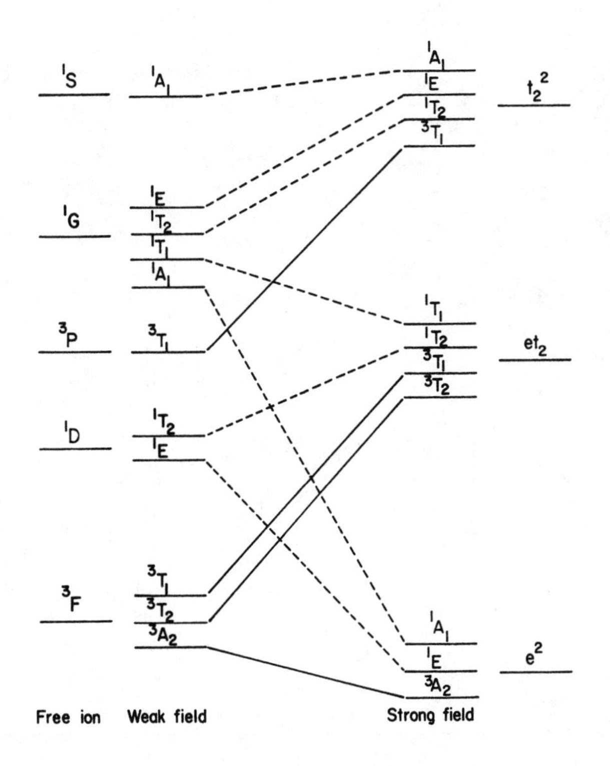

5-39.

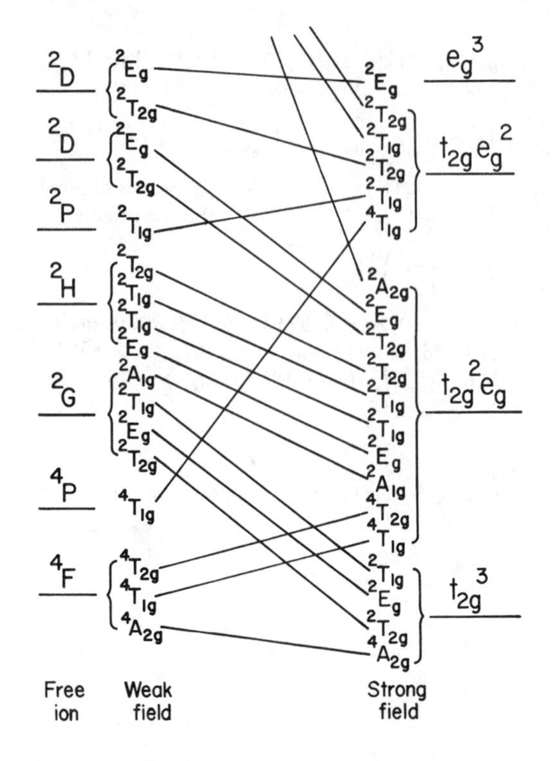

5-40. Any drawing you try for the excited state will not have five unpaired electrons.

5-41. Yes, because H_2O lies lower in the spectrochemical series than ethylenediamine.

5-42.

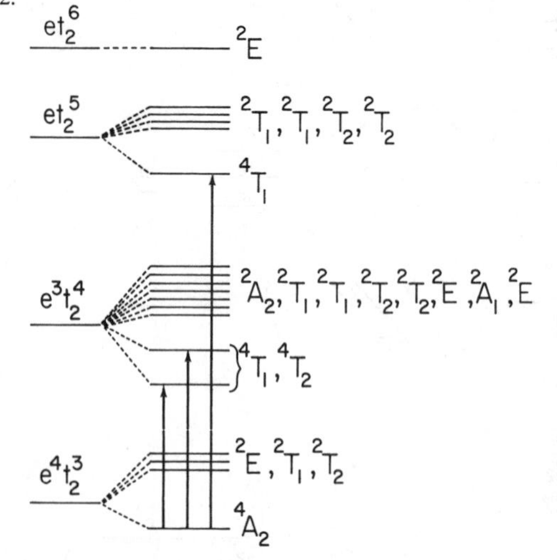

(The order of the $^4T_1$ and $^4T_2$ infrared transitions cannot be assigned without some kind of calculation.) An octahedral d^3 complex such as $V(H_2O)_6^{2+}$ will have the same state diagram as a tetrahedral d^7 complex.

5-43. High spin: CoF_6^{3-}; low spin: $Co(H_2NCH_2CH_2NH_2)_3^{3+}$. The discontinuity occurs at the point where the spin pairing energy is equal to Δ_0. To the left, $\Delta_0 < P$ and complexes are high spin. To the right, $\Delta_0 > P$ and complexes are low spin.

5-44. (a) Cl(3p) electrons are held more tightly than Br(4p) electrons.
(b) It is easier to add electrons to Os(IV) ($OsCl_6^{2-}$) than to Os(III) ($OsCl_6^{3-}$).

5-45.

5-46.

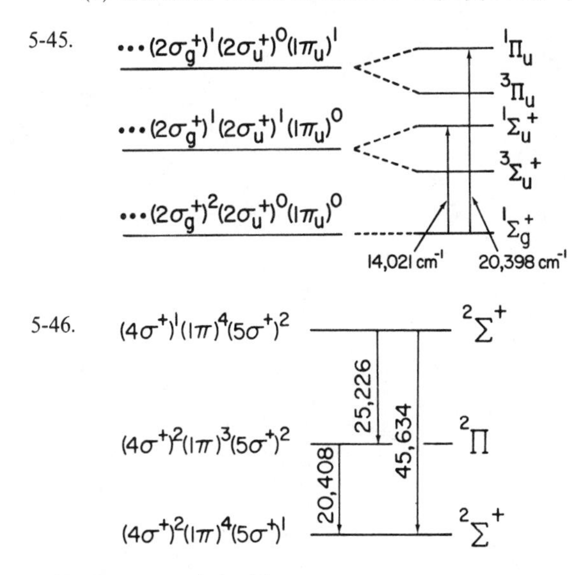

5-47. Quartet and doublet.

5-48. Ground state: $^3\Sigma^-$. Excited states: $^3\Pi$ and $^1\Pi$. Assignments: $^3\Sigma^- \rightarrow {}^1\Delta$(10 kK), $^1\Sigma^+$(22 kK), $^3\Pi$(30 kK), $^1\Pi$(31 kK).

5-49. ^5A_g, $^3T_{1g}$, $^3T_{2g}$, ^3G_g, ^3H_g, ^1A_g, ^1G_g, (3) ^1H_g.

5-50.

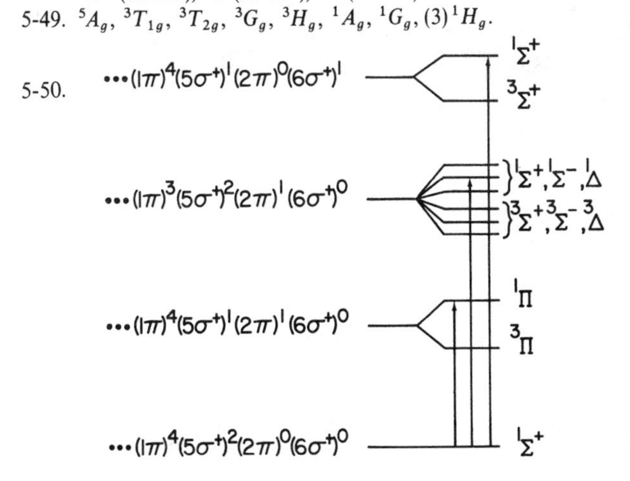

(The ordering of the two highest configurations is not certain.)

5-51. $\chi(E) = 7$; $\chi(C_5) = \dfrac{\sin (7/2)(2\pi/5)}{\sin (\pi/5)} = -1.61803399\ldots$; $\chi(C_5{}^2) = \dfrac{\sin (7/2)(4\pi/5)}{\sin (2\pi/5)}$ $= 0.61803399\ldots$; $\chi(C_3) = \dfrac{\sin (7/2)(2\pi/3)}{\sin (\pi/3)} = 1$; $\chi(C_2) = \dfrac{\sin (7/2)(\pi)}{\sin (\pi/2)} = -1.$

I	E	$12C_5$	$12C_5{}^2$	$20C_3$	$15C_2$
Γ(f)	7	−1.61803399...	0.61803399...	1	−1
T_2	3	−0.61803399...	1.61803399...	0	−1
G	4	−1	−1	1	0
$T_2 + G$	7	−1.61803399...	0.61803399...	1	−1

In I_h, the f orbitals will transform as $T_{2u} + G_u$.

5-52.

v	IE(eV)	$IE(v) - IE(0)$(eV)	$IE(v) - IE(0)$(cm^{-1})	ΔIE(cm^{-1})/v
0	16.53	—	—	—
1	16.73	0.20	1613	1613
2	16.92	0.39	3146	1573
3	17.11	0.58	4678	1559
4	17.29	0.76	6130	1532
5	17.47	0.94	7582	1516
6	17.64	1.11	8953	1492
7	17.81	1.28	10324	1475
8	17.98	1.45	11695	1462
9	18.15	1.62	13066	1452
10	18.31	1.78	14356	1436
11	18.47	1.94	15647	1422

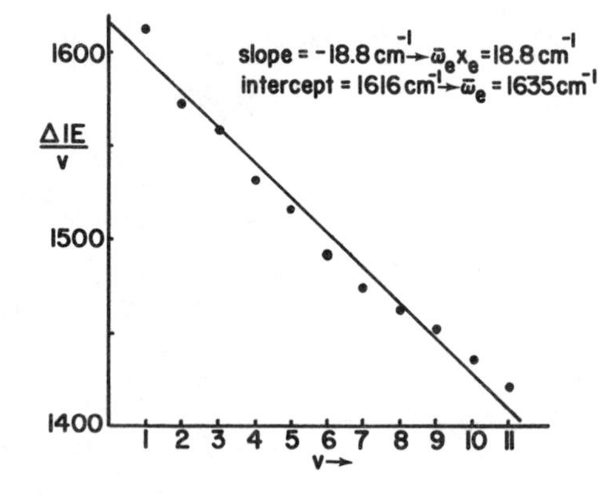

5-53. It will probably be a weaker Lewis base since the density of n electrons is decreased. The dipole moment should decrease in magnitude (and perhaps change sign). The $n \to \pi^*$ transition of NNH^+ will probably be lower in energy than that of HCN since the π^* orbital will have more N(2p) character and will probably be lower in energy than the HCN π^* orbital.

5-54. (a) (Case 5, Section 5.3-D) ^3P, ^1S, ^1D. (b) (Case 3) ^1S, ^1P, ^1D, ^3S, ^3P, ^3D. (c) (Case 3) ^1D, ^3D. (d) (Case 3) ^1P, ^1D, ^1F, ^3P, ^3D, ^3F. (e) ^3F, ^3P, ^1G, ^1D, ^1S. (f) (Case 6) We will work this out in detail, using the arbitrary angle $\phi = 28.7°$, as described in the text of Appendix B. We transform the K character table as follows:

K	E	$\infty C_\infty^{\phi = 28.7°}$	
S	1	1	
P	3	2.754292...	
D	5	3.831834...	
F	7	3.967864...	

	R	E	$C_\infty^{\phi = 28.7°}$	
	R^2	E	$C_\infty^{2\phi = 57.4°}$	
	R^3	E	$C_\infty^{3\phi = 86.1°}$	
	$\chi(R)$	3	2.754292...	
	$[\chi(R)]^3$	27	20.89440...	
	$\chi(R^2)$	3	2.077542...	
	$\chi(R^3)$	3	1.136031...	
eq. 5-23	χ(doublet)	8	6.586123...	$= {}^2P + {}^2D$
	χ(quartet)	1	0.9999999...	$= {}^4S$

5-55. (a) allowed, $\psi_{e'}\hat{\mu}\psi_e \sim P(P)S \sim S + P + D$. (b) allowed, $P(P)S \sim S + P + D$. (c) forbidden, $D(P)S \sim P + D + F$. (d) allowed, $D(P)P \sim S + 2P + 3D + 2F + G$. (e) forbidden, $F(P)P \sim P + 2D + 3F + 2G + H$. (f) forbidden, $S(P)S \sim P$.

5-56. (a) $\ldots(b_{2g})^2 \Rightarrow {}^1A_{1g}$ ground state. $\ldots(b_{2g})^1(b_{1u})^1 \Rightarrow {}^{1,3}A_{2u}$ excited state. The transition moment integral for the ${}^1A_{1g} \rightarrow {}^1A_{2u}$ transition transforms as $A_{2u}\begin{pmatrix} e_u \\ a_{2u} \end{pmatrix} A_{1g} = \begin{pmatrix} e_g \\ a_{1g} \end{pmatrix}$, which says that the transition is allowed and z-polarized. (b) The i_0^1 transition moment integral will have the symmetry $A_{2u}(v_i)\begin{pmatrix} e_u \\ a_{2u} \end{pmatrix} A_{1g}a_{1g}$. Only if $v_i = \mathrm{a}_{1g}$ is the integral not zero. (c) There will be no hot bands at 5 K. (d) One possible assignment is the following: A, 3_0^1; B, 2_0^1; C, $2_0^1 3_0^1$; D, 2_0^2; E, $2_0^2 3_0^1$; F, 2_0^3; G, $2_0^3 3_0^1$; H, 2_0^4; I, $2_0^4 3_0^1$; J, 2_0^5; K, $2_0^5 3_0^1$; L, 2_0^6. The average value of v_3' is 132 cm^{-1}, which is higher than v_3. The average value of v_2' is 244 cm^{-1}. Peak C might also contain the 1_0^1 transition. Peaks F, I, and L might contain 1_0^2, 1_0^3, and 1_0^4, respectively, giving $v_1' \approx 369$ cm^{-1}.

5-57. $3\sigma_u^+$ is weakly bonding. $1\pi_u$ is strongly bonding. $2\sigma_u^+$ is weakly antibonding.

5-58. The octahedral d^8 species will have the same diagram as a tetrahedral d^2 species (Problem 5-38).

5-59. From the graph we find the following: At 330 nm, slope = -1.03×10^{-2} $\text{min}^{-1} \Rightarrow k = 2.38 \times 10^{-2}\ \text{min}^{-1} = 3.97 \times 10^{-4}\ \text{s}^{-1}$. Intercept = -0.517 $= \log[c_0(\epsilon_j - \epsilon_i)] \Rightarrow \epsilon_i = 667$ (since $\epsilon_j = 1277$). At 340 nm, slope = $-9.60 \times 10^{-3}\ \text{min}^{-1} \Rightarrow k = 3.52 \times 10^{-4}\ \text{s}^{-1}$. Intercept = $-0.514 \Rightarrow \epsilon_i =$ 505.

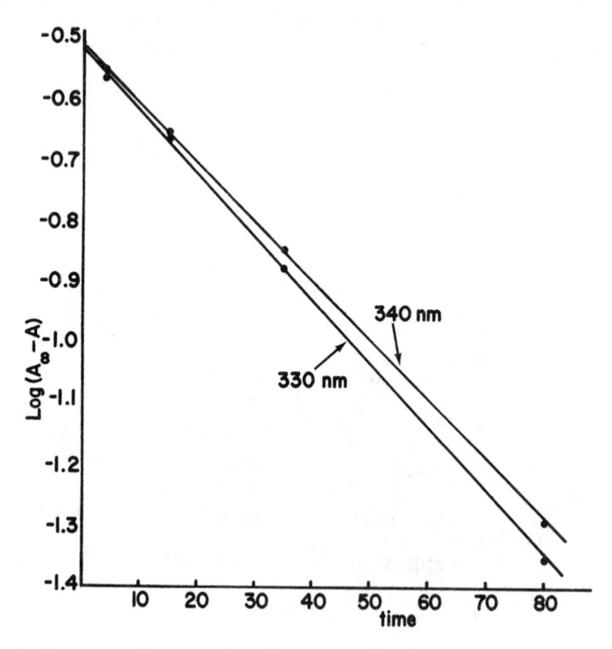

5-60. For any composition of the solution, the ratio $[M-XY]/[M-YX] = K_1/K_2$ is constant. Therefore the spectrum of the product will always be a multiple of the spectrum of this ratio ($[M-XY]/[M-YX]$), and we are effectively looking only at two superimposed spectra: that of free XY and that of bound XY. Isosbestic points will be exhibited by such a system despite the presence of three chromophores (XY, $M-XY$, $M-YX$), since two are always present in a constant ratio.

Index

A CATALOG OF SELECTED
DOVER BOOKS
IN SCIENCE AND MATHEMATICS

Mathematics

FUNCTIONAL ANALYSIS (Second Corrected Edition), George Bachman and Lawrence Narici. Excellent treatment of subject geared toward students with background in linear algebra, advanced calculus, physics and engineering. Text covers introduction to inner-product spaces, normed, metric spaces, and topological spaces; complete orthonormal sets, the Hahn-Banach Theorem and its consequences, and many other related subjects. 1966 ed. 544pp. 6⅛ x 9¼. 0-486-40251-7

DIFFERENTIAL MANIFOLDS, Antoni A. Kosinski. Introductory text for advanced undergraduates and graduate students presents systematic study of the topological structure of smooth manifolds, starting with elements of theory and concluding with method of surgery. 1993 edition. 288pp. 5⅜ x 8½. 0-486-46244-7

VECTOR AND TENSOR ANALYSIS WITH APPLICATIONS, A. I. Borisenko and I. E. Tarapov. Concise introduction. Worked-out problems, solutions, exercises. 257pp. 5⅜ x 8¼. 0-486-63833-2

AN INTRODUCTION TO ORDINARY DIFFERENTIAL EQUATIONS, Earl A. Coddington. A thorough and systematic first course in elementary differential equations for undergraduates in mathematics and science, with many exercises and problems (with answers). Index. 304pp. 5⅜ x 8½. 0-486-65942-9

FOURIER SERIES AND ORTHOGONAL FUNCTIONS, Harry F. Davis. An incisive text combining theory and practical example to introduce Fourier series, orthogonal functions and applications of the Fourier method to boundary-value problems. 570 exercises. Answers and notes. 416pp. 5⅜ x 8½. 0-486-65973-9

COMPUTABILITY AND UNSOLVABILITY, Martin Davis. Classic graduate-level introduction to theory of computability, usually referred to as theory of recurrent functions. New preface and appendix. 288pp. 5⅜ x 8½. 0-486-61471-9

AN INTRODUCTION TO MATHEMATICAL ANALYSIS, Robert A. Rankin. Dealing chiefly with functions of a single real variable, this text by a distinguished educator introduces limits, continuity, differentiability, integration, convergence of infinite series, double series, and infinite products. 1963 edition. 624pp. 5⅜ x 8½. 0-486-46251-X

METHODS OF NUMERICAL INTEGRATION (SECOND EDITION), Philip J. Davis and Philip Rabinowitz. Requiring only a background in calculus, this text covers approximate integration over finite and infinite intervals, error analysis, approximate integration in two or more dimensions, and automatic integration. 1984 edition. 624pp. 5⅜ x 8½. 0-486-45339-1

INTRODUCTION TO LINEAR ALGEBRA AND DIFFERENTIAL EQUATIONS, John W. Dettman. Excellent text covers complex numbers, determinants, orthonormal bases, Laplace transforms, much more. Exercises with solutions. Undergraduate level. 416pp. 5⅜ x 8½. 0-486-65191-6

RIEMANN'S ZETA FUNCTION, H. M. Edwards. Superb, high-level study of landmark 1859 publication entitled "On the Number of Primes Less Than a Given Magnitude" traces developments in mathematical theory that it inspired. xiv+315pp. 5⅜ x 8½. 0-486-41740-9

Physics

OPTICAL RESONANCE AND TWO-LEVEL ATOMS, L. Allen and J. H. Eberly. Clear, comprehensive introduction to basic principles behind all quantum optical resonance phenomena. 53 illustrations. Preface. Index. 256pp. 5⅜ x 8½.
0-486-65533-4

QUANTUM THEORY, David Bohm. This advanced undergraduate-level text presents the quantum theory in terms of qualitative and imaginative concepts, followed by specific applications worked out in mathematical detail. Preface. Index. 655pp. 5⅜ x 8½.
0-486-65969-0

ATOMIC PHYSICS (8th EDITION), Max Born. Nobel laureate's lucid treatment of kinetic theory of gases, elementary particles, nuclear atom, wave-corpuscles, atomic structure and spectral lines, much more. Over 40 appendices, bibliography. 495pp. 5⅜ x 8½.
0-486-65984-4

A SOPHISTICATE'S PRIMER OF RELATIVITY, P. W. Bridgman. Geared toward readers already acquainted with special relativity, this book transcends the view of theory as a working tool to answer natural questions: What is a frame of reference? What is a "law of nature"? What is the role of the "observer"? Extensive treatment, written in terms accessible to those without a scientific background. 1983 ed. xlviii+172pp. 5⅜ x 8½.
0-486-42549-5

AN INTRODUCTION TO HAMILTONIAN OPTICS, H. A. Buchdahl. Detailed account of the Hamiltonian treatment of aberration theory in geometrical optics. Many classes of optical systems defined in terms of the symmetries they possess. Problems with detailed solutions. 1970 edition. xv + 360pp. 5⅜ x 8½. 0-486-67597-1

PRIMER OF QUANTUM MECHANICS, Marvin Chester. Introductory text examines the classical quantum bead on a track: its state and representations; operator eigenvalues; harmonic oscillator and bound bead in a symmetric force field; and bead in a spherical shell. Other topics include spin, matrices, and the structure of quantum mechanics; the simplest atom; indistinguishable particles; and stationary-state perturbation theory. 1992 ed. xiv+314pp. 6⅛ x 9¼.
0-486-42878-8

LECTURES ON QUANTUM MECHANICS, Paul A. M. Dirac. Four concise, brilliant lectures on mathematical methods in quantum mechanics from Nobel Prize-winning quantum pioneer build on idea of visualizing quantum theory through the use of classical mechanics. 96pp. 5⅜ x 8½.
0-486-41713-1

THIRTY YEARS THAT SHOOK PHYSICS: THE STORY OF QUANTUM THEORY, George Gamow. Lucid, accessible introduction to influential theory of energy and matter. Careful explanations of Dirac's anti-particles, Bohr's model of the atom, much more. 12 plates. Numerous drawings. 240pp. 5⅜ x 8½. 0-486-24895-X

ELECTRONIC STRUCTURE AND THE PROPERTIES OF SOLIDS: THE PHYSICS OF THE CHEMICAL BOND, Walter A. Harrison. Innovative text offers basic understanding of the electronic structure of covalent and ionic solids, simple metals, transition metals and their compounds. Problems. 1980 edition. 582pp. 6⅛ x 9¼.
0-486-66021-4

HYDRODYNAMIC AND HYDROMAGNETIC STABILITY, S. Chandrasekhar. Lucid examination of the Rayleigh-Benard problem; clear coverage of the theory of instabilities causing convection. 704pp. 5⅝ x 8¼. 0-486-64071-X

INVESTIGATIONS ON THE THEORY OF THE BROWNIAN MOVEMENT, Albert Einstein. Five papers (1905–8) investigating dynamics of Brownian motion and evolving elementary theory. Notes by R. Fürth. 122pp. 5⅜ x 8½. 0-486-60304-0

THE PHYSICS OF WAVES, William C. Elmore and Mark A. Heald. Unique overview of classical wave theory. Acoustics, optics, electromagnetic radiation, more. Ideal as classroom text or for self-study. Problems. 477pp. 5⅜ x 8½. 0-486-64926-1

GRAVITY, George Gamow. Distinguished physicist and teacher takes reader-friendly look at three scientists whose work unlocked many of the mysteries behind the laws of physics: Galileo, Newton, and Einstein. Most of the book focuses on Newton's ideas, with a concluding chapter on post-Einsteinian speculations concerning the relationship between gravity and other physical phenomena. 160pp. 5⅜ x 8½. 0-486-42563-0

PHYSICAL PRINCIPLES OF THE QUANTUM THEORY, Werner Heisenberg. Nobel Laureate discusses quantum theory, uncertainty, wave mechanics, work of Dirac, Schroedinger, Compton, Wilson, Einstein, etc. 184pp. 5⅜ x 8½. 0-486-60113-7

ATOMIC SPECTRA AND ATOMIC STRUCTURE, Gerhard Herzberg. One of best introductions; especially for specialist in other fields. Treatment is physical rather than mathematical. 80 illustrations. 257pp. 5⅜ x 8½. 0-486-60115-3

AN INTRODUCTION TO STATISTICAL THERMODYNAMICS, Terrell L. Hill. Excellent basic text offers wide-ranging coverage of quantum statistical mechanics, systems of interacting molecules, quantum statistics, more. 523pp. 5⅜ x 8½. 0-486-65242-4

THEORETICAL PHYSICS, Georg Joos, with Ira M. Freeman. Classic overview covers essential math, mechanics, electromagnetic theory, thermodynamics, quantum mechanics, nuclear physics, other topics. First paperback edition. xxiii + 885pp. 5⅜ x 8½. 0-486-65227-0

PROBLEMS AND SOLUTIONS IN QUANTUM CHEMISTRY AND PHYSICS, Charles S. Johnson, Jr. and Lee G. Pedersen. Unusually varied problems, detailed solutions in coverage of quantum mechanics, wave mechanics, angular momentum, molecular spectroscopy, more. 280 problems plus 139 supplementary exercises. 430pp. 6½ x 9¼. 0-486-65236-X

THEORETICAL SOLID STATE PHYSICS, Vol. 1: Perfect Lattices in Equilibrium; Vol. II: Non-Equilibrium and Disorder, William Jones and Norman H. March. Monumental reference work covers fundamental theory of equilibrium properties of perfect crystalline solids, non-equilibrium properties, defects and disordered systems. Appendices. Problems. Preface. Diagrams. Index. Bibliography. Total of 1,301pp. 5⅜ x 8½. Two volumes. Vol. I: 0-486-65015-4 Vol. II: 0-486-65016-2

WHAT IS RELATIVITY? L. D. Landau and G. B. Rumer. Written by a Nobel Prize physicist and his distinguished colleague, this compelling book explains the special theory of relativity to readers with no scientific background, using such familiar objects as trains, rulers, and clocks. 1960 ed. vi+72pp. 5⅜ x 8½. 0-486-42806-0

CATALOG OF DOVER BOOKS

A TREATISE ON ELECTRICITY AND MAGNETISM, James Clerk Maxwell. Important foundation work of modern physics. Brings to final form Maxwell's theory of electromagnetism and rigorously derives his general equations of field theory. 1,084pp. 5⅜ x 8½. Two-vol. set. Vol. I: 0-486-60636-8 Vol. II: 0-486-60637-6

MATHEMATICS FOR PHYSICISTS, Philippe Dennery and Andre Krzywicki. Superb text provides math needed to understand today's more advanced topics in physics and engineering. Theory of functions of a complex variable, linear vector spaces, much more. Problems. 1967 edition. 400pp. 6½ x 9¼. 0-486-69193-4

INTRODUCTION TO QUANTUM MECHANICS WITH APPLICATIONS TO CHEMISTRY, Linus Pauling & E. Bright Wilson, Jr. Classic undergraduate text by Nobel Prize winner applies quantum mechanics to chemical and physical problems. Numerous tables and figures enhance the text. Chapter bibliographies. Appendices. Index. 468pp. 5⅜ x 8½. 0-486-64871-0

METHODS OF THERMODYNAMICS, Howard Reiss. Outstanding text focuses on physical technique of thermodynamics, typical problem areas of understanding, and significance and use of thermodynamic potential. 1965 edition. 238pp. 5⅜ x 8½. 0-486-69445-3

THE ELECTROMAGNETIC FIELD, Albert Shadowitz. Comprehensive undergraduate text covers basics of electric and magnetic fields, builds up to electromagnetic theory. Also related topics, including relativity. Over 900 problems. 768pp. 5⅝ x 8¼. 0-486-65660-8

GREAT EXPERIMENTS IN PHYSICS: FIRSTHAND ACCOUNTS FROM GALILEO TO EINSTEIN, Morris H. Shamos (ed.). 25 crucial discoveries: Newton's laws of motion, Chadwick's study of the neutron, Hertz on electromagnetic waves, more. Original accounts clearly annotated. 370pp. 5⅜ x 8½. 0-486-25346-5

EINSTEIN'S LEGACY, Julian Schwinger. A Nobel Laureate relates fascinating story of Einstein and development of relativity theory in well-illustrated, nontechnical volume. Subjects include meaning of time, paradoxes of space travel, gravity and its effect on light, non-Euclidean geometry and curving of space-time, impact of radio astronomy and space-age discoveries, and more. 189 b/w illustrations. xiv+250pp. 8⅜ x 9¼. 0-486-41974-6

THE VARIATIONAL PRINCIPLES OF MECHANICS, Cornelius Lanczos. Philosophic, less formalistic approach to analytical mechanics offers model of clear, scholarly exposition at graduate level with coverage of basics, calculus of variations, principle of virtual work, equations of motion, more. 418pp. 5⅜ x 8½. 0-486-65067-7